The Environmental Law and
Compliance Handbook

The Environmental Law and Compliance Handbook

James F. Berry, J.D., Ph.D.

Mark S. Dennison, J.D.

McGraw-Hill
New York   San Francisco   Washington, D.C.   Auckland   Bogotá
Caracas   Lisbon   London   Madrid   Mexico City   Milan
Montreal   New Delhi   San Juan   Singapore
Sydney   Tokyo   Toronto

Library of Congress Cataloging-in-Publication Data

Berry, James F.
    The environmental law and compliance handbook / James F. Berry, Mark S. Dennison.
      p.   cm.
    Includes index.
    ISBN 0-07-134094-7
    1. Environmental law—United States. I. Dennison, Mark S. II. Title.
  KF3775.B42 2000
  344.73'046—dc21
                                                         00-022357
                                                         CIP

McGraw-Hill

A Division of The McGraw·Hill Companies

   3 4 5 6 7 8 9 0 DOC/DOC 0 6 5 4 3 2 1

ISBN 0-07-134094-7

The sponsoring editor for this book was Robert Esposito, the editing supervisor was Frank Kotowski, Jr., and the production supervisor was Pamela A., Pelton. It was set in Times New Roman by James F. Berry.

Printed and bound by R. R. Donnelley & Sons, Inc.

McGraw-Hill books are available at special quantity discounts to use as premiums and sales promotions, or for use in corporate training programs. For more information, please write to the Director of Special Sales, Professional Publishing, McGraw-Hill, Two Penn Plaza, New York, NY 10121-2298. Or contact your local bookstore.

This book was printed on recycled, acid-free paper containing a minimum of 50% recycled, de-inked fiber.

For Lynn,

Who makes it all worthwhile

J.F.B.

For Tracey,

*Who makes even the cloudy days
seem sunny and bright*

M.S.D.

Contents in Brief

Contents

x   Contents

Preface

Environmental professionals–whether they are working as in-house environmental managers or outside environmental consultants–must possess a full understanding of the seemingly unmanageable body of environmental laws and regulations, and know how those regulations apply to particular types of business operations, manufacturing processes, hazardous substances and wastes, and so forth. Even though most environmental laws have been in existence for quite some time, the broad system of environmental regulation remains complex, and its understanding is further confounded by periodic statutory amendments, ongoing changes to existing regulations, and the continuing introduction of new ones. *The Environmental Law and Compliance Handbook* is intended to serve the needs of environmental professionals by providing comprehensive guidance on the complex array of federal environmental laws and implementing regulations that environmental professionals must know and master in order to be successful in their positions as environmental managers, compliance officers, engineers, and consultants.

The book has been organized into eight distinct parts, each of which essentially corresponds to a particular type of environmental medium (i.e., water, land, and air) or specialized area governed by the environmental laws presented within that part. Part I, *Introduction*, serves to explain the purpose and coverage of the book, as well as introduce some basic principles of administrative law which are necessary to understand fully the nature and scope of authority granted to governmental agencies charged with the administration and enforcement of the environmental laws and implementing regulations. Part II, *Origins of U.S. Environmental Regulation*, first explains the common-law theories of trespass, strict liability for ultrahazardous activities, nuisance, and negligence, which may form the basis for claims of environmental liability and which are separate and distinct from the statutory claims that may arise under environmental laws enacted by federal and state legislatures. Also included in this part is a full discussion of the National Environmental Policy Act of 1969 and its system of environmental impact review. NEPA is included here because it was the first major federal environmental law enacted in the United

States, and laid a foundation for other environmental legislation that was soon to follow. Part III, *Water Pollution Control*, focuses on federal laws that control water quality, and includes chapters dealing with the regulation of water pollution under the Clean Water Act and Safe Drinking Water Act. Part IV, *Air Pollution Control*, explains the detailed regulatory scheme for controlling air pollutant emissions under the Clean Air Act. Part V, *Hazardous and Toxic Substance Regulation*, is the largest part of the book and covers the primary federal laws governing the generation, storage, handling, disposal, and cleanup of hazardous substances and wastes, as required by the Resource Conservation and Recovery Act, the Comprehensive Environmental Response, Compensation and Liability Act, the Federal Insecticide, Fungicide, and Rodenticide Act, and the Toxic Substances Control Act. In Part VI, *Pollution Prevention*, the authors turn to a full discussion of the important subject of pollution prevention, explaining the policy and goals of the Pollution Prevention Act of 1990, and examine federal and state programs that focus on waste minimization and recycling. Part VII, *Environmental Control of Land Use*, explains environmental laws designed to control the adverse impacts of certain types of land use activities on the environment, in particul regulation of land use activities in wetlands, coastal areas, and endangered species habitats under Section 404 of the Clean Water Act, the Coastal Zone Management Act, and the Endangered Species Act. Finally, Part VIII, *The International Perspective*, rounds out the coverage of the book with a discussion of international environmental law and policy.

The authors have endeavored to write the book in such a way that it will serve the needs of the environmental professional regardless of the level of his or her experience. For the environmental professional in pursuit of an educational degree or license, the book functions as a detailed textbook of environmental law and regulation. For the experienced environmental manager, engineer, compliance officer, or consultant, the book is a professional desk reference that can be consulted to refresh one's knowledge or obtain answers to questions about a specific environmental law or set of regulations. For the less experienced environmental professional, or one who has specialized knowledge in a particular area of environmental regulation but only a general understanding of other areas, the book serves as a useful handbook for determining and evaluating regulatory compliance requirements.

James F. Berry
Elmhurst, Illinois

Mark S. Dennison
Westwood, New Jersey

Acknowledgments

The authors wish to express their appreciation to everyone who played a role in the publication of this book. First, we thank the editorial and production staff at McGraw-Hill, especially our Sponsoring Editor, Bob Esposito, who invited us to write the book and guided it through several stages of the publication process, our Editor Scott Grillo, and our Editing Supervisor, Frank Kotowski, Jr. who did a wonderful job editing the manuscript.

In addition, Jim Berry expresses his gratitude to everyone who has contributed to this volume, either directly or indirectly, including Professors Fred P. Bosselman and A. Dan Tarlock (IIT Chicago–Kent College of Law), Dean Stuart L. Deutsch (Rutgers University–Newark School of Law), Nancy E. Stroud (Weiss, Serota, Helfman, Pastoriza & Guedes, P.A., Fort Lauferdale, FL), Dean Zia Hassan and Dr. George P. Nassos (Stuart Graduate School of Business, Illinois Institute of Technology); to Elmhurst College for support in the preparation of this book; and to his wife Lynn for her patience and unquestioning support.

Mark Dennison also gives thanks to family and friends who provided support and inspiration, including Erin Carrather, Bobbie Waits, and Keith Dennison. Most of all, he wishes to express his appreciation to Tracey Huff, whose constant love and companionship means more than anything else.

J.F.B.
M.S.D.

Part

I

Introduction

Overview

1.1 Introduction; Purpose of this Book

The climate in which business and industry operate today no longer requires embarking on a voyage through uncharted waters while attempting to navigate through the snags, snares, and changing currents known as environmental law and regulation. The environmental laws that have formed the basis for a system of environmental regulation have been around for the better part of three decades, commencing with enactment of the National Environmental Policy Act (NEPA) in 1969, continuing through the 1970s with enactment of the Clean Water Act, the Clean Air Act, the Resource Conservation and Recovery Act, the Endangered Species Act, and other federal environmental laws. Perhaps the peak of this flurry of environmental legislation was reached in 1980 with passage of the Comprehensive Environmental Response, Compensation, and Liability Act (CERCLA). By now, it would indeed be rare to find a company--large or small--that has not in some way found its day-to-day business operations influenced, changed, or otherwise subject to compliance with the regulatory requirements of one or more of these environmental laws. Over the years, the steady proliferation of federal and state environmental laws and regulations has forced business, industry, public enterprises, and even governmental agencies to change the manner in which various types of commercial and industrial operations are performed in order to come into compliance with environmental regulations.

Despite the fact that most environmental laws have been in existence for quite some time, the broad system of environmental regulation remains complex, and its understanding is ever confounded by periodic statutory amendments, ongoing changes to existing regulations, and the continuing introduction of new regulations. Much in the way that companies must employ specialists in such areas as computer analysis, financial management, or taxes, companies have learned that only well-trained environmental professionals can properly determine what is necessary to achieve and maintain environmental compliance. Consequently, in order to minimize, if not eliminate, the risk of incurring costly liability and penalties for noncompliance with environmental laws, most companies have created distinct environmental management departments staffed by environmental professionals who specialize in the evaluation of day-to-day operations for compliance with the extensive network of environmental regulatory requirements. These environmental professionals--whether working as in-house environmental managers or outside environmental consultants--obviously must possess a full understanding of the seemingly unmanageable body of environmental laws and regulations, know how those regulations apply to particular types of business operations, manufacturing processes, hazardous substances and wastes, and so forth, and stay abreast of any changes in the law and regulations.

This book is intended to serve the needs of environmental professionals by providing comprehensive guidance on the complex array of environmental compliance issues that confront environmental professionals in their positions as environmental managers, engineers, compliance officers, and consultants to private business/industry and government agencies. The authors have endeavored to write the book in such a way that it will serve the needs of the environmental professional regardless of the level of his or her experience. For the environmental professional in pursuit of an educational degree or license, the book functions as a detailed textbook of environmental law and regulation. For the experienced environmental manager, engineer, compliance officer, or consultant, the book is a professional desk reference that can be consulted to refresh one's knowledge or obtain answers to questions about a specific environmental law or set of regulations. For the less experienced environmental professional, or one who has specialized knowledge in a particular area of environmental regulation (e.g., National Pollutant Discharge Elimination System or storm water permitting) but only a general understanding of other areas, the book serves as a useful handbook for determining and evaluating regulatory compliance requirements.

1.2 Scope and Organization of the Book

The book provides a detailed discussion of the major federal environmental laws and implementing regulations that environmental professionals need to know and master in order to be successful in their positions as environmental managers, compliance officers, engineers, and consultants. The authors have organized the book into distinct parts, each of which corresponds to a particular type of environmental medium (i.e., water, land, and air) or specialized area governed by the environmental laws presented within that part. Although this approach is certainly less than perfect given the broad scope of many environmental laws, the authors determined that this form of organization would enable environmental professionals to locate most easily the laws and regulations governing a particular compliance issue. The general purpose of most federal environmental laws is to promote the protection of human health and the environment through regulation of particular types of pollutants, substances, or processes. For example, in general terms, the Resource Conservation and Recovery Act (RCRA) sets forth a "cradle-to-grave" regulatory framework for management of hazardous and solid wastes; the Clean Air Act places technology-based controls on emissions of air pollutants; the Clean Water Act seeks to protect the physical integrity of U.S. waters; and the Comprehensive Environmental Response, Compensation, and Liability Act (CERCLA) governs the cleanup of hazardous substances that have been released, or pose a threat of release, into the environment.

The book has been divided into eight separate parts. Part I, Introduction, consists of this chapter and serves to explain the purpose and scope of the book, as well as introduce some basic principles of administrative law which is necessary for a full understanding of the nature and scope of authority granted to governmental agencies to administer and enforce the environmental laws and implementing regulations. Part II, Origins of U.S. environmental Regulation, first explains the common-law theories of trespass, strict liability for ultrahazardous activities, nuisance, and negligence, which may form the basis for claims of environmental liability and which are separate and distinct from the statutory claims that may arise under environmental laws enacted by federal and state legislatures.[1] Also included in this part is a full discussion of the National Environmental Policy Act of 1969 (NEPA) and its system of environmental impact review since NEPA was the first major federal environmental law enacted in the United States and laid a foundation for other environmental legislation that was soon to follow.[2] Part III, Water Pollution Control, focuses on federal laws that control water quality, and includes chapters dealing with the regulation of water pollution under the Clean Water Act (CWA) and Safe Drinking Water Act (SDWA).[3] Part IV, Air Pollution Control, as the title suggests, explains the detailed regulatory scheme for controlling air pollutant emissions under the Clean Air Act (CAA).[4] Part V, Hazardous and Toxic Substance Regulation, is the largest Part of the book and covers the primary federal laws governing the generation, storage, handling, disposal, and cleanup of hazardous

---

[1] See chapter 2 for discussion of common law principles.
[2] See chapter 3 for discussion of NEPA.
[3] See chapter 4 for discussion of CWA regulation of water pollutant discharges (focusing on NPDES permitting); chapter 5 for discussion of CWA regulation of storm water discharges; and chapter 6 for discussion of the SDWA.
[4] See chapter 7 for discussion of the CAA.

substances and wastes, as required by the Resource Conservation and Recovery Act (RCRA), the Comprehensive Environmental Response, Compensation, and Liability Act (CERCLA), the Federal Insecticide, Fungicide, and Rodenticide Act (FIFRA), and the Toxic Substances Control Act (TSCA).[5] In Part VI, Pollution Prevention, the authors turn to a full discussion of the important subject of pollution prevention, explaining the policy and goals of the Pollution Prevention Act of 1990 (PPA), and examining federal and state programs that focus on waste minimization and recycling.[6] Part VII, Environmental Control of Land Use, explains environmental laws designed to control the adverse impacts of certain types of land use activities on the environment, in particular the regulation of land use activities in wetlands, coastal areas, and endangered species habitats under Section 404 of the CWA, the Coastal Zone Management Act (CZMA), and the Endangered Species Act (ESA).[7] Finally, Part VIII, The International Perspective, rounds out the coverage of the book with a discussion of international environmental law and policy.[8]

1.3 State Environmental Law

This book focuses on the regulatory requirements of the federal environmental laws. Unfortunately, it must be emphasized that knowing and understanding the federal requirements will not alone suffice to ensure regulatory compliance. Business, industry, public enterprises, and governmental agencies must be in compliance with state-specific environmental laws and regulations as well. Although many state environmental laws are modeled after the federal laws, and may even mirror their federal counterparts in certain respects, the state laws often have some additional and unique requirements of their own. Thus, it is imperative that the environmental professional also review all applicable state environmental laws and regulations and contact state environmental authorities, information hotlines, and trade organizations for additional guidance on state-specific environmental laws and regulations. A discussion of state environmental laws is obviously beyond the scope of this book;[9] however, the authors have provided some examples of state

---

[5] See chapter 8 for discussion of RCRA; chapter 9 for discussion of CERCLA; chapter 10 for discussion of FIFRA; and chapter 11 for discussion of TSCA.
[6] See chapter 12 for discussion of the PPA and federal and state pollution prevention programs.
[7] See chapter 13 for discussion of wetlands regulation; chapter 14 for discussion of coastal zone management; and chapter 15 for discussion of sensitive areas and endangered species regulation.
[8] See chapter 16 for discussion of international environmental law and policy.
[9] For a good presentation of state environmental laws, see Selmi and Manaster, *State Environmental Law* (West Group, 1989 & Supps.)

regulatory requirements wherever possible.

In addition, it must be noted that some states may have authority to operate certain state environmental programs in place of corresponding federal programs. Although federal agencies, such as the U.S. EPA, the Army Corps of Engineers, and the U.S. Fish & Wildlife Service, are empowered to administer and enforce the federal environmental laws discussed throughout this book, some of the federal laws also allow federal agencies to delegate authority to the states to administer their own environmental programs in lieu of the federal ones, conditioned on approval by the federal agency. Such is frequently the case with RCRA solid and hazardous waste management programs, RCRA underground storage tank programs, and CWA National Pollutant Discharge Elimination System (NPDES) permitting programs. Federal agency approval of these state programs basically indicates that a state has passed legislation and promulgated regulations to administer a state regulatory program that is fully equivalent to, and at least as stringent as, the corresponding federal program. As such, it must be kept in mind that although a state's environmental program must be equivalent to the federal program to receive federal approval, equivalent does not mean identical and states are not prohibited from establishing regulatory regimes that are more stringent than those delineated in the federal law and regulations.

1.4 Introduction to the Legal System

There are several fundamental questions about the legal system that must be answered before we can begin our analysis of environmental laws. First, what are the differences between environmental "laws" and "regulations," and how does each affect environmental professionals? We can answer this question by recalling that the government of the United States of America consists of three branches, with the expectation that each branch will function to balance any excesses by the others.[10]

The Legislative Branch consists of the Senate and the House of Representatives. It is the task of the Legislative Branch to pass laws. Typically, a bill is introduced in either the Senate or the House. The bill is then assigned to a committee that may hold hearings, and/or conduct investigations or studies, and ultimately issues a report to the Senate or House with a recommendation on approval. If the bill passes a vote, it is then sent to the other house (either House of Representatives or Senate) for its consideration. If different versions of the same bill are approved by each house, then a joint committee will attempt to reach a compromise that

---

[10] As was explained in section 1.3 above, our discussion will focus on the federal or national government, but it is important to remember that each *state* has a government that parallels the federal system (most states are essentially identical to the federal system).

satisfies both houses. If members of both houses pass the bill, it becomes an act. The act becomes a law if the president signs it (or fails to veto it within ten days). Once an act has become a law, it is codified within the United States Code (usually abbreviated "U.S.C."), and is often considered as a "statute."[11]

The executive branch of government consists of the president and the cabinet. The president is elected by the people, and the members of the cabinet are nominated by the president and approved by the Senate. Examples of cabinet-level positions of greatest importance in environmental laws are the Administrator of the U.S. Environmental Protection Agency, and the Secretary of the Department of the Interior. Generally speaking, the function of the executive branch is to "execute," or administer, the laws. Most environmental laws contain a provision in which the legislature "delegates" to the executive branch the authority to conduct the day-to day administration of the laws by way of "regulations." If Congress has failed to delegate the authority to create regulations, then the president may do so via an "Executive Order."[12]

Regulations, then, are the rules by which executive agencies administer the laws. The process of "rule making" begins when an executive agency publishes a proposed regulation in the *Federal Register*. This publication invites comments from all interested persons, either through the submission of written comments or through public hearings. Once the agency has scrutinized the comments (and made whatever changes seem proper), a final rule is published in the *Federal Register*. This final rule, or "regulation," has the force and effect of law once it reaches its effective date. Regulations are codified in the *Code of Federal Regulations* (usually abbreviated CFR).[13]

For most environmental professionals, it is the regulations that guide our everyday activities, rather than the statutes. Nevertheless, it is important to remember the relationship between the statutes themselves and the regulations that are used to administer them.

The third branch of government in the United States is the judicial branch, composed of the federal district courts, the Circuit Courts of Appeal, and the U.S. Supreme Court. The purpose of the federal courts is,

---

[11] Once a law has been codified, it is still frequently known by its title and section number within the original Act. For example, the section of the Clean Water Act that applies to the filling of wetlands is usually called "Clean Water Act §404," rather than "33 U.S.C. § 1344 " [volume 33 of the U.S. Code, Section 1344].

[12] For example, in 1977 President Carter signed Executive Order 11,990 (42 *Fed. Reg.* 26,967 (1977)) authorizing the Council on Environmental Quality to promulgate regulations under NEPA, because the statute contained no rule making authority. This situation is discussed more fully in chapter 3.

[13] For example, the regulation that determines what organisms are "pests" under the Federal Insecticide, Fungicide, and Rodenticide Act (FIFRA, see chapter 10) is "40 CFR § 152.15" [volume 40 of the Code of Federal Regulations, Section 152, Subsection 15].

in general, to make decisions regarding disputes among parties over issues that are national in scope.[14] In deciding the cases that are brought before them, judges interpret the relevant statutes and regulations, apply constitutional principles, and (when asked to do so) follow precedents.

The geographically delineated district courts are the "trial courts" of the federal system; that is, the first court to which a plaintiff turns to ask that a decision be made.[15] The district courts examine evidence and hear arguments, then render their decision based on the record established at trial. The loser in the district court may then appeal to a regional Circuit Court of Appeals. There are eleven regional Circuit Courts of Appeals, plus the Court of Appeals for the Federal Circuit in Washington, D.C., that hears appeals only in a few types of cases. It is important to note that very few cases are automatically appealed as a matter of right. Instead, the Circuit Courts of Appeals pick the cases they will hear on appeal based on their importance. If the Circuit Court of Appeals denies an appeal, that is usually the end of the case for the parties. If the Circuit Court does agree to hear an appeal, that appeal is limited to the record established in the District Court (that is, no new evidence is allowed). In an appeal, it is usually only the attorneys representing the parties that are invited to speak to the court.

In fairly unusual circumstances, the loser in the Circuit Court of Appeals may then appeal to the U.S. Supreme Court. Given the large number of cases appealed to the Supreme Court every year, it is only a small percentage that are actually heard.[16] Once again, if an appeal is heard, it is limited to the record established in the district court, and it is usually only the attorneys representing the parties that are invited to speak to the Court. There is no opportunity for appeal within the judicial system beyond the U.S. Supreme Court. Given the small number of environmental cases heard by the Supreme Court each year, it is not surprising that they take on extreme importance.

1.5 Administrative Law

The practice of environmental law in the United States is primarily regulatory and public. As a consequence, federal agencies are largely

---

[14] This includes disputes under most (but not all) federal environmental laws, but may also include disputes among parties located across state boundaries from each other.

[15] The District Court for the District of Columbia, and the U.S. Claims Court are not truly geographical.

[16] In actual practice, the loser in the Circuit Court of Appeals petitions the Supreme Court for "writ of certiorari." Following some cases in this book you will see "cert. denied" before the Supreme Court reference, indicating that the Supreme Court did not grant certiorari, and the appeal was denied.

responsible for enforcing environmental laws. The field of law that controls the actions of federal agencies and the enforcement of regulatory controls is known as administrative law, and is generally governed by the federal Administrative Procedures Act (APA).[17]

In fact, the role of federal agencies is in many ways a hybrid of the activities of all three branches of government. The executive, or administrative, role is the most obvious, since agencies clearly function to administer the various environmental laws. However, federal agencies also have a distinctive "legislative" function when they participate in the rule-making process and draft regulations as described in Section 1.4 above.

Federal agencies also have a "judicial" function in that they are often authorized to make findings of fact and conclusions of law in matters related to the activities they regulate. The primary mechanism for these actions is the administrative hearing process. An administrative hearing is held before an administrative law judge (ALJ), who determines whether a particular defendant (or the defendant's activities) is within the purview of a particular regulation, and whether that regulation has been violated. If the defendant is found guilty of violating the provisions of a regulation, then enforcement actions specified by the regulations begin.

If a disappointed party wishes to appeal the ALJ's decision following an administrative hearing, then an appeal to the agency is appropriate if an administrative appeal is specified in the statute. An administrative appeal involves another administrative judge, who makes a decision based on the record at the original administrative hearing.[18] Some members of the regulated community feel uncomfortable with this process, since they are appealing to the very agency that rebuked them in the first place.

If a statute does not specify an administrative appeal process, or if a defendant is disappointed with the result of an administrative appeal, then appeal is usually to the local federal district court. As with other appeals, the appeal to the district court is usually restricted to the record developed in the administrative hearing (that is, no new evidence is allowed). From here, the administrative process proceeds as in any other judicial procedure.

There are many variations on the administrative process, and they are far too numerous to discuss here. Many variations will be discussed throughout the book.

One recent administrative innovation worth mention is the Environmental Appeals Board (EAB), which was created in response to concerns by the regulated community that agency decisions (especially those of the EPA) on environmental matters could not be appealed except by expensive litigation in the federal courts. Established by EPA regulations in 1992,[19] the EAB was designed to inspire confidence in the

---

[17] 5 U.S.C. §§ 551–559.
[18] See APA at 5 U.S.C. § 702.
[19] Scattered sections of 40 CFR.

fairness of agency adjudication in administrative hearings.

The EAB consists of three senior-level executive attorneys plus a small staff. It allows immediate appeals of agency decisions without the time/expense of a formal trial. The EAB has published an EAB Practice Manual with EAB rules for the conduct of hearings, and other useful information.[20]

Over 500 appeals have been heard since 1992, and over 150 written opinions have been published. To date, most EAB appeals have been under CERCLA, RCRA, TSCA, the Clean Air Act, and the Clean Water Act. It appears that the EAB and its administrative process is being grudgingly accepted by the regulated community as an improvement on the old system.

---

[20] EPA Publication No. 100-B-94-002 (1994).

Origins of U.S. Environmental Regulation

2

Common-Law Principles

2.1 Introduction

In addition to the federal and state environmental laws discussed throughout this book, it is important to examine various common-law theories that may be asserted as a basis for environmental liability claims. Common law, in contrast to statutory law, refers to those general principles and rules pertaining to persons and property that have developed over time from custom and usage, and from court decisions that have ruled on those customs and usages. A more general way of conceptualizing the distinction between common law and statutory law is to think of the common law as nonstatutory law. Statutory law pertains to all fedcral and state laws specifically created by legislative enactment. Common law, on the other hand, refers to all remaining rules, principles, and customs not derived from express legislative authority.

Common-law claims have increasingly taken on environmental implications. Liability may stem from personal injuries and property damage caused by environmental conditions under common-law theories of negligence, nuisance, trespass, and strict liability for abnormally dangerous activities. For example, an individual or company may be sued by environmental authorities to clean up contaminated property, and neighboring property owners may assert statutory and common-law claims for damage caused to their properties. Except for negligence, common-law liability for environmental conditions is imposed *without regard to fault*. Although the statutory claims, such as liability under the Comprehensive Environmental Response, Compensation, and Liability Act (CERCLA), may be more typical, common-law claims for environmental injury to persons and property are more prevalent today. Therefore, this chapter explores and explains each of these common-law theories.

2.2 Trespass

Trespass is an common-law action to remedy an unauthorized invasion of a person's right to exclusive possession of the land.[1] By contrast, an actionable invasion of a person's right to the use and enjoyment of land is a nuisance.[2] For example, one court equated knowledge on the part of a defendant concerning release of sludge from its industrial waste dump onto the plaintiff's land with an intentional invasion of the landowner's property interest.[3] Since the defendants had been notified that their sludge was entering the plaintiffs' mine and the defendants did nothing to stop it, plaintiffs had a valid common-law action for trespass.[4]

Because trespass is a strict liability claim, plaintiffs in such common-law actions may succeed even in the absence of proof of actual harm or injury.[5] Section 158 of the Restatement (Second) of Torts states that "one is subject to liability to another for trespass, irrespective of whether he thereby causes harm."[6] However, in the case of a negligent trespass, actual damage is an essential element of the claim.[7]

2.3 Strict Liability for Ultrahazardous Activities

Under the common-law theory of strict liability for ultrahazardous or abnormally dangerous activities, liability is imposed without regard to fault for injuries caused from an ultrahazardous activity. This longstanding theory, which originated in 1866 with the famous English case of *Rylands v. Fletcher*,[8] has been applied in the modern context of injuries caused by toxic, hazardous, and radioactive substances.[9]

---

[1] See, for example, Cassinos v. Union Oil of Cal., 14 Cal.App.4th 1770, 18 Cal.Rptr.2d 574 (1993) ("[t]he essence of a cause of action for trespass is an 'unauthorized entry' onto the land of another.")

[2] See section 2.4 for discussion of nuisance.

[3] Curry Coal Co. v. M.C. Arnoni Co., 266 A.2d 678 (Pa. 1970).

[4] See also Regan v. Cherry Corp. 706 F. Supp. 145 (R.I. 1985) (knowing deposit of toxic wastes on another's property is a continuing trespass).

[5] See, for example, Ream v. Keen, 112 Or. App. 197, 828 P.2d 1038, aff'd, 314 Or. 370 (1992) (under Oregon law, actual damage unnecessary to establish liability for an intentional trespass).

[6] Restatement (Second) of Torts § 158.

[7] See, for example, Cereghino v. Boeing Co., 873 F. Supp. 398 (D.Or. 1994) (plaintiffs claims of negligent trespass and nuisance dismissed for failure to show actual damage as a result of Boeing's TCA contamination).

[8] Rylands v. Fletcher, L.R. 1 Ex. 265 (1866), aff'd, House of Lords, 3 H.L. 330 (1868).

[9] The *Rylands* case actually involved property loss caused by water escaping from a reservoir that had been created in an abandoned coal mine. The water flooded an operating coal mine on the property of an adjoining landowner.

Section 519 of the Restatement (Second) of Torts establishes the elements of strict liability for harm caused by abnormally dangerous activities as follows:

1. One who carries on an abnormally dangerous activity is subject to liability for harm to the person, land or chattels of another resulting from the activity, although he has exercised the utmost care to prevent the harm.

2. This strict liability is limited to the kind of harm, the possibility of which makes the activity abnormally dangerous.[10]

Further, the Restatement (Second) of Torts lists several factors upon which the courts often rely when determining whether an activity is abnormally dangerous:

1. Existence of a high degree of risk of some harm to the person, land or chattels of others

2. Likelihood that the harm that results from it will be great

3. Inability to eliminate the risk by exercise of reasonable care

4. Extent to which the activity is not a matter of common usage

5. Inappropriateness of the activity to the place where it is carried on; and

6. Extent to which its value to the community is outweighed by its dangerous attributes.[11]

Although the vast majority of cases involving injury to person or property from hazardous substances are based on violations of various federal and state statutes and regulations, the common-law cause of action for ultrahazardous or abnormally dangerous activities may be applicable in certain toxic tort situations.[12] For example, in a New Jersey case,[13] the owners of land used as a dump for toxic wastes from a mercury processing plant were held strictly liable for harm caused to others, including the costs of cleanup. Some 625 corporations that had generated wastes hauled to the site also found strictly liable at common law for abnormally dangerous activities.

---

[10] Restatement (Second) of Torts § 519.
[11] Restatement (Second) of Torts § 520.
[12] See Section 2.5.3 for discussion of toxic torts.
[13] State of New Jersey v. Ventron, 468 A.2d 150 (N.J. 1983).

Similarly, a federal district court in Connecticut adopted the view seldom endorsed by other courts that the disposal of wastes at a landfill may constitute an ultrahazardous activity, thus subjecting the defendant-owner of the landfill to strict liability for damage to the land of neighboring property owners.[14] In that case, the plaintiffs were owners of land adjacent to a landfill that was owned, operated, and managed by the defendant, Town of Bristol, Connecticut. The plaintiffs alleged that the defendant had caused and allowed the disposal of hazardous wastes and substances in the landfill, resulting in the pollution and contamination of the soil, ground, and water beneath the plaintiffs' land. The plaintiffs asserted that the disposal of wastes at the landfill constituted an ultrahazardous activity for which the defendant should be held strictly liable. The defendant moved for dismissal of plaintiffs' strict liability claim, arguing that operation of the landfill was not an unreasonably dangerous activity under Connecticut law.

The court began by noting that successful assertion of a strict liability claim requires proof of three factors: "(1) an instrumentality capable of producing harm; (2) circumstances and conditions in its use which, irrespective of a lawful purpose or due care, involve a risk of probable injury to such a degree that the activity fairly can be said to be intrinsically dangerous to the person or property of others; (3) and a causal relation between the activity and the injury for which damages are claimed." Further, the court stated that "[a] person who uses an intrinsically dangerous means to accomplish a lawful end, in such a way as will necessarily or obviously expose the person of another to the danger of probable injury, is liable if such injury results, even though he [or she] uses all proper care."[15]

Reviewing Connecticut law, the court found that the Connecticut Supreme Court had recognized ultrahazardous activity only with respect to blasting, pile driving, and conducting research with highly volatile chemicals. In the absence of any controlling Connecticut Supreme Court precedent as to whether disposal of hazardous substances in a landfill constituted an ultrahazardous activity, the court was thus required to determine what it believed the state's highest court would find if the same issue were before it. The court found that in one case, the Connecticut Superior Court had addressed the issue, concluding that disposal of hazardous and toxic wastes at a landfill constitutes an abnormally dangerous or ultrahazardous activity sufficient to maintain a cause of

---

[14] Albahary v. City and Town of Bristol, Connecticut, 1997 WL 22084 (D. Conn. Mar. 24, 1997).
[15] Id., 1997 WL 220284, at *3.

action for strict liability.[16]

Finding that hazardous materials are an instrumentality capable of producing harm, and because the circumstances and conditions of its disposal into a municipal landfill, irrespective of a lawful purpose or due care, involve a substantial risk of probable injury to the person or property of others, the federal district court concluded that disposal of hazardous and toxic wastes at a landfill may constitute an abnormally dangerous or ultrahazardous activity sufficient to maintain a cause of action for strict liability. In light of plaintiffs' allegation that defendant allowed the discharge into the landfill of hazardous substances, the court held that the plaintiffs were entitled to put forth evidence of the extent to which disposal of hazardous materials was allowed and encouraged at the landfill, as well as the toxic nature of the hazardous wastes plaintiffs alleged were disposed of there. Accordingly, the court denied the defendant's motion to dismiss the strict liability claim.

2.4 Nuisance

Nuisance represents another form of common-law action for an unreasonable interference with the use and enjoyment of property.[17] The law of nuisance recognizes two conflicting rights: (1) property owners have a right to control their land and use it to benefit their best interests and (2) the public and neighboring landowners have a right to prevent unreasonable use that substantially impairs the peaceful use and enjoyment of other land. Nuisance law is based on the principle that "[o]ne must use his own property so that his neighbor's comfortable and reasonable use and enjoyment of his estate will not be unreasonably interfered with or disturbed."[18]

The legal remedy for a nuisance may be in the form of monetary damages and/or injunctive relief. In determining whether a challenged activity constitutes a nuisance, the courts generally weigh the utility and public acceptability of the activity with the extent of harm or cost of

---

[16] Barnes v. General Electric Co., No. CV 930529354, 1995 WL 337904 (Conn. Super. July 20, 1995) (applying the factors provided in Section 520 of the Restatement (Second) of Torts, the court concluded that the defendants' storage, burial, and disposal of hazardous and toxic wastes at a municipal landfill involved a high degree of risk of harm to property and persons and was thus abnormally dangerous).

[17] See, for example, Frank v. Environmental Sanitation Management, Inc., 687 S.W.2d 876 (Mo. 1985) (en banc).

[18] Patz v. Farmegg Products, Inc., 196 N.W.2d 557, 560 (Iowa 1972), quoting Bates v. Quality Ready Mix Co., 261 Iowa 696, 702, 154 N.W.2d 852, 857 (1967).

compensation for the injury to the complaining party's property.[19] A court may choose to deny injunctions when the resulting losses to the polluters would greatly exceed the benefits to their victims.[20]

Two types of nuisance actions are possible: public nuisance and private nuisance. Although the two categories often overlap, there is an important, and often misunderstood, distinction between the two. Public nuisance actions arise when there has been an unreasonable interference with a property right "common to the general public," such as when an improperly managed public landfill violates the public's right to a safe environment.[21] Private nuisance actions, on the other hand, involve unreasonable interferences with an individual's use and enjoyment of his or her property, such as when a neighboring landowner discharges hazardous substances onto the property of the complaining landowner.[22] In either case, the claim of nuisance requires that the pollution-causing activity be unreasonable. The unreasonable use element of nuisance balances the rights of adjoining property owners.[23]

2.4.1 Public Nuisance

For a private person to recover damages for a public nuisance, the individual must show that the injury suffered was in some way different from that suffered by other members of the general public.[24] A private party generally has no right of action for a public nuisance, because "[i]t is the province of the public authorities to procure redress for public wrongs."[25] Still, an aggrieved landowner may bring a successful private action to abate or enjoin a public nuisance if the aggrieved party demonstrates special and peculiar injury apart from the injury suffered by

---

[19] See, for example, Stevinson v. Deffenbaugh Indus., 870 S.W.2d 851 (Mo. App. 1993).

[20] See, for example, Pate v. City of Martin, 614 S.W.2d 46 (Tenn. 1981).

[21] See, for example, Blair v. Anderson, 570 N.E.2d 1337 (Ind. App. 1991).

[22] See, for example, O'Neal v. Department of the Army, 852 F. Supp. 327 (M.D. Pa. 1994); Mel Foster Co. Properties v. American Oil Co., 427 N.W.2d 171 (Iowa 1988).

[23] See, for example, Stevinson v. Deffenbaugh Indus., 870 S.W.2d 851 (Mo. App. 1993).

[24] Compare Leo v. General Electric Co., 538 N.Y.S.2d 844 (App. Div. 1989) (commercial fishermen successfully sued defendant for a public nuisance caused by PCB pollution of the Hudson River, which tortious act deprived them of their livelihood) with Burgess v. M/V Tamano, 380 F. Supp. 247 (Me. 1972), aff'd, 559 F.2d 1200 (1st Cir. 1977) (holding that commercial fishermen could recover for their special losses caused by an oil spill, but local businesspersons could not recover since their losses were remote and of the kind common to all businesses in the area).

[25] Blair v. Anderson, 570 N.E.2d 1337 (Ind. App. 1991).

the general public,[26] but the injury must be different in kind and not merely different in degree.[27]

2.4.2 Private Nuisance

A private nuisance is the unreasonable, unusual, or unnatural use of one's property so that it substantially impairs the right of another to peacefully enjoy his property.[28] It is anything that annoys or disturbs the free use of one's property, or which renders its ordinary use or physical occupation uncomfortable.[29] Section 822 of the Restatement (Second) of Torts defines the elements of a private nuisance as follows:

> One is subject to liability for a private nuisance if, but only if, his conduct is a legal cause of an invasion of another's interest in the private use and enjoyment of land, and the invasion is either (a) intentional and unreasonable, or (b) unintentional and otherwise actionable under the rules controlling liability for negligent or reckless conduct, or for abnormally dangerous conditions or activities.[30]

In a Florida case,[31] for example, the court concluded that an oil company's failure to prevent noise, vibrations, and emissions from its plant from unreasonably interfering with the peaceful enjoyment, use, and occupation of the neighboring landowners' property constituted a nuisance. The court specifically found that the vibrations and noise through the air and ground from the plant caused severe structural damage which rendered the neighboring homes unsafe for human habitation. The court awarded the plaintiff-homeowners damages totaling $304,750 for the full market value of their house, which would exist in the absence of the nuisance, finding that it would be unreasonable to require them to repair the house in light of the fact that the structure could not be warranted against future deterioration.

It is important to note that a claim of private nuisance generally only

---

[26] See, for example, Town of Rome City v. King, 450 N.E.2d 72 (Ind. App. 1983). See also Restatement (Second) of Torts § 821C.

[27] See, for example, Blair v. Anderson, 570 N.E.2d 1337 (Ind. App. 1991) (water flow blockage to creek on plaintiff's property caused by public landfill constituted sufficient special injury to give standing to bring a private action to abate and enjoin the nuisance).

[28] See, for example, Frank v. Environmental Sanitation Management, Inc., 687 S.W.2d 876, 880 (Mo. en banc 1985); Mel Foster Co. Properties v. American Oil Co., 427 N.W.2d 171 (Iowa 1988); Patz v. Farmegg Products, Inc., 196 N.W.2d 557 (Iowa 1972).

[29] See, for example, Pate v. City of Martin, 614 S.W.2d 46 (Tenn. 1981).

[30] Restatement (Second) of Torts § 822.

[31] Exxon Corp., U.S.A. v. Dunn, 474 So.2d 1269 (Fla. App. 1985).

applies when a landowner's use and enjoyment of his or her property is unduly infringed by an unreasonable activity conducted on neighboring property. Private nuisance claims brought by a buyer against a seller for cleanup costs are generally dismissed on grounds that nuisance does not apply to pollution of the property by former owners.[32] Likewise, courts have dismissed claims of private nuisance by current owners or lessees of contaminated property against former lessees who polluted the site. For example, in one case,[33] a Maryland court declined to extend common-law liability to cover economic losses of a lessee of commercial property against a prior lessee whose underground gasoline storage tanks had contaminated the site. The court ruled that, under Maryland law, commercial tenants cannot maintain claims against prior tenants based on trespass, private nuisance, negligence, or strict liability for abnormally dangerous activities. The court held that these common-law tort actions were only available to recover damages sustained by occupants of neighboring property or others to whom the polluter owed a duty of care, not subsequent lessees of the contaminated site. Still, although most courts refuse to allow common-law nuisance actions against prior owners or lessees by subsequent occupiers of the land, a handful of courts have broadened the doctrine's application beyond the claims of neighboring landowners.[34]

2.4.3 Temporary vs. Permanent Nuisance

Whether a court characterizes a particular nuisance as "temporary" or "permanent" will be a key factor in determining the proper measure of damages for the injury inflicted to the plaintiff's property. Whether a particular nuisance is permanent or temporary is often a confusing area of the law. "The terms are, in reality, often only short-hand conclusions to determine the outcome of a particular case or the legal effects of certain defenses, such as the statute of limitations."[35] As a general rule, however, a nuisance is considered temporary if it may be abated and permanent if

---

[32] See, for example, Pinole Point Properties, Inc. v. Bethlehem Steel Corp., 596 F. Supp. 283, n. 283 (N.D. Cal. 1984) (allegation that "defendant disposed of waste on its own land and then sold the land ... would not ordinarily state a claim in nuisance.").

[33] Rosenblatt v. Exxon Co., USA, 642 A.2d 180 (Md. 1994).

[34] See, for example, Mangini v. Aerojet-General Corp., 230 Cal. App. 3d 1125, 281 Cal. Rptr. 827 (1991) (current landowner allowed to sue former lessee of property under theories of public and private nuisance), rev'd, 31 Cal. App. 4th 945, 31 Cal.Rptr.2d 696 (1994) (plaintiffs failed to present substantial evidence of continuing nuisance or continuing trespass). See also Wilson Auto Enterprises v. Mobil Oil Corp., 778 F. Supp. 101 (D.R.I. 1991) (purchaser of real estate has a cause of action in trespass against a former lessee of the seller for damage caused by the lessee's chemical contamination of the property).

[35] Spain v. City of Cape Girardeau, 484 S.W.2d 498 (Mo.App. 1972).

abatement is impracticable or impossible.[36]

The character of the source of injury often distinguishes temporary and permanent nuisances.[37] For example, in a Missouri case,[38] the plaintiff-landowners sued the defendant-owner of a landfill for damages arising from its maintenance of a permanent nuisance. The plaintiffs alleged that the owner of the landfill allowed contaminated water to run across their property. The owner argued that the jury should have been instructed on damages for a temporary rather than a permanent nuisance. However, because the evidence showed that the landfill's expensive and sophisticated leachate control plans had failed to stop the contamination, the court ruled that permanent damages were warranted.[39] The court also held that a stigma had attached to the land from the repeated leachate outbreaks and that the stigma permanently devalued the plaintiff's property, further justifying an award of permanent damages.

How the court chooses to characterize the nuisance (as either permanent or temporary) will also determine the applicable statute of limitations for the claim.[40] The period of limitations for a temporary nuisance runs anew from the accrual of injury from every successive invasion of interest. On the other hand, for a permanent nuisance, the period of limitations runs immediately upon creation of the permanent nuisance and bars all claims of damage, present and future, after lapse of the statutory period.[41]

If a nuisance is deemed temporary, damages to property affected by the nuisance are recurrent and may be recovered from time to time until the nuisance is abated.[42] Thus, when a nuisance is "temporary, continuing or abatable," an injured party can bring a subsequent action for injuries sustained by the continuation of a temporary nuisance.[43] The recovery is for the damage actually sustained to the commencement of suit, but not for

---

[36] See, for example, Stevinson v. Deffenbaugh Indus., 870 S.W.2d 851 (Mo.App. 1993).

[37] See, for example, Mel Foster Co. Properties v. American Oil Co., 427 N.W.2d 171 (Iowa 1988).

[38] Frank v. Environmental Sanitation Management, Inc., 687 S.W.2d 876 (Mo. 1985) (en banc).

[39] Id., 687 S.W.2d at 883.

[40] See, for example, Kohler v. Germain Investment Co., 934 P.2d 867 (Colo. App. 1996).

[41] See, for example, Rebel v. Big Tarkio Drainage Dist., 602 S.W.2d 787, 792 (Mo.App. 1980).

[42] See, for example, Stevinson v. Deffenbaugh Indus., 870 S.W.2d 851 (Mo. App. 1993) (permanent damages, including damages for reduction in fair market value of landowners' properties, were not available in temporary nuisance action).

[43] Spain v. City of Cape Girardeau, 484 S.W.2d 498 (Mo. App. 1972).

prospective injury.[44] The theory is that a temporary nuisance may be abated at any time by a reasonable effort or by an order of the court, but if not, then the injured party can bring a successive action for the continuance of damage.[45]

In one case,[46] the Massachusetts Supreme Judicial Court dismissed the continuing nuisance and trespass claims of plaintiff-landowners whose property had been damaged by a leaking underground gasoline storage tank on an adjacent parcel. The court concluded that the landowners' claims, based on the continued presence of gasoline on their property, were time-barred because the source of the offending condition had been abated outside of the applicable statute of limitations period. The court held that actions for continuing trespass or nuisance must be based on recurring tortious or unlawful conduct, not a continuation of harm caused by tortious conduct that the defendant has stopped.

The measure of damages from a temporary nuisance is the decrease in rental or use value of the land during continuance of injury, as well as any special costs.[47] In contrast, damages for a permanent nuisance are measured by the difference in the land's market value immediately before and after injury. In one case,[48] for example, an Indiana appellate court upheld the summary judgment dismissal of a plaintiff-landowner's permanent damage claims based on polychlorinated biphenyl (PCB) contamination of his property. Although the court concluded that PCB contamination should generally be treated as a temporary injury capable of remediation or repair, for which the proper measure of damages would be the cost of restoration, the court nevertheless ruled that in the context of environmental contamination a plaintiff may also recover any proven reduction in the fair market value of real property remaining after remediation. The defendant had completed remediation of the site at a cost of $25 million. However, because the plaintiff failed to produce evidence establishing any "remaining loss" or reduction in fair market value of the property after remediation, the court concluded that the landowner had already been compensated for its proven loss.

---

[44] See, for example, Sundell v. Town of New London, 119 N.H. 839, 409 A.2d 1315 (1979).

[45] See, for example, Rebel v. Big Tarkio Drainage Dist., 602 S.W.2d 787 (Mo. App. 1980).

[46] Carpenter v. Texaco, Inc., 419 Mass. 581, 646 N.E.2d 398 (1995).

[47] See Prosser, *Law of Torts* § 90, at 602 (1971).

[48] Terra-Products, Inc. v. Kraft General Foods, Inc., 653 N.E.2d 89 (Ind. App. 1995).

2.4.4 Environmental Conditions on Neighboring Land

Some environmental problems originating on neighboring land may physically invade and cause harm to a landowner's property, such as groundwater contamination.[49] In contrast, an off-site environmental condition may merely stigmatize and devalue the property of a nearby landowner without actually causing any physical damage. Some property owners are bringing a relatively new brand of environmental lawsuit for diminished property values caused by the close proximity of off-site environmental hazards to their property, such as nearby public landfills,[50] hazardous waste disposal sites,[51] oil spills,[52] polluted industrial sites,[53] overhead power lines,[54] and other environmental conditions on neighboring land.[55] Property owners have used various common-law and statutory grounds to form the basis of a claim for property devaluation caused by off-site environmental conditions. Nuisance has been the most prevalent cause of action for recovering damages for diminished market values although other common-law and statutory-based claims have formed the basis for property devaluation damages.

In some cases, plaintiffs have supported their claims for damages for diminished market value with evidence of market stigma resulting from the location of their property in close proximity to off-site contamination. Although some courts may allow recovery of property devaluation damages based on allegations of market stigma,[56] expert witness testimony is required to establish that a property is stigmatized in the marketplace. The expert witness must show that the stigma exists in the minds of potential buyers and not mere personal belief. For example, some courts

---

[49] See, for example, Cereghino v. Boeing Co., 873 F. Supp. 398 (D.Or. 1994) (groundwater contamination from adjacent industrial facilities); Ayers v. Township of Jackson, 202 N.J. Super. 106, 493 A.2d 1314 (App. Div. 1985) (contamination of groundwater from negligent operation of landfill).
[50] See, for example, City of Atlanta v. Murphy, 194 Ga. App. 652, 391 S.E.2d 474 (1990); Strawn v. Canuso, 271 N.J. Super. 88, 638 A.2d 141 (App. Div. 1994).
[51] See, for example, Frank v. Environmental Sanitation Management, Inc., 687 S.W.2d 876 (Mo. 1985) (en banc).
[52] See, for example, Adams v. Star Enterprise, 851 F. Supp. 770 (E.D. Va. 1994).
[53] See, for example, In re Paoli R.R. Yard Litigation, 35 F.3d 717 (3d Cir. 1994); Terra-Products, Inc. v. Kraft General Foods, Inc., 653 N.E.2d 89 (Ind.App. 1995).
[54] See, for example, Criscuola v. Power Auth. of New York, 81 N.Y.2d 649, 602 N.Y.S.2d 588 (1993).
[55] See, for example, Crawford v. National Lead Co., 784 F. Supp. 439 (S.D. Ohio 1989) (neighboring uranium metals production plant).
[56] See, for example, F.D.I.C. v. Jackson-Shaw Partners No. 46, Ltd., 850 F. Supp. 839 (N.D. Cal. 1994); Frank v. Environmental Sanitation Management, Inc., 687 S.W.2d 876 (Mo. en banc 1985).

allow testimony about fear in the marketplace affecting the value of property located near electrical transmission lines, as long as the testimony is not based on a personal fear.[57]

Landowners have, however, not faired well with property devaluation claims based on the market stigma caused by off-site conditions.[58] In one case,[59] for example, a group of homeowners brought a nuisance action against the owners of a nearby tank farm based on petroleum contamination that had leached into the groundwater, forming an underground plume that extended under a portion of the homeowners' residential development. The homeowners sought property devaluation damages for an alleged market stigma in the minds of the buying public. The federal district court ruled that an alleged diminution in property values based solely on "fear of future contamination" from a nearby oil spill was not a compensable injury.

In dismissing the plaintiffs' claim for damages, the court reviewed Virginia's state nuisance law to conclude that "nuisance ... does not extend to diminution of property values due to the buying public's attitude towards that property as a result of the nuisance."[60] Although the court found that a nuisance did exist because the oil spill clearly was not a reasonable use of the defendants' property, the plaintiffs had no compensable injury because the only present risk posed by the contamination plume was "mere mental distress." The court agreed with the defendants that "negative publicity or stigma resulting from unfounded fear about the dangers in the vicinity of the property 'does not constitute a significant interference with the use and enjoyment of land.'"[61]

Likewise, a California appellate court dismissed a plaintiff-landowner's "continuing" nuisance action for "stigma" damages based on property devaluation associated with petroleum contamination from leaking underground storage tanks located on adjacent land.[62] The defendant, owner of a gasoline service station, undertook cleanup action to abate the contamination and the plaintiff did not incur any expenses or suffer any loss of use of his property during the cleanup. Since the statute of limitations barred assertion of a permanent nuisance claim, the plaintiff brought suit on a continuing nuisance theory, alleging its entitlement to

---

[57] See Criscuola v. Power Auth. of New York, 81 N.Y.2d 649, 602 N.Y.S.2d 588 (1993); Ryan v. Kansas Power & Light Co., 249 Kan. 1, 815 P.2d 528 (1991).
[58] See, for example, Adams v. Star Enterprise, 851 F. Supp. 770 (E.D. Va.), aff'd, 51 F.3d 417 (4th Cir. 1994); In re Paoli Railroad Yard PCB Litigation, 811 F. Supp. 1071 (E.D. Pa. 1992).
[59] Adams v. Star Enterprise, 851 F. Supp. 770 (E.D. Va.), aff'd, 51 F.3d 417 (4th Cir. 1994).
[60] Id., 851 F. Supp. at 773 n.4.
[61] Id. (citations omitted).
[62] Santa Fe Partnership v. ARCO Products Co., 54 Cal.Rptr.2d 214 (Ct. App. 1996).

property devaluation damages premised on the market stigma associated with historically contaminated property. The court ruled that property devaluation damages are only recoverable in "permanent" nuisance cases where, as a practical matter, the offending activity is considered unabatable. Although acknowledging that some authority from other jurisdictions existed to support an award of stigma damages in environmental contamination cases where property suffers permanent physical injury despite remediation efforts, the California appellate court was bound to follow the decisions of the state's highest court which did not allow recovery of damages for diminution in property value in cases where the nuisance was continuing and abatable.

Similarly, in a New York case,[63] the court refused to recognize a landowner's claim of permanent damages to his property based on the market stigma associated with an oil spill. The court concluded that since the petroleum contamination had been fully remediated, the landowner was only entitled to temporary damages measured by the cost of restoration and loss of rental income from the time the property was damaged until the time it was restored. The court found that the evidence failed to support the landowner's contention that the oil spill had permanently stigmatized the property and had thereby rendered it unmarketable and of no value.

On the other hand, in a Missouri case,[64] the court held that in an action for permanent nuisance, evidence of devaluation of property as a result of market stigma is substantial evidence for an award of permanent damages. The plaintiffs alleged that the landfill owner had allowed contaminated water to run across their property. Evidence showed that expensive and sophisticated leachate control plans had failed to stop the contamination. The court ruled that a stigma had attached to the land from repeated leachate outbreaks and that this stigma permanently devalued the plaintiff's property.

2.4.5 Noise as a Nuisance

A number of common-law theories may be used to establish claims for compensatory damages or injunctive relief from excessive noise. Common-law claims may be based on nuisance, trespass, and inverse condemnation theories, but nuisance is the most common and successful claim for relief from noise on neighboring land. The noise emitted from a number of different types of activities have been the subject of nuisance actions by neighboring landowners. Common-law nuisance claims have been recognized for noise produced by activities and machineries as varied

---

[63] Putnam v. State of New York, 636 N.Y.S.2d 473 (App. Div. 1996).
[64] Frank v. Environmental Sanitation Management, Inc., 687 S.W.2d 876 (Mo. 1985) (en banc).

as air conditioners,[65] barking dogs,[66] bars and restaurants,[67] exhaust fans,[68] wind generators,[69] gun and shooting clubs,[70] manufacturing plants,[71] racetracks,[72] and trucks.[73]

Measurement of Noise Levels. Loud, harsh, nonharmonious sounds or vibrations that are unpleasant and uncomfortable to the ear comprise noise. There is no objective standard to measure the annoyance level of noise because individual sensitivity to noise varies. Noise is the most subtle of pollutants because it leaves no residue accumulations. Loss of sleep is one of the most disruptive effects of noise pollution, and other physiological and psychological effects have been recognized, including hearing loss, ulcers, chronic fear, and anxiety.[74]

In general, the louder the noise and the higher the pitch of its components, the greater will be the level of discomfort produced. Other

---

[65] See, for example, Massey v. Long, 608 S.W.2d 547 (Mo.App. 1980); Mandel v. Geloso, 614 N.Y.S.2d 645 (App.Div. 1994).

[66] See, for example, Brewton v. Young, 596 So.2d 577 (Ala. 1991); Parker v. Reaves, 505 So.2d 323 (Ala. 1987); Higgs v. Anderson, 685 S.W.2d 521 (Ark.App. 1985); State v. Olson, 511 A.2d 379 (Conn.App. 1986).

[67] See, for example, People v. Mason, 177 Cal.Rptr. 284 (App. 1981); King v. Western Club, Inc., 587 So.2d 122 (La.App. 1991).

[68] See, for example, Daugherty v. Ashton Feed and Grain Co., Inc., 303 N.W.2d 64 (Neb. 1981).

[69] See, for example, Rose v. Chaikin, 187 N.J.Super. 210, 453 A.2d 1378 (Ch.Div. 1982); Rassier v. Houim, 488 N.W.2d 635 (N.D. 1992).

[70] See, for example, Shepard v. Pollution Control Board, 651 N.E.2d 555 (Ill.App. 1995); Hinsdale Golf Club v. Kochanski, 555 N.E.2d 31 (Ill. App. 1990); Anne Arundel County Fish & Game Conservation Ass'n, Inc. v. Carlucci, 573 A.2d 847 (Md.App. 1990); Racine v. Glendale Shooting Club, Inc., 755 S.W.2d 369 (Mo. App. 1988); Christensen v. Hilltop Sportsman Club, Inc., 573 N.E.2d 1183 (Ohio App. 1990); Gray v. Barnhart, 601 A.2d 924 (Pa.Commonw. 1992).

[71] See, for example, Exxon Corp., U.S.A. v. Dunn, 474 So.2d 1269 (Fla. App. 1985); Roberts v. Southern Wood Piedmont Co., 328 S.E.2d 391 (Ga.App. 1985).

[72] See, for example, Patterson v. Robinson, 620 So.2d 609 (Ala. 1993); Renz v. 33rd District Agricultural Ass'n, 46 Cal.Rptr.2d 67 (App. 1995); McCombs v. Joplin Slavant v. Calhoun Motor Speedway, 626 So.2d 771 (La. App. 1993); 66 Fairgrounds, Inc., 925 S.W.2d 946 (Mo. App. 1996); Hoover v. Durkee, 622 N.Y.S.2d 348 (App. Div. 1995).

[73] See, for example, Garner v. Walker, 577 So.2d 1276 (Ala. 1991); Village of Caseyville v. Cunningham, 484 N.E.2d 499 (Ill. App. 1985); Sherk v. Indiana Waste Systems, Inc., 495 N.E.2d 815 (Ind. App. 1986); Escobar v. Continental Baking Co., 596 N.E.2d 394 (Mass. App. 1992).

[74] See Hildebrand, Noise Pollution: An Introduction to the Problem and an Outline for Future Research, 70 Colum. L. Rev. 652 (1970).

factors include the characteristics of the sound and modulation of loudness and pitch. Higher-frequency sounds are more disturbing and harmful than those of lower frequencies. Sound is measured in decibels (dB) and also in A-weighted decibels (dBA). Decibel measurements are logarithmic. The intensity of sound multiplies by 10 with every 10 decibel increase. Therefore, 20 dBA is 10 times more intense than 10 dBA, 30 dBA is 100 times more intense than 10 dBA, and 40 dBA is 1,000 times more intense than 10 dBA. The following sound levels are emitted from certain familiar machines: a washing machine (above 70 dBA), a food blender (80 dBA), a power lawn mower (95 dBA), and a motorcycle (110 dBA under maximum acceleration). The sound levels experienced by drivers of various automobiles cruising at 70 miles per hour has been calculated as 67 dBA for a Lexus SC300, 72 dBA for a Chevrolet Impala SS, and 71 dBA for a Chrysler LHS.[75]

Nuisance Factors. Noise may constitute a nuisance and be enjoined under certain circumstances.[76] Generally, noise is not a nuisance per se, but it may be of such a character as to constitute a nuisance in fact, which may serve as the basis of an action at law or in equity, even though it arises from the operation of a factory, industrial plant, or other lawful business or occupation.[77] The mere fact that a business is operated in accordance with various rules and regulations does not require a finding that the use is reasonable. A determination of reasonableness of use in an action for nuisance depends upon the effect of the activity upon one's neighbors in the particular circumstances and locality, not merely upon whether one operates within the confines of particular authority. A lawful business may be of such a nature, so situated, or so conducted as to constitute or become a nuisance.[78]

Noise is an actionable private nuisance if two elements are present: (1) injury to the health and comfort of ordinary people in the vicinity, and (2) unreasonableness of that injury under all the circumstances.[79] The plaintiff

---

[75] See, for example, Anne Arundel County Fish & Game Conservation Ass'n, Inc. v. Carlucci, 573 A.2d 847 (Md. App. 1990) (expert witness testimony on sound measurements); McCombs v. Joplin 66 Fairgrounds, Inc., 925 S.W.2d 946 (Mo. App. 1996) (expert witness testimony on various sound levels).

[76] See, for example, Friendship Farms Camps v. Parsons, 172 Ind.App. 73, 359 N.E.2d 280 (1977); Rose v. Chaikin, 187 N.J.Super. 210, 453 A.2d 1378, 36 A.L.R.4th 1148 (Ch. Div. 1982).

[77] See, for example, Racine v. Glendale Shooting Club, Inc., 755 S.W.2d 369 (Mo.App. 1988); Lee v. Rolla Speedway, Inc., 668 S.W.2d 200 (Mo. App. 1984).

[78] See, for example, Yeager & Sullivan, Inc. v. O'Neill, 324 N.E.2d 846 (Ind. App. 1975).

[79] See, for example, Malhame v. Demarest, 162 N.J.Super. 248, 392 A.2d 652 (Law Div. 1978).

carries the burden of proving that the defendant's use of its property created a nuisance.[80]

The character, volume, frequency, duration, time, and locality are all relevant factors in determining whether the complained of annoyance materially interferes with the ordinary comfort of human existence. In determining whether a nuisance exists, the character of the locality is a circumstance of great importance. Courts have stated that people who reside in a neighborhood with businesses close by must compromise their comfort to some degree to accommodate the commercial necessities of business.[81] Further, where the plaintiff comes to the area with knowledge of the existence of the offending activity, the courts consider this a relevant factor when balancing the equities of the case. For example, where the plaintiff purchases property near the alleged nuisance and the purchase price reflects the property's location near the offending activity, the courts are less sympathetic to the plaintiff's nuisance claim.[82]

Whether noise is sufficient to constitute a nuisance depends upon its effect upon an ordinary, reasonable person, that is, a normal person of ordinary habits and sensibilities.[83] Relief cannot be based solely upon the subjective likes and dislikes of a particular plaintiff; it must be based upon an objective standard of reasonableness.[84]

State Noise Nuisance Statutes. Traditionally, most state and local governments have regulated noise as a nuisance[85] under general nuisance statutes[86] and local ordinances. Some state statutes exempt certain noise-producing activities from the purview of state and local noise control regulations. For example, Pennsylvania's noise pollution statute grants immunity to owners of pistol, silhouette, skeet, trap, blackpowder, and other shooting ranges from civil and criminal prosecution for noise

---

[80] See, for example, Burgess v. Omahawks Radio Control Organization, 363 N.W.2d 27 (Neb. 1985).

[81] See, for example, Karpiak v. Russo, 676 A.2d 270 (Pa.Super. 1996).

[82] See, for example, Escobar v. Continental Baking Co., 596 N.E.2d 394 (Mass. App. 1992).

[83] See, for example, Kaiser v. Western R/C Flyers, Inc., 477 N.W.2d 557 (Neb. 1991).

[84] See, for example, Daugherty v. Ashton Feed and Grain Co., Inc., 303 N.W.2d 64 (Neb. 1981).

[85] See, for example, Keane v. Patcher, 598 N.E.2d 1067 (Ind. App. 1992) (noise produced from marble floors installed in condominum unit located above plaintiff's unit constituted a nuisance under Indiana Code § 34-1-52-1).

[86] For example, the Indiana nuisance statute states: "Whatever is injurious to health, or indecent, or offensive to the senses, or an obstruction to the free use of property, so as essentially to interfere with the comfortable enjoyment of life or property, is a nuisance, and the subject of anaction." Ind. Code § 34-1-52-1. See also Iowa Code § 657.1, which contains a virtually identical definition of nuisance.

pollution[87] and private actions for nuisance[88] provided that a local noise control ordinance was in effect at the time the shooting range was constructed and the range was in compliance with the provisions of that ordinance.[89] However, unless the statute expressly provides otherwise, a landowner is usually not precluded from bringing a common-law nuisance action for abatement of noise.[90]

Some states have enacted comprehensive noise control acts which regulate noise from a variety of sources under the authority of a specially created or designated state agency.[91] These statutes are separate and distinct from statutes designed to control transportation noise, particularly the noise generated by motor vehicles.[92]

Local Noise Control Ordinances. Many municipalities have sought to control noise through local ordinances. Pursuant to a police power statute, a local ordinance may legitimately protect public health and welfare by proscribing excessive noise.[93] To be a valid exercise of the police power a municipal ordinance must bear a reasonable relationship to the public interest sought to be protected, and the means adopted must be a reasonable method of accomplishing the chosen objective.[94] "Due process principles operate to limit the police power only to the extent that the power is arbitrarily or unreasonably used."[95] Further, local noise control

---

[87] 35 Pa.Stat. § 4501.
[88] 35 Pa.Stat. § 4502.
[89] See, for example, Gray v. Barnhart, 144 Pa.Commonw. 474, 601 A.2d 924 (1992) (shooting range owner was not entitled to statutory exemption since no ordinance was in effect at the time the range was constructed).
[90] See, for example, Anne Arundel County Fish & Game Conservation Ass'n, Inc. v. Carlucci, 83 Md.App. 121, 573 A.2d 847 (1990) (Maryland Environmental Code did not preclude action for common law nuisance against gun club); Rassier v. Houim, 488 N.W.2d 635 (N.D. 1992) (common law remains relevant where there is no conflict between the common law and state statute); Christensen v. Hilltop Sportsman Club, Inc., 61 Ohio App.3d 807, 573 N.E.2d 1183 (1990) (state nuisance statute did not supersede common law nuisance actions).
[91] See, for example, Cal. Health & Safety Code §§ 46000-46080; Conn. Gen. Stat. §§ 22a-67 to -76; Del. Code Ann. §§ 7101-7124; 415 ILCS § 5/24; Ky. Rev. Stat. Ann. §§ 224.710-.800; N.J. Stat. Ann. § 13:1G-1 to -23.
[92] See, for example, Cal. Veh. Code § 23130; Colo. Rev. Stat. § 42-4-222; Minn. Stat. Ann. § 163.693.
[93] See, for example, Rose v. Chaikin, 187 N.J.Super. 210, 453 A.2d 1378, 36 A.L.R.4th 1148 (Ch.Div. 1982) ("Limiting noise from windmills indisputably advances that legitimate purpose and does so in a reasonable way.").
[94] See, for example, City of Carbondale v. Brewster, 78 Ill.2d 111, 115, 398 N.E.2d 829, 831 (1979).
[95] Crocker v. Finley, 99 Ill.2d 444, 456, 459 N.E.2d 1346, 1352 (1984).

ordinances must provide adequate notice to persons of common understanding concerning the level, degree, and type of noise prohibited; otherwise it may not withstand constitutional challenges based on overbreadth or vagueness.[96]

"A municipality has the power to declare anything a nuisance, which is either a nuisance per se, a nuisance at common law or statute. A municipality also has the authority to regulate as a nuisance anything in which there could be an honest difference of opinion if, in the municipality's opinion, such item constitutes a nuisance."[97] In doubtful cases, where it is questionable whether something is a nuisance, the municipality's decision will be conclusive unless its judgment and use of discretion is clearly erroneous.[98]

2.5 Negligence

Negligence occurs when someone fails to use such care as a reasonably prudent person would use to avoid causing a foreseeable harm to someone else. This common-law doctrine can be used in the context of environmental liability when someone negligently releases a hazardous substance into the environment that causes personal injury or property damage to others. Negligence delineates conduct falling below the standard of a reasonably prudent person under similar circumstances. As stated in Section 282 of the Restatement (Second) of Torts, negligence encompasses all conduct "which falls below the standard established by law for the protection of others against unreasonable risk of harm."

2.5.1 Elements of Negligence

To sustain a claim based on negligence, the plaintiff must prove each of the following elements:

1. A *duty* of care owed to the plaintiff by the defendant

2. Negligent conduct by the defendant resulted in a *breach* of that duty

---

[96] See, for example, Easy Way of Lee County v. Lee County, 674 So.2d 863 (Fla. App. 1996) (night club owner successfully challenged constitutionality of local noise control ordinance on grounds of overbreath and vagueness).
[97] Village of Riverwoods v. Untermyer, 54 Ill. App.3d 816, 822, 369 N.E.2d 1385, 1390 (1977).
[98] Id.

3. Legal *causation* of the plaintiff's harm by the defendant's conduct

4. Actual *harm* to plaintiff[99]

Proving each of these elements of negligence may be difficult.[100] For example, an Iowa court dismissed a plaintiff-buyer's common-law negligence claim against the defendant-seller of a gasoline service station property later found to have petroleum contamination.[101] The court concluded that since the evidence failed to support a finding that the defendant's conduct had, in fact, produced the contamination, the plaintiff could not prove the necessary causation element of his case. The expert witnesses for both the plaintiff and the defendant agreed that the contaminant on the property was gasoline and that the source of the gasoline was underground storage tanks and associated equipment; however, none of the experts could determine the precise location of the leak. The experts testified that there could be several potential causes, including system leaks, customer overfills, and accidents at the pump island. Most importantly, none of the experts, including the plaintiff's expert witness, could pinpoint the time when the contamination occurred. The court stated that without evidence to establish when the contamination occurred, testimony identifying possible sources of contamination would still leave the jury to speculate as to who was responsible for those sources at the time of the contamination. Thus, the court dismissed the negligence claim, holding that the plaintiff had failed to offer sufficient proof that the defendant had caused the petroleum discharge during his former ownership and operation of the gas station.

The duty of care component of negligence is dependent on proof that it was reasonably foreseeable that a particular act or omission by the defendant would cause injury to the plaintiff.[102] No duty of care can be imposed on the defendant to take precautions against events that could not be reasonably foreseen. In determining whether the plaintiff's injury from a hazardous substance was reasonably foreseeable by the defendant, courts will generally employ a standard of reasonable care. Under this standard, the courts will consider what a reasonably prudent person should have done or the applicable standard of care for a particular industry.

---

[99] See, for example, O'Neil v. Department of the Army, 852 F. Supp. 327 (M.D. Pa. 1994); Jacques v. First National Bank, 515 A.2d 756 (Md. 1986).
[100] See, for example, Exxon Corp. v. Amoco Oil Co., 875 F.2d 1085 (4th Cir. 1989) (jury found that plaintiff had not been damaged by defendant's leaking gasoline storage tanks); Malone v. Ware Oil Co., 534 N.E.2d 1003 (Ill. App. 1989) (evidence was insufficient to support the jury finding that any invasion of property from a leaking underground gasoline storage tank was intentional, negligent, or unreasonable).
[101] Gerst v. Marshall, 1996 WL 333149 (Iowa, June 19, 1996).
[102] See, for example, Elam v. Alocolac, Inc., 765 S.W.2d 42 (Mo. App. 1988).

2.5.2 Negligence per se

In the absence of an applicable industry standard, Section 286 of the Restatement (Second) of Torts provides that the reasonable person standard may be defined by legislative enactment if the party injured by the violation is a member of the class of persons the statute or regulation is designed to protect. Thus, a plaintiff may establish *negligence per se* by the fact that a defendant violated a federal or state statute or regulation.[103] Still, although negligence per se allows a shortcut to establishing a duty of care (and a breach of that duty), the plaintiff must still establish causation and damages.

Obviously, assertion of a claim premised on the theory of negligence per se depends on the language and requirements of the applicable statute or regulation. Thus, in property damage cases based on injury from environmental contamination originating on neighboring land, the evidence necessary to sustain a claim based on negligence per se will require a showing that the defendant violated the requirements of a statute or regulation that was designed to protect the plaintiff's property from environmental harm. Plaintiffs have found it difficult to make this showing in cases involving environmental injury to property because the language of most environmental statutes is not supportive of their claims of negligence per se.

For example, in one case,[104] the plaintiff-landowners sought to recover damages for personal injury and property damage resulting from allegedly negligent contamination of their drinking water by toxic chemicals that had migrated from an Army aircraft maintenance facility. The federal district court stated that:

> Negligence per se may be demonstrated by proof that a defendant has violated a law or regulation whose purpose is found to be, at least in part (a) to protect a class of persons which includes the one whose interest is invaded, (b) to protect the particular interest which is invaded, (c) to protect that interest against the kind of harm that has resulted, and (d) to protect that interest against the particular hazard from which the harm results.[105]

The plaintiffs pointed to several environmental statutes that they alleged were violated by the Army's conduct. However, because the regulations at issue were designed to guard against hazards unrelated to well

---

[103] See, for example, Sanford Street Local Development Corp. v. Textron, Inc., 768 F. Supp. 1218 (W.D. Mich. 1991) (defendant's violation of state toxic substance control act constituted negligence per se).

[104] O'Neal v. Department of the Army, 852 F. Supp. 327 (M.D. Pa. 1994).

[105] Id. 852 F. Supp. at 335, citing Centolanza v. Lehigh Valley Dairies, 430 Pa.Super. 463, 635 A.2d 143 (1993).

contamination, the court concluded the evidence did not support a claim for negligence per se. The environmental laws in question only referred to protection of aquatic life.

2.5.3 Toxic Torts

An environmental toxic tort is a claim for damages arising from exposure to a harmful chemical or substance.[106] The typical toxic tort arises when a defendant's negligence causes a plaintiff to sustain personal injuries as a result of exposure to a toxic substance. Common-law claims of nuisance, trespass, and strict liability for abnormally dangerous activities are usually asserted for recovery of damage to property. Personal injury damages may be sought under these theories, in addition to property damage claims; however, recovery of personal injury-related damages are generally based on negligence.[107]

Increasingly, environmental torts are now related to injuries and potential injuries from a wide variety of substances, including polychlorinated biphenyls (PCBs), pesticides, benzene, heavy metals, and other contaminants in the air, soil, or water. When bringing a toxic tort action, the plaintiff must prove that the substance to which she was exposed is toxic. She must prove that there was sufficient exposure to cause harm, such as ingestion of a sufficient dosage and concentration of the substance to bring about a harmful effect. The plaintiff must also prove that exposure to the substance was caused by the defendant's wrongful conduct and that the exposure was not from a substance that occurs naturally in the environment.

Claims for Emotional Distress. Emotional distress claims resulting from exposure to toxic substances have increased dramatically in recent years, not only in number but in the size of verdicts.[108] Courts are increasingly faced with claims for emotional distress based on mental anguish, cancerphobia, fear of future disease, and other related psychological

---

[106] Toxic torts will be discussed in the context of toxic substances in section 11.7 of this book.

[107] See, for example, McGregor v. Barton Sand & Gravel, 62 Or.App. 24, 660 P.2d 175 (1983) ("No reason occurs to us why the same kind of injury resulting from the same kind of act should not be compensable solely because the actor's conduct is deliberate rather than negligent.").

[108] See Krochock & Solheim, Psychological Damages from Toxic Substances: Problems and Solutions, Def. Couns. J 80-87 (Jan. 1993).

injury.[109] Infliction of emotional distress may be intentional or negligent.[110]

The legal standard for recovery of emotional distress damages generally follows Section 456 of the Restatement (Second) of Torts, which states: "If the actor's negligent conduct has so caused any bodily harm to another as to make him liable for it, the actor is also subject to liability for (a) fright, shock, or other emotional disturbance resulting from the bodily harm or from the conduct which causes it, and (b) further bodily harm resulting from such emotional disturbance."[111]

Although psychological damage claims, without accompanying physical injury or harm, have traditionally been rejected,[112] some courts allow an award of damages for emotional distress associated with the fear of contracting a future disease even without a showing of present physical injury when the evidence shows a reasonable basis for the fear itself.[113]

Although a majority of jurisdictions still adhere to the traditional rule that some form of physical injury must be present before damages can be awarded for negligent infliction of emotional distress,[114] the modern trend has been to abolish the physical injury requirement as a guarantee of the genuineness of claims for mental distress. These jurisdictions place greater emphasis on general tort law principles and the ability of jurors and the medical profession to determine the genuineness of the plaintiff's emotional distress and fear of future disease.

Emotional distress claims based on the fear of contracting some future disease are difficult to sustain for several reasons. First, fear of future disease claims are generally premised on emotional distress that may or may not be accompanied by physical injury. Furthermore, unlike more traditional claims for damages stemming from another person's

---

[109] See, for example, Sterling v. Velsicol Chemical Corp., 855 F.2d 1188 (6th Cir. 1988).

[110] Intentional infliction of emotional distress is illustrated by the case of Ashland Oil Co. v. Miller Oil Purchasing Co., 678 F.2d 1293 (5th Cir. 1982), wherein an unscrupulous toxic waste hauler deliberately dumped toxic materials on or near the plaintiff's property in order to compel the plaintiff to move. See also Wisniewski v. Johns-Manville Corp., 759 F.2d 271 (3d Cir. 1985) (intentional infliction of emotional distress); O'Neal v. Department of the Army, 852 F. Supp. 327 (M.D. Pa. 1994) (rejecting landowners' claim of negligent infliction of emotional distress allegedly caused by Army's contamination of groundwater).

[111] Restatement (Second) of Torts § 456.

[112] See Section 436A of Restatement (Second) of Torts which states that: "If the actor's conduct is negligent as creating an unreasonable risk of causing either bodily harm or emotional disturbance to another, and it results in such emotional disturbance alone, without bodily harm or other compensable damage, the actor is not liable for such emotional disturbance."

[113] See, for example, Potter v. Firestone Tire and Rubber Co., 863 P.2d 795 (Cal. 1993).

[114] See, for example, Ferrara v. Gallucchia, 5 N.Y.2d 16, 176 N.Y.S.2d 996, 152 N.E.2d 249 (1958).

negligence, fear of future disease claims are based on the fear of a future event that may never occur. When no visible physical injury is present, the genuineness of the plaintiff's emotional distress can be difficult to substantiate.[115]

Damages. In toxic tort lawsuits, an injured plaintiff is entitled to recover for all the natural and proximate consequences of a defendant's tortious act or omission. Plaintiffs may seek damages for both present and prospective injuries. The elements of compensatory damages are not always easy to identify, although property damages, medical expenses, lost income, and lost earnings or profits are relatively easy to document. Intangible losses, such as pain and suffering, emotional distress, and mental anguish also fall into the category of compensatory damages.

Courts have increasingly allowed recovery of the costs of medical monitoring to detect the onset of disease related to exposure to hazardous and toxic substances, even without evidence of any present physical injury.[116] The purpose of medical monitoring compensation is to enable the plaintiff to obtain information about future diseases as early as possible. That information, in turn, enables the plaintiff to seek early treatment so that the injuries will be minimized.[117]

Punitive damages may be awarded in toxic tort cases when the evidence shows wantonly reckless, intentional, or egregious misconduct by the defendant. Punitive damages may be assessed in addition to compensatory damages in order to punish a defendant who commits an aggravated act of misconduct against a plaintiff and in order to deter the defendant and others from similar misbehavior. However, punitive damages are generally disfavored by the courts and awarded only upon proof that the defendant's tortious conduct contains elements of intentional wrongdoing or grossly negligent disregard for the plaintiff's welfare.[118]

---

[115] See Comment, Emotional Distress Damages in Toxic Tort Litigation: The Move Towards Foreseeability, 3 Vill. Envtl. L.J. 113 (1992).
[116] See, for example, Miranda v. Shell Oil Co., 26 Cal.Rptr.2d 655 (Ct. App. 1993).
[117] See, for example, Paoli R.R. Yard PCB Litig., 916 F.2d 829, 852 (3d Cir. 1990); Ayers v Jackson Township, 525 A.2d 287, 311-312 (N.J. 1987).
[118] See Pacific Mutual Life Insurance Co. v. Haslip, 111 S.Ct. 1032 (1992) (rejecting a constitutional attack on punitive damages, leaving juries with broad discretion over damage awards).

3

Environmental Impact Review

3.1 The National Environmental Policy Act of 1969

The National Environmental Policy Act of 1969 (NEPA)[1] was the United States' first significant acknowledgment of the relationship between human welfare and the environment. In fact, NEPA has been described as "the most famous statute of its kind on the planet."[2] While this statement may seem to be an exercise in legal hyperbole, there can be little question that NEPA introduced a new emphasis on environmental quality in virtually every action taken by the federal government.

At the heart of NEPA is the mandate that all "major Federal actions significantly affecting the quality of the human environment" require the preparation of an Environmental Impact Statement (EIS).[3] But perhaps most important was the recognition by Congress that legislation was necessary to encourage productive and enjoyable harmony between humans and their environment.

3.1.1 History and Origins of NEPA

NEPA resulted from a series of major policy changes in the 1960s. The first "environmental decade" was fueled by post-World War II affluence, and focused on increased public apprehension over scientific evidence of

---

[1] 42 U.S.C. § 4331 et seq.
[2] Rodgers, *Environmental Law*, 2d ed. (1994) at § 9.1.
[3] NEPA § 102(2)(C), 42 U.S.C. § 4332(2)(C).

rapid, irreversible environmental damage from anthropogenic pollution, the squandering of natural resources, and abrupt increases in human population growth rates. But perhaps the greatest consequence of the environmental movement was the increase in demands for major political and legal reform at all levels.

It became clear during the 1960s that major federal legislation would be necessary to renovate a government bureaucracy whose actions frequently encouraged environmental damage. Federal agencies at the time tended to be program- or mission-oriented, with little regard for environmental effects. For example, the U.S. Forest Service, an agency of the Department of Agriculture, managed the nation's forest resources in a way that optimized timber production with little regard for the environmental impacts of modifications or destruction of forest ecosystems.

Advocates of legal reform did not attempt to amend existing federal laws in order to force federal agencies to consider the environmental consequences of their actions, nor did they attempt (initially, at least) to adopt standards to regulate pollution. Instead, they opted for a single piece of legislation that would intrude into every aspect of federal decision making.

The approach actually employed by NEPA was not, of course, the only action Congress could have taken. Professor Tarlock suggested three alternative forms that environmental legislation might take:[4]

(a) agencies could be required to consider environmental consequences in their decision making (the approach adopted in NEPA);

(b) designated natural resource areas could be withdrawn from development, or environmental agencies would have a veto (the approach used for resource areas such as floodplains and wetlands);

(c) agencies could be authorized to adopt environmental standards and to prohibit development that violates these standards (other environmental laws such as the Clean Air Act, Clean Water Act, and Resource Conservation and Recovery Act use this approach).

In NEPA, the shortest of all environmental statutes, Congress opted for a single piece of legislation that would intrude into every aspect of federal decision making. Its language was intentionally short and vague, which encouraged the development of a "common law" of NEPA. The idea was that the courts would provide the details where Congress had merely set national policy. Unfortunately, this approach has had several important drawbacks. First, it led to "forum shopping," in which litigants would search for the court with jurisdiction that would provide them with the

---

[4] Tarlock, Balancing Environmental Considerations and Energy Demands: A Comment on Calvert Cliff's Coordinating Committee, Inc. v. AEC, 47 Ind. L.J. 645, 649 (1972).

most favorable decision. A second drawback was inconsistent court decisions based on the inclination of judges. A final important drawback to NEPA's common law approach was that the U.S. Supreme Court at the time was trying to extend *agency* power, while lower courts extended *judicial* power.

3.2 Overview of NEPA Policies and Goals

NEPA was passed late in 1969 after relatively little congressional conflict. It was a unique piece of legislation at the time because it was environmentally proactive, requiring all federal agencies to consider the environmental consequences of their actions before carrying them out.

The legislative history does not make clear whether Congress intended for NEPA to have the far-reaching effects on environmental protection that it has had, although comments suggest that Congress was well aware that it was considering a major revision in existing policy.[5] The following joint statement from the Senate Committee on Interior and Insular Affairs and House Committee on Science and Aeronautics in 1968 is suggestive:

> Alteration and use of the environment must be planned and controlled rather than left to arbitrary decisions. Technological development, introduction of new factors affecting the environment, and modifications of the landscape must be planned to maintain diversity of plants and animals. Furthermore, such activities should proceed only after an ecological analysis and projection of probable effects. Irreversible or difficult reversible changes should be accepted only after the most thorough study.[6]

At the core of NEPA are three important sections that reflect the conceptual approaches of the statute. Section 101,[7] often called "substantive NEPA," states the national environmental policy in emphatic terms. Section 102, called "procedural NEPA," requires that a "detailed statement" (now known as an "environmental impact statement" or EIS) must be prepared for all "major Federal actions significantly affecting the

---

[5] See Bear, The National Environmental Policy Act: Its Origins and Evolution, 10(2) Natural Resources & Env't 3 (1995).
[6] Senate Committee on Interior and Insular Affairs, and House Committee on Science and Aeronautics, *Congressional White Paper on a National Policy for the Environment*, 90th Congress.
[7] 42 U.S.C. § 4331.

quality of the human environment."[8] Title II of NEPA establishes the Council on Environmental Quality (CEQ), an Executive Office of the President, with responsibility for promulgating regulations to implement NEPA.[9] These sections will be discussed in detail below.

3.2.1 General Approach of NEPA

The general approach of NEPA is quite different from that of other environmental laws. Most environmental statutes are quite lengthy, often including volumes of information on the applicability of the statute to specific situations and entities, procedures for judicial review, and so on. By contrast, NEPA is quite short and is drafted only in very general terms. It is not known precisely what Congress intended, but this approach was soundly criticized by two Supreme Court justices who argued in their dissent in the landmark NEPA case *Kleppe v. Sierra Club* that "[t]his vaguely worded statute seems designed to serve as no more than a catalyst for development of a 'common law' of NEPA."[10]

The result has been precisely the one predicted by the dissenting Justices in *Kleppe*. The courts have produced many decisions interpreting the various sections of NEPA, and have created what amounts to a judge-made "common law" of NEPA. However, for its part, the U.S. Supreme Court has been reluctant to deal with lower court interpretations; there have been only twelve Supreme Court decisions specifically on NEPA since its passage.[11] This has created a situation in which lower courts have often disagreed on how to interpret NEPA, creating conflicting interpretations of the statute in different courts. The unfortunate result has been that litigants often engage in various forms of "forum shopping," or attempting to bring a NEPA lawsuit in the court that is most likely to give the desired interpretation of the statute.

Despite these criticisms, there is little question that NEPA has introduced a new layer into federal agency decision making and action. NEPA makes environmental protection an integral part of the mandate of

---

[8] 42 U.S.C. § 4332.
[9] 42 U.S.C. § 4321.
[10] Kleppe v. Sierra Club, 427 U.S. 390, 420 (1976), Justices Marshall and Brennan dissenting.
[11] These 12 Supreme Court decisions are known as the "dirty dozen." See Blumm, The National Environmental Policy Act at Twenty: A Preface, 20 Envt'l L. 447 (1990); Rodgers, NEPA at Twenty: Mimicry and Recruitment in Environmental Law, 20 Envt'l L. 485 (1990); and Sheldon, NEPA in the Supreme Court, 25 Land & Water L. Rev. 83 (1990).

every federal agency, requiring the agency to consider environmental concerns just as it considers other matters within its mandates.

It is important to note that NEPA Section 104 contains a federal "nonderogation clause" that states that "[n]othing in Section 102 or 103 [of NEPA] shall in any way affect the specific statutory obligations of any Federal agency . . . to comply with criteria or standards of environmental quality . . . ."[12] This section was apparently added to prevent NEPA mandates from overruling environmental standards contained in other federal laws.[13] In a leading decision on Section 104 of NEPA, *Calvert Cliff's Coordinating Committee, Inc. v. Atomic Energy Commission*,[14] the court held that the specific mandates of NEPA remain in force, unless they are clearly overruled by the requirements of another law (in the *Calvert Cliff's* case, the Water Quality Improvement Act of 1971 did not overrule NEPA).

From the outset, it is important to note that NEPA applies specifically to federal agency actions. Except under very limited circumstances, it does not apply to private individuals or entities, or even to state and local governments.[15] Nevertheless, the federal NEPA statute has inspired similar statutes in many states and local governments. These state or "little NEPAs" will be discussed in section 3.9.

3.2.2 Declaration of Purpose

Perhaps nowhere else are the lofty goals of NEPA more succinctly expressed than in the statement of purpose in Section 2 of NEPA, which states:

The Purposes of this Act are: To declare a national policy which will encourage productive and enjoyable harmony between man and his environment; to promote efforts which will prevent or eliminate

---

[12] 42 U.S.C. § 4334.
[13] See Mandelker, *NEPA Law and Litigation*, 2d ed. (1997) (hereafter "Mandelker") at § 2.09.
[14] Calvert Cliff's Coordinating Committee, Inc. v. Atomic Energy Commission, 449 F.2d 1109 (D.C. Cir. 1971). Subsequent decisions by the U.S. Supreme Court have been in agreement, see United States v. Students Challenging Regulatory Agency Procedures (SCRAP I), 412 U.S. 669 (1973); and Vermont Yankee Nuclear Power Corp. v. NRDC, 435 U.S. 519 (1978).
[15] See Brooklyn Bridge Park Coalition v. Port Auth. of N.Y. and N.J., 951 F. Supp. 383 (E.D.N.Y. 1997) [Port Authority of N.Y. and N.J. is not a "federal agency" subject to NEPA jurisdiction].

damage to the environment and biosphere and stimulate the health and welfare of man; to enrich the understanding of the ecological systems and natural resources important to the Nation; and to establish a Council on Environmental Quality.[16]

What is less clear is how Congress intended to implement and enforce these purposes, since NEPA does not contain the kinds of compliance-forcing language found in many other statutes.

3.2.3 Substantive Provisions

Section 101 of NEPA[17] sets out a series of general purposes, goals and policies of NEPA. These have come to be known as the "substantive provisions" of NEPA, and they at least allow the argument that NEPA creates a series of substantive responsibilities that federal agencies are required to meet.

Section 101(a) begins by recognizing the problem:

> The Congress, recogniz[es] the profound impact of man's activity on the interrelationships of all components of the natural environment, particularly the profound influences of population growth, high density urbanization, industrial expansion, natural resource exploitation, and new and expanding technological advances and recogniz[es] further the critical importance of restoring and maintaining environmental quality to the overall welfare and development of man . . .[18]

It is difficult to express adequately the profound significance of this statement. It recognizes for the first time in U.S. history both the nature and the extent of the impact of anthropogenic activity on the environment, and succinctly ties environmental vitality to human welfare. After centuries of thoughtless environmental exploitation, the nation was now poised to begin a period of new environmental sensitivity. This simple statement has been the catalyst for environmental protection actions at all levels.

Section 101(a) continues by stating the policies by which the federal government intends to rectify the problem:

---

[16] 42 U.S.C. § 4321.
[17] 42 U.S.C. § 4331.
[18] 42 U.S.C. § 4331(a).

[The Congress] . . . declares that it is the continuing policy of the Federal Government, in cooperation with State and local governments, and other concerned public and private organizations, to use all practicable means and measures, including financial and technical assistance, in a manner calculated to foster and promote the general welfare, to create and maintain conditions under which man and nature can exist in productive harmony, and fulfill the social, economic, and other requirements of present and future generations of Americans.[19]

It is significant that Congress intended from the outset that state and local governments, as well as "concerned public and private organizations," be included in the new environmental protection mandate. As we will see, these governmental bodies and private groups have not always approached their participation with enthusiasm.

It is also notable that Section 101(a) directs the government to use "all practicable means and measures" to attain environmental protection. It is less clear what those means and measures were meant to be. One early, influential federal court decision, *Calvert Cliffs' Coordinating Committee v. U.S. Atomic Energy Commission*, interpreted this phrase as stating that "Congress did not establish environmental protection as an exclusive goal; rather, it desired a reordering of priorities, so that environmental costs and benefits will assume their proper place along with other considerations."[20] The following sections of this chapter will explore some of the ways in which the government has approached this problem.

Section 101(b) imposes explicit duties on federal officials, providing that:

In order to carry out the policy set forth in this Act, it is the continuing responsibility of the Federal Government to use all practicable means, consistent with other essential considerations of national policy, to improve and coordinate Federal plans, functions, programs, and resources to the end that the nation may-

(1) fulfill the responsibilities of each generation as trustee of the environment for succeeding generations;

(2) assure for all Americans safe, healthful, productive, and esthetically and culturally pleasing surroundings;

(3) attain the widest range of beneficial uses of the environment

---

[19] 42 U.S.C. § 4331(a).
[20] Calvert Cliff's Coordinating Committee, Inc. v. Atomic Energy Commission, 449 F.2d 1109 (D.C. Cir. 1971).

without degradation, risk to health or safety, or other undesirable or unintended consequences;

(4) preserve important historic, cultural, and natural aspects of our national heritage, and maintain, wherever possible, an environment which supports diversity and variety of individual choice;

(5) achieve a balance of between population and resource use which will permit high standards of living and a wide sharing of life's amenities; and

(6) enhance the quality of renewable resources and approach the maximum attainable recycling of depletable resources.[21]

Section 101(b) includes references to several popular environmental themes. For example, the popular idea of environmental stewardship is clearly seen in (1). Paragraph (6) emphasizes the importance of recycling, a concept that has become extremely popular. Most communities have now made some provision for recycling of materials such as paper, glass, metals, and lawn waste. Some communities require recycling under their local ordinances.[22]

In the final section of the substantive provisions, Section 101(c) states that "Congress recognizes that each person should enjoy a healthful environment and each person has a responsibility to contribute to the preservation and enhancement of the environment."[23] Interestingly, this section does not create a judicially enforceable right to a healthful environment, although an earlier version of the statute would have done so. The form in which NEPA was passed could be used by a court as the basis to refuse standing to individuals seeking to require agencies to comply with NEPA, but courts have generally interpreted NEPA as creating an interest capable of judicial enforcement.[24] The issue of "standing to sue" in NEPA cases is discussed in section 3.7.1.

3.2.4 Procedural Provisions

As it was originally drafted, NEPA was little more than a statement of environmental policy. When it became apparent to the drafters of the

---

[21] 42 U.S.C. § 4331(b).

[22] For example, effective January 1, 1995 the City of Chicago, Illinois *requires* the recycling of many materials by businesses and residences. Such actions are likely to become more popular in the future, particularly in congested urban areas. See also chapter 12 on pollution prevention.

[23] 42 U.S.C. § 4331(c).

[24] See Anderson, Mandelker, and Tarlock, *Environmental Protection: Law and Policy*, 2d ed. (1990) at 686.

original statute that policies alone would probably result in little if any change in agency actions, Section 102(2) was added. This controversial section of NEPA imposes several procedural duties on federal agencies. The most prominent of these obligations is that of Section 102(2)(C), which states that:

The Congress authorizes and directs that, to the fullest extent possible:
. . . (2) all agencies of the Federal Government shall -
. . . (c) include in every recommendation or report on proposals for legislation and other major Federal actions significantly affecting the quality of the human environment, a detailed statement by the responsible official on -
(i) the environmental impact of the proposed action,
(ii) any adverse environmental effects which cannot be avoided should the proposal be implemented,
(iii) alternatives to the proposed action,
(iv) the relationship between local short-term uses of man's environment and the maintenance and enhancement of long-term productivity, and
(v) any irreversible and irretrievable commitments of resources which would be involved in the proposed action should it be implemented.[25]

The detailed statement of the environmental impact of an action that must be prepared by each federal agency has come to be called the *environmental impact statement* (EIS). The EIS requirement is the cornerstone of NEPA. Clauses (i) through (v) enumerate the contents of the EIS, which are discussed in more detail in section 3.6.4.

Clauses (iv) and (v) address the problems of long-term commitments of resources and their effects on future generations. Unfortunately, these clauses generally do not receive adequate attention either in EISs or in reviewing courts.[26]

The EIS requirement obliges federal agencies to consider carefully the environmental consequences of their actions. Furthermore, the EIS is on public display, and public scrutiny of the results of the agency's analysis places a heavy burden on the agency to fulfill its responsibility.

NEPA does not define environmental "impact" in the context of an EIS. However, the Council on Environmental Quality regulations encompass

---

[25] 42 U.S.C. § 4332(2)(C).
[26] See Mandelker, § 2.04.

effects or impacts that are direct, indirect (i.e., "reasonably foreseeable"), as well as cumulative. Effects or impacts include ecological, esthetic, historic, cultural, economic, social, or health.[27]

3.3 Council on Environmental Quality

Title II of NEPA creates the Council on Environmental Quality (CEQ),[28] and requires the president to provide an annual Environmental Quality Report (EQR) to Congress.[29] The CEQ is composed of three members, who are appointed by the president and confirmed by the Senate. The CEQ members are to be "exceptionally well qualified" to interpret environmental information, appraise federal programs, assess the environmental needs of the nation, and to recommend policies.[30]

NEPA gives the CEQ responsibilities to study and analyze environmental information, make policy recommendations to the president, and assist the president in preparing the EQR. Initially, the CEQ had policy-making responsibilities only as specified by the statute, but in 1970 President Nixon issued an executive order giving the CEQ the authority to issue "guidelines" for the EIS process.[31] In 1977 President Carter signed an executive order[32] authorizing the CEQ to promulgate regulations for the EIS process, which it has done.

The CEQ has primary responsibility for interpreting NEPA, and courts have generally given deference to the CEQ's interpretations. In fact, the U.S. Supreme Court has stated that the CEQ's opinion is to be given "substantial deference" in cases that depend on interpretations of NEPA.[33]

3.4 Role of the Environmental Protection Agency

The U.S. Environmental Protection Agency (EPA) has limited responsibility for reviewing and evaluating EISs. It should be noted that the EPA does not necessarily produce the EIS; in fact, many EPA

---

[27] 40 CFR § 1508.8.
[28] 42 U.S.C. §§ 4342-4347.
[29] 42 U.S.C. § 4341.
[30] 42 U.S.C. § 4342.
[31] Exec. Order No. 11,514, 35 *Fed. Reg.* 4247 (1970).
[32] Exec. Order No. 11,990, 42 *Fed. Reg.* 26967 (1977).
[33] Andrus v. Sierra Club, 442 U.S. 347, 358 (1979). The Supreme Court reiterated its deference to CEQ in Robertson v. Methow Valley Citizens Council, 490 U.S. 332 (1989).

activities are exempt from the EIS process.[34]

The EPA mandate to review and evaluate EISs is found in Section 309 of the Clean Air Act (CAA), which was passed in 1970 a short time after the passage of NEPA. CAA Section 309 authorizes the EPA to "review and comment in writing on the environmental impact of . . . any major Federal action . . . to which . . . § 102(2)(C) of NEPA applies."[35] Since NEPA Section 102(2)(C) requires all agencies preparing an EIS to "obtain the comments of any Federal agency which has jurisdiction by law or special expertise with respect to any environmental impact involved," the EPA has review and commenting authority over every EIS prepared by a federal agency.[36]

It is significant that CAA Section 309 also requires that "[s]uch written comment [by the EPA with respect to an EIS] shall be made public at the conclusion of any such review."[37] In drafting CAA Section 309, Congress apparently intended that the results of the EPA's environmental analysis be made available to all interested parties prior to the time a federal agency makes a final decision on an action.[38] Public access to EPA's EIS review has had tremendous impact on the ability of interested parties to participate in (and challenge) agency actions.

CAA Section 309(a) requires the EPA to determine if a proposed action is unsatisfactory from the standpoint of "public health or welfare," or "environmental quality" in its review of an EIS. The EPA then refers its EIS determination to CEQ for further action.

The EPA has responded to its responsibilities under CAA Section 309(a) by constructing a formal set of responses to an EIS submitted by a federal agency.[39] Typically, the EPA reviews a "draft" EIS (DEIS), making comments that permit the federal agency to make additions or changes prior to issuing a "final" EIS (FEIS). The EPA first responds to the DEIS by issuing one of four ratings of the environmental impacts of the proposed action. A proposed action may be rated as "LO" (lack of objections), "EC" (environmental concerns, meaning some impacts have been identified that should be avoided or mitigated), "EO" (environmental objections, identifying significant impacts that may require substantial

---

[34] For example, EPA is exempt from the EIS process under the Clean Air Act (15 U.S.C. § 793(c)(1)), and most of the Clean Water Act (33 U.S.C. § 1371(c)(2)). See Mandelker, § 2.10.
[35] 42 U.S.C. § 7609.
[36] 42 U.S.C. § 4332(2)(C).
[37] 42 U.S.C. § 7609.
[38] See Mandelker, § 2.08[1].
[39] See U.S.E.P.A., Office of Enforcement, *EPA's Section 309 Review: The Clean Air Act and NEPA* (Dec. 1992).

changes to the proposed action), or "EU" (environmentally unsatisfactory, with impacts so severe that the action must not proceed as proposed). In addition, the EPA rates the adequacy of the EIS as "1" (adequate, requiring no additional information), "2" (insufficient information, requiring more information or an evaluation of alternatives), or "3" (inadequate, seriously lacking information or analysis). Of course, the EPA (and the EIS) must comply with all CEQ regulations.[40]

The EPA's rating of the proposed action is made available in the form of a comment letter. In the event of an unsatisfactory EIS, these comments are forwarded to the CEQ, and the agency is given an opportunity to respond. The CEQ may then take several steps, including mediation or a finding that the concerns expressed by the EPA have been resolved.[41] While the CEQ may choose to publish a finding that an agency's proposed action remains unsatisfactory, it has no authority to prevent an unsatisfactory agency action.

An interesting question arises when there is a negative review of an EIS: What, if anything, must the agency do with a negative EIS review? This question has been only partly answered. In the federal circuit court decision in *Alaska v. Andrus*,[42] the court noted that an agency could proceed in the face of a negative EIS, but that it has a heightened obligation to explain fully its reasons for doing so. Surprisingly, this issue has arisen only rarely.[43]

3.5 Environmental Impact Statements

A unique characteristic of NEPA is it's requirement that a "detailed statement" (now known as an environmental impact statement, or EIS) be prepared for all "proposals for legislation and other major Federal action significantly affecting the quality of the human environment."[44] The purpose of the EIS is to guarantee that no federal agency will conduct its business without thoroughly considering the adverse environmental consequences of its actions. Unfortunately, NEPA gives little guidance as

---

[40] 40 CFR Part 6. These regulations apply whenever an agency is required to complete an EIS.
[41] 40 CFR Part 1504.
[42] Alaska v. Andrus, 580 F.2d 465 (D.C. Cir. 1978), vacated in part and remanded on other grounds, 439 U.S. 922 (1979).
[43] See Concerned About Trident v. Schlesinger, 400 F. Supp. 454 (D.D.C. 1975), modified on other grounds sub nom. Concerned About Trident v. Rumsfeld, 555 F.2d 817 (D.C. Cir. 1977); and Massachusetts v. Andrus, 594 F.2d 872 (1st Cir. 1979).
[44] NEPA § 102(2)(C), 42 U.S.C. § 102(2)(C).

to when and how an EIS must be prepared, and circumstances under which an agency is excused from the requirement. Policies implemented by the CEQ, EPA, and the courts have largely filled in the gaps where NEPA itself is silent.[45]

3.5.1 Action Triggering the EIS Requirement

A critical question whenever a federal agency contemplates action is whether an EIS must be prepared. Under NEPA Section 102(2)(C) an EIS must be prepared for proposed "major Federal actions significantly affecting the quality of the human environment."[46] However, this simple statement defies a simple interpretation. Instead, it is best to approach the statement as a series of threshold questions. Referring to a proposed action by a federal agency, these questions are: (a) Is the proposed action *major*? (b) Is it *federal*? (c) Is it an *action*? (d) Is it *significant*? and (e) Will it affect the *quality of the human environment*? We will consider these elements separately.

Is the proposed action federal? CEQ regulations define several categories of federal actions. In general, federal actions tend to fall within one of the following categories:

(1) Adoption of official policy . . . [and] formal documents establishing an agency's policies which will result in or substantially alter agency programs.

(2) Adoption of formal plans . . . upon which future agency actions will be based.

(3) Adoption of programs . . . ; systematic and connected agency decisions allocating agency resources to implement a specific statutory program or executive directive.

(4) Approval of specific projects . . . Projects include actions approved by permit or other regulatory decisions as well as federal and federally assisted activities.[47]

Clearly, proposed actions such as private housing projects receiving federal subsidies or projects that receive federal agency licenses would be considered federal under these regulations.

---

[45] CEQ's general regulations regarding the preparation of EISs are found at 40 CFR §§ 6.200-6.205.
[46] 42 U.S.C. § 4332(2)(C).
[47] 40 CFR § 1508.18(b).

Federal courts have established that even minimal federal involvement in an action may be enough to trigger the EIS requirement. For example, federal block grants, funding for state projects, and federal-state partnerships normally trigger the requirement for an EIS.[48] On the other hand, federal involvement is generally insufficient in federal revenue-sharing programs, funding for preliminary studies, unrestricted subsidies, federal technical assistance, and situations where the actions of the federal agency are ministerial (i.e., not involving federal agency discretion).[49]

Federal funds are often made available to state, regional, and local governments for various planning activities. Interestingly, one court held that federal funding for a state coastal zone management program did not require an EIS.[50] Federal courts have held that an EIS is generally not required when federal approval of the plan is not required.[51] One court held that a decision to pursue a course of inaction, in contrast to a course of action, does not trigger NEPA's requirement that an EIS be prepared. The court held that the U.S. Forest Service's decision to terminate

---

[48] See Ely v. Veld, 451 F.2d 1130 (4th Cir. 1971) [block grant for law enforcement assistance subject to NEPA]. Some courts have held that indirect use of block grants does not trigger the EIS requirement, for example, Citizens for Better St. Clair Co. v. James, 648 F.2d 246 (5th Cir. 1981) [block grant supporting state prison system indirectly related to purchase of prison site, and no EIS required]. See also Monarch Chemical Works, Inc. v. Ebon, 452 F. Supp. 493 (D. Neb. 1978), *vacated*, 466 F. Supp. 639, *aff'd*, 604 F.2d 1083 (8th Cir. 1979) [state, acting as federal agency, contracted with city to build correctional facility]; Blue Ocean Preservation Society v. Watkins, 754 F. Supp. 1450 (D. Haw. 1991) [federal funding for a geothermal project triggers the EIS requirement]; and Ross v. Federal Highway Admin., 972 F. Supp. 552 (D.Kan. 1997) [all four segments of 14-mile highway demonstration project constituted a "major federal action"].

[49] See Kings County Economic Community Development Ass'n v. Hardin, 478 F.2d 478 (9th Cir. 1973) [unrestricted subsidies]; Carolina Action v. Simon, 389 F. Supp. 1244 (M.D.N.C.), *aff'd per curiam*, 522 F.2d 295 (4th Cir. 1975) [revenue sharing]; State of South Dakota v. Andrus, 614 F.2d 1190, *cert. denied*, 449 U.S. 822 (1980) [issuance of mineral patent by U.S. Dept. of the Interior was ministerial, and did not come under NEPA]; National Organization for Reform of Marijuana Laws v. USDEA, 545 F. Supp. 981 (D.D.C. 1982) [federal technical assistance for marijuana spraying project]; Almond Hill School v. Dept. of Agriculture, 768 F.2d 1030 (9th Cir. 1985) [involvement of federal officials in state beetle eradication program]; and Macht v. Skinner, 916 F.2d 13 (D.C. Cir. 1990) [preliminary studies].

[50] Save Our Dunes v. Pegues, 642 F. Supp. 393 (M.D. Ala. 1985), *rev'd on other grounds*, 834 F.2d 984 (11th Cir. 1987).

[51] See Atlanta Coalition on Transp. Crisis v. Atlanta Regional Comm'n, 599 F.2d 1333 (5th Cir. 1979) [federal funding of Regional Development Plan not sufficient federal involvement to trigger EIS requirement]; and Bradley v. U.S. Dept. of Housing & Urban Development, 658 F.2d 290 (5th Cir. 1981) [redevelopment plan with federal funding not sufficiently federal].

herbicide application in a national forest was not a federal "action" requiring the preparation of an EIS.[52]

As a general rule, an action is "federal" and triggers NEPA if the federal agency has the power to influence the outcome materially.

Is it major? CEQ regulations state that the word "major" means "reinforces but does not have a meaning independent of significant."[53] This formulation provides little guidance to individuals attempting to determine if an EIS is required for a specific project, and the federal courts have been of little assistance. The most often quoted definition is from *NRDC v. Grant*,[54] which states that a "major" action is one "that requires substantial planning, time, resources or expenditure." Nevertheless, federal courts treat the determination of "major" federal actions on a case-by-case basis.[55]

Does it significantly affect the quality of the human environment? Judged by the sheer volume of litigation, this requirement is the most important when considering whether an agency must complete an EIS. In most of these cases, it has already been determined that the proposed action is both "federal" and "major," leaving the "significance" of the action as the remaining issue.

CEQ regulations define *significantly* as based on "considerations of both context and intensity."[56] *Context* means that the action must be analyzed with respect to factors such as its affect on society as a whole, the affected region and the locality.[57] Significance varies with the setting of the proposed action, and includes consideration of both short and long-term effects. *Intensity* refers to the "severity of impact" of the action, and includes factors such as the "highly controversial" nature of the impact, effects on public health and safety, the cumulative effects of the action, and the degree to which the effects are "highly uncertain or involve unique

---

[52] Minnesota Pesticide Info. & Educ., Inc. v. Espy, 29 F.3d 442 (8th Cir. 1994).
[53] 40 CFR § 1508.18. See, for example, Sierra Club v. Babbitt, 65 F.3d 1502 (9th Cir. 1995) [Bureau of Lan Management did not have to prepare an EIS for right-of-way agreement with private logging company because the agency had discretion to influence the private activity for the benefit of endangered species].
[54] NRDC v. Grant, 341 F. Supp. 356 (E.D.N.Y. 1972).
[55] See Fund for Anmals, Inc. v. Thomas, 127 F.3d 80 (D.C. Cir. 1997) [U.S. Forest Service's adoption of new policy concerning game baiting was not a major federal action requiring preparation of an EIS]. See also Mandelker, § 8.06[3] for examples of actions held by federal courts to be major and minor.
[56] 40 CFR § 1508.27.
[57] 40 CFR § 1508.27(a).

or unknown risks."[58]

Some agencies define the term *significant* in their regulations, and these are used by the agency and the courts as guidance. For example, Federal Highway Administration NEPA regulations[59] contains a definition of *significant* similar to the CEQ definition, but applicable to highway projects. Some courts collapse the terms *major* and *significant* into one standard, although other courts have disagreed.[60]

Probably the most important discussion of "significance" has come to be known as the "Hanly II" factors (from *Hanly v. Kleindienst*[61]). The Hanly II court stated that a consideration of the "significance" of an action must consider: (1) the extent to which the action will cause adverse environmental effects in excess of those created by existing uses; and (2) the absolute quantitative adverse environmental effects of the action itself. Many courts have followed the formula in Hanly II, although a few courts have followed Judge Friendly's dissent in Hanly II, which would require an EIS any time "when there are 'arguably' cases of true significance."[62] In deciding what factors may be considered, courts have allowed inclusion of esthetic values, but not psychological damage.[63]

3.5.2 Exceptions to the EIS Requirement

There are several circumstances when a major federal action significantly affects the quality of the human environment, but which nevertheless does not fall within the EIS requirement of NEPA. These are variously known as "exceptions," "exclusions," or "exemptions," but in each case they shield the agency from NEPA's otherwise comprehensive reach.

In some cases, there is direct conflict between NEPA and some other statute. For example, under the Interstate Land Sales Full Disclosure Act, developers must submit disclosure statements to HUD that become

---

[58] 40 CFR § 1508.27(b).

[59] 23 CFR § 771.3(d).

[60] See Minnesota Public Interest Research Group v. Butz (I), 498 F.2d 1314 (8th Cir. 1974); and NAACP v. Medical Center, Inc., 584 F.2d 619 (3rd Cir. 1978).

[61] Hanly v. Kleindienst, 471 F.2d 823 (2d Cir. 1972).

[62] See Maryland-Nat'l Capital Park Planning Comm'n v. Postal Service, 487 F.2d 1029 (D.C. Cir. 1973). Another case argued that the *Friendly* dissent is no longer applicable because federal agencies are now more environmentally sensitive. River Road Alliance v. Corps of Engineers, 764 F.2d 445, *cert. denied*, 475 U.S. 1055 (1986).

[63] See Maryland-Nat'l Capital Park Planning Comm'n v. Postal Service, 487 F.2d 1029 (D.C. Cir. 1973); and Metro. Edison Co. v. People Against Nuclear Energy, 460 U.S. 766 (1983).

effective in 30 days.[64] Since it would be virtually impossible to complete an EIS in that time period (and because the environmental consequences would be slight) it is considered to be an exception to EIS requirement.[65]

In other cases, there is an express statutory exemption for a particular action. For example, EPA actions are exempt from the EIS process under Section 511(c) of the Clean Water Act.[66] In the context of the CWA, the EPA must prepare an EIS only when federal funds are used for public treatment works, or for a permit for a new pollution source. EPA actions are exempt from Clean Air Act requirements under the Energy Supply and Environmental Coordination Act of 1974.[67] Other agencies have limited exemptions.

Yet another exemption from the EIS requirement is for "emergencies" where, presumably, time and circumstances do not permit the usual EIS process. However, before an agency can exploit this exemption, it must consult with the CEQ, which limits the exemption "to actions necessary to control the immediate impacts of the emergency."[68] For example, emergency exemptions have been authorized to permit emergency use of a pesticide and to allow the capture of endangered California condors in the wild.[69]

In some circumstances, an agency may rely on a "categorical exclusion," defined by CEQ regulations as "a category of actions which do not individually or cumulatively have a significant effect on the human environment and which have been found to have no such effect in procedures adopted by a Federal agency in implementation of these regulations ([40 CFR] Sec. 1507.3) and for which, therefore, neither an environmental assessment nor an environmental impact statement is required."[70] For the most part, categorical exemptions have been upheld by the courts when they are used for agency activities that are routine and

---

[64] 15 U.S.C. §§ 1701-1720.

[65] See Flint River Development Co. v. Scenic Rivers Assoc., 426 U.S. 776 (1976).

[66] 33 U.S.C. § 1371(c)(1).

[67] 15 U.S.C. § 793(c)(1).

[68] 40 CFR § 1506.11.

[69] Environmental Defense Fund v. Blum, 458 F. Supp. 650 (D.D.C. 1978) [pesticide]; and National Audubon Society v. Hester, 801 F.2d 405 (D.C. Cir. 1986) [endangered species].

[70] 40 CFR § 1508.4.

inconsequential.[71] However, courts have not hesitated to reverse an agency's categorical exclusion where an EIS (or, at least, an environmental assessment) are required.[72]

Other exemptions have been "implied" by reviewing courts. For example, the EPA's adherence to statutes it administers is often considered to be the "functional equivalent" of NEPA compliance. The federal circuit court in *Portland Cement Associates v. Ruckelshaus* [73] held that there is a "workable balance" between Section 111 of the CAA and NEPA, so no EIS is required when EPA establishes new source performance standards.[74] In *Texas Committee on Natural Resources v. Bergland*[75] the court held that the Forest Service is exempt under the National Forest Management Act[76] under certain circumstances where the Act contained the necessary guidelines which precluded an EIS requirement. In *Douglas County v. Babbitt*, the ninth Circuit Court of Appeals held that the U.S. Fish and Wildlife Service need not prepare an EIS for designation of the critical habitat of the northern spotted owl under the Endangered Species Act.[77]

However, other courts have not implied an exemption. For example, *Foundations on Economic Trends v. Heckler*[78] held that a National Institute of Health review of a recombinant DNA release was not the equivalent of an EIS.

Some actions are exempt from the EIS requirement. For example,

---

[71] See, for example, Town of Beverly Shores v. Lujan, 736 F. Supp. 934 (N.D. Ind. 1989) [paving a parking lot along a lakeshore national park]; and Oregon Natural Resources Council v. Bureau of Reclamation, 49 F.3d 1441 (9th Cir. 1995) [dredging of a channel]. See also Mandelker, § 7.04[2] for additional discussion.

[72] See, for example, Washington Trails Ass'n v. U.S. Forest Service, 935 F. Supp. 1117 (W.D. Wash. 1996) [proposal to reconstruct and relocate off-road vehicle trails not categorically exempt under NEPA]; Runway 27 Coalition v. Engen, 679 F. Supp. 95 (D. Mass. 1995) [airport flight path changes over residential neighborhoods]; and State of Mississippi v. Marsh, 710 F. Supp. 1488 (S.D. Miss. 1989) [removing dredge spoil from riverbank]. See also Mandelker, § 7.04[2].

[73] Portland Cement Assoc. v. Ruckelshaus, 486 F.2d 375 (D.C. Cir. 1973), *cert. denied*, 417 U.S. 921 (1974).

[74] EPA was specifically exempted under the CAA, 15 U.S.C. § 793(c)(1).

[75] Texas Committee on Natural Resources v. Bergland, 573 F.2d 201 (5th Cir.), *cert. denied*, 439 U.S. 966 (1978).

[76] 16 U.S.C. §§ 1600-1687.

[77] Douglas County v. Babbitt, 48 F.3d 1495 (9th Cir. 1995). No EIS is necessary for critical habitat designation because: (a) the Endangered Species Act displaces the EIS requirement; and (b) no EIS is required for actions that do not alter the natural physical environment.

[78] Foundations on Economic Trends v. Heckler, 587 F. Supp. 753 (D.D.C. 1984), aff'd in part and vacated in part, 756 F.2d 143 (D.C. Cir. 1985).

appropriation requests do not require an EIS.[79] There may be a national security exemption as well. In *Weinberger v. Catholic Action of Hawaii*[80] information on a weapons storage facility was considered a national security issue exempt from a Freedom of Information Act (FOIA) disclosure, so an EIS (with its ample public disclosure requirements) was not required. The 9th Circuit Court of Appeals in *Kasza v. Browner*[81] upheld the national security exemption in disallowing a suit under the Resource Conservation and Recovery Act[82] (not NEPA) at the U.S. Air Forces' Groom Lake experimental facility (well known to UFO enthusiasts as "Area 51"). While the issue of a national security exemption under NEPA has not been resolved, it appears likely that a federal court would support such an exemption.

3.6 EIS Requirements Under NEPA Section 102(2)(c)

NEPA Section 102 offers only general assistance to federal agencies in the preparation of an EIS. CEQ regulations[83] are considerably more helpful, but the need for judicial interpretation of NEPA has left abundant room for confusion and misunderstanding.

It is important to understand that the process by which a federal agency complies with the mandates of NEPA Section 102(2)(C) seldom begins with the preparation of an EIS. Instead, there is a relatively complex process involving several kinds of EISs, EPA and CEQ review, and public disclosure.

3.6.1 The EIS Process

If an agency is unsure whether an EIS is required or not, it prepares an Environmental Assessment (EA) to determine if a full EIS is required.[84]

---

[79] Andrus v. Sierra Club, 442 U.S. 347 (1979).

[80] Weinberger v. Catholic Action of Hawaii, 454 U.S. 139 (1981).

[81] Kasza v. Browner, 133 F.3d 1159 (9th Cir. 1998), cert. denied, 525 U.S. 967 (1998).

[82] "RCRA," 42 U.S.C. §§ 6901-6992.

[83] See 40 CFR Part 6, and 40 CFR §§ 1502-1506, 1508.

[84] The regulations specifically state that an EA "is not necessary if the agency has decided to prepare an [EIS]," 40 CFR § 1501.3(a). See Missouri Mining, Inc. v. Interstate Commerce Comm'n, 33 F.3d 980 (8th Cir. 1994) [an agency can waive the EA requirement and proceed directly to the remainder of the EIS process].

CEQ regulations describe an EA as: "a concise public document for which a Federal agency is responsible that serves to: (1) Briefly provide sufficient evidence and analysis for determining whether to prepare an environmental impact statement or a finding of no significant impact; (2) Aid an agency's compliance with the Act when no environmental impact statement is necessary; [and] (3) Facilitate preparation of [an EIS] when one is necessary."[85]

EAs differ dramatically from EISs.[86] Most EAs are 10 to 15 pages in length (as compared to EISs that are hundreds or thousands of pages long). Compared to an EIS, an EA is short and relatively simple. CEQ regulations state simply that an EA "[s]hall include brief discussions of the need for the proposal, of alternatives as required by Section 102(2)(E) [of NEPA], of the environmental impacts of the proposed action and alternatives, and a listing of agencies and persons consulted."[87] To the extent that it is practicable to do so, the agency is required to "involve environmental agencies, applicants, and the public" in the preparation of the EA.[88]

Approximately 30,000 EAs are produced by federal agencies each year. The vast majority of these determine that a full EIS is not necessary. In some cases, an EIS is not required because the proposed agency action is covered by one of the exemptions or exceptions discussed in section 3.5.2 above. In many more cases, however, there is no EIS because the agency determines that the proposed action will not significantly affect the quality of the human environment. Such a determination is called a "Finding of No Significant Impact," or "FONSI." FONSIs are a frequent source of litigation.[89]

CEQ regulations define a FONSI as "a document by a Federal agency briefly presenting the reasons why an action, not otherwise excluded ([40 CFR] Sec. 1508.4), will not have a significant effect on the human environment and for which an environmental impact statement therefore

[85] 40 CFR § 1508.9(a).

[86] For discussions of the differences between EAs and EISs, see Sierra Club v. Marsh, 769 F.2d 868 (1st Cir. 1985); and Sierra Club v. Hodel, 848 F.2d 1068 (10th Cir. 1988).

[87] 40 CFR § 1508.9(b). For discussions of the contents and preparation of EAs, see Sierra Club v. Hodel, 848 F.2d 1068 (10th Cir. 1988); Cronin v. USDA, 919 F.2d 439 (7th Cir. 1990); Friends of Fiery Gizzard v. Farmers Home Admin., 61 F.3d 501 (6th Cir. 1995); and Kelley v. Selin, 42 F.3d 1501 (6th Cir. 1995).

[88] 40 CFR § 1501.4(b).

[89] See, for example, National Audubon Soc'y v. Hoffman, 132 F.3d 7 (2d Cir. 1997); Presidio Golf Club v. National Park Serv., 155 F.3d 1153 (9th Cir. 1998); Brandon v. Pierce, 725 F.2d 555 (10th Cir. 1984); and Surfrider Found. v. Dalton, 989 F. Supp. 1309 (S.D. Cal. 1998). See also Mandelker, Chapter 4.

will not be prepared."[90] A FONSI must include the EA (or a summary of the EA), as well as mention of any other EAs or EISs on related projects.[91]

If an EA determines that a full EIS is required, then the agency prepares a "notice of Intent" (NOI) that is published in the *Federal Register*.[92] CEQ regulations state that the NOI "shall briefly: (a) describe the proposed action and possible alternatives; (b) describe the agency's proposed scoping process including whether, when, and where any scoping meeting will be held; and (c) state the name and address of a person within the agency who can answer questions about the proposed action and the environmental impact statement."[93]

It is not unusual for several federal agencies to share responsibility for the preparation of an EIS. CEQ regulations necessitate the designation of a "lead agency" to supervise the preparation of the EIS.[94] The lead agency is typically the one with the greatest involvement, expertise, or experience. If there is disagreement among the agencies as to which will be the lead agency, CEQ regulations require the use of the following factors (listed in order of descending importance) to be used in determining which agency is the lead agency:

(1) Magnitude of agency's involvement.
(2) Project approval/disapproval authority.
(3) Expertise concerning the action's environmental effects.
(4) Duration of agency's involvement.
(5) Sequence of agency's involvement.[95]

3.6.2 Preparation of the EIS

Once it has been determined by a federal agency that an EIS is required, several important and controversial issues arise. An important initial issue is *when* an EIS should be prepared. There is tension between preparing an EIS late enough to get the best possible information, but early enough to ensure that the information is available for the decision-making process. CEQ regulations state that the EIS should be prepared "as close as possible to the time the agency is developing or is presented with a proposal."[96]

---

[90] 40 CFR § 1508.13.
[91] 40 CFR § 1501.7(a)(5).
[92] 40 CFR § 1501.7.
[93] 40 CFR § 1508.22.
[94] 40 CFR § 1501.5.
[95] 40 CFR § 1501.5(c). If the agencies cannot agree on a lead agency, there are rules by which the CEQ will choose one at 40 CFR § 1501.5(e).
[96] 40 CFR § 1502.5.

Where there has been an application to a federal agency, the EIS must commence "no later than immediately after the application is received [by the agency]."[97]

Several federal courts have considered the timing issue. In *Aberdeen & Rockfish R.R. Co. v. Students Challenging Regulatory Agency Procedures*[98] (SCRAP II), the U.S. Supreme Court held that an EIS is not required until a final agency "proposal" is prepared.[99] According to a Ninth Circuit decision in *Conner v. Burford*,[100] an EIS is probably not required prior to leases of federal land.

A second issue is *who* prepares the EIS? Without question, NEPA states that the EIS is to be prepared by the agency. However, there is disagreement among the federal courts as to when, if ever, an agency may "delegate" to another agency or private entity the responsibility of preparing the EIS.

May responsibility for preparing an EIS be delegated to a state? Amendments to NEPA Section 102(2)(D) made in the 1980s allow such delegation. These amendments were made in 1975 as a congressional response to federal court decisions that disallowed delegation to state agencies because it was too self-serving. The most famous of these decisions was *Conservation Society of Southern Vermont v. Secretary of Transportation (I)*, in which the inability of New England states to prepare EISs prior to the 1975 amendments brought a halt to federal highway funding, and a brief halt to all major road construction.[101]

Delegation of EIS preparation to a private applicant is probably not acceptable. In *Greene County Planning Board v. Federal Power Commission*,[102] the FPC had delegated to applicants the authority to prepare EISs for high-voltage transmission line permits. The court held that the FPC had violated NEPA by "abdicating . . . its authority," and creating a situation in which an EIS might be based on "self-serving

---

[97] 40 CFR § 1502.5(b).
[98] Students Challenging Regulatory Agency Procedures (SCRAP II), 422 U.S. 289 (1975).
[99] See also Kleppe v. Sierra Club, 427 U.S. 390 (1976).
[100] Conner v. Burford, 848 F.2d 1441 (9th Cir. 1988), cert. denied, 489 U.S. 1012 (1989).
[101] Conservation Society of Southern Vermont v. Secretary of Transportation (I), 508 F.2d 927 (2d Cir. 1974), vacated and remanded, 423 U.S. 809 (1975). See also Idaho Public Utilities Comm'n v. Interstate Commerce Comm'n, 35 F.3d 585 (D.C. Cir. 1994) [ICC, in approving application to abandon a railroad line, could not abrogate its responsibility to prepare an EIS by directing railroad to consult with other state and fewderal agencies to identify environmental probelms].
[102] Greene County Planning Board v. Federal Power Commission, 455 F.2d 412 (2d Cir.), cert. denied, 409 U.S. 849 (1972).

assumptions" by the applicant. On the other hand, *Sierra Club v. Lynn* held that it is acceptable for the applicant to assist the agency by submitting environmental information or participating in environmental studies used in preparation of the EIS.[103]

It is quite common for federal agencies to delegate the responsibility for EIS preparation to private consultants with the scientific and technical qualifications often missing in agency personnel. In *Natural Resources Defense Council v. Callaway*,[104] the U.S. Navy had delegated to a private consultant the responsibility for preparation of an EIS for the deposit of polluted spoil on Long Island Sound. The court held that this delegation was acceptable because the consultant served only the Navy's interests. Other cases have emphasized the importance of agency supervision and control over the preparation of the EIS in deciding if EIS preparation can be delegated to a consultant.[105]

3.6.3 Scope of the EIS

Once the lead agency has made the determination that an EIS is required, the process known as "scoping" begins. Scoping is described in the regulations as "an early and open process for determining the scope of issues to be addressed and for identifying the significant issues related to a proposed action."[106] Scoping is an important part of the EIS process because it begins communication between appropriate federal agencies, state and local agencies, and the public.[107] Scoping also allows the lead agency to set boundaries on the EIS process with respect to such issues as timing, assignments of the lead and cooperating agencies in preparing the EIS, identifying and narrowing the discussion of nonsignificant issues, and identifying other environmental review and consultation requirements so the agencies can prepare required analyses and studies.

---

[103] Sierra Club v. Lynn, 502 F.2d 43 (5th Cir. 1973), cert. denied, 421 U.S. 994, 422 U.S. 1049 (1975). See also Friends of Endangered Species v. Jantzen, 596 F. Supp. 518 (N.D. Cal. 1984); and Friends of Earth v. Hintz, 800 F.2d 822 (9th Cir. 1986).
[104] Natural Resources Defense Council v. Callaway, 524 F.2d 79 (2d Cir. 1975).
[105] See, for example, Life of Land v. Brinegar, 485 F.2d 460 (9th Cir. 1973), cert. denied, 416 U.S. 961 (1974); Save Our Wetlands, Inc. v. Sands, 711 F.2d 634 (5th Cir. 1983); and Brandon v. Pierce, 725 F.2d 555 (10th Cir. 1984).
[106] 40 CFR § 1501.7.
[107] 40 CFR § 1501.7(a). This early public participation probably invites public confidence in the EIS process by identifying potential problem areas. See Mandelker, § 7.04[4].

The scope of an EIS has been an important concern since NEPA was passed in 1969. The primary reason is concern regarding the "segmentation" problem, or the tendency to break up large projects into several smaller ones thus avoiding "cumulative impact" problems. In order to avoid segmentation problems as much as possible, CEQ regulations require an EIS to discuss:

> Actions (other than unconnected single actions) which may be:
> (1) Connected actions, which means that they are closely related and therefore should be discussed in the same impact statement. Actions are connected if they:
>   (i) Automatically trigger other actions which may require environmental impact statements.
>   (ii) Cannot or will not proceed unless other actions are taken previously or simultaneously.
>   (iii) Are interdependent parts of a larger action and depend on the larger action for their justification.
> (2) Cumulative actions, which when viewed with other proposed actions have cumulatively significant impacts and should therefore be discussed in the same impact statement.
> (3) Similar actions, which when viewed with other reasonably foreseeable or proposed agency actions, have similarities that provide a basis for evaluating their environmental consequences together, such as common timing or geography. An agency may wish to analyze these actions in the same impact statement. It should do so when the best way to assess adequately the combined impacts of similar actions or reasonable alternatives to such actions is to treat them in a single impact statement."[108]

Despite CEQ's specific requirements, the issue of scope of an EIS has been frequently litigated. Federal courts have asked several probative questions of a particular action in considering whether the action should be viewed as part of a larger project. In *Daly v. Volpe*[109] the court asked whether the segment of a larger project "fulfills important state and local needs," which might make it stand alone as a separate project. The court in *Hawthorn Environmental Preservation Ass'n v. Coleman*[110] the court asked whether it was "an extension or a connecting link" to a larger

---

[108] 40 CFR § 1508.25.
[109] Daly v. Volpe, 514 F.2d 1106 (9th Cir. 1975).
[110] Hawthorn Environmental Preservation Ass'n v. Coleman, 417 F. Supp. 1091 (N.D. Ga. 1976), aff'd per curiam, 551 F.2d 1055 (5th Cir. 1977).

project. In *Sierra Club v. Stamm*[111] the court asked if a particular project was a "unit unto itself," or an increment of a larger plan. In *Patterson v. Ebon*[112] asked whether the commitment of resources in one section made further construction more likely.

CEQ regulations have solved some segmentation problems by implementing Program EISs (PEISs), which permit broad evaluation of the environmental impact of a group of related actions.[113] Examples of PEISs are coal leasing by the Bureau of Land Management, the Central Arizona Project by the Bureau of Reclamation, and oil and gas leasing on the outer continental shelf by the Bureau of Land Management. In *Kleppe v. Sierra Club*[114] the U.S. Supreme Court considered a PEIS for coal production in the northern Great Plains. The Court held that a PEIS was not necessary in this case, but it noted that PEISs might be required for "interrelated actions" at the agency's discretion. In a later case, *Environmental Defense Fund v. Adams*[115] a federal district court interpreted Kleppe as requiring PEISs to be defined with "geographical, temporal and subject matter limits."

The Supreme Court's decision in *Kleppe* went much farther than discussing PEISs, however. It actually seemed to encourage segmentation by suggesting that an agency could prepare an EIS on one action, but postpone EISs that might be required on later, related actions. However, a subsequent Fifth Circuit decision, *Environmental Defense Fund v. Marsh*, interpreted *Kleppe* as permitting a court to require a comprehensive PEIS if the agency has "egregiously or arbitrarily violated the underlying purpose of NEPA."[116]

3.6.4 Contents of the EIS

CEQ regulations state that the purpose of the NEPA EIS requirement is to "insure that environmental information is available to public officials

---

[111] Sierra Club v. Stamm, 507 F.2d 788 (10th Cir. 1974).

[112] Patterson v. Ebon, 415 F. Supp. 1276 (D. Neb. 1976). See also Ross v. Federal Highway Admin., 172 F. Supp. 552 (D. Kan. 1997) [FHA could not divide highway project into four segments, complete three then fail to allocate federal funding to final controversial segment so as to avoid preparing a supplemental EIS for the remaining segment].

[113] 40 CFR §§ 1508.18(b) and 1502.4(b).

[114] Kleppe v. Sierra Club, 427 U.S. 390 (1976).

[115] Environmental Defense Fund v. Adams, 434 F. Supp. 403 (D.D.C. 1977).

[116] Environmental Defense Fund v. Marsh, 651 F.2d 983 (5th Cir. 1981). For more discussion of Kleppe and its impact, see Mandelker, §§ 9.02–9.04.

and citizens before decisions are made and before actions are taken."[117] The information contained in the EIS "must be of high quality . . . [a]ccurate scientific analysis, expert agency comments, and public scrutiny are essential to implementing NEPA."[118]

NEPA Section 102(2)(C) provides the basic framework for the contents of an EIS by stating that the "detailed statement" by the agency must discuss:

(i) the environmental impact of the proposed action,

(ii) any adverse environmental effects which cannot be avoided should the proposal be implemented,

(iii) alternatives to the proposed action,

(iv) the relationship between local short-term uses of man's environment and the maintenance and enhancement of long-term productivity, and

(v) any irreversible and irretrievable commitments of resources which would be involved in the proposed action should it be implemented.[119]

The details of the EIS process are contained in CEQ regulations rather than in NEPA itself.[120] On their face, the regulatory requirements under NEPA do not appear to be especially onerous. For simple agency actions with little environmental impact, the EIS process is generally quite simple. However, for larger projects (or those with serious environmental impacts), the process can be tedious indeed.

Environmental professionals have developed numerous techniques for assessing the environmental impacts of proposed actions. There are several excellent sources of information for environmental professionals that describe these techniques, although a detailed discussion is beyond the scope of this book.[121]

Several courts have entered the discussion of how environmental impacts are to be assessed in an EIS. The range of impacts that must be

---

[117] 40 CFR § 1500.1(b).

[118] Id.

[119] 42 U.S.C. § 4332(2)(C).

[120] Primarily at 40 CFR § 1502.

[121] For example, see Rao and Wooten (eds.), *Environmental Impact Analysis Handbook* (McGraw-Hill, 1980); and Camougis, *Environmental Biology for Engineers: a Guide to Environmental Assessment* (McGraw-Hill, 1981). An exhaustive list of sources is in Mandelker, § 10.02[1].

discussed is broad.[122] They include crime, emotional and physical isolation, and esthetic considerations.[123] However, remote, highly speculative impacts need not be considered.[124] The discussion of environmental impacts must be explanatory, and written in layperson's language.[125] Responsible opposing views must be presented.[126]

The EIS must also discuss all adverse environmental effects of the proposed action.[127] The EIS must have sufficient information regarding environmental costs and consequences that a well reasoned decision can be made.[128]

NEPA Section102(2)(C)(iii) requires a discussion of "alternatives" to the proposed action in the EIS.[129] Section 102(2)(E) specifies further that the agency must "study, develop, and describe appropriate alternatives to recommended courses of action in any proposal which involves unresolved conflicts concerning alternative uses of available resources."[130] CEQ regulations state that the discussion of alternatives is "the heart of the environmental impact statement," and specifies the types of alternatives that the agency must consider in developing the EIS:

(a) Rigorously explore and objectively evaluate all reasonable alternatives, and for alternatives which were eliminated from detailed study, briefly discuss the reasons for their having been eliminated.

(b) Devote substantial treatment to each alternative considered in detail including the proposed action so that reviewers may evaluate their comparative merits.

(c) Include reasonable alternatives not within the jurisdiction of the lead agency.

(d) Include the alternative of no action.

---

[122] 40 CFR § 1502.2(b) states that " Impacts shall be discussed in proportion to their significance. There shall be only brief discussion of other than significant issues."

[123] See Maryland Nat'l Cap. Park v. U.S. Postal Serv., 487 F.2d 1029 (D.C. Cir. 1973); and Chelsea Neighborhood Assoc. v. U.S. Postal Serv., 516 F.2d 378 (2d Cir. 1975).

[124] See Concerned About Trident v. Rumsfeld, 555 F.2d 817 (D.C. Cir. 1977).

[125] 40 CFR § 1502.8. See Minnesota Public Int. Group v. Butz, 541 F.2d 1292 (8th Cir. 1976).

[126] 40 CFR § 1502.9(b). See Committee for Nuclear Responsibility v. Seaborg, 463 F.2d 783 (D.C. Cir. 1971).

[127] 40 CFR § 1502.16.

[128] See Johnston v. Davis, 698 F.2d 1088 (10th Cir. 1983); and Coalition for Canyon Preservation v. Bowers, 632 F.2d 774 (9th Cir. 1980).

[129] 42 U.S.C. § 4332(2)(C)(iii).

[130] 42 U.S.C. § 4332(2)(E).

(e) Identify the agency's preferred alternative or alternatives, if one or more exists, in the draft statement and identify such alternative in the final statement unless another law prohibits the expression of such a preference.

(f) Include appropriate mitigation measures not already included in the proposed action or alternatives.[131]

It should come as no surprise that challenges to EISs have frequently been based on the argument that alternatives were not explored, were inadequately explored, or were erroneously rejected by the agency.[132] The U.S. Supreme Court ruled in *Vermont Yankee Nuclear Power Corp. v. Natural Resources Defense Council (NRDC)* that consideration of alternatives must be based on "feasibility," but that the agency need not include "every alternative device and thought conceivable by the mind of man."[133] On the other hand, courts have not hesitated to find an EIS inadequate if the discussion of alternatives is not addressed adequately.[134] It should be noted that a party challenging an EIS has the burden of proving that alternatives ignored by the agency are "reasonable."[135]

One confusing and often contradictory line of cases suggests that the agency need not discuss those alternatives that it feels lie outside the projects narrowly defined parameters.[136] By interpreting a project very narrowly, agencies have been able to avoid consideration of some alternatives. In *Northwest Coalition for Alternatives to Pesticides (NCAP) v. Lyng*, the court allowed the agency to ignore alternatives to weed herbicides, including changes in grazing practices that might have

---

[131] 40 CFR § 1502.14.

[132] For excellent reviews of the issues, see Mandelker, §§ 9.05[5], and 10.09; and Rodgers, at § 9.8(C).

[133] Vermont Yankee Nucl. Power Corp. v. NRDC, 435 U.S. 519, 551 (1978) [AEC need not consider alternative of "energy conservation," because there are limitless possibilities].

[134] See Neighbors of Cuddy Mtn. v. U.S. Forest Serv., 137 F.3d 1372 (9th Cir. 1998); DuBois v. U.S. Dept. of Agric., 102 F.3d 1273 (1st Cir. 1996); Massachusetts v. Clark, 594 F. Supp. 1373 (D. Mass. 1984); Aurora v. Hunt, 749 F.2d 1457 (10th Cir. 1984); and NRDC v. Hodel, 865 F.2d 288 (D.C. Cir. 1988).

[135] See Alaska Ctr. for Env't v. Armbruster, 131 F.3d 1285 (9th Cir. 1997); Mt. Lookout-Mt. Nebo Prop. Protection Ass'n v. FERC, 154 F.3d 165 (4th Cir. 1998); and Texas Comm. on Natural Resources v. Marsh, 736 F.2d 262 (5th Cir. 1984).

[136] See City of Angoon v. Hodel, 803 F.2d 1016 (9th Cir. 1986), cert. denied, 484 U.S. 870 (1987). However, if an agency defines an impermissibly narrow purpose for the contemplated project, thereby eliminating consideration of the full range of reasonable alternatives, the EIS will be deficient. See Simmons v. U.S. Army Corps of Eng'rs, 120 F.3d 664 (7th Cir. 1997).

encouraged weed growth. [137] A surprising number of courts have followed this "project purposes" approach, although an EIS still runs the risk of rejection if it fails reasonably to discuss alternatives.[138]

Related to the question of alternatives is the question of how an agency should deal with the issue of unavailable or uncertain scientific information in the preparation of an EIS.[139] From a practical perspective, this question has come to mean: Must an EIS contain a worst-case analysis? CEQ regulations prior to 1986 required that each EIS contain such an analysis, even though it seemed to fly in the face of early case law stating that the "alternatives" an agency must explore are merely those that are "feasible," not those that are possible however improbable. During the 1980s, several federal courts (especially the 9th Circuit) began to enforce the requirement for worst-case analysis in all EISs.[140] For its part, the U.S. Supreme Court in *Baltimore Gas & Electric Co. v. Natural Resources Defense Council*[141] allowed the Nuclear Regulatory Commission to adopt a "zero release" assumption based on risk analysis in licensing nuclear power plants, rather than adopting a worst-case analysis.

Following a politically charged attack on CEQ policies in the mid-1980s, CEQ amended its regulations to the current form:

> When an agency is evaluating reasonably foreseeable significant adverse effects on the human environment in an environmental impact statement and there is incomplete or unavailable information, the agency shall always make clear that such information is lacking.
>
> (a) If the incomplete information relevant to reasonably foreseeable significant adverse impacts is essential to a reasoned choice among

---

[137] Northwest Coalition for Alternatives to Pesticides (NCAP) v. Lyng, 844 F.2d 588 (9th Cir. 1988).

[138] See, for example, Roosevelt Campobello Int'l Park Comm'n v. EPA, 684 F.2d 1041 (1st Cir. 1982); National Wildlife Federation v. Federal Energy Regulatory Comm'n, 912 F.2d 1471 (D.C. Cir. 1990); and North Buckhead Civic Ass'n v. Skinner, 903 F.2d 1533 (11th Cir. 1990).

[139] See Gelpe and Tarlock, The Uses of Scientific Information in Environmental Decisionmaking, 48 S. Cal. L. Rev. 371 (1974).

[140] See, for example, Save Lake Washington v. Frank, 641 F.2d 1330 (9th Cir. 1981) [risks from lake navigation]; and Southern Oregon Citizens Against Toxic Sprays, Inc. v. Clark, 720 F.2d 1475 (9th Cir. 1983), cert. denied, 469 U.S. 1028 (1984) [herbicide spraying]. Some 9th Circuit cases disapproved of worst-case analysis, see Save Our Ecosystems v. Clark, 747 F.2d 1240 (9th Cir. 1984). Some other federal Circuit Courts also enforced the worst case analysis requirement, see City of New York v. USDOT, 715 F.2d 732 (2d Cir. 1983), appeal dismissed, 465 U.S. 1055 (1984) [transport of nuclear material].

[141] Baltimore Gas & Electric Co. v. Natural Resources Defense Council, 462 U.S. 87 (1983).

alternatives and the overall costs of obtaining it are not exorbitant, the agency shall include the information in the environmental impact statement.

(b) If the information relevant to reasonably foreseeable significant adverse impacts cannot be obtained because the overall costs of obtaining it are exorbitant or the means to obtain it are not known, the agency shall include within the environmental impact statement: (1) A statement that such information is incomplete or unavailable; (2) a statement of the relevance of the incomplete or unavailable information to evaluating reasonably foreseeable significant adverse impacts on the human environment; (3) a summary of existing credible scientific evidence which is relevant to evaluating the reasonably foreseeable significant adverse impacts on the human environment, and (4) the agency's evaluation of such impacts based upon theoretical approaches or research methods generally accepted in the scientific community.[142]

"Reasonably foreseeable" impacts are defined as those "which have catastrophic consequences, even if their probability of occurrence is low, provided that the analysis of the impacts is supported by credible scientific evidence, is not based on pure conjecture, and is within the rule of reason."[143]

The U.S. Supreme Court upheld CEQ's revocation of the worst case analysis requirement in *Robertson v. Methow Valley Citizens Council*.[144] In doing so, the Supreme Court reversed a series of 9th Circuit cases that had rejected the revised regulations.[145] However, the Supreme Court offered little guidance as to the form that analysis *should* take under the current regulations. Agencies are apparently still required to perform an analysis of low probability but disastrous impacts, even though the

---

[142] 40 CFR § 1502.22. For discussions of the events surrounding the withdrawal of worst-case analysis, see Rosenbaum, Amending CEQ's Worst Case Analysis Rule: Towards Better Decisionmaking? 15 Envtl. L. Rep. 10275 (1985); and Fitzgerald, The Rise and Fall of Worst Case Analysis, 18 U. Dayton L. Rev. 1 (1992).

[143] Id.

[144] Robertson v. Methow Valley Citizens Council, 490 U.S. 332 (1989).

[145] See Methow Valley Citizens Council v. Regional Forester, 833 F.2d 810 (9th Cir. 1987), reversed, 490 U.S. 332 (1989); Oregon Natural Resources Council v. Marsh, 832 F.2d 1489 (9th Cir. 1987); and Animal Defense Council v. Hodel, 840 F.2d 1432 (9th Cir. 1988), corrected, 867 F.2d 1244 (1989). For discussion of the Methow decision and its impact, see Mandelker, NEPA Alive and Well: The Supreme Court Takes Two, 19 Envtl. L. Rep. 10385 (1989); Robertson v. Methow Valley Citizens Council and the New "Worst Case" Regulation, 8 U.C.L.A. J. Envtl. L. & Pol'y 287 (1989); and Mandelker, § 10.07.

requirement is not well described.[146]

NEPA Section 102(2)(B) states that agencies must develop methods "which will insure that presently unquantified environmental amenities and values may be given appropriate consideration in decision making along with economic and technical considerations."[147] One way that agencies can comply with this requirement is to engage in "cost-benefit" analysis in their decision making.[148]

Cost-benefit analysis is an analytical technique by which all benefits of a proposed action are summed in the numerator of a fraction, while the costs are summed in the denominator. If the resulting ratio exceeds 1.0, then the project is said to be justified.[149] CEQ regulations are somewhat cautious, however, in requiring the following when preparing a cost-benefit analysis for an EIS:

> To assess the adequacy of compliance with Section 102(2)(B) of [NEPA] the [EIS] shall, when a cost-benefit analysis is prepared, discuss the relationship between that analysis and any analyses of unquantified environmental impacts, values, and amenities. For purposes of complying with the Act, the weighing of the merits and drawbacks of the various alternatives need not be displayed in a monetary cost-benefit analysis and should not be when there are important qualitative considerations.[150]

Federal courts have followed the CEQ's lead. In a leading case, *Trout Unlimited v. Morton*,[151] the court refused to find that an EIS for a dam project was inadequate just because the EIS did not contain a formal,

---

[146] See Sierra Club v. Marita, 46 F.3d 606 (7th Cir. 1995); and Salmon River Concerned Citizens v. Robertson, 32 F.3d 1346 (9th Cir. 1994).

[147] 42 U.S.C. § 4332(2)(B).

[148] See Anderson, Mandelker, and Tarlock, *Environmental Protection: Law and Policy*, 2d ed. (1990) at 44-40, 869-872; and Fraas, The Role of Economic Analysis in Shaping Environmental Policy, 54 Law & Contemp. Prob. 113 (1991).

[149] See discussion in Mandelker, § 10.08[1]; Farber, Revitalizing Regulation, 91 Mich. L. Rev. 1278 (1993); and Farmer & Hemmersbaugh, The Shadow of the Future: Discount Rates, Later Generations, and the Environment, 46 Vand. L. Rev. 267 (1993).

[150] 40 CFR § 1502.23.

[151] Trout Unlimited v. Morton, 509 F.2d 1276 (9th Cir. 1974). See also Columbia Basin Land Protection Assoc. v. Schlesinger, 643 F.2d 585 (9th Cir. 1981); NRDC v. Hodel, 435 F. Supp. 590, affirmed on other grounds sub nom. NRDC v. Munro, 626 F.2d 134 (9th Cir. 1980); and Stow v. United States, 696 F. Supp. 857 (W.D.N.Y. 1988).

quantitative cost-benefit analysis. The problem faced by the *Trout Unlimited* court was that it is often difficult (or impossible) to place quantitative values on the environment to determine either costs or benefits. As a result, courts have generally upheld cost-benefit analyses that were not quantified.[152]

A final source of concern comes from the requirement under NEPA Section 102(2)(C)(v) that the EIS discuss "any irreversible and irretrievable commitments of resources which would be involved in the proposed action should it be implemented."[153] This requirement has been interpreted to mean that an EIS must discuss "mitigation" of any environmental damage.[154] CEQ regulations define the term "mitigation" in an EIS as:

(a) Avoiding the impact altogether by not taking a certain action or parts of an action.

(b) Minimizing impacts by limiting the degree or magnitude of the action and its implementation.

(c) Rectifying the impact by repairing, rehabilitating, or restoring the affected environment.

(d) Reducing or eliminating the impact over time by preservation and maintenance operations during the life of the action.

(e) Compensating for the impact by replacing or providing substitute resources or environments.[155]

Despite the guidance from CEQ regulations, it is still difficult for an agency to determine the degree to which mitigation measures must be discussed in an EIS. The U.S. Supreme Court dealt with the mitigation issue in *Robertson v. Methow Valley Citizens Council*.[156] Citizens had sued the U.S. Forest Service for permitting the construction of a ski resort in a national forest without a detailed mitigation plan. The Ninth Circuit Court

---

[152] See Friends of Southeast's Future v. Morrison, 153 F.3d 1959 (9th Cir. 1998); Izaak Walton League of Am. v. Marsh, 655 F.2d 346 (D.C. Cir. 1981), cert denied, 454 U.S. 1092 (1981); South Louisiana Environmental Council v. Sand, 629 F.2d 1005 (D.C. Cir. 1981); and Gerosa v. Dole, 576 F. Supp. 344 (S.D.N.Y. 1983).
[153] 42 U.S.C. § 4332(2)(C)(v).
[154] See Mandelker, NEPA Alive and Well: The Supreme Court Takes Two, 19 Envtl. L. Rep. 10385 (1989); and Rossmann, NEPA: Not So Well at Twenty, 20 Envtl. L. 10174 (1990).
[155] 40 CFR § 1508.20.
[156] Robertson v. Methow Valley Citizens Council, 490 U.S. 332 (1989). See Mandelker, § 10.13; and Note, Robertson v. Methow Valley Citizens Council: The Gray Area of Environmental Impact Statement Mitigation. 10 J. Energy L. & Pol'y 217 (1990).

of Appeals had reversed the forest service decision because its EIS did not contain "a detailed explanation of specific measures which will be employed to mitigate the adverse impacts of a proposed action."[157] The Supreme Court reversed the Ninth Circuit, however, holding that "it would be inconsistent with NEPA's reliance on procedural mechanisms — as opposed to substantive, result-based standards — to demand the presence of a fully developed plan that will mitigate environmental harm before an agency can act."[158] In other words, the Supreme Court felt that "[b]ecause NEPA imposes no substantive requirement that mitigation measures actually be taken, it should not be read to require agencies to obtain an assurance that third parties will implement particular measures."[159] The Court then reminds us that "NEPA merely prohibits uninformed — rather than unwise — agency action."[160]

Unfortunately, the Court in *Methow Valley* did not provide the instruction that environmental professionals had anticipated. It is possible, however, to draw some important conclusions by reading *Methow Valley* together with other federal court cases. First, an EIS that does not discuss mitigation measures at all will be found to be inadequate.[161] It is also likely that an EIS will be found inadequate if it merely lists possible alternatives without explanation, if the mitigation measures are too vague, or if the information on which the mitigation measures are based is inadequate or delinquent.[162] On the other hand, EISs will probably withstand judicial scrutiny if mitigation measures are adequate, even if mitigation needs are not yet known, or if the agency is inexperienced.[163]

---

[157] Methow Valley Citizens Council v. Regional Forester, 833 F.2d 810, 819 (9th Cir. 1987), reversed, 490 U.S. 332 (1989).

[158] Robertson v. Methow Valley Citizens Council, 490 U.S. at 353.

[159] Id., at footnote 16.

[160] Id. at 351.

[161] See City of Carmel-by-the-Sea v. USDOT, 123 F.3d 1142 (9th Cir. 1997) [but finding that Federal Highway Administration's discussion of wetlands and wetland mitigation plan was "reasonably thorough"].

[162] See Neighbors of Cuddy Mtn. v. U.S. Forest Serv., 137 F.3d 1372 (9th Cir. 1998) [EIS for timber sales was inadequate, in part because description of mitigation measures to offset adverse impacts on redband trout habitat was insufficient]; Northwest Indian Cemetery Protective Ass'n v. Peterson, 565 F. Supp. 586 (N.D. Cal. 1983) [road through national forest]; Sierra Club v. Marsh, 769 F.2d 868 (1st Cir. 1985); Stein v. Barton, 740 F. Supp. 743 (D. Alaska 1990); and Jones v. Gordon, 792 F.2d 821 (9th Cir. 1986); and Friends of the Earth v. Hall, 693 F. Supp. 904 (W.D. Wash. 1988).

[163] Citizens Against Burlington v. Busey, 938 F.2d 190 (D.C. Cir.), cert. denied, 502 U.S. 994 (1991) [expansion of airport]; Town of Norfolk v. EPA, 761 F. Supp. 867 (D. Mass. 1991), affirmed, 960 F.2d 143 (1st Cir. 1992) [groundwater mitigation for sewage plant]; and Northern Crawfish Frog v.

3.6.5 Agency Review of EIS

The first version of the EIS produced by the lead agency is known as a "draft EIS," or DEIS. NEPA Section 102(2)(C) requires that the lead agency "consult with and obtain the comments of any Federal agency which has jurisdiction by law or special expertise with respect to any environmental impact involved."[164] In addition, CEQ regulations require that the lead agency:

(2) Request the comments of:
   (i) Appropriate State and local agencies which are authorized to develop and enforce environmental standards;
   (ii) Indian tribes, when the effects may be on a reservation; and
   (iii) Any agency which has requested that it receive statements on actions of the kind proposed.
. . .
(3) Request comments from the applicant, if any.
(4) Request comments from the public, affirmatively soliciting comments from those persons or organizations who may be interested or affected."[165]

Federal agencies that have jurisdiction "by law" are specified in other federal statutes, and require consultation with specific agencies under certain circumstances. Examples are the Fish and Wildlife Coordination Act,[166] the Endangered Species Act,[167] the Department of Transportation Act,[168] and the Clean Air Act.[169]

It is important to note that NEPA and CEQ regulations make numerous attempts to involve as many interested parties as possible in the EIS process. NEPA's "environmental full disclosure" is intended to inspire public confidence in agency decision making.[170]

Once the comments on the DEIS have been obtained by the lead agency, the comments are made available to the President, CEQ, and to the

---

Federal Highway Administration, 858 F. Supp. 1503 (D. Kan. 1994) [wetlands mitigation for highway]. For additional examples of adequate mitigation, see Mandelker, § 10.13. See also NRDC v. Hodel, 819 F.2d 888 (9th Cir. 1987) [need for grazing restrictions not determined].

[164] 42 U.S.C. § 4332(2)(C).
[165] 40 CFR § 1503.1(a).
[166] 16 U.S.C. § 661 ff., U.S. Fish and Wildlife Service.
[167] 16 U.S.C. § 1531 ff., U.S. Fish and Wildlife Service.
[168] 49 U.S.C. § 1653(f), U.S. Department of Transportation.
[169] 42 U.S.C. § 7609, U.S. Environmental Protection Agency.
[170] See Rodgers, at § 9.3(B).

public.[171] Any member of the public may obtain access to comments made on a DEIS through the Freedom of Information Act.[172] The comments must also "accompany the [EIS] through the existing agency review processes."[173]

Once the lead agency has received and reviewed the comments on the DEIS, it prepares a "final EIS," or FEIS. The FEIS incorporates and responds to all comments on the draft, including "any responsible opposing view which was not adequately discussed in the draft statement and shall indicate the agency's response to the issues raised."[174] The entire FEIS is then sent to: "(a) Any Federal agency which has jurisdiction by law or special expertise with respect to any environmental impact involved and any appropriate Federal, State or local agency authorized to develop and enforce environmental standards; (b) The applicant, if any; (c) Any person, organization, or agency requesting the entire environmental impact statement; (d) . . . any person, organization, or agency which submitted substantive comments on the draft."[175] The FEIS must then accompany the proposal at every subsequent step where balances of environmental and non-environmental factors are appropriate, or where alterations might be made to minimize costs.[176]

Even after an agency has filed a DEIS or a FEIS, there will not be a formal decision on the proposed action by the agency "until the later of the following dates: (1) Ninety (90) days after publication of the notice . . . for a draft environmental impact statement; [or] (2) Thirty (30) days after publication of the notice . . . for a final environmental impact statement," absent a showing of compelling reason.[177]

If the FEIS differs substantially from the DEIS (e.g., if the proposed environmental impacts have changed), then the process begins again with the DEIS.[178]

An agency must file a "supplemental EIS" (SEIS) if: "(i) The agency makes substantial changes in the proposed action that are relevant to environmental concerns; or (ii) There are significant new circumstances or information relevant to environmental concerns and bearing on the

---

[171] 42 U.S.C. § 4332(2)(C).
[172] 5 U.S.C. § 552.
[173] 42 U.S.C. § 4332(2)(C).
[174] 40 CFR § 1502.9(b).
[175] 40 CFR § 1502.19(d).
[176] See Calvert Cliffs' Coordinating Comm., Inc. v. U.S. Atomic Energy Comm'n, 449 F.2d 1109 (D.C. Cir. 1971).
[177] 40 CFR § 1506.10.
[178] See City of Carmel-by-the-Sea v. USDOT, 95 F.3d 892 (9th Cir. 1996), superseded by 123 F.3d 1142 (9th Cir. 1997).

proposed action or its impacts.[179] The U.S. Supreme Court in *Marsh v. Oregon Natural Resources Department* stated further that an SEIS may be required where the original EIS is inadequate in some details, a court requires follow-up procedures, or "new information" is made available.[180] SEISs follow the same process of review as the original DEIS or FEIS.[181]

3.7 Judicial Review of Agency Decisions

Recall that Congress created NEPA with language that is intentionally short and vague, which encouraged the development of a "common law" of NEPA. The development of NEPA's common law has given it a rich history of judicial review.

This chapter has already discussed many forms of judicial review under NEPA. However, it is important to review the mechanisms by which judicial review proceeds. Key questions include: (1) Who, exactly, may sue under NEPA, and how is it done? (2) When must a NEPA lawsuit be brought? and (3) How do NEPA's substantive and procedural provisions lend themselves to lawsuits?

3.7.1 Who May Sue? ... The Issue of Standing

Article III of the U.S. Constitution limits federal judicial involvement to "cases and controversies." In other words, a person who wishes to bring a lawsuit in any federal (or most state) court must demonstrate that they have "standing" to sue. The Administrative Procedures Act clarifies the situation as it applies to federal agencies by granting standing to "persons suffering legal wrong . . . or adversely affected or aggrieved by agency action."[182]

In order to assert standing in a NEPA challenge, a plaintiff must first claim "injury in fact." The U.S. Supreme Court initially made the burden of proving injury in fact a relatively small hurdle in NEPA cases.[183] In

---

[179] 40 CFR § 1502.9(c).
[180] Marsh v. Oregon Natural Resources Council, 490 U.S. 360 (1989).
[181] 40 CFR § 1502.9(c)(4).
[182] 5 U.S.C. § 702.
[183] For reviews of the evolution of the standing issue in NEPA cases, see Mandelker, § 4.06; Lawrence, Standing for Environmental Groups: An Overview and Recent Developments in the D.C. Circuit, 19 Envtl. L. Rep. 10289 (1989); and Steuer & Juni, Court Access for Environmental Plaintiffs: Standing Doctrine in Lujan v. National Wildlife Federation, 15 Harv. Envtl. L. Rev. 187 (1991).

Sierra Club v. Morton,[184] the U.S. Supreme Court allowed a claim of injury to the "esthetic and environmental well-being" of club members to establish standing in a challenge to a U.S. Forest Service action under NEPA. One year later, the Supreme Court in *U.S. v. Students Challenging Regulatory Agency Procedures (SCRAP I)*[185] allowed standing to students and an environmental group, even though the injury (plaintiffs argued that an ICC action would discourage recycling) was "far less direct and perceptible" than was the case in *Sierra Club v. Morton* and followed a "more attenuated line of causation." A few years later in *Duke Power Co. v. Carolina Environmental Study Group*,[186] the Court let stand a lower court holding that an environmental group had standing to protest a limitation on liability for nuclear accidents because nuclear power plants cause environmental harm.

This line of cases freely giving standing in NEPA cases may have come to a halt in 1990, however. In *Lujan v. National Wildlife Federation*,[187] members of an environmental group sued the Bureau of Land Management to prevent mining and oil and gas leases on federal land, arguing that it would interfere with their recreational use and enjoyment of the land. The Supreme Court denied standing to the environmentalists. Although the Supreme Court based its decision to deny standing on the related issue of "ripeness" (discussed below), the Supreme Court in *Lujan* made suggestions that the days of open-ended standing may be over.[188] In an interesting case involving standing under the Endangered Species Act,[189] the Supreme Court seems to have allowed standing for "procedural injury" (where a person's procedural rights to protect their interests have been impaired) by specifically noting a NEPA example.[190]

---

[184] Sierra Club v. Morton, 405 U.S. 727 (1972).
[185] U.S. v. Students Challenging Regulatory Agency Procedures (SCRAP I), 412 U.S. 669 (1973).
[186] Duke Power Co. v. Carolina Environmental Study Group, 438 U.S. 59 (1978).
[187] Lujan v. National Wildlife Federation, 497 U.S. 871 (1990).
[188] See references in note 182 above, and Sheldon, N.W.F. v. Lujan: Justice Scalia Restricts Government Standing to Constrain the Courts, 10 Envtl. L. 10557 (1990).
[189] 16 U.S.C. §§ 1531-1544.
[190] Lujan v. Defenders of Wildlife, 540 U.S. 555, 572 n. 7 (1992). See also Bennett v. Spear, 520 U.S. 154 (1997) [broadly worded standing requirements of the Endangered Species Act's citizen suit provision permits "any person" to commence a civil suit since the overall subject matter of the ESA is the environment, a matter in which all persons have an interest]. See also Sunstein, What's Standing After Lujan? Of Citizen Suits, "Injuries," and Article III, 91 Mich. L. Rev. 163 (1992).

In addition to establishing "injury in fact," a NEPA plaintiff must also be among those injured,[191] and the injury must be real.[192]

Governmental entities (like local and state governments) must demonstrate injury in fact to obtain standing in a NEPA challenge just like private individuals and organizations. A state may have standing if it is located in the geographical proximity of a proposed agency action such that it must be consulted for an EIS, or if the state's environmental resources would be damaged by agency action.[193] A local government may have standing if its environmental resources or property might be damaged by federal actions, or if federal agency actions might interfere with local environmental projects.[194]

Environmental groups may have standing if their members have suffered injury to their esthetic, conservation, or economic interests, if members use a wilderness area for recreation, or if noise and air pollution would harm members' health or conservational interests.[195] On the other hand,

---

[191] See Sierra Club v. Morton, 405 U.S. 727 (1972); and Warth v. Seldon, 422 U.S. 490 (1975).

[192] See Presidio Golf Club v. National Park Serv., 155 F.3d 1153 (9th Cir. 1998) [National Park Service's construction of a public clubhouse could constitute sufficient future injury-in-fact to members of a private clubhouse since members might stop paying dues as a result of construction of a public clubhouse]; Hiatt Grain & Feed v. Bergland, 446 F. Supp. 457 (D. Kan. 1978), affirmed, 602 F.2d 939 (10th Cir. 1979), cert. denied, 444 U.S. 1073 (1980) [increased price supports by USDA too speculative and attenuated for argument that air pollution would increase]; and Nevada Land Ass'n v. U.S. Forest Svc., 8 F.3d 713 (9th Cir. 1994) [ranchers who suffered a "lifestyle" loss due to forest management practices lacked standing].

[193] See California v. Block, 690 F.2d 753 (9th Cir. 1982) [State had standing to challenge Forest Service EIS allocating national forest lands]; and Idaho v. ICC, 35 F.3d 585 (D.C. Cir. 1994) [railroad activities that pollute state land]; but see Michigan v. United States, 994 F.2d 1197 (6th Cir. 1993) [no injury to state].

[194] See City of Coleman v. Davis, 521 F.2d 661 (9th Cir. 1975) [highway construction would affect city's waste supply and population]; and Catron County Board of Commissioners v. USFWS, 75 F.3d 1429 (10th Cir. 1996) [county property damaged by flood]. But see City of Evanston v. Regional Transportation Authority, 825 F.2d 1121 (7th Cir. 1987), cert. denied, 484 U.S. 1005 (1988) [city failed to demonstrate that conversion of steel business to garage would increase environmental damage]. See also City of Klamath Falls v. Babbitt, 947 F. Supp. 1 (D.D.C. 1996) [river designation prevented dam construction which would solve environmental problems].

[195] See Minnesota Public Int. Research Group v. Butz (I), 498 F.2d 1314 (8th Cir. 1974) [wilderness area used for recreation]; Cady v. Morton, 527 F.2d 786 (9th Cir. 1975) [esthetic and conservation interests]; and Committee for Auto Responsibility v. Solomon, 603 F.2d 992 (D.C.Cir. 1979) [health and conservation interests].

some courts have held that environmental groups do not have standing if their interests are too abstract or remote.[196]

3.7.2 Ripeness, and Exhaustion of Administrative Remedies

Two other potential impediments to a plaintiff in a NEPA challenge are the related issues of "ripeness" and "exhaustion of administrative remedies." Both issues are related to the issue of "standing" discussed above.

The issue in "ripeness" obstacles is whether or not a case has matured or "ripened" into a case or controversy worthy of adjudication under Article III of the Constitution.[197] From a practical standpoint, this means that a plaintiff wishing to challenge an agency action (or proposed action) under NEPA must wait until the administrative process has matured to the point that there is a "final agency action for which there is no adequate remedy."[198]

As would be expected, most cases in which ripeness is an issue involve a challenge at some point to the EIS process. Most courts have held that the EIS process is "final" once a FEIS has been prepared.[199] However, there is considerable variation among federal courts in their willingness to find that a case is ripe. Many ripeness cases are extremely fact-specific, such that a decision on an issue in one court may be carefully distinguished by another court that wishes to reach a different result. Nonetheless, a court will likely find that a case is not ripe for adjudication if the agency has not reached a final decision on an action, or if only

---

[196] See Florida Audubon Soc'y v. Bentsen, 94 F.3d 658 (D.C. Cir. 1996); Broadened Horizons Riverkeepers v. U.S. Army Corps of Eng'rs, 8 F. Supp. 2d 730 (E.D. Tenn. 1998); Continued Action on Transportation v. Adams, 9 ELR 20648 (D.D.C. 8/10/79), aff'd, 618 F.2d 1078; 1980 (4th Cir. 1980).

[197] See Hust, Ripeness Doctrine in NEPA Cases: A Rotten Jurisdictional Barrier, 11 Law & Ineq. J. 505 (1993).

[198] Administrative Procedures Act, 5 U.S.C. § 704.

[199] See Resources Ltd., Inc. v. Robertson, 35 F.3d 1300 (9th Cir. 1994); Northern Alaska Environmental Center v. Hodel, 803 F.2d 466 (9th Cir. 1986); Public Service Co. of Colo. v. Andrus, 825 F. Supp. 475 (D. Idaho 1993); and Sierra Club v. Marita, 46 F.3d 606 (7th Cir. 1995). In Friends of the Wild Swan, Inc. v. U.S. Forest Service, 910 F. Supp. 1500 (D. Or. 1995), the court allowed a claim of unreasonable delay because the agency had failed to respond to information and concerns that required immediate action.

preliminary activities have taken place.[200] Other courts have disagreed.[201] Some courts invoke the ripeness doctrine if agency determinations are necessary to judge a plaintiff's claim.[202]

An issue related both to standing and to ripeness is the requirement of "exhaustion of administrative remedies." Here, a plaintiff wishing to challenge in court a proposed agency action under NEPA must first exhaust all administrative, nonjudicial procedures available.[203]

The intent of the "exhaustion" requirement is twofold; it assures that the administrative process is allowed to progress without interruption, and it helps to prevent the waste of judicial time and energy on issues that are best resolved by the agencies.

The exhaustion rule works very differently under NEPA than it does under other environmental laws. Under a more typical, non-NEPA scenario, a decision from a federal agency decision must first be appealed internally (i.e., to the agency that made the decision, since it has expertise). For example, denial of a NPDES permit by the EPA under the federal Clean Water Act must be appealed to an EPA hearing officer (and rejected there) before it can be appealed to a federal district court.[204] In a NEPA case, however, the court will typically read NEPA and the non-NEPA statute together. The court will then defer to the agency with special expertise in the area of the non-NEPA statute for administrative

---

[200] See Eastern Connecticut Citizens Action Group v. Dole, 638 F. Supp. 1297 (D. Conn.), aff'd per curiam, 804 F.2d 804 (2d Cir. 1986), cert. denied, 481 U.S. 1068 (1987); Coconut Grove House, Inc. v. U.S. Dept. of Health and Human Servs, 805 F. Supp. 39 (S.D. Fla. 1992); and Oregon Natural Resources Council v. Bureau of Reclamation, 49 F.3d 1441 (9th Cir. 1995). See also Environmental Defense Fund v. Johnson, 629 F.2d 239 (2d Cir. 1980 [study of river project prior to FEIS]; and Lujan v. National Wildlife Fed'n, 497 U.S. 871 (1990) [BLM "program" withdrawing protected land not ripe]. See also Note, Preserving Review of Undeclared Programs: A Statutory Redefinition of Final Agency Action, 101 Yale L.J. 643 (1991).
[201] See Friedman Brothers Investment Co. v. Lewis, 676 F.2d 1317 (9th Cir. 1982) [grant obtained for site purchase and design, but no contracts as yet]; Save Barton Creek Assoc. v. Federal Highway Admin., 950 F.2d 1129 (5th Cir. 1991) [highway construction had begun in some instances]; and Blue Ocean Preservation Society v. Watkins, 754 F. Supp. 1450 (D. Haw. 1991) [appropriations for final project, but no contracts as yet]. See also Comment, Opening the Door to Early Judicial Review of Environmental Impact Statements, 55 U. Colo. L. Rev. 99 (1983).
[202] See Izaak Walton League of America v. St. Clair, 497 F.2d 849 (8th Cir. 1974).
[203] See Mandelker, § 4.08[1]; and Gelpe, Exhaustion of Administrative Remedies: Lessons from Environmental Cases, 53 Geo. Wash. L. Rev. 1 (1983-1984).
[204] This particular scenario will be discussed more fully in chapter 4.

process. Because NEPA applies equally to all federal agencies, and because no agency has special expertise in EIS preparation, the administrative process under the non-NEPA statute is followed.[205] Due primarily to the dual nature of NEPA in these instances, courts have usually adopted a "flexible balancing test" in determining the applicability of the exhaustion requirement.[206]

Courts generally have not applied the exhaustion of administrative remedies requirement where there is no administrative remedy under the non-NEPA statute, or where an administrative appeal would be pointless or futile.[207] Some courts have not required exhaustion of administrative remedies where it appeared that the agency had used the requirement to delay or avoid its responsibilities under NEPA.[208]

3.7.3 Substantive and Procedural Review

The U.S. Supreme Court in *Aberdeen & Rockfish R.R. Co. v. Students Challenging Regulatory Agency Procedures*[209] (SCRAP II) held that private citizens may sue to enforce federal agency compliance with NEPA. If a hopeful plaintiff overcomes the obstacles of standing, ripeness, and exhaustion of administrative remedies, the next key question is whether a federal court can reverse an agency's proposed action (or inaction) if the action is offensive to either NEPA's substantive or procedural requirements.

---

[205] See Park City Resource Agency v. U.S. Dep't of Agric., 817 F.2d 609 (10th Cir. 1987). Note that administrative authority must be specifically delegated to the federal agency by the non-NEPA statute; see Darby v. Cisneros, 509 U.S. 137 (1993).

[206] In Foundation on Economic Trends v. Heckler, 756 F.2d 143 (D.C. Cir. 1985), the court held that the exhaustion requirement was inapplicable due to the importance of the first genetic engineering case. See also Ayers v. Espy, 873 F. Supp. 455 (D. Colo. 1994) [statutory issues are an exception to the exhaustion rule]. Some courts have required exhaustion after the balancing test was applied: see Sierra Club v. U.S. Forest Serv., 878 F. Supp. 1295 (D. S.D. 1993).

[207] See Park City Resource Agency v. U.S. Dep't of Agric., 817 F.2d 609 (10th Cir. 1987); and Cornell Village Tower Condominium v. Department of Housing & Urban Development, 750 F. Supp. 909 (N.D. Ill. 1990) [no administrative remedy]. See also Sierra Club v. Espy, 822 F. Supp. 356 (E.D. Tex. 1993), reversed, 18 F.3d 1202 (5th Cir. 1994) [appeal futile]; and Wright v. Inman, 923 F. Supp. 1295 (D. Nev. 1996).

[208] See Jette v. Bergland, 579 F.2d 59 (10th Cir. 1978); and Sierra Club v. Robertson, 764 F. Supp. 546 (W.D. Ark. 1991).

[209] Aberdeen & Rockfish R.R. Co. v. Students Challenging Regulatory Agency Procedures (SCRAP II), 422 U.S. 289 (1975).

Substantive Review. NEPA Section 101 is usually designated "substantive NEPA," as was discussed in section 3.2.3 above. Many early federal court decisions held that an agency's violations of NEPA's substantive provisions was reversible.[210] However, the U.S. Supreme Court put substantive agency review to rest in *Vermont Yankee Nuclear Power Corp. v. NRDC*, holding that "NEPA does set forth significant substantive goals for the Nation, but its mandate to the agencies is essentially procedural."[211] In other decisions the Supreme Court has stated that an agency's responsibility under NEPA Section 101 is merely to take a "hard look" at the environmental consequences of its actions, and that a reviewing court cannot "substitute its judgment for that of the agency as to the environmental consequences of its actions."[212] Today, there is little chance that a reviewing court will find that an agency decision violates NEPA's substantive provisions.[213]

Procedural Review. As was discussed in section 3.2.3 above, NEPA Section 102 is usually designated "procedural NEPA." By holding that NEPA is "essentially procedural,"[214] the U.S. Supreme Court apparently determined that a federal court *can* reverse a decision in favor of an agency action for failure to follow NEPA's procedural requirements. Unlike the relatively few cases that have argued for substantive review of agency decisions, procedural review under NEPA has been an extremely active area of litigation.

Most challenges to agency action under NEPA Section 102 are those that challenge an EIS, or some aspect of the EIS process. The first issue a reviewing court must address is the question of the scope of judicial

---

[210] See Environmental Defense Fund v. Corps of Eng'rs, 470 F.2d 289 (8th Cir. 1972), cert. denied, 412 U.S. 931 (1973); Sierra Club v. Froehlke, 486 F.2d 946 (7th Cir. 1973); and Conservation Council v. Froehlke, 473 F.2d 664 (4th Cir. 1973).
[211] Vermont Yankee Nuclear Power Corp. v. NRDC, 435 U.S. 519, 558 (1978).
[212] Kleppe v. Sierra Club, 427 U.S. 390, 410 (1976). See also Strycker's Bay Neighborhood Council v. Karlen, 444 U.S. 223, 229 (1980), and Robertson v. Methow Valley Citizens Council, 490 U.S. 332, 351 (1989) [NEPA "simply prescribes the necessary process for preventing uninformed-rather than unwise-agency action."].
[213] See Cabinet Mountains Wilderness v. Peterson, 685 F.2d 678 (D.C.Cir. 1982); Sierra Club v. Army Corps of Eng'rs 701 F.2d 1011 (2d Cir. 1983); and Citizens Against Burlington v. Busey, 938 F.2d 190 (D.C. Cir. 1991), cert. denied, 112 S.Ct. 616 (1991). See also Weinberg, It's Time to Put NEPA Back on Course, 3 N.Y.U. Envtl. L. J. 99 (1994); Ferester, Revitalizing the National Environmental Policy Act: Substantive Law Adaptations from NEPA's Progeny, 16 Harv. Envtl. L. Rev. 207 (1992).
[214] Vermont Yankee Nuclear Power Corp. v. NRDC, 435 U.S. 519, 558 (1978).

inquiry. In other words, in what depth should the court review the actions (or inactions) by the agency? In *Citizens to Preserve Overton Park v. Volpe*, the U.S. Supreme Court held that a reviewing court must employ a "thorough, probing, in depth review" of the agency decision that goes beyond the "bare record" in the case.[215] The *Overton Park* standard is occasionally called the "hard look" doctrine (that is, the court must take a "hard look" at the agency's decision), although this terminology more properly refers to the agency's requirement to take a "hard look" at the environmental consequences of its actions.[216]

The next issue a reviewing court must address is the standard used by the court in reviewing the agency's decision (known as a "standard of review"). In other words, to what level must the agency have erred in its decision making for a court to reverse that decision? Unfortunately, the Administrative Procedures Act[217] (APA) as well as court decisions have applied at least three standards depending on the type of NEPA case involved. For convenience, the following discussion will address three types of NEPA EIS cases: (1) failure to prepare an Environmental Assessment, where the agency determined that the proposed project did not trigger NEPA; (2) FONSI cases, in which the agency prepared an EA, but determined that the proposed project had no significant environmental impact; and (3) adequacy of the EIS cases, where an EIS was prepared, but there is question as to its adequacy.

First, in cases where an EA was not prepared, there is a split of authority among the courts as to the standard to be applied. Most courts have held that these cases are reviewable only under the APA "arbitrary and capricious" standard.[218] In other words, a reviewing court may only reverse the decision of an agency when the agency decision was "arbitrary, capricious, an abuse of [agency] discretion, or otherwise not in accordance with law."[219] However, a few courts have held that absence of EIS cases are reviewable under the "rational basis" test, in which case the agency's decision must be accepted if it has a rational basis in law.[220] Neither of

---

[215] Citizens to Preserve Overton Park v. Volpe, 401 U.S. 402, 415-420 (1971). The Supreme Court's mandate in the Overton Park decision has come to be known as the "substantial inquiry" requirement.

[216] Kleppe v. Sierra Club, 427 U.S. 390, 410 (1976). The two usages of the "hard look" doctrine have become so thoroughly intertwined that Professor Mandelker suggested "courts must take a hard look to ensure that the agency took a hard look." Mandelker, § 3.04[4].

[217] 5 U.S.C. § 706(2).

[218] See Hanly v. Kleindienst (Hanly II), 471 F.2d 823 (2d Cir. 1972); and Providence Rd. Community Ass'n v. EPA, 683 F.2d 80 (4th Cir. 1982).

[219] 5 U.S.C. § 706(2)(A).

[220] See Brandon v. Pierce, 725 F.2d 555 (10th Cir. 1984).

these standards is particularly favorable to a plaintiff wishing to challenge agency action (or inaction). It is debatable whether the burden is greater on a plaintiff to demonstrate that an agency decision was "arbitrary and capricious," or that there was not a "rational basis" in law for that decision.

The standard of review in FONSI cases (where the agency found that there was no significant impact from the proposed action) was given by the U.S. Supreme Court's three-part test in *Baltimore Gas & Electric Co. v. NRDC*:[221] (1) Did the agency take a "hard look" at the environmental consequences of the proposed action? (2) Did the agency adequately consider and disclose the environmental impacts of the proposed action? and (3) Is the agency decision arbitrary and capricious?

In an "adequacy of EIS" case (where the issue is whether or not the EIS that was prepared was adequate), there has been another split among the courts. The most common standard is the "reasonableness" standard (i.e., was the EIS "reasonably" adequate).[222] Some courts use the "arbitrary and capricious" standard,[223] while a few courts have allowed only a "limited review" of NEPA compliance (very deferential to the agencies).[224] The U.S. Supreme Court *seems* to support the "arbitrary and capricious" standard, but the issue is still unresolved.[225]

3.8 Remedies Under NEPA

Should a plaintiff prevail after having survived the foregoing, there is another interesting revelation still to come. What, if anything, is the plaintiff's "remedy" under NEPA? In other words, what can a plaintiff expect to get out of a NEPA case?

As a first order of business, it is important to note that NEPA has no provision for traditional damages. If a person has experienced damage at the hands of a federal agency for which they desire compensation, then the person must seek relief under some other federal law such as the Federal Tort Claims Act,[226] or under a common law action.

More realistically, NEPA plaintiffs wish either to delay a proposed agency action until the proper environmental safeguards can be taken, or

---

[221] Baltimore Gas & Electric Co. v. NRDC, 462 U.S. 87 (1983).
[222] See Warm Springs Dam Task Force v. Gribble, 565 F.2d 549 (9th Cir. 1977).
[223] See Chelsea Neighborhood Assoc. v. U.S. Postal Svc., 516 F.2d 378 (2d Cir. 1975).
[224] See Lathan v. Brinegar, 506 F.2d 677 (9th Cir. 1974).
[225] See Marsh v. Oregon Natural Resources Council, 490 U.S. 360 (1989).
[226] 28 U.S.C. §§ 1291 ff.

stop it altogether. They do this typically by requesting "injunctive relief" as well as "declaratory relief." "Injunctive relief" attempts to enjoin (i.e., prohibit) the agency's proposed action, while "declaratory relief" forces the agency to discharge their legal obligations under NEPA. Most NEPA plaintiffs request both injunctive and declaratory relief.

The most common form of relief in NEPA cases is the preliminary injunction (PI), which temporarily stops (or limits) the actions by the defendant agency until it complies with its responsibilities under NEPA. Unlike preliminary injunctions in other kinds of cases, NEPA plaintiffs rarely have to post a cash bond (or the bond amount is minimal) to protect the defendant against damages in the event that the injunction is lifted.[227]

The decision to grant a PI lies with the federal district court. The district court's decision to grant (or not to grant) a PI can be appealed to the appropriate circuit court of appeal, but the circuit courts rarely reverse the district court's decision.[228]

Most courts will examine at least three factors in determining if a PI is appropriate.[229] First, does the plaintiff have at least a reasonable chance of success on the merits of the claim? Second, does the harm to the environment if the PI is not granted outweigh the harm to the agency if the PI is granted? Third, is there a public interest that would be served by granting the PI? A few courts have required a demonstration of irreparable harm to the environment coupled with probable success by the plaintiff before they will issue a PI.[230] On the other hand, some courts make an exception to the traditional balancing of equities in NEPA challenges on policy grounds, making a PI much easier to accomplish.[231]

---

[227] See Morgan v. Walter, 728 F. Supp. 1483 (D. Idaho 1989). However, a few courts have required substantial bonds (see Monarch Chemical Works v. Exon, 452 F. Supp. 493 (D. Neb. 1987)) or an injunction is refused (see Conservation Law Foundation v. Air Force, 26 Env't Rep. Cas. 2146 (D. Mass. 1987)). See also Mandelker, § 4.10[2][a].

[228] See Lakeshore Terminal & Pipeline Co. v. Defense Fuel Supply Center, 777 F.2d 1171 (6th Cir. 1985); and National Wildlife Fed'n v. Coston, 773 F.2d 1513 (9th Cir. 1985). However, if the district court has clearly abused its discretion, the circuit court will reverse its decision, see Scherr v. Volpe, 466 F.2d 1027 (7th Cir. 1972).

[229] See Smith v. Soil Conservation Serv., 563 F. Supp. 843 (W.D. Okla. 1982); and Half Moon Bay Fisherman's Marketing Assn. v. Carlucci, 857 F.2d 505 (9th Cir. 1988); Sierra Club v. Marsh, 714 F. Supp. 539 (D. Me. 1989).

[230] See Northern Alaska Environmental Center v. Hodel, 803 F.2d 466 (9th Cir. 1986).

[231] See Puna Peaks v. Edwards, 554 F. Supp. 117 (D. Haw. 1982); and Foundation on Economic Trends v. Heckler, 610 F. Supp. 829 (D. D.C. 1985). See also Comment, Injunctions for NEPA Violations: Balancing the Equities, 59 U. Chi. L. Rev. 1263 (1992).

A PI is more likely to be granted if the project is in its early stages, and there is a real threat of environmental harm.[232] A PI allows an agency to prepare an EIS, or correct an EIS if it is defective. If an EIS is inadequate, a PI may spell out corrective action.[233] Some courts have allowed work on a project to continue during the corrective period.[234]

A "permanent injunction" goes beyond a PI, and stops an agency action completely. Most courts use an equity balancing process similar to the PI process to determine if a permanent injunction is appropriate, although many cases have not considered harm to the defendant as a factor.[235] Courts have granted permanent injunctions where irreparable injury or strong public interest outweigh factors to the contrary.[236] However, most courts will refuse to grant a permanent injunction if continuing irreparable harm is unlikely or other factors don't favor it.[237]

Many NEPA plaintiffs request "declaratory relief" in addition to their request for an injunction. If a court issues a "declaratory judgment" in a NEPA case, it is specifying the legal obligations of the defendant agency. This assists the agency by specifying at an early point what the court has determined are its obligations, and it aids the plaintiff by avoiding the necessity of proving irreparable harm as is the case with a PI.[238] In *Seattle Audubon Society v. Moseley*, the Ninth Circuit Court of Appeals held that declaratory relief is appropriate when there is a "substantial controversy" between the parties.[239] While some courts have been willing to grant

---

[232] See National Wildlife Fed'n v. Adams, 629 F.2d 687 (9th Cir. 1980) [highway through wetlands]).
[233] See Cape Henry Bird Club v. Laird, 359 F. Supp. 404 (W.D.Va. 1973), aff'd, 484 F.2d 453 (4th Cir. 1973).
[234] See Env'l Defense Fund v. Froehlke (Truman Dam), 477 F.2d 1033 (8th Cir. 1973).
[235] See Minnesota Public Interest Research Group v. Butz (I), 358 F. Supp. 584 (D. Minn. 1973), affirmed, 498 F.2d 1314 (8th Cir. 1974); Friends of the Earth v. Hall, 693 F. Supp. 904 (W.D. Wash. 1988); and Sierra Club v. U.S. Army Corps of Engineers, 935 F. Supp. 1556 (S.D. Ala. 1996).
[236] See Minnesota Public Interest Research Group v. Butz (II), 541 F.2d 1292 (8th Cir. 1976), cert. denied, 430 U.S. 922 (1976); Seattle Audubon Soc'y v. Mosely, 798 F. Supp. 1484 (W.D. Wash. 1992), aff'd sub nom., Resources Ltd. v. Robertson, 8 F.3d 1394 (9th Cir. 1993).
[237] See Conservation Soc'y of Vt. v. Secretary of Transportation, 508 F.2d 927 (2d Cir. 1974), vacated and remanded, 423 U.S. 809 (1975); Sierra Club v. U.S. Army Corps of Eng'rs, 772 F.2d 1043 (2d Cir. 1985); and Sierra Club v. U.S. Army Corps of Eng'rs, 935 F. Supp. 1556 (S.D. Ala. 1996).
[238] See Mandelker, § 4.10[4].
[239] Seattle Audubon Soc'y v. Moseley, 80 F.3d 1401 (9th Cir. 1996).

declaratory relief, others have refused it if there is uncertainty whether the project would proceed.[240]

3.9 Federal vs. State EISs

Over a dozen states have adopted legislation that is comparable to NEPA. These state statutes, sometimes called "little NEPAs," require state agencies to prepare impact statements for proposed actions affecting the state's environment. The issues encountered by states in implementing these little NEPAs are often the same issues that are encountered at the federal level.[241]

As might be expected, there is variation among states in the application of the impact statement requirement. A few states (e.g., Maryland and Massachusetts) require a discussion of mitigation and alternatives to a proposed state action, whereas most do not. In most states the preparation of an impact statement is required, whereas in others (e.g., South Dakota[242]) it is optional. In some states (e.g., Connecticut, Georgia, Indiana, Maryland, Montana, North Carolina, Virginia, and Wisconsin[243]), the requirement applies only to governmental agencies at the state level. In other states (e.g., California, Hawaii, Massachusetts, Minnesota, New York, and Washington[244]), the requirement applies to local governments as well as to state agencies. In states for which the requirement applies to local governments, it requires impact statements on at least some development of private land (a major difference from the federal EIS requirement).

---

[240] For cases favorable to declaratory relief, see D'Agnillo v. Department of Housing and Urban Development, 738 F. Supp. 1454 (S.D.N.Y. 1990); and NRDC v. Lujan, 768 F. Supp. 870 (D.D.C. 1991). For cases less favorable, see Atchison, Topeka & Santa Fe Ry. Co. v. Callaway, 459 F. Supp. 188 (D. D.C. 1978); and Utah v. Andrus, 636 F.2d 276 (10th Cir. 1980).

[241] Excellent reviews of state "little NEPAs" can be found in Mandelker, Chapter 12; Renz, The Coming of Age of State Environmental Policy Acts, 5 Pub. Land L. Rev. 31 (1984); and Weinberg, A Powerful Mandate: NEPA and State Environmental Review Acts in the Courts, 5 Pace Envtl. L. Rev. 1 (1987).

[242] S.D. Codified Laws § 34A-9.

[243] Conn. Gen. Stat. § 22a; Ga. Code Ann. § 12-16; Ind. Code Ann. § 13-1-10; Md. Nat. Res. Code §§ 1-301 to 1-305; Mont. Code Ann. §§ 75-1-101 to 75-1-105, 75-1-201 to 75-1-207; N.C. Gen. Stat. § 113A; Va. Code §§ 3.1-18.8, 10.1-1200 to 10.1-1212; and Wis. Stat. § 1.11.

[244] Cal. Pub. Res. Code §§ 21000-21177; Haw. Rev. Stat. § 343; Mass. Gen. Laws ch. 30, §§ 61-62; Minn. Stat. Ann. § 116D; N.Y. Envtl. Conserv. Law §§ 8-0101 to 8-0117; and Wash. Rev. Code § 43.21C.

Two states, California and New York, have particularly well-known "little NEPA" statutes with an extensive record of judicial review. These are discussed in more detail.

California's version of NEPA is known as the California Environmental Quality Act (CEQA).[245] CEQA is the most comprehensive and the most famous of the little NEPA statutes. It requires that an Environmental Impact Report (EIR) be prepared by any state or local governmental agency for "any project they intend to carry out or approve that may have a significant effect on the environment."[246] CEQA delineates the process for EIR preparation, including requirements that the agency consult with other agencies and persons with special expertise.[247] If a proposed state or local agency action does not require an EIR, the agency must produce a "negative declaration" (similar to a FONSI under the federal NEPA).[248] If an EIS is prepared under the federal NEPA, then the EIS may be accepted "in lieu of all or any part" of the EIR under CEQA.[249]

A California state court may reverse an agency decision under CEQA if the agency's decision is not supported by "substantial evidence,"[250] although the courts are now directed to balance the environment against economic considerations.[251] California courts have often looked to federal court decisions for guidance in interpreting CEQA, although one court found that CEQA is more supportive of the environment than NEPA.[252]

The State of New York's version of NEPA is the State Environmental Quality Review Act (SEQRA).[253] Under SEQRA, state and local agencies must prepare an impact statement for "any action they propose or approve which may have a significant effect on the environment."[254] SEQRA does

---

[245] Cal. Pub. Res. Code §§ 21000-21177. For a history of CEQA, see Selmi, The Judicial Development of the California Environmental Quality Act, 18 U.C. Davis L. Rev. 197 (1984).
[246] Cal. Pub. Res. Code § 21000.
[247] Cal. Pub. Res. Code §§ 21104, and 21153.
[248] Cal. Pub. Res. Code § 21064.
[249] Cal. Pub. Res. Code § 21083.5.
[250] Cal. Pub. Res. Code § 21168.5. In general, this is a standard more favorable to plaintiffs than the federal "arbitrary and capricious" standard.
[251] Cal. Pub. Res. Code § 21003(f).
[252] San Francisco Ecology Center v. City of San Francisco, 122 Cal. Rptr. 100 (Cal. App. 1975). See No Oil v. City of Los Angeles, 529 P.2d 66 (Cal. 1975); and Karlson v. City of Camarillo, 161 Cal. Rptr. 260 (Cal. App. 1980).
[253] N.Y. Envtl. Conserv. Law §§ 8-0101 to 8-0117. See Bowers, New York's SEQRA in the Courts, 5 Pace Envtl. L. Rev. 25 (1987); Snider & Levine, A Prolegomenon to Understanding the Developer's True Statutory Responsibilities Under SEQRA, 5 Touro L. Rev. 255 (1989); and Regulatory Compliance, Environmental Impact Review of Development Projects, 1(7) Envt'l Strategies for Real Estate 4 (1994).
[254] N.Y. Envtl. Conserv. Law §§ 8-0109(2).

not require that the proposed action be "major" as is the case in the federal NEPA.

Similar to NEPA, the SEQRA process includes the preparation of both draft and final impact statements, and the collection of comments from other agencies and the public.[255] Unlike NEPA, however, SEQRA requires discussions of energy conservation, and the population growth effects of a proposed action.[256]

3.10 NEPA In the International Arena

An issue of increasing importance in an expanding global economy is whether NEPA's mandates extend beyond the limits of the U.S. into other countries. The issue has received a great deal of discussion in recent years, but is still largely unsettled.[257]

NEPA Section 102(2)(F) states that all agencies of the federal government must "recognize the worldwide and long-range character of environmental problems and, where consistent with the foreign policy of the United States, lend appropriate support to initiatives, resolutions, and programs that are designed to maximize international cooperation in anticipating and preventing a decline in the quality of mankind's world environment."[258] However, this statement seems to make participation by federal agencies voluntary, and does not appear to require an EIS for major federal actions outside the United States.

In 1979, President Carter issued an executive order that purported to clarify NEPA's international application.[259] The executive order does not require the preparation of an EIS, however, but requires a less demanding procedure (such as an environmental assessment) for most federal

---

[255] N.Y. Envtl. Conserv. Law § 8-0109(2).

[256] N.Y. Envtl. Conserv. Law §§ 8-0105(7), and 8-109(4).

[257] See Mandelker, § 5.04; Burnhans, Exporting NEPA: The Export-Import Bank and the National Environmental Policy Act, 7 Brook. J. Int'l L. 1 (1981); Millan, Wanted NEPA Dead or Alive - Reward: Our Global Environment, 24 Env't Rep. (Curr. Dev.) 2081 (1991); and Recent Development, Application of the National Environmental Policy Act to the Extraterritorial Activities of United States Agencies, 2 Tul. J. Int'l & Comp. L. 337 (1994).

[258] 42 U.S.C. § 4332(2)(F).

[259] Environmental Effects Abroad of Major Federal Actions, Executive Order No. 12114, 44 *Fed. Reg.* 1957. See discussions in Mandelker, § 5.04 [3]; Gaines, Environmental Effects Abroad of Major Federal Actions: An Executive Order Ordains a National Policy, 3 Harv. Envtl. L. Rev. 136 (1979); and Note, The Extra-Territorial Application of NEPA Under Executive Order 12,114, 13 Vand. J. Transnational L. 173 (1980).

actions.[260] Many agency actions are specifically exempted from the review process, such as any actions by the president, national security actions and those during armed conflict, all nuclear activities (except those involving production and waste management facilities), disaster and emergency relief activities, and any actions determined by the agency that will not have "a significant effect on the environment outside the United States."[261] In addition, the executive order specifically does *not* allow for a private "cause of action" by plaintiffs wishing to challenge a federal agency's decision.[262]

Given the limitations within NEPA itself as well as the Carter executive order, it appears that NEPA's international reach is limited at best. A few federal courts have dealt with the issue. One court held that NEPA applies to Trust Territories, largely because foreign policy issues are not raised.[263] The D.C. Circuit Court held that no impact statement was required when the Nuclear Regulatory Commission licensed the export of nuclear power plant components to the Phillippines near a U.S. military base due to the short time frame involved, and because foreign policy issues were raised.[264]

In 1991 the U.S. Supreme Court held in a non-NEPA case that there is a presumption against the effectiveness of a federal statute outside the boundaries of the United States.[265] This holding might seem to mean that NEPA could not be applied outside the U.S. However, the D.C. Circuit Court held in *Environmental Defense Fund v. Massey* that NEPA applied to a decision by the National Science Foundation to incinerate wastes at research facilities in Antarctica.[266] The *Massey* court stated that NEPA would apply to those situations where, as in Antarctic research stations, the U.S. retains some measure of legislative control. Exactly what measure of legislative control must be retained by the United States for NEPA to

---

[260] See Greenpeace U.S.A. v. Stone, 748 F. Supp. 749 (D. Haw. 1990), appeal dismissed, 924 F.2d 175 (9th Cir. 1991).

[261] Executive Order 12114 § 2-5(a).

[262] Executive Order 12114 § 3-1. See Environmental Defense Fund v. Massey, 986 F.2d 528 (D.C. Cir. 1993).

[263] People of Enewetak v. Laird, 353 F. Supp. 811 (D. Haw. 1973). Enewetak is a Pacific Island Trust Territory under U.S. supervision.

[264] Natural Resources Defense Council v. NRC, 647 F.2d 1345 (D.C. Cir. 1981). Only two judges participated in the decision, and they disagreed on the reason for the decision! See Note, Nuclear Power Plant Licensing — Jurisdiction to Consider Foreign Impacts, 23 Nat. Resources J. 225 (1983).

[265] Equal Employment Opportunity Comm'n v. Arabian American Oil Co., 499 U.S. 244 (1991).

[266] Environmental Defense Fund v. Massey, 986 F.2d 528 (D.C. Cir. 1993).

apply remains to be determined.[267]

Although NEPA's international reach remains a question at this point, there is ample evidence that NEPA has been an important international role model. Several countries have adopted procedures for environmental impact analysis, including Canada, Australia, Germany, France, and the United Kingdom.[268] In 1985 the European Economic Union adopted requirements that environmental impact assessments be completed for proposed projects "which are likely to have significant effects on the environment," whether they are public or private.[269] Other countries in the Americas, Europe, Asia, and Africa (and some multilateral agencies) are currently considering similar requirements.

3.11 The Status of NEPA

An appropriate end to this chapter is a brief examination of NEPA's effectiveness and status. To be sure, there has been no scarcity of commentators willing to evaluate NEPA.[270] Critics of NEPA have argued that agencies tend to be politically and economically motivated in their decision making, that alternatives and mitigation receive insufficient attention, and that NEPA has become little more than a delaying tactic.[271] Other commentators have been more optimistic, finding that NEPA has

---

[267] See Maragia, Defining the Jurisdictional Reach of NEPA: An Analysis of the Extraterritorial Application of NEPA in *Environmental Defense Fund v. Massey*, 4 Widener J. Pub. L. 129 (1994).

[268] See Wood, *Environmental Impact Assessment: A Comparative Review* (1996); Canadian Environmental Assessment Agency and International Association for Impact Assessment, *International Study of the Effectiveness of Environmental Assessment, Final Report: Environmental Assessment in a Changing World* (1996).

[269] EEC Council Directive 337 (June 27, 1985). For more complete discussions of the EEC process, see Mandelker, § 13.02; Visek, Implementation and Enforcement of EC Environmental Law; and McHugh, The European Community Directive - An Alternative Environmental Impact Assessment Procedure? 34 Nat. Resources J. 629 (1994).

[270] For an excellent review of the legal literature evaluating NEPA, see Mandelker, Chapter 11.

[271] See Sax, The (Unhappy) Truth About NEPA, 26 Okla. L. Rev. 239 (1973); Cramton & Berg, On Leading a Horse to Water: NEPA and the Federal Bureaucracy, 71 Mich. L. Rev. (1973); Funk, NEPA at Energy: An Exercise in Legal Narrative, 20 Envtl. L. 759 (1990); and Comment, NEPA: Business as Usual: The Weaknesses of the National Environmental Policy Act, 59 J. Air L. & Com. 709 (1994).

encouraged public participation, and has improved agency decision making.[272]

In the future, NEPA may have impact beyond the expectations of its drafters. For example, NEPA and the EIS requirement may be a tool to protect air quality where the Clean Air Act does not.[273] NEPA may prove to be an invaluable tool in protecting the nation's biodiversity.[274]

It remains true that NEPA offers less substantive protection than other environmental laws, but it remains the *only* statute that reaches all federal agency actions. NEPA is still the chief statute for citizens protecting the environment.

---

[272] See Liroff, *a National Policy for the Environment* (1976); Liroff, NEPA - Where Have We Been and Where Are We Going?, 46 J. Amer. Planning Assn. 154 (1980); Bear, NEPA at 19: A Primer on an "Old" Law with Solutions to New Problems, 19 Envtl. L. Rep. 10060 (1989); Culhane, NEPA's Impacts on Federal Agencies, Anticipated and Unanticipated, 20 Envtl. L. 681 (1990); Ackerman, Observations on the Transformation of the Forest Service: The Effects of the National Environmental Policy Act on U.S. Forest Service Decision-Making, 20 Envtl. L. 703 (1990); Bear, The National Environmental Policy Act: Its Origins and Evolution, 10(2) Natural Resources & Env't 3 (1995).

[273] See Kite, Air Quality Regulation through NEPA: A Southwest Wyoming Experience, 12(1) Natural Resources & Environment 25 (1997).

[274] See Bear, Using the National Environmental Policy Act to Protect Biological Diversity, 76 Tul. Envtl. L. J. 77 (1994).

III

Water Pollution Control

4

The Clean Water Act —
Water Pollutant Discharges

4.1 Background

It has been estimated that over 70 percent of the surface of the earth is water. Of all the earth's water, 97 percent is in the oceans and seas, and only 3 percent is fresh water in glaciers, lakes, groundwater, rivers, and the atmosphere.[1]

The five Great Lakes represent about 95 percent of all fresh water above ground in the United States In addition, there are:

- 3.5 million miles of rivers and streams
- 41 million acres of lakes
- 34,400 square miles of estuaries (excluding Alaska)
- 101 million acres of wetlands (excluding Alaska)
- 170-200 million acres of wetlands in Alaska

Water offers many valuable uses to individuals and communities. The beaches, whitewater rivers, and lakes found throughout most of the United States contribute to a thriving recreation and tourism industry. Other sectors of the economy rely on clean water to grow, process, or deliver products and services.

Americans use rivers, lakes and aquifers for drinking water. About half

---

[1] Source of much of the following information is U.S.E.P.A., Office of Water, *Water Factsheet* (1997).

of the population drinks from rivers and lakes and the other half taps underground water sources.

Water is an important resource for commerce:

● The nation's $45 billion commercial fishing and shellfishing industry relies on clean water to deliver products safe to eat. The average American now eats 15 pounds of fish and shellfish every year.
● In the southeastern United States, over 90 percent of the commercial catch of fish and shellfish depends on clean, coastal wetland systems.
● Manufacturers use about 13 trillion gallons of water every year-more than nine times the volume that flows through the Mississippi River into the Gulf of Mexico daily.
● The soft drink industry alone uses over 12 billion gallons of clean water annually to produce products valued at more than $50 billion.

Water is an important resource for agriculture:
● Farmers irrigate about 15 percent of American farm lands to grow food and fiber.
● Crops grown on irrigated lands are valued at nearly $70 billion a year about 40 percent of the total value of all crops sold.

Water is an important resource for tourism:
● Beaches, rivers, and lakes are the primary vacation choice for Americans, helping to support a flourishing recreation and tourism industry.
● Each year, Americans take over 1.8 billion trips (or about seven trips per person) to go fishing, swimming or boating, or just to relax around their favorite water destinations.

Water is an important resource for quality of life:

● A *Money* magazine survey found clean water and clean air rank among the top factors Americans consider in choosing a place to live.
● Proximity to clean water is a neighborhood attribute that has significant impact on real estate values, according to the National Association of Home Builders. A clean body of water in the vicinity increases the value of a home by 22 percent.

4.1.1 Early Approaches to Water Pollution Control

Pollution of water resources was recognized as a problem in the early 1800s, due to degradation of sport and commercial fisheries. In 1899, Congress passed the Rivers and Harbors Act, known as the "Refuse Act."[2]

---

[2] 33 U.S.C. § 401 *et seq.*

The Refuse Act required a permit from the U.S. Army Corps of Engineers to discharge "refuse" into navigable waters. Its purpose was to keep rivers and harbors open, *not* to prevent pollution.

Congress adopted a new strategy in 1948 with the Federal Water Pollution Control Act (FWPCA).[3] The FWPCA required states to develop "water quality standards," which generally determined how polluted a body of water was permitted to become. Legal action was taken against violators. Unfortunately, problems of proof were overwhelming and the 1948 FWPCA failed.

In 1965, the Water Quality Act (WQA) was enacted to charge states with setting water quality standards for interstate navigable waters.[4]

The Refuse Act and the FWPCA represent the two approaches the federal government has taken toward water pollution control. The first approach is the "effluent limitations" approach, which regulates how much pollutant you can be present in the water (like the Refuse Act), with "end-of-the-pipe" restrictions on discharge. The second approach is the "water quality standards" approach, regulates how polluted the water can be (like the Federal Water Pollution Act).

4.1.2 Federal Water Pollution Control Act of 1972

In 1972, Congress enacted the first comprehensive national clean water legislation in response to growing public concern for serious and widespread water pollution. During the late 1960s, the general public became aware that tha nation's water was at risk: Lake Erie was dying, the Potomac River was clogged with blue-green algae blooms that were a nuisance and a threat to public health, many of the nation's rivers were little more than open sewers and sewage frequently washed up on shore, fish kills were a common sight, and wetlands were disappearing at a rapid rate.[5]

The Federal Water Pollution Control Act of 1972 (FWPCA) was the first "clean water act."[6] Congress stated the general goals of the FWPCA in Section 101(a):

The objective of this chapter is to restore and maintain the chemical, physical, and biological integrity of the Nation's waters. In order to

---

[3] Pub.L. No. 80-845, 62 Stat. 1155.
[4] Pub.L. No. 80-845, 79 Stat. 903.
[5] U.S.E.P.A., Office of Water, *History of the Clean Water Act* (1997).
[6] 33 U.S.C. Chap. 26. Although technically a series of amendments to the FWPCA of 1948, the FWPCA of 1972 virtually replaced the 1948 Act.

achieve this objective it is hereby declared that, consistent with the provisions of this chapter -

(1) it is the national goal that the discharge of pollutants into the navigable waters be eliminated by 1985;

(2) it is the national goal that wherever attainable, an interim goal of water quality which provides for the protection and propagation of fish, shellfish, and wildlife and provides for recreation in and on the water be achieved by July 1, 1983;

(3) it is the national policy that the discharge of toxic pollutants in toxic amounts be prohibited;

(4) it is the national policy that Federal financial assistance be provided to construct publicly owned waste treatment works;

(5) it is the national policy that areawide waste treatment management planning processes be developed and implemented to assure adequate control of sources of pollutants in each State;

(6) it is the national policy that a major research and demonstration effort be made to develop technology necessary to eliminate the discharge of pollutants into the navigable waters, waters of the contiguous zone, and the oceans; and

(7) it is the national policy that programs for the control of nonpoint sources of pollution be developed and implemented in an expeditious manner so as to enable the goals of this chapter to be met through the control of both point and nonpoint sources of pollution.[7]

CWA Section 101(a)(1) makes the elimination of the discharge of "pollutants" a priority. Under CWA Section 502(6), pollutants are defined:

The term "pollutant" means dredged spoil, solid waste, incinerator residue, sewage, garbage, sewage sludge, munitions, chemical wastes, biological materials, radioactive materials, heat, wrecked or discarded equipment, rock, sand, cellar dirt and industrial, municipal, and agricultural waste discharged into water. This term does not mean:

(A) "sewage from vessels or a discharge incidental to the normal operation of a vessel of the Armed Forces" within the meaning of [13 U.S.C. Section 1322]; or

(B) water, gas, or other material which is injected into a well to facilitate production of oil or gas, or water derived in association with oil or gas production and disposed of in a well, if the well used either to facilitate production or for disposal purposes is approved by authority of the State in which the well is located, and if such State

---

[7] 33 U.S.C. § 1251(a).

determines that such injection or disposal will not result in the degradation of ground or surface water resources.[8]

In 1977 the FWPCA was substantially amended, and formally renamed the Clean Water Act (CWA).[9] The 1977 CWA emphasized control of toxic pollutants and established a program to transfer the responsibility of federal clean water programs to the states.

Toxic pollutants are defined as pollutants or combinations of pollutants which, after discharge and upon exposure, ingestion, inhalation, or assimilation into any organism, will cause harm to public health or death. Toxic pollutants include 63 chemicals and classes of chemicals listed under CWA Section 307.[10]

With the passage of the CWA, Congress opted for the "effluent limitations" approach, although "water quality standards" is an important backup. "Planning" is an integral part of the CWA. The CWA's mandates are enforced by the National Pollutant Discharge Elimination System (NPDES) permit program.

There have been additional amendments to the CWA in an effort to gradually implement increasingly stricter standards in its various programs. The Water Quality Act of 1987 focused on stricter regulation of toxic chemicals from industry, acid rain, and reduction of "non-point source" pollution discharges like agricultural runoff and urban stormwater runoff. The CWA was substantially amended in 1981 and 1987. It was amended and reauthorized in 1994.

4.2 Point Source Discharges into U.S. Waters

4.2.1 Regulation Under the Clean Water Act

CWA Section 301 applies to all point sources of water pollution, *except* Publicly Owned Treatment Works (POTW), which are discussed separately in CWA Title II. A "point source" is defined at CWA Section 502(14):

The term "point source" means any discernible, confined and discrete conveyance, including but not limited to any pipe, ditch, channel, tunnel, conduit, well, discrete fissure, container, rolling stock,

---

[8] 33 U.S.C. § 1362(6).
[9] Section 1 of Pub. L. 95-217 provided: "That this Act . . . may be cited as the 'Clean Water Act of 1977'."
[10] 33 U.S.C § 1317.

concentrated animal feeding operation, or vessel or other floating craft, from which pollutants are or may be discharged. This term does not include agricultural stormwater discharges and return flows from irrigated agriculture.[11]

The fact that the definition of "point source" specifically excludes "agricultural stormwater discharges and return flows from irrigated agriculture" probably represents Congress' view that agriculture requires special consideration.

CWA Section 301 is enforced and regulated by the Section 402 NPDES program (see below). Interestingly, the original CWA of 1972 did not contain an express authorization for The EPA to issue regulations. However, The U.S. Supreme Court in *E.I. du Pont de Nemours & Co. v. Train* held that the EPA could promulgate regulations under CWA Section 301 even though regulations are not expressly authorized.[12]

Section 301 uses the "effluent limitations" approach to regulate point sources. An effluent limitation is a limit on the quantity, discharge rates, and concentration of each pollutant a facility may discharge into waters of the United States. Under the CWA, all municipal and industrial point sources of water pollution are subject to effluent limitations unique for each discharger.[13] CWA Section 301 established limits and guidelines on effluents, and requires states to issue NPDES permits based on effluent limitations.

4.2.2 Point versus Nonpoint Source Pollution

The CWA divides pollution into "point source" and "non-point source." While this makes little scientific sense, Congress apparently felt that point sources would be much easier to monitor than would nonpoint sources. Congress apparently felt that engineers were more likely to find ways to measure and control "end of pipe" pollution than would be the case with water quality standards.

Nonpoint source regulation is primarily at CWA Section 319,[14] and will be discussed below. For the most part, nonpoint source regulation is left to the states (see Section 319 State Plans). The relationship between nonpoint source regulation and the NPDES permit process remains a matter of some debate.

---

[11] 33 U.S.C. § 1362(14).
[12] E.I. du Pont de Nemours & Co. v. Train, 430 U.S. 112 (1977).
[13] U.S.E.P.A., Office of Water, *Clean Water Act Fact Sheet* (1997).
[14] 33 U.S.C. § 1329.

For most environmental professionals, a central question is: When is it a point source? If it is a point source, it requires a NPDES permit under CWA Section 402; but applicants generally want the lesser controls under CWA Section 319. Some examples of point sources are runoff from a mining operation that passes through ditches into navigable waters,[15] runoff from a mine that entered non-navigable waters,[16] and a trap and skeet shooting range overlooking Long Island Sound.[17]

On the other hand, examples of nonpoint sources include return flow from irrigated agricultural land,[18] and a hydroelectric dam which reduces dissolved oxygen downstream.[19]

4.3 Point Source Effluent Standards

CWA Section 301 contains a set standards known best by their acronyms. In general, these "technology forcing" standards consider a series of factors, including the technological capabilities of the discharger and the cost of the pollution controls. The idea seems to be that polluters will be compelled to upgrade their operations to meet a specific technological standard. While these standards are largely responsible for the dramatic improvement in much of the nation's surface waters, they have been criticized as a "blunt instrument" that forces expensive technological advances that accomplish little. Because technology-based standards are set without reference to the waters receiving the effluent, these standards have also been criticized as permitting unacceptably high

---

[15] See United States v. Earth Sciences; see also Beartooth Alliance v. Crown Butte Mines, 904 F. Supp. 1168 (D.Mont. 1995).

[16] Because pollutants could enter navigable waters via an aquifer (see Quivira Mining Co. v. United States, 765 F.2d 126 (10th Cir. 1985); but see Friends of Santa Fe Cty. v. LAC Minerals, Inc., 892 F. Supp. 1333 (D.N.M. 1995) [subsurface mine seepage is not a point source].

[17] Shooting platforms "discharged" lead into coastal waters. Long Island Soundkeeper Fund, Inc. v. N.Y. Athletic Club, 1996 WL 131863 (S.D.N.Y. March 22, 1996). See also Stone v. City of Naperville Park Dist., 38 F. Supp. 2d 651 (N.D. Ill. 1999) [discharge of lead shot into navigable waters from trap shooting facility in city park was a point source].

[18] Exempted under § 502(14), 33 U.S.C. § 1362(14). United States v. Frezzo Bros., 642 F.2d 59 (3d Cir. 1981).

[19] National Wildlife Fed'n v. Gorsuch, 530 F. Supp. 1291 (D.D.C. 1982) [holding point source], rev'd, 693 F.2d 156 (D.C. Cir. 1982) [reversing district court, holding EPA can determine dams are non-point sources].

levels of pollutants to enter sensitive aquatic systems.[20]

The CWA gave the EPA the authority to implement the effluent limitations. Individual point sources then choose any control technology, as long as it meets the technology-based standards. Presumably, this allows the point source to choose a control technology that is effective for the particular conditions, but is also cost effective.

In its 1972 formulation, CWA Section 301(b)(1)(A) required point sources of water pollution to meet a "generic" effluent standard (i.e., one standard that applied to all pollutants) of "Best practicable control technology currently available" (BPT) by 1977.[21] CWA Section 304(b)(1)(B) characterized the BPT standard:

> Factors relating to the assessment of best practicable control technology currently available to comply with subsection (b)(1) of section 1311 of this title shall include consideration of the total cost of application of technology in relation to the effluent reduction benefits to be achieved from such application, and shall also take into account the age of equipment and facilities involved, the process employed, the engineering aspects of the application of various types of control techniques, process changes, non-water quality environmental impact (including energy requirements), and such other factors as the Administrator deems appropriate.[22]

The BPT standard is clearly one based on "cost-benefit" relationships. In other words, it carefully balances the effective pollution control against the cost to the polluter. Presumably, a technological standard that exceeded the ability of the affected point sources to implement them would be replaced with another, more cost effective standard.[23]

In addition, CWA Section 306 required that "new" point sources constructed after 1972 achieve a more stringent standard of "best available control technology" (BAT).[24] In devising BAT standards, the EPA must consider several factors:

---

[20] See generally Wardzinski, Sandalow, Burgin, Ginsberg, and McGaffey, Water Pollution Control under the National Pollutant Discharge Elimination System, Chap. 2 in Evans (ed.) *The Clean Water Act Handbook* (Amer. Bar Ass'n, 1994).

[21] 33 U.S.C. § 1311(b)(1)(A). See Weyerhauser Co. v. Costle, 590 F.2d 1011 (D.C. Cir. 1978), [discussing factors EPA may consider in determining the extent of BPT].

[22] 33 U.S.C. § 1314(b)(1)(B).

[23] See Rodgers, *Environmental Law*, 2nd ed. (West, 1994) at § 4.1 (hereafter, "Rodgers").

[24] 33 U.S.C. § 1316(b)(2)(B).

Factors relating to the assessment of best available technology shall take into account the age of equipment and facilities involved, the process employed, the engineering aspects of the application of various types of control techniques, process changes, the cost of achieving such effluent reduction, non-water quality environmental impact (including energy requirements), and such other factors as the Administrator deems appropriate.[25]

CWA Section 307 placed effluent limitations on pretreatment of wastes that discharge into municipal sewer systems, and charged the EPA with setting effluent limitations for toxic pollutants (these will be discussed below).[26] Under Section 402, POTWs were required to adopt "Best practicable waste treatment over the life of the works" by 1983.[27]

Over time, the "generic" effluent standards gradually gave way to standards that are specific for each of three categories of pollutants: conventional, toxic, and nonconventional pollutants.

4.3.1 Conventional Pollutants

CWA Section 304(a)(4) requires the Administrator of the EPA to identify "conventional pollutants:"

including but not limited to, pollutants classified as biological oxygen demanding, suspended solids, fecal coliform, and pH.[28]

These include pollutants traditionally regulated as discharges from treatment plants.

With respect to these conventional pollutants, CWA Section 301(a)(2)(E) sets the effluent standard:

[There shall be achieved] as expeditiously as practicable but in no case ... later than March 31, 1989, compliance with effluent limitations for categories and classes of point sources, other than publicly owned treatment works, which in the case of [conventional] pollutants identified pursuant to section 1314(a)(4) of this title shall require application of the best conventional pollutant control technology as

---

[25] Id.
[26] 33 U.S.C. § 1317(b)(1).
[27] 33 U.S.C. § 1281.
[28] 33 U.S.C. § 1314(a)(4).

determined in accordance with regulations issued by the Administrator [of the EPA] pursuant to section 1314(b)(4) of this title.

This "best conventional pollutant control technology" (BCT) standard is described in CWA Section 304(b)(4(B):

Factors relating to the assessment of best conventional pollutant control technology (including measures and practices) shall include consideration of the reasonableness of the relationship between the costs of attaining a reduction in effluents and the effluent reduction benefits derived, and the comparison of the cost and level of reduction of such pollutants from the discharge from publicly owned treatment works to the cost and level of reduction of such pollutants from a class or category of industrial sources, and shall take into account the age of equipment and facilities involved, the process employed, the engineering aspects of the application of various types of control techniques, process changes, non-water quality environmental impact (including energy requirements), and such other factors as the Administrator deems appropriate.

In addition to the CWA Section 304(b)(4)(B) "industry cost effectiveness" standard, the EPA applies a "POTW cost-comparison test," which examines candidate BCT technologies in relation to additional costs over BPT technology.[29]

The BCT standard was apparently applied by Congress to conventional pollutants to make sure that technologies beyond BPT be cost effective. In fact, BCT standards have rarely been more strict than BPT.[30]

4.3.2 Toxic Pollutants

The CWA has dealt separately with toxic pollutants since the 1977 amendments. CWA Section 307(a)(1) Identifies toxic pollutants:

The Administrator in publishing any [list of toxic pollutants], including the addition or removal of any pollutant from such list, shall take into account toxicity of the pollutant, its persistence,

---

[29] See 51 *Fed. Reg.* 24,974-24,976 (July 9, 1986).
[30] See Wardzinski, Sandalow, Burgin, Ginsberg, and McGaffey, Water Pollution Control under the National Pollutant Discharge Elimination System, Chap. 2 in Evans (ed.), *The Clean Water Act Handbook* (Amer. Bar Ass'n, 1994).

degradability, the usual or potential presence of the affected organisms in any waters, the importance of the affected organisms, and the nature and extent of the effect of the toxic pollutant on such organisms.[31]

The list of toxic pollutants includes 65 chemicals and classes of chemicals at this writing.[32]

CWA Section 307(a)(2) sets the effluent standard for toxic pollutants:

Each toxic pollutant listed in accordance with paragraph (1) of this subsection shall be subject to effluent limitations resulting from the application of the best available technology economically achievable for the applicable category or class of point sources established in accordance with sections 1311(b)(2)(A) and 1314(b)(2) of this title.[33]

The "best available technology" (BAT) standard represents the very best performance achieved within a category or subcategory of points sources. In developing BAT standards, the EPA considers "the optimally operating plant, the pilot plant that acts as a beacon to show what is possible."[34]

In setting BAT standards, CWA Section 304(b)(2)(B) requires the EPA to consider the following factors:

Factors relating to the assessment of best available technology shall take into account the age of equipment and facilities involved, the process employed, the engineering aspects of the application of various types of control techniques, process changes, the cost of achieving such effluent reduction, non-water quality environmental impact (including energy requirements), and such other factors as the Administrator deems appropriate.[35]

Although the BAT standard requires a cost-benefit analysis, costs are less important than they were with the BPT and BCT standards. Rather than requiring the EPA to compare costs and effluent reduction benefits as in the BPT and BCT standards, the BAT standard requires only that the EPA consider whether costs are "economically achievable."[36]

---

[31] 33 U.S.C. § 1317(a)(1).
[32] 40 CFR § 401.15.
[33] 33 U.S.C. § 1317(a)(2).
[34] Kennecott v. EPA, 780 F.2d 445, 448 (4th Cir. 1985).
[35] 33 U.S.C. § 1314(b)(2)(B).
[36] Compare 33 U.S.C. § 1314(b)(2)(B) [BAT] with 33 U.S.C. § 1314(b)(1)(B) [BPT] and 33 U.S.C. § 1314(b)(4)(B) [BCT]. See also Wardzinski, Sandalow, Burgin, Ginsberg, and McGaffey, Water Pollution Control under the National Pollutant Discharge Elimination System, Chap. 2 in Evans (ed.), *The Clean*

4.3.3 Nonconventional Pollutants

"Nonconventional pollutants" are all pollutants that are neither toxic nor conventional,[37] including such pollutants as ammonia, chlorides, nitrates, and color. For nonconventional pollutants, CWA Section 301(b)(2)(F) requires BAT technology by July 1, 1984.[38]

CWA Section 301(g) contains an interesting statement regarding modification of the BAT standard with regard to nonconventional pollutants:

The Administrator [of the EPA], with the concurrence of the State, may modify the requirements of subsection (b)(2)(A) of this section with respect to the discharge from any point source of ammonia, chlorine, color, iron, and total phenols (4AAP) (when determined by the Administrator to be a pollutant covered by subsection (b)(2)(F) [nonconventional pollutants] of this section) and any other pollutant which the Administrator lists under paragraph (4) of this subsection.[39]

In other words, the EPA may modify the BAT standard if the state agrees.

4.3.4 New Source Performance Standards

Congress determined that "new" sources of water pollution should be subject to the most rigid performance standards under the CWA. The idea seems to be that it is far less expensive for a "new" source to install state-of-the-art pollution controls during the construction process than it would be for an existing source to retrofit the same controls.[40]

The New Source Performance Standard (NSPS) established by CWA Section 306(a)(1) is:

The term [new source] "standard of performance" means a standard for the control of the discharge of pollutants which reflect the greatest degree of effluent reduction which the Administrator determines to be achievable through application of the *best available demonstrated control technology*, processes, operating methods, or other

---

Water Act Handbook (Amer. Bar Ass'n, 1994).
[37] 33 U.S.C. § 1311(b)(2)(F).
[38] Id.
[39] 33 U.S.C. § 1311(g).
[40] See Maloney, Assessing NEPA's Effect on NPDES New Source Permit Issuance: Do the New NPDES Regulations Strike the Proper Balance? 38 Sw. L. J. 1231 (1985).

alternatives, including, where practicable, a standard permitting no discharge of pollutants.[41]

The standard "Best Available Demonstrated Control Technology" (BDT) applied to "new sources" is at least as stringent as BAT, and may be more stringent.[42]

The EPA considers several factors in determining the BDT standard. CWA Section 306(b)(1)(B) states that: "the Administrator shall take into consideration the cost of achieving such effluent reduction, and any non-water quality, environmental impact and energy requirements."[43] Section 306(b)(2) of the CWA states that: "the Administrator may distinguish among classes, types, and sizes within categories of new sources for the purpose of establishing such standards and shall consider the type of process employed (including whether batch or continuous)."[44] Costs are even less significant with the BDT standard than they are with BAT because entire categories of costs (e.g., the cost of retrofitting technologies) are not relevant in NSPS.

Of course, a critical question in determining NSPS is when is a source a "new source"? CWA Section 306(a)(2) defines a "source:"

The term "source" means any building, structure, facility, or installation from which there is or may be the discharge of pollutants.[45]

And a "new source:"

The term "new source" means any source, the construction of which is commenced after the publication of proposed regulations prescribing a standard of performance under this section which will be applicable to such source, if such standard is thereafter promulgated in accordance with this section.[46]

However, EPA regulations define a "new discharger:"

New discharger means any building, structure, facility, or installation:
(a) From which there is or may be a "discharge of pollutants;"
(b) That did not commence the "discharge of pollutants" at a particular

---

[41] 33 U.S.C. § 1316(a)(1), emphasis added.
[42] See American Iron & Steel Inst. v. EPA, 526 F.2d 1027 (3d Cir. 1974).
[43] 33 U.S.C. § 1316(b)(1)(B).
[44] 33 U.S.C. § 1316(b)(2).
[45] 33 U.S.C. § 1316(a)(3).
[46] 33 U.S.C. § 1316(a)(2).

"site" prior to August 13, 1979;
(c) Which is not a "new source;" and
(d) Which has never received a finally effective NDPES permit for discharges at that "site."[47]

This has come to mean that if the EPA has not issued a NSPS for a particular category of dischargers, then the less strict BCT or BAT standards will apply to that source. Courts have been mixed in their willingness to accept this definition.[48]

4.3.5 Indirect Discharges

Given the CWA's prohibition of discharges of pollutants into the nation's waters, it is not surprising that some unscrupulous point sources might wish to avoid detection by discharging their pollutants into the local sewer system. Besides the obvious violations of the CWA, this practice has a serious negative effect on Publicly Owned Treatment Works (POTW) or any other treatment facility that is unable to cope with the load pollutants. To discourage this practice, Congress passed CWA Section 307(b)(1) requires the EPA to promulgate regulations requiring "pretreatment" of pollutants by those point sources that discharge into a POTW rather than directly into navigable waters.[49]

More specifically, CWA Section 307(b)(1) states:

The Administrator shall, within one hundred and eighty days after October 18, 1972, and from time to time thereafter, publish proposed regulations establishing pretreatment standards for introduction of pollutants into treatment works (as defined in section 1292 of this title) which are publicly owned for those pollutants which are determined not to be susceptible to treatment by such treatment works or which would interfere with the operation of such treatment works.[50]

---

[47] 40 CFR § 122.2.
[48] Contrast NRDC v. EPA, 822 F.2d 104 (D.C. Cir. 1987) [upholding 40 CFR § 122.2 definition of "new source"]; with National Ass'n of Metal Finishers v. EPA, 719 F.2d 624 (3d. Cir. 1983) rev'd on other grounds sub nom. Chemical Mfrs. Ass'n v. NRDC 470 U.S. 116 (1985) [rejecting regulatory definition].
[49] 33 U.S.C. § 1317(b)(1). See United States v. City of Detroit, 940 F. Supp. 1097 (E.D. Mich. 1996) [pretreatment must be part of an NPDES permit to be enforceable].
[50] 33 U.S.C. § 1317(b)(1).

The EPA has established a Pretreatment Program with National Pretreatment Standards (NPS). The objectives of the program are:

(a) To prevent the introduction of pollutants into POTWs which will interfere with the operation of a POTW, including interference with its use or disposal of municipal sludge;
(b) To prevent the introduction of pollutants into POTWs which will pass through the treatment works or otherwise be incompatible with such works; and
(c) To improve opportunities to recycle and reclaim municipal and industrial wastewaters and sludges.[51]

Elsewhere in the regulations, "indirect discharge" is defined as "the introduction of pollutants into a POTW from any non-domestic source regulated under section 307 [POTWs] of the {CWA].[52]
The EPA has promulgated two kinds of pretreatment regulations.[53] "General" pretreatment regulations apply to all nondomestic sources that discharge to POTWs. The regulations contain the "General prohibitions," which state that "a User may not introduce into a POTW any pollutant(s) which cause Pass Through or Interference."[54] "Pass through" and "Interference" are defined as:

The term Pass Through means a Discharge which exits the POTW into waters of the United States in quantities or concentrations which, alone or in conjunction with a discharge or discharges from other sources, is a cause of a violation of any requirement of the POTW's NPDES permit (including an increase in the magnitude or duration of a violation).[55]

and:

The term Interference means a Discharge which, alone or in conjunction with a discharge or discharges from other sources, both: (1) Inhibits or disrupts the POTW, its treatment processes or operations, or its sludge processes, use or disposal; and (2) Therefore is a cause of a violation of any requirement of the POTW's NPDES permit, Section 405 of the Clean Water Act, [RCRA], [the Clean Air

---

[51] 40 CFR § 403.2.
[52] 40 CFR § 403.3(g).
[53] See discussion in Goldstein, Pretreatment and Indirect Dischargers, Chap. 5 in Evans (ed.), *The Clean Water Act Handbook* (Amer. Bar Ass'n 1994).
[54] 40 CFR § 403.5(a)(1).
[55] 40 CFR § 403.3(n).

Act] the Toxic Substances Control Act, and the Marine Protection, Research and Sanctuaries Act.[56]

POTWs with pretreatment programs must develop local limits to implement the prohibitions. These limits must be adequate to guarantee that the POTW will comply with its permit (including such things as sludge use and disposal practices).[57]

In addition to the general pretreatment regulations, the EPA has also promulgated "National categorical" pretreatment standards. These are technology-based standards that the EPA applies to categories of industrial dischargers. Categorical pretreatment standards are based on BAT technology, and apply whether the discharging facility is existing or new.[58] However, Pretreatment Standards for Existing Sources (PSES) are somewhat less strict than Pretreatment Standards for New Sources (PSNS) because the latter have an additional set of standards.[59] There are significant monitoring, reporting, and record keeping obligations, and the discerning discharger or POTW will be careful to comply with them.[60]

Under CWA Section 309(f), the EPA may, after 30 day's notice to the POTW and the state, commence an enforcement action an indirect discharger "commence a civil action for appropriate relief, including but not limited to, a permanent or temporary injunction, against the owner or operator of such treatment works."[61] Penalties can include all civil and criminal penalties available under the CWA. Furthermore, a private citizen can bring action against an indirect discharger under CWA Section 505(f)(4) for violations of pretreatment standards.[62]

4.3.6 Variances

The CWA allows for three types of "variances" from the technology-based standards discussed above, although these are rarely granted by the EPA. The first, and most controversial is the CWA Section 301(n) variance for "Fundamentally Different Factors (the "FDF" variance).[63] To

---

[56] 40 CFR § 403.3(j).
[57] 40 CFR § 403.5(c)(1) and (2).
[58] See Goldstein, Pretreatment and Indirect Dischargers, Chap. 5 in Evans (ed.), *The Clean Water Act Handbook* (Amer. Bar Ass'n, 1994).
[59] 33 U.S.C. §§ 1317(c) and 1316(a). See also 40 CFR § 122.2.
[60] See discussion in Goldstein, Pretreatment and Indirect Dischargers, Chap. 5 in Evans (ed.) *The Clean Water Act Handbook* (Amer. Bar Ass'n, 1994).
[61] 33 U.S.C. § 1319(f).
[62] 33 U.S.C. § 1365(f)(4).
[63] 33 U.S.C. § 1311(n).

qualify for the FDF variance, a facility must demonstrate that its processes are "fundamentally different" from those on which the effluent standards are based. Cost of controlling pollutant discharges is not a consideration in determining a FDF variance.[64] It is important to note that the FDF variance does not excuse a facility from the obligation to meet technology based standards, but only allows a variance from the particular standard specified in EPA guidelines. The FDF variance does not apply to New Source Performance Standards (NSPS) or Pretreatment Standards for New Sources (PSNS).

Based on the Supreme Court's holding in *Chemical Manufacturers Association v. NRDC*, Section 301(n) allows a variance for all pollutants, including toxics, for point sources that are subject to the FDF variance.[65]

A second variance is the CWA Section 301(c) variance for "economic incapability."[66] Although these variances are rarely given, discharger can be granted a variance from BAT standards if it can be demonstrated that the discharger will utilize the "maximum use of technology within the capability" of the discharger, and that "reasonable further progress" toward elimination of pollutant discharges is being met.[67] The discharger is relieved only from the BAT standard, but must still comply with BPT. The Supreme Court has held that a similar variance must be available for BPT deadlines,[68] although the variance from BPT may not be based exclusively on economic hardship.[69]

Finally, a variance is available from BAT limits for five conventional pollutants under CWA Section 301(g).[70] Dischargers must demonstrate that they can still meet effluent standards.

4.4 Water Quality Standards

Water quality standards differ from effluent limitations in focusing on the impact that discharges have on the receiving water body (rather than on the pollutants present in the effluent).

Recall that a goal of the CWA as stated in CWA Section 101(a)(2):

---

[64] 33 U.S.C. § 1311(n)(1)(A).
[65] Chemical Manufacturers Ass'n v. NRDC, 470 U.S. 116 (1985).
[66] 33 U.S.C. § 1311(c).
[67] Id.
[68] See E.I. du Pont de Nemours & Co. v. Train, 430 U.S. 112 (1977).
[69] See EPA v. National Crushed Stone Ass'n, 449 U.S. 64 (1980).
[70] 33 U.S.C. § 1311(g). The five pollutants are ammonia, chlorine, color, iron, and phenols.

The objective of this chapter is to restore and maintain the chemical, physical, and biological integrity of the Nation's waters. In order to achieve this objective it is hereby declared that, consistent with the provisions of this chapter . . .

(2) it is the national goal that wherever attainable, an interim goal of water quality which provides for the protection and propagation of fish, shellfish, and wildlife and provides for recreation in and on the water be achieved by July 1, 1983.[71]

This standard is usually called the "fishable/swimmable" standard.

In order to implement these goals, CWA Section 303 requires that each state produce water quality standards that are approved by the EPA. States designate a use for each specific body of water (e.g., municipal water supply, recreation, etc.), and designate criteria for allowable concentrations of pollutants. In designating uses for a water body, states must consider use as a public water supply, propagation of fish and wildlife, recreation, agriculture, industry, and other uses.[72]

The U.S. Supreme Court held that a state can use water quality criteria to protect a specific species of fish, even though the protection was not directly related to the conditions that gave rise to the need for a CWA Section 303 permit.[73]

The CWA Section 303 water quality standards are non-technology based, and provide important support to effluent limitations. If the state's standards do not comply with EPA standards, then the EPA must designate standards with which the state must comply.[74] CWA Section 302 allows the EPA to establish stringent effluent limitations if combined sources fail to meet water quality criteria.[75]

Under CWA Section 303(d)(1)(C), each state must develop a Total Maximum Daily Load (TMDL) for each of several identified pollutants in waters for which existing effluent limitations are not stringent enough.[76] Where a state has failed to issue TMDLs for impaired waters, the EPA must do so.[77]

---

[71] 33 U.S.C. § 1311(a)(2).

[72] 33 U.S.C. § 1313(c)(2)(A). See also EPA regulations at 40 CFR § 131.2.

[73] PUD No. 1 v. Washington Dept. of Ecology, 114 S.Ct. 1900 (1994).

[74] 33 U.S.C. § 1313(c)(3). See Defenders of Wildlife v. Browner, 909 F. Supp. 1342 (D. Ariz. 1995); and Raymond Proffitt Found. v. EPA, 930 F. Supp. 1088 (E.D. Pa. 1996).

[75] 33 U.S.C. § 1312.

[76] 33 U.S.C. § 1313(d)(1)(C). See Dioxin/Organochloride Center v. Clarke, 57 F.3d 1517 (9th Cir. 1995) [TMDL for dioxin in the Columbia River].

[77] See Sierra Club v. Hankinson, 939 F. Supp. 865 and 872 (N.D. Ga. 1996).

The courts have espoused differing views on whether private citizens can file suit under Section 505 of the CWA (the "citizen suit" provision) to enforce water quality standards that are incorporated into a permit.[78]

4.5 National Pollutant Discharge Elimination System

The National Pollutant Discharge Elimination System (NPDES) is the national program for attaining and maintaining effluent limitations and water quality standards across the nation. The NPDES program is fundamentally a permitting program, which makes the EPA responsible for issuing, modifying, revoking and reissuing, terminating, monitoring, and enforcing permits, and imposing and enforcing pretreatment requirements.[79] The permit system makes point source regulations easier to administer.

The NPDES program requires permits for the discharge of pollutants from any point source into waters of the United States.[80] Discharges must comply with all terms and conditions of an EPA, state, or local permit. Two kinds of permits are issued. "Individual permits" are issued to one particular facility or source based on site-specific information.[81] "Storm water discharge" permits are discussed in chapter 5.

In February 1996 the EPA announced a new program by which a point source with an NPDES permit would be allowed to achieve greater pollution reductions and "trade" (or sell) credits for its excess reduction to another source. This "effluent trading" model is based on similar provisions in the federal Clean Air Act. Some environmentalists have argued that a problem with this program is that there is no net improvement in water quality.

---

[78] 33 U.S.C. § 1365. Compare Northwest Env'l Advocates v. City of Portland, 56 F.3d 979 (9th Cir. 1995) [environmental group could bring CWA citizen suit to enforce state water quality standards included as condition in city's NPDES permit], and Upper Chattahoochie River Keeper Fund, Inc. v. City of Atlanta, 953 F. Supp. 1541 (N.D. Ga. 1996) [private citizens could bring suit to enforce water quality standards imposed pursuant to state law as condition of city's NPDES permit] with Atlantic Legal States Found. v. Eastman Kodak Co., 12 F.3d 353 (2d Cir. 1993) [CWA citizen suit provision could not be used to enforce state wastewater effluent standards that were stricter than, and applied in lieu of, federal law].
[79] Source: U.S.E.P.A., Office of Water. *History of the Clean Water Act* (1997).
[80] See, generally, Wardzinski, Sandalow, Burgin, Ginsberg, and McGaffey, Water Pollution Control under the National Pollutant Discharge Elimination System, Chap. 2 in Evans (ed.), *The Clean Water Act Handbook* (Amer. Bar Ass'n, 1994).
[81] 40 CFR § 122.4(m).

Most permitting authority has been delegated to the states, so long as the state program demonstrates "adequate authority to carry out the [NPDES] program."[82] Nevertheless, the EPA retains a "veto" over state NPDES permits by issuing a letter of disapproval.[83] Reasons for the EPA to disapprove an NPDES permit are listed in the regulations:

(1) The permit fails to apply, or to ensure compliance with, any applicable requirement of this part;

(2) . . . written recommendations of an affected State have not been accepted by the permitting State and the Regional Administrator finds the reasons for rejecting the recommendations are inadequate;

(3) The procedures followed in connection with formulation of the proposed permit failed in a material respect to comply with procedures required by CWA or by regulations . . .;

(4) Any finding made by the State Director in connection with the proposed permit misinterprets CWA or any guidelines or regulations under CWA, or misapplies them to the facts;

(5) Any provisions of the proposed permit relating to the maintenance of records, reporting, monitoring, sampling, or the provision of any other information by the permittee are inadequate, in the judgment of the Regional Administrator, to assure compliance with permit conditions, including effluent standards and limitations or standards for sewage sludge use and disposal required by CWA, by the guidelines and regulations issued under CWA, or by the proposed permit;

(6) In the case of any proposed permit with respect to which applicable effluent standards and limitations or standards for sewage sludge use and disposal under . . . the CWA have not yet been promulgated by the Agency, the proposed permit, in the judgment of the Regional Administrator, fails to carry out the provisions of CWA or of any regulations issued under CWA . . . ;

(7) Issuance of the proposed permit would in any other respect be outside the requirements of CWA, or regulations issued under CWA.

(8) The effluent limits of a permit fail to satisfy the requirements of 40 CFR 122.44(d) (limitations on NPDES permits).[84]

---

[82] 40 CFR § 123.

[83] See United States v. City of Menominee, 727 F. Supp. 1110 (W.D. Mich. 1989).

[84] 40 CFR § 123.44(c).

State programs have been unsuccessfully challenged in court.[85] States normally have wastewater discharge regulations similar to the NPDES program.[86]

NPDES permits are often joint permits issued pursuant to both the federal CWA and state legislation. Frequently, the state will not administer the NPDES program but will issue a state permit even though the EPA has issued an NPDES permit. The states and the EPA generally cooperate in the permit issuance process to ensure that the two permits are consistent, but there may be differences in monitoring requirements and the number of pollutants limited. These requirements normally do not conflict but may require additional sampling and dual reporting.[87]

Under Section 402(b)(1)(B) of the CWA, the EPA can issue permits for up to 5 years, but most permits are subject to modification or revocation for cause.[88]

CWA Section 511(c)(1) exempts NPDES permits from the EIS requirements of NEPA, except POTW grants and new source discharge permits.[89]

Issuance of an NPDES permit has been held to be a quasi-judicial proceeding requiring an on-the-record hearing.[90] Violations of an NPDES permit may result in civil and, for "knowing" violations, criminal penalties under CWA Sections 309(b) and (c).[91]

Section 505 of the CWA has a "citizen suit" provision that allows "any citizen" to commence a civil action against "any person" (including the United Stated) or against the EPA to enforce any effluent limitation or standard.[92] During the early 1990s, the question arose whether a defendant in a CWA § 505 citizen suit case for violation of an NPDES permit would prevail if they could prove that the activity that gave rise to the suit had stopped, and that further permit violations could not reasonably be expected. Several federal court decisions concluded that defendants in CWA Section 505 citizen suits would prevail under these conditions.[93] However, this line of cases was reversed by the U.S. Supreme Court in

---

[85] For example, see NRDC v, USEPA, 859 F.2d 156 (D.C. Cir. 1988) [upholding state program requirements].

[86] See chapter 5.

[87] Source: USEPA, Office of Water. History of the Clean Water Act (1997).

[88] 33 U.S.C. § 1342(b)(1)(B).

[89] 33 U.S.C. § 1371(c)(1).

[90] Marathon Oil Co. v. EPA, 564 F.2d 1253 (9th Cir. 1977).

[91] 33 U.S.C. § 1319(b) and (c). See United States v. Hopkins, 53 F.3d 533 (2d Cir. 1995).

[92] 33 U.S.C. § 1365(a).

[93] See Friends of the Earth, Inc. v. Chevron Chem. Co., 900 F. Supp. 67 (E.D. Tex. 1995); and Molokai Chamber of Commerce v. Kukui (Molokai), Inc., 891 F. Supp. 1389 (D.Haw. 1995).

early 2000 in *Friends of the Earth, Inc. v. Laidlaw Environmental Services (TOC), Inc.*[94] In the *Laidlaw* case, Friends of the Earth (FOE) and other environmental groups sued a private wastewater treatment plant (Laidlaw, Inc.) under CWA § 505(a) for repeatedly violating its NPDES permit by discharging mercury and other pollutants into a nearby waterway. The District Court assessed a civil penalty of $405,800, but declined to order injunctive relief because Laidlaw, after the lawsuit began, had achieved substantial compliance with the terms of its permit.[95] FOE appealed to the Fourth Circuit Court of Appeals as to the amount of the District Court's civil penalty judgment. The Fourth Circuit vacated the District Court's order and remanded with instructions to dismiss the action, holding that the case had become moot once Laidlaw complied with the terms of its permit and the plaintiffs failed to appeal the denial of equitable relief.[96] The Supreme Court reversed the Fourth Circuit, holding that a CWA Section 505(a) citizen suit's claim for civil penalties need not be dismissed as moot when the defendant, after commencement of the litigation, has come into compliance with its NPDES permit.[97] The Supreme Court held that a defendant's voluntary cessation of a challenged practice ordinarily does not deprive a federal court of its power to determine the legality of the practice unless subsequent events make it absolutely clear that the allegedly wrongful behavior could not reasonably be expected to recur.[98]

4.5.1 General Permits

"General permits" cover an entire category or group of similar facilities with similar conditions but separate locations. EPA regulations state that general permits are appropriate for:

A category of point sources other than storm water point sources, or a category of "treatment works treating domestic sewage," [POTWs] if the sources or "treatment works treating domestic sewage" all:
(A) Involve the same or substantially similar types of operations;

---

[94] Friends of the Earth, Inc. v. Laidlaw Environmental Services (Toc), Inc., 120 S.Ct. 693 (2000).
[95] Friends of the Earth, Inc. v. Laidlaw Envtl. Servs. (TOC), Inc., 956 F. Supp. 588 (D.S.C., 1997).
[96] Friends of the Earth v. Laidlaw Environ Ser., 49 F.3d 303 (4th Cir., 1998).
[97] Friends of the Earth, Inc. v. Laidlaw Environmental Services (Toc), Inc., 120 S.Ct. 693 (2000).
[98] If such voluntary cessation rendered a case moot, "courts would be compelled to leave the defendant free to return to its old ways." Id., citing United States v. Phosphate Export Assn., 393 U.S. 199 (1968).

(B) Discharge the same types of wastes or engage in the same types of sludge use or disposal practices;

(C) Require the same effluent limitations, operating conditions, or standards for sewage sludge use or disposal;

(D) Require the same or similar monitoring; and

(E) In the opinion of the Director, are more appropriately controlled under a general permit than under individual permits.[99]

Unlike individual permits, general permits require only the submission to the EPA of a "notice of intent" rather than a full application.

4.5.2 The Individual Permit Process

EPA regulations contain the details of the process for individual permit applications. "Any person" who discharges or proposes to discharge pollutants into the waters of the United States must obtain a permit from the EPA (or the appropriate state agency).[100] When a facility or activity is owned by one person but is operated by another person, it is the operator's duty to obtain a permit.[101]

A person proposing a new discharge must submit an application at least 180 days before the date on which the discharge is to commence, unless permission for a later date has been granted by the EPA.[102] Certain construction activities require that an application be submitted at least 90 days before the date on which construction is to commence.[103] Persons proposing a new discharge are encouraged to submit their applications well in advance of the 90 or 180 day requirements to avoid delay.

In general, applicants for NPDES individual permits must provide the following information to the EPA, using the application form provided by the EPA or the appropriate state agency:

(1) The activities conducted by the applicant which require it to obtain an NPDES permit.

(2) Name, mailing address, and location of the facility for which the application is submitted.

(3) Up to four SIC codes which best reflect the principal products or services provided by the facility.

---

[99] 40 CFR § 122.28(a)(2).
[100] 49 CFR § 122.21(a).
[101] 40 CFR § 122.21(b). See Eckenfelder, *Industrial Water Pollution Control*, 3rd Ed. (McGraw-Hill, 1999).
[102] 40 CFR § 122.21(c).
[103] Id., referring to 40 CFR § 122.26(b)(14)(x)

(4) The operator's name, address, telephone number, ownership status, and status as Federal, State, private, public, or other entity.

(5) Whether the facility is located on [Native American] lands.

(6) A listing of all permits or construction approvals received or applied for under any of the following programs:

(i) Hazardous Waste Management program under RCRA.

(ii) UIC program under SDWA.

(iii) NPDES program under CWA.

(iv) Prevention of Significant Deterioration (PSD) program under the Clean Air Act.

(v) Nonattainment program under the Clean Air Act.

(vi) National Emission Standards for Hazardous Pollutants (NESHAPS) preconstruction approval under the Clean Air Act.

(vii) Ocean dumping permits under the Marine Protection Research and Sanctuaries Act.

(viii) Dredge or fill permits under section 404 of CWA.

(ix) Other relevant environmental permits, including State permits.

(7) A topographic map (or other map if a topographic map is unavailable) extending one mile beyond the property boundaries of the source, depicting the facility and each of its intake and discharge structures; each of its hazardous waste treatment, storage, or disposal facilities; each well where fluids from the facility are injected underground; and those wells, springs, other surface water bodies, and drinking water wells listed in public records or otherwise known to the applicant in the map area.

(8) A brief description of the nature of the business.[104]

Specific types of point sources have specific requirements, and some of these will be discussed below.

All permit applications must be complete before the EPA (or the appropriate state agency) can issue an individual NPDES permit. An application for a permit is complete when the EPA receives an application form and any supplemental information which are completed to the EPA's satisfaction. The completeness of any application for a permit is judged independently of the status of any other permit application or permit for the same facility or activity. For EPA administered NPDES programs, an application which is reviewed under 40 CFR Section 124.3 is complete when the EPA receives either a complete application, or the information listed in a notice of deficiency.[105]

---

[104] 40 CFR § 121.21(f).
[105] 40 CFR § 122.21(e).

Exemptions

Some persons are specifically exempted from the requirement for an individual NPDES permit. These include the following:

- Persons covered by general permits (see section 4.5.1).[106]
- Any discharge of sewage from vessels, effluent from properly functioning marine engines, laundry, shower, and galley sink wastes, or any other discharge incidental to the normal operation of a vessel. This exclusion does not apply to rubbish, trash, garbage, or other such materials discharged overboard; nor to other discharges when the vessel is operating in a capacity other than as a means of transportation such as when used as an energy or mining facility, a storage facility or a seafood processing facility, or when secured to a storage facility or a seafood processing facility, or when secured to the bed of the ocean, contiguous zone or waters of the United States for the purpose of mineral or oil exploration or development.[107]
- Discharges of dredged or fill material into waters of the United States which are regulated under Section 404 of the CWA.[108]
- The introduction of sewage, industrial wastes or other pollutants into publicly owned treatment works by indirect dischargers. Plans or agreements to switch to this method of disposal in the future do not relieve dischargers of the obligation to have and comply with permits until all discharges of pollutants to waters of the United States are eliminated.[109] This exclusion does not apply to the introduction of pollutants to privately owned treatment works or to other discharges through pipes, sewers, or other conveyances owned by a state, municipality, or other party not leading to treatment works.[110]
- Any discharge in compliance with the instructions of an On-Scene Coordinator under the National Oil and Hazardous Substances Pollution Contingency Plan under CERCLA,[111] or by the U.S. Coast Guard under the Oil Pollution Act.[112]
- Any introduction of pollutants from nonpoint-source agricultural and silvicultural activities, including storm water runoff from orchards, cultivated crops, pastures, range lands, and forest lands, but not

---

[106] 40 CFR § 122.28.
[107] 49 CFR § 122.21(a).
[108] 40 CFR § 122.3(b). Referring to 33 U.S.C. § 1344, the wetlands "dredge and fill" permit section, which is described at length in chapter 13.
[109] See also 40 CFR § 122.47(b).
[110] 40 CFR § 122.3(c).
[111] 40 CFR Part 300. CERCLA will be described in detail in chapter 9.
[112] 40 CFR § 122.3(d). See 33 CFR 153.10(e) (Pollution by Oil and Hazardous Substances).

discharges from concentrated animal feeding operations,[113] discharges from concentrated aquatic animal production facilities as defined in Sec. 122.24, discharges to aquaculture projects as defined in Sec. 122.25, and discharges from silvicultural point sources as defined in Sec. 122.27.[114]

- Return flows from irrigated agriculture.[115]
- Discharges into a privately owned treatment works, except as directed by the EPA.[116]

Application Requirements for New Sources and New Discharges

New manufacturing, commercial, mining and silvicultural dischargers applying for NPDES permits (except for new discharges of manufacturing, commercial, mining and silvicultural facilities which discharge only non-process wastewater, or new discharges of storm water associated with industrial activity) must provide the following information to the EPA, using the application forms provided by the EPA or the appropriate state agency:

(1) Expected outfall location. The latitude and longitude to the nearest 15 seconds and the name of the receiving water.

(2) Discharge dates. The expected date of commencement of discharge.

(3) Flows, sources of pollution, and treatment technologies:

(i) Expected treatment of wastewater. Description of the treatment that the wastewater will receive, along with all operations contributing wastewater to the effluent, average flow contributed by each operation, and the ultimate disposal of any solid or liquid wastes not discharged.

(ii) Line drawing. A line drawing of the water flow through the facility with a water balance as described in [40 CFR Section 122.21(g)(2)].

(iii) Intermittent flows. If any of the expected discharges will be intermittent or seasonal, a description of the frequency, duration and maximum daily flow rate of each discharge occurrence (except for stormwater runoff, spillage, or leaks).

(4) Production. If a new source performance standard promulgated under section 306 of CWA or an effluent limitation guideline applies to the applicant and is expressed in terms of production (or other

---

[113] Defined in 40 CFR § 122.23.
[114] 40 CFR § 122.3(e).
[115] 40 CFR § 122.3(f).
[116] Except as otherwise required by the EPA under Sec. 122.44(m). 40 CFR § 122.3(g).

measure of operation), a reasonable measure of the applicant's expected actual production reported in the units used in the applicable effluent guideline or new source performance standard as required by [40 CFR Section 122.45(b)(2)] for each of the first three years. Alternative estimates may also be submitted if production is likely to vary.

(5) Effluent characteristics. The requirements in [40 CFR Sections (h)(4)(i), (ii), and (iii)] that an applicant must provide estimates of certain pollutants expected to be present do not apply to pollutants present in a discharge solely as a result of their presence in intake water; however, an applicant must report such pollutants as present. Net credits may be provided for the presence of pollutants in intake water if the requirements of [40 CFR Section 122.45(g)] are met. All levels (except for discharge flow, temperature, and pH) must be estimated as concentration and as total mass.

(i) Each applicant must report estimated daily maximum, daily average, and source of information for each outfall for the following pollutants or parameters. . .

(A) Biochemical Oxygen Demand (BOD).
(B) Chemical Oxygen Demand (COD).
(C) Total Organic Carbon (TOC).
(D) Total Suspended Solids (TSS).
(E) Flow.
(F) Ammonia (as N).
(G) Temperature (winter and summer).
(H) pH.

(ii) Each applicant must report estimated daily maximum, daily average, and source of information for each outfall for [a series of conventional, conconventional, and toxic pollutants discussed in EPA regulations].

(6) Engineering Report. Each applicant must report the existence of any technical evaluation concerning his wastewater treatment, along with the name and location of similar plants of which he has knowledge.

(7) Other information. Any optional information the permittee wishes to have considered.

(8) Certification. Signature of certifying official under [40 CFR Section 122.22].[117]

If a new source is located in a state without an EPA approved NPDES program, it must comply with special regulatory provisions. Before

---

[117] 40 CFR § 122.21(k).

beginning any on-site construction, the owner or operator of any facility which may be a new source must submit information to the Regional Administrator of the EPA so that the EPA can determine if the facility is a new source. The Regional Administrator may request any additional information needed to determine whether the facility is a new source, and must make an initial determination whether the facility is a new source within 30 days of receiving all necessary information.[118] The Regional Administrator then issues a public notice of a new source determination, stating that the applicant must comply with all environmental review requirements.[119] Any interested person can challenge the Regional Administrator's initial new source determination by requesting an evidentiary hearing within 30 days of issuance of the public notice.[120]

Application Requirements for Existing Manufacturing, Commercial Mining, and Silvicultural Dischargers

Existing manufacturing, commercial mining, and silvicultural dischargers applying for NPDES permits must generally provide the following information to the EPA or the appropriate state agency, using application forms provided by the EPA:

(1) Outfall location. The latitude and longitude to the nearest 15 seconds and the name of the receiving water.

(2) Line drawing. A line drawing of the water flow through the facility with a water balance, showing operations contributing wastewater to the effluent and treatment units. Similar processes, operations, or production areas may be indicated as a single unit, labeled to correspond to the more detailed identification under [40 CFR Section 122.21(g)(3)]. The water balance must show approximate average flows at intake and discharge points and between units, including treatment units. If a water balance cannot be determined (for example, for certain mining activities), the applicant may provide instead a pictorial description of the nature and amount of any sources of water and any collection and treatment measures.

(3) Average flows and treatment. A narrative identification of each type of process, operation, or production area which contributes wastewater to the effluent for each outfall, including process wastewater, cooling water, and stormwater runoff; the average flow which each process contributes; and a description of the treatment the wastewater receives, including the ultimate disposal of any solid or

---

[118]   40 CFR § 122.21(l)(2).
[119]   See 40 CFR §§ 6.600 et seq.
[120]   40 CFR § 122.21(l)(3).

fluid wastes other than by discharge. Processes, operations, or production areas may be described in general terms (for example, "dye-making reactor," "distillation tower"). For a privately owned treatment works, this information shall include the identity of each user of the treatment works. The average flow of point sources composed of storm water may be estimated. The basis for the rainfall event and the method of estimation must be indicated.

(4) Intermittent flows. If any of the discharges described in [40 CFR Section] (g)(3) . . . are intermittent or seasonal, a description of the frequency, duration and flow rate of each discharge occurrence (except for stormwater runoff, spillage or leaks).

(5) Maximum production. If an effluent guideline promulgated under Section 304 of CWA applies to the applicant and is expressed in terms of production (or other measure of operation), a reasonable measure of the applicant's actual production reported in the units used in the applicable effluent guideline. The reported measure must reflect the actual production of the facility as required by [40 CFR Section 122.45(b)(2)].

(6) Improvements. If the applicant is subject to any present requirements or compliance schedules for construction, upgrading or operation of waste treatment equipment, an identification of the abatement requirement, a description of the abatement project, and a listing of the required and projected final compliance dates.

(7) Effluent characteristics. Information on the discharge of pollutants specified in this paragraph (except information on storm water discharges).[121]

When quantitative data for a pollutant are required, the applicant must collect a sample of effluent and analyze it for the pollutant in accordance with analytical methods approved under EPA regulations.[122] When no analytical method is approved, the applicant may use any suitable method but must provide a description of the method. When an applicant has two or more outfalls with substantially identical effluents, the EPA may allow the applicant to test only one outfall and report that the quantitative data also apply to the substantially identical outfalls. Additional reporting requirements are contained in EPA regulations.[123] Under EPA regulations, every applicant must report quantitative data for every outfall for the following pollutants:

---

[121] 40 CFR § 122.21(g).
[122] 40 CFR part 136.
[123] See 40 CFR § 122.21(g).

Biochemial Oxygen Demand (BOD_5)
Chemical Oxygen Demand
Total Organic Carbon
Total Suspended Solids
Ammonia (as N)
Temperature (both winter and summer)
pH[124]

The EPA may waive the reporting requirements for individual point sources or for a particular industry category for one or more of the pollutants if the applicant has demonstrated that such a waiver is appropriate because information adequate to support issuance of a permit can be obtained with less stringent requirements.[125]

Any applicant with processes in one or more primary industry category contributing to a discharge[126] must report quantitative data for toxic pollutants in each outfall containing process wastewater.[127]

Each applicant must indicate whether it knows or has reason to believe that certain conventional and nonconventional pollutants is discharged from each outfall.[128] Likewise, each applicant must indicate whether it knows or has reason to believe that certain toxic pollutants and total phenols for which quantitative data are not otherwise required is discharged from each outfall.[129] For every pollutant expected to be discharged in concentrations of 10 ppb or greater the applicant must report quantitative data. In addition, each applicant must indicate whether it knows or has reason to believe that certain hazardous substances and asbestos are discharged from each outfall.[130]

Certain small businesses are exempt from the requirement to submit quantitative data for organic toxic pollutants.[131] These include coal mines with a probable total annual production of less than 100,000 tons per year; and, for all other applicants, gross total annual sales averaging less than $100,000 per year.[132]

---

[124] 40 CFR § 122.21(g)(i)(A).
[125] 40 CFR § 122.21(g)(i)(B).
[126] See appendix A to 40 CFR Part 122.
[127] 40 CFR § 122.21(g)(ii).
[128] 40 CFR § 122.21(g)(6)(iii)(A). The conventional and nonconventional pollutants are list in Table IV of appendix D to 40 CFR Part 122.
[129] 40 CFR § 122.21(g)(6)(iii)(B). The toxic chemicals and phenols are listed in Tables II and III of appendix D to 40 CFR Part 122.
[130] 40 CFR § 122.21(g)(6)(iii)(C). These chemicals are listed in Table V of appendix D to 40 CFR Part 122.
[131] That is, those listed in Table II of appendix D of 40 CFR Part 122.
[132] 40 CFR § 122.21(g)(8).

Certain additional requirements may also apply to a permit application, including:

(1) Biological toxicity tests. An identification of any biological toxicity tests which the applicant knows or has reason to believe have been made within the last 3 years on any of the applicant's discharges or on a receiving water in relation to a discharge.

(2) Contract analyses. If a contract laboratory or consulting firm performed any of the analyses required by [40 CFR Section 122.21(g)(7)], the identity of each laboratory or firm and the analyses performed.

(3) Additional information. In addition to the information reported on the application form, applicants shall provide to the Director [of the EPA], at his or her request, such other information as the Director may reasonably require to assess the discharges of the facility and to determine whether to issue an NPDES permit. The additional information may include additional quantitative data and bioassays to assess the relative toxicity of discharges to aquatic life and requirements to determine the cause of the toxicity.[133]

Application Requirements for Manufacturing, Commercial, Mining and Silvicultural Facilities Which Discharge Only Non-process Wastewater

Except for stormwater discharges, all manufacturing, commercial, mining and silvicultural dischargers applying for NPDES permits which discharge only non-process wastewater not regulated by an effluent limitations guideline or new source performance standard shall provide the following information to the EPA, using application forms provided by the EPA or the appropriate state agency:

(1) Outfall location. Outfall number, latitude and longitude to the nearest 15 seconds, and the name of the receiving water.

(2) Discharge date (for new dischargers). Date of expected commencement of discharge.

(3) Type of waste. An identification of the general type of waste discharged, or expected to be discharged upon commencement of operations, including sanitary wastes, restaurant or cafeteria wastes, or noncontact cooling water. An identification of cooling water additives (if any) that are used or expected to be used upon commencement of operations, along with their composition if existing composition is available.

---

[133] 40 CFR §§ 122.21(g)(11)-(14).

(4) Effluent characteristics.

(i) Quantitative data for the pollutants or parameters listed below, unless testing is waived by the Director [of the EPA]. The quantitative data may be data collected over the past 365 days, if they remain representative of current operations, and must include maximum daily value, average daily value, and number of measurements taken. The applicant must collect and analyze samples in accordance with 40 CFR part 136. Grab samples must be used for pH, temperature, oil and grease, total residual chlorine, and fecal coliform. For all other pollutants, 24-hour composite samples must be used. New dischargers must include estimates for the pollutants or parameters listed below instead of actual sampling data, along with the source of each estimate. All levels must be reported or estimated as concentration and as total mass, except for flow, pH, and temperature.

(A) Biochemical Oxygen Demand (BOD_5).

(B) Total Suspended Solids (TSS).

(C) Fecal Coliform (if believed present or if sanitary waste is or will be discharged).

(D) Total Residual Chlorine (if chlorine is used).

(E) Oil and Grease.

(F) Chemical Oxygen Demand (COD) (if non-contact cooling water is or will be discharged).

(G) Total Organic Carbon (TOC) (if non-contact cooling water is or will be discharged).

(H) Ammonia (as N).

(I) Discharge Flow.

(J) pH.

(K) Temperature (Winter and Summer).

(5) Flow. A description of the frequency of flow and duration of any seasonal or intermittent discharge (except for stormwater runoff, leaks, or spills).

(6) Treatment system. A brief description of any system used or to be used.

(7) Optional information. Any additional information the applicant wishes to be considered, such as influent data for the purpose of obtaining "net" credits pursuant to [40 CFR Section 122.45(g)].

(8) Certification. Signature of certifying official under [40 CFR Section 122.22].[134]

The EPA may waive the testing and reporting requirements for any of the pollutants or flow listed in [40 CFR Section 122.21(h)(4)(i)] if the

---

[134] 40 CFR § 122.21(h).

applicant submits a request for such a waiver, before or with the application, which demonstrates that information adequate to support issuance of a permit can be obtained through less stringent requirements.

If the applicant is a new discharger, then they must complete and submit Item IV of Form 2e [see 40 CFR Section 122.21(h)(4)] by providing quantitative data in accordance with that section no later than two years after commencement of discharge.

The above requirements for quantitative data or estimates of certain pollutants do not apply to pollutants present in a discharge solely as a result of their presence in intake water. However, an applicant must report such pollutants as present.

Application Requirements for New and Existing POTWs

All POTWs with design influent flows equal to or greater than one million gallons per day, those with approved pretreatment programs, and those required to develop a pretreatment program must provide the results of valid whole effluent biological toxicity testing to the EPA.[135] The EPA may require other POTWs to submit the results of toxicity tests with their permit applications, based on consideration of the following factors:

(1) The variability of the pollutants or pollutant parameters in the POTW effluent (based on chemical-specific information, the type of treatment facility, and types of industrial contributors);

(2) The dilution of the effluent in the receiving water (ratio of effluent flow to receiving stream flow);

(3) Existing controls on point or nonpoint sources, including total maximum daily load calculations for the waterbody segment and the relative contribution of the POTW;

(4) Receiving stream characteristics, including possible or known water quality impairment, and whether the POTW discharges to a coastal water, one of the Great Lakes, or a water designated as an outstanding natural resource; or

(5) Other considerations (including but not limited to the history of toxic impact and compliance problems at the POTW), which the Director [of the EPA] determines could cause or contribute to adverse water quality impacts.[136]

For those POTWs required to conduct toxicity testing, POTWs must use the EPA's methods or other established protocols which are scientifically defensible and sufficiently sensitive to detect aquatic toxicity. Such testing

---

[135] 40 CFR § 122.21(j).
[136] Id.

must have been conducted since the last NPDES permit reissuance or permit modification under 40 CFR Section 122.62(a), whichever occurred later. All POTWs with approved pretreatment programs must provide a written technical evaluation of the need to revise local limits under 40 CFR Section 403.5(c)(1).[137]

Application Requirements for POTWs Treating Domestic Sewage

EPA regulations under Section 405(f) of the CWA require POTWs treating domestic sewage, and which request site specific pollutant limitations, to submit a permit application within 180 days after publication of a standard applicable to its sewage sludge use or disposal practice(s).[138] After this 180 day period, the POTW may only apply for site-specific pollutant limits for good cause, and such requests must be made within 180 days of becoming aware that good cause exists. For all other new or existing POTWs treating domestic sewage, there must generally be an application for a permit (or permit renewal) 180 days before operations commence, or an existing permit expires.[139] The POTW's permit application must contain the following:

(A) Name, mailing address and location of the [POTW];

(B) The operator's name, address, telephone number, ownership status, and status as Federal, State, private, public or other entity;

(C) A description of the sewage sludge use or disposal practices (including, where applicable, the location of any sites where sewage sludge is transferred for treatment, use, or disposal, as well as the name of the applicator or other contractor who applies the sewage sludge to land, if different from the [POTW], and the name of any distributors if the sewage sludge is sold or given away in a bag or similar enclosure for application to the land, if different from the [POTW]);

(D) Annual amount of sewage sludge generated, treated, used or disposed (dry weight basis); and

(E) The most recent data the [POTW] may have on the quality of the sewage sludge.[140]

---

[137] Id.

[138] 40 CFR § 122.21(c)(2)(i).

[139] 40 CFR §§ 122.21(c)(2)(ii) and (iii). However, the EPA may require permit applications from any POTW at any time if the Director of the EPA determines that a permit is necessary to protect public health and the environment from any potential adverse effects that may occur from toxic pollutants in sewage sludge. 40 CFR § 122.21(c)(2)(iv),

[140] 40 CFR § 122.21(c)(2)(iii). Additional information requirements are at 40 CFR § 501.15(a)(2).

Requests for Variances

As discussed in section 4.3.6, the CWA allows variances of several kinds from the requirements of the NPDES permit system. EPA regulations provide guidance on the process of applying for these variances, depending on whether the source is a POTW or not.

A discharger that is not a POTW can request a variance from otherwise applicable effluent limitations under any of the following statutory or regulatory provisions within specified times:

(1) Fundamentally different factors. A request for a variance based on the presence of "fundamentally different factors" from those on which the effluent limitations guideline was based shall be filed as follows:

(A) For a request from best practicable control technology currently available (BPT), by the close of the public comment period under [40 CFR Section 124.10].

(B) For a request from best available technology economically achievable (BAT) and/or best conventional pollutant control technology (BCT), by no later than: (1) July 3, 1989, for a request based on an effluent limitation guideline promulgated before February 4, 1987, to the extent July 3, 1989 is not later than that provided under previously promulgated regulations; or (2) 180 days after the date on which an effluent limitation guideline is published in the Federal Register for a request based on an effluent limitation guideline promulgated on or after February 4, 1987.

The request shall explain how the requirements of the applicable regulatory and/or statutory criteria have been met.

(2) Non-conventional pollutants. A request for a variance from the BAT requirements for CWA Section 301(b)(2)(F) pollutants (commonly called "non-conventional" pollutants) pursuant to Section 301(c) of CWA because of the economic capability of the owner or operator, or pursuant to Section 301(g) of the CWA (provided however that a Sec. 301(g) variance may only be requested for ammonia; chlorine; color; iron; total phenols (4AAP) (when determined by the Administrator [of the EPA] to be a pollutant covered by Section 301(b)(2)(F) [of the CWA]) and any other pollutant which the Administrator lists under Section 301(g)(4) of the CWA) must be made as follows:

(i) For those requests for a variance from an effluent limitation based upon an effluent limitation guideline by:

(A) Submitting an initial request to the Regional Administrator [of the EPA], as well as to the State Director if applicable, stating the name of the discharger, the permit number, the outfall number(s), the applicable effluent guideline, and whether the discharger is requesting a Section 301(c) or Section 301(g) [of the CWA] modification or both. . . ; and

(B) Submitting a completed request no later than the close of the public comment period under [40 CFR Section 124.10] demonstrating that the requirements of [40 CFR Section 124.13] and the applicable requirements of [40 CFR Part 125] have been met. Notwithstanding this provision, the complete application for a request under Section 301(g) shall be filed 180 days before EPA must make a decision (unless the Regional Division Director establishes a shorter or longer period).

(ii) For those requests for a variance from effluent limitations not based on effluent limitation guidelines, the request need only comply with [40 CFR Section 122.21(m)(2)(i)(B)] and need not be preceded by an initial request under [40 CFR Section 122.21 (m)(2)(i)(A)]. . .

(5) Water quality related effluent limitations. A modification under Section 302(b)(2) [of the CWA] of requirements under Section 302(a) for achieving water quality related effluent limitations may be requested no later than the close of the public comment period under [40 CFR Section 124.10] on the permit from which the modification is sought.

(6) Thermal discharges. A variance under CWA Section 316(a) for the thermal component of any discharge must be filed with a timely application for a permit under this section, except that if thermal effluent limitations are established under CWA Section 402(a)(1) or are based on water quality standards the request for a variance may be filed by the close of the public comment period under [40 CFR Section 124.10]. A copy of the request as required under 40 CFR part 125, subpart H, shall be sent simultaneously to the appropriate State or interstate certifying agency as required under 40 CFR part 125. (See [40 CFR Section 124.65] for special procedures for [CWA] Section 316(a) thermal variances.)[141]

A discharger that is a POTW may request a variance from otherwise applicable effluent limitations under any of the following statutory provisions:

(1) Discharges into marine waters. A request for a modification under CWA Section 301(h) of requirements of CWA Section 301(b)(1)(B) for discharges into marine waters must be filed in accordance with the requirements of 40 CFR part 125, subpart G. . .

(3) Water quality based effluent limitation. A modification under

---

[141] 40 CFR § 122.21(m).

CWA Section 302(b)(2) of the requirements under Section 302(a) for achieving water quality based effluent limitations shall be requested no later than the close of the public comment period under [40 CFR Section 124.10] on the permit from which the modification is sought.[142]

Despite the time limitations for variance requests, the EPA may notify a permit applicant before a draft permit is issued that the draft permit will likely contain limitations which are eligible for variances. In the notice the EPA may require the applicant as a condition of consideration of any potential variance request to submit a request explaining how the requirements of EPA regulations applicable to the variance have been met, and may require its submission within a specified reasonable time after receipt of the notice. The notice may be sent before the permit application has been submitted. The draft or final permit may contain the alternative limitations which may become effective upon final grant of the variance. This is known as an "expedited variance."[143]

If an applicant cannot file a complete request in the time required, then the applicant may request an extension. The extension is for no more than 6 months, and may be granted or denied at the discretion of the EPA.[144]

4.6 Nonpoint Source Discharges

Nonpoint source (NPS) pollution remains the nation's largest source of water quality problems. NPS pollution is the primary reason that approximately 40 percent of our surveyed rivers, lakes, and estuaries are not clean enough to meet basic uses such as fishing or swimming.

In discussing NPS regulation, Congress was apparently thinking of "disparate runoff caused primarily by rainfall."[145] However, NPS pollution occurs when rainfall, snowmelt, or irrigation runs over land or through the ground, picks up pollutants, and deposits them into rivers, lakes, and coastal waters or introduces them into groundwater. NPS pollution also includes adverse changes to the vegetation, shape, and flow of streams and other aquatic systems.

NPS pollution is widespread because it can occur any time activities disturb the land or water. Agriculture, forestry, grazing, septic systems, recreational boating, urban runoff, construction, physical changes to

---

[142] 40 CFR § 122.21(n).
[143] 40 CFR § 122.21(o)(1).
[144] 40 CFR § 122.21(o)(2).
[145] United States v. Earth Sciences, Inc., 599 F.2d 368 (10th Cir. 1979).

stream channels, and habitat degradation are potential sources of NPS pollution. Careless or uninformed household management also contributes to NPS pollution problems.[146]

A 1996 National Water Quality Inventory indicates that agriculture is the leading contributor to water quality impairments, degrading 60 percent of the impaired river miles and half of the impaired lake acreage surveyed by states, territories, and tribes. Runoff from urban areas is the largest source of water quality impairments to surveyed estuaries.[147]

The most common NPS pollutants are sediment and nutrients, which wash into water bodies from agricultural land, small and medium-sized animal feeding operations, construction sites, and other areas of disturbance. Other common NPS pollutants include pesticides, pathogens (bacteria and viruses), salts, oil, grease, toxic chemicals, and heavy metals.

NPS problems are regulated at several levels and by several agencies. Recent federal NPS control programs include the Nonpoint Source Management Program established by the 1987 Clean Water Act Amendments, and the Coastal Nonpoint Pollution Program established by the 1990 Coastal Zone Act Reauthorization Amendments. Other recent federal programs, as well as state, territorial, tribal and local programs also tackle NPS problems. In addition, public and private organizations have developed and used pollution prevention and pollution reduction initiatives and NPS pollution controls known as management measures.[148]

4.6.1 Federal NPS Programs

The EPA administers CWA Section 319, also known as the Nonpoint Source Management Program (NSMP).[149] Under Section 319, states, territories, and tribes apply for and receive grants from the EPA to implement NPS pollution programs. The EPA had awarded more than $370 million under Section 319 to address NPS pollution problems as of 1995.[150]

Under the NSMP, the EPA administers other sections of the Clean Water Act to help states, territories, and tribes to plan for and implement water pollution programs, which can include measures for NPS control. These

---

[146] U.S.E.P.A., *Nonpoint Source Pollution: The Nation's Largest Water Quality Problem* (1996).
[147] Id.
[148] U.S.E.P.A., *Managing Nonpoint Source Pollution: Final Report to Congress on Section 319 of the Clean Water Act* (EPA-506/9-90, 1990).
[149] 33 U.S.C. § 1329.
[150] U.S.E.P.A., *Nonpoint Source Pollution: The Nation's Largest Water Quality Problem* (1996).

include CWA Section 104(b)(3), Water Quality Cooperative Agreements;[151] CWA Section 104(g), Small Community Outreach;[152] CWA Section 106, Grants for Pollution Control Programs;[153] CWA Section 314, Clean Lakes Program;[154] CWA Section 320, National Estuary Program;[155] and CWA Section 604(b), Water Quality Management Planning.[156] Together with the National Oceanic Atmospheric Administration, the EPA helps administer section 6217 of the 1990 Coastal Zone Act Reauthorization Amendments, a program that deals with nonpoint source pollution affecting coastal waters.[157]

The National Oceanic and Atmospheric Administration (NOAA) administers Coastal Zone Management Act (CZMA) Section 306 that provides funds for water pollution control projects, including NPS management activities, in states with coastal zones.[158] Together with the EPA, NOAA also helps administer section 6217 of the Coastal Zone Act Reauthorization Amendments. This requires the 29 states with approved Coastal Zone Management Programs to establish and implement Coastal Nonpoint Pollution Control Programs.

The U.S. Department of Agriculture (USDA) administers incentive-based conservation programs through the Consolidated Farm Services Agency, the Natural Resources Conservation Service, and the U.S. Forest Service to help control NPS pollution from agriculture, forestry, and urban sources.

The U.S. Department of Transportation (DOT) and Federal Highway Administration developed erosion control guidelines for federally funded construction projects on roads, highways, and bridges under the Intermodal Surface Transportation Efficiency Act of 1991 (ISTEA).[159]

The Bureau of Reclamation, the Bureau of Land Management, and the Fish and Wildlife Service, all agencies within the U.S. Department of the Interior, administer several programs to help states manage NPS pollution by providing technical assistance and financial support. For example, the Fish and Wildlife Service administers the Clean Vessel Act, which provides grants to construct sewage pumpout stations at marinas.[160]

---

[151] 33 U.S.C. § 1254(c).
[152] 33 U.S.C. § 1254(g).
[153] 33 U.S.C. § 1256.
[154] 33 U.S.C. § 1324.
[155] 33 U.S.C. § 1330.
[156] 33 U.S.C. § 1384(b).
[157] 16 U.S.C. §§ 1451–1465. See chapter 14.
[158] 16 U.S.C. § 1455.
[159] 49 U.S.C §§ 101–112.
[160] 33 U.S.C. § 1322. See U.S.E.P.A., *Guidance Specifying Management Measures for Sources of Nonpoint Pollution in Coastal Waters* (EPA-840-B-92-002, 1993).

4.6.2 Coastal Waters and NPS Pollution

The U.S. EPA found that high levels of pollution prevented people from swimming safely at coastal beaches on more than 12,000 occasions from 1988 through 1994, and the latest National Water Quality Inventory reports that one-third of surveyed estuaries (areas near the coast where seawater and freshwater mixing occurs) are damaged. Rapidly increasing population growth and development in coastal regions could be a source of even more coastal water quality problems in the future. A significant portion of the threats to coastal waters are caused by NPS pollution. This is caused primarily by agriculture and urban runoff, but also from faulty septic systems, forestry, marinas and recreational boating, physical changes to stream channels, and habitat degradation, especially the destruction of wetlands and vegetated areas near streams.[161]

The Coastal Zone Act Reauthorization Amendments (CZARA) of addresses NPS pollution problems in coastal waters.[162] The CZARA requires the 29 states and territories with approved Coastal Zone Management Programs to develop Coastal Nonpoint Pollution Control Programs. In its program, a state or territory describes how it will implement nonpoint source pollution controls, known as management measures, that conform with those described in Guidance Specifying Management Measures for Sources of Nonpoint Pollution in Coastal Waters.[163] If these original management measures fail to produce the necessary coastal water quality improvements, a state or territory then must implement additional management measures to address remaining water quality problems.[164]

Under Section 306 of the Coastal Zone Management Act, approved programs must update and expand upon NPS Management Programs developed under Section 319 of the Clean Water Act and Coastal Zone Management Programs developed.[165] Coastal states submitted their coastal NPS programs to the EPA and NOAA for review and approval in 1995. States and territories are scheduled to implement the first phase of their

---

[161] U.S.E.P.A., *Nonpoint Source Pollution: The Nation's Largest Water Quality Problem* (1996).
[162] 33 U.S.C. § 1321.
[163] U.S.E.P.A., *Coastal Nonpoint Pollution Control Program Development and Approval Guidance* (EPA-841-B-93-003, 1993).
[164] See Wiltshire, Nonpoint Source Pollution Control, Chap. 10 in Evans (ed.), *The Clean Water Act Handbook* (Amer. Bar Ass'n, 1994).
[165] 16 U.S.C. § 1455.

approved program by 2004 and, if necessary, the second phase by 2009. Approved programs include several key elements:[166]

● **Boundary.** The boundary defines the region where land and water uses have a significant impact on a states or territorys coastal waters. It also includes areas where future land uses reasonably can be expected to impair coastal waters. To define the boundary, a state or territory may choose a region suggested by NOAA or may propose its own boundary based on geologic, hydrologic, and other scientific data.

● **Management Measures.** The state or territory coastal nonpoint program describes how a state or territory plans to control NPS pollution within the boundary. To help states and territories identify appropriate technologies and tools, the EPA issued Guidance Specifying Management Measures for Sources of Nonpoint Pollution in Coastal Waters. This technical guidance describes the best available, economically achievable approaches used to control NPS pollution from the major categories of land management activities that can degrade coastal water quality. States or territories may elect to implement alternative measurement measures as long as the alternative measures will achieve the same environmental results as those described in the guidance.

● **Enforceable Policies and Mechanisms.** States and territories should ensure the implementation of the management measures. Mechanisms may include, for example, permit programs, zoning, bad actor laws, enforceable water quality standards, and general environmental laws and prohibitions. States and territories may also use voluntary approaches like economic incentives if they are backed by appropriate regulations.

4.6.3 NPS Pollution from Agriculture

The United States has over 330 million acres of agricultural land, but, when improperly managed, agricultural activities produce NPS pollution. The most recent National Water Quality Inventory reports that agricultural NPS pollution is the leading source of water quality impacts to surveyed rivers and lakes, the third largest source of impairments to surveyed estuaries, and also a major contributor to groundwater contamination and wetlands degradation. Agricultural activities that cause NPS pollution

---

[166] U.S.E.P.A., *Nonpoint Source Pollution: The Nation's Largest Water Quality Problem* (1996).

include confined animal facilities, grazing, plowing, pesticide spraying, irrigation, fertilizing, planting, and harvesting. The major agricultural NPS pollutants that result from these activities are sediment, nutrients, pathogens, pesticides, and salts. Agricultural activities also can damage habitat and stream channels.[167]

Agricultural impacts on surface water and groundwater can be minimized by properly managing activities that can cause NPS pollution. A variety of government programs are available to help design and pay for management approaches to prevent and control NPS pollution. For example, over 40 percent of CWA Section 319 grants were used to control agricultural NPS pollution. Several U.S. Department of Agriculture and state-funded programs provide cost-share, technical assistance, and economic incentives to implement NPS pollution management practices.[168]

Excessive sedimentation has been identified as a particularly harmful form of agricultural NPS pollution. Sedimentation occurs when wind or water runoff carries soil particles from an area, such as a farm field, and transports them to a water body, such as a stream or lake. Excessive sedimentation clouds the water, which reduces the amount of sunlight reaching aquatic plants. Excessive sedimentation also covers fish spawning areas and food supplies, and clogs the gills of fish. Erosion and sedimentation can be reduced by 20 to 90 percent by applying management measures to control the volume and flow rate of runoff water, keep the soil in place, and reduce soil transport.[169]

Nutrients such as phosphorus, nitrogen, and potassium in the form of fertilizers, manure, sludge, irrigation water, legumes, and crop residues are applied to enhance production. When applied in excess of plant needs, nutrients can wash into aquatic ecosystems where they can cause excessive plant growth, which inhibits swimming and boating, creates unpleasant tastes and odors in drinking water, and kills fish. Farmers can implement nutrient management plans which help maintain high yields and save money on the use of fertilizers while reducing NPS pollution.[170]

Pesticides, herbicides, and fungicides are used to kill pests and control the growth of weeds and fungus.[171] These chemicals can enter and contaminate water through direct application, runoff, wind transport, and atmospheric deposition. They can kill fish and wildlife, poison food sources, and destroy the habitat that animals use for protective cover.

---

[167] Id.
[168] See Wiltshire, Nonpoint Source Pollution Control, Chap. 10 in Evans (ed.) *The Clean Water Act Handbook* (Amer. Bar Ass'n, 1994).
[169] National Research Council, *Soil And Water Quality: An Agenda for Agriculture* (National Academy Press, 1993).
[170] Id.
[171] See chapter 10

Integrated Pest Management (IPM) techniques based on the specific soils, climate, pest history, and crop for a particular field can be used to reduce NPS contamination from pesticides. IPM helps limit pesticide use and manages necessary applications to minimize pesticide movement from the field.[172]

4.6.4 Urban Runoff and Nonpoint Source Pollution

The most recent National Water Quality Inventory reports that runoff from urban areas is the leading source of damage to estuaries and the third largest source of water quality impairments to surveyed lakes. Population and development trends indicate that by 2010 more than half of the Nation will live in coastal towns and cities. Runoff from these rapidly growing urban areas will continue to degrade coastal waters.[173] In addition, urbanization increases the variety and amount of pollutants transported to receiving waters.

Sediment from development and new construction; oil, grease, and toxic chemicals from automobiles; nutrients and pesticides from turf management and gardening; viruses and bacteria from failing septic systems; road salts; and heavy metals are examples of pollutants generated in urban areas. Sediments and solids constitute the largest volume of pollutant loads to receiving waters in urban areas.[174]

Urban NPS pollution is covered by nonpoint source management programs developed by states and territories under CWA Section 208(a). Under CWA Section 208(a), each state must produce an Area-Wide Waste Treatment Management (AWTM) plan. In states and territories with coastal zones, programs to protect coastal waters from nonpoint source pollution also are required by the Coastal Zone Act Reauthorization Amendments.[175]

4.6.5 Wetlands and NPS Pollution

Wetlands will be discussed in more detail in Chapter 13, but it is worth mentioning that wetlands can help prevent NPS pollution from degrading water quality. Wetlands can intercept runoff and transform and store NPS

---

[172] Id. See also National Research Council, *Alternative Agriculture* (National Academy Press, 1989).
[173] U.S.E.P.A., *Nonpoint Source Pollution: The Nation's Largest Water Quality Problem* (1996).
[174] Id.
[175] 33 U.S.C. § 1321.

pollutants like sediment, nutrients, and certain heavy metals without being degraded. In addition, wetlands vegetation can keep stream channels intact by slowing runoff and by evenly distributing the energy in runoff.[176]

It is important that healthy wetlands be maintained. Degraded wetlands do not provide effective water quality benefits and become significant sources of NPS pollution. Excessive amounts of decaying wetlands vegetation, for example, can increase biochemical oxygen demand, making habitat unsuitable for fish and other aquatic life. Degraded wetlands also release stored nutrients and other chemicals into surface water and groundwater.

The U.S. EPA recommends three management strategies to maintain the water quality benefits provided by wetlands: preservation, restoration, and construction of engineered systems that pretreat runoff before it reaches receiving waters and wetlands.

Preservation of wetlands is largely the funtion of CWA Section 404, and is discussed at length in Chapter 13.[177] Preservation protects the full range of wetlands functions by discouraging development activity. At the same time, this strategy encourages proper management of upstream watershed activities, such as agriculture, forestry, and urban development.

Restoration of degraded wetlands creates valuable wetlands and riparian zones with NPS pollution control potential. Restoration activities include replanting degraded wetlands with native plant species and, depending on the location and the degree of degradation, using structural devices to control water flows. Restoration projects encompass ecological principles, such as habitat diversity and the connections between different aquatic and riparian habitat types, which distinguish these kinds of projects from wetlands that are constructed for runoff pretreatment.[178]

Finally, construction of engineered systems promotes the use of engineered vegetated treatment systems (VTS), which are especially effective at removing suspended solids and sediment from NPS pollution before the runoff reaches natural wetlands. One type of VTS, the vegetated filter strip (VFS), is a swath of land planted with grasses and trees that intercepts uniform sheet flows of runoff, before the runoff reaches wetlands. VFSs are most effective at sediment removal, with removal rates usually greater than 70 percent. Constructed wetlands, another type of VTS, are typically engineered complexes of water, plants, and animal life that simulate naturally occurring wetlands. Studies indicate that

---

[176] See Dennison and Berry, *Wetlands: Guide to Science, Law, and Technology* (Noyes, 1993).

[177] 33 U.S.C. § 1344.

[178] See discussion of wetlands in chapter 13. See Dennison and Berry, *Wetlands: Guide to Science, Law, and Technology* (Noyes, 1993); and U.S.E.P.A., *The Quality of Our Nation's Water* (EPA-841-S-95-004, 1994).

constructed wetlands can achieve sediment removal rates greater than 90 percent. Like VFS, constructed wetlands offer an alternative to other systems that are more structural in design.[179]

4.7 State Water Pollution Prohibitions

Most states have some enforceable statutory authority to deal generally with the subject of water pollution and activities that may lead to water pollution. These laws come in a variety of forms. Many are parts of states' water pollution control laws, others are in public health and penal codes, which prohibit specific kinds of discharges and substances that impair public waters. Still other states use statutory nuisance and public health laws provide additional authorities where certain adverse effects can be proven. State fish and game protection laws frequently contain general provisions prohibiting pollution harmful to fish; or imposing liability for fish kills due to pollution events, not limited to point source pollution.[180] In most states, however, agricultural runoff in particular, or NPS pollution in general, is exempted from state regulation.

For example, in the northeast New Jersey prohibits discharge of pollutants without a permit or as otherwise authorized,[181] and also prohibits the placement of "deleterious" substances into the waters or where they can find their way into such waters, but exempts from the latter provision chemicals used in agriculture, forestry, horticulture, and livestock if done in an approved manner.[182] New York prohibits the direct or indirect discharge of any substance that "shall cause or contribute to" a condition in violation of water quality standards.[183] Vermont prohibits the discharge of any substance without a permit, but exempts the "proper application of fertilizer to fields and crops."[184] Delaware requires a permit for any activity "which may cause or contribute to a discharge of a pollutant into any surface or groundwater."[185] The adopted implementing regulations appear limited to point source discharges to water and land, but the statute is not so limited and Delaware maintains that this authority also applies to nonpoint sources; indeed, Delaware's nonpoint programs rely in

---

[179] Id.
[180] See Environmental Law Institute, *Nonpoint Source Pollution Control Program* (1997). Probably the best recent review of state laws affecting nonpoint regulation.
[181] N.J. Stat. Ann. § 58:10-6.
[182] N.J. Stat. Ann. § 23:5-28.
[183] N.Y. Env. Cons. L. § 17-0501.
[184] 10 Vt. Stat. Ann. § 1259.
[185] 7 Del. Code § 6003.

part upon this authority.[186]

In the southeast, Virginia law prohibits the discharge of wastes or any "noxious or deleterious substances" or the pollution of waters without a permit,[187] as well as the placement of any substance which may contaminate or impair the lawful use or enjoyment of waters of the state except as permitted by law.[188] Alabama requires a permit for discharges of "pollution,"[189] but the regulations provide that a permit is not required for discharges "from non-point source agricultural and silvicultural activities."[190] Florida law provides that causing pollution except as provided by law is prohibited,[191] and requires permits for discharges of waste that contribute to violation of water quality standards,[192] but further provides that agricultural activities (including all "normal and customary" farming and forestry operations), and agricultural water management systems, are authorized and do not require permits.[193] Tennessee has a general prohibition against any discharge causing "pollution" except as properly authorized,[194] but the law does not apply to any nonpoint source discharges from any agricultural or forestry activity.[195]

In the midwest, Illinois prohibits any person from causing, threatening, or allowing the discharge of any "contaminants" that would cause or tend to cause water pollution, or that would violate regulations or standards adopted by the Pollution Control Board.[196] This provision is not limited to point sources, but a second provision prohibits the unpermitted discharge of contaminants (without requiring evidence of water pollution) and is expressly limited to point source discharges.[197] Michigan prohibits the direct or indirect discharge of any substance that may be injurious to health, safety or welfare, uses of waters, riparian lands, and fish and wildlife.[198] Minnesota has a general requirement of notice to the state of water pollution events and requires reasonable attempts by the discharger

---

[186] See Environmental Law Institute, *Nonpoint Source Pollution Control Program* (1997).
[187] Va. Code § 62.1-44.5.
[188] Va. Code § 62.1-194.1.
[189] Ala. Code § 22-22-9(I)(3).
[190] Ala. Admin. Code § 335-6-6-.03(a).
[191] Fla. Stat. § 403.161.
[192] Fla. Stat. § 403.088.
[193] Fla. Stat. § 403.927.
[194] Tenn. Code Ann. § 69-3-114.
[195] Tenn. Code Ann. § 69-3-120(g). See Environmental Law Institute, *Nonpoint Source Pollution Control Program* (1997).
[196] 415 ILCS § 5/12(a).
[197] 415 ILCS § 5/12(f).
[198] Mich. Comp. Laws § 324.3109(1).

to minimize or abate pollution.[199] By regulation, Minnesota has stated that "no sewage, industrial waste or other wastes shall be discharged from either a point or nonpoint source into the waters of the state in such quantity or in such a manner alone or in combination with other substances as to cause pollution."[200] Ohio's water pollution law prohibits causing pollution, or placing any wastes where they cause pollution except in accordance with a permit, but exempts agricultural and silvicultural runoff and earthmoving activities subject to regulation under Ohio's nonpoint source control programs administered by soil and water conservation districts and local governments.[201]

In the southwest, New Mexico's water pollution law authorizes the Water Quality Control Commission to adopt regulations "to prevent or abate water pollution in the state" and to require permits.[202] Texas law prohibits the discharge of waste, including agricultural waste, into or adjacent to any waters, and prohibits any other act which causes pollution of any waters, except as authorized.[203] The law exempts agricultural and silvicultural discharges in compliance with a certified water quality management plan.[204] Nebraska law prohibits water pollution or to placing wastes in a location where they are likely to cause water pollution, or discharging wastes that reduce the water quality in the receiving waters below adopted water quality standards.[205]

In the west, Colorado's water pollution control law authorizes the Water Quality Control Commission to adopt regulations relating to any "activity" that "does or could reasonably be expected to cause pollution of any state waters in violation of control regulations or...any applicable water quality standard."[206] However, "control regulations related to agricultural practices shall be promulgated only if incentive, grant, and cooperative programs are determined by the commission to be inadequate and such regulations are necessary to meet state law or the federal act."[207] Utah prohibits causing pollution that constitutes a menace to public health and welfare, is harmful to fish or wildlife, or impairs beneficial uses of water, and

---

[199] Minn. Stat. § 115.061.
[200] Minn. Rules § 7050.0210(13). See Environmental Law Institute, *Nonpoint Source Pollution Control Program* (1997).
[201] Ohio Rev. Stat. § 6111.04.
[202] N.M. Stat. Ann. § 74-6-4.
[203] Tex. Water Code Ann. § 26.121(a).
[204] Tex. Agric. Code Ann. § 201.026.
[205] Neb. Rev. Stat. § 81-1506.
[206] Colo. Rev. Stat. § 25-8-205.
[207] Colo. Rev. Stat. § 25-8-205(5).

prohibits placement of waste where there is "probable cause" to believe it will cause pollution.[208] California law requires a "report of waste discharge" from any person proposing to discharge "waste." The regional water quality control board must then issue waste discharge requirements (WDRs), which is essentially a permit.[209] However, these requirements may be conditionally waived by the regional board.[210] California uses these requirement by first seeking to abate nonpoint source pollution through nonregulatory means, but reserves the power to either grant a conditional waiver (to secure operational changes in a discharger) or to require the report of waste discharge and issue a WDR. Alaska law provides that "a person may not pollute or add to the pollution of the...water of the state."[211] Washington prohibits the discharge of "any organic or inorganic matter that shall cause or tend to cause" water pollution,[212] and permits are required for disposal of material into the waters of the state.[213] However, the law does not authorize the adoption of a permit system for nonpoint sources or imposition of penalties for pollution arising from forest practices conducted in compliance with the state's forest practices law.[214]

4.8 Publicly Owned Treatment Works

Publicly owned treatment works (POTWs) have been discussed in several contexts in this chapter. For example, indirect discharges and pretreatment standards under CWA Section 307(b)(1) were discussed in section 4.3.5 above.

CWA Title II applies specifically to POTWs. Title II standards originally required "Best practicable waste treatment over the life of the works" by 1983.

Title II contains EPA grant provisions to encourage responsible waste treatment. A number of problems with local planning agencies has led to much litigation. For example, EPA grants are subject to NEPA, which opens the door to NEPA litigation.[215] The EPA may refuse a grant where

---

[208] Utah Code Ann. § 19-5-107.
[209] Cal. Water Code § 13260.
[210] Cal. Water Code § 13269.
[211] Alaska Stat. § 46.03.710.
[212] Wash. Rev. Code § 90.48.080.
[213] Wash. Rev. Code § 90.48.160.
[214] Wash. Rev. Code § 90.48.420.
[215] See Bosco v. Beck, 475 F. Supp. 1029 (D.N.J. 1979), aff'd, 614 F.2d 769 (3d Cir. 1980).

the POTW has too much reserve capacity, which might lead to overbuilding in the area.[216]

Funds may not be used to treat combined sewer-stormwater systems. Funds for collection systems are limited to a major rehabilitation of an existing system, and new systems in existing communities. Presumably, funds are not available for a new system in a new community because it might lead to overbuilding.

4.9 Oil Pollution

Prevention and response to oil spills, particularly those at sea, has been an environmental concern in the United States for over a century. On March 24, 1989, the supertanker Exxon Valdez ran aground on the coast of Prince William Sound in Alaska, spilling over 250,000 barrels (11 million gallons) of crude oil in one of the nation's most biologically productive waters. While the Exxon Valdez spill is the largest spill in U.S. history, it is just one of the many oil spill incidents that have been reported. Over 10,000 oil spills were reported in the United States between 1973 and 1985. Between 1984 and 1988, over 30 million gallons of oil were discharged into the United States waters by tankers, nearly all of which resulted in at least some environmental damage.[217]

4.9.1 The Spill Prevention Act and Related Regulation

The 1989 Exxon Valdez incident was instrumental in inspiring the passage of the Oil Pollution Act of 1990 (OPA, discussed below),[218] but the OPA is not the only federal statute that deals with oil pollution. The Spill Prevention, Control and Countermeasures (SPCC) amendments to CWA Section 311(j) charges the president with the responsibility to regulate discharges of oil from vessels and on-shore and off-shore facilities.

By executive order, the president delegated the responsibility to develop and enforce regulations establishing procedures, methods, and equipment to prevent discharges of oil and to contain such discharges to the U.S. EPA and the Department of Transportation (DOT). A memorandum of

---

[216] See State ex rel. Burch v. Costle, 452 F. Supp. 1154 (D.D.C. 1978); but see Cape May Greene, Inc. v. Warren, 698 F.2d 179 (3d Cir. 1983) [EPA cannot decide where growth will occur].

[217] See J. Burger, *Oil Spills* (Rutgers Univ. Press, 1997).

[218] 33 U.S.C. §§ 2701-2706.

understanding (MOU) between the EPA and DOT established their respective jurisdictions. DOT has responsibility for transportation-related on-shore and off-shore facilities, while the EPA has responsibility for all non-transportation-related facilities.

The EPA Oil Pollution Prevention regulations, better known as the "SPCC Regulations," became effective during January, 1974.[219] The principal requirement is that owners or operators of non-transportation related facilities, subject to this regulation, must prepare and implement an oil Spill Prevention, Control and Countermeasures Plan (SPCC Plan) in accordance with guidelines prescribed in the regulations.[220] Additional EPA regulation of oils spills (and spill prevention) include the following:

● Criteria for State, Local and Regional Oil Removal Contingency Plans — These establish the minimum criteria for the development and implementation of state, local and regional oil removal contingency plans. The intended purpose is to foster and ensure coordination among state and local governments and private interests so that actions taken in response to a major discharge of oil are timely, efficient, coordinated and effective in minimizing damage. Depending on the requirements of the situation, the federal government may advise or assist. Should the other three entities fail to take appropriate containment or cleanup action, the federal government can assume control of the response activity.[221]

● Discharge of Oil — These define and prohibit discharges of "harmful" quantities of oil, as required by CWA Sections 311 (b)(3) and (4).[222]

● Liability Limits for Small On-Shore Storage Facilities — These establish the limits of liability for an owner or operator, as a function of the amount of storage capacity, for Federal government removal actions resulting from an oil discharge from the facility. These apply to all on-shore oil storage facilities of any kind having 1,000 barrels or less of oil storage capacity. Does not limit liability under any State or local law or other Federal law, or for discharges resulting from the willful negligence or misconduct of the owners or operators.[223]

---

[219] 40 CFR Part 112.
[220] The MOU establishing the respective jurisdictions of EPA and DOT is included as an appendix to the SPCC regulations, Id. at Appendix B.
[221] 40 CFR Part 109.
[222] 33 U.S.C. § 1321(b)(3) and (4). 40 CFR Part 110.
[223] 40 CFR Part 113.

● Civil Penalties for Violation of Oil Pollution Prevention Regulations — These define a violation for purposes of 40 CFR Part 112, outline enforcement and appeal procedures, limit the maximum penalty, and list criteria for consideration of a lesser penalty.[224]

4.9.2 The Oil Pollution Act of 1990

The 1989 Exxon Valdez disaster forced the United States to reconsider how we have dealt with the marine environment. Congress responded with the Oil Pollution Act of 1990 (OPA).

The OPA increased significantly the liability of the responsible party of an oil spill in U.S. waters and the Exclusive Economic Zone, including liability for all environmental damage.

OPA Section 1002(a) provides that the responsible party for a vessel or facility from which oil is discharged, or which poses a substantial threat of a discharge, is liable for: (1) certain specified damages resulting from the discharged oil; and (2) removal costs incurred in a manner consistent with the National Contingency Plan (NCP).[225]

OPA Section 1002(c) lists exceptions to the CWA liability provisions, which include: (1) discharges of oil authorized by a permit under federal, state, or local law; (2) discharges of oil from a public vessel; or (3) discharges of oil from onshore facilities covered by the liability provisions of the Trans-Alaska Pipeline Authorization Act.

If a responsible party can establish that the removal costs and damages resulting from an incident were caused solely by an act or omission by a third party, then OPA Section 1002(d) provides that the third party will be held liable for such costs and damages.[226]

Section 1004 of the OPA establishes liability for tank vessels larger than 3,000 gross tons is $1,200 per gross ton or $10 million ($2 million for vessels smaller than 3,000 gross tons), whichever is greater. Responsible parties at onshore facilities and deepwater ports are liable for up to $350 million per spill; holders of leases or permits for offshore facilities, except deepwater ports, are liable for up to $75 million per spill, plus removal costs. The federal government has the authority to adjust, by regulation, the $350 million liability limit established for onshore facilities.[227]

---

[224] 40 CFR Parts 112 and 114.
[225] 33 U.S.C. § 2702(a). The "NCP" is the "National Oil and Hazardous Substances Contingency Plan" under CERCLA (see chapter 9.)
[226] 33 U.S.C. § 2702(d).
[227] 33 U.S.C. § 2704(a).

Under OPA Section 4301(a) and (c) the fine for failing to notify the appropriate federal agency of a discharge is a maximum of $250,000 for an individual or $500,000 for an organization. The maximum prison term is five years. The penalties for violations have a maximum fine of $250,000 and 15 years in prison.[228]

Section 1016 of the OPA requires offshore facilities to maintain evidence of financial responsibility of $150 million and vessels and deepwater ports must provide evidence of financial responsibility up to the maximum applicable liability amount.[229] Claims for removal costs and damages may be asserted directly against the guarantor providing evidence of financial responsibility.

States have the authority under Section 1019 of the OPA to enforce, on the navigable waters of the State, OPA requirements for evidence of financial responsibility. States are also given access to Federal funds (up to $250,000 per incident) for immediate removal, mitigation, or prevention of a discharge, and may be reimbursed by the Trust fund for removal and monitoring costs incurred during oil spill response and cleanup efforts that are consistent with the National Contingency Plan (NCP).[230]

OPA Section 4202 strengthens planning and prevention activities by: (1) providing for the establishment of spill contingency plans for all areas of the United States (2) mandating the development of response plans for individual tank vessels and certain facilities for responding to a worst case discharge or a substantial threat of such a discharge; and (3) providing requirements for spill removal equipment and periodic inspections.[231]

OPA Section 1006 addresses the issue of damage to natural resources on U.S. federal lands, state land, land belonging to Indian tribes, or lands of foreign countries. Either the president or a representative of the state, Indian tribe, or foreign country can be designated as a "trustee" and present a claim to recover damages.[232]

In assessing natural resources damages, the measure of damages is —

(A) the cost of restoring, rehabilitating, replacing, or acquiring the equivalent of, the damaged natural resources;
(B) the diminution in value of those natural resources pending restoration; plus
(C) the reasonable cost of assessing those damages.[233]

---

[228] 33 U.S.C. § 2716a.
[229] 33 U.S.C. § 2716.
[230] 33 U.S.C. § 2719. The "NCP" is the "National Oil and Hazardous Substances Contingency Plan" under CERCLA (see chapter 9.)
[231] 33 U.S.C. § 2720.
[232] 33 U.S.C. § 2706(b).
[233] 33 U.S.C. § 2706(d).

4.9.3 Oil Spill Liability Trust Fund

Under the OPA of 1990, the owner or operator of a facility from which oil is discharged (also known as the *responsible party*) is liable for the costs associated with the containment or cleanup of the spill and any damages resulting from the spill. The EPA's first priority is to ensure that responsible parties pay to clean up their own oil releases. However, when the responsible party is unknown or refuses to pay, funds from the Oil Spill Liability Trust Fund[234] can be used to cover removal costs or damages resulting from discharges of oil.[235]

The primary source of revenue for the Fund is a five-cents-per-barrel fee on imported and domestic oil. Collection of this fee ceased on December 31, 1994 due to a sunset provision in the statute, but other revenue sources for the Fund include interest on the Fund, cost recovery from the parties responsible for the spills, and any fines or civil penalties collected. The Fund is administered by the U.S. Coast Guard's National Pollution Funds Center (NPFC).

The Fund can provide up to $1 billion for any one oil pollution incident, including up to $500 million for the initiation of natural resource damage assessments and claims in connection with any single incident. The main uses of Fund expenditures are: (1) state access for removal actions, and payments to federal, state, and Indian tribe trustees to carry out natural resource damage assessments and restorations; (2) payment of claims for uncompensated removal costs and damages; and (3) research and development and other specific appropriations.

4.9.4 Limitations and Problems with the OPA

The OPA's rather broad damage and liability provisions have been controversial from the start, and there have been some unexpected repercussions. There is concern for potential liability for oil spills will discourage international oil companies from risking potentially catastrophic liability by shipping oil to the United States

Several shipping companies threatened to boycott the United States and cease all oil shipments.[236] For example, Shell Oil Company withdrew its fleet from U.S. trade waters, opting instead to use chartered vessels for the transport of its oil to the U.S. mainland. In addition, Shell decreased the

---

[234] See 26 U.S.C. § 9509.
[235] 33 U.S.C. § 2736.
[236] See Price, U.S. Oil Spill Law to Cause Growing Tanker Problem, 27 Oil & Gas J. 21 (1991).

size of its tanker fleet from ninety to fifty vessels and phased out its third-party transportation business.[237]

Several international shipping companies have removed tankers from U.S. waters as well, including A.P. Moiler, Petrofina, Teekay Shipping, and Maersk.[238] Chevron Oil Company and Amoco Oil Company have significantly reduced shipments in U.S waters. Furthermore, a number of tanker owners, including Shell, have refused to carry heavy crude oil and certain fuels that are considered to be "dirty" because they are costly and difficult to clean up in the event of a spill.[239]

Despite these concerns, it does not appear that the flow of imported oil into the United States has decreased. At the same time, oil spills have decreased. Most tanker owners have continued trading with the United States at a rate about the same as that before the passage of the OPA in 1990.[240] To date, the only important results of the OPA have been changes in operational procedures, safety provisions, and inspection routines implemented by the oil trades.[241]

---

[237] Id.

[238] See Burger, *Oil Spills* (Rutgers Univ. Press, 1997); and Bamber, Oil Tanker Market Never Up Long, August, 1991 Petroleum Economist 15 (1991).

[239] Id.

[240] Petroleum Industry Research Foundation, *Transporting U.S. Oil Imports: The Impact of Oil Spill Legislation on the Tanker Market* (Office of Domestic & Int'l Energy Policy, U.S. Dept. of Energy, 1992).

[241] Id.

The Clean Water Act— Regulation of Storm Water Discharges

5.1 Introduction

Storm water discharges have been a major contributing source of the degradation of navigable waterways. According to the U.S. Environmental Protection Agency (EPA), "pollution from diffuse sources, such as runoff from agricultural, urban areas, construction sites, land disposal and resource extraction, is cited by the States as the leading cause of water quality impairment."[1] The problem is difficult to solve primarily because storm water pollution is easily created: a discharge occurs whenever rainwater falls on contaminated soils or piles of materials containing pollutants; this storm water then carries the contaminants into waterways.

The volume and nature of storm water discharges depend on a number of factors, including the types of industrial activities occurring at a facility, the nature of precipitation, and the degree of surface imperviousness. The sources of pollutants in storm water discharges differ with the type of industry operation and specific facility features. Storm water discharges from industrial facilities may contain toxics and conventional pollutants when material management practices allow exposure to storm water. Rainwater may pick up pollutants from structures and other surfaces as it drains from the land.

Besides storm water, other sources of pollutants may increase the pollutant loads discharged from separate storm sewers, such as illicit connections, spills, and improperly disposed wastes. Illicit connections are

---

[1] 55 *Fed. Reg.* 47,990, 47,991 (Nov. 16, 1990).

contributions of unpermitted nonstorm water discharges to storm sewers from any of a number of sources including sanitary sewers, industrial facilities, commercial establishments, or residential dwellings. In some municipalities, illicit connections of sanitary, commercial, and industrial discharges to storm sewer systems have had a significant impact on the water quality of receiving waters. Studies have shown that illicit connections to storm sewers can create severe, widespread contamination problems.[2] Removal of these discharges presents opportunities for dramatic improvements in the quality of storm water discharges.

Six activities can be identified as major potential sources of pollutants in storm water discharges associated with industrial activities:
1. Loading or unloading of dry bulk materials or liquids
2. Outdoor storage of raw materials or products
3. Outdoor process activities
4. Dust or particulate generating processes
5. Illicit connections or inappropriate management practices and
6. Waste disposal practices.
The potential for pollution from many of these activities may be influenced by the presence and use of toxic chemicals.

Loading and Unloading Operations. Loading and unloading operations typically are performed along facility access roads and railways and at loading/unloading docks and terminals. These operations include pumping of liquids or gases from a truck or rail car to a storage facility (or vice versa), pneumatic transfer of dry chemicals to or from the loading or unloading vehicle, transfer by mechanical conveyor systems, and transfer of bags, boxes, drums, or other containers from vehicle by forklift trucks or other materials handling equipment. Material spills or losses may discharge directly to the storm drainage systems or may accumulate in soils or on surfaces and be washed away during a storm or facility washdown.

Outdoor storage of raw materials or products. Outdoor storage activities include the storage of fuels, raw materials, byproducts, intermediates, final products, and process residuals. Methods of material storage include using storage containers (e.g., drums or tanks), platforms or pads, bins, silos, boxes, or piles. Materials, containers, and material storage areas that are exposed to rainfall and/or runoff may contribute

---

[2] For example, the Huron River Pollution Abatement Program inspected 660 businesses, homes, and other buildings located in Washtenaw County, Michigan, and found that 14% of the buildings had improper storm drain connections. Illicit discharges were detected at a higher rate of 60% for automobile-related businesses, including service stations, automobile dealerships, car washes, body shops, and light industrial facilities. 57 *Fed. Reg.* 41236, 41238 (Sept. 9, 1992).

pollutants to storm water when solid materials wash off or materials dissolve into solution.

Outdoor process activities. Other outdoor activities include certain types of manufacturing and commercial operations and land-disturbing operations. Although many manufacturing activities are performed indoors, some activities, such as equipment maintenance and/or cleaning, timber processing, rock crushing, vehicle maintenance and/or cleaning, and concrete mixing, typically occur outdoors. Processing operations may result in liquid spillage and losses of material solids to the drainage system or surrounding surfaces, or creation of dusts or mists, which can be deposited locally. Some outdoor industrial activities cause substantial physical disturbance of land surfaces that result in soil erosion by storm water. For example, disturbed land occurs in construction and mining. Disturbed land may result in soil losses and other pollutant loadings associated with increased runoff rates. Facilities whose major process activities are conducted indoors may still apply chemicals such as herbicides, pesticides, and fertilizer outdoors for a variety of purposes.

Dust or particulate generating processes. Dust or particulate generating processes include industrial activities with stack emissions or process dusts that settle on plant surfaces. Localized atmospheric deposition is a particular concern with heavy manufacturing industries. For example, monitoring of areas surrounding smelting industries has shown much higher levels of metals at sites nearest the smelter. Other industrial sites, such as mines, cement manufacturing, and refractories, generate significant levels of dusts.

Illicit connections or inappropriate management practices. Illicit connections or inappropriate management practices result in improper nonstorm water discharges to storm sewer systems. The likelihood of illicit discharges to storm water collection systems is expected to be higher at older facilities, due to past practices, as well as for facilities that use high volumes of process water or dispose of significant amounts of liquid wastes, including process wastewaters, cooling waters, and rinse waters. Pollutants from nonstorm water discharges to the storm sewer system of individual facilities are caused typically by a combination of improper connections, spills, improper dumping, and the belief that the absence of visible solids in a discharge is equivalent to the absence of pollution. Illicit connections are often associated with floor drains that are connected to separate storm sewers. Rinse waters used to clean or cool objects discharge to floor drains connected to separate storm sewers. Large amounts of rinse waters may originate from industries that use regular washdown procedures; for example, bottling plants use rinse waters for removing waste products, debris, and labels. Rinse waters can be used to cool materials by dipping, washing, or spraying objects with cool water; for example, rinse water is sometimes sprayed over the final products of a metal plating facility for cooling purposes. Condensate return lines of heat exchangers often discharge to floor drains. Heat exchangers, particularly those used under stressed conditions (such as exposure to corrosive fluids) such as in the metal finishing and electroplating industry, may develop pinhole leaks that result in contamination of condensate by process wastes. These and other nonstorm water discharges to a storm sewer may be intentional, based on the belief that the discharge does not contain pollutants, or they may be inadvertent, if the operator is unaware that a floor drain is connected to the storm sewer.

Waste disposal practices. Waste management practices include temporary storage of waste materials, operating landfills, waste piles, and land application sites that involve land disposal. Outdoor waste treatment operations also include wastewater and solid waste treatment and disposal processes, such as waste pumping, additions of treatment chemicals, mixing, aeration, clarification, and solids dewatering. Industrial facilities often conduct some waste management on site.

5.2 Scope of Storm Water Program

Due to the complexity of the storm water problem, government regulation of storm water discharges was debated for many years. The Clean Water Act (CWA),[3] specifically, its National Pollutant Discharge Elimination System (NPDES) permit scheme, was the obvious means for regulating storm water discharges.[4] For various reasons, however, the EPA and NPDES-authorized states failed to issue NPDES permits for the majority of point source discharges of storm water. Recognizing this, Congress added Section 402(p) to the CWA in 1987 to establish a comprehensive framework for addressing storm water discharges under the NPDES program.[5]

Section 402(p) of the CWA establishes a comprehensive two-phased

---

[3] 33 U.S.C. §§ 1251 through 1386.

[4] CWA § 402, 33 U.S.C. § 1342. In 1972, the Clean Water Act was amended to provide that the discharge of any pollutant to waters of the United States from any point source is unlawful, except if the discharge is in compliance with a National Pollutant Discharge Elimination System (NPDES) permit. See chapter 4 for full discussion of NPDES permit requirements.

[5] 33 U.S.C. § 1342(p). Congress amended the Clean Water Act with the Water Quality Act of 1987, Pub. L. No. 100-4, 101 Stat. 7 (codified as amended in scattered Sections of 33 U.S.C. § 1251 through 1386).

approach for EPA to address storm water discharges under its NPDES permit program.[6] In November 1990, the EPA issued "Phase I" storm water regulations, which subject certain types of facilities to storm water discharge permitting requirements.[7] Under these Phase I regulations, storm water discharges associated with municipal separate storm sewers (MS4s) serving large or medium-sized populations (greater than 250,000 or 100,000 people, respectively), and storm water discharges associated with industrial activity must be authorized by a NPDES storm water permit. One of the categories of "industrial activities" is construction activities affecting five or more acres of land.[8] Permits are also to be issued, on a case-by-case basis, if the EPA or NPDES-authorized state determines that a storm water discharge contributes to the violation of a water quality standard or is a significant contributor of pollutants to U.S. waters.

Under "Phase II" of the storm water program, the EPA was required to assess remaining storm water discharges not covered under the Phase I regulations. The Phase II regulations were to have been issued by the EPA not later than October 1, 1992; however, because it took so long to finalize the Phase I program, the EPA did not complete and issue appropriate regulations for the Phase II program until December 1999. The Phase II regulations are discussed in section 5.2.6.

5.2.1 Phase I Permit Regulations

The EPA promulgated the Phase I permit regulations for storm water discharges on November 16, 1990.[9] These regulations established the scope of the Phase I storm water program by defining two major classes of storm water discharges identified under Section 402(p)(2)(B), (C), and (D) of the CWA: storm water discharges associated with industrial activity; and discharges from municipal separate storm sewer systems (MS4s) serving a population of 100,000 or more. Five classes of storm

---

[6] The term "storm water" is defined in the regulations as "storm water runoff, snow melt runoff, and surface runoff and drainage." 40 CFR § 122.26(b)(13). Only storm water discharged from a "point source" is subject to regulation. The regulatory definition of "point source" is quite broad and encompasses "any discernible, confined, and discrete conveyance, including but not limited to, any pipe, ditch, channel, tunnel, conduit, well, discrete fissure, container, rolling stock, concentrated animal feeding operation, landfill leachate collection system, vessel or other floating craft from which pollutants are or may be discharged. 40 CFR § 122.2. Specifically excluded from this definition are return flows from irrigated agriculture and agricultural storm water runoff.
[7] 55 *Fed. Reg.* 47990 (Nov. 16, 1990).
[8] See Section 5.2.4 for discussion of construction activities subject to storm water discharge regulation.
[9] 55 *Fed. Reg.* 47990 (Nov. 16, 1990).

water discharges are specifically listed under Section 402(p)(2) which are covered under Phase I of the storm water program:

1. A discharge with respect to which a permit has been issued prior to February 4, 1987

2. A discharge associated with industrial activity

3. A discharge from a municipal separate storm sewer system serving a population of 250,000 or more

4. A discharge from a municipal separate storm sewer system serving a population of 100,000 or more, but less than 250,000 or

5. A discharge for which the EPA Administrator or the NPDES-authorized state, as the case may be, determines that the storm water discharge contributes to a violation of a water quality standard or is a significant contributor of pollutants to U.S. waters.

The regulations define a MS4 serving a population of 100,000 or more to include MS4s within the boundaries of 173 incorporated cities, and within unincorporated portions of 47 counties that were identified as having populations of 100,000 or more in unincorporated, urbanized portions of the county.[10] In addition, the regulations allow for additional MS4s to be designated by the Director of the NPDES program as being part of a large or medium MS4. The regulations also defined the term "storm water discharges associated with industrial activity" to include 11 categories of industrial facilities subject to permitting requirements.[11] One of the categories includes discharges from construction sites larger than five acres, which are presumed to discharge storm water.[12]

5.2.2 Municipal Separate Storm Sewer Systems

With respect to storm water permits for large and medium MS4s, the EPA has defined the role of municipalities in a flexible manner that allows local governments to assist in defining priority pollutant sources within the municipality, and to develop and implement appropriate controls for such discharges. Municipal programs address the control of pollutants in storm water from all areas within the boundaries of the MS4 that discharge to the system, including privately-owned lands, as well as modifying municipal activities (e.g., road deicing and maintenance, flood control efforts, maintenance of municipal lands, etc.) to address storm water quality concerns. The regulations establish comprehensive two-part permit

---

[10] See Appendices F, G, H, and I to 40 CFR § pt. 122 for a listing of municipalities.

[11] 40 CFR § 122.26(b)(14)(i)-(xi).

[12] 40 CFR § 122.26(b)(14)(x).

applications for discharges from large or medium MS4s.[13] Permits for discharges from MS4s may be issued on a system- or jurisdiction-wide basis. The permit application requirements for large and medium MS4s, among other things, require municipal applicants to propose municipal storm water management programs to control pollutants to the maximum extent practicable (MEP) and to effectively prohibit nonstorm water discharges to the MS4.[14]

5.2.3 Industrial Activities

Section 402(p)(3) confirms that, like all other "point source" discharges under the CWA,[15] discharges of storm water associated with industrial activity must meet all applicable provisions of Sections 402 and 301 of the CWA, including technology-based requirements and any necessary water quality-based requirements. The EPA has defined the term "storm water discharge associated with industrial activity" in a comprehensive manner to address over 100,000 industrial facilities. All storm water discharges associated with industrial activity that discharge directly to U.S. waters or through MS4s are required to obtain NPDES storm water permits, including those which discharge through systems located in municipalities with populations of less than 100,000. Discharges of storm water to a combined sewer system or to Publicly Owned Treatment Works (POTW) are excluded. Facilities with storm water discharges associated with industrial activity include: manufacturing/industrial facilities; hazardous waste treatment, storage, or disposal facilities; landfills; certain sewage treatment plants; recycling facilities; power plants; mining operations; some oil and gas operations; airports; and certain other transportation facilities. Operators of industrial facilities that are federally, state, or municipally owned or operated (with the exception of certain facilities owned or operated by a municipality of less than 100,000 people)[16] that meet the description of the facilities listed in 40 CFR Section 122.26(b) (14)(i)-(xi) must also submit applications.

---

[13] See Section 5.6 for discussion of permit requirements for MS4s.

[14] See 40 CFR § 122.26(d)(2)(iv).

[15] A "point source" is defined as "any discernible, confined, and discrete conveyance, including (but not limited to) any pipe, ditch, channel, tunnel, conduit, well, discrete fissure, container, rolling stock, concentrated animal feeding operation, landfill leachate collection system, vessel, or other floating craft from which pollutants are or may be discharged." 40 CFR § 122.2.

[16] In the Intermodal Surface Transportation Efficiency Act of 1991, Congress provided that industrial activities owned or operated by municipalities with a population of less than 100,000 be placed into Phase II of the storm water program with the exception of airports, power plants, and controlled sanitary landfills.

5.2.4 Construction Activities

One of the 11 categories of industrial facilities listed in the storm water regulations is construction sites larger than five acres, which are presumed to discharge storm water.[17] Construction activities impacting less than five acres were exempted from the Phase I permitting requirements. It should be noted that the Court of Appeals for the Ninth Circuit invalidated the EPA's exemption for construction sites of less than five acres;[18] nevertheless, the EPA is unlikely to require permits for these construction sites until the Phase II application deadline set for March 10, 2003.

The calculation of the five-acre threshold for purposes of triggering storm water permitting requirements has been somewhat problematic. The EPA adopted a broad approach under its regulations which considers the cumulative acreage impacted by the construction activity. The regulations and a 1992 EPA Memorandum take this cumulative approach in order to prevent developers from avoiding permit requirements by undertaking development in a piecemeal fashion. Thus, if a project, when taken as a whole, will disturb a total of five acres, each individual phase or segment of the project is subject to the permit requirements, even though a separate segment is less than five acres. Likewise, the subdivision of a larger parcel cannot be used to circumvent the permit requirements where there a common plan of development or sale is contemplated.

Storm water discharges from construction sites that will result in the disturbance of five or more acres of land are normally authorized under general permits.[19] However, dischargers who do not obtain coverage under a general permit must submit an individual permit application.[20]

---

[17] 40 CFR § 122.26(b)(14)(x).

[18] See section 5.2.5 for discussion of the Ninth Circuit decision.

[19] See section 5.5 for full discussion of general permit requirements. The general permits were originally published in the Federal Register at 57 *Fed. Reg.* 41176 (Sept. 9, 1992); 57 *Fed. Reg.* 44412 (Sept. 25, 1992). In 1998, the EPA reissued the general permits for storm water discharges associated with construction activities, which were published in the Federal Register at 63 *Fed. Reg.* 7858 (Feb. 17, 1998) (reissuance of general permits for construction activities in EPA Regions 1-3, 7-10); 63 *Fed. Reg.* 15622 (Mar. 31, 1998) (reissuance of general permits for construction activities in EPA Region 4); 64 *Fed. Reg.* 39136 (July 21, 1999) (draft modifications to reissued general permits for construction activities in EPA Region 4); 63 *Fed. Reg.* 36490 (July 6, 1998) (reissuance of general permits for construction activities in EPA Region 6).

[20] See section 5.7 for discussion of individual permits.

5.2.5 Significant Legal Challenges

Several legal actions have been filed against the EPA to challenge various aspects of the storm water regulations. Of those cases that were successful, particular mention should be made of rulings that struck down the EPA's "light industries" exemption, and the exemption for construction sites comprising less than five acres.

Challenge to exemption for oil and gas activities. In 1992, the Ninth Circuit Court of Appeals issued an important decision in a case filed by the Natural Resources Defense Council (NRDC) to challenge several provisions of the EPA's storm water regulations.[21] One of the challenged provisions was the EPA's exemption for oil and gas activities.[22] This challenge focused on the disparity of treatment between oil and gas operations and mining operations. The regulations require owners of oil and gas facilities to apply for a permit only if the facility has a discharge of a reportable quantity,[23] whereas mining operators must submit applications whenever storm water comes into contact with overburden or other listed materials. The court upheld the exemption for oil and gas operations, stating that "the determination of whether storm water is contaminated is within the [EPA] Administrator's discretion."

Challenge to light industries exemption. In the same case, the NRDC also challenged the EPA's so-called "light industries" exemption from the definition of "associated with industrial activity." For these discharges, the EPA presumed that there would be no storm water discharge associated with industrial activity, and did not require permit applications unless there was actual exposure of industrial pollutants to storm water at the facility.[24] "Light industries" include businesses with certain Standard Industrial Classification (SIC) numbers, such as manufacturers of pharmaceuticals, paints, varnishes, lacquers, enamels, machinery, computers, electrical equipment, transportation equipment, glass products, fabrics, furniture, paper board, food processors, printers, jewelry, toys, and tobacco products. These facilities must acquire a permit for storm water runoff only if there are areas "where material-handling equipment or activities, raw materials, intermediate products, final products, waste

---

[21] Natural Resources Defense Council v. EPA, 966 F.2d 1292 (9th Cir. 1992).
[22] Section 402(l)(2) of the CWA, 33 U.S.C. § 1342(l)(2), specifically exempts storm water discharges from mining and oil and gas operations if the discharges are "composed entirely of flows . . . which are not contaminated by contact with, or do not come in contact with," any materials or waste products.
[23] The EPA relied on the use of reportable quantities (RQs) in CWA § 311(b)(4), 33 U.S.C. § 1321(b)(4).
[24] 40 CFR § 122.26(b)(14)(xi).

materials, by-products, or industrial machinery at these facilities are exposed to storm water."

The EPA justified this exclusion for light industries because "most of the activity at these types of manufacturers takes place indoors, and that emissions from stacks, use of unhoused manufacturing equipment, outside material storage or disposal, and generation of large amounts of dust and particles will all be minimal." The EPA determined that "these industries are more akin or comparable to businesses, such as retail, commercial, or service industries . . . and storm water discharges from these facilities are not 'associated with industrial activity.'"[25]

The Ninth Circuit struck down the EPA's exclusion of certain SIC categories from the definition of "associated with industrial activity." The court held that the EPA did not provide any facts in the record to support its determination; therefore, it was arbitrary for the EPA to include this added criteria in the case of "light industries" when other industries are presumed to be discharging contaminated storm water.

Challenge to exemption for construction sites less than five acres. The NRDC also challenged the EPA's exemption for construction sites under five acres. The NRDC contended that the EPA was without authority to exempt any discharges associated with industrial activity. Further, the NRDC maintained that the five-acre threshold was arbitrary, especially in light of an acknowledgement by the EPA that "[e]ven small construction sites may have a significant negative impact on water quality in localized areas."[26] The EPA said that it chose the five-acre threshold because the agency wanted to limit the amount of permit applications and because it felt that this acreage limitation was most appropriate for identifying sites that would constitute industrial activity. The Ninth Circuit agreed with the NRDC and struck down this part of the regulations. The court noted that the EPA conceded that construction was an industrial activity, and therefore, the EPA could not create exemptions from the permitting requirements. Furthermore, the court ruled that a five-acre limit was arbitrary due to the lack of factual findings by the EPA and due to the admission that even small sites may have a significant impact. Despite the court ruling, the EPA is unlikely to require permits for construction activities impacting less than five acres until the Phase II application deadline of March 10, 2003. Still, it is important to note that some states may impose their own, more stringent storm water permit requirements for construction activities involving less than five acres.

---

[25] 55 *Fed. Reg.* 47990, at 48008
[26] 55 *Fed. Reg.* 47990, at 48033.

Challenge to designation of abandoned mining sites as industrial category. Also in 1992, the Ninth Circuit issued an important decision with regard to a challenge by the American Mining Congress to the EPA's designation of abandoned mining sites as a major industrial category under the Phase I permitting regulations.[27] The court held that the EPA was justified in its determination that inactive sites should be included "because some mining sites represent a significant source of contaminated storm water runoff." Further, the court found EPA's regulation of inactive mines reasonable because it was limited to "those sites at which storm water discharge is likely to have become contaminated through association with industrial activity."[28]

5.2.6 Phase II Permit Regulations

Sections 402(p)(5) and (6) of the CWA require the EPA to identify storm water discharges not covered under Phase I which should be regulated to protect water quality. As originally adopted, Section 402(p)(1) specified that the "moratorium" on permitting of these storm water sources, which the EPA refers to as "Phase II" sources, would expire on October 1, 1992. Section 312 of the Water Resource Development Act of 1992, which was signed by the President on October 31, 1992, extended the moratorium to October 1, 1994. In 1992, 1995, and 1998, the EPA issued proposed regulations for Phase II identifying additional categories of storm water activities in need of control and establishing a sequential application process for all Phase II storm water discharges.[29] However, the final Phase II regulations were not promulgated until December 1999.[30]

The Phase II regulations expand Phase I of the storm water program to include regulation of storm water discharges from construction sites between one and five acres and municipal separate storm sewer systems in urbanized areas serving populations of less than 100,000. In addition, the final regulations contain "safety valves" that (1) allow certain sources to be excluded from the storm water program based on the lack of impact on water quality, and (2) pull in other sources not regulated on a national

---

[27] American Mining Congress v. EPA, 965 F.2d 759 (9th Cir. 1992).
[28] EPA excluded certain inactive mines that may not be regulated efficiently, including: (1) inactive mines without an identifiable owner or operator; (2) those that have been "reclaimed" under the Surface Mining Control and Reclamation Act or other state laws; (3) those with minimal disturbances (such as those that are undisturbed or where mining operations are solely for the purpose of maintaining a claim); and (4) those where storm water does not come in contact with overburden or other materials. 40 CFR § 122.26(b)(14)(iii).
[29] See 57 *Fed. Reg.* 41344 (Sept. 9, 1992); 60 *Fed. Reg.* 17950 (Apr. 7, 1995); 60 *Fed. Reg.* 40230 (Aug. 7, 1995); 63 *Fed. Reg.* 1536 (Jan. 9, 1998).
[30] 64 *Fed. Reg.* 68722 (Dec. 8, 1999).

basis based on localized adverse impact on water quality. Further, the Phase II regulations conditionally exclude from the storm water program, industrial facilities that have "no exposure" of industrial activities to storm water, thereby reducing application of the program to many industrial activities currently covered by the program that have no industrial storm water discharges. The final rule took effect on February 7, 2000 and sets March 10, 2003 as the deadline by which Phase II regulated facilities must obtain coverage under a storm water permit.[31]

Prior to promulgation of the final Phase II regulations, the EPA submitted a mandatory report to Congress[32] that addresses the following issues with respect to the Phase II rule: (1) an analysis of the impact of the rule on local governments, (2) an explanation of the rationale for lowering the threshold for regulation of construction sites from five acres to one acre, (3) an explanation of why the coverage of the regulation is based on a census-determined population instead of a water quality threshold and documentation that storm water runoff is generally a problem in communities with populations of 50,000 to 100,000, and (4) information that supports the position of the EPA Administrator that the Phase II storm water program should be administered as part of the NPDES permit program.[33]

5.3 Types of Storm Water Permits

The EPA essentially provides two options for obtaining a permit for storm water discharges: (1) file a Notice of Intent (NOI) to be covered by a General Permit or (2) submit an application for an Individual Permit. If coverage is not available under one of the General Permits, the discharger is required to submit an Individual Permit application.

It is important to note at the outset that the EPA permits only authorize storm water discharges in states without approved NPDES programs. Permit applicants in states without NPDES permitting authority must submit the EPA forms discussed in this chapter. However, most states have been granted EPA approval to administer their own storm water

---

[31] Those with questions regarding the applicability of the final Phase II regulations to a particular facility may contact George Utting, Office of Wastewater Management, Environmental Protection Agency, Mail Code 4203, 401 M Street, S.W., Washington, D.C. 20460; telephone (202) 260-5816; email: sw2@epa.gov.

[32] See 64 *Fed. Reg.* 68852 (Dec. 8, 1999).

[33] The report is available through the Internet on the EPA Office of Wastewater Management web site at http://www.epa.gov/owm/sw/phase2.

programs in place of the federal program.[34] Thus, in states where the EPA has delegated NPDES permitting authority to the state, the permit applicant will need to contact the state agency responsible for administering the storm water permit program to obtain the appropriate application forms and instructions. Although the state forms and requirements are similar, if not identical, in most respects to the EPA forms and requirements--and the EPA encourages the states to model their storm water permits after the EPA permits--state permitting authorities may impose some additional or unique requirements of their own.

Following the EPA's issuance of the final regulations for Phase I of the storm water program in 1990,[35] the following permit application options were available to facilities applying for storm water discharge permits:

1. File a Notice of Intent (NOI) to be covered by a general permit for storm water discharges associated with industrial activities,[36]

2. File a NOI to be covered by a general permit for storm water discharges associated with construction activities,[37]

3. Submit an individual permit application consisting of Forms 1 and 2F.

4. Become a participant in a group permit application.

In addition, operators of large and medium MS4s were required to submit a twopart general permit application designed to facilitate development of site-specific permit conditions.

Significantly, in September 1995, the EPA also provided an additional permit option, known as a multi-sector general permit.[38] The EPA

---

[34] Through the end of 1999, only the states of Arizona, Florida, Idaho, Louisiana, Maine, Massachusetts, New Hampshire, New Mexico, Oklahoma, and Texas did *not* have EPA approval to administer their own storm water permit programs in place of the federal program.

[35] 55 *Fed. Reg.* 47990 (Nov. 16, 1990).

[36] The general permits for storm water discharges associated with industrial activities were originally published in the Federal Register in September 1992. 57 *Fed. Reg.* 41236 (Sept. 9, 1992); 57 *Fed. Reg.* 44438 (Sept. 25, 1992). As discussed below, the EPA decided not to reissue these general permits and, in 1998, required all facilities with unexpired permits to transfer coverage to a multi-sector general permit or submit an application for an individual permit. 63 *Fed. Reg.* 52430 (Sept. 30, 1998).

[37] The general permits were originally published in the Federal Register at 57 *Fed. Reg.* 41176 (Sept. 9, 1992); 57 *Fed. Reg.* 44412 (Sept. 25, 1992). In 1998, the EPA reissued the general permits for storm water discharges associated with construction activities, which were published in the Federal Register at 63 *Fed. Reg.* 7858 (Feb. 17, 1998) (reissuance of general permits for construction activities in EPA Regions 1-3, 7-10); 63 *Fed. Reg.* 15622 (Mar. 31, 1998) (reissuance of general permits for construction activities in EPA Region 4); 64 *Fed. Reg.* 39136 (July 21, 1999) (draft modifications to reissued general permits for construction activities in EPA Region 4); 63 *Fed. Reg.* 36490 (July 6, 1998) (reissuance of general permits for construction activities in EPA Region 6).

[38] 60 *Fed. Reg.* 50804 (Sept. 29, 1995).

designed the multi-sector general permit in response to the submission of group permit applications by over 1,200 groups with over 60,000 member facilities. In drafting the multi-sector permit, the EPA reviewed both parts of the applications and formulated permit language. To facilitate the process of developing permit conditions for each of the 1,200 group applications submitted, the EPA classified groups into 29 industrial sectors where the nature of industrial activity, type of materials handled, and material management practices employed were sufficiently similar for the purposes of developing permit conditions. Each of the industrial sectors were represented by one or more groups which participated in the group application process.

Enamored by the multi-sector general permit concept, the EPA decided not to reissue the baseline general permit for industrial activities upon the five-year expiration date of the permit. The first of those general permits expired in September 1997. Those industrial facilities originally covered by the baseline general permit for industrial activities were thus required, within 180 days prior to expiration of their general permits, to submit an application for coverage under a multi-sector general permit or to submit an application for an individual permit. Moreover, in September 1998, the EPA announced the termination of all unexpired baseline general permits for industrial activities, with a few limited exceptions, and required those facilities to apply for a transfer of coverage to a multi-sector general permit or submit an individual permit application.[39]

For dischargers covered by a general permit for construction activities, the EPA did reissue those general permits in 1998, with some modifications.[40]

As a result of the changes to the EPA's storm water permit program, the following permit options are now available:

1. File a Notice of Intent (NOI) to be covered by a multi-sector general permit for storm water discharges associated with industrial activities (discussed in section 5.4);

2. File a NOI to be covered by a general permit for storm water discharges associated with construction activities (discussed in Section 5.5); or

3. Submit an individual permit application consisting of Forms 1 and 2F (discussed in section 5.7).

---

[39] 63 *Fed. Reg.* 52430 (Sept. 30, 1998).

[40] 63 *Fed. Reg.* 7858 (Feb. 17, 1998) (reissuance of general permits for construction activities in EPA Regions 1-3, 7-10); 63 *Fed. Reg.* 15622 (Mar. 31, 1998) (reissuance of general permits for construction activities in EPA Region 4); 64 *Fed. Reg.* 39136 (July 21, 1999) (draft modifications to reissued general permits for construction activities in EPA Region 4); 63 *Fed. Reg.* 36490 (July 6, 1998) (reissuance of general permits for construction activities in EPA Region 6).

In addition, general permit requirements for MS4s are summarized in section 5.6.

5.4 Multi-Sector General Permit for Industrial Activities

In September 1992, the EPA issued "baseline" general permits to cover the majority of storm water discharges associated with industrial activities in states and territories without authorized NPDES programs.[41] Subsequently, in September 1995, the EPA issued a special type of general permit for industrial activities known as a "multi-sector" general permit.[42] The requirements of the two types of permits are almost the same, except that the baseline general permits only included generic best management practices (BMPs) for controlling storm water from a wide variety of industries, whereas the multi-sector general permits contain sector-specific BMPs tailored to specific types of facilities.

In 1997, upon expiration of the first of the baseline general permits issued in 1992, the EPA decided not to reissue those permits--instead requiring that those permittees seek coverage under a multi-sector permit or submit an individual permit application. Furthermore, in 1998, the EPA terminated all of the remaining baseline general permits for industrial activities, with a few limited exceptions, and required those facilities to transfer coverage to a multi-sector general permit or submit an individual permit application.[43] Thus, the baseline general permit for storm water discharges associated with industrial activities, first issued in September 1992, was completely eliminated and replaced by the multi-sector general permit as of January 1, 1999. This section discusses the requirements of the EPA's multi-sector general permits for storm water discharges associated with industrial activities.

5.4.1 Industrial Sectors Covered by Permit

Multi-sector general permits were originally designed to cover storm water discharges associated with industrial activities from 29 industrial sectors, including discharges through large and medium MS4s, and through other MS4s. The EPA first issued the final multi-sector general permit in 1995,[44] and subsequently made modifications to the permit in

---

[41] See 57 *Fed. Reg.* 41236 (Sept. 9, 1992); 57 *Fed. Reg.* 44438 (Sept. 25, 1992).
[42] 60 *Fed. Reg.* 50804 (Sept. 29, 1995).
[43] 63 *Fed. Reg.* 52430 (Sept. 30, 1998).
[44] 60 *Fed. Reg.* 50804 (Sept. 29, 1995).

1996[45] and 1998,[46] which included the addition of a new Sector AD to authorize discharges from Phase I facilities which did not fall into one of the original 29 sectors of the permit, and selected Phase II discharges which are designated for permitting in accordance with 40 CFR Section 122.26(g)(1)(i); the addition of a new Addendum I to provide guidance and information to assist applicants in determining permit eligibility concerning protection of historic properties; and an update of the list of endangered and threatened species found in Addendum H, and a listing of additional sources to reference for future updates to the list.

The EPA's general permits cover the majority of storm water discharges associated with industrial activity. The permit is intended to cover storm water discharges from the following industrial sectors:

Sector A: Timber products facilities;

Sector B: Paper and allied products manufacturing facilities;

Sector C: Chemical and allied products manufacturing facilities;

Sector D: Asphalt paving and roofing materials manufacturers and lubricant manufacturers;

Sector E: Glass, clay, cement, concrete and gypsum product manufacturing facilities;

Sector F: Primary metals facilities;

Sector G: Metal mining (ore mining and dressing) facilities;

Sector H: Coal mines and coal mining-related facilities;

Sector I: Oil and gas extraction facilities;

Sector J: Mineral mining and processing facilities;

Sector K: Hazardous waste treatment, storage, or disposal facilities;

Sector L: Landfills and land application sites;

Sector M: Automobile salvage yards;

Sector N: Scrap recycling and waste recycling facilities;

Sector O: Steam electric power generating facilities, including coal handling areas;

Sector P: Motor freight transportation facilities, passenger transportation facilities, petroleum bulk oil stations and terminals, rail transportation facilities, and United States Postal Service transportation facilities;

Sector Q: Water transportation facilities that have vehicle maintenance shops and/or equipment cleaning operations;

Sector R: Ship and boat building or repairing yards;

Sector S: Vehicle maintenance areas, equipment cleaning areas, or deicing Areas located at air transportation facilities;

Sector T: Treatment works;

---

[45] 61 *Fed. Reg.* 5248 (Feb. 9, 1996); 61 *Fed. Reg.* 6412 (Feb. 20, 1996); 61 *Fed. Reg.* 50020 (Sept. 24, 1996).
[46] 63 *Fed. Reg.* 52430 (Sept. 30, 1998).

Sector U: Food and kindred products facilities;

Sector V: Textile mills, apparel, and other fabric product manufacturing facilities;

Sector W: Wood and metal furniture and fixture manufacturing facilities;

Sector X: Printing and publishing facilities;

Sector Y: Rubber, miscellaneous plastic products, and miscellaneous manufacturing industries;

Sector Z: Leather tanning and finishing facilities;

Sector AA: Fabricated metal products industry;

Sector AB: Facilities that manufacture transportation equipment, industrial, or commercial machinery;

Sector AC: Facilities that manufacture electronic equipment and components, photographic and optical goods.

Addition of Sector AD. Upon reissuance of the multi-sector permit in 1998, the EPA added Sector AD to cover discharges from Phase I facilities which may not fall into one of the sectors of the final modified permit, and to provide a readily available means for covering many of the Phase II storm water facilities which are designated for permitting prior to the permit application deadline for Phase II sources of March 10, 2003. For cases where Sector AD is inappropriate, individual permits or an alternate general permit are required. No analytical monitoring requirements are included for the Sector AD; however, quarterly visual examinations are required as in most other sectors. In addition, the requirements common to all sectors set forth in Parts I-X and XII of the multi-sector general permit apply to Sector AD.

The EPA has further divided some of the sectors into subsectors in order to establish more specific and appropriate permit conditions, including best management practices (BMPs) and monitoring requirements. It should be noted that, in 1997, a new North American Industry Classification System (NAICS) was adopted by the Office of Management and Budget,[47] which replaces the 1987 Standard Industrial Classification (SIC) code system for the collection of statistical economic data. However, the use of the new system for nonstatistical purposes is optional. The EPA considered the use of NAICS for the modified multi-sector permit, but elected to retain the 1987 SIC code system since the storm water regulations reference the existing system and this system has generally proven to be adequate.[48]

---

[47] 62 *Fed. Reg.* 17288 (Apr. 9, 1997).

[48] Readers should, however, watch for future EPA rulemakings with regard to possible adoption of the NAICS system.

5.4.2 Discharges Not Eligible for Coverage

The following storm water discharges associated with industrial activity are **not** eligible for coverage under the EPA's multi-sector general permits:

- Certain storm water discharges subject to an existing effluent limitations guideline. Storm water discharges subject to effluent guidelines under 40 CFR § part 436 or for mine drainage under 40 CFR Part 440 are not covered under the permit, nor are discharges subject to effluent guidelines for acid or alkaline mine drainage under 40 CFR Part 434.[49]
- Storm water discharges that are mixed with nonstorm water, with a few exceptions (discussed in Section 5.4.4 below), unless the nonstorm water discharges are in compliance with a different NPDES permit;
- Storm water discharges subject to an existing NPDES individual or general permit;
- Storm water discharges that were subject to a NPDES permit that was terminated by the permitting authority;
- Storm water discharges from construction activities;
- Storm water discharges that will affect a property that either is listed on, or is eligible for listing, on the National Historic Register, unless the applicant has obtained and is in compliance with a written agreement signed by the State Historic Preservation Officer (SHPO) that outlines measures to be undertaken by the applicant to mitigate or prevent adverse effects to the historic property.
- Storm water discharges that are, or may reasonably be expected to be, contributing to a violation of a water quality standard;
- Storm water discharges that are likely to adversely affect a listed, or proposed to be listed, endangered or threatened species or its critical habitat;
- Storm water discharges from inactive mines, inactive landfills, or inactive oil and gas operations, that are located on federal lands where an operator cannot be identified.

---

[49] On the other hand, four types of storm water discharges subject to effluent limitation guidelines may be covered under the multi-sector general permit if they are not already subject to an existing or expired NPDES permit. These discharges include contaminated storm water runoff from phosphate fertilizer manufacturing facilities, runoff associated with asphalt paving or roofing emulsion production, runoff from material storage piles at cement manufacturing facilities, and coal pile runoff at steam electric generating facilities.

5.4.3 Notice of Intent Requirements

Storm water dischargers that submit a Notice of Intent (NOI) to be covered by the multi-sector general permit are not required to submit an individual permit application, provided the discharger is eligible for the general permit and an individual permit application is not required. Submitting an NOI is significantly less burdensome than submitting an individual permit application.[50] The NOI requirements for general permits usually address only general information and typically do not require the collection of monitoring data.[51]

NOIs to be covered under the multi-sector general permit must be sent to: Storm Water Notice of Intent (4203), 401 M Street, S.W., Washington, D.C. 20460. New facilities seeking permit coverage for storm water discharges associated with industrial activity must submit a NOI at least 2 days prior to the commencement of the industrial activity. Facilities that discharge to a large or medium MS4 must also submit signed copies of the NOI to the operator of the municipal system.[52]

Contents of NOIs. The specific information required of industrial facilities that are submitting a NOI for coverage under the multi-sector general permit are as follows:
1. The operator's name, address, telephone number, and status as federal, state, private, public, or other entity.
2. Street address of the facility for which the notification is submitted. Where a street address for the site is not available, the location can be described in terms of the latitude and longitude of the facility to the nearest 15 seconds, or the quarter, section, township, and range (to the nearest quarter section) of the approximate center of the site.
3. An indication of whether the facility is located on Federal Indian Reservations.
4. Up to four 4-digit Standard Industrial Classification (SIC) codes that best represent the principal products or activities provided by the facility. For hazardous waste treatment, storage, or disposal facilities, land disposal

---

[50] See Section 5.7 for full discussion of individual permit application requirements.
[51] A copy of the NOI form can be found in Addendum B to the multi-sector general permit published at 60 *Fed. Reg.* 51265 (Sept. 29, 1995). The same NOI form is also reproduced at the end of the September 1998 modification of the multi-sector general permit published at 63 *Fed. Reg.* 52430 (Sept. 30, 1998).
[52] The terms large and medium municipal separate storm sewer systems (systems serving a population of 100,000 or more) are defined at 40 CFR § 122.26(b)(4) & (7). Most of the cities and counties in which these systems are found are listed in Appendices F, G, H, and I to 40 CFR Part 122. Other municipal systems have been designated by EPA on a case-by-case basis.

facilities that receive or have received any industrial waste, steam electric power generating facilities, or treatment works treating domestic sewage, a two-character code must be provided.

5. The permit number of any NPDES permit for any discharge (including nonstorm water discharges) from the site that is currently authorized by an NPDES permit.

6. The name of the receiving water(s), or if the discharge is through a municipal separate storm sewer, the name of the municipal operator of the storm sewer and the receiving water(s) for the discharge through the municipal separate storm sewer.

7. The analytical monitoring status of the facility (monitoring or not).

8. For a co-permittee, if a storm water general permit number has been issued, it should be included.

9. A certification that the operator of the facility has read and understands the eligibility requirements for the permit and that the operator believes the facility to be in compliance with those requirements.

10. Identify type of permit requested (multi-sector); longitude and latitude; indication of presence of endangered species; indication of historic preservation agreement; signed certification stating compliance with the National Historic Preservation Act, Endangered Species Act, and the new source performance standard requirements.

11. A certification that a storm water pollution prevention plan (SWPPP) has been prepared for the facility in accordance with Part IV of the permit. A copy of the SWPPP should **not** be included with the NOI submission.

The NOI must be signed in accordance with the signatory requirements of 40 CFR Section 122.22. A complete description of the signatory requirements is provided in the instructions accompanying the NOI.

The EPA may deny coverage under the permit and require submittal of an individual permit application based on a review of the completeness and/or content of the NOI or other information (e.g., Endangered Species Act compliance, National Historic Preservation Act compliance, water quality information, compliance history, history of spills, etc.). If the EPA requires a discharger to apply for an individual permit, the EPA will notify the discharger in writing that a permit application is required by an established deadline. Coverage under the multi-sector general permit will automatically terminate if the discharger fails to submit the required permit application in a timely manner. If the discharger does submit a requested permit application, coverage under the multi-sector general permit will automatically terminate on the effective date of the issuance or denial of the individual permit.

5.4.4 Special Conditions

The conditions of the multi-sector general permit are designed to comply with the technology-based standards of the Clean Water Act--i.e., Best Available Technology (BAT) and Best Control Technology (BCT). The permit includes numeric effluent limitations for coal pile runoff, contaminated runoff from fertilizer manufacturing facilities, runoff from asphalt emulsion manufacturing facilities, and material storage pile runoff located at cement manufacturing facilities or cement kilns. For other discharges covered by the permit, the permit conditions reflect the EPA's decision to identify a number of best management practices (BMPs) and traditional storm water management practices which prevent pollution in storm water discharges as the BAT/BCT level of control for the majority of storm water discharges covered by the permit. The permit conditions applicable to these discharges are not numeric effluent limitations, but rather are flexible requirements for developing and implementing site-specific plans to minimize and control pollutants in storm water discharges associated with industrial activity.

The following special conditions are applicable to the multi-sector general permits:

- Prohibition on most types of nonstorm water discharges as a component of discharges authorized by the permit. (These discharges should already have a NPDES permit.) However, the multi-sector general permits do authorize certain types of nonstorm water discharges.
- In the event there is a release of a hazardous substance in excess of reportable quantities, as established under the CWA or CERCLA (see 40 CFR Section 117.3; 40 CFR Section 302.4), the discharger must notify the National Response Center and modify its Storm Water Pollution Prevention Plan (SWPPP).
- Co-located industrial activities are authorized under the general permit provided that the industrial facility complies with the SWPPP and the monitoring requirements for each co-located activity.

Prohibition of Nonstorm Water Discharges. The multi-sector general permit does not authorize nonstorm water discharges that are mixed with storm water except under limited circumstances. The only nonstorm water discharges that are intended to be authorized under the permit include discharges from fire fighting activities; fire hydrant flushings; potable water sources, including waterline flushings; irrigation drainage; lawn watering; routine external building washdown without detergents; pavement washwaters where spills or leaks of toxic or hazardous materials have not occurred (unless all spilled material has been removed) and where detergents are not used; air conditioning condensate; compressor condensate; springs; uncontaminated groundwater; and foundation or footing drains where flows are not contaminated with process materials such as solvents that are combined with storm water discharges associated with industrial activity. To be authorized under the general permit, these sources of nonstorm water (except flows from firefighting activities)[53] must be identified in the SWPPP prepared for the facility. Where such discharges occur, the plan must also identify and ensure the implementation of appropriate pollution prevention measures for the nonstorm water component(s) of the discharge.

Where a storm water discharge is mixed with nonstorm water that is not authorized by the multi-sector general permit or another NPDES permit, the discharger should submit the appropriate application forms (Forms 1, 2C, and/or 2E) to gain permit coverage of the nonstorm water portion of the discharge.

Releases of Reportable Quantities of Hazardous Substances and Oil. The multi-sector general permit provides that the discharge of hazardous substances or oil from a facility must be eliminated or minimized in accordance with the SWPPP developed for the facility. Where a permitted storm water discharge contains a hazardous substance or oil in an amount equal to or in excess of a reporting quantity established under 40 CFR Part 117, or 40 CFR Part 302 during a 24-hour period, the following actions must be taken:

1. Any person in charge of the facility that discharges hazardous substances or oil is required to notify the National Response Center at 800-424-8802 (202-426-2675 in the Washington, D.C. metropolitan area) in accordance with the requirements of 40 CFR Part 117 and 40 CFR Part 302 as soon as they have knowledge of the discharge.

2. The SWPPP for the facility must be modified within 14 calendar days of knowledge of the release to provide a description of the release, an account of the circumstances leading to the release, and the date of the release. In addition, the plan must be reviewed to identify measures to prevent the reoccurrence of such releases and to respond to such releases, and it must be modified where appropriate.

3. The permittee must also submit to the EPA within 14 calendar days of knowledge of the release a written description of the release (including the type and estimate of the amount of material released), the date that such release occurred, the circumstances leading to the release, and steps to be taken to modify the SWPPP for the facility.

---

[53] The permit does not require pollution prevention measures to be identified and implemented for nonstorm water flows from firefighting activities because these flows will generally be unplanned emergency situations where it is necessary to take immediate action to protect the public.

Anticipated discharges containing a hazardous substance in an amount equal to or in excess of reportable quantities are those caused by events occurring within the scope of the relevant operating system. Facilities that have more than one anticipated discharge per year containing a hazardous substance in an amount equal to or in excess of a reportable quantity are required to:

1. Submit notifications of the first release that occurs during a calendar year (or for the first year of the permit, after submittal of a NOI); and
2. Provide a written description in the SWPPP of the dates on which such releases occurred, the type and estimate of the amount of material released, and the circumstances leading to the releases. In addition, the SWPPP plan must address measures to minimize such releases.

Where a discharge of a hazardous substance or oil in excess of reporting quantities is caused by a nonstorm water discharge (e.g., a spill of oil into a separate storm sewer), that discharge is not authorized by the permit and the discharger must report the discharge as required under 40 CFR Part 110, 40 CFR Part 117, or 40 CFR Part 302. In the event of a spill, the requirements of Section 311 of the CWA and other applicable provisions of Sections 301 and 402 of the CWA continue to apply.

Co-located Industrial Facilities. The multi-sector general permit addresses storm water discharges from industrial activities co-located at an industrial facility described in the coverage section of the permit. Co-located industrial activities occur when activities being conducted onsite meet more than one of the descriptions in the coverage sections of Part XI. of the permit (e.g., a landfill at a wood treatment facility or a vehicle maintenance garage at an asphalt batching plant). Co-located industrial activities are authorized under the permit provided that the industrial facility complies with the SWPPP and the monitoring requirements for each co-located activity. Authorizing co-located discharges allows industrial facilities to develop SWPPPs that fully address all industrial activities at the site. For example, if a wood treatment facility has a landfill, the SWPPP requirements for the wood treatment facility will differ greatly from those needed for a landfill. Therefore, by authorizing co-located industrial activities, the wood treatment facility will develop a SWPPP to meet the requirements addressing the storm water discharges from the wood treatment facility and the landfill. The facility is also subject to applicable monitoring requirements for each type of industrial activity as described in the applicable sections of the permit. By monitoring the discharges from the different industrial activities, the facility can better determine the effectiveness of the SWPPP requirements for controlling storm water discharges from all activities.

5.4.5 Historic Preservation Certification

The National Historic Preservation Act (NHPA)[54] requires federal agencies to take into account the effects of "federal undertakings" on historic properties that are either listed on, or eligible for listing on, the National Register of Historic Places.[55] Federal undertakings include the EPA's issuance of NPDES general permits.[56] In light of NHPA requirements, the EPA included a provision in the eligibility requirements of the multi-sector general permit for the consideration of the effects to historic properties. That provision provides that an applicant is eligible for permit coverage only if:

1. The applicant's storm water discharges and best management practices (BMPs) to control storm water runoff do not affect a historic property, or
2. The applicant has obtained, and is in compliance with, a written agreement between the applicant and the State Historic Preservation Officer (SHPO) or Tribal Historic Preservation Officer (THPO) which outlines all measures to be taken by the applicant to mitigate or prevent adverse effects to the historic property.[57]

When applying for permit coverage, applicants are required to certify in the NOI that they are in compliance with the historic preservation eligibility requirements. Provided there are no other factors limiting permit eligibility, permit coverage is then granted 48 hours after the postmark on the envelope used to mail the NOI. Facilities that cannot certify compliance with the NHPA requirements must submit individual permit applications to the permitting authority.

---

[54] 16 U.S.C. § 470 et seq.

[55] Historic properties are defined in the NHPA regulations to include prehistoric or historic districts, sites, buildings, structures, or objects that are included in, or are eligible for inclusion in, the National Register of Historic Places. See 36 CFR § 802(e).

[56] The term "federal undertaking" is defined in the NHPA regulations to include any project, activity, or program under the direct or indirect jurisdiction of a federal agency that can result in changes in the character or use of historic properties, if any such historic properties are located in the area of potential effects for that project, activity, or program. See 36 CFR § 802(o).

[57] In the September 1998 modification of the multi-sector general permit, the EPA added a new Addendum I, which provides guidance and a list of SHPO and THPO addresses to assist applicants with the certification process for permit eligibility under this condition.

5.4.6 Endangered Species Certification

The Endangered Species Act (ESA) requires federal agencies to ensure, in consultation with the U.S. Fish and Wildlife Service (FWS) and the National Marine Fisheries Service (NMFS) that any actions authorized, funded, or carried out by the agency (e.g., EPA-issued NPDES permits authorizing discharges to waters of the United States) are not likely to jeopardize the continued existence of any federally listed endangered or threatened species or adversely modify or destroy critical habitat of such species.[58] This consultation resulted in a joint FWS/NMFS biological opinion issued by the FWS on March 31, 1995, and by the NMFS on April 5, 1995, which concluded that the issuance and operation of the multi-sector general permits was not likely to jeopardize the existence of any listed endangered or threatened species, or result in the adverse modification or destruction of any critical habitat.

The multi-sector general permit contains a number of conditions to protect listed species and critical habitat. All dischargers applying for coverage under the multi-sector general permit must provide in the application information on the Notice of Intent form:

1. A determination as to whether there are any species identified in Addendum H of the permit in proximity to the storm water discharges and BMP construction areas, and

2. A certification that their storm water discharges and the construction of BMPs to control storm water are not likely to adversely affect species identified in Addendum H of the permit, or are otherwise eligible for coverage due to a previous authorization under the ESA.

Species List. Addendum H of the multi-sector permit contains a list of proposed and listed endangered and threatened species that could be affected by the discharges and measures to control pollutants in the discharges. The Addendum also provides instructions to assist applicants in determining whether they met the eligibility requirements.

Because the EPA determined that its 1998 modification of the multi-sector general permit constituted an action that may affect listed endangered and threatened species, the EPA reinitiated consultation with the FWS and NMFS. The FWS and NMFS provided written concurrences on the EPA's findings that the permit modification was not likely to result in adverse effects to listed species or critical habitat. As a result of this consultation, the EPA updated the species list in Addendum H to include species that were listed or proposed for listing since the Addendum H list was first compiled on March 31, 1995. The EPA also decided to expand

---

[58] See 16 U.S.C. § 1536(a)(2); 50 CFR § 402 and 40 CFR § 122.49(c). See also Chapter 15 for full discussion of the ESA.

the list to include all of the terrestrial (i.e., non-aquatic) listed and proposed species in recognition that those species may be impacted by permitted activities such as the construction and operation of the BMPs.[59] The Addendum H list will be updated on a regular basis and an electronic copy of that list will be made available at the Office of Wastewater Management website at http://www.epa.gov/owm. Information on the availability of an electronic list is also being added to the Addendum H instructions.

Eligibility Determination. To be eligible for coverage under the multi-sector general permit, facilities must review the list of species and their locations, which are contained in Addendum H of the permit and which are described in the instructions for completing the application requirements under the permit. If an applicant determines that none of the species identified in the Addendum are found in the county in which the facility is located, then there is no likelihood of an adverse effect and they are eligible for permit coverage. If species identified in Addendum H are found to be located in the same county as the facility seeking permit coverage, then the applicant must determine whether the species are in proximity to the storm water discharges at the facility, or any BMPs to be constructed to control storm water runoff. A species is in proximity to a storm water discharge when the species is located in the path or down gradient area through which or over which point source storm water flows from industrial activities to the point of discharge into the receiving water, and once discharged into the receiving water, in the immediate vicinity of, or nearby, the discharge point. A species is also in proximity if a species is located in the area of a site where storm water BMPs are planned to be constructed. If an applicant determines there are no species in proximity to the storm water discharge, or the BMPs to be constructed, then there is no likelihood of adversely affecting the species and the applicant is eligible for permit coverage.

If species are in proximity to the storm water discharges or areas of BMP construction, as long as they have been considered as part of a previous ESA authorization of the applicant's activity, and the environmental baseline established in that authorization is unchanged, the applicant may be covered under the permit. The environmental baseline generally includes the past and present impacts of all federal, state, and private actions that were occurring at the time the initial NPDES authorization and current ESA section 7 action by the EPA was taken. Therefore, if a permit applicant has received previous authorization and nothing has changed or been added to the environmental baseline

---

[59] 63 *Fed. Reg.* 52430 (Sept. 30, 1998).

established in the previous authorization, then coverage under the multi-sector permit will be provided.

In the absence of such previous authorization, if species identified in Addendum H are in proximity to the discharges or construction areas for BMPs, then the applicant must determine whether there is any likely adverse effect upon the species. This is done by the applicant conducting a further examination or investigation, or an alternative procedure, as described in the instructions in Addendum H of the permit. If the applicant determines that there is no likely adverse effect upon the species, then the applicant is eligible for permit coverage. If the applicant determines that there likely is, or will likely be an adverse effect, then the applicant is not eligible for permit coverage.

5.4.7 Storm Water Pollution Prevention Plan (SWPPP)

A Storm Water Pollution Prevention Plan (SWPPP) must be developed for each facility covered by the multi-sector general permit. The SWPPP is considered to be the most important requirement of the general permit. Facilities must implement the provisions of the SWPPP as a condition of permit issuance. SWPPPs must be prepared in accordance with good engineering practices and in accordance with the factors outlined in 40 CFR Sections 125.3(d) (2) or (3), as appropriate. The plan must identify potential sources of pollution that may reasonably be expected to affect the quality of storm water discharges from the facility. In addition, the plan must describe and ensure the implementation of practices to reduce the pollutants in storm water discharges associated with industrial activity at the facility and to ensure compliance with the terms and conditions of the permit. The following checklist outlines the primary elements of the SWPPPs required by the EPA's multi-sector general permits for storm water discharges associated with industrial activities.

SWPPP Requirements for Multi-Sector General Permit.
- **Pollution Prevention Team** — Each facility will select a Pollution Prevention Team from its staff, and the team will be responsible for developing and implementing the plan.
- **Components of the Plan** — The permit requires that the plan contain a description of potential pollutant sources and a description of the measures and controls to prevent or minimize pollution of storm water. The description of potential pollutant sources must include:
—A map of the facility indicating the areas which drain to each storm water discharge point.
—An indication of the industrial activities which occur in each drainage area.

—A prediction of the pollutants which are likely to be present in the storm water.

—A description of the likely source of pollutants from the site.

—An inventory of the materials which may be exposed to storm water.

—The history of spills or leaks of toxic or hazardous materials for the past 3 years.

●**Measures and Controls.** The measures and controls to prevent or minimize pollution of storm water must include the following basic Best Management Practices (BMPs):

—Good housekeeping or upkeep of industrial areas exposed to storm water.

—Preventive maintenance of storm water controls and other facility equipment.

—Spill prevention and response procedures to minimize the potential for and the impact of spills.

—Test all outfalls to ensure there are no cross connections (only storm water is discharged).

—Training of employees on pollution prevention measures and controls, and recordkeeping.

●**Sector-specific BMPs.** In addition, the permit contains sector-specific BMPs which are unique to the types of facilities in the various sectors. The applicant will need to review the sector-specific BMPs outlined in the multi-sector general permit to ensure compliance with these SWPPP requirements.

●**Inspection/Site Compliance Evaluation:**

Facility personnel must inspect the plant equipment and industrial areas on a regular basis. At least once every year a more thorough site compliance evaluation must be performed by facility personnel to:

—Look for evidence of pollutants entering the drainage system.

—Evaluate the performance of pollution prevention measures.

—Identify areas where the plan should be revised to reduce the discharge of pollutants.

—Document both the routine inspections and the annual site compliance evaluation in a report.

●**Consistency** — The plan can incorporate other plans that a facility may have already prepared for other permits including Spill Prevention Control and Countermeasure (SPCC) Plans, or BMP Programs for other NPDES permits issued under the CWA.

●**Signature** — The plan must be signed by a responsible corporate official, such as the president, vice-president, or general partner.

● **Plan Review** — The plan is to be kept at the permitted facility at all times. The plan should be submitted for review only when requested by the EPA.

Each of the SWPPP requirements of the multi-sector general permit are discussed in the sections that follow.

5.4.8 Pollution Prevention Team

The SWPPP provisions of the multi-sector general permit require that the permittee choose a pollution prevention team. Each SWPPP must identify a specific individual or individuals within the facility organization as members of a storm water Pollution Prevention Team that are responsible for developing the SWPPP and assisting the facility or plant manager in its implementation, maintenance, and revision. The plan must clearly identify the responsibilities of each team member. The activities and responsibilities of the team must address all aspects of the facility's SWPPP.

Evaluation of Related Environmental Management Plans. Many industrial facilities may have already incorporated storm water management practices into day—to—day operations as a part of an environmental management plan required by other regulations. In some cases, it may be possible to build on elements of these plans that are relevant to storm water pollution prevention. Potentially relevant elements of a number of different types of plans may be incorporated into the SWPPP, including RCRA Preparedness, Prevention and Contingency Plans;[60] OSHA Emergency Action Plans;[61] and Spill Prevention Control and Countermeasure (SPCC) Plans.[62] The pollution prevention plan provisions of the EPA's multi-sector general permit specifically state that the permittee is allowed to incorporate provisions of SPCC plans or BMPs from other NPDES permits into SWPPPs.

It is the responsibility of the pollution prevention team to evaluate these other plans to determine which, if any, provisions may be incorporated into the SWPPP. For example, if the facility already has an effective SPCC plan in place, elements of that spill prevention strategy may be relevant to the approach taken for storm water pollution prevention. More specifically, lists of potential pollutants or constituents of concern may provide a starting point for a list of potential storm water pollutants.

---

[60] See 40 CFR pts. 264 and 265.
[61] See 29 CFR § 1910.
[62] See 40 CFR § 112.

However, although the facility should try to build on relevant portions of other environmental plans, it is important to remember that the facility's SWPPP must be a comprehensive, stand—alone document.

5.4.9 Description of Potential Pollutant Sources

After identifying who is responsible for developing and implementing the facility's SWPPP, a pollutant source assessment must be performed. The assessment is used to evaluate what materials or practices are or may be a source of storm water contaminants at the facility site. The SWPPP provisions of the EPA's multi-sector general permit require that the permittee provide a description of potential sources of storm water pollution. Each SWPPP must describe activities, materials, and physical features of the facility that may contribute significant amounts of pollutants to storm water runoff or, during periods of dry weather, result in pollutant discharges through the separate storm sewers or storm water drainage systems that drain the facility. This assessment of storm water pollution risk will support subsequent efforts to identify and set priorities for necessary changes in materials, materials management practices, or site features, as well as aid in the selection of appropriate structural and nonstructural control techniques. The SWPPP provisions of the EPA's multi-sector general permit require that the permittee complete the following tasks to complete the pollutant source assessment:
1. Provide drainage and site map.
2. Complete inventory of exposed materials.
3. Evaluate significant spills and leaks.
4. Identify nonstorm water discharges and illicit connections.
5. Collect or evaluate sampling data on storm water quality.

A discussion of each of these components of the pollutant source assessment is provided here. Upon completion of the pollutant source assessment, there should be enough information to determine which areas, activities, or materials may contribute pollutants to storm water runoff from the facility site. At the conclusion of the pollutant source assessment, the EPA's multi-sector general permit requires that the permittee provide a narrative description of potential pollutant sources that pose a risk to storm water quality. The description must specifically list any significant potential source of pollutants at the site and, for each potential source, any pollutant or pollutant parameter (e.g., biochemical oxygen demand, etc.) of concern must be identified.

Drainage and Site Map. The plan must contain a map of the site that shows the location of outfalls covered by the permit (or by other NPDES permits), the pattern of storm water drainage, an indication of the types of discharges contained in the drainage areas of the outfalls, structural features that control pollutants in runoff,[63] surface water bodies (including wetlands), places where significant materials are exposed to rainfall and runoff, and locations of major spills and leaks that occurred in the three years prior to the date of the submission of a NOI to be covered under the permit. The map also must show areas where the following activities take place: fueling, vehicle and equipment maintenance and/or cleaning, loading and unloading, material storage (including tanks or other vessels used for liquid or waste storage), material processing, and waste disposal. For areas of the facility that generate storm water discharges with a reasonable potential to contain significant amounts of pollutants, the map must indicate the probable direction of storm water flow and the pollutants likely to be in the discharge. Flows with a significant potential to cause soil erosion also must be identified. In order to increase the readability of the map, the inventory of the types of discharges contained in each outfall may be kept as an attachment to the site map.

Inventory of Exposed Materials. Facility operators are required to carefully conduct an inspection of the site and related records to identify significant materials that are or may be exposed to storm water.[64] The inventory must address materials, that within three years prior to the date of the submission of a NOI to be covered under the permit, have been handled, stored, processed, treated, or disposed of in a manner to allow exposure to storm water. Findings of the inventory must be documented in detail in the SWPPP. At a minimum, the plan must describe the method and location of onsite storage or disposal; practices used to minimize contact of materials with rainfall and runoff; existing structural and nonstructural controls that reduce pollutants in runoff; and any treatment the runoff receives before it is discharged to surface waters or a separate storm sewer system. The description must be updated whenever there is a significant change in the types or amounts of materials, or material management practices, that may affect the exposure of materials to storm water.

---

[63] Nonstructural features such as grass swales and vegetative buffer strips also should be shown.
[64] Significant materials include, but are not limited to the following: raw materials; fuels; solvents, detergents, and plastic pellets; finished materials, such as metallic products; raw materials used in food processing or production; hazardous substances designated under Section 101(14) of CERCLA; any chemical the facility is required to report pursuant to Section 313 of EPCRA; fertilizers; pesticides; and waste products, such as ashes, slag, and sludge that have the potential to be released with storm water discharges. 40 CFR § 122.26(b)(8).

Significant Spills and Leaks. The plan must include a list of any significant spills and leaks of toxic or hazardous pollutants that occurred in the three years prior to the date of the submission of a NOI to be covered under the permit. Significant spills include, but are not limited to, releases of oil or hazardous substances in excess of quantities that are reportable under Section 311 of the CWA (see 40 CFR Section 110.10 and 40 CFR Section 117.21) or Section 102 of CERCLA (see 40 CFR Section 302.4). Significant spills may also include releases of oil or hazardous substances that are not in excess of reporting requirements and releases of materials that are not classified as oil or a hazardous substance. The listing should include a description of the causes of each spill or leak, the actions taken to respond to each release, and the actions taken to prevent similar such spills or leaks in the future. This effort will aid the facility operator as she or he examines existing spill prevention and response procedures and develops any additional procedures necessary to fulfill the requirements of Part XI. of the permit.

Nonstorm Water Discharges. The SWPPP provisions of the permit require that the plan include a certification that all storm water outfalls have been tested or evaluated for the presence of nonstorm water discharges. The certification must include:
- Identification of potential nonstorm water discharges.
- A description of the results of any test and/or evaluation for the presence of nonstorm water discharges.
- The evaluation criteria or test method used.
- The date of testing and/or evaluation.
- The on-site drainage points that were directly observed during the test and/or evaluation.

If this certification is not feasible because the facility does not have access to an outfall, manhole, or other point of access to the final storm water discharge point(s), the permit applicant should describe why the certification was not feasible. The facility also must notify the permitting authority, within 180 days after submitting the NOI, of any potential sources of nonstorm water discharges to the storm water discharge and explain why the facility could not perform the test for nonstorm water discharges.

Sampling Data. The multi-sector general permit requires that the permittee provide a summary of existing discharge sampling data describing pollutants in storm water discharges from the facility, including a summary of sampling data collected during the term of the permit. During the pollutant source assessment, permit applicants should collect and summarize any storm water sampling data that were collected in the past. Historical storm water monitoring data may be very useful in locating areas which have previously contributed pollutants to storm water discharges and identifying problem pollutants. When summarizing these data, the applicant must describe the sample collection procedures used. In addition, the particular storm water outfall sampled should be cross-referenced to one of the outfalls designated on the site map.

5.4.10 Measures and Controls

After completion of each of the steps in the pollutant source assessment, there should be enough information to determine which areas, activities, or materials may contribute pollutants to storm water runoff from the facility site. With this information, the facility can select the most appropriate measures and controls for pollutants from these areas. Best Management Practices (BMPs) are recognized as an important part of the Clean Water Act's NPDES permitting process to prevent the release of toxic and hazardous chemicals.[65] The SWPPP provisions of the EPA's multi-sector general permit require that the permittee develop a description of BMPs for controlling storm water discharges at the facility, and implement such practices. The appropriateness and priorities of BMPs in a plan must reflect identified potential sources of pollutants at the facility. The description of storm water management controls must address the following general or "baseline" BMPs, including a schedule for implementing such BMPs:

- *Good housekeeping*: Practices designed to maintain the facility in a clean and orderly fashion.
- *Preventive maintenance:* Practices focused on preventing releases caused by equipment problems, rather than repair of equipment after problems occur.
- *Visual inspections:* Practices established to oversee facility operations and identify actual or potential problems.
- *Spill prevention and response:* Practices designed to avoid releases due to accidental or intentional entry.
- *Sediment and erosion control:* Practices designed to identify structural, vegetative, and/or stabilization measures to control significant soil erosion caused by topography, site activities, or other factors.
- *Management of runoff:* Practices designed to evaluate the appropriateness of traditional storm water management practices used to divert, infiltrate, reuse, or otherwise manage storm water runoff in a

---

[65] See chapter 12, section 12.6 for full discussion of BMP plans for NPDES permits issued under the Clean Water Act.

manner that reduces pollutants in storm water discharges from the site.
- *Employee training:* Practices developed to instill in employees an understanding of the BMP plan.
- *Recordkeeping and reporting:* Practices designed to maintain relevant information and foster communication.

Each of these BMPs are discussed below. In addition to these baseline BMPs, which all permittees must incorporate into their SWPPPs, the multi-sector general permit requires sector-specific BMPs which are unique to the types of facilities in the various sectors. The applicant will need to review the sector-specific BMPs outlined in the multi-sector general permit to ensure compliance with those SWPPP requirements.

Good Housekeeping Measures. Good housekeeping requires maintenance in a clean, orderly manner of areas that may contribute pollutants to storm water discharges. Maintaining an orderly facility means that materials and equipment are neat and well kept to prevent releases to the environment. Maintaining a clean facility also involves the expeditious remediation of releases to the environment.

Good housekeeping measures can be easily and simply implemented. Some examples of commonly implemented good housekeeping measures include the orderly storage of bags, drums, and piles of chemicals; prompt cleanup of spilled liquids to prevent significant runoff to receiving waters; expeditious sweeping, vacuuming, or other cleanup of accumulations of dry chemicals to prevent them from reaching receiving waters; and proper disposal of toxic and hazardous wastes to prevent contact with and contamination of storm water runoff.

Maintaining good housekeeping is the heart of a facility's overall pollution control effort. Some of the benefits that may result from a good housekeeping program include ease in locating materials and equipment; improved employee morale; improved manufacturing and production efficiency; lessened raw, intermediate, and final product losses due to spills, waste, or releases; fewer health and safety problems arising from poor materials and equipment management; environmental benefits resulting from reduced releases of pollution; and overall cost savings.

Preventive Maintenance Practices. Preventive maintenance (PM) is a method of periodically inspecting, maintaining, and testing storm water management devices and facility equipment and systems to uncover conditions that could cause breakdowns or failures resulting in discharges of pollutants to surface waters. Most facilities have existing PM programs. It is not the intent of the SWPPP to require development of a redundant PM program. Instead, the objective is to expand the current PM program to include storm water considerations, especially the upkeep and maintenance of storm water management devices. The pollution prevention team should evaluate the existing PM program for the facility and recommend any necessary changes.

A PM program accomplishes its goals by shifting the emphasis from a repair maintenance system to a PM system. It should be noted that in some cases, existing PM programs are limited to machinery and other moving equipment. The PM program prescribed to meet the goals of the SWPPP includes all other items (human-made and natural) used to contain and prevent releases of toxic and hazardous materials. Ultimately, the well-operated PM program devised to support the SWPPP should produce environmental benefits of decreased releases to the environment, as well as reduce total maintenance costs and increase the efficiency and longevity of equipment, systems, and structures.

In terms of pollution prevention plans, the PM program should prevent breakdowns and failures of equipment, containers, systems, structures, or other devices used to handle the toxic or hazardous chemicals or wastes. To meet this goal, a PM program should include a suitable system for evaluating equipment, systems, and structures; recording results; and facilitating corrective actions.

The multi-sector general permit contains additional preventive maintenance inspection requirements for facilities subject to EPCRA Section 313 reporting for water priority chemicals. For these facilities, all areas of the facility must be inspected for the following at appropriate intervals as specified in the plan:
- Leaks or conditions that would lead to discharges of Section 313 water priority chemicals.
- Conditions that could lead to direct contact of storm water with raw materials, intermediate materials, waste materials, or products.
- Examine piping, pumps, storage tanks and bins, pressure vessels, process and material-handling equipment, and material bulk storage areas for leaks, wind blowing, corrosion, support or foundation failure, or other deterioration or noncontainment.

These inspections must occur at intervals based on facility design and operational experience, and the timing must be specified in the plan. When a leak or other threatening condition is found, corrective action must be taken immediately or the facility unit or process must be shut down until the problem is repaired.

Visual Inspections. As part of the SWPPP requirements of the permit, the facility must perform visual inspections. Inspections provide an ongoing method to detect and identify sources of actual or potential environmental releases. Qualified facility personnel must be identified to inspect designated equipment and areas of the facility at appropriate intervals specified in the plan. A set of tracking or follow-up procedures must be used to ensure that appropriate actions are taken in response to the inspections. Records of inspections must also be maintained.

Every facility is different, so it is up to the facility owner/operator to determine which areas of the facility could potentially contribute pollutants to storm water runoff, and to devise and implement an inspection program based on this information. The following list identifies some types of equipment and plant areas to include in the facility's inspection plan:

- Areas around equipment, including around pipes, pumps, storage tanks and bins, pressure vessels, pressure release valves, process and material-handling equipment, and storm water management devices.
- Areas where spills and leaks have occurred in the past.
- Material storage areas.
- Outdoor material-processing areas.
- Material-handling areas (loading, unloading, transfer).
- Waste generation, storage, treatment, and disposal areas.

Inspection records should note when inspections were done, who conducted the inspection, what areas were inspected, what problems were found, and steps taken to correct any problems, including who has been notified. Many facilities will already have some sort of incident reporting procedure in place. Existing incident reporting and security surveillance procedures could easily be incorporated into the SWPPP. These records should be kept with the plan.

Spill Prevention and Response Measures. As part of the SWPPP requirements, the facility must implement spill prevention and response procedures. Areas where potential spills can occur and their accompanying drainage points must be identified clearly in the SWPPP. Where appropriate, material-handling procedures, storage requirements, and use of equipment should be spelled out in the plan. Procedures for cleaning up spills must be identified in the plan and made available to the appropriate personnel. The necessary equipment to implement a cleanup should be available to personnel.

Spills and leaks together comprise one of the largest sources of storm water pollutants and, in most cases, are avoidable. Establishing standard operating procedures, such as safety and spill prevention procedures, along with proper employee training, can reduce these accidental releases. Avoiding spills and leaks is preferable to cleaning them up after they occur, not only from an environmental standpoint, but also because spills cause increased operating costs and lower productivity.

Development of spill prevention and response procedures is a very important element of an effective SWPPP. A spill prevention and response plan may have already been developed in response to other environmental regulatory requirements. If the facility already has a spill prevention and response plan, it should be evaluated and revised if necessary to address the objectives of the SWPPP.

When developing the SWPPP, the facility should have created a list or inventory of materials that are handled, used, and disposed of. A site map indicating the drainage area of each storm water outfall should also be created. Overlay the drainage area map with the locations of these areas and activities with high material spill potential to determine where spills are likely to occur. Spill potential also depends on how materials are handled, the types and volumes of materials handled, and how materials are stored at the site. These factors must be described in the plan.

Also evaluate the possibility of storm water contamination from underground sources, such as tanks and pipes. Leaking underground storage tanks are often a source of storm water contamination. In addition to identifying these and other potential spill areas, projecting the possible spill volume and type of material is critical to developing the correct response procedures for a particular area.

At all times during operation of a facility, personnel should be available who have appropriate training and authority to respond to spills. The response plan should describe the following:

- Identification of the spill response "team" responsible for implementing the spill response plan.
- Safety measures.
- Procedures to notify appropriate authorities providing assistance, such as police, fire, hospital, POTWs, and so on.
- Spill containment, diversion, isolation, cleanup.
- Spill response equipment, including safety equipment and cleanup equipment.

In addition, the general permit sets forth more specific requirements for facilities that are subject to EPCRA Section 313 reporting for water priority chemicals.[66] Whenever a leak or spill of a Section 313 water priority chemical occurs, the contaminated soil, material, or debris must be removed promptly and disposed of in accordance with federal, state, and local requirements and as described in the SWPPP. These facilities are also required to designate a person responsible for spill prevention, response, and reporting procedures.

---

[66] The multi-sector general permit also provides an exemption from the EPCRA Section 313 requirements for situations where an operator certifies that all water priority chemicals which are handled and/or stored on-site are only in gaseous or non-soluble liquid or solid forms (at atmospheric pressure and temperature).

Sediment and Erosion Control Measures. As part of the SWPPP requirements, the facility must identify areas which, due to topography, activities, or other factors, have a high potential for significant soil erosion, and identify structural, vegetative, and/or stabilization measures to be used to limit erosion. There may be areas at the facility site that are prone to soil erosion due to construction activities, steep slopes, sandy soils, or other reasons. Construction activities typically remove grass and other protective ground covers, resulting in exposure of the underlying soil to wind and rain. Similarly, steep slopes or sandy soils may not be able to hold plant life so that these soils become exposed. Because the soil surface is unprotected, dirt and sand particles are easily picked up by wind and/or washed away by rain. Erosion can be controlled or prevented with the use of certain BMPs, including buffer zones, vegetated filter strips, stream bank stabilization, interceptor dikes, pipe slope drains, silt fences, and sediment traps.

Management of Runoff. The facility's SWPPP must contain a narrative evaluation of the appropriateness of traditional storm water management practices (practices other than those which control the generation or source(s) of pollutants) used to divert, infiltrate, reuse, or otherwise manage storm water runoff in a manner that reduces pollutants in storm water discharges from the site. The potential of various sources at the facility to contribute pollutants to storm water discharges must be considered when determining reasonable and appropriate measures. Measures that the permittee identifies in the plan as reasonable and appropriate must be implemented and maintained. Appropriate measures may include vegetative swales, reuse of collected storm water (such as for a process or as an irrigation source), inlet controls (such as oil/water separators), snow management activities, infiltration devices, and wet detention/retention devices.

Employee Training. As part of the SWPPP requirements, the facility must perform employee training. Employee training programs must inform personnel responsible for implementing activities identified in the SWPPP, or otherwise responsible for storm water management, of the components and goals of the SWPPP. Training should address topics such as spill response, good housekeeping, and material management practices. The plan must identify periodic dates for such training.

The multi-sector general permit contains additional training requirements for employees and contractor personnel that work in areas where EPCRA Section 313 water priority chemicals are used or stored.

These individuals must be trained in the following areas at least once per year:

● Preventive measures, including spill prevention and response and preventive maintenance.
● Pollution control laws and regulations.
● The facility's SWPPP.
● Features and operations of the facility which are designed to minimize discharges of Section 313 water priority chemicals, particularly spill prevention procedures.

Recordkeeping and Internal Reporting Procedures. As part of the SWPPP requirements of the permit, the facility must implement recordkeeping and reporting procedures. A description of incidents (such as spills or other discharges) along with other information describing the quality and quantity of storm water discharges must be included in the plan. Inspections and maintenance activities must be documented and records of such activities shall be incorporated into the plan. Records of spills, leaks, or other discharges, inspections, and maintenance activities must be retained for at least one year after coverage under the permit expires.

Records should include the following, as appropriate:

● The date and time of the incident, weather conditions, duration, cause, environmental problems, response procedures, parties notified, recommended revisions of the BMP program, operating procedures, and/or equipment needed to prevent recurrence.
● Formal written reports. These are helpful in reviewing and evaluating the discharges and making revisions to improve the BMP program. Document all reports called in to the National Response Center in the event of a reportable quantity discharge.[67]
● A list of the procedures for notifying the appropriate plant personnel and the names and telephone numbers of responsible employees. This enables more rapid reporting of and response to spills and other incidents.

A recordkeeping system set up for documenting spills, leaks, and other discharges, including discharges of hazardous substances in reportable quantities, should help the facility minimize incident recurrence, correctly respond with appropriate cleanup activities, and comply with legal requirements.

---

[67] For more information on reporting spills or other discharges, refer to 40 CFR § 117.3 and 40 CFR § 302.4.

5.4.11 Comprehensive Site Compliance Evaluation

After the SWPPP has been put into action, it must be kept up-to-date by regularly evaluating the information collected in the pollutant source assessment phase and the controls selected in the measures and controls identification phase. Regular site evaluations must be conducted and the plan must be revised as needed. The permit requires that the SWPPP describe the scope and content of the comprehensive site evaluations that qualified personnel will conduct to (1) confirm the accuracy of the description of potential pollution sources contained in the plan, (2) determine the effectiveness of the plan, and (3) assess compliance with the terms and conditions of the permit. Note that the comprehensive site evaluations are not the same as periodic or other routine inspections described in the preceding discussion of baseline BMPs. However, in the instances when frequencies of inspections and the comprehensive site compliance evaluation overlap they may be combined allowing for efficiency, as long as the requirements for both types of inspections are met. The plan must indicate the frequency of comprehensive evaluations which must be at least once a year, except where comprehensive site evaluations are shown in the plan to be impractical for inactive mining sites, due to remote location and inaccessibility.[68] The individual or individuals who will conduct the comprehensive site evaluation must be identified in the plan and should be members of the pollution prevention team. Material handling and storage areas and other potential sources of pollution must be visually inspected for evidence of actual or potential pollutant discharges to the drainage system. Inspectors also must observe erosion controls and structural storm water management devices to ensure that each is operating correctly. Equipment needed to implement the SWPPP, such as that used during spill response activities, must be inspected to confirm that it is in proper working order.

The results of each comprehensive site evaluation must be documented in a report signed by an authorized company official. The report must describe the scope of the comprehensive site evaluation, the personnel making the comprehensive site evaluation, the date(s) of the comprehensive site evaluation, and any major observations relating to implementation of the SWPPP. Comprehensive site evaluation reports must be retained for at least three years after the date of the evaluation. Based on the results of each comprehensive site evaluation, the description in the plan of potential pollution sources and measures and controls must be revised as appropriate within two weeks after each comprehensive site

---

[68] Where annual site inspections are shown in the plan to be impractical for inactive mining sites, due to remote location and inaccessibility, site inspections must be conducted at least once every three years.

evaluation, unless indicated otherwise in Section XI of the permit. Changes in procedural operations must be implemented on the site in a timely manner for non-structural measures and controls, not more than 12 weeks after completion of the comprehensive site evaluation. Procedural changes that require construction of structural measures and controls are allowed up to three years for implementation. In both instances, an extension may be requested from the EPA Director.

5.4.12 Monitoring and Reporting Requirements

The multi-sector general permit contains three basic types of monitoring requirements: (1) analytical monitoring or chemical monitoring; (2) compliance monitoring for effluent guidelines compliance; and (3) visual examinations of storm water discharges. Analytical monitoring requirements involve laboratory chemical analyses of samples collected by the permittee. The results of the analytical monitoring are quantitative concentration values for different pollutants, which can be easily compared to the results from other sampling events, other facilities, or to National benchmarks. Compliance monitoring requirements are imposed under the permit to insure that discharges subject to numerical effluent limitations under the storm water effluent limitations guidelines are in compliance with those limitations. Visual examinations of storm water discharges are the least burdensome type of monitoring requirement under the permit. Almost all of the industrial activities are required to perform visual examinations of their storm water discharges when they are occurring on a quarterly basis.

This section provides a general description of each of these types of monitoring. Actual monitoring requirements for a given facility under the permit will vary depending upon the industrial activities that occur at a facility and the criteria for determining monitoring used to develop the permit.

5.4.13 Analytical Monitoring Requirements

The multi-sector general permit requires analytical monitoring for discharges from certain classes of industrial facilities. Analytical monitoring is a means by which to measure the concentration of a pollutant in a storm water discharge. Analytical results are quantitative and therefore can be used to compare results from discharge to discharge and to quantify the improvement in storm water quality attributable to the SWPPP, or to identify a pollutant that is not being successfully controlled by the plan. The permit only requires analytical monitoring for the industry sectors or subsectors that demonstrated a potential to discharge pollutants at concentrations of concern.

To conduct a comparison of the results of the statistical analyses to determine when analytical monitoring would be required, the EPA established "benchmark" concentrations for the pollutant parameters on which monitoring results had been received. The benchmarks are the pollutant concentrations above which the EPA determined would represent a level of concern. The level of concern is a concentration at which a storm water discharge could potentially impair water quality or affect human health from ingestion of water or fish.[69] The benchmarks are also viewed by the EPA as a level, that if below, a facility represents little potential for water quality concern. As such, the benchmarks also provide an appropriate level to determine whether a facility's storm water pollution prevention measures are successfully implemented. The benchmark concentrations are not effluent limitations and should not be interpreted or adopted as such. These values are merely levels which the EPA has used to determine if a storm water discharge from any given facility merits further monitoring to ensure that the facility has been successful in implementing its SWPPP. As such, these levels represent a target concentration for a facility to achieve through implementation of pollution prevention measures at the facility. Table 5.1 lists the parameter benchmark values.

Table 5.1. Parameter Benchmark Values

Parameter name	Benchmark level	Source
Biochemical Oxygen Demand(5)	30 mg/L	4
Chemical Oxygen Demand	120 mg/L	5
Total Suspended Solids	100 mg/L	7
Oil and Grease	15 mg/L	8
Nitrate + Nitrite Nitrogen	0.68 mg/L	7
Total Phosphorus	2.0 mg/L	6

---

[69] The primary source of benchmark concentrations is the EPA's National Water Quality Criteria, published in 1986 (often referred to as the "Gold Book"). For the majority of the benchmarks, the EPA chose to use the acute aquatic life, fresh water ambient water quality criteria.

Table 5.1. Parameter Benchmark Values (*Continued*)

Parameter name	Benchmark level	Source
pH	6.0–9.0 stand. units	4
Acrylonitrile (c)	7.55 mg/L	2
Aluminum, Total (pH 6.5–9)	0.75 mg/L	1
Ammonia	19 mg/L	1
Antimony, Total	0.636 mg/L	9
Arsenic, Total (c)	0.16854 mg/L	9
Benzene	0.01 mg/L	10
Beryllium, Total (c)	0.13 mg/L	2
Butylbenzyl Phthalate	3 mg/L	3
Cadmium, Total (H)	0.0159 mg/L	9
Chloride	860 mg/L	1
Copper, Total (H)	0.0636 mg/L	9
Dimethyl Phthalate	1.0 mg/L	11
Ethylbenzene	3.1 mg/L	3
Fluoranthene	0.042 mg/L	3
Fluoride	1.8 mg/L	6
Iron, Total	1.0 mg/L	12
Lead, Total (H)	0.0816 mg/L	1
Manganese	1.0 mg/L	13
Mercury, Total	0.0024 mg/L	1
Nickel, Total (H)	1.417 mg/L	1
PCB-1016 (c)	0.000127 mg/L	9
PCB-1221 (c)	0.10 mg/L	10
PCB-1232 (c)	0.000318 mg/L	9
PCB-1242 (c)	0.00020 mg/L	10
PCB-1248 (c)	0.002544 mg/L	9
PCB-1254 (c)	0.10 mg/L	10
PCB-1260 (c)	0.000477 mg/L	9
Phenols, Total	1.0 mg/L	11
Pyrene (PAH,c)	0.01 mg/L	10
Selenium, Total (*)	0.2385 mg/L	9
Silver, Total (H)	0.0318 mg/L	9
Toluene	10.0 mg/L	3
Trichloroethylene (c)	0.0027 mg/L	3
Zinc, Total (H)	0.117 mg/L	1

Table 5.1. Parameter Benchmark Values (*Continued*)

Sources
1. "EPA Recommended Ambient Water Quality Criteria." Acute
   Aquatic Life Freshwater
2. "EPA Recommended Ambient Water Quality Criteria." LOEL
   Acute Freshwater
3. "EPA Recommended Ambient Water Quality Criteria." Human
   Health Criteria for Consumption of Water and Organisms
4. Secondary Treatment Regulations (40 CFR Section 133)
5. Factor of 4 times BOD5 concentration--North Carolina benchmark
6. North Carolina storm water benchmark derived from NC Water
   Quality Standards
7. National Urban Runoff Program (NURP) median concentration
8. Median concentration of Storm Water Effluent Limitation Guideline
   (40 CFR Part 419)
9. Minimum Level (ML) based upon highest Method Detection Limit
   (MDL) times a factor of 3.18
10. Laboratory derived Minimum Level (ML)
11. Discharge limitations and compliance data
12. "EPA Recommended Ambient Water Quality Criteria." Chronic
    Aquatic Life Freshwater
13. Colorado--Chronic Aquatic Life Freshwater--Water Quality Criteria

Notes
(*) Limit established for oil and gas exploration and production
facilities only.
(c) carcinogen
(H) hardness dependent
(PAH) Polynuclear Aromatic Hydrocarbon

Assumptions
Receiving water temperature--20^0 C
Receiving water pH 7.8
Receiving water hardness $CaCO_3$ 100 mg/L
Receiving water salinity 20 g/kg
Acute to Chronic Ratio (ACR)--10

Where the EPA could identify a source of a potential pollutant that is
directly related to industrial activities of the industry sector or subsector,
the permit identifies that parameter for analytical monitoring. If the EPA

could not identify a source of a potential pollutant that was associated with the sector/subsector's industrial activity, the permit does not require monitoring for the pollutant in that sector/subsector. Industries with no pollutants for which the median concentrations are higher than the benchmark levels are not required to perform analytical monitoring under the permit, with the exceptions of Sector K (hazardous waste treatment storage and disposal facilities), and Sector S (airports which use more than 100,000 gallons per year of glycol-based fluids or 100 tons of urea for deicing). These industries are required to perform analytical monitoring under the permit due to the high potential for contamination of storm water discharge.

All facilities within an industry sector or subsector identified for analytical monitoring must, at a minimum, monitor their storm water discharges during the second year of permit coverage, unless the facility exercises the alternative certification discussed below in Section 5.4.16. At the end of the second year of permit coverage, a facility must calculate the average concentration for each parameter for which the facility is required to monitor. If the permittee collects more than four samples in this period, then an average concentration must be calculated for each pollutant of concern for all samples analyzed. Monitoring must be conducted for the same storm water discharge outfall in each sampling period. Where a given storm water discharge is addressed by more than one sector/subsector's monitoring requirements, then the monitoring requirements for the applicable sector's/subsector's activities are cumulative. Therefore, if a particular discharge fits under more than one set of monitoring requirements, the facility must comply with all sets of sampling requirements. Monitoring requirements must be evaluated on an outfall-by-outfall basis.

If the average concentration for a pollutant parameter is less than or equal to the benchmark value, then the permittee is not required to conduct analytical monitoring for that pollutant during the fourth year of the permit. If, however, the average concentration for a pollutant is greater than the benchmark value, then the permittee is required to conduct quarterly monitoring for that pollutant during the fourth year of permit coverage. Analytical monitoring is not required during the first, third, and fifth years of the permit. The exclusion from analytical monitoring in the fourth year of the permit is conditional on the facility maintaining industrial operations and BMPs that will ensure a quality of storm water discharges consistent with the average concentrations recorded during the second year of the permit.

5.4.14 Compliance Monitoring

In addition to the analytical monitoring requirements for certain sectors, the permit contains monitoring requirements for discharges that are subject to effluent limitations. These discharges must be sampled annually and tested for the parameters that are limited by the permit. Discharges subject to compliance monitoring include: coal pile runoff, contaminated runoff from phosphate fertilizer manufacturing facilities, runoff from asphalt paving and roofing emulsion production areas, material storage pile runoff from cement manufacturing facilities, and mine dewatering discharges from crushed stone, construction sand and gravel, and industrial sand mines located in Texas, Louisiana, Oklahoma, New Mexico, and Arizona. All samples are to be grab samples taken within the first 30 minutes of discharge where practicable, but in no case later than the first hour of discharge. Where practicable, the samples must be taken from the discharges subject to the numeric effluent limitations prior to mixing with other discharges. Monitoring for these discharges is required to determine compliance with numeric effluent limitations. Furthermore, discharges covered under the permit which are subject to numeric effluent limitations are not eligible for the alternative certification discussed below in Section 5.4.16.

5.4.15 Quarterly Visual Examination of Storm Water Quality

The multi-sector general permit requires quarterly visual examinations of storm water discharges for all sectors except Sector S, which covers air transportation. The visual examination of storm water outfalls should include any observations of color, odor, clarity, floating solids, settled solids, suspended solids, foam, oil sheen, or other obvious indicators of storm water pollution. No analytical tests are required to be performed on these samples. The visual examination is not required if there is insufficient rainfall or snow-melt to runoff or if hazardous conditions prevent sampling. Grab samples for the examination must be collected within the first 30 minutes (or as soon thereafter as practical, but not to exceed 1 hour) of when the runoff begins discharging. Reports of the visual examination include the examination date and time, examination personnel, visual quality of the storm water discharge, and probable sources of any observed storm water contamination. The sampling must be conducted quarterly during the following time periods: January-March, April–June, July–September, and October–December of each year. The reports summarizing these quarterly visual storm water examinations must be maintained on-site with the SWPPP.

5.4.16 Alternate Certification

Throughout the permit, the EPA has included monitoring requirements for facilities that the agency believes have the potential for contributing significant levels of pollutants to storm water discharges. The alternative certification described here is included in the permit to ensure that monitoring requirements are only imposed on those facilities which do, in fact, have storm water discharges containing pollutants at concentrations of concern. The EPA has determined that if there are no sources of a pollutant exposed to storm water at the site then the potential for that pollutant to contaminate storm water discharges does not warrant monitoring. Therefore, a discharger is not subject to the analytical monitoring requirements provided the discharger makes a certification for a given outfall, on a pollutant-by-pollutant basis, that material handling equipment or activities, raw materials, intermediate products, final products, waste materials, by-products, industrial machinery or operations, significant materials from past industrial activity that are located in areas of the facility that are within the drainage area of the outfall are not presently exposed to storm water and will not be exposed to storm water for the certification period. Such certification must be retained in the SWPPP, and submitted to the EPA in lieu of monitoring reports required under Part XI of the permit. The permittee is required to complete any and all sampling until the exposure is eliminated. If the facility is reporting for a partial year, the permittee must specify the date exposure was eliminated. If the permittee is certifying that a pollutant was present for part of the reporting period, nothing relieves the permittee from the responsibility to sample that parameter up until the exposure was eliminated and it was determined that no significant materials remained. This certification is not to be confused with the low concentration sampling waiver. The test for the application of this certification is whether the pollutant is exposed, or can be expected to be present in the storm water discharge. If the facility does not use a parameter, or if exposure is eliminated and no significant materials remain, then the facility can exercise this certification. The permit does not allow facilities with discharges subject to numeric effluent limitations to submit alternative certification in lieu of the compliance monitoring requirements. The permit also does not allow air transportation facilities subject to the analytical monitoring requirements to exercise an alternative certification.

A facility is not precluded from exercising the alternative certification in lieu of analytical monitoring requirements in the fourth year of permit coverage, even if that facility failed to qualify for a low concentration waiver in year two. The EPA encourages facilities to eliminate exposure of industrial activities and significant materials where practicable.

5.4.17 Reporting and Retention Requirements

Permittees are required to submit all analytical monitoring results obtained during the second and fourth year of permit coverage within three months of the conclusion of the second and fourth year of coverage of the permit. For each outfall, one Discharge Monitoring Report Form must be submitted per storm event sampled. For facilities conducting monitoring beyond the minimum requirements an additional Discharge Monitoring Report Form must be filed for each analysis. The permittee must include a measurement or estimate of the total precipitation, volume of runoff, and peak flow rate of runoff for each storm event sampled. Permittees subject to compliance monitoring requirements are required to submit all compliance monitoring results annually on the 28th day of the month following the anniversary of the publication of the multi-sector general permit. Compliance monitoring results must be submitted on signed Discharge Monitoring Report Forms. For each outfall, one Discharge Monitoring Report form must be submitted for each storm event sampled.

Permittees are not required to submit records of the visual examinations of storm water discharges unless specifically asked to do so by the EPA Director. Records of the visual examinations must be maintained at the facility. Records of visual examination of storm water discharge need not be lengthy. Permittees may prepare typed or handwritten reports using forms or tables which they may develop for their facility. The report need only document: the date and time of the examination; the name of the individual making the examination; and any observations of color, odor, clarity, floating solids, suspended solids, foam, oil sheen, and other obvious indicators of storm water pollution. The location for submittal of all reports is contained in the permit. Permittees are required to retain all records for a minimum of three years from the date of the sampling, examination, or other activity that generated the data.[70]

5.4.18 Sample Type

Grab samples may be used for all monitoring unless otherwise stated. All such samples must be collected from the discharge resulting from a storm event that is greater than 0.1 inches in magnitude and that occurs at least 72 hours from the previously measurable (greater than 0.1 inch rainfall) storm event. The required 72-hour storm event interval may be waived by the permittee where the preceding measurable storm event did not result in a measurable discharge from the facility. The 72-hour

---

[70] 40 CFR § 122.41(j).

requirement may also be waived by the permittee where the permittee documents that less than a 72-hour interval is representative for local storm events during the season when sampling is being conducted. The grab sample must be taken during the first 30 minutes of the discharge. If the collection of a grab sample during the first 30 minutes is impracticable, a grab sample can be taken during the first hour of the discharge, and the discharger must submit with the monitoring report a description of why a grab sample during the first 30 minutes was impracticable. A minimum of one grab is required. Where the discharge to be sampled contains both storm water and nonstorm water, the facility must sample the storm water component of the discharge at a point upstream of the location where the nonstorm water mixes with the storm water, if practicable.

5.4.19 Representative Discharge

The permit allows permittees to use the substantially identical outfalls to reduce their monitoring burden. This representative discharge provision provides facilities with multiple storm water outfalls with a means for reducing the number of outfalls that must be sampled and analyzed. This may result in a substantial reduction of the resources required for a facility to comply with analytical monitoring requirements. When a facility has two or more outfalls that, based on a consideration of industrial activity, significant materials, and management practices and activities within the area drained by the outfall, the permittee reasonably believes discharge substantially identical effluents, the permittee may test the effluent of one of such outfalls and report that the quantitative data also applies to the substantially identical outfalls provided that the permittee includes in the SWPPP a description of the location of the outfalls and explaining in detail why the outfalls are expected to discharge substantially identical effluent.[71] In addition, for each outfall that the permittee believes is representative, an estimate of the size of the drainage area (in square feet) and an estimate of the runoff coefficient of the drainage area [e.g., low (under 40 percent), medium (40 to 65 percent) or high (above 65 percent)] must be provided in the plan. Facilities that select and sample a representative discharge are prohibited from changing the selected discharge in future monitoring periods unless the selected discharge ceases to be representative or is eliminated. Permittees do not need EPA approval to claim discharges are representative, provided they have documented their rationale within the SWPPP. However, the EPA Director may determine the discharges are not

---

[71] "Substantially identical effluents" are defined as discharges from drainage areas undergoing similar activities where the discharges are expected to be of similar quantity and quality, and indistinguishable in expected composition. 40 CFR § 122.21(g)(7).

representative and require sampling of all non-identical outfalls. The representative discharge provision in the permit is available to almost all facilities subject to the analytical monitoring requirements (not including compliance monitoring for effluent guideline limit compliance purposes) and to facilities subject to visual examination requirements.

5.4.20 Sampling Waiver

The permit allows for temporary waivers from sampling based on adverse climatic conditions. This temporary sampling waiver is only intended to apply to insurmountable weather conditions such as drought or dangerous conditions such as lightning, flash flooding, or hurricanes. The sampling waiver is not intended to apply to difficult logistical conditions, such as remote facilities with few employees or discharge locations which are difficult to access. When a discharger is unable to collect samples within a specified sampling period due to adverse climatic conditions, the discharger must collect a substitute sample from a separate qualifying event in the next sampling period, as well as a sample for the routine monitoring required in that period. Both samples should be analyzed separately and the results of that analysis submitted to the EPA. Permittees are not required to obtain advance approval for sampling waivers.

The permit also allows for a waiver from sampling and/or visual examinations for facilities that are both inactive and unstaffed. This waiver is only intended to apply to these types of facilities when the ability to conduct sampling and/or perform visual examinations would be severely hindered and result in the inability to meet the time and representative rainfall sampling specifications. This waiver is not intended to apply to remote facilities that are active and staffed, or typical difficult logistical conditions. When a discharger is unable to collect samples and/or perform visual examinations as specified in the permit, the discharger must certify to the EPA director that the facility is unstaffed and inactive and the ability to conduct samples within the specifications is not possible. When a discharger is unable to perform visual examinations as specified in the permit, the discharger must maintain a certification onsite with the pollution prevention plan stating that the facility is unstaffed and inactive and the ability to perform visual examinations within the specifications is not possible. Permittees are not required to obtain advance approval for these sampling or visual examination waivers.

5.4.21 EPCRA Section 313 Facilities

The multi-sector general permit does not contain special monitoring requirements for facilities subject to the Toxic Release Inventory (TRI) reporting requirements under Section 313 of EPCRA. The EPA has reviewed data and determined that storm water monitoring requirements are more appropriately based upon the industrial activity or significant material exposed than upon a facility's status as a TRI reporter under Section 313 of EPCRA.

5.4.22 Numeric Effluent Limitations

Part XI of the multi-sector general permit contains numeric effluent limitations for phosphate fertilizer manufacturing facilities, asphalt emulsion manufacturers, cement manufacturers, coal pile runoff from steam electric power generating facilities, and sand, gravel, and crushed stone quarries. These limitations are required under the EPA's storm water effluent limitation guidelines in the Code of Federal Regulations at 40 CFR Part 411, Part 418, Part 423, Part 436, and Part 443.

The permit also establishes effluent limitations of 50 mg/L total suspended solids and a pH range of 6.0 to 9.0 for coal pile runoff. Any untreated overflow from facilities designed, constructed, and operated to treat the volume of coal pile runoff associated with a 10-year, 24-hour rainfall event is not subject to the 50 mg/L limitation for total suspended solids. Steam electric generating facilities must comply with these limitations upon submittal of the NOI. The EPA has adopted these technology-based pH limitations in the general permit in accordance with setting limits on a case-by-case basis as allowed under 40 CFR Section 125.3, and Section 402 of the Clean Water Act. These case-by-case limits are derived by transferring the known achievable technology from an effluent guideline to a similar type of discharge. When developing these technology-based limitations, variables such as rainfall pH, sizes of coal piles, pollutant characteristics, and runoff volume were considered. Therefore, these variables need not be considered again.

5.5 General Permit for Construction Activities

In most instances, storm water discharges associated with construction activities are authorized under general permits. These general permits basically cover storm water discharges from construction sites that will result in the disturbance of five or more acres of land. The EPA's regulations for Phase I of the storm water program specifically exempt construction sites of less than five acres from permit requirements.[72] However, it is important to note that the EPA will likely require permits for these construction sites under Phase II of the storm water program, with an application deadline of March 10, 2003 for construction sites of one to five acres.[73]

The EPA originally issued the general permits for construction activities in September 1992.[74] In 1998, the EPA reissued the general permits for storm water discharges associated with construction activities.[75] The most significant changes from the 1992 permits are:

● New conditions to protect listed endangered and threatened species and critical habitats;
● Expanded coverage to construction sites under five acres of disturbed land which are not part of a larger common plan of development or sale when an operator has been designated by the EPA Regional Director to obtain coverage.
● A requirement to post at the construction site the confirmation of permit coverage (the permit number or copy of the Notice of Intent (NOI) if a permit number has not yet been assigned) including a brief description of the project;
● The addition of certain storm water pollution prevention plan performance objectives.

This section discusses the requirements of the EPA's general permit for storm water discharges associated with construction activities.

5.5.1 Unauthorized Discharges

The following discharges are not authorized by the construction general permit:

● Storm water discharges associated with industrial activity that originate from the site after construction activities have been completed and the site has undergone final stabilization;
● Nonstorm water discharges (except certain nonstorm water discharges specifically listed in the general permit). However, the permit can

---

[72] See section 5.2.4 for additional discussion of the five-acre threshold.
[73] See section 5.2.6 for discussion of the Phase II regulations.
[74] 57 *Fed. Reg.* 41176 (Sept. 9, 1992); 57 *Fed. Reg.* 44412 (Sept. 25, 1992).
[75] 63 *Fed. Reg.* 7858 (Feb. 17, 1998) (reissuance of general permits for construction activities in EPA Regions 1-3, 7-10); 63 *Fed. Reg.* 15622 (Mar. 31, 1998) (reissuance of general permits for construction activities in EPA Region 4); 64 *Fed. Reg.* 39136 (July 21, 1999) (draft modifications to reissued general permits for construction activities in EPA Region 4); 63 *Fed. Reg.* 36490 (July 6, 1998) (reissuance of general permits for construction activities in EPA Region 6).

authorize storm water discharges from construction activities where such discharges are mixed with nonstorm water discharges that are authorized by a different NPDES permit;

- Storm water discharges from construction sites that are covered by an existing NPDES individual or general permit. However, storm water discharges associated with industrial activity from a construction site that are authorized by an existing permit may be authorized by the general permit after the existing permit expires, provided the expired permit did not establish numeric limitations for such discharges;
- Storm water discharges from construction sites that the EPA director has determined to be, or may reasonably be expected to be, contributing to a violation of a water quality standard; and
- Storm water discharges from construction sites if the discharges are likely to adversely affect a listed endangered or threatened species, or a species that is proposed to be listed as endangered or threatened, or its critical habitat.

5.5.2 Notice of Intent Requirements

The following information is required of applicants submitting a Notice of Intent (NOI) for storm water discharges associated with construction activities:

- The mailing address of the construction site for which the notification is submitted. Where a mailing address for the site is not available, the location of the approximate center of the site must be described in terms of the latitude and longitude to the nearest 15 seconds, or the section, township, and range to the nearest quarter;
- The site owner's name, address, and telephone number;
- The name, address, and telephone number of the operator(s) with day-to-day operational control who have been identified at the time of the NOI submittal, and their status as a federal, state, private, public, or other entity. Where multiple operators have been selected at the time of the initial NOI submittal, NOIs must be attached and submitted in the same envelope. When an additional operator submits a NOI for a site with a pre-existing NPDES permit, the NOI of the additional operator must indicate the pre-existing NPDES permit number for discharge(s) from the site;
- The name of the receiving water(s), or if the discharge is through a municipal separate storm sewer, the name of the municipal operator of the storm sewer and the ultimate receiving water(s);
- The permit number of any NPDES permit(s) for any other discharge(s) (including any other storm water discharges or any nonstorm water discharges) from the site;

- An indication of whether the operator has existing sampling data that describe the concentration of pollutants in storm water discharges. Existing data should not be included as part of the NOI and should not be submitted unless and until requested by the EPA; and
- An estimate of project start date and completion dates, estimates of the number of acres of the site on which soil will be disturbed, and a certification that a storm water pollution prevention plan has been prepared for the site in accordance with the permit and that such plan complies with approved state and/or local sediment and erosion plans or permits and/or storm water management plans or permits. A copy of the plans or permits should not be included with the NOI submission, and should not be submitted unless and until requested by the EPA.

In addition to submitting the NOI to the EPA, facilities operating under approved state or local sediment and erosion plans, grading plans, or storm water management plans are required to submit signed copies of the NOI to the state or local agency approving such plans. Failure to do so constitutes a violation of the permit. The NOI must be signed in accordance with the signatory requirements of 40 CFR Section 122.22. A complete description of these signatory requirements is provided in the instructions accompanying the NOI.

5.5.3 Special Conditions

The construction general permits contain special conditions with regard to nonstorm water discharges and releases of reportable quantities of hazardous substances and oil, each of which are discussed in this subsection.

Prohibition on Nonstorm Water Discharges. The construction general permit does not authorize nonstorm water discharges that are mixed with storm water except for specific classes of nonstorm water discharges specified in the permit. Nonstorm water discharges that can be authorized under the permit include:
- Discharges from firefighting activities;
- Fire hydrant flushings;
- Waters used to wash vehicles or control dust in accordance with permit requirements;
- Potable water sources including waterline flushings;
- Irrigation drainage;
- Routine external building washdown that does not use detergents;
- Pavement washwaters where spills or leaks of toxic or hazardous

materials have not occurred (unless all spilled material has been removed) and where detergents are not used;
- Air conditioning condensate;
- Springs; and
- Foundation or footing drains where flows are not contaminated with process materials such as solvents.[76]

To be authorized under the general permit, sources of nonstorm water (except flows from firefighting activities) must be specifically identified in the storm water pollution prevention plan (SWPPP) prepared for the facility. Where such discharges occur, the SWPPP must also identify and ensure the implementation of appropriate pollution prevention measures for the nonstorm water components of the discharge. The permit does not require pollution prevention measures to be identified and implemented for nonstorm water flows from firefighting activities since these flows will usually occur as unplanned emergency situations where it is necessary to take immediate action to protect the public.

Where a storm water discharge is mixed with process wastewaters or other sources of nonstorm water prior to discharge, and the discharge is currently not authorized by an NPDES permit, the discharge cannot be covered by the permit and the discharger should (1) submit the appropriate application forms (Forms 1 and 2C) to obtain permit coverage or (2) discontinue the discharge.

The permit authorizes a storm water discharge associated with industrial activity from a construction site that is mixed with a storm water discharge from an industrial source other than construction, only if (1) the industrial source other than construction is located on the same site as the construction activity; and (2) storm water discharges from where the construction activities are occurring are in compliance with the terms of the construction general permit.

Releases of Reportable Quantities of Hazardous Substances and Oil.
The construction general permit provides that the discharge of hazardous substances or oil from a facility must be eliminated or minimized in accordance with the SWPPP developed for the facility. Where a permitted storm water discharge contains a hazardous substance or oil in an amount equal to or in excess of a reporting quantity established under 40 C.F.R part 110, 40 CFR § part 117, or 40 C.F.R part 302, during a 24-hour period, the permit requires the following actions:
- The permittee must notify the National Response Center at 800-424-

---

[76] These discharges are consistent with the allowable classes of nonstorm water discharges to municipal separate storm sewer systems. 40 CFR § 122.26(d)(iv)(D).

8802 in accordance with the requirements of 40 CFR Part 110, 40 CFR Part 117, and 40 CFR Part 302, upon knowledge of the discharge;
- The permittee must modify the SWPPP for the facility within 14 calendar days of knowledge of the release to provide (1) a description of the release, (2) the date of the release and (3) the circumstances leading to the release. In addition, the permittee must modify the plan, as appropriate, to identify measures to prevent the reoccurrence of such releases and to respond to such releases.
- Within 14 calendar days of the knowledge of the release, the permittee must submit to the EPA (1) a written description of the release (including the type and estimated amount of material released), (2) the date that such release occurred, (3) the circumstances leading to the release, and (4) any steps to be taken to modify the SWPPP for the facility.

Where a discharge of a hazardous substance or oil in excess of reporting quantities is caused by a nonstorm water discharge (e.g., a spill of oil into a separate storm sewer), the spill is not authorized by the permit. The discharger must report the spill as required under 40 CFR § part 110. In the event of a spill, the requirements of section 311 of the CWA and otherwise applicable provisions of sections 301 and 402 of the CWA continue to apply.

5.5.4 Endangered Species Protection

Based on consultation with the U.S. Fish and Wildlife Service (FWS) and the National Marine Fisheries Service (NMFS), the EPA placed conditions in the construction general permit to ensure the activities regulated by it are protective of species that are listed under the Endangered Species Act (ESA) as endangered or threatened, and listed species habitat that is designated under the ESA as critical habitat.[77] Coverage under the permit is available for construction projects only if:
1. The storm water discharges and storm water discharge-related activities are not likely to adversely affect listed species or critical habitat (Part I.B.3.e.(2)(a)); or
2. Formal or informal consultation with the FWS and NMFS under Section 7 of the ESA has been concluded which addresses the effects of the applicant's storm water discharges and storm water discharge-related activities on listed species and critical habitat and the consultation results in either a no jeopardy opinion or a written concurrence by the FWS

---

[77] See chapter 16 for a complete discussion of the Endangered Species Act and consultation requirements.

and/or NMFS on a finding that the applicant's storm water discharges and storm water discharge-related activities are not likely to adversely affect listed species or critical habitat;[78] or

3. The applicant's construction activities are covered by a permit under Section 10 of the ESA and that permit addresses the effects of the applicant's storm water discharges and storm water discharge-related activities on listed species and critical habitat (Part I.B.3.e.(2)(c)); or

4. The applicant's storm water discharges and storm water discharge-related activities were already addressed in another operator's certification of eligibility under Part I.B.3.e.(2)(a), (b), or (c) which included the applicant's project area.[79]

If adverse effects are likely, the applicant would have to meet one of the eligibility requirements of Part I.B.3.e.(2)(b)–(d) to receive permit coverage (items 2, 3 or 4 listed above).

The permit also requires that applicants consider effects to listed species and critical habitat when developing SWPPPs and require that those plans include measures, as appropriate, to protect those resources. Failure by permittees to abide by measures in the SWPPPs to protect species and critical habitat may invalidate permit coverage.

Addendum A of the permit provides procedures for making the determination of permit eligibility with regard to the endangered species protection requirements. The EPA Director may also require any existing permittee or applicant to provide documentation of eligibility for the permit using the procedures in Addendum A where the EPA or the FWS determines that there is a potential impact on endangered or threatened species or a critical habitat. The instructions in Addendum A require that the applicant ascertain:

1. Whether the construction activities would occur in critical habitat;

2. Whether listed species are in the project area; and

3. Whether the applicant's storm water discharges and discharge-related activities are likely to adversely affect listed species or critical habitat.

"Discharge-related activities" include activities which cause point source storm water pollutant discharges including but not limited to excavation, site development, and other surface disturbing activities, and measures to control, reduce or prevent storm water pollution including the siting, construction and operation of BMPs. The "project area" includes:

1. Area(s) on the construction site where storm water discharges originate and flow towards the point of discharge into the receiving waters (this

---

[78] Consultation under Section 7 of the ESA may occur in the context of another federal action (e.g., a Section 7 consultation was performed for issuance of a wetlands dredge and fill permit for the project, or as part of a NEPA review).

[79] By certifying eligibility under Part I.B.3.e.(2)(d) of the permit, the applicant agrees to comply with any measures or controls upon which the other operator's certification under Part I.B.3.e.(2)(a), (b) or (c) was based.

includes the entire area or areas where excavation, site development, or other ground disturbance activities occur), and the immediate vicinity;

2. Area(s) where storm water discharges flow from the construction site to the point of discharge into receiving waters;

3. Area(s) where storm water from construction activities discharges into the receiving waters and the area(s) in the immediate vicinity of the point of discharge; and

4. Area(s) where storm water BMPs will be constructed and operated, including any area(s) where storm water flows to and from BMPs.

The project area will vary with the size and structure of the construction activity, the nature and quantity of the storm water discharges, the measures (including BMPs) to control storm water runoff, and the type of receiving waters.

Addendum A also contains information on where to find information on listed and proposed species organized by state and county to assist applicants in determining if further inquiry is necessary as to whether listed species are present in the project area. Applicants can check the Office of Wastewater Management's website (http://www.epa.gov/owm). Permit applicants can also get updated species information for their county by contacting the appropriate FWS or NMFS office, or by calling the EPA Storm Water Hotline (1-800-245-6510) or EPA Regional storm water coordinator.

5.5.5 Historic Property Protection

The National Historic Preservation Act (NHPA)[80] establishes a national historic preservation program for the identification and protection of historic properties and resources. Under the NHPA, identification of historic properties is coordinated by the State Historic Preservation Officers (SHPOs), Tribal Historic Preservation Officers (THPOs), or other Tribal Representatives (in the absence of a THPO). Section 106 of the NHPA requires federal agencies to take into account the effects of their actions on historic properties that are listed or eligible for listing on the National Register of Historic Places. In order to be eligible for the construction general permit, the permittee must not adversely affect a property that is listed or is eligible for listing on the National Historic Register maintained by the Secretary of the Interior.[81] When applying for

---

[80] 16 U.S.C. § 470 et seq.
[81] Note that the construction general permit reissued for Region 6 did not impose the historic preservation requirements contained in the construction general permits issued for the other EPA Regions. See 63 *Fed. Reg.* 36490 (July 6, 1998).

permit coverage, applicants are required to certify in the NOI that they are in compliance with the historic preservation requirements. If another operator has certified eligibility for the project in his NOI (or at least the portion of the project that the applicant will be working on), the applicant will usually be able to rely on the other operator's certification of project eligibility and not have to repeat the process. The EPA created this "coat tail" eligibility option for protection of historic places to allow the site developer/owner to obtain upfront "clearance" for a project, thereby avoiding duplication of effort by his or her contractors and unnecessary delays in construction.

5.5.6 Storm Water Pollution Prevention Plan (SWPPP)

All construction permit applicants must develop a storm water pollution prevention plan (SWPPP) that focuses on two major tasks: (1) providing a site description that identifies sources of pollution to storm water discharges associated with industrial activity from the facility and (2) identifying and implementing appropriate measures to reduce pollutants in storm water discharges to ensure compliance with the terms and conditions of the permit. The plan must be completed prior to the submittal of a NOI to be covered under the permit and updated as appropriate.

5.5.7 Site Description

The first part of the SWPPP requires an evaluation of the sources of pollution at a specific construction site. The plan must identify potential sources of pollution that may reasonably be expected to affect the quality of storm water discharges from the construction site. In addition, the source identification components for SWPPPs must provide a description of the site and the construction activities. At a minimum, SWPPPs must include the following information:
- A description of the nature of the construction activity. This would typically include a description of the ultimate use of the project (e.g., low-density residential, shopping mall, highway).
- A description of the intended sequence of major activities that disturb soils for major portions of the site (e.g., grubbing, excavation, grading).
- Estimates of the total area of the site and the total area of the site that is expected to be disturbed by excavation, grading, or other activities. Where the construction activity is to be staged, it may be appropriate to describe areas of the site that will be disturbed at different stages of the construction process.

- Estimates of the runoff coefficient of the site after construction activities are completed as well as existing data describing the quality of any discharge from the site or the soil. The runoff coefficient is defined as the fraction of total rainfall that will appear at the conveyance as runoff. Runoff coefficients can be estimated from site plan maps, which provide estimates of the area of impervious structures planned for the site and estimates of areas where vegetation will be precluded or incorporated. Runoff coefficients are one tool for evaluating the volume of runoff that will occur from a site when construction is completed. These coefficients assist in evaluating pollutant loadings, potential hydraulic impacts to receiving waters, and flooding impacts. They are also used for sizing of post-construction storm water management measures.
- A site map indicating drainage patterns and approximate slopes anticipated after major grading activities, areas of soil disturbance; an outline of areas that will not be disturbed; the location of major structural and nonstructural controls identified in the plan; the location of areas where stabilization practices are expected to occur; the location of surface waters (including wetlands); and locations where storm water is discharged to a surface water. Site maps should also include other major features and potential pollutant sources, such as the location of impervious structures and the location of soil piles during the construction process.
- The name of the receiving water(s), and areal extent of wetland acreage at the site.

5.5.8 Controls to Reduce Pollutants

The SWPPP must describe and ensure the implementation of practices that will be used to reduce the pollutants in storm water discharges from the site and assure compliance with the terms and conditions of the permit. Permittees are required to develop a description of four classes of controls appropriate for inclusion in the facility's SWPPP, and implement those controls in accordance with the plan. The description of controls must address:
1. Erosion and sediment controls;
2. Storm water management;
3. A specified set of other controls; and
4. Any applicable procedures and requirements of state and local sediment and erosion plans or storm water management plans.
The SWPPP must also clearly describe the intended sequence of major activities and when, in relation to the construction process, the control will be implemented. The description of the intended sequence of major

activities will typically describe the intended staging of activities on different parts of the site.

Erosion and Sediment Controls. The requirements for erosion and sediment controls for construction activities in the permit have three goals: (1) To divert upslope water around disturbed areas of the site; (2) to limit the exposure of disturbed areas to the shortest duration possible; and (3) to remove sediment from storm water before it leaves the site. Erosion and sediment controls include both stabilization practices and structural practices.

Stabilization Practices. The SWPPP must include a description of interim and permanent stabilization practices, including site-specific scheduling of the implementation of the practices. The plan should ensure that existing vegetation is preserved where attainable and that disturbed portions of the site are stabilized as quickly as possible. Stabilization practices are the first line of defense for preventing erosion. They include temporary seeding, permanent seeding, mulching, geotextiles, sod stabilization, vegetative buffer strips, protection of trees, preservation of mature vegetative buffer strips, as well as other appropriate measures. Temporary stabilization practices may be the single most important factor in reducing erosion at construction sites.

Stabilization also involves preserving and protecting selected trees that were on the site prior to development. Mature trees have extensive canopy and root systems, which help to hold soil in place. Shade trees also keep soil from drying rapidly and becoming susceptible to erosion. Measures taken to protect trees can vary significantly, from simple measures such as installing tree fencing around the drip line and installing tree armoring, to more complex measures such as building retaining walls and tree wells.

Since stabilization practices play such an important role in preventing erosion, it is critical that they are rapidly employed in appropriate areas. The construction general permit requires, except in certain situations, that stabilization measures be initiated on disturbed areas as soon as practicable, but no more than 14 days after construction activity on a particular portion of the site has temporarily or permanently ceased. Exceptions to this requirement are situations where construction activities will resume on a portion of the site within 21 days from when the construction activities ceased; and the initiation of stabilization measures is precluded by snow cover, in which case, stabilization measures must be initiated as soon as practicable.

Structural Practices. The SWPPP must include a description of structural practices to the degree economically attainable, to divert flows from exposed soils, store flows, or otherwise limit runoff and the discharge of pollutants from exposed areas of the site. Structural controls are necessary because vegetative controls cannot be employed at areas of the site that are continually disturbed and because a finite time period is required before vegetative practices are fully effective. Options for such controls include silt fences, earth dikes, drainage swales, check dams, subsurface drains, pipe slope drains, level spreaders, storm drain inlet protection, rock outlet protection, sediment traps, rock outlet protection, reinforced soil retaining systems, gabions, and temporary or permanent sediment basins. Structural measures should be placed on upland soils to the degree possible.

For sites with more than 10 disturbed acres at one time that are served by a common drainage location, a temporary or permanent sediment basin providing 3,600 cubic feet of storage per acre drained, or equivalent control measures (such as suitably sized dry wells or infiltration structures), must be provided where economically attainable until final stabilization of the site has been accomplished. Flows from offsite areas and flows from onsite areas that are either undisturbed or have undergone final stabilization may be diverted around both the sediment basin and the disturbed area. The requirement to provide 3,600 cubic feet of storage area per acre drained does not apply to such diverted flows.

For the drainage locations which serve more than 10 disturbed acres at one time and where a sediment basin providing storage or equivalent controls for 3,600 cubic feet per acre drained is not economically attainable, smaller sediment basins or sediment traps should be used. At a minimum, silt fences, or equivalent sediment controls are required for all sideslope and downslope boundaries of the construction area. Diversion structures should be used on upland boundaries of disturbed areas to prevent runon from entering disturbed areas.

For drainage locations serving 10 or fewer acres, smaller sediment basins or sediment traps should be used and at a minimum, silt fences, or equivalent sediment controls are required for all sideslope and downslope boundaries of the construction area. Alternatively, the permittee may provide a sediment basin providing storage for 3,600 cubic feet of storage per acre drained. Diversion structures should be used on upland boundaries of disturbed areas to prevent runon from entering disturbed areas.

Storm Water Management. The plan must include a description of "storm water management" measures.[82] The permit addresses only the installation of storm water management measures and not the ultimate operation and maintenance of such structures after the construction

---

[82] For the purpose of the special requirements for construction activities, the term "storm water management" measures refers to controls that will primarily reduce the discharge of pollutants in storm water from sites after completion of construction activities.

activities have been completed and the site has undergone final stabilization. Permittees are responsible only for the installation and maintenance of storm water management measures prior to final stabilization of the site and are not responsible for maintenance after storm water discharges associated with construction activities have been eliminated from the site. However, this does not release a facility from responsibilities to operate and maintain storm water management systems in perpetuity after final stabilization in accordance with the requirements set forth by local environmental permitting actions.

Storm water management measures that are installed during the construction process can control the volume of storm water discharged and peak discharge velocities, as well as reduce the amount of pollutants discharged after the construction operations have been completed. Reductions in peak discharge velocities and volumes can also reduce pollutant loads, as well as reduce physical impacts such as stream bank erosion and stream bed scour. Storm water management measures that mitigate changes to predevelopment runoff characteristics assist in protecting and maintaining the physical and biological characteristics of receiving streams and wetlands.

Options for storm water management measures that are to be evaluated in the development of plans include infiltration of runoff on site; flow attenuation by use of open vegetated swales and natural depressions; storm water retention structures and storm water detention structures (including wet ponds); and sequential systems that combine several practices. Structural measures should be placed on upland soils to the degree attainable. The installation of such devices may be subject to Section 404 of the CWA if the devices are placed in wetlands (or other waters of the United States).

The SWPPP must include an explanation of the technical basis used to select the practices to control pollution where flows exceed pre-development levels. The explanation of the technical basis for selecting practices should address how a number of factors were evaluated, including the pollutant removal efficiencies of the measures, the costs of the measure, site-specific factors that will affect the application of the measures, the economic achievability of the measure at a particular site, and other relevant factors. Proper selection of a technology depends on site factors and other conditions.[83]

Other Controls. Other controls to be addressed in the SWPPP for construction activities require that no nonstorm water solid materials, including building material wastes shall be discharged at the site, except

---

[83] See U.S.E.P.A., *Storm Water Management for Construction Activities: Developing Pollution Prevention Plans and Best Management Practices* (1992).

as authorized by a Section 404 permit. The permit requires that offsite vehicle tracking of sediments and the generation of dust be minimized. This can be accomplished by measures such as providing gravel or paving at access entrance and exit drives, parking areas, and unpaved roads on the site carrying significant amounts of traffic; providing entrance wash racks or stations for trucks; and/or providing street sweeping. In addition, the permit requires that the SWPPP ensure and demonstrate compliance with applicable state and/or local sanitary sewer, septic system, and waste disposal regulations.

State and Local Controls. Many states and municipalities have developed sediment and erosion control requirements for construction activities. A significant number of states and municipalities have also developed storm water management controls. The general permit requires that SWPPPs for facilities that discharge storm water associated with industrial activity from construction sites include procedures and requirements of state and local sediment and erosion control plans or storm water management plans. Permittees are required to provide a certification that their SWPPPs reflect requirements related to protecting water resources that are specified in state or local sediment and erosion plans or storm water management plans. In addition, permittees are required to amend their SWPPPs to reflect any change in a sediment and erosion site plan or site permit or storm water management site plan or site permit approved by state or local officials for which the permittee receives written notice. Where such amendments are made, the permittee must provide a recertification that the SWPPP has been modified. This provision does not apply to provisions of master plans, comprehensive plans, nonenforceable guidelines, or technical guidance documents, but rather to site-specific state or local permits or plans.

5.5.9 Maintenance

Erosion and sediment controls can become ineffective if they are damaged or not properly maintained. Maintenance of controls has been identified as a major part of effective erosion and sediment programs. SWPPPs must contain a description of prompt and timely maintenance and repair procedures addressing all erosion and sediment control measures (e.g., sediment basins, traps, silt fences), vegetation, and other measures identified in the site plan to ensure that such measures are kept in good and effective operating condition.

5.5.10 Inspections

Procedures in the SWPPP must provide that specified areas on the site are inspected by qualified personnel provided by the discharger a minimum of once every seven calendar days and within 24 hours after any storm event of greater than 0.25 inches. Areas of the site that must be observed during such inspections include disturbed areas, areas used for storage of materials that are exposed to precipitation, structural control measures, and locations where vehicles enter or exit the site. Where sites have been temporarily or finally stabilized, the inspection must be conducted at least once every month.

Disturbed areas and areas used for storage of materials that are exposed to precipitation must be inspected for evidence of, or the potential for, pollutants entering the runoff from the site. Erosion and sediment control measures identified in the plan must be observed to ensure that they are operating correctly. Observations can be made during wet or dry weather conditions. Where discharge locations or points are accessible, they must be inspected to ascertain whether erosion control measures are effective in preventing significant impacts to receiving waters. This can be done by inspecting receiving waters to see whether any signs of erosion or sediment are associated with the discharge location. Locations where vehicles enter or exit the site must be inspected for evidence of offsite sediment tracking. Based on the results of the inspection, the site description and the pollution prevention measures identified in the plan must be revised as soon as possible after an inspection that reveals inadequacies.

The inspection and plan review process must provide for timely implementation of any changes to the plan within 7 calendar days following the inspection. An inspection report that summarizes the scope of the inspection, name(s) and qualifications of personnel conducting the inspection, the dates of the inspection, major observations relating to the implementation of the SWPPP, and actions taken must be retained as part of the plan for at least three years after the date of inspection. The report must be signed in accordance with the signatory requirements in the Standard Conditions section of the permit.

5.5.11 Nonstorm Water Discharges

The construction general permit may authorize storm water discharges from construction activities that are mixed with discharges from firefighting activities, fire hydrant flushings, waters used to wash vehicles or control dust in accordance with efforts to minimize offsite sediment tracking, potable water sources including waterline flushings, irrigation drainage from watering vegetation, routine exterior building washdown that does not use detergents, pavement washwaters where spills or leaks of toxic or hazardous materials have not occurred (unless all spilled material has been removed) and where detergents are not used, air conditioning condensate, springs, and foundation or footing drains where flows are not contaminated with process materials such as solvents, provided the nonstorm water component of the discharge is specifically identified in the SWPPP. In addition, the plan must identify and ensure the implementation of appropriate pollution prevention measures for each of the nonstorm water component(s) of the discharge.

5.5.12 Signature and Plan Review

The signature and plan review requirements of the SWPPP are as follows:
- The plan must be signed by all permittees for a site in accordance with the signatory requirements in the Standard Permit Conditions section of the permit, and must be retained on site at the facility that generates the storm water discharge.
- The permittee must make plans available, upon request, to the EPA, and state or local agency approved sediment and erosion plans, grading plans, or storm water management plans. In the case of a storm water discharge associated with industrial activity that discharges through a municipal separate storm sewer system with an NPDES permit, permittees must make plans available to the municipal operator of the system upon request.
- The EPA may notify the permittee at any time that the plan does not meet one or more of the minimum requirements. Within 7 days of such notification from the EPA (or as otherwise requested by EPA), the permittee must make the required changes to the plan and submit to the EPA a written certification that the requested changes have been made.

5.5.13 Amendments to Plan

The permittee must amend the SWPPP whenever there is a change in design, construction, operation, or maintenance, that has a significant effect on the potential for the discharge of pollutants to waters of the United States or to municipal separate storm sewer systems. The plan must also be amended if it proves to be ineffective in eliminating or significantly minimizing pollutants in the storm water discharges from the construction activity. In addition, the plan must be amended to identify any new contractor and/or subcontractor that will implement a measure of the storm water pollution prevention plan.

5.5.14 Contractors

The SWPPP must clearly identify for each measure identified in the plan, the contractor(s) and/or subcontractor(s) that will implement the measure. All contractors and subcontractors identified in the plan must sign a copy of the following certification statement before conducting any professional service at the site identified in the SWPPP:

"I certify under penalty of law that I understand the terms and conditions of the general National Pollutant Discharge Elimination System (NPDES) permit that authorizes the storm water discharges associated with industrial activity from the construction site identified as part of this certification."

All certifications must be included in the SWPPP.

5.5.15 Record Retention

The permittee is required to retain records or copies of all reports required by the permit, including SWPPPs and records of all data used to complete the NOI to be covered by the permit, for a period of at least three years from the date of final stabilization. This period may be extended by request of the Director.

5.5.16 Notice of Termination

A discharger may submit a Notice of Termination (NOT) to the EPA in two sets of circumstances: (1) after a site has undergone final stabilization and the facility no longer discharges storm water associated with industrial activity from a construction site and (2) when the permittee has transferred operational control to another permittee and is no longer an operator for the site. NOTs must be submitted using the form provided by the EPA Regional Director (or a photocopy thereof).

The permit defines final stabilization for the purpose of submitting an NOT as occurring when all soil-disturbing activities are completed and a uniform perennial vegetative cover with a density of 70 percent for the unpaved areas and areas not covered by permanent structures has been established or equivalent stabilization measures have been employed. Equivalent stabilization measures include permanent measures other than establishing vegetation, such as the use of rip-rap, gabions, and/or geotextiles. A copy of the NOT, and instructions for completing the NOT, are provided the construction general permit. Notices of Termination are to be sent to the following address: Storm Water Notice of Termination (4203), 401 M Street, S.W., Washington, D.C. 20460. The NOT must be signed by the appropriate individual in accordance with the signatory requirements of 40 CFR Section 122.22.

5.5.17 Answers to Commonly Asked Questions

In the "Summary of Responses to Comments on the Proposed Permit" section of the reissuance of the general permit for construction activities,[84] the EPA provided answers to some of the more common questions on the storm water permitting program for discharges from construction sites. Some of these questions and EPA responses are provided here. However, be aware these answers are fairly broad and may not take into account all possible scenarios at a particular construction site.

How Do I Know If I Need a Permit?
You need a storm water permit if you can be considered an "operator" of the construction activity that would result in the "discharge of storm water associated with construction activity." You must become a permittee if you meet either of the following two criteria:
1. You have operational control of construction project plans and specifications, including the ability to make modifications to those plans and specifications; or
2. You have day-to-day operational control of those activities at a project which are necessary to ensure compliance with a SWPPP for the site or other permit conditions (e.g., you are authorized to direct workers at a site to carry out activities required by the SWPPP or comply with other permit conditions).

There may be more than one party at a site performing the tasks relating to "operational control" as defined above. Depending on the site and the relationship between the parties (e.g., owner, developer), there can either be a single party acting as site operator and consequently be responsible for obtaining permit coverage, or there can be two or more operators with all needing permit coverage. The following are three general operator scenarios (variations on any of the three are possible as the number of "owners" and contractors increases):
1. *Owner as sole permittee.* The property owner designs the structures for the site, develops and implements the SWPPP, and serves as general

---

[84] 63 *Fed. Reg.* 7858 (Feb. 17, 1998).

contractor (or has an on-site representative with full authority to direct day-to-day operations). He or she may be the only party that needs a permit, in which case everyone else on the site may be considered subcontractors and not need permit coverage.

2. *Contractor as sole permittee.* The property owner hires a construction company to design the project, prepare the SWPPP, and supervise implementation of the plan and compliance with the permit (e.g., a "turnkey" project). Here, the contractor would be the only party needing a permit. It is under this scenario that an individual having a personal residence built for his or her own use (e.g., not those to be sold for profit or used as rental property) would not be considered an operator. The EPA believes that the general contractor--being a professional in the building industry--should be the entity, rather than the individual, who is better equipped to meet the requirements of both applying for permit coverage and developing and properly implementing a SWPPP. However, individuals would meet the definition of "operator" and require permit coverage in instances where they perform general contracting duties for construction of their personal residences.

3. *Owner and contractor as co-permittees.* The owner retains control over any changes to site plans, SWPPPs, or storm water conveyance or control designs; but the contractor is responsible for overseeing actual earth-disturbing activities and daily implementation of SWPPP and other permit conditions. In this case, both parties may need coverage. However, you are probably not an operator and subsequently do not need permit coverage if:

—you are a subcontractor hired by, and under the supervision of, the owner or a general contractor (i.e., if the contractor directs your activities onsite, you probably are not an operator); or

—your activities onsite result in earth disturbance and you are not legally a subcontractor, but a SWPPP specifically identifies someone other than you (or your subcontractor) as the party having operational control to address the impacts your activities may have on storm water quality (i.e., another operator has assumed responsibility for the impacts of your construction activities). In addition, for purposes of the construction general permit and determining who is an operator, "owner" refers to the party that owns the structure being built. Ownership of the land where construction is occurring does not necessarily imply the property owner is an operator (e.g., a landowner whose property is being disturbed by construction of a gas pipeline). Likewise, if the erection of a structure has been contracted for, but possession of the title or lease to the land or structure is not to occur until after construction, the would-be owner may not be considered an operator (e.g., having a house built by a residential homebuilder).

My Project Will Disturb Less Than Five Acres, But It May Be Part of a "Larger Common Plan of Development or Sale." How Can I Tell and What Must I Do?

If your smaller project is part of a larger common plan of development or sale that collectively will disturb five or more acres (e.g., you are building on six half-acre residential lots in a 10-acre development or are putting in a parking lot in a large retail center) you need permit coverage. The "plan" in a common plan of development or sale is broadly defined as any announcement or piece of documentation (including a sign, public notice or hearing, sales pitch, advertisement, drawing, permit application, zoning request, computer design, etc.) or physical demarcation (including boundary signs, lot stakes, surveyor markings, etc.) indicating construction activities may occur on a specific plot. You must still meet the definition of operator in order to be required to get permit coverage, regardless of the acreage you personally disturb. As a subcontractor, it is unlikely you would need a permit.

For some situations where less than five acres of the original common plan of development remain undeveloped, a permit may not be needed for the construction projects "filling in" the last parts of the common plan of development. A case in which a permit would not be needed is where several empty lots totaling less than five acres remain after the rest of the project had been completed, provided that stabilization had also been completed for the entire project. However, if the total area of all the undeveloped lots in the original common plan of development was more than five acres, a permit would be needed.

When Can You Consider Future Construction on a Property to Be Part of a Separate Plan of Development or Sale?

In many cases, a common plan of development or sale consists of many small construction projects that collectively add up to five or more acres of total disturbed land. For example, an original common plan of development for a residential subdivision might lay out the streets, house lots, and areas for parks, schools and commercial development that the developer plans to build or sell to others for development. All these areas would remain part of the common plan of development or sale until the intended construction occurs. After this initial plan is completed for a particular parcel, any subsequent development or redevelopment of that parcel would be regarded as a new plan of development, and would then be subject to the five-acre cutoff for storm water permitting.

Must Every Permittee Have His or Her Own Separate SWPPP or Is a Joint Plan Allowed?

The only requirement is that there be at least one SWPPP for a site which incorporates the required elements for all operators, but there can be separate plans if individual permittees so desire. The EPA encourages permittees to explore possible cost savings by having a joint SWPPP for several operators. For example, the prime developer could assume the inspection responsibilities for the entire site, while each homebuilder shares in the installation and maintenance of sediment traps serving common areas.

If a Project Will Not Be Completed Before the Permit Expires, How Can I Keep Permit Coverage?

If the permit is reissued or replaced with a new one before the current one expires, you will need to comply with whatever conditions the new permit requires in order to transition coverage from the old permit. This usually includes submitting a new NOI. If the permit expires before a replacement permit can be issued, the permit will be administratively "continued." You will be required to submit an NOI for coverage under the continued permit, until the earliest of:
1. The permit being reissued or replaced;
2. Submittal of a Notice of Termination (NOT);
3. Issuance of an individual permit for your activity; or
4. The EPA Director issues a formal decision not to reissue the permit, at which time you must seek coverage under an alternate permit.

5.6 General Permit for Municipal Separate Storm Sewer Systems

The EPA or NPDES-authorized states may issue system-wide or jurisdiction-wide permits covering all discharges from a municipal separate storm sewer system. Operators of large and medium municipal separate storm sewer systems are required to submit a two-part permit application designed to facilitate development of site-specific permit conditions. The general permit applications requirements for covered municipalities are as follows:[85]

---

[85] See also *Guidance Manual for the Preparation of Part 1 of the NPDES Permit Application for Discharges from Municipal Separate Storm Sewer Systems* (available from NTIS at (703) 487—4650, order number PB 92—114578); *Guidance Manual for the Preparation of Part 2 of the NPDES Permit Applications for Discharges from Municipal Separate Storm Sewers Systems* (available from the Storm Water Hotline by calling (800) 245-6510).

5.6.1 Part 1 Application Requirements

1. General information (name, address, etc.).
2. Existing legal authority and any additional authorities needed.
3. Source identification information.
4. Discharge characterization, including:
—Monthly mean rain and snowfall estimates.
—Existing quantitative data on volume and quality of storm water discharges.
—A list of receiving water bodies and existing information on the impacts of receiving waters.
—Field screening analysis for illicit connections and illegal dumping.
5. Characterization plan identifying representative outfalls for further sampling in Part 2.
6. Description of existing management programs to control pollutants from the municipal separate storm sewer and to identify illicit connections.
7. Description of financial budget and resources currently available to complete Part 2.

5.6.2 Part 2 Application Requirements

1. Demonstration of adequate legal authority to control discharges, prohibit illicit discharges, require compliance, and carry out inspections, surveillance, and monitoring.
2. Source identification indicating the location of any major outfalls and identifying facilities that discharge storm water associated with industrial activity through the municipal separate storm sewer.
3. Discharge characterization data including:
—Quantitative data from five to 10 representative locations in approved sampling plans.
—For selected conventional pollutants and heavy metals, estimates of the annual pollutant load and event mean concentration of system discharges.
—Proposed schedule to provide estimates of seasonal pollutant loads and the mean concentration for certain detected constituents in a representative storm event.
—Proposed monitoring program for representative data collection.
4. Proposed management program including descriptions of:
—Structural and source control measures that are to be implemented to reduce pollutants in runoff from commercial and residential areas.

—Program to detect and remove illicit discharges.

—Program to monitor and control pollutants from municipal landfills; hazardous waste treatment, disposal, and recovery facilities; EPCRA Section 313 facilities; and other priority industrial facilities.

—Program to control pollutants in construction site runoff.

5. Estimated reduction in loadings of pollutants as a result of the management program.

6. Fiscal analysis of necessary capital and operation and maintenance expenditures.

The permit application requirements provide municipal applicants an opportunity to propose appropriate management programs to control pollutants in discharges from their municipal systems. This increases flexibility to develop appropriate permit conditions and ensures input from municipalities in developing appropriate controls.

5.6.3 Sampling Requirements for Municipal Separate Storm Sewer Systems

Both parts of the two-part application for operators of large and medium municipal separate storm sewer systems contain sampling requirements. Part 1 requires information characterizing discharges from the separate storm sewer system, including field screening sample data for identifying illicit/illegal connections. Part 2 requires sampling at representative locations and estimates of pollutant loadings for those sites. These sampling data are to be used to design a long-term storm water monitoring plan that will be implemented during the term of the permit. The sampling data that must be submitted in Parts 1 and 2 of municipal applications are as follows:

Part 1
- Monthly mean rainfall and snowfall estimates.
- Existing quantitative data on the depth and quality of storm water discharges.
- A list of receiving water bodies and existing information concerning known water quality impacts.
- Field screening analysis for illicit connections and illegal dumping.
- Identification of representative outfalls for further sampling in Part 2.

Part 2
- Quantitative data from five to 10 representative locations in approved sampling plans.

- Estimates of the annual pollutant load and event mean concentration (EMC) of system discharges.
- Proposed schedule to provide estimates of seasonal pollutant loads and the EMC for certain detected constituents in a representative storm event during the term of the permit.
- Proposed monitoring program for representative data collection during the term of the permit.

Municipal applicants are required to conduct sampling for both Parts 1 and 2 of their applications. In Part 1, municipalities must conduct a field screening analysis to detect illicit connections and illegal dumping into their storm sewer system. Where flow is observed during dry weather, two grab samples must be collected during a 24-hour period with a minimum of four hours between samples. These samples must be analyzed for pH, total chlorine, total copper, total phenol, and detergents (surfactants). Note that these are dry weather samples, rather than storm water samples. EPA's Guidance Manual for the Preparation of Part 1 of the NPDES Permit Applications for Discharges from Municipal Separate Storm Sewer Systems presents a description of conducting field screening sampling and provides a data sheet.

For Part 2 of the application, municipalities must submit grab (for certain pollutants) and flow-weighted sampling data from selected sites (five to 10 outfalls) for three representative storm events at least one month apart. The flow-weighted composite sample must be taken for either the entire discharge or the first three hours (if the event lasts longer than three hours). Municipal facilities are not required to collect grab samples within the first 30 minutes of a storm event.

In addition to submitting quantitative data for the application, municipalities must also develop programs for future sampling activities that specify sampling locations, frequency, pollutants to be analyzed, and sampling equipment. Where necessary (as determined by the municipality or if required by the permitting authority), responsibilities may also include monitoring industries connected to the municipality's storm sewers for compliance with their facility-specific NPDES permits. Refer to EPA's Guidance Manual for the Preparation of Part 1 of the NPDES Permit Applications for Discharges from Municipal Separate Storm Sewer Systems for information on how to develop municipal sampling programs.

5.7 Individual Permits

Operators of facilities with storm water discharges associated with industrial activity who do not obtain coverage under a general permit must submit an individual permit application. The permit applicant must first determine whether the state in which the discharging facility is located has an authorized NPDES program. If so, obtain a copy of the permit application from the state permitting agency. If not, obtain a copy of the permit application from the EPA Regional Office. This section describes the requirements for individual permit applications submitted to the EPA. It is important to note that because the requirements for individual permits issued by states with authorized NPDES programs may not, in all respects, be the same as the requirements for individual permits issued by the EPA, the permit applicant must contact the state permitting agency for further information.

5.7.1 Individual Permit Application Forms

The EPA requirements for individual permit application for most types of discharges composed of storm water associated with industrial activity are incorporated into Form 1 and Form 2F. Form 1 (EPA Form 3510-1) requires the applicant to provide the name and address of the facility; the facility type (SIC code); a map showing specified features, and other general information about the facility. Form 2F (EPA Form 3510-2F) requests information that can be used to evaluate the pollution potential of storm water discharges associated with industrial activity, including the following requirements:

- Site map showing topography and/or drainage areas and site characteristics.
- Estimate of impervious surface area and the total area drained by each outfall.
- Description of significant materials exposed to storm water, including current materials management practices.
- Certification that outfalls have been tested or evaluated for the presence of nonstorm water discharges that are not covered by a NPDES permit.
- Information on significant leaks and spills in the last 3 years.
- Quantitative testing data for the following parameters:
—Any pollutants limited in an effluent guideline to which the facility is subject
—Any pollutant listed in the facility's NPDES permit for process wastewater

—Oil and grease, pH, biochemical oxygen demand (BOD_5), chemical oxygen demand (COD), total suspended solids (TSS), total phosphorus, nitrate plus nitrite nitrogen, and total Kjeldahl nitrogen
—Certain pollutants known to be in the discharge
—Flow measurements or estimates
—Date and duration of storm event.

Additional Forms for Nonstorm Water Discharges. Where a storm water discharge associated with industrial activity is mixed with a nonstorm water component prior to discharge, an additional application form must be submitted. A complete permit application for a storm water discharge associated with industrial activity mixed with process wastewater includes Form 1, Form 2F, and Form 2C. Process wastewater is water that comes into direct contact with, or results from, the production or use of any raw material, intermediate product, finished product, by-product, waste product or wastewater. A complete permit application for a storm water discharge associated with industrial activity mixed with new sources or new discharges of nonstorm water (non—NPDES permitted discharges commencing after August 13, 1979) includes Form 1, Form 2F, and Form 2D. A complete permit application for a storm water discharge associated with industrial activity mixed with nonprocess wastewater includes Form 1, Form 2F, and Form 2E. Nonprocess wastewater includes noncontact cooling water and sanitary wastes which are not regulated by effluent guidelines or a new source performance standard, except discharges by educational, medical, or commercial chemical laboratories.

Application deadline/certification. Individual permit applications for a new discharge of storm water associated with industrial activity must be submitted 180 days before that facility commences industrial activity which may result in a discharge of storm water associated with that industrial activity. Applications submitted by industrial facilities must be certified by a responsible corporate officer as described in 40 CFR Section 122.22 (e.g., president, secretary, treasurer, vice-president of the corporation in charge of a principal business function).

5.7.2 Special Provisions for Construction Activities

The application requirements for operators of storm water discharges associated with industrial activity from construction sites include Form 1 and a narrative description of the following:

1. The location (including a map) and the nature of the construction activity;

2. The total area of the site and the area of the site that is expected to undergo excavation during the life of the permit;

3. Proposed measures, including best management practices, to control pollutants in storm water discharges during construction, including a brief description of applicable state and local erosion and sediment control requirements;

4. Proposed measures to control pollutants in storm water discharges that will occur after construction operations have been completed, including a brief description of applicable state and local storm water management controls;

5. An estimate of the runoff coefficient of the site and the increase in impervious area after the construction addressed in the permit application is completed, the nature of fill material and existing data describing the soil or the quality of the discharge; and

6. The name of the receiving water.

Submission of Form 2F is not required. The EPA has not developed a standardized form for the narrative information accompanying Form 1 that is required in individual applications for storm water discharges associated with industrial activity from construction sites. Permit applications for a new discharge of storm water associated with industrial activity from a construction site must be submitted at least 90 days before the date on which construction is to commence.

5.7.3 Exemptions for Mining; Oil and Gas Operations

The storm water regulations provide certain exemptions for storm water discharges associated with industrial activity from mining and oil and gas operations.[86]

Mining Operations. The permitting authority may not require a permit for discharges of storm water runoff from active or inactive mining operations composed entirely of flows which are from conveyances or systems of conveyances (including but not limited to pipes, conduits, ditches, and channels) used for collecting and conveying precipitation runoff unless the discharge has come into contact with any overburden, raw material, intermediate or finished products, by-products, or waste products located onsite. Inactive coal mining operations released from Surface Mining Coal Reclamation Act (SMCRA) performance bonds and noncoal mining

---

[86] 40 CFR § 122.26(a)(2); 40 CFR § 122.26(c)(1)(iii).

operations released from applicable state or federal reclamation requirements after December 17, 1990 are not required to submit permit applications.

Oil and Gas. The permitting authority may not require a permit for discharges of storm water runoff from oil and gas exploration, production, processing, or treatment operations or transmission facilities composed entirely of flows which are from conveyances or systems of conveyances (including but not limited to pipes, conduits, ditches, and channels) used for collecting and conveying precipitation runoff and which are not contaminated by contact with or that has not come into contact with any overburden, raw material, intermediate products, finished product, by-product, or waste products located on the site of such operations. In addition, the operator of an existing or new discharge composed entirely of storm water from an oil or gas exploration, production, processing, or treatment operation or transmission facility is not required to submit a permit application, unless the facility:

—-has had a discharge of storm water resulting in the discharge of a reportable quantity for which notification is or was required pursuant to 40 CFR Section 117.21 or 40 CFR Section 302.6 at any time since November 16, 1987; or

—has had a discharge of storm water resulting in the discharge of a reportable quantity for which notification is or was required pursuant to 40 CFR Section 110.6 at any time since November 16, 1987; or

—contributes to a violation of a water quality standard.

5.7.4 Sampling Requirements for Individual Permits

Storm water sampling provides a means for evaluating the environmental risk of the storm water discharge by identifying the types and amounts of pollutants present. Evaluating these data helps to determine the relative potential for the storm water discharge to contribute to water quality impacts or water quality standard violations. Further, storm water sampling data can be used to identify potential sources of pollutants. These sources can then be either eliminated or controlled more specifically by the permit.

Operators of large- and medium-size municipalities must submit storm water sampling data with their two-part permit applications for municipal separate storm sewer systems (MS4s).[87] Operators of facilities that have

---

[87] See section 5.6 for discussion of the two-part permit application for municipal separate storm sewer systems.

storm water discharges associated with industrial activity are also required to conduct storm water sampling as part of their applications for individual permits. Sampling data generally will not be required for a Notice of Intent (NOI) to be covered by a general permit; however, the general permit may require sampling during the term of the permit. State permitting authorities may also require sampling information for a NOI at their discretion and should, therefore, be consulted prior to submittal.

This section explains the sampling information required by Form 2F of individual permit applications. The types of information required for each section of Form 2F are outlined in Table 5.2 below. Sampling requirements for the two-part application for MS4s is explained in section 5.6.3.

Table 5.2. Form 2F Sampling Data

Facilities discharging storm water associated with industrial activity and submitting individual permit applications must supply the following information and sampling data on a completed application Form 2F:

Section	Requirement
2F—I	Outfall location(s), including longitude and latitude and receiving water(s)
2F—II	Facility improvements which may affect the discharges described in the application
2F—III	Site drainage map
2F—IVA	Estimates of impervious area within each outfall drainage area
2F—IVB	A narrative description of pollutant sources (i.e., onsite materials which maycome in contact with storm water runoff)
2F—IVC	Location and description of existing structural and nonstructural pollutant control measures
2F—VA	Certification that outfalls have been tested or evaluated for nonstorm water discharges
2F—VB	Description of method used for testing/evaluating presence of nonstorm water discharges
2F—VI	History of significant leaks or spills of toxic or hazardous pollutants at the facility within the last 3 years

Table 5.2. Form 2F Sampling Data (*Continued*)

Section	Requirement
2F—VII	Discharge characterization for all required pollutants
2F—VIII	Statement of whether biological testing for acute or chronic toxicity was performed and list of pollutants for which it was performed
2F—IX	Information on contract laboratories or consulting firms
2F—X	Certification that information supplied is accurate and complete

5.7.5 Site Drainage Map

Section III of Form 2F requires that a site drainage map be attached to the application. The site drainage map must show either topography or a delineation of the drainage area served by each outfall which discharges storm water associated with industrial activity if a topographic base map is not used. The delineation of the drainage area for each outfall that discharges storm water associated with industrial activity can be based on site observations which identify drainage patterns. Drainage patterns should be shown on the site drainage map so that runoff from each drainage area drains to a separate outfall.

The site drainage map must show the location and size (approximate for earthen structures) of all drainage conveyances or natural channels that convey or drain storm water off the applicant's property. The map must indicate whether the drainage system receiving the discharge is a natural water body, part of a municipal or nonmunicipal drainage system, or other system as applicable.

The following information must be provided and recorded on the map where appropriate:

1. Paved areas and buildings at the facility.
2. Past and present outdoor areas used for storage or disposal of significant materials.
3. Hazardous waste treatment, storage or disposal facilities, or accumulation areas (including those not requiring a RCRA permit).

4. Injection wells.

5. Material loading and access areas (e.g., loading docks and main truck routes on the facility property).

6. Areas where pesticides, herbicides, soil conditioners, and fertilizers are applied.

7. Structural control measures to reduce pollutants in storm water runoff.

8. Surface water bodies which receive storm water discharges from the facility.

During the preparation of a site drainage map, or the review of an existing one, emphasis should be placed on the identification of all inflow sources to ensure that inappropriate sources of nonstorm water entry are not present. The map should identify points of entry to the facility site storm water drain system, including catch basins, floor drains, and roof leaders.

The site drainage map required in Form 2F should show the location and an identifying number or name for each storm water outfall at the facility.

5.7.6 Outfalls To Be Monitored

Form 2F requires that applicants provide quantitative data for samples of storm water discharges associated with industrial activity. If a facility discharges storm water associated with industrial activity to a municipal separate storm sewer, then the facility should sample the storm water from the site prior to discharging to the municipal separate storm sewer. Storm water runoff from employee parking lots, administration buildings, and landscaped areas that is not mixed with storm water associated with industrial activity, or storm water discharges to municipal sanitary sewers, are not defined as storm water associated with industrial activity and hence do not need to be sampled.

Individual permit applicants must collect and analyze a grab sample taken within the first 30 minutes of the storm event and flow-weighted composite samples collected during the first three hours of discharge (or the entire discharge, if it is less than three hours) from each of the industrial storm water "point source" outfalls identified on the site drainage map submitted for Section III of Form 2F. Information from both types of samples is critical to fully evaluate the types and concentrations of pollutants present in the storm water discharge.

Storm water samples should be taken at a storm water point source. A "point source" is defined as "any discernible, confined, and discrete conveyance, including (but not limited to) any pipe, ditch, channel, tunnel, conduit, well, discrete fissure, container, rolling stock, concentrated animal feeding operation, landfill leachate collection system, vessel, or other floating craft from which pollutants are or may be discharged."[88] Included in the definition of storm water "point sources" is storm water from an industrial facility that enters, and is discharged through, a municipal separate storm sewer system (MS4). In short, most storm water discharges can be defined as point source discharges, since they ultimately flow into some kind of conveyance (e.g., a channel or swale).

The ideal sampling location would be the lowest point in the drainage area where a conveyance discharges storm water to U.S. waters or to an MS4. A sample point also should be easily accessible on foot in a location that will not cause hazardous sampling conditions. Ideally, the sampling site should be on the applicant's property or within the municipality's easement; if not, the field personnel should obtain permission from the owner of the property where the discharge outfall is located. Typical sampling locations may include the discharge at the end of a pipe, a ditch, or a channel.

However, logistical problems with sample locations may arise (e.g., nonpoint discharges, inaccessibility of discharge point, etc.). In many cases, it may be necessary to locate a sampling point further upstream of the discharge point (e.g., in a manhole or inlet). If the storm water at a selected location is not representative of a facility's total runoff, the facility may have to sample at several locations to best characterize the total runoff from the site. In situations where discharge points are difficult to sample for various reasons, the applicant should take the best sample possible and explain the conditions in the application.

5.7.7 Nonstorm Water Discharges

Form 2F requires applicants to certify that all outfalls that discharge storm water associated with industrial activity have been tested or evaluated for the presence of nonstorm water discharges. Applicants do not have to test for the presence of nonstorm water discharges already subject to a NPDES permit. This testing should be conducted during dry weather to avoid any flows of storm water through the conveyance. A narrative description of the method used to conduct dry weather evaluations and the date and the drainage points must be included in Section V.A of Form 2F.

The applicant should make every attempt to halt nonstorm water discharges to the storm sewer system unless the discharge is covered by a NPDES permit. If it is not feasible to halt the discharge of nonstorm water to the storm sewer system, and the discharge is not authorized by a

---

[88] 40 CFR § 122.2.

process wastewater or storm water permit, the applicant must submit either Form 2C (for a process water discharge) or Form 2E (for a nonprocess water discharge), which must accompany the individual storm water discharge permit application (Form 1 and Form 2F). The applicant should also check with state officials to see if alternate forms are required.

5.7.8 Storm Event Criteria

Individual permit applicants must include sampling data from at least one representative storm event. The permit application requirements establish specific criteria for the type of storm event that must be sampled:
1. The depth of the storm must be greater than 0.1 inch accumulation.
2. The storm must be preceded by at least 72 hours of dry weather.
3. Where feasible, the depth of rain and duration of the event should not vary by more than 50 percent from the average depth and duration.
These criteria were established to ensure that adequate flow would be discharged, allow some build—up of pollutants during the dry weather intervals; and ensure that the storm would be "representative," (i.e., typical for the area in terms of intensity, depth, and duration).

Collection of samples during a storm event meeting these criteria ensures that the resulting data will accurately portray the most common conditions for each site.

All outfalls should be sampled during the same representative storm event if possible. If this is not feasible, outfalls may be sampled during different representative storm events upon approval by the permitting authority. Descriptions of each storm event and which outfalls were sampled during each event must be included in the application.

If samples from more than one storm are analyzed and the results are representative of the discharge, the data representing each event must be reported. The facility must provide a description of each storm event tested. The average of all values within the last year must be determined and the concentration, mass, and total number of storm events sampled must be reported on Form 2F.

If an applicant has two or more outfalls with "substantially identical effluents," the facility may petition the permitting authority to sample and analyze only one of the identical outfalls and submit the results as representative of the other. "Substantially identical effluents" are defined as discharges from drainage areas undergoing similar activities where the discharges are expected to be of similar quantity and quality, and indistinguishable in expected composition.[89]

---

[89] 40 CFR § 122.21(g)(7).

5.7.9 Representative Storm Event

In determining whether a storm is representative, there are two important steps to take. First, data on local weather patterns should be collected and analyzed to determine the range of representative storms for a particular area. Second, these results should be compared to measurements of duration, intensity, and depth to ensure that the storm to be sampled fits the representativeness criteria.

Obtaining Rainfall Data. Several sources provide accurate local weather information for both: (1) determining what a representative storm event is for a particular area and (2) assessing expected storm events to determine whether a predicted rainfall will be "representative," and thus, meet the requirements for storm water sampling. The National Climatic Data Center (NCDC) of the National Oceanic and Atmospheric Administration (NOAA) is responsible for collecting precipitation data. Data on hourly, daily, and monthly precipitation for each measuring station (with latitude and longitude) are available to the public on computer diskette, microfiche, or hard copy.[90]

The National Weather Service (NWS) of the NOAA can also provide information on historic, current, and future weather conditions. Local NWS telephone numbers can be obtained from the NWS Public Affairs Office at (301) 713-0622. Telephone numbers are also usually in local phone directory listings under "National Weather Service" or "Weather."

Logistical Problems with When to Sample. Applicants may encounter weather conditions that may not meet minimum "representative" storm criteria. These conditions may prevent adequate collection of storm water samples prior to application submission deadlines. For instance, sampling may be problematic in parts of the country that experience drought or near-drought conditions or areas that are under adverse weather conditions such as freezing and flooding. Events with false starts and events with stop-start rains can also cause problems.

Where the timing of storm event sampling poses a problem, it may be appropriate for the applicant to petition the permitting authority for a sampling protocol/procedure modification either prior to sampling or after sampling is conducted (if the storm event is not acceptable). When the applicant requests a sampling protocol/procedure modification, a narrative

---

[90] Orders can be placed by calling (704) 259-0682, or by writing to: NCDC, Climate Services Branch, The Federal Building, Asheville, NC 28071-2733.

justification should be attached. This justification should be certified by a corporate official in accordance with 40 CFR Section 122.22.

Arid Areas. For arid or drought-stricken areas where a storm event does not occur prior to the time the applicant must sample and submit data with the application form, the applicant should submit the application, complete to the extent possible, with a detailed explanation of why sampling data are not provided and an appraisal of when sampling will be conducted. This explanation must be certified by the appropriate party (as required by 40 CFR Section 122.22). The applicant should also contact the permitting authority for further direction. Where the applicant can anticipate such problems, approval for an extension to submit sampling data should be acquired prior to the deadline.

False Starts and Stop/Start Rains. False start and stop/start rains can also cause problems. False starts may occur when weather conditions are unpredictable and it appears that a storm event may be representative, collection begins, and then the rain stops before an adequate sample volume is obtained. Some latitude may be given for the 0.1-inch rainfall requirement as long as the sample volume is adequate; the permitting authority may accept the results with applicant justification and certification. During stop/start rains (those in which rainfall is intermittent), samples should be taken until an adequate sample volume is obtained.

Use of Historical Data. Data from storm water samples analyzed in the past can be submitted with applications in lieu of new sampling data if:
1. All data requirements in Form 2F are met.
2. Sampling was performed no longer than 3 years prior to submission of the permit application.
3. All water quality data are representative of the present discharge.
The historical data may be unacceptable if there have been significant changes since the time of that storm event in production level, raw materials, processes, or final products. Significant changes that may also impact storm water runoff include construction or installation of treatment or sedimentation/erosion control devices, buildings, roadways, or parking lots. Applicants should assess any such changes to determine whether they have altered storm water runoff since the time of the storm event chosen for use in the permit application. Historical data can be used only in applications. Historical data cannot be used for fulfilling permit requirements.

5.7.10 Pollutants to be Analyzed

Section VII of Form 2F requires that several common pollutants must be analyzed in both the grab sample and the flow—weighted composite sample, while additional analyses are dependent upon existing NPDES permit conditions or whether the discharger has reason to believe other pollutants may be present in the storm runoff discharge. A separate table should be completed for each outfall. A grab sample must be used (rather than a flow-weighted composite sample) for quantitative data for pH, temperature, cyanide, total phenols, residual chlorine, oil and grease, fecal coliform, and fecal streptococcus.[91]

Part A of Section VII of Form 2F requires that both grab samples and flow-weighted composite samples be analyzed for:
- Biochemical oxygen demand (BOD_5)
- Chemical oxygen demand (COD)
- Total suspended solids (TSS)
- Total Kjeldahl Nitrogen (TKN)
- Nitrate plus nitrite nitrogen
- Total phosphorus

Part B of Section VII of Form 2F requires that each pollutant limited in an effluent guideline to which the facility is subject or any pollutant listed in the facility's NPDES permit for its process wastewater (if the facility is operating under an existing permit) be analyzed for and reported separately for each outfall in Part B.

Part C of Section VII requires the listing of any pollutant shown in Tables 2F-2, 2F-3, and 2F-4 that the discharger knows or has reason to believe is present in the discharge and was not already identified above. Table 2F-2 includes conventional and nonconventional pollutants. For any pollutant from this table listed in Part C, the applicant is required to either report quantitative data or briefly describe the reason the pollutant is expected to be discharged. Table 2F-3 lists toxic pollutants. For every pollutant listed in Table 2F-3 that is expected to be discharged in concentrations of 10 parts per billion (ppb) or greater, the applicant is required to submit quantitative data. For acrolein; acrylonitrile; 2,4-dinitrophenol; and 2-methyl-4, 6-dinitrophenol, the applicant must submit quantitative data if these four pollutants (collectively) are expected to be discharged in concentrations of 100 ppb or greater. For every other pollutant listed in Table 2F-3 that is expected to be discharged in concentrations less than 10 ppb (or 100 ppb total for the four pollutants

---

[91] 40 CFR § 122.21(g)(7).

listed above), then the applicant must either submit quantitative data or briefly describe the reasons the pollutant is expected to be discharged. Table 2F-4 lists hazardous substances. For each outfall, the applicant must list any pollutant from Table 2F-4 that is known or believed to be present in the discharge and explain why they believe it to be present. No analysis is required, but if the applicant has analytical data, it must be reported.

5.7.11 Discharge Flow Rates And Volumes

Form 2F requires applicants to provide quantitative data based on samples collected during storm event(s). One set of parameters that must be provided for such storm event(s) are flow estimates or flow measurements and an estimate of the total volume of the discharge. The method of flow estimation or measurement must be described in the application. The EPA intends that applicants need only provide rough estimates of flows in Form 2F.

Estimating Flows and Volumes. Runoff flow rates and volumes can be estimated by using the total rainfall amount for the storm event and estimated runoff coefficients for the facility. Runoff coefficients represent the fraction of total rainfall that will be transmitted as runoff from the facility. As such, the coefficients reflect the ground surface or cover material. To estimate runoff volume and rates, it can be assumed that paved areas and other impervious structures such as roofs have a runoff coefficient of 0.90 and, therefore, 90% of the rainfall is conveyed from the facility as runoff. For unpaved surfaces, it can be assumed that the runoff coefficient is about 0.50. The total volume of discharge for the event is then estimated by:

total runoff volume (cubic ft) = total rainfall (ft) x [facility paved area x 0.90 + facility unpaved area x 0.50]

The facility areas used in this calculation should be in units of square feet and should include only those areas drained by the outfall sampled. To estimate an average flow rate, divide the volume by the duration of the rainfall event. If desired, a more accurate estimate can be made by using more specific runoff coefficients for different parts of the facility based on the type of ground cover.

To estimate flow rates in units of volume per time, such as cubic feet per second, information on flow velocities and depth of flow are required. Flow rate estimates may be obtained by measuring depth of flow and velocity in a pipe of known diameter or other conveyance structure at frequent intervals during a storm runoff event. For a pipe or other structure of known size, the cross—sectional area of flow can be calculated for any depth of flow using geometric relationships. Flow velocities can be measured by using suitable units (e.g., propeller-operated devices) attached to a portable current meter. Flow velocity measurements should be obtained from representative locations throughout the flow cross—section. Flow velocities can also be estimated using simpler methods, such as measuring the time of passage of an object (e.g., an orange) between two points a known distance apart (e.g., manholes).

Estimation of Flow Rates Based on Flow Velocity. If the measurements of flow depth are recorded and converted to cross-sectional areas (in square feet), and the corresponding velocities for each depth are recorded (in feet per minute), then the flow rate (Q) in cubic feet per minute (cfm) is: Q = (area)(velocity). The maximum flow rate is the highest value recorded during the storm event. The time-weighted average flow rate for the storm event can be estimated by the average of the individual values recorded.

Estimation of Volumes Based on Flow Rate Estimates. The total volume of discharge can be estimated by first multiplying each of the flow rates determined above by a time interval that represents the portion of the total storm duration associated with the measurement, and then adding all such partial volumes. If the time intervals used are seconds, then the total flow of runoff will be in units of cubic feet.

5.7.12 Sampling at Retention Ponds

Retention ponds with greater than a 2-hour holding time for a representative storm event may be sampled by grab sample. Composite sampling is not necessary because the water is held for at least 24 hours and, thus, a thorough mixing occurs within the pond. Therefore, a single grab sample of the effluent from the discharge point of the pond accurately represents a composite of the storm water contained in the pond. If the pond does not thoroughly mix the discharge, thereby compositing the sample, then a regular grab and composite sample should be taken at the inflow to the pond. Since each pond may vary in its capability to "composite" a sample, applicants must carefully evaluate whether the pond is thoroughly mixing the discharge. Such factors as pond design and maintenance are important in making this evaluation. Poor pond design, for example, where the outfall and inflow points are too closely situated, may cause inadequate mixing. In addition, poor maintenance may lead to excessive resuspension of any deposited silt and sediment during heavy inflows. Because of factors such as these, the applicant should determine the best location to sample the pond (e.g., at the outfall, at the outfall structure, in the pond) to ensure that a representative composite sample is taken. If adequate compositing is not occurring within the pond, the applicant should conduct routine grab and flowweighted composite sampling.

A grab sample and a flow—weighted sample must be taken for storm water discharges collected in holding ponds with less than a 24-hour retention period. The applicant must sample the discharge in the same manner as for any storm water discharge, as described in 40 CFR Section 122.21(g)(7).

5.8 Storm Water Enforcement and Penalties

The EPA or state permitting authority may seek to impose administrative,[92] civil,[93] or criminal penalties[94] on violators of storm water permit conditions and limitations. For example, the CWA provides that any person who violates a permit condition is subject to a civil penalty not to exceed $25,000 per day of violation.[95] Under the criminal penalty provisions, any person who negligently violates a permit is subject to a fine of not less than $2,500 or more than $25,000 per day of violation, or imprisonment for not more than one year, or both.[96] Any person who "knowingly" violates any permit condition or limitation is subject to a fine of not less than $5,000 or more than $50,000 per day of violation, or by imprisonment of not more than three years, or both.[97] The CWA also provides that any person who knowingly falsifies any record or document, tampers with or renders inaccurate any monitoring device, will be punished by a fine of not more than $10,000 per violation, or by imprisonment for not more than 2 years, or both.[98]

In addition, under Section 505 of the CWA,[99] private citizens may bring suits to enforce the Act and its implementing regulations, including water

---

[92] 33 U.S.C. § 1319(g).
[93] 33 U.S.C. § 1319(d).
[94] 33 U.S.C. § 1319(c).
[95] The amount of the civil penalty depends on "the seriousness of the violation or violations, the economic benefit (if any) resulting from the violation, any history of such violations, any good-faith efforts to comply with the applicable requirements, the economic impact of the penalty on the violator, and such other matters as justice may require." 33 U.S.C. § 1319(d).
[96] 33 U.S.C. § 1319(c)(1). See also 40 CFR § 122.41(a).
[97] 33 U.S.C. § 1319(c)(2).
[98] 33 U.S.C. § 1319(c)(4). See also 40 CFR § 122.41(j)(5) & (k)(2).
[99] 33 U.S.C. § 1365.

quality standards,[100] NPDES permit conditions,[101] and storm water discharge requirements.[102] Citizens may sue the EPA to compel enforcement of a nondiscretionary duty of the agency[103] or may sue a violator directly to enjoin a violation[104] and/or seek the assessment of civil penalties.[105] Commencement of a citizen suit against an alleged violator is not allowed, however, if the government is already diligently prosecuting the alleged violation.[106] In addition, the alleged violation must be a continuing one and not a wholly past violation.[107]

---

[100] See, for example, Northwest Envtl. Advocates v. City of Portland, 56 F.3d 979 (9th Cir. 1995); Upper Chattachoochee Riverkeeper Fund, Inc. v. City of Atlanta, 953 F. Supp. 1541 (N.D. Ga. 1996); Community Ass'n for Restoration of Environment (CARE) v. Sid Koopman Dairy, 54 F.Supp.2d 976 (E.D. Wash. 1999).

[101] See, for example, Committee to Save Mokelumne River v. East Bay Util. Dist., 13 F.3d 305 (9th Cir. 1993); Mancuso v. New York Thruway Auth., 909 F. Supp. 133 (S.D.N.Y. 1995).

[102] See, for example, Molokai Chamber of Commerce v. Kukui (Molokai), Inc., 891 F. Supp. 1389 (D. Hawaii 1995); Beartooth Alliance v. Crown Butte Mines, 904 F. Supp. 1168 (D. Mont. 1995).

[103] See, for example, National Wildlife Fed. v. Browner, 127 F.3d 1126 (D.C. Cir. 1997); Alaska Clean Water Alliance v. Clarke, 45 Env't Rep. Cas. (BNA) 1664, 1997 WL 446499 (W.D. Wash. July 8, 1997).

[104] See, for example, Atlantic States Legal Found. v. Stroh Die Casting Co. 116 F.3d 814 (7th Cir. 1997). However, an injunction does not automatically follow from the existence of a violation. Bucholz v. Dayton Int'l Airport, 1995 WL 811897 (S.D. Ohio Oct. 30, 1995) (court found that runoff from an airport violated a permit, but stressed that equitable relief should be fashioned from traditional principles, including a balancing of the equities).

[105] See, for example, Friends of the Earth v. Laidlaw Envtl. Serv. (TOC), Inc. 120 S.Ct. 693 (U.S. 2000) (environmental groups had standing to bring citizen suit seeking both injunctive relief and criminal penalties for defendant-company's violation of mercury discharge limits in NPDES permit); and Old Timer, Inc. v. Blackhawk-Central City Sanitation Dist., 51 F.Supp.2d 1109 (D. Colo. 1999) (in citizen suit against sanitation district and operators of its sewage treatment plant for effluent limit violations, claim for injunctive relief, but not claim for civil penalties, was rendered moot by district's showing that it was in compliance with discharge permit).

[106] 33 U.S.C. § 1365(b)(2). See, for example, Comfort Lake Ass'n, Inc. v. Dresel Contracting, Inc., 138 F.3d 351 (8th Cir. 1998) (citizen suit claims for injunctive relief and civil penalties against construction company that violated NPDES permit precluded by state environmental agency's diligent prosecution of enforcement action); Old Timer, Inc. v. Blackhawk-Central City Sanitation Dist., 51 F.Supp.2d 1109 (D. Colo. 1999) (citizen suit against sanitation district and operators of its sewage treatment plant for effluent limit violations not barred by state's issuance of administrative compliance order).

[107] Gwaltney of Smithfield v. Chesapeake Bay Found., Inc., 484 U.S. 49 (1987). See also Molokai Chamber of Commerce v. Kukui (Molokai), Inc., 891 F. Supp. 1389 (D. Hawaii 1995) (finding that commencement of construction without a permit constituted a continuing violation even after the construction ceased because storm water discharges continued).

Prior to initiating an action, a citizen suit plaintiff must give 60 days' notice of its intent to sue to the alleged violator, the EPA, and the affected state.[108] The notice must set forth the details of the alleged violation and the regulations allegedly violated.[109] In addition to statutory notice requirements, a citizen suit plaintiff must also establish standing to pursue the action, by demonstrating that it has an interest that is or may be adversely affected by the alleged violation.[110]

At the beginning of 2000, the U.S. Supreme Court handed down an important decision on the issue of standing in Clean Water Act citizen suits. In *Friends of the Earth v. Laidlaw Environmental Services*,[111] various environmental groups filed a citizen suit seeking both injunctive relief and civil penalties for the defendant-company's violation of mercury discharge limits in its NPDES permit. The federal district court denied injunctive relief because the defendant had come into compliance with the effluent limitations for mercury after the lawsuit was filed but assessed a civil penalty of $405,800 to deter the defendant from future violations.[112] The environmental groups did not appeal the denial of injunctive relief but did appeal the amount of the civil penalty. The Court of Appeals for the Fourth Circuit dismissed the appeal on grounds that the only remedy available--civil penalties payable to the government--would not redress any injury suffered by the environmental groups.[113] The Court of Appeals held that even if the environmental groups initially had standing, the case had become moot once the defendant complied with the terms of its permit and the plaintiffs failed to appeal the denial of equitable relief. The Supreme Court reversed, holding that the environmental groups had standing notwithstanding that the appeal only sought judicial review of the amount of the civil penalty. The Court rejected the Fourth Circuit's reasoning, explaining that because civil penalties are a form of sanction that effectively abates illegal conduct ongoing at the time of suit and prevents its recurrence, it provides a form of redress to a citizen suit plaintiff who is injured or threatened with injury as a result of the illegal conduct. The Court stated that it need not explore the outer limits of the principle that civil penalties provide sufficient deterrence to support

---

[108] 33 U.S.C. § 1365(b)(1).

[109] See 40 CFR pt. 135 (regulations governing citizen suit requirements).

[110] See, for example, Public Interest Research Group of New Jersey, Inc. v. Magnesium Elektron, Inc., 123 F.3d 111 (3d Cir. 1997); Friends of the Earth, Inc. v. Chevron Chem. Co., 129 F.3d 826 (5th Cir. 1997); San Francisco Baykeeper v. Vallejo Sanitation and Flood Control Dist., 36 F.Supp.2d 1214 (E.D. Cal. 1999)

[111] Friends of the Earth, Inc. v. Laidlaw Envtl. Servs. (TOC), Inc., 120 S.Ct. 693 (U.S. 2000).

[112] Friends of the Earth, Inc. v. Laidlaw Envtl. Servs. (TOC), Inc., 956 F. Supp. 588 (D.S.C. 1997).

[113] Friends of the Earth, Inc. v. Laidlaw Envtl. Servs. (TOC), Inc., 149 F.3d 303 (4th Cir. 1998).

redressability, because the civil penalties sought here carried a deterrent effect that made it likely, as opposed to merely speculative, that the penalties would redress the environmental groups' injuries--as the district court reasonably found when it assessed a penalty of $405,800.

6

The Safe Drinking Water Act

6.1 Introduction

From our earliest days as a nation until the early 1960s, most Americans had assumed that we had an inexhaustible supply of drinking water. However, information began to emerge during the 1960s suggesting that water supplies in the United States were in jeopardy due to contamination from a variety of pollutants. Faced with a potential crisis, Congress enacted the Safe Drinking Water Act (SDWA) in 1974 to manage potential contamination threats to groundwater.[1]

The SDWA authorizes the U.S. Environmental Protection Agency (EPA) to protect the nations drinking water supply in three ways: First, to develop national primary and secondary drinking water regulations; second, to promulgate underground injection control regulations to protect underground sources of drinking water; and third, to develop groundwater protection grant programs. The SDWA permits these activities to be implemented by the states.

However, the SDWA does *not* regulate discharge of pollutants into waterways, even though these activities might eventually affect drinking water supplies. Such activities are regulated by the Clean Water Act,[2] and are discussed in chapter 4. In addition, the SDWA has been held not to permit action against a polluter of a municipal water supply with toxic chemicals.[3] At least four other federal statutes regulate groundwater at one level or another; these are the Solid Waste Disposal Act [known more generally as the Resource Conservation and Recovery Act (RCRA)],[4] the

---

[1] 42 U.S.C. §§ 300f-j.
[2] 42 U.S.C. § 300f et seq.
[3] See City of Evansville v. Kentucky Liquid Recycling, 604 F.2d 1008 (7th Cir. 1979); and U.S. v. Price, 523 F.Supp. 1055 (D.N.J. 1981).
[4] 42 U.S.C. § 6901 et seq., which is discussed in chapter 8.

Comprehensive Environmental Response, Compensation and Liability Act (CERCLA),[5] the Federal Insecticide, Fungicide and Rodenticide Act (FIFRA),[6] and the Toxic Substances Control Act (TSCA).[7]

6.2 Drinking Water Supplies

One of the primary functions of the SDWA is to protect the nation's drinking water supplies. Based on evidence that has accumulated through the years, contaminants of drinking water can be placed into several categories.[8]

Pathogens. Among the most common drinking water contaminants are pathogens such as bacteria, parasites (e.g., *Giardia* and *Cryptosporidium*), and viruses. These pathogens are considered a public health problem in their own right, but they often indicate water contamination by human or animal wastes. While microbial contaminants have largely been controlled in U.S. public water supplies, they continue to be the most common cause of water-related diseases in the country. Since 1971, the EPA and the Centers for Disease Control (CDC) have maintained a program for collecting and reporting data on waterborne disease outbreaks. For the two-year period 1993–1994, CDC reports that 17 states identified 30 disease outbreaks associated with drinking water, and that the outbreaks caused an estimated 405,366 persons to become ill, including 403,000 from an outbreak of cryptosporidiosis in Milwaukee, Wisconsin (the largest waterborne disease outbreak ever documented in the United States).

Organic chemicals. Although data were inadequate at the time, concern about synthetic organic chemicals in drinking water supplies of some cities was a significant force in the passage of the 1974 Safe Drinking Water Act. In 1981, the EPA conducted the Ground Water Supply Survey to determine the occurrence of volatile organic chemicals (VOCs) in public drinking water supplies drawing on groundwater. The survey showed detectable levels of these chemicals in 28.7 percent of public water systems serving more than 10,000 people and in 16.5 percent of smaller systems. Other EPA and state surveys also revealed VOCs in public water

---

[5] 42 U.S.C. § 9601 et seq., which is discussed in chapter 9.
[6] 7 U.S.C. § 136 et seq., which is discussed in chapter 10.
[7] 15 U.S.C. § 2601 et seq., which is discussed in chapter 11.
[8] Much of the following is based on the Committee for the National Institute for the Environment, *Safe Drinking Water Act: Implementation and Reauthorization* (Congressional Research Service 91041, 1999).

supplies. The EPA has used these surveys to support regulation of numerous organic chemicals, many of which are suspected to be carcinogenic.[9]

Pesticides. Several kinds of agricultural chemicals (mainly fertilizers and pesticides) have been detected in ground and surface water in recent years. There has been heightened public and governmental concern, even though concentrations of most of the detected pesticides have been very low.[10] In areas of heavy agricultural-chemical use, pesticides have been detected more frequently and at higher levels. In 1992, EPA issued the Pesticides in Ground Water Database (1971–1991) which showed that nearly 10,000 of 68,824 tested wells contained pesticides at levels that exceeded drinking water standards or health advisory levels. Almost all the data were from drinking water wells. EPA has placed restrictions on 54 pesticides found in groundwater, 28 of which are no longer registered for use in the United States but may still be present in soils and groundwater.[11]

Under the SDWA, EPA sets the maximum permissible levels of contaminants in drinking water supplies.[12] This is done with two levels of controls. National primary drinking water regulations (NPDWR) protect human health to the extent feasible, taking technology, treatment techniques, and costs into consideration. In addition, EPA sets national secondary drinking water regulations (NSDWR) which specify the maximum contaminant levels necessary to protect public welfare. NSDWRs are concerned primarily with contaminants affecting drinking water odor and appearance. These mainly "aesthetics" standards are not federally enforceable and are issued only as guidelines for the states.

Neither NPDWR nor NSDWR pertain to every drinking water source, but specifically to the protection of "public drinking water" supplies. SDWA Section 1401(4)(a) states that:

The term "public water system" means a system for the provision to the public of water for human consumption through pipes or other constructed conveyances, if such system has at least fifteen service connections or regularly serves at least twenty-five individuals. Such term includes (i) any collection, treatment, storage, and distribution facilities under control of the operator of such system and used primarily in connection with such system, and (ii) any collection or

---

[9] Id.

[10] Little is known about the long-term health effects of low-level exposures to pesticides and other chemicals. See chapters 10 and 11.

[11] Committee for the National Institute for the Environment, *Safe Drinking Water Act: Implementation and Reauthorization* (Congressional Research Service 91041, 1999).

[12] Regulations are at 40 CFR Parts 141–143.

pretreatment storage facilities not under such control which are used primarily in connection with such system.[13]

In other words, water supply systems that service fewer than fifteen regular customers are exempt from NPDWR and NSDWR. On the other hand, all water treatment or storage facilities, whether they are independent of the control of the public water supply system or not, are covered.

6.2.1 National Primary Drinking Water Regulations

In order to institute National Primary Standards, EPA establishes a Maximum Contaminant Level (MCL) for each of several chemicals. These MCLs are the maximum concentration of a chemical that is allowed in public drinking water systems. Currently there are fewer than 100 chemicals for which MCLs have been established, but these represent chemicals that are thought to pose the most serious risk.

The EPA guidance for establishing MCLs states that MCLs are enforceable standards and are to be set as close to the maximum contaminant level goals (MCLGs) (Health Goals) as is feasible. In addition, MCLs are based upon treatment technologies, costs (affordability) and other feasibility factors, such as availability of analytical methods, treatment technology and costs for achieving various levels of removal.[14]

The process of determining an MCL starts with an evaluation of the adverse effects caused by the chemical in question and the doses needed to cause such effects. The final result of this process is a safe dose (the dose thought to provide protection against adverse effects including a margin of safety), now called a Reference Dose (RfD) by the EPA. This evaluation is based on the results of animal experiments and the research results are extrapolated to humans using standard EPA methods.

For chemicals that do not cause cancer, a MCLG is established by first converting the safe dose (RfD) to a water concentration. Then, this number is divided by five based on the assumption that exposure to the chemical through drinking water represents only one-fifth of the possible exposure to this substance. Other sources of exposure may be air, soil, and food. In almost all cases, the MCLG value is the same one that is used as the MCL.

For chemicals believed to cause cancer (known or probable human carcinogens, EPA Class A or B), the MCLG is set at zero (i.e., no amount of chemical is considered acceptable). However, since zero cannot be

---

[13] 42 U.S.C. § 300f(4)(a).
[14] 42 U.S.C. § 300g-1(b).

measured, the MCL is based on the lowest concentration that can be measured on a routine basis. This is known as the Practical Quantitation Limit (PQL). Thus for known or probable carcinogens, the MCL is not a safe level but instead is the lowest measurable level.

For chemicals that are possible cancer-causing agents (EPA Class C); i.e., there is some evidence that they may cause cancer but this is not very convincing, a value equivalent to the MCLG is calculated as if they were not carcinogens. Then this value is divided by a factor of 10 to give the final MCL. This provides an additional margin of safety in case the chemical is later determined to be a carcinogen.

For lead and copper, the MCL approach is not used. In these two cases, water treatment programs are required.

6.2.2 National Secondary Drinking Water Standards

Unlike primary drinking water regulations, secondary drinking water regulations are not designed to protect public health. Instead, they are intended to help protect public welfare by establishing concentration limits for chemicals that cause undesirable taste, odor, and color of the water or that cause staining or corrosion of fixtures that come into contact with the water.

6.2.3 Setting Drinking Water Standards

For all drinking water standards, the SDWA directs EPA to conduct a thorough cost-benefit analysis and provide comprehensive, informative, and understandable information to the public. SDWA Section 1412(b)(3)(A) requires top-level scientific evaluation:

Use of science in decisionmaking. — In carrying out this section, and, to the degree that an Agency action is based on science, the Administrator [of the EPA] shall use —

(i) the best available, peer-reviewed science and supporting studies conducted in accordance with sound and objective scientific practices; and

(ii) data collected by accepted methods or best available methods (if the reliability of the method and the nature of the decision justifies use of the data).[15]

---

[15] 42 U.S.C. § 300g-1(b)(3)(A).

The standard-setting process first defines a maximum contaminant level (MCL) or treatment technique standard based on affordable technology.[16] Then the costs of that standard are compared to the expected health benefits to determine if they would be justified by the benefits. If not, then EPA may adjust an MCL to a level that "maximizes health risk reduction benefits at a cost that is justified by the benefits."[17] MCLs must be set as close to MCLGs "as is feasible," except when EPA determines that the cost of a standard at that level are not justified by the benefits, or when certain "risk-risk" considerations apply. SDWA Section 1412(b)(5)(B) states that these "risk-risk" balancing situations require flexibility to "minimize the overall risk of adverse health effects" so that controlling the level of one contaminant does not increase the risk from another contaminant.[18]

In addition, EPA (after consultation with the Department of Health and Human Services) may issue interim regulations for any contaminant which poses an urgent threat to human health without making the usual "determination to regulate" and completing the cost-benefit analysis. However, a cost-benefit analysis and the required determination (to regulate or not) must be done within 3 years after the interim regulation, and the rule must be repromulgated or revised if necessary.[19]

EPA is required to issue regulations that establish requirements for water systems to provide annual reports to all customers. These regulations are developed in consultation with environmental groups, public interest groups, risk communication experts, and the states. The regulations must include a plainly worded explanation of the definition of MCLs and MCLGs, as well as plain language explanations of the health concerns associated with contaminants. The reports must contain information on the source of a water systems supply, the level of detected contaminants, information on the health effects of contaminants found in violation of the standard, and information on unregulated contaminants.[20]

The 1996 Amendments to the SDWA specify a number of changes to the current law that will take effect in the next few years. One is a new risk-based contaminant selection process. The EPA must use three criteria to determine whether or not to regulate a contaminant: (1) that the contaminant adversely affects human health; (2) it is known or substantially likely to occur in public water systems with a frequency and at levels of public health concern; and (3) regulation of the contaminant

---

[16] 42 U.S.C. § 300f(1)(C)(i).
[17] 42 U.S.C. § 300g-1(b)(3)(C).
[18] 42 U.S.C. § 300g-1(b)(5)(B).
[19] 42 U.S.C. § 300g-1(b)(1)(ii).
[20] 42 U.S.C. § 300g-3(c)(3).

presents a meaningful opportunity for health risk reduction. In short, the new provision makes risk prioritization dominant in selecting which contaminants to regulate.

In 1998, and every 5 years thereafter, EPA will publish a list of contaminants which are not subject to any proposed or final national primary drinking water regulation but which are known or anticipated to occur in public water systems and may require regulation. In developing the list of potential contaminants, EPA must consult with the scientific community, allow for public comment, and consider the occurrence database.

Starting in 2001, and every 5 years thereafter, EPA is required to determine whether or not to regulate at least five of the contaminants listed as potential contaminants for regulation. EPA is directed to make determinations for contaminants that present the greatest public health concern. In selecting such contaminants, EPA must take into consideration the effect of contaminants upon sensitive subpopulations, such as infants, children, pregnant women, the elderly, and individuals with a history of serious illness. After EPA makes a determination to regulate a contaminant it must publish an MCLG and final national primary drinking water regulation within 3 ½ years.

6.3 Wellhead Protection Area Program

SDWA Section 1428 creates a special "Wellhead Protection Area" program that is specifically designed to protect wells that serve as public drinking water supplies. SDWA Section 1428(e) defines "Wellhead protection area" as:

> [T]he surface and subsurface area surrounding a water well or wellfield, supplying a public water system, through which contaminants are reasonably likely to move toward and reach such water well or wellfield. The extent of a wellhead protection area, within a State, necessary to provide protection from contaminants which may have any adverse effect on the health of persons is to be determined by the State in the program submitted under subsection (a) of this section.[21]

---

[21] 42 U.S.C. § 399h-7(e).

The EPA provides technical guidance which states may use in making these determinations.[22] EPA guidance reflects factors such as the radius of influence around a well or wellfield, the depth of drawdown of the water table by the well or wellfield at any given point, the time or rate of travel of various contaminants in various hydrologic conditions, distance from the well or wellfield, or other factors affecting the likelihood of contaminants reaching the well or wellfield, taking into account available engineering pump tests or comparable data, field reconnaissance, topographic information, and the geology of the formation in which the well or wellfield is located.[23]

SDWA Section 1428(a) contains instructions for states on the requirements for state wellhead programs.[24] Each state is required to submit to EPA a program that will, at a minimum:

(1) specify the duties of State agencies, local governmental entities, and public water supply systems with respect to the development and implementation of programs required by this section;

(2) for each wellhead, determine the wellhead protection area . . . based on all reasonably available hydrogeologic information on ground water flow, recharge and discharge and other information the State deems necessary to adequately determine the wellhead protection area;

(3) identify within each wellhead protection area all potential anthropogenic sources of contaminants which may have any adverse effect on the health of persons;

(4) describe a program that contains, as appropriate, technical assistance, financial assistance, implementation of control measures, education, training, and demonstration projects to protect the water supply within wellhead protection areas from such contaminants;

(5) include contingency plans for the location and provision of alternate drinking water supplies for each public water system in the event of well or wellfield contamination by such contaminants; and

(6) include a requirement that consideration be given to all potential sources of such contaminants within the expected wellhead area of a new water well which serves a public water supply system.[25]

---

[22] See U.S.E.P.A., Office of Water, *Protecting Local Ground Water Supplies Through Wellhead Protection* (570-09-91/007, 1991); and U.S.E.P.A., Office of Ground Water Protection, *Guidelines for Delineation of Wellhead Protection Areas* (440-5-93/001, 1993).

[23] 42 U.S.C. § 399h-7(e)..

[24] For additional instructions, see U.S.E.P.A., Office of Ground Water Protection, *Developing a State Wellhead Protection Program: a User's Guide to Assist State Agencies under the Safe Drinking Water Act.* (440-6-88/003, 1988).

[25] 42 U.S.C. § 300h-7(a).

6.4 Underground Injection Control Program

Sections 1421-1428 of the SDWA regulates the injection of pollutants (often toxic, hazardous, or radioactive wastes) into very deep injection wells below groundwater levels. The idea is that they are so deep that they will never contaminate drinking water. SDWA Section 1421(d)(2) contains Congress' concern about the potential dangers of this practice:

Underground injection endangers drinking water sources if such injection may result in the presence in underground water which supplies or can reasonably be expected to supply any public water system of any contaminant, and if the presence of such contaminant may result in such system's not complying with any national primary drinking water regulation or may otherwise adversely affect the health of persons.[26]

The Underground Injection Control (UIC) program has separate regulations for each of five types of underground injection wells. Class I wells are used to dispose of hazardous wastes by generators or treatment, storage and disposal facilities regulated under the Resource Conservation and Recovery Act (RCRA),[27] and some other municipal and industrial injection wells.[28] Class II wells are used by oil and gas companies for conventional oil and gas extraction.[29] Class III wells are used in mining and power generation.[30] Class IV injection wells are those used for hazardous or radioactive waste injection within ¼ mile of a drinking water source, which are prohibited under RCRA Section 3020(a).[31] Class V wells are cesspools and septic systems serving multifamily or industrial structures, drainage wells, and assorted other wells that must be identified, studied, and regulated under the state's EPA-approved UIC program.

Under the SDWA's UIC program, each state establishes an EPA-approved UIC program. The state then issues permits for injection wells. In general, these permits must demonstrate that the well will not endanger any drinking water supply. Under EPA regulations, a states may *exempt* a specific drinking water supply from UIC restrictions (i.e., allow a well to be approved that would otherwise be restricted or prohibited):

---

[26] 42 U.S.C. § 300h(d)(2).
[27] 42 U.S.C. § 6939b.
[28] 40 CFR § 146.5(a)(2).
[29] 40 CFR § 146.5(b)(1).
[30] 40 CFR § 146.5(c).
[31] 42 U.S.C. § 6939b(a).

An aquifer or a portion thereof which meets the criteria for an "underground source of drinking water" in [40 CFR] Sec. 146.3 may be determined under 40 CFR 144.8 to be an "exempted aquifer" if it meets the following criteria:

(a) It does not currently serve as a source of drinking water; and

(b) It cannot now and will not in the future serve as a source of drinking water because:

(1) It is mineral, hydrocarbon or geothermal energy producing, or can be demonstrated by a permit applicant as part of a permit application for a Class II or III operation to contain minerals or hydrocarbons that considering their quantity and location are expected to be commercially producible.

(2) It is situated at a depth or location which makes recovery of water for drinking water purposes economically or technologically impractical;

(3) It is so contaminated that it would be economically or technologically impractical to render that water fit for human consumption; or

(4) It is located over a Class III well mining area subject to subsidence or catastrophic collapse; or

(c) The total dissolved solids content of the ground water is more than 3,000 and less than 10,000 mg/l and it is not reasonably expected to supply a public water system.[32]

In addition, SDWA Section 1425(a) states that EPA will delegate all regulation to the state for:

[A]ny State underground injection control program which relates to —

(1) the underground injection of brine or other fluids which are brought to the surface in connection with oil or natural gas production or natural gas storage operations, or

(2) any underground injection for the secondary or tertiary recovery of oil or natural gas.[33]

States may regulate these wells so long as they demonstrate to the EPA that the affected portion of the state program "represents an effective program (including adequate recordkeeping and reporting) to prevent underground injection which endangers drinking water sources."[34]

---

[32] 40 CFR § 146.4.
[33] 42 U.S.C. § 300h-4(a).
[34] Id.

6.5 Sole Source Aquifer Protection Program

Under circumstances where an aquifer is the sole source of drinking water for an area, Congress felt that special precautions should be taken to protect the aquifer. The Sole Source Aquifer Demonstration Project provides grants to states for up to 50 percent of the costs for the state to identify and protect "Critical Aquifer Protection Areas,"[35] which are defined as:

> [T]the term "critical aquifer protection area" means either of the following:
> (1) All or part of an area located within an area for which an application or designation as a sole or principal source aquifer pursuant to section 300h-3(e) of this title, has been submitted and approved by the Administrator and which satisfies the criteria established by the Administrator under subsection (d) of this section.
> (2) All or part of an area which is within an aquifer designated as a sole source aquifer as of June 19, 1986, and for which an areawide ground water quality protection plan has been approved under section 208 of the Clean Water Act (33 U.S.C. § 1288) prior to June 19, 1986.[36]

State and local governments can apply for the funds under SDWA Section 1427(e). The application must include, among other things, a hydrogeologic assessment of surface and groundwater resources within the critical protection area, and a "comprehensive management plan," the objective of which is "to maintain the quality of the ground water in the critical protection area in a manner reasonably expected to protect human health, the environment and ground water resources."[37] SDWA Section 1427(f) continues:

> In order to achieve such objective, the plan may be designed to maintain, to the maximum extent possible, the natural vegetative and hydrogeological conditions. Each of the following elements shall be included in such a protection plan:
> (A) A map showing the detailed boundary of the critical protection area.
> (B) An identification of existing and potential point and nonpoint sources of ground water degradation.
> (C) An assessment of the relationship between activities on the land

---

[35] 42 U.S.C. § 300h-6(f).
[36] 42 U.S.C. § 300h-6(b).
[37] 42 U.S.C. § 300h-6(f).

surface and ground water quality.

(D) Specific actions and management practices to be implemented in the critical protection area to prevent adverse impacts on ground water quality.

(E) Identification of authority adequate to implement the plan, estimates of program costs, and sources of State matching funds.

(2) Such plan may also include the following:

(A) A determination of the quality of the existing ground water recharged through the special protection area and the natural recharge capabilities of the special protection area watershed.

(B) Requirements designed to maintain existing underground drinking water quality or improve underground drinking water quality if prevailing conditions fail to meet drinking water standards, pursuant to this chapter and State law.[38]

The local or state must establish procedures for public participation in the development of the plan, for review, approval, and adoption of the plan, and for assistance to municipalities and other public agencies with authority under state law to implement the plan.

While the infusion of federal money into the protection of sole source aquifers is very attractive to many state and local governments, the total funding available for any one aquifer may not exceed $4 million.[39]

6.6 State Enforcement

States have primary enforcement authority (called "primacy") for the SDWA. A state's standards can be more stringent than the federal standards, but a state can lose primacy if state standards (or enforcement) are less stringent.[40] However, if a state does not properly enforce SDWA requirements, EPA will assume the authority to do so.[41] Once EPA makes a determination that a public water system is not in compliance, then it must act.[42]

The federal government provides funds to assist in state enforcement of the SDWA. In certain circumstances states may consider cost, benefits, alternatives, public interest, and the protection of human health and the environment in granting variances and exemptions from the national regulation. For example, many of the regulations affecting ground water

---

[38] Id.

[39] 42 U.S.C. § 300h-6(j).

[40] 42 U.S.C. § 300h(b).

[41] 42 U.S.C. § 300g-2.

[42] See National Wildlife Federation v. EPA, 980 F.2d 765 (D.C. Cir. 1992).

provide for exemptions, variances, or alternate concentration limits.[43]

State drinking water programs vary in their approaches and coverage, although all must meet EPA standards. Several states have representative programs.

6.6.1 Florida's Wellhead Protection Program

Florida's groundwater resource is the primary source of drinking water in the State, supplying over 90 percent of all public water supply. Florida uses wellhead protection to protect potable water wells, and to prevent the need for their replacement or restoration due to contamination.[44] "Wellhead Protection Area" is defined as "an area designated by the Department [of Environmental Protection; DEP] consisting of a 500-foot radial setback distance around a potable water well where groundwater is provided the most stringent protection measures to protect the groundwater source for a potable water well and includes the surface and subsurface area surrounding the well."[45]

Florida's Wellhead Protection Program sets performance standards, and prohibits a long list of activities that might contaminate drinking water supplies. The DEP requires new installations to meet a series of restrictions within a wellhead protection area:

(a) New domestic wastewater treatment facilities shall be provided with Class I reliability . . . and flow equalization. New wastewater ponds, basins, and similar facilities shall be lined or sealed to prevent measurable seepage. Unlined reclaimed water storage systems are allowed for [certain permitted] reuse projects.

(b) New reuse and land application projects shall be prohibited except for new [permitted] projects.

(c) New domestic wastewater residuals land application sites . . . shall be prohibited.

(d) New discharges to ground water of industrial wastewater . . . shall be prohibited except as provided below: (1) All non-contact cooling water discharges (without additives); and (2) Discharges specifically allowed-within a wellhead protection area . . .

(e) New phosphogypsum stack systems . . . are prohibited.

(f) New Class I and Class III underground injection control wells . . . are prohibited.

(g) New Class V underground injection control wells . . . are

---

[43] 40 CFR § 141.4.
[44] Fla. Admin. Code § 62-521.100.
[45] Id.

prohibited except as provided below: (1) Thermal exchange process wells (closed-loop without additives) for use at single family residences; and (2) Aquifer storage and recovery systems wells, where the injected fluid meets the applicable drinking water quality standards . . .

(h) New solid waste disposal facilities . . . are prohibited.

(i) New generators of hazardous waste, . . . which excludes household hazardous waste . . . shall comply with the secondary containment requirements of [federal regulations].

(j) New hazardous waste treatment, storage, disposal, and transfer facilities requiring permits . . . are prohibited.

(k) New aboveground and underground tankage of hazardous wastes . . . is prohibited.

(l) Underground storage tanks . . . shall not be installed 90 days after the effective date of this rule [although most replacement of existing underground storage tanks is allowed].

(m) Aboveground storage tanks . . . shall not be installed 90 days after the effective date of this rule [although most replacement of existing aboveground storage tanks is allowed].

(n) Storage tanks which meet the auxiliary power provisions . . . for operation of a potable water well and storage tanks for substances used for the treatment of potable water are exempt from the provisions of this rule.

(o) To prevent the vertical migration of fluids, a construction permit may be required from the appropriate water management district for new water wells, and shall meet the applicable construction standards for wells . . .[46]

The Florida plan has exemptions for equipment to supply emergency power for sewer service, telephone service, and the like,[47] and for DEP approved remedial corrective
actions for contaminated sites located within wellhead protection areas.[48]

6.6.2 Illinois' Protection of Public Water Supplies

Illinois recognizes the value of public water supplies in the policy statement in the Illinois Environmental Protection Act, Public Water Supplies:

---

[46] Fla. Admin. Code § 62-521.400(1) Ground Water Protection Measures in Wellhead Protection Areas.
[47] Fla. Admin. Code § 62-521.400(2).
[48] Fla. Admin. Code § 62-521.400(3).

The General Assembly finds that state supervision of public water supplies is necessary in order to protect the public from disease and to assure an adequate supply of pure water for all beneficial uses.

It is the purpose of this Title to assure adequate protection of public water supplies.[49]

Illinois attempts to meet this policy in three basic ways. First, "setback zones" are created for variable distances around wells. A "setback zone" is defined as "a geographic area . . . containing a potable water supply well or a potential route, having a continuous boundary, and within which certain prohibitions or regulations are applicable in order to protect ground waters."[50] No new public water supply well may be located within from 200 to 400 feet from any potential source of groundwater pollution, depending on conditions.[51] Conversely, no non-drinking water well or source of groundwater contamination may be located within 200 feet of any public water supply well or other potable water supply well.[52] Waivers are available for certain wells from the Illinois EPA (IEPA).

Second, Illinois creates two groundwater protection agencies. The Interagency Coordinating Committee on Ground Water (ICCG) is composed of members of a variety of state agencies.[53] The ICCG reviews and coordinates the state's policy, laws, regulations, and procedures that relate to groundwater. The Groundwater Advisory Council (GAC) is composed of nine members from the general public who are appointed by the Governor. The GAC has functions very similar to those of the ICCG.

Third, Illinois turns to the ICCG and the GAC to propose regulations to the Illinois Pollution Control Board for a series of activities:

(1) landfilling, land treating, surface impounding or piling of special waste and other wastes which could cause contamination of groundwater and which are generated on the site, other than hazardous, livestock and landscape waste, and construction and demolition debris;

(2) storage of special waste in an underground storage tank for which federal regulatory requirements for the protection of groundwater are not applicable;

(3) storage and related handling of pesticides and fertilizers at a facility for the purpose of commercial application;

(4) storage and related handling of road oils and de-icing agents at

---

49 415 Ill. Comp. Stat. § 5/14.
50 415 Ill. Comp. Stat. § 5/3.61.
51 415 Ill. Comp. Stat. § 5/14.1.
52 415 Ill. Comp. Stat. § 5/14.2.
53 415 Ill. Comp. Stat. § 55/4.

a central location; and

(5) storage and related handling of pesticides and fertilizers at a central location for the purpose of distribution to retail sales outlets.[54]

Fourth, Illinois creates special setback zones related to storage, handling, and application of agrichemicals (including pesticides and fertilizers) used for commercial purposes.[55] The IEPA and Illinois Department of Agriculture jointly administer the program.

Finally, the IEPA creates a regiional groundwater protection planning program which designates priority groundwater protection planning regions.[56] A Regional Planning Committee is created for each region, which functions to:

(1) identification of and advocacy for region-specific groundwater protection matters;

(2) monitoring and reporting the progress made within the region regarding implementation of protection for groundwaters;

(3) maintaining a registry of instances where the Agency has issued an advisory of groundwater contamination hazard within the region;

(4) facilitating informational and educational activities relating to groundwater protection within the region; and

(5) recommending to the Agency whether there is a need for regional protection pursuant to Section 17.3. Prior to making any such recommendation, the regional planning committee shall hold at least one public meeting at a location within the region. Such meeting may be held after not less than 30 days notice is provided, and shall provide an opportunity for public comment.

6.6.3 Montana's Wellhead Protection Program

The Montana Wellhead Protection Program was approved by EPA in 1994, and provides the foundation for these methods and criteria for wellhead and drinking water protection in the state. Source water protection areas are divided into regions to distinguish areas where it takes only days for groundwater to reach a well from areas where it takes years for groundwater to travel to the well. This division allows for differential management of potential sources of pollution. Differential management imposes stricter controls on potential sources of contamination close to a well than those further away from the well. Delineating regions allows a

---

[54] 415 Ill. Comp. Stat. § 5/14.4.
[55] 415 Ill. Comp. Stat. § 5/14.6.
[56] 415 Ill. Comp. Stat. § 5/17.2.

community to focus limited resources on the regions closest to the well or intake. The methods and criteria of Montana's plan are tailored to the unique character of Montana's public water supplies and the nature of the source waters available to them. Sixty percent of community public water supplies in Montana serve 100 or fewer people. These small supplies have very limited financial and staff resources, so methods and criteria are designed to be cost-effective so resources can be directed toward effective management of source water protection areas.

Montana Department of Environmental Quality regulations prohibit construction of wells that allow deleterious interflow between aquifers.[57] Under these rules an aquifer is defined as any discrete water-bearing unit with a specific water chemistry, temperature, or hydrostatic head.[58] Deleterious interflow is deemed to occur if any of these parameters are changed in an aquifer because a well provides a conduit for flow from another aquifer.

In establishing Control Regions for wellhead protection, Montana establishes a circular area within a fixed radius from a water supply well. For groundwater systems determined to be under direct influence of surface water the control region should encompass all flowing water including flowing drain ditches within 1 mile of the intake. Also, a vegetated land-buffer is desirable to prevent contaminant spills or contaminated runoff from flowing directly into the water source.

Montana also establishes Inventory Regions associated with the control regions that focus on pollution prevention activities where water is expected to contribute to a public water supply. Management actions may address specific contaminants such as microbes, nitrates, volatile organic compounds, pesticides and herbicides, or specific metals. Regulations may be implemented to prohibit potential sources of contamination or to require leak detection monitoring or secondary containment for chemical storage tanks. Houses utilizing septic systems can be hooked up to public sewage treatment systems, parkways or greenways can be dedicated to filter runoff and increase infiltration, and containment barriers can be constructed to prevent accidental chemical spills on roads or railways adjacent to surface water sources.

In addition, Recharge Regions are created in which potential sources of contamination can be limited or controlled, best management practices can be implemented, and public education programs can be organized. Land use agreements, surface water monitoring, and site plan reviews are additional tools of protective management. Areas where recharge to groundwater originates are managed to maintain or improve the quality of the water that replenishes the aquifer. Recharge regions correspond to

---

[57] Admin. Rules Mont. § 36.21.650.
[58] Admin. Rules Mont. § 36.21.634.

areas bounded by physical and hydrologic limits of unconfined or semi-confined aquifers. The method recommended for delineating recharge regions for these groundwater sources is hydrogeologic mapping, which identifies groundwater flow boundaries, physical and hydrologic features that limit groundwater flow to a well.

Part

IV

Air Pollution Controls

7

The Clean Air Act

7.1 History and Policy of Air Pollution Controls

Air pollution is not a new concern, but was recognized in Europe as early as the Industrial Revolution when the burning of fossil fuels created lethal conditions when combined with congested living conditions. The first "smoke abatement" ordinances were used to forbid coal burning in London in the year 1273 during the reign of Edward I. These early ordinances were used to convict (and execute) violators.[1]

7.1.1 Early U.S. Actions

Similar circumstances occurred in the United States during the nineteenth century, and led to the first smoke abatement ordinances in several larger, industrial U.S. cities (e.g., Chicago and Cincinnati). These early ordinances required furnaces to consume their own smoke or forbid burning high-sulfur coal, and levied fines of $10–100 for offenders.[2] These early smoke abatement ordinances were based on common law nuisance (public or private), and were required to be "reasonably necessary" to protect public welfare, and not "unduly oppressive" to the regulated community.[3] Despite challenges by a regulated community that was used to getting its own way, these early controls were held by the Supreme Court to be a valid exercise of the "police power" of the municipality to

---

[1] See Anderson, Mandelker, and Tarlock, *Environmental Protection: Law and Policy*, 2d ed. (Little, Brown & Co., 1990), Chap. 3.
[2] Laitos, Legal Institutions and Pollution: Some Intersections Between Law and History, 15 Nat. Resources J. 423 (1975).
[3] See Northwestern Laundry v. City of Des Moines, 239 U.S. 486 (1916).

protect public health and welfare.[4]

Early ordinances based on common law were problematic, however, because of the difficulty of proving actual harm, and difficulties in establishing and measuring ambient concentrations of pollutants. In *Boomer v. Atlantic Cement Co.*,[5] the New York Court of Appeals (New York's "supreme court") stated that a cement plant was guilty of a nuisance by seriously polluting the air, but permitted the plant to continue polluting indefinitely (while paying modest damages) because the monetary damages were small compared to the value of the plant's operation to the community.

Some particularly dangerous activities that produced air pollution were considered to be "ultrahazardous," which triggered a "strict liability" standard on the part of the polluter. As discussed more fully in chapter 2, this common law standard means that a plaintiff need not prove that the damage was intentional or even negligent (a very favorable standard for the plaintiff). Some examples of ultrahazardous activities are blasting,[6] crop dusting,[7] factory emissions,[8] and oil wells and refineries.[9]

7.1.2 Actions After World War II

Along with other environmental problems, concerns about air pollution increased during the period of relative tranquility and prosperity following World War II. Common law actions were generally viewed as inadequate to stop widespread air pollution. In 1955, the federal government first entered the air pollution field with the Air Pollution Control Act, which authorized a modest research and technical assistance program.[10] During this time, motor vehicle emissions were first linked conclusively to smog problems in cities like Los Angeles. In dealing with its problems of severe air pollution, California's pollution controls became the model for federal efforts.

Eight years later, the Clean Air Act of 1963[11] directed the Department of Health, Education, and Welfare (HEW, now the Department of Health

---

[4] Huron Cement Co. v. City of Detroit, 362 U.S. 440 (1960).
[5] Boomer v. Atlantic Cement Co., 257 N.E.2d 870 (1970).
[6] Caporale v. C.W. Blackeslee & Sons, Inc., 175 A.2d 561 (Conn. 1961).
[7] Young v. Dater, 363 P.2d 829 (Okla. 1961), and Loe v. Lenhardt, 362 P.2d 3112 (Ore. 1961).
[8] Dutton v. Rocky Mountain Phosphates, 438 P.2d 674 (Mont. 1968).
[9] Berry v. Shell Petroleum Co., 33 P.2d 953 (Kan. 1934), *reh'g denied*, 40 P.2d 359 (Kan. 1935).
[10] Pub. L. No. 84-159, 69 Stat. 322.
[11] Pub. L. No. 88-206, 77 Stat. 392.

and Human Services) to provide scientific information called "criteria documents" to states on the effects of air pollutants. The individual states were then required to establish abatement programs based on the criteria documents. The Act also authorized HEW to monitor interstate pollution hot spots, but abatement remained the responsibility of state and local governments. If pollution "endangered the health or welfare" of the public, HEW could theoretically begin a lengthy enforcement process under the U.S. Attorney General, along with a series of conferences. Unfortunately, the system was so cumbersome that only 11 conferences and one enforcement action ever took place.

In 1965, the Motor Vehicle Air Pollution Control Act required HEW to promulgate emission standards.[12] Automobile manufacturers were given 2 years notice and a reasonable time to comply with new standards. Enforcement was intermittent. At about the same time, the 1967 Air Quality Act[13] required HEW to designate national geographic air quality control regions that defined levels of pollution that would maintain health and welfare. Within these regions, states adopted numerical air quality standards for each major pollutant. State standards were based on HEW criteria documents, and were subject to HEW approval. Moreover, HEW could establish its own standards if the state failed to do so. Each state was also required to develop an "implementation plan" which set standards for individual sources of pollution.

The 1967 Air Quality Act was a failure, however, despite its commendable intentions. The Act failed because: (a) the scientific and jurisdictional problems faced by state and federal agencies were insurmountable; (b) the preparation of implementation plans proved to be an enormous task; (c) enforcement was extremely difficult; and (d) at the time it was less clear that the constitution's Commerce Clause allowed federal regulation of air pollution.[14]

7.2 The Nature and Sources of Air Pollution

"Air pollution" can be defined as substances or particles that are undesirable in ambient (outdoor) air. Some air pollution occurs naturally, such as when the decay of natural organic materials releases hydrocarbons and H_2S (hydrogen sulfide) into the air. The air pollutants of greatest concern are anthropogenic (man-made) pollutants.

Section 302 of the federal Clean Air Act (CAA) states:

---

[12] Pub. L. No. 89-272, 79 Stat. 992.
[13] Pub. L. No. 90-148, 81 Stat. 485.
[14] See Pennsylvania v. EPA, 500 F.2d 246 (3d Cir. 1974).

The term "air pollutant" means any air pollution agent or combination of such agents, including any physical, chemical, biological, radioactive (including source material, special nuclear material, and byproduct material) substance or matter which is emitted into or otherwise enters the ambient air. Such term includes any precursors to the formation of any air pollutant, to the extent the Administrator [of the EPA] has identified such precursor or precursors for the particular purpose for which the term "air pollutant" is used.[15]

There are, of course, many pollutants that may reach the ambient air that we breathe. The Clean Air Act currently recognizes six "criteria pollutants" that receive special attention. These six criteria pollutants were selected because their health affects are known; and relatively simple technology is available to measure them. As will be discussed below, the original list included "hydrocarbons," but these were dropped because causation of health effects was extremely difficult to prove. However, "lead" was subsequently added following several court decisions.[16] The six criteria pollutants will be discussed in section 7.3.2 below.

7.2.1 Monitoring Air Pollution

Monitoring is a major problem in air quality controls because laws and regulations set precise numerical standards and assume monitoring technology is adequate (often it is not). TSPs can be measured using the "Ringelmann Smoke Chart," devised in the 1890s, consists of progressively shaded gray spots on a chart that are compared to the plume of smoke to estimate opacity. It cannot be used at night, in rain, or in high winds, and is highly subjective. Other simple, much more accurate techniques are available (e.g., particulate filtration devices). The Ringelmann chart is still used today, although it is a frequent source of controversy.[17] Gases such as SO_2, NO_x, CO, and ozone must be measured using more complicated absorption and adsorption techniques. These techniques are relatively reliable and inexpensive, but often fail to give the precise, long-term values anticipated by clean air legislation and regulations. Direct physical measurement techniques such as chemiluminescence is gaining in popularity, but the techniques remain expensive or unreliable for many pollutants.

---

[15] 42 U.S.C. § 7602(g).
[16] See NRDC v. Train, 545 F.2d 320 (2d Cir. 1976); and Lead Indust. Ass'n, Inc. v. EPA, 647 F.2d 1130 (D.C. Cir. 1980).
[17] See Chemithon Corp. v. Puget Sound Air Pollution Control Agency, 18 Envt. Rep. Cas. (BNA) 1647 (Wash. Ct. App. 1983).

Measurements from monitoring devices vary tremendously depending on the distance from the source, height above the ground, weather conditions, time of day, and ground configuration. Although it appears increasingly irrelevant, monitoring ambient air quality remains the major form of air quality regulation. Some progress has been made in modeling air pollution using sophisticated computer technologies that incorporate a variety of factors that affect air quality.

7.2.2 Global Issues in Air Pollution

Accumulations of chlorofluorocarbons (CFCs) in the upper atmosphere has resulted in the depletion of ozone in the earth's upper atmosphere. As a result, harmful radiation is not filtered as effectively. Continuous exposure to such radiation can increase the incidence of skin cancers, increase birth defects, and has a deleterious effect on many plants and animals (e.g., it may be responsible for worldwide, large scale extinction of many species of amphibians).

Global warming, also known as the "greenhouse effect," results from an increase in the atmosphere of carbon dioxide (CO_2) from the burning of fossil fuels and other "greenhouse gasses." Global warming is leading to a gradual elevation of sea levels, and is the likely cause of changes in weather patterns (which may ultimately lead to massive crop failures) worldwide. Acid rain is caused when droplets of acids form in clouds and fall to earth as rain. The usual source of the acids is industrial air pollution, such as SO_2 and NO_x. Pollutants from industries in the Ohio River valley in the United States are thought to be responsible for extensive acid rain damage in New England, and in eastern Canada. Acid rain from western Europe is responsible for extensive acid rain damage to the Black Forest and other areas in Bavaria.

7.3 Federal Clean Air Act

Although it technically consisted of "amendments" to the 1967 Air Quality Act, the entire Clean Air Act (CAA) was completely rewritten in 1970.[18] The 1970 Act is based on a series of federal/state partnerships. The federal EPA would set air quality and emission standards, but states would determine how the federal standards were to be met, and issue permits to sources of air pollution. The new CAA recognized that stationary and mobile air pollution sources should be regulated separately.

---

[18] The CAA originally appeared at 42 U.S.C. § 1857.

In general, Title I deals with stationary sources, Title II with mobile sources, and Title III with administration and judicial review.

Title I of the CAA requires EPA to set up three different kinds of nationwide standards. First, EPA must create Air Quality Control Regions and develop National Ambient Air Quality Standards (NAAQS), which determine the maximum concentrations of designated "criteria" pollutants (e.g., CO and SO_2) in ambient air. Second, EPA must designate New Source Performance Standards (NSPS), which establish allowable emission limitations for various kinds of new stationary sources of air pollution. Third, EPA must set National Emissions Standards for Hazardous Air Pollutants (NESHAP) for which no ambient air quality standards exist.[19]

In the 1977 amendments, a pollutant subject to NSPS which is not listed as either "hazardous" or as a "criteria" pollutant is called a "designated" pollutant.[20] Some examples of designated pollutants are fluorides from aluminum plants and sulfuric acid mist. These receive little attention from the federal government.

Once EPA had developed the NAAQS, states were required to determine how they can be attained and maintained by developing a State Implementation Plan (SIP).

7.3.1 Ambient Air Quality Standards

The 1970 CAA required EPA to prepare National Ambient Air Quality Standards for criteria pollutants that have an adverse effect on public health and welfare, and result from diverse mobile and stationary sources. "Primary standards" were to be developed "allowing an adequate margin of safety, [that] are requisite to protect the public health."[21] These are, in essence, health-based standards.

"Secondary standards" were to be developed that "specify a level of air quality the attainment and maintenance of which . . . is requisite to protect the public welfare from any known or anticipated adverse effects associated with the presence of such air pollutants in the ambient air."[22] Secondary standards are welfare-based standards.

Primary and secondary standards for each of the criteria pollutants must "reflect the latest scientific knowledge useful in indicating the kinds and

---

[19] "Hazardous" air pollutants are defined as those contained in an extensive list at 42 U.S.C. § 7412(b).
[20] 42 U.S.C. § 7411(d)(1).
[21] 42 U.S.C. § 7409(b)(1).
[22] 42 U.S.C. § 7409(b)(2).

extent of all identifiable effects on public health and welfare."[23] These standards must establish a minimally acceptable level of ambient air quality that protects humans and the human environment from all known effects, plus any unknown effects that are legitimate sources of concern.

An important question to economists as well as policy interests is whether EPA should consider costs of compliance in regulating criteria pollutants? Interestingly, under CAA Section 108 and Section 109,[24] EPA must regulate criteria pollutants that "endanger the public health or welfare" without considering costs. However, under CAA Section 211 EPA may regulate fuel additives that "endanger the public health or welfare" but has discretion whether to consider compliance costs or not.[25]

The CAA amendments of 1990 established that attainment of primary NAAQSs in effect at the time (for attainment and unclassified areas) was to be reached by November 15, 1993, or within five years of a finding that a SIP is inadequate, whichever is later.[26] Secondary NAAQS are to be the subject of a report by EPA that will discuss the effects on welfare and the environment of criteria pollutants, among other things.

NAAQS standards are implemented in three ways. First, NAAQS are implemented by nationwide, technology-forcing emission limitations on mobile sources such as automobiles. These standards include motor vehicle inspection and maintenance programs, the requirement for catalytic convertors on the exhaust systems of automobiles, and the use of low-lead gasolines. Second, NAAQS are implemented by nationwide, technology-forcing emission limitations on new or modified stationary sources of pollution. Emission limitations under several circumstances will be discussed at length below. Finally, NAAQS are implemented by state SIPs that implement the NAAQSs through emission limitations on stationary and, to a limited extent, mobile sources.

States are given latitude as to the mix of control technologies they impose on existing sources under the SIP,[27] but the CAA requires that the NAAQS be met through the use of "continuous" controls. Continuous controls include full-time technologies such as scrubbers, catalytic converters, and filters. Dispersion techniques (e.g., tall smoke stacks that disperse pollution over a wider area) and intermittent controls (reduce

---

[23] 42 U.S.C. § 7408(a)(2).

[24] 42 U.S.C. §§ 7408-7409.

[25] 42 U.S.C. § 7545. See Small Refiners Lead Phase-Down Task Force v. U.S. EPA, 705 F.2d 506 (D.C. Cir. 1983).

[26] 42 U.S.C. § 7410(n)(1).

[27] See Train v. NRDC, 421 U.S. 60 (1975).

emissions only periodically) are disfavored, and may be used *only* when continuous controls are technologically or economically unfeasible.[28]

7.3.2 Criteria Pollutants

As noted above, six "criteria pollutants" were originally designated by EPA: CO, SO_2, ozone, NO_x, hydrocarbons, and TSPs. If the EPA finds that an additional pollutant has an adverse effect on health and welfare, it must be included on the list of criteria pollutants. Lead was subsequently added following several court decisions.[29] Hydrocarbons were eventually dropped from the original list when proving causation became a problem.

Once a substance is listed as a criteria pollutant, EPA must establish a NAAQS standard for it. Under CAA Section 109(d)(1), EPA must review and revise NAAQS air quality criteria by December 31, 1980, and do so again every five years.[30] Several attempts have been made to use the judicial system to force EPA to revise its standards more often, although none have been successful.[31]

The six current criteria pollutants are as follows:

Total Suspended Particulates (TSPs, PM_{10}). Solid particles or liquid droplets that remain suspended in the air, with diameters of 10μ or larger. About two-thirds of TSPs are from stationary industrial sources, while the remainder comes from from motor vehicles (mainly diesel engines) and from waste disposal. EPA requires that PM_{10} levels be monitored for annual levels, as well as 24-hour levels.[32] The U.S. EPA has proposed TSP regulations that would regulate particles as small as $2.5\mu m$ (the "$PM_{2.5}$" standard), since it has been demonstrated that small particles are a greater health risk than larger particles. Many industries and trade groups opposed the proposed standard, in part because they felt that it would be prohibitively expensive to implement. A three-judge panel of the D.C. Circuit Court of Appeals invalidated EPAs $PM_{2.5}$ standard (along with its

---

[28] See Kennecott Copper Corp. v. Train, 526 F.2d 1149 (9th Cir. 1975), cert. denied, 425 U.S. 935 (1976). For what constitutes a "dispersion technique," see Kamp v. Hernandez, 752 F.2d 144, opinion modified, 778 F.2d 527 (9th Cir. 1985).
[29] See NRDC v. Train, 545 F.2d 320 (2d Cir. 1976); and Lead Indust. Assoc., Inc. v. EPA, 647 F.2d 1130 (D.C. Cir. 1980).
[30] 42 U.S.C. § 7409(d)(1).
[31] For example, see American Petroleum Inst. v. Costle, 665 F.2d 1176 (D.C. Cir. 1981) [seeking revision of the ozone standard]; and Environmental Defense Fund v. EPA, 27 Env't Rep. Cas. (BNA) 2008 (S.D.N.Y. 1988) [seeking revision of the SO_2 standard].
[32] See Ober v. EPA, 84 F.3d 304 (9th Cir. 1996).

proposed ozone standard) in *American Truckers Associations, Inc. v EPA*,[33] holding that EPA failed to articulate an "intelligible principle" for balancing factors in making its decision, such that the regulations violated the nondelegation clause of the U.S. Constitution.[34] At this writing, it is unclear whether EPA will appeal the decision, or take further action to implement the $PM_{2.5}$ standard.

Sulfur Dioxide (SO_2). Most (80 percent) atmospheric SO_2 results from fossil fuel combustion (especially electric utilities), with the remainder originating primarily from the smelting of ores (e.g., lead and copper). SO_2 (along with NO_x) is a major source of acid rain, which results from the reaction of SO_2 with water. Water droplets containing sulfuric acid molecules then fall to earth as rain. This acid rain has caused structural damage to buildings, and significantly lowered the pH of many ecosystems particularly in the eastern U.S. and northern Europe. Acid rain problems were addressed in the late 1980s by the federal National Acid Precipitation Abatement Program (NAPAP), but little has been done to rectify the problem.

Nitrogen Oxides (NO, NO_2 = "No_x"). About half of NO_x emissions comes from motor vehicles, while the other half is primarily from power plants. NO_x emissions alone can exacerbate heart, lung, and cardiovascular diseases, as well as damaging materials and contributing to acid rain. No_x also combine with VOCs (volatile organic comounds) in the presence of sunlight to form ozone (O_3).

Carbon Monoxide (CO). Most CO production results from incomplete combustion of fossil fuels, and can be reduced with the use of catalytic converters or thermal exhaust conversion on motor vehicles. CO has pronounced health effects, binding to hemoglobin in the blood and displacing oxygen. This causes damage to the respiratory and cardiovascular systems, reduces mental functions, and alters fetal development.

Ozone (O_3). Ozone is a "photochemical oxidant," which results from photochemical reactions between NO_2 and volatile organic compounds (VOCs). Ozone in the upper atmosphere serves to filter harmful radiation, but high concentrations in the lower atmosphere cause health and economic damage. At ground level, most ozone production results from

---

[33] American Truckers Ass'n, Inc. v EPA, 175 F.3d 1027 (D.C. Cir. 1999).
[34] See U.S. Const. art. I, § 1 ("All legislative powers herein granted shall be vested in a Congress of the United States."). The court felt that EPA had attempted to exercise legislative powers not delegated by Congress.

industrial sources and from motor vehicles. Ozone is controlled indirectly, by decreasing emissions of NO_x, SO_2, and VOCs (for example, gasoline pumps that recapture evaporated gasoline).

Lead (Pb). Lead was not recognized as a dangerous air pollutant until the 1970s. Until the late 1970s, most airborne lead resulted from motor vehicle emissions due to leaded gasoline. Today, most lead comes from industrial sources. Lead can be reduced by using unleaded fossil fuels, and by catalytic conversion and scrubbers. Lead causes nervous system disorders, particularly in children, and can cause death with prolonged exposure.

Following the 1970 amendments, the EPA classified the entire country into "attainment" and "nonattainment" areas with respect to each criteria pollutant. Under CAA Section 107(d)(1)(A), each state must designate all areas within the state as one of the following:

(i) nonattainment, any area that does not meet (or that contributes to ambient air quality in a nearby area that does not meet) the national primary or secondary ambient air quality standard for the pollutant,

(ii) attainment, any area (other than an area identified in clause (i)) that meets the national primary or secondary ambient air quality standard for the pollutant, or

(iii) unclassifiable, any area that cannot be classified on the basis of available information as meeting or not meeting the national primary or secondary ambient air quality standard for the pollutant.[35]

The 1990 amendments began a special "acid rain" program, which specifically targets SO_2 emissions by limiting tonnage in areas of concern.[36]

7.3.3 State Implementation Plans

A centerpiece of the CAA is the requirement that each state develop a State Implementation Plan (SIP), which illustrates how the state will attain the NAAQS by the applicable attainment deadlines. SIPs must be approved by the EPA as containing sufficient measures to timely attain NAAQS and meet other requirements described below. SIPs must contain air pollution measures in adopted, "regulatory" form within one year after

---

[35] 42 U.S.C. § 7407(d)(1)(A).
[36] CAA §§ 401 et seq., 42 U.S.C. §§ 7651 et seq.; see Indianapolis Power & Light Co. v. EPA, 58 F.3d 643 (D.C. Cir. 1995).

approval by the EPA. Upon approval by the EPA, SIP requirements can be enforced against regulated sources by EPA and by any citizen.

CAA Section 110(a)(2) spells out the contents of the SIP, some of which are:

Each implementation plan submitted by a State under this chapter shall be adopted by the State after reasonable notice and public hearing. Each such plan shall —

(A) include enforceable emission limitations and other control measures, means, or techniques (including economic incentives such as fees, marketable permits, and auctions of emissions rights), as well as schedules and timetables for compliance, as may be necessary or appropriate to meet the applicable requirements of this chapter;

(B) provide for establishment and operation of appropriate devices, methods, systems, and procedures necessary to -

(i) monitor, compile, and analyze data on ambient air quality, and

(ii) upon request, make such data available to the Administrator [of the EPA];

(C) include a program to provide for the enforcement of the measures described in subparagraph (A), and regulation of the modification and construction of any stationary source within the areas covered by the plan as necessary to assure that national ambient air quality standards are achieved, including a permit program as required in parts C and D of this subchapter;

(D) contain adequate provisions —

(i) prohibiting, consistent with the provisions of this subchapter, any source or other type of emissions activity within the State from emitting any air pollutant in amounts which will — (I) contribute significantly to nonattainment in, or interfere with maintenance by, any other State with respect to any such national primary or secondary ambient air quality standard, or (II) interfere with measures required to be included in the applicable implementation plan for any other State under part C of this subchapter to prevent significant deterioration of air quality or to protect visibility,

(ii) insuring compliance with the applicable requirements of sections 7426 and 7415 of this title (relating to interstate and international pollution abatement); . . .

(F) require, as may be prescribed by the Administrator —

(i) the installation, maintenance, and replacement of equipment, and the implementation of other necessary steps, by owners or operators of stationary sources to monitor emissions from such sources,

(ii) periodic reports on the nature and amounts of emissions and emissions-related data from such sources, and

(iii) correlation of such reports by the State agency with any emission limitations or standards established pursuant to this chapter, which reports shall be available at reasonable times for public inspection; . . .

(K) provide for —

(i) the performance of such air quality modeling as the Administrator may prescribe for the purpose of predicting the effect on ambient air quality of any emissions of any air pollutant for which the Administrator has established a national ambient air quality standard, and

(ii) the submission, upon request, of data related to such air quality modeling to the Administrator; . . .[37]

Among the numerous other SIP requirements are: a mandate that the region achieve a three percent annual reduction in emissions of ozone precursors (VOC an NO_x); a requirement that new sources over 10 tons per year of VOC or NO_x, and modifications to such sources, achieve lowest achievable emission rate and offset their emission increases by equal reductions elsewhere in the region; transportation control measures to reduce vehicle trips; and measures to increase average vehicle occupancy of commuters to employers of over 100 employees.[38]

States can be forced to adopt or revise SIPs or SIP provisions, either through direct legal challenge or via the CAA Section 304 "citizen suit"provision.[39] On the other hand, states can challenge SIP requirements, or sanctions imposed against the state for failing to meet SIP standards.[40]

7.4 Nonattainment Areas

Nonattainment areas are those areas that have failed to meet primary or secondary NAAQS for a criteria pollutant, or those that contribute to the failure of a nearby area to meet primary or secondary NAAQS.[41] A key provision of the 1990 amendments was a restructuring of the sections

---

[37] 42 U.S.C. § 7410(a)(2).
[38] Id.
[39] 42 U.S.C. § 7604. See NRDC v. N.Y. State Dep't of Environmental Conservation, 668 F. Supp. 848 (S.D.N.Y. 1987), 834 F.2d 1987 (2d Cir. 1987); and Citizens for a Better Environment v. Costle, 515 F. Supp. 264 (N.D. Ill. 1981).
[40] See Virginia v. United States, 74 F.3d 517 (4th Cir. 1996).
[41] 42 U.S.C. § 7407(d)(1)(A)(I).

related to nonattainment areas, designed to bring them into compliance with NAAQS.

Nonattainment areas are further subdivided into "low," "moderate," or "severe" nonattainment for each criteria pollutant, with increasingly stringent standards for each subdivision.

Section 107 of the CAA generally gives states a choice in dealing with nonattainment area; they can either ban all major new sources or modifications in nonattainment areas, or prepare a SIP that meets more stringent than usual statutory requirements for the area. For most cities, this is no choice at all! The harshness of the SIP depends on the severity of the pollutant(s) in the area, and the specific pollutant.[42]

SIPs in nonattainment areas must comply with nine statutory requirements that generally restrict local and state decision making:[43]

(1) In general . . . [SIP] provisions shall provide for the implementation of all reasonably available control measures [RACM] as expeditiously as practicable (including such reductions in emissions from existing sources in the area as may be obtained through the adoption, at a minimum, of reasonably available control technology [RACT]) and shall provide for attainment of the national primary ambient air quality standards.

(2) . . . provisions shall require reasonable further progress [toward attainment].

(3) . . . provisions shall include a comprehensive, accurate, current inventory of actual emissions from all sources of the relevant pollutant or pollutants.

(4) . . . provisions shall expressly identify and quantify the emissions, if any, of any such pollutant or pollutants which will be allowed . . . from the construction and operation of major new or modified stationary sources in each such area. The plan shall demonstrate to the satisfaction of the Administrator that the emissions quantified for this purpose will be consistent with the achievement of reasonable further progress and will not interfere with attainment of the applicable national ambient air quality standard by the applicable attainment date.

(5) . . . provisions shall require permits for the construction and operation of new or modified major stationary sources anywhere in the nonattainment area, in accordance with section 7503 of this title.

(6) . . . provisions shall include enforceable emission limitations, and such other control measures, means or techniques (including economic incentives such as fees, marketable permits, and auctions of

---

[42] 42 U.S.C. § 7502(c)(1).
[43] 42 U.S.C. § 7502(c)(1)–(9).

emission rights), as well as schedules and timetables for compliance, as may be necessary or appropriate to provide for attainment of such standard in such area by the applicable attainment date specified in this part.

(7) . . . provisions shall also meet the applicable provisions of section 7410(a)(2) of this title [discussed in section 7.3.4 above].

(8) Upon application by any State, the Administrator [of the EPA] may allow the use of equivalent modeling, emission inventory, and planning procedures, unless the Administrator determines that the proposed techniques are, in the aggregate, less effective than the methods specified by the Administrator.

(9) Such [SIP] plan shall provide for the implementation of specific measures to be undertaken if the area fails to make reasonable further progress, or to attain the national primary ambient air quality standard by the attainment date applicable under this part. Such measures shall be included in the plan revision as contingency measures to take effect in any such case without further action by the State or the Administrator.[44]

7.4.1 Emission Standards for Stationary Sources

As noted above, Section 172(c)(5) of the CAA, the SIP must contain a program that requires permits "for the construction and operation of new or modified major stationary sources anywhere in the nonattainment area."[45] A "major" source is defined generally in Section 302(j) of the CAA:

[T]he terms "major stationary source" and "major emitting facility" mean any stationary facility or source of air pollutants which directly emits, or has the potential to emit, one hundred tons per year or more of any air pollutant (including any major emitting facility or source of fugitive emissions of any such pollutant, as determined by rule by the Administrator).

EPA regulations define "major" sources more specifically:

A major stationary source as defined in part D of title I of the [CAA], including:

(i) For ozone nonattainment areas, sources with the potential to emit 100 tpy or more of volatile organic compounds or oxides of

---

[44]   42 U.S.C. § 7502(c)(1)–(9).
[45]   42 U.S.C. § 7502(c)(5).

nitrogen in areas classified as "marginal" or "moderate," 50 tpy or more in areas classified as "serious," 25 tpy or more in areas classified as "severe, and 10 tpy or more in areas classified as "extreme"; except that the references in this paragraph to 100, 50, 25 and 10 tpy of nitrogen oxides shall not apply with respect to any source for which the Administrator has made a finding, under section 182(f) (1) or (2) of the [CAA], that requirements under section 182(f) of the [CAA] do not apply;

    (ii) For ozone transport regions established pursuant to section 184 of the [CAA], sources with the potential to emit 50 tpy or more of volatile organic compounds;

    (iii) For carbon monoxide nonattainment areas: (A) That are classified as "serious," and (B) in which stationary sources contribute significantly to carbon monoxide levels as determined under rules issued by the Administrator, sources with the potential to emit 50 tpy or more of carbon monoxide; and

    (iv) For particulate matter (PM-10) nonattainment areas classified as "serious," sources with the potential to emit 70 tpy or more of PM-10.

Existing major stationary sources in a nonattainment area must meet the standard of "Reasonably Available Control Technology" (RACT).[46] EPA regulations define RACT as devices, systems, process modifications, etc. that permit attainment of limits for stationary sources established by EPA.[47] This loose interpretation of RACT was upheld by the federal courts.[48] However, within the Lake Michigan Ozone Study area (LMOS, an ozone nonattainment area), the EPA is granting exemptions from the RACT requirements for NO_x effective February 26, 1996. Apparently, stationary sources will be subject to a "reasonable further progress" standard where exemptions are granted.

All new major stationary sources or modifications of existing stationary sources must obtain a permit and demonstrate that it will use the Lowest Achievable Emission Rate (LAER) for each nonattainment pollutant. A "stationary source" to which the LAER standard applies is:

[A]ny building, structure, facility, or installation which emits or may emit any air pollutant. Nothing in subchapter II of this chapter

---

[46] 42 U.S.C. § 7502(c)(1).
[47] 40 CFR § 51.1(o).
[48] See Bethlehem Steel Corp v. EPA (723 F.2d 1303 (7th Cir. 1983); National Steel Corp. v. Gorsuch, 700 F.2d 314 (6th Cir. 1983); and Michigan v. Thomas, 805 F.2d 176 (6th Cir. 1986).

relating to nonroad engines shall be construed to apply to stationary internal combustion engines.[49]

A "modified" stationary source is:

The term "modification" means any physical change in, or change in the method of operation of, a stationary source which increases the amount of any air pollutant emitted by such source or which results in the emission of any air pollutant not previously emitted.[50]

The LAER standard applied to new or modified stationary sources is defined in CAA Section 171(3) as:

The term "lowest achievable emission rate" means for any source, that rate of emissions which reflects —
  (A) the most stringent emission limitation which is contained in the implementation plan of any State for such class or category of source, unless the owner or operator of the proposed source demonstrates that such limitations are not achievable, or
  (B) the most stringent emission limitation which is achieved in practice by such class or category of source, *whichever is more stringent*. In no event shall the application of this term permit a proposed new or modified source to emit any pollutant in excess of the amount allowable under applicable new source standards of performance.[51]

SIPs or SIP revisions for nonattainment areas must be submitted to EPA within three years of the time EPA designates the area as nonattainment, after which EPA has one year to take action on the SIP.[52] The deadline for attainment is five years from the date of designation, although EPA can extend the date by ten years based on the severity of nonattainment and the difficulty of pollution controls.[53] EPA must also set an attainment date for secondary NAAQS "as expeditiously as practicable."[54]

If EPA disapproves all or part of a SIP, if a state fails to submit all or part of a SIP, or if the state fails to implement a SIP requirement, then the state may be subject to sanctions under CAA Sections 110(k) and 110(m).[55] For a nonattainment area, there are two such sanctions, both of

---

[49] 42 U.S.C. § 7411(a)(3).
[50] 42 U.S.C. § 7411(a)(4).
[51] 42 U.S.C. § 7501(3). Emphasis added.
[52] 42 U.S.C. §§ 7502(b) and 7410(k).
[53] 42 U.S.C. § 7502(a)(2)(A).
[54] 42 U.S.C. §§ 7502(a)(2)(B).
[55] 42 U.S.C. § 7410(k) and (m).

which must be applied if there is a lack of good faith by the state. One such sanction permits EPA to impose an offset ratio of at least 2 to 1 on the state for all new stationary sources. The second is that EPA, with the approval of the Secretary of Transportation, may prevent the state from receiving federal funding for its highways.

7.4.2 "Offsets" and Emission Trading

New, major stationary sources or modifications of existing major sources within a nonattainment area for a criteria pollutant must meet other requirements as well. For example, emissions from a new or modified source may be offset by reductions in pollution elsewhere in the area, so long as "reasonable further progress" toward implementation is achieved.[56] This technique would allow a new producer of air pollution to enter what would appear to be a closed market.

The idea of "offsets" has been extremely controversial since it actually allows a source to *increase* its level of pollution, so long as the increase is offset by decreases elsewhere. EPA uses a concept known as the air pollution "bubble" to explain offsets. Imagine that all sources of particular criteria pollutant were placed under a large "bubble." It should be possible, then, to measure and regulate the air pollution within the bubble. While real "air pollution bubbles" do not exist, they at least allow EPA to represent how a single source can increase pollution while total pollution within the area (i.e., within the bubble) decreases.

The amount of offset required depends on the severity of nonattainment within the bubble, and the particular pollutant involved. Offsets are always greater than 1:1, and may be increased to 2:1 if a state has been subject to CAA Section 179 sanctions for failures in submission or implementation of a SIP.[57]

Offsets can be traded among polluters within the same bubble. If one polluter is able to reduce emissions to levels below what is allowed under its permit, it can trade the excess to new polluter. Pollution "trading" allows a new polluter to purchase (or trade) "credits" which represent the other polluter's pollution reduction. EPA regulations under CAA Section 179 govern the emissions trading process.[58]

Offsets have also allowed for emissions "banking" and a healthy trade in "pollution futures" by polluters who traded away their rights to pollute to other polluters. The U.S. Supreme Court approved the bubble concept

---

[56] 42 U.S.C. §§ 7503(a)(1) and 7503(c).
[57] 42 U.S.C. § 7509. See discussion in Wooley, *Clean Air Act Handbook: A Practical Guide to Compliance*, 8th ed. (West Group, 1998) at § 1.05[5][a].
[58] 40 CFR Part 51.

in *Chevron, USA, Inc. v. NRDC.*[59]

Advocates of pollution trading argue that this approach saves money, promotes innovative technology, and continuously reduces pollution through market incentives. They claim that technology-based regulations, commonly referred to as "command and control," are economically inefficient and rigidly over-prescriptive.[60] Critics argue that these programs have resulted in adverse public health impacts, fraud, and manipulation of the trading market to reward the worst polluters.[61]

7.5 Prevention of Significant Deterioration

A series of court decisions in the early 1970s led EPA to designate many relatively unpolluted areas as "prevention of significant deterioration" (PSD) areas.[62] The 1977 amendments required by statute that PSD areas be designated.[63]

At any time, the air in a particular area may be either cleaner or dirtier than the NAAQS. For the parts of the country with cleaner air, CAA Part C seeks to "prevent the significant deterioration" of the air quality, particularly in areas of special natural, recreational, scenic, or historic value.[64] It also seeks to protect the special visibility values of certain clean air areas, for example, the scenic qualities of many parks.[65]

The purposes of the PSD program under CAA Part C are:

(1) to protect public health and welfare from any actual or potential adverse effect which in the Administrator's judgment may reasonably be anticipate[d] to occur from air pollution or from exposures to pollutants in other media, which pollutants originate as emissions to the ambient air, notwithstanding attainment and maintenance of all national ambient air quality standards;

(2) to preserve, protect, and enhance the air quality in national parks,

---

[59] Chevron, USA, Inc. v. NRDC, 467 U.S. 837 (1984). For a case involving a trade in pollution rights, see Citizens Against Refinery's Effects, Inc. v. EPA, 643 F.2d 183 (4th Cir. 1981).
[60] See Driesen, Is Emissions Trading an Economic Incentive Program? Replacing the Command and Control/Economic Incentive Dichotomy, 55 Wash.& Lee L. Rev. 289 (1998).
[61] See Drury, Belliveau, Kuhn, and Bansal, Pollution Trading and Environmental Injustice: Los Angeles' Failed Experiment in Air Quality Policy, 9 Duke Envtl. L. & Pol'y F. 231 (1997).
[62] See Sierra Club v. Ruckelshaus, 344 F. Supp. 253 (D.D.C. 1972), *aff'd sub nom.*, Fri v. Sierra Club, 412 U.S. 541 (1973).
[63] 42 U.S.C. §§ 7470-7479.
[64] 42 U.S.C. Part C.
[65] 42 U.S.C. § 7491.

national wilderness areas, national monuments, national seashores, and other areas of special national or regional natural, recreational, scenic, or historic value;

(3) to [e]nsure that economic growth will occur in a manner consistent with the preservation of existing clean air resources;

(4) to assure that emissions from any source in any State will not interfere with any portion of the applicable implementation plan to prevent significant deterioration of air quality for any other State; and

(5) to assure that any decision to permit increased air pollution in any area to which this section applies is made only after careful evaluation of all the consequences of such a decision and after adequate procedural opportunities for informed public participation in the decisionmaking process.[66]

It should be noted that paragraph 3 demonstrates that the PSD program does not attempt to stop all construction in PSD areas, but rather seeks to balance development and additional air pollution against economic considerations.

The PSD program addresses resource protection through the establishment of ceilings on additional amounts of air pollution over base-line levels in clean air areas, the protection of the air quality-related values of certain special areas, and additional protection for the visibility values of certain special areas. The PSD title reserves an important resource protection role to the federal land manager, which the CAA defines as the secretary of the department with authority over the affected lands. For example, the Secretary of the Interior has delegated his authority as federal land manager to the Assistant Secretary for Fish and Wildlife and Parks where parklands are concerned.[67]

Within PSD areas, an additional designation of Class I, Class II, or Class III for an area reflects Congress' judgment that certain areas deserve an even higher level of air-quality protection than others (Class I areas are the cleanest). Under CAA Section 162(a), 158 areas were originally designated as Class I areas, including national parks larger than 6,000 acres and national wilderness areas larger than 5,000 acres, in existence on August 7, 1977. These "mandatory" Class I areas may not be redesignated to a less protective classification.[68]

The PSD program also includes measures that can protect many remaining clean areas, known as Class II areas. All areas in either

---

[66] 42 U.S.C. § 7470.
[67] See Ross, The Clean Air Act, Chap. 4, in Mantell (ed.), *Managing National Park System Resources: A Handbook of Legal Duties, Opportunities, and Tools* (The Conservation Foundation, 1990).
[68] See Id. 42 U.S.C. § 7472(a).

attainment or unclassifiable which are not established as Class I are designated as Class II areas unless redesignated under the provisions of CAA Section 164.[69] States and Indian governing bodies can redesignate Class II (and Class III) areas to Class I on their own authority.[70]

7.5.1 Preconstruction Permit

The preconstruction permit program is the primary mechanism of the CAA for implementing the special protection for PSD areas. To obtain a permit, major new and modified sources proposing to locate in clean air areas must:

(1) a permit has been issued for such proposed facility in accordance with this part setting forth emission limitations for such facility which conform to the requirements of this part;

(2) the proposed permit has been subject to a review in accordance with this section, the required analysis has been conducted in accordance with regulations . . ., and a public hearing has been held with opportunity for interested persons . . . to appear and submit written or oral presentations on the air quality impact of such source, alternatives thereto, control technology requirements, and other appropriate considerations;

(3) the owner or operator of such facility demonstrates . . . that emissions from construction or operation of such facility will not cause, or contribute to, air pollution in excess of any (A) maximum allowable increase or maximum allowable concentration for any pollutant in any area to which this part applies more than one time per year, (B) national ambient air quality standard in any air quality control region, or (C) any other applicable emission standard or standard of performance under this chapter;

(4) the proposed facility is subject to the best available control technology [BACT] for each pollutant subject to regulation under this chapter emitted from, or which results from, such facility;

(5) the provisions of subsection (d) of this section with respect to protection of class I areas have been complied with for such facility;

(6) there has been an analysis of any air quality impacts projected for the area as a result of growth associated with such facility;

(7) the person who owns or operates, or proposes to own or operate, a major emitting facility for which a permit is required under this part agrees to conduct such monitoring as may be necessary to determine

---

[69] 42 U.S.C. § 7474.
[70] 42 U.S.C. § 7472.

the effect which emissions from any such facility may have, or is having, on air quality in any area which may be affected by emissions from such source; and

(8) in the case of a source which proposes to construct in a class III area, emissions from which would cause or contribute to exceeding the maximum allowable increments applicable in a class II area and where no standard . . . has been promulgated . . . for such source category, the Administrator [of the EPA] has approved the determination of best available technology as set forth in the permit.[71]

In addition, all major new and modified sources with the potential to affect the visibility of a mandatory Class I area must obtain a new source permit that assures no adverse impact on the Class I area's visibility.[72] However, the limitation in paragraph (3) above (pertaining to maximum allowable increases) does not apply to maximum allowable increases for class II areas in the case of an expansion or modification of a major emitting facility which was already in existence on August 7, 1977, if the plants total allowable emissions of air pollutants is less than fifty tons per year, and if the owner or operator of the facility demonstrates that emissions of particulate matter and sulfur oxides will not cause or contribute to ambient air quality levels in excess of the NAAQS for either pollutant.[73]

A "major emitting facility" is any one of 28 types of plants that might emit 100 tons or more of any regulated pollutant per year, or *any* plant that might emit 250 tons per year. Some nonprofit institutions are exempt.[74] EPA's decision not to consider "fugitive" emissions from surface mining for PSD purposes was upheld.[75] However, even "minor source violations" can carry heavy penalties.[76]

7.5.2 Class I Areas and the Adverse Impact Test

A "baseline" is first established in a Class I area when the first preconstruction permit application is submitted for a major new or modified source. PSD provisions the allow only a small increment of pollutants (initially, only sulfur dioxide and particulate matter) to be added

---

[71] 42 U.S.C. § 7475(a).
[72] 42 U.S.C. § 7475(d)(2)(B).
[73] 42 U.S.C. § 7475(b).
[74] For example, see Town of Brookline v. Gorsuch, 667 F.2d 215 (1st Cir. 1981) [exemption for Harvard University's power plant].
[75] NRDC v. EPA, 937 F.2d 641 (D.C. Cir. 1991).
[76] See United States v. Marine Shale Processors, 81 F.3d 1329 (5th Cir. 1996) [$2.5 milliion penalty].

to the air. CAA Section 163(a) requires EPA to promulgate increments or equivalent protective measures for all pollutants that have national ambient air quality standards.[77] As a result of the decision in *Sierra Club v. Thomas*, EPA promulgated nitrogen oxide increments in 1988.[78]

The PSD program also establishes a site-specific resource test, known as the "adverse impact test," to determine whether emissions from major new and modified sources will cause an "adverse impact" on the "air quality related values" of the Class I area.[79] "Air quality related values" include all values of an area dependent upon and affected by air quality, such as scenic, cultural, biological, and recreational resources, including visibility itself.[80] The adverse impact test imposes an "affirmative responsibility" on federal land managers "to protect the air quality related values (including visibility)" of Class I areas. EPA has further defined adverse impact on visibility to mean perceptible visibility changes that "interfere with the management, protection, preservation, or enjoyment of the visitor's visual experience."

7.5.3 Class I Areas and Visibility Protection

In addition to increment ceilings and the adverse impact test, visibility is also regulated in Class I areas under CAA Section 169A(a):

> Congress hereby declares as a national goal the prevention of any future, and the remedying of any existing, impairment of visibility in mandatory class I Federal areas which impairment results from manmade air pollution.[81]

Protection from visibility impairment due to man-made air pollution is available for the 156 (of 158) statutory Class I areas where visibility is an important value.

CAA Section 169A's goals have been held to apply to "regional haze."[82] With the 1990 amendments, Congress added CAA Section 169B which specifically regulates haze.[83]

---

[77] 42 U.S.C. § 7473(a). See Wooley, *Clean Air Act Handbook: A Practical Guide to Compliance*, 8th ed. (West Group, 1998) at § 1.03.
[78] Sierra Club v. Thomas, 658 F. Supp. 165 (N.D. Cal. 1987).
[79] See Ross, The Clean Air Act. Chapter 4, in Mantell (ed.), *Managing National Park System Resources: a Handbook of Legal Duties, Opportunities, and Tools* (The Conservation Foundation, 1990).
[80] 42 U.S.C. § 7475(d)(2)(D).
[81] See 42 U.S.C. § 7491(a)(1).
[82] Maine v. Thomas, 874 F.2d 883 (1st Cir. 1989).
[83] 42 U.S.C. § 7492.

The EPA is still developing the regulatory program to assure "reasonable progress" toward the national visibility goal. Prodded by lawsuits, EPA has issued "Phase I visibility regulations" that address "plume blight" and other visibility impairment "reasonably attributable" to a specific source or sources.[84] At the time of this writing, EPA has not yet proposed "Phase II" regulations to address visibility impairment from "regional haze."

Under EPA regulations, the thirty-five states and one territory (Virgin Islands) containing Mandatory Class I areas must submit a SIP revision meeting the requirements of the regulations.[85] The SIP revision must:

- Contain a list of integral vistas that are to be listed by the State for the purpose of implementing the regulations, and identification of impairment of visibility in any mandatory Class I Federal area(s).
- Include an assessment of visibility impairment and a discussion of how each element of the plan relates to the preventing of future or remedying of existing impairment of visibility in any mandatory Class I Federal area within the State.
- Adopt emission limitations representing BART and schedules for compliance with BART for each existing stationary facility identified according to regulations.
- Require each source to maintain control equipment required by this subpart and establish procedures to ensure such control equipment is properly operated and maintained.
- Identify and analyze for BART each existing stationary facility which may reasonably be anticipated to cause or contribute to impairment of visibility in any mandatory Class I federal area where the impairment in the mandatory Class I federal area is reasonably attributable to that existing stationary facility.
- BART must be determined for fossil-fuel fired generating plants having a total generating capacity in excess of 750 megawatts.[86]
- Require that each existing stationary facility required to install and operate BART do so as expeditiously as practicable but in no case later than five years after plan approval.
- Provide for a BART analysis of any existing stationary facility that might cause or contribute to impairment of visibility in any mandatory Class I Federal area if: (A) The pollutant is emitted by that existing stationary facility, (B) Controls representing BART for the pollutant

---

[84] See 40 CFR § 51.301.
[85] 40 CFR § 51.302.
[86] See U.S.E.P.A., *Guidelines for Determining Best Available Retrofit Technology for Coal-fired Power Plants and Other Existing Stationary Facilities* (450/3-80-009b, 1980).

have not previously been required under this subpart, and (C) The impairment of visibility in any mandatory Class I Federal area is reasonably attributable to the emissions of that pollutant.

"BART," or "Best Available Retrofit Technology," is defined in the regulations as:

[A]n emission limitation based on the degree of reduction achievable through the application of the best system of continuous emission reduction for each pollutant which is emitted by an existing stationary facility. The emission limitation must be established, on a case-by-case basis, taking into consideration the technology available, the costs of compliance, the energy and nonair quality environmental impacts of compliance, any pollution control equipment in use or in existence at the source, the remaining useful life of the source, and the degree of improvement in visibility which may reasonably be anticipated to result from the use of such technology.

7.5.4 Class II and Class III Areas

Under CAA Section 162(b), all attainment or unclassifiable areas within a state that are not designated Class I areas are automatically classifies as "Class II" areas.[87]

Class II increment ceilings on additional pollution over base-line concentrations allow for moderate development in Class II areas. Class II increments constitute an absolute ceiling on additional pollution in these areas, however, because Congress did not qualify the Class II increment with a variance procedure similar to the adverse impact test for Class I areas.[88]

Class III areas are the dirtiest areas within an attainment or unclassifiable area. There are no Class III areas at the present time. Class III designation could allow for substantial air pollution increases over base-line concentrations, subject (as with all increments) to the ceiling imposed by the national ambient air quality standards. The redesignation process itself, as well as subsequent new source reviews and implementation proceedings, provide opportunities to seek protection of park values.

---

[87] 42 U.S.C. § 7472(b).
[88] See Ross, The Clean Air Act, Chap. 4, in Mantell (ed.), *Managing National Park System Resources: a Handbook of Legal Duties, Opportunities, and Tools* (The Conservation Foundation, 1990).

Redesignation of certain Class II areas to the dirtier Class III classification is prohibited.[89] However, under CAA Section 164(a), reclassification from Class II to Class I (or from Class III to Class II) is permitted for:

(1) an area which exceeds ten thousand acres in size and is a national monument, a national primitive area, a national preserve, a national recreation area, a national wild and scenic river, a national wildlife refuge, a national lakeshore or seashore, and

(2) a national park or national wilderness area established after August 7, 1977, which exceeds ten thousand acres in size.

An area can be reclassified as Class III only if:

(A) such redesignation has been specifically approved by the Governor of the State, after consultation with the appropriate Committees of the legislature . . . and if general purpose units of local government representing a majority of the residents of the area so redesignated enact legislation (including for such units of local government resolutions where appropriate) concurring in the State's redesignation;

(B) such redesignation will not cause, or contribute to, concentrations of any air pollutant which exceed any maximum allowable increase or maximum allowable concentration permitted under the classification of any other area; and

(C) such redesignation otherwise meets the requirements of this part. Subparagraph (A) of this paragraph shall not apply to area redesignations by Indian tribes.[90]

Before an area can be redesignated, the redesignating authority must describe and analyze the health, environmental, economic, social, and energy effects of the redesignation. If the redesignation includes federal land, the redesignating authority must provide notice and opportunity for a conference with the federal land manager. If the federal land manager responds with written comments, the state must explain any inconsistency between those comments and the state's redesignation decision.

---

[89] 42 U.S.C. § 7474(a).
[90] 42 U.S.C. § 7474(a).

7.6 Vehicle Emission Controls

The 1970 CAA required EPA to adopt emission limitations for motor vehicles. The 1990 amendments required EPA to adopt regulations to achieve further reductions in emissions from motor vehicles, as well as from other mobile sources such as locomotives. States are preempted from adopting emission limitations for motor vehicles and certain other mobile sources.

Under CAA Section 177, there is an exception for the state of California.[91] California is free to adopt its own (more stringent) motor vehicle standards, and standards for some (but not all) other mobile sources. Other states can adopt the California standards if they wish.

The 1990 amendments to the CAA put strict tailpipe standards on cars, buses, and trucks, and have expanded Inspection and Maintenance (I/M) programs to include more areas within the United States, and allow for more stringent tests.

The act mandates that improved gasoline formulations be sold in some polluted cities to reduce emissions of carbon monoxide or ozone-forming hydrocarbons.[92] Other programs set low vehicle emission standards to stimulate the introduction of cleaner cars and fuels.[93]

The 1990 amendments require EPA to consider emissions from off-highway vehicles as well as from highway vehicles such as cars and trucks.[94] The so-called "nonroad" category includes boats, farm equipment, bulldozers, lawn and garden devices, and construction machinery. EPA has determined that emissions from nonroad engines are a significant source of urban air pollution and is working with industry and the public to develop effective control strategies.

In cities that are in ozone nonattainment areas, growth in vehicle travel must be limited by encouraging alternatives to solo driving. In areas where ozone levels exceed criteria, employers of 100 or more will be asked to find ways to increase the average number of passengers in each vehicle for commutes to work and during work-related driving trips. New 1994 and later model cars must be equipped with "onboard diagnostic systems."[95]

Tailpipe (exhaust) standards for cars have been reduced under the 1990 amendments. The previous standards of 0.41 gram per mile (gpm) total hydrocarbons, 3.4 gpm carbon monoxide, and 1.0 gpm nitrogen oxides have been replaced with standards of 0.25 gpm nonmethane hydrocarbons and 0.4 gpm nitrogen oxides (the 3.4 gpm standard for carbon monoxide

---

[91] 42 U.S.C § 7507.
[92] 42 U.S.C § 7545(j) and (k).
[93] 42 U.S.C § 7543.
[94] 42 U.S.C § 7547.
[95] 42 U.S.C § 7521(m)(3).

does not change).[96] These standards will be fully phased in with 1996 models. EPA is required to study whether even tighter standards are needed, technologically feasible, and economical. If EPA determines by 1999 that lower standards are warranted, the standards will be cut in half beginning with 2004 model year vehicles.

7.7 The Acid Rain Program

Acid rain was first identified as an environmental problem in the mid-1970s. It was given a grim name by reporters and commentators. The northeastern United States was especially hard hit. In 1980, the National Academy of Science issued a report suggesting that a major culprit was emissions of sulfur dioxide (SO_2) from electricity generating power plants power plants in the Midwest that were being carried prevailing winds. The emissions were then transformed in the atmosphere into sulfuric acid, which then fell to the surface of the earth as acid precipitation, or "acid rain." It has subsequently been suggested that both SO_2 and NO_x can form acid rain.

In addition to causing acidification of lakes and streams, acid rain contributes to damage of trees at high elevations (e.g., red spruce trees above 2,000 feet in elevation). In addition, acid rain accelerates the decay of building materials and paints, including irreplaceable buildings, statues,and sculptures that are part of our nation's cultural heritage. Prior to falling to the earth, SO_2 and NO_x gases and their particulate matter derivatives, sulfates and nitrates, contribute to visibility degradation and impact public health.[97]

The primary source of SO_2 pollution is coal burning power plants. SO_2 is created when the sulfur in coal is released during combustion and reacts with oxygen in the air. The amount of sulfur dioxide created depends on the amount of sulfur in the coal. While all coal contains some sulfur, but the amount varies significantly depending on where the coal is mined. and NO_x are from coal and/or oil burning plants, and nitrogen oxides from the car and truck exhaust and from the oil burning plants.

Congress responded to concerns about acid rain with the Acid Precipitation Act of 1980,[98] which authorizes EPA to develop an Acid Rain Program. The overall goal of the Acid Rain Program is to achieve significant environmental and public health benefits through reductions in emissions of SO_2 and NO_x. To achieve this goal at the lowest cost to

---

[96] 42 U.S.C § 7582.
[97] U.S.E.P.A., *Acid Rain Overview* (1992).
[98] 42 U.S.C. § 8901 et seq.

society, the program employs both traditional and innovative, market-based approaches for controlling air pollution. In addition, the program encourages energy efficiency and pollution prevention.

Title IV of the Clean Air Act sets as its primary goal the reduction of annual SO_2 emissions by 10 million tons below 1980 levels. To achieve these reductions, the CAA requires a two-phase tightening of the restrictions placed on fossil fuel-fired power plants. Phase I began in 1995 and affects 263 units at 110 mostly coal-burning electric utility plants located in 21 eastern and midwestern states. An additional 182 units joined Phase I of the program as substitution or compensating units, bringing the total of Phase I affected units to 445.[99] Emissions data indicate that 1995 SO_2 emissions at these units nationwide were reduced by almost 40 percent below their required level.[100]

Phase II will begin in the year 2000, and tightens the annual emissions limits imposed on these large, higher emitting plants and also sets restrictions on smaller, cleaner plants fired by coal, oil, and gas, encompassing over 2,000 units in all.[101] The program affects existing utility units serving generators with an output capacity of greater than 25 megawatts and all new utility units.

The Act also calls for a 2 million ton reduction in NO_x emissions by the year 2000. A significant portion of this reduction will be achieved by coal-fired utility boilers that will be required to install low NO_x burner technologies and to meet new emissions standards.

7.8 Hazardous Air Pollutants

"Hazardous" air pollutants (HAPs), also known as "air toxics," are defined as any of the 189 pollutants list in CAA Section 112.[102] These 189 HAPs are listed in Table 7.1, and include a variety of chemical substances including organic chemicals, pesticides, metals, mineral fibers, and radionuclides.[103] Section 112 was originally added in the 1970 amendments, but the program was a failure because EPA refused to list many hazardous pollutants due to the extreme cost to industry of compliance.

---

[99] 42 U.S.C. § 7651(c).
[100] U.S.E.P.A., *Acid Rain Overview* (1992).
[101] 42 U.S.C. § 7651(d).
[102] 42 U.S.C. § 7412.
[103] Wooley, *Clean Air Act Handbook*, 8th ed. (West, 1998) at Chap. 3.

Table 7.1 Hazardous Air Pollutants Listed in Section 112 of the Clean Air Act

CAS no.	Chemical name	CAS no.	Chemical name
75070	Acetaldehyde	7782505	Chlorine
60355	Acetamide	79118	Chloroacetic acid
75058	Acetonitrile	532274	2-Chloroacetophenone
98862	Acetophenone	108907	Chlorobenzene
53963	2-Acetylaminofluorene	510156	Chlorobenzilate
107028	Acrolein	67663	Chloroform
79061	Acrylamide	107302	Chloromethyl methyl
79107	Acrylic acid		ether
107131	Acrylonitrile	126998	Chloroprene
107051	Allyl chloride	1319773	Cresols/Cresylic acid
92671	4-Aminobiphenyl		(isomers and mixture)
62533	Aniline	95487	o-Cresol
90040	o-Anisidine	108394	m-Cresol
1332214	Asbestos	106445	p-Cresol
71432	Benzene (including	98828	Cumene
	benzene from gasoline)	94757	2,4-D, salts and esters
92875	Benzidine	3547044	DDE
98077	Benzotrichloride	334883	Diazomethane
100447	Benzyl chloride	132649	Dibenzofurans
92524	Biphenyl	96128	1,2-Dibromo-3-
117817	Bis(2-ethylhexyl)-		chloropropane
	phthalate (DEHP)	84742	Dibutylphthalate
542881	Bis(chloromethyl)ether	106467	1,4-Dichlorobenzene(p)
75252	Bromoform	91941	3,3-Dichlorobenzidene
106990	1,3-Butadiene	111444	Dichloroethyl ether (Bis
156627	Calcium cyanamide		(2- chloro-ethyl) ether)
105602	Caprolactam	542756	1,3-Dichloro propene
133062	Captan	62737	Dichlorvos
63252	Carbaryl	111422	Diethanolamine
75150	Carbon disulfide	121697	N,N-Diethyl aniline
56235	Carbon tetrachloride		(N,N-Dimethylaniline)
463581	Carbonyl sulfide	64675	Diethyl sulfate
120809	Catechol	119904	3,3-Dimethoxy
133904	Chloramben		benzidine
57749	Chlordane		

Table 7.1 Hazardous Air Pollutants Listed in Section 112 of the Clean Air Act (*Continued*)

CAS no.	Chemical name	CAS no.	Chemical name
60117	Dimethyl amino-azobenzene	76448	Heptachlor
119937	3,3 -Dimethyl benzidine	118741	Hexachloro-benzene
79447	Dimethyl carbamoyl chloride	87683	Hexachloro-butadiene
68122	Dimethyl formamide	77474	Hexachloro-cyclopentadiene
57147	1,1-Dimethyl hydrazine	67721	Hexachloroethane
131113	Dimethyl phthalate	822060	Hexamethylene-1,6-diisocyanate
77781	Dimethyl sulfate	680319	Hexamethylphosphor-amide
534521	4,6-Dinitro-o-cresol, and salts	110543	Hexane
51285	2,4-Dinitrophenol	302012	Hydrazine
121142	2,4-Dinitro toluene	7647010	Hydrochloric acid
123911	1,4-Dioxane (1,4-Diethylene oxide)	7664393	Hydrogen fluoride (Hydrofluoric acid)
122667	1,2-Dipheny hydrazine	123319	Hydroquinone
106898	Epichlorohydrin (l-Chloro-2,3-epoxypropane)	78591	Isophorone
		58899	Lindane (all isomers)
		108316	Maleic anhydride
106887	1,2-Epoxybutane	67561	Methanol
140885	Ethyl acrylate	72435	Methoxychlor
100414	Ethyl benzene	74839	Methyl bromide (Bromomethane)
51796	Ethyl carbamate (Urethane)	74873	Methyl chloride (Chloromethane)
75003	Ethyl chloride (Chloroethane)	71556	Methyl chloroform (1,1,1-Trichlorothane)
106934	Ethylene dibromide (Dibromoethane)	78933	Methyl ethyl ketone (2-Butanone)
107062	Ethylene dichloride (1,2-Dichloroethane)	60344	Methyl hydrazine
107211	Ethylene glycol	74884	Methyl iodide (Iodomethane)
151564	Ethylene imine (Aziridine)	108101	Methyl isobutyl ketone (Hexone)
75218	Ethylene oxide	624839	Methyl isocyanate
96457	Ethylene thiourea	80626	Methyl methacrylate
75343	Ethylidene dichloride (1,1- Dichloroethane)	1634044	Methyl tert butyl ether
50000	Formaldehyde		

Table 7.1 Hazardous Air Pollutants Listed in Section 112 of the Clean Air Act (*Continued*)

CAS no.	Chemical name	CAS number	Chemical name
101144	4,4-Methylene bis(2-chloroaniline)	75558	1,2-Propylenimine (2-Methyl aziridine)
75092	Methylene chloride (Dichloromethane)	91225	Quinoline
101688	Methylene diphenyl diisocyanate (MDI)	106514	Quinone
		100425	Styrene
101779	4,4 -Methylene-dianiline	96093	Styrene oxide
91203	Naphthalene	1746016	2,3,7,8-Tetra-chloro-dibenzo-*p*-dioxin
98953	Nitrobenzene	79345	1,1,2,2-Tetra-chloro-ethane
92933	4-Nitrobiphenyl		
100027	4-Nitrophenol	127184	Tetrachloro-ethylene (Perchloro-ethylene)
79469	2-Nitropropane	7550450	Titanium tetrachloride
684935	N-Nitroso-N-methyl-urea	108883	Toluene
62759	N-Nitrosodimethyl-amine	95807	2,4-Toluene diamine
59892	N-Nitroso-morpholine	584849	2,4-Toluene diisocyanate
56382	Parathion	95534	*o*-Toluidine
82688	Pentachloro-nitrobenzene (Quintobenzene)	8001352	Toxaphene (chlorinated camphene)
87865	Pentachloro-phenol	120821	1,2,4-Trichloro-benzene
108952	Phenol	79005	1,1,2-Trichloro-ethane
106503	*p*-Phenylenedi-amine	79016	Trichloroethylene
75445	Phosgene	95954	2,4,5-Trichloro-phenol
7803512	Phosphine	88062	2,4,6-Trichloro-phenol
7723140	Phosphorus	121448	Triethylamine
85449	Phthalic anhydride	1582098	Trifluralin
1336363	Polychlorinated biphenyls (Aroclors)	540841	2,2,4-Trimethyl-pentane
1120714	1,3-Propane sultone	108054	Vinyl acetate
57578	beta-Propiol-actone	593602	Vinyl bromide
123386	Propionaldehyde	75014	Vinyl chloride
114261	Propoxur (Baygon)	75354	Vinylidene chloride (1,1- Dichloroethylene)
78875	Propylene dichloride (1,2- Dichloropropane)	1330207	Xylenes (isomers and mixture)
75569	Propylene oxide	95476	*o*-Xylenes
		108383	*m*-Xylenes
		106423	*p*-Xylenes

Table 7.1 Hazardous Air Pollutants Listed in Section 112 of the
Clean Air Act (*Continued*)

--------------------------------------- --------------------------------------

CAS number	Chemical name	CAS no.	Chemical name
0	Antimony Compounds	0	Lead Compounds
0	Arsenic Compounds	0	Manganese Compounds
	(inorganic including arsine)	0	Mercury Compounds
0	Beryllium Compounds	0	Fine mineral fibers *3
0	Cadmium Compounds	0	Nickel Compounds
0	Chromium Compounds	0	Polycylic Organic Matter *4
0	Cobalt Compounds	0	Radionuclides (including
0	Coke Oven Emissions		radon) *5
0	Cyanide Compound *1	0	Selenium Compounds
0	Glycol ethers *2		

NOTE: For all listings above which contain the word "compounds"
and for glycol ethers, the following applies: Unless otherwise
specified, these listings are defined as including any unique chemical
substance that contains the named chemical (i.e., antimony, arsenic,
etc.) as part of that chemical's infrastructure.
*1: X CN where X = H or any other group where a formal dissociation
     may occur. For example KCN or $Ca(CN)_2$.
*2: Includes mono- and di-ethers of ethylene glycol, diethylene glycol,
     and triethylene glycol $R-(OCH_2CH_2)_n-OR$ where $n = 1, 2,$ or 3
     R = alkyl or aryl groups
     R = R, H, or groups which, when removed, yield glycol ethers with
     the structure: $R-(OCH_2CH)_n-OH$. Polymers are excluded from the
     glycol category.
*3: Includes mineral fiber emissions from facilities manufacturing or
     processing glass, rock, or slag fibers (or other mineral derived
     fibers) of average diameter 1 micrometer or less.
*4: Includes organic compounds with more than one benzene ring, and
     which have a boiling point greater than or equal to 100° C.
*5: A type of atom which spontaneously undergoes radioactive decay.

Congress established the original list of 189 hazardous air pollutants,
although the EPA (or a petitioner) can add to the list any substances it
finds pose a "threat of adverse human health effects" or environmental

effects, including any toxic, carcinogenic, mutagenic, teratogenic, or neurotoxic substance. It is sufficient for listing purposes that a *possibility* exists that a substance poses a health threat.[104] A substance can be withdrawn from the list if EPA (or a petitioner) shows that enough information on the substance exists to demonstrate that it does not pose unreasonable risk of adverse health or environmental effects.[105] However, Congress abandoned the "complete elimination of risk to public health" approach with the 1990 amendments, in favor of a technology-based system.

The EPA published a broader Section 112 program in 1984 based on risk assessment and risk management practices, but this cost/benefit approach was even more controversial. After 20 years, EPA had issued only 8 National Emission Standards for Hazardous Air Pollutants (NESHAP), and the HAP program was completely revised in 1990.[106]

Under the 1990 CAA Amendments, the EPA is required to publish a list of area source categories and subcategories of listed hazardous air pollutants. This list must include enough source categories by November 15, 1995 to ensure that 90 percent of the area source emissions of the 30 hazardous pollutants posing the greatest threat to public health in the most cities are subject to Section 112 regulation.[107] This technology-based system largely replaced the health-risk-based program in effect prior to 1990. Nevertheless, Section 112(f) of the CAA retains some health-based standards. The EPA must assess residual risk of HAPs after the implementation of technology-based standards, and may impose additional regulation to reduce such risks.[108]

Under CAA Section 112(a)(1),[109] a "major source" is any stationary source (or combination of sources) that produce 10 tons/year of any hazardous air pollutant, or 25 tons/year of any combination of hazardous pollutants.[110] An "area source" is any source of hazardous air pollutants that is not a major source. EPA is in the process of establishing special regulatory programs that apply to major and area sources, under which major sources will have special limitations that do not apply to area sources. Obviously, a source would rather be regulated as an "area source" than as a "major source." One court held that EPA could include "fugitive emissions" in deciding if a source is a major source.[111]

---

[104] See 42 U.S.C. § 7412(b)(3)(B).
[105] 42 U.S.C. § 7412(b)(3)(C).
[106] See Wooley, *Clean Air Act Handbook*, 8th ed. (West, 1998) Chap. 3.
[107] 42 U.S.C. § 7412(c)(3). See National Mining Assoc. v. USEPA, 59 F.3d 1351 (D.C. Cir. 1995) [challenge to source definitions].
[108] 42 U.S.C. § 7412(f).
[109] 42 U.S.C. § 7412(a)(1).
[110] See Sierra Club v. Larson, 2 F.3d 462 (1st Cir. 1993).
[111] National Mining Assoc. v. USEPA, 59 F.3d 1351 (D.C. Cir. 1995).

Technology-based emission standards for hazardous pollutants (NESHAP) must require the maximum emissions reduction that EPA determines to be achievable, including a prohibition on any emission (MACT).[112] For a new source, emission standards must be at least as stringent as the emission control achieved by the best controlled similar source.[113] These are based on the Standard Industry Code (SIC).

At least some health-based standards for hazardous pollutants remain. In late 1996, the EPA reported to Congress that health risks remain from hazardous pollutants after the application of the technology-based standards.

7.9 Permits

Unlike other environmental statutes, the CAA did not originally contain any provisions for permits. In the 1990 amendments, however, Congress added a new "Permits" section at CAA Title V,[114] which was modeled after the NPDES system of the Clean Water Act. Unfortunately, Congress left most of the important issues out of the legislation (e.g., exemptions, permit revisions, EPA review and veto powers, the relationship between SIPs and permits, the degree of flexibility under a permit, etc.). The following discussion is based on the Title V permit program as it has evolved following the 1990 CAA amendments, although the Title V program regulations remain in a period of rapid change.

All major sources of air pollutants are covered by the "generic permit program" of CAA Title V. This includes all covered sources under the NSPS standards (Section 111), PSD areas (Section 165), nonattainment areas (Sections 172 and 173), hazardous air pollutants (Subsection 112(j)), acid rain (Section 408), and so on.[115]

Section 504 of the CAA authorizes states to issue three kinds of permits: individual permits, general permits, and permits for temporary sources.[116]

General permits. The [state] permitting authority may, after notice and opportunity for public hearing, issue a general permit covering numerous similar sources. Any general permit shall comply with all requirements applicable to permits under this subchapter. No source

---

[112] 42 U.S.C. § 7412(d)(2).
[113] 42 U.S.C. § 7412(d)(3).
[114] 42 U.S.C. § 7661.
[115] See Commonwealth of Virginia v. United States of America, 80 F.3d 869 (4th Cir. 1996) [rejecting challenges to Title V of the CAA]. See also discussion in Wooley, *Clean Air Act Handbook: a Practical Guide to Compliance*, 8th ed. (West Group, 1998) at § 5.01.
[116] 42 U.S.C. § 7661c.

covered by a general permit shall thereby be relieved from the obligation to file an application [for a CAA permit] under section 7661b of this title.[117]

Temporary sources. The [state] permitting authority may issue a single permit authorizing emissions from similar operations at multiple temporary locations. No such permit shall be issued unless it includes conditions that will assure compliance with all the requirements of this chapter at all authorized locations, including, but not limited to, ambient standards and compliance with any applicable increment or visibility requirements under part C of subchapter I of this chapter. Any such permit shall in addition require the owner or operator to notify the permitting authority in advance of each change in location. The permitting authority may require a separate permit fee for operations at each location.[118]

Any source required to obtain a permit under Section 503 of the CAA that is neither a general source nor a temporary source is an individual source, and will be the focus of the discussion to follow. It should be noted that a state may generate as list of "insignificant activities" and emission levels that may be excluded from the state's permit program.[119]

Clearly the CAA contemplates that permits will be issued by the states under U.S. EPA approved programs. There are, however, two situations in which the EPA rather than the state will receive a permit application, issue the permit, or both. First, if a state has obtained Title V program approval from the EPA, or if the EPA has withdrawn approval from a state's Title V program, then Sections 502(d)(3) and (i)(4) of the CAA authorize the EPA to become the permitting authority and issue permits.[120] The EPA may subsequently delegate its permitting authority to the state under specified conditions, the state must comply with all regulatory requirements in exercising that authority, and appeals of such permit decisions are brought in accordance with Part 71 procedures.[121] Second, if the EPA objects to a proposed permit and the state permitting authority fails to revise and submit a proposed permit in response to the objection

---

[117] 42 U.S.C. § 7661c(d).
[118] 42 U.S.C. § 7661c(e).
[119] 40 CFR § 70.5(c).
[120] 40 CFR Part 71. See 42 U.S.C. §§ 7661a(d)(3) and (i)(4).
[121] 40 CFR § 71.10(a) and (i). Appeals will be discussed further below.

within 90 days, the EPA is authorized to issue or deny that permit in accordance with EPA regulations under 40 CFR Part 71.[122]

State Permit Application Programs

Many of the details of permitting procedures are left to states, and it is to state agencies that a source must ordinarily apply. The state agency's permit application program requirements must contain the following elements as specified in Section 503 of the CAA and EPA regulations:[123]

- A timely application. For a source applying for a permit for the first time, this must be submitted within 12 months after the source becomes subject to the permit program. Sources producing hazardous air pollutants under Section 112(g) of the CAA,[124] or those that require preconstruction review under part C or D of title I of the CAA,[125] must file a complete application within 12 months after commencing operation (or earlier at the permitting authority's discretion).[126] Permit renewals must be submitted at least 6 months prior to the date of permit expiration, although the EPA can extend the time to 18 months to ensure that the term of the permit will not expire before renewal.[127]
- A complete application. The permitting authority (normally the appropriate state agency) must provide criteria and procedures for determining in a timely fashion when applications are complete. To be deemed complete, an application must provide all required information in the state's standard application form, which includes sufficient information to evaluate the source and its application, and to determine all applicable requirements.[128] However, applications for permit revision need supply the information if it is related to the proposed change. Information required in the application will be discussed below.
- Timely review. Unless the permitting authority determines that an application is not complete within 60 days of receipt, the application is deemed to be complete. If additional information is required, the agency can request the information in writing and set a deadline for a response.

---

[122] 40 CFR § 70.8(c)(4). See Morgan, Lewis & Bockius LLP, *EPA's Operating Permit Regulations: Title V Permit Appeals and Strategic Considerations* (1996), found at http://envinfo.com/caain; and Wooley, *Clean Air Act Handbook: a Practical Guide to Compliance*, 8th ed. (West Group, 1998) at § 5.02.
[123] 42 U.S.C.§ 7661b; and 40 CFR Part 70.
[124] 42 U.S.C. § 7412(g), related to hazardous air pollutants (see section 7.8).
[125] Related to PSD areas and nonattainment areas, see sections 7.4 and 7.5 above.
[126] Where an existing permit would prohibit such construction or change in operation, the source must obtain a permit revision before commencing operation. 40 CFR § 70.5(a)(1)(ii).
[127] 40 CFR § 70.5(a)(1)(iii).
[128] 40 CFR § 70.5(a)(2).

An applicant must promptly submit such supplementary facts or corrected information.
- Standard application form and required information. The permitting authority provides a standard application form for CAA Title V permits.

The forms and attachments include the following elements:[129]

(1) Identifying information, including company/plant name and address, owner's name and agent, and telephone number and names of plant site manager/contact.

(2) A description of the source's processes and products (by Standard Industrial Classification Code).

(3) Emission-related information, including: all emissions of pollutants for which the source is major, and all emissions of regulated air pollutants; identification and description of all points of emissions in sufficient detail to establish the basis for fees and applicability of requirements; emissions rate in tons per year (tpy), and in terms to establish compliance consistent with the applicable standard reference test method; all pertinent information on fuels, fuel use, raw materials, production rates, and operating schedules; identification and description of air pollution control equipment and compliance monitoring devices or activities; all calculations on which emission information is based; and any other information required by any applicable requirement.

(4) Air pollution control requirements, including citation and description of all applicable requirements, and description or reference to any applicable test method for determining compliance with each applicable requirement.

(5) Other specific information that may be necessary to implement and enforce other applicable requirements of the CAA or to determine the applicability of such requirements.

(6) An explanation of any proposed exemptions from otherwise applicable requirements.

(7) Additional information as determined to be necessary by the permitting authority to define alternative operating scenarios identified by the source[130] or to define permit terms and conditions implementing CAA regulations.[131]

(8) A compliance plan for each source containing: A description of the compliance status of the source with respect to all applicable requirements; a statement that the source has met all requirements, or

---

[129] 40 CFR § 70.5(c).
[130] As defined in 40 CFR § 70.6(a)(9).
[131] As defined in 40 CFR §§ 70.4(b)(12) and 70.6(a)(10).

will do so in a timely fashion; a compliance schedule,[132] including a
statement that the source is in compliance and will continue to do so,
or will meet any new requirements in a timely fashion, including a
schedule of any necessary remedial measures; a schedule for
submission of certified progress reports (at least every 6 months for
sources required to have a schedule of compliance to remedy a
violation); a compliance plan applicable to acid rain emissions
limitations for an affected source,[133] unless specifically superseded by
regulations under title IV of the CAA.

(9) Requirements for compliance certification, including the
following: a certification of compliance with all applicable
requirements by a responsible state official; a statement of methods
used for determining compliance, including a description of
monitoring, record keeping, and reporting requirements and test
methods; a schedule for submission of compliance certifications
during the permit term, submitted at least annually, or more frequently
if specified by the requirement or by the permitting authority; and a
statement indicating the source's compliance status with any
applicable enhanced monitoring and compliance certification
requirements of the CAA.

(10) The use of nationally standardized forms for acid rain portions
of permit applications and compliance plans, as required by
regulations promulgated under title IV of the CAA.[134]

- Any application form, report, or compliance certification must contain
  certification by a responsible state official of truth, accuracy, and
  completeness. This certification must state that, based on information
  and belief formed after reasonable inquiry, the statements and
  information in the document are true, accurate, and complete.[135]
- Unless the permitting authority requests additional information or
  otherwise notifies the applicant of incompleteness within 60 days of
  receipt of an application, the application is deemed to be complete.[136]
- Of course, state agencies are authorized to collect reasonable fees.[137]

---

[132] "This compliance schedule shall resemble and be at least as stringent as that
contained in any judicial consent decree or administrative order to which the
source is subject." 40 CFR § 70.5(c)(8)(iii)(C).

[133] See section 7.7 for a discussion of the CAA acid rain program.

[134] See section 7.7.

[135] 40 CFR § 70.5(d).

[136] 40 CFR § 70.7(a)(4).

[137] 42 U.S.C. § 7410((a)(2)(A). EPA regulations state that "The State program
shall require that the owners or operators of part 70 sources pay annual fees, or
the equivalent over some other period, that are sufficient to cover the permit
program costs and shall ensure that any fee required by this section will be used
solely for permit program costs." 40 CFR § 70.9(a).

Content of CAA Permits

Prior to the 1990 amendments to the CAA, the contents of CAA permits was left largely to the EPA. The resulting EPA guidelines were only moderately effective in requiring stationary sources to install emission controlling and monitoring equipment. The 1990 amendments to the CAA provided guidance in several sections for the EPA to promulgate regulations related to emissions controlling and monitoring. However, due to debates on the stringency and scope of the rules, final regulations are still incomplete at this writing.[138] Nevertheless, current EPA regulations provide considerable guidance as to the contents of CAA permits.[139]

● Standard permit requirements. Each permit must include the following elements:

(1) Emission limitations and standards, including those operational requirements and limitations that assure compliance with all applicable requirements at the time of permit issuance. The permit must specify the origin of and authority for each term or condition, and identify any difference in form as compared to the applicable requirement upon which the term or condition is based; where an applicable requirement of the CAA is more stringent than an applicable requirement of regulations promulgated under title IV of the Act, both provisions must be incorporated into the permit; if a SIP allows an alternative emission limit at a stationary source, then the permit for that source must contain provisions to ensure that any resulting emissions limit has been demonstrated to be quantifiable, accountable, enforceable, and based on replicable procedures.

(2) Permit duration. The permitting authority issues permits for a fixed term of 5 years in the case of "affected sources" (that is, those sources regulated under the CAA Title IV acid rain program), and for a term not to exceed 5 years in the case of all other sources. Notwithstanding this requirement, the permitting authority must issue permits for solid waste incineration units combusting municipal waste subject to standards under section 129(e) of the CAA[140] for a period not to exceed 12 years and must review such permits at least every 5 years.

(3) Monitoring and related recordkeeping and reporting requirements. Each permit must contain the following requirements with respect to monitoring:

---

[138] Wooley, *Clean Air Act Handbook: a Practical Guide to Compliance*, 8th ed. (West Group, 1998) at § 5.04[1][a]. See also Martineau and Novello (eds.) *Clean Air Act Handbook* (Amer. Bar Assn., 1997).

[139] 40 CFR § 70.6.

[140] 42 U.S.C. § 7429(e).

(A) All emissions monitoring and analysis procedures or test methods required under the applicable requirements, including any procedures and methods promulgated pursuant to sections 114(a)(3) or 504(b) of the CAA;[141]

(B) Where the applicable requirement does not require periodic testing or instrumental or noninstrumental monitoring, periodic monitoring sufficient to yield reliable data from the relevant time period that are representative of the source's compliance with the permit; and

(C) As necessary, requirements concerning the use, maintenance, and, where appropriate, installation of monitoring equipment or methods.

With respect to recordkeeping, the permit must incorporate all applicable recordkeeping requirements and require, where applicable, the following:

(A) Records of required monitoring information that include the following: the date, place as defined in the permit, and time of sampling or measurements; the date(s) analyses were performed; the company or entity that performed the analyses; the analytical techniques or methods used; the results of the analyses; and the operating conditions as existing at the time of sampling or measurement;

(B) Retention of records of all required monitoring data and support information for a period of at least 5 years from the date of the monitoring sample, measurement, report, or application. Support information includes all calibration and maintenance records and all original strip-chart recordings for continuous monitoring instrumentation, and copies of all reports required by the permit.

With respect to reporting, the permit must incorporate all applicable reporting requirements and require the following:

(A) Submittal of reports of any required monitoring at least every 6 months. All instances of deviations from permit requirements must be clearly identified in the reports. All required reports must be certified by a responsible official consistent with EPA regulations.[142]

(B) Prompt reporting of deviations from permit requirements, including those attributable to upset conditions as defined in the permit, the probable cause of the deviations, and any corrective actions or preventive measures taken. The permitting authority must define "prompt" in relation to the degree and type of deviation likely to occur and the applicable requirements.

(4) A permit condition prohibiting emissions exceeding any

---

[141] 42 U.S.C. §§ 7414(a)(3) or 7661c(b).
[142] 40 CFR § 70.5(d).

allowances that the source lawfully holds under title IV of the CAA or CAA regulations. No permit revision can be required for increases in emissions that are authorized by allowances acquired pursuant to the CAA Title IV acid rain program, provided that the increases do not require a permit revision under any other applicable requirement. No limit can be placed on the number of allowances held by the source. The source may not, however, use allowances as a defense to noncompliance with any other applicable requirement. Any allowance must be accounted for according to the procedures established in regulations promulgated under Title IV of the CAA.

(5) A severability clause to ensure the continued validity of the various permit requirements in the event of a challenge to any portions of the permit.

(6) Provisions stating the following: The permittee must comply with all conditions of the 40 CFR Part 70 permit. Any permit noncompliance constitutes a violation of the Act and is grounds for enforcement action; for permit termination, revocation and reissuance, or modification; or for denial of a permit renewal application. Need to halt or reduce activity is not a defense. It shall not be a defense for a permittee in an enforcement action that it would have been necessary to halt or reduce the permitted activity in order to maintain compliance with the conditions of this permit. The permit may be modified, revoked, reopened, and reissued, or terminated for cause. The filing of a request by the permittee for a permit modification, revocation and reissuance, or termination, or of a notification of planned changes or anticipated noncompliance does not stay any permit condition. The permit does not convey any property rights of any sort, or any exclusive privilege. The permittee must furnish to the permitting authority, within a reasonable time, any information that the permitting authority may request in writing to determine whether cause exists for modifying, revoking and reissuing, or terminating the permit or to determine compliance with the permit. Upon request, the permittee shall also furnish to the permitting authority copies of records required to be kept by the permit or, for information claimed to be confidential, the permittee may furnish the records directly to the EPA, along with a claim of confidentiality.

(7) A provision to ensure that a 40 CFR Part 70 source pays fees to the permitting authority consistent with the fee schedule approved pursuant to EPA regulations.[143]

(8) Emissions trading. A provision stating that no permit revision can be required, under any approved economic incentives, marketable permits, emissions trading and other similar programs or processes for

---

[143] 40 CFR § 70.9.

changes that are provided for in the permit.

(9) Terms and conditions for reasonably anticipated operating scenarios identified by the source in its application as approved by the permitting authority. The terms and conditions require the source, contemporaneously with making a change from one operating scenario to another, to record in a log at the permitted facility a record of the scenario under which it is operating. The source may extend the permit shield described in EPA regulations[144] to all terms and conditions under each operating scenario; and ensure that the terms and conditions of each alternative scenario meet all applicable requirements and the requirements of this part.

(10) Terms and conditions, if the permit applicant requests them, for the trading of emissions increases and decreases in the permitted facility, to the extent that the applicable requirements provide for trading such increases and decreases without a case-by-case approval of each emissions trade. Such terms and conditions must include all terms required under EPA regulations to determine compliance,[145] may extend the permit shield described in EPA regulations to all terms and conditions that allow such increases and decreases in emissions,[146] and must meet all applicable requirements and requirements EPA regulations.

● Federally enforceable requirements.

(1) All terms and conditions in a 40 CFR Part 70 permit, including any provisions designed to limit a source's potential to emit, are enforceable by the EPA and citizens under the CAA.

(2) The permitting authority must specifically designate as not being federally enforceable under the CAA any terms and conditions included in the permit that are not required under the CAA or under any of its applicable requirements.[147]

● Compliance requirements. All 40 CFR Part 70 permits must contain the following elements with respect to compliance:

(1) Consistent with EPA regulation,[148] compliance certification, testing, monitoring, reporting, and recordkeeping requirements sufficient to assure compliance with the terms and conditions of the permit. Any document (including reports) required by a 40 CFR Part

---

[144]   40 CFR § 70.6(f). See discussion below.

[145]   40 CFR §§ 70.6(a) and (c).

[146]   40 CFR § 70.6(f). See discussion below.

[147]   Terms and conditions so designated are not subject to the requirements of 40 CFR §§ 70.6, 70.7, or 70.8, other than those contained in this paragraph 40 CFR § 70.6(b).

[148]   40 CFR § 70.6(a)(3).

70 permit must contain a certification by a responsible official that meets the requirements of EPA regulations.[149]

(2) Inspection and entry requirements that require that, upon presentation of credentials and other documents as may be required by law, the permittee must allow the state permitting authority or an authorized representative to perform the following: enter upon the permittee's premises where a source is located or emissions-related activity is conducted, or where records must be kept under the conditions of the permit; have access to and copy, at reasonable times, any records that must be kept under the conditions of the permit; inspect at reasonable times any facilities, equipment (including monitoring and air pollution control equipment), practices, or operations regulated or required under the permit; and, as authorized by the CAA, sample or monitor at reasonable times substances or parameters for the purpose of assuring compliance with the permit or applicable requirements.

(3) A schedule of compliance consistent with EPA regulations.[150]

(4) Progress reports consistent with an applicable schedule of compliance and EPA regulations[151] to be submitted at least semiannually, or at a more frequent period if specified in the applicable requirement or by the state permitting authority. Progress reports must contain the following: dates for achieving the activities, milestones, or compliance required in the schedule of compliance, and dates when such activities, milestones or compliance were achieved; and an explanation of why any dates in the schedule of compliance were not or will not be met, and any preventive or corrective measures adopted.

(5) Requirements for compliance certification with terms and conditions contained in the permit, including emission limitations, standards, or work practices. Permits must include the frequency (not less than annually or such more frequent periods as specified in the applicable requirement or by the permitting authority) of submissions of compliance certifications; and, in accordance with EPA regulations,[152] a means for monitoring the compliance of the source with its emissions limitations, standards, and work practices. There is a requirement that the compliance certification include the following: the identification of each term or condition of the permit that is the basis of the certification; the compliance status; whether compliance was continuous or intermittent; the method(s) used for determining the

---

[149] 40 CFR § 70.5(d).

[150] 40 CFR § 70.5(c)(8).

[151] Id.

[152] 40 CFR § 70.6(a)(3).

compliance status of the source, currently and over the reporting period consistent with EPA regulations;[153] and such other facts as the permitting authority may require to determine the compliance status of the source. There is a requirement that all compliance certifications be submitted to the EPA as well as to the permitting authority. Finally, any additional requirements as may be specified by CAA Sections 114(a)(3) and 504(b)[154] must be included.

(6) Any other provisions required by the permitting authority.

Permit Shield

Prior to the 1990 amendments, well-meaning sources of air pollution often had great difficulty determining what behaviors were unlawful under the CAA. Permits were often vague and imprecise, and SIP provisions often contradicted the permits. Sources who followed their permits occasionally found themselves in potential violation of the CAA.[155]

Under the 1990 CAA amendments, a permit issued to a source by the appropriate permitting authority or the EPA can shield permit holders from some additional enforcement risk. Section 504(f) of the CAA states the following regarding permit shields:

(1) [T]he [state] permitting authority may expressly include in a [40 CFR] part 70 permit a provision stating that compliance with the conditions of the permit shall be deemed compliance with any applicable requirements as of the date of permit issuance, provided that:

(i) Such applicable requirements are included and are specifically identified in the permit; or

(ii) The [state] permitting authority, in acting on the permit application or revision, determines in writing that other requirements specifically identified are not applicable to the source, and the permit includes the determination or a concise summary thereof.

(2) A [40 CFR] part 70 permit that does not expressly state that a permit shield exists shall be presumed not to provide such a shield.

(3) Nothing in this paragraph or in any [40 CFR] part 70 permit shall alter or affect the following:

(i) The provisions of section 303 of the Act (emergency orders), including the authority of the Administrator under that section;

(ii) The liability of an owner or operator of a source for any

---

[153] 40 CFR § 70.6(a)(3).
[154] 42 U.S.C. §§ 7414(a)(3), and 7661c(b).
[155] See Wooley, *Clean Air Act Handbook: A Practical Guide to Compliance*, 8th ed. (West Group, 1998) at § 5.04[6]. See also Martineau and Novello (eds.), *Clean Air Act Handbook* (Amer. Bar Assn., 1997).

violation of applicable requirements prior to or at the time of permit issuance;

(iii) The applicable requirements of the acid rain program, consistent with section 408(a) of the [Clean Air] Act; or

(iv) The ability of EPA to obtain information from a source pursuant to section 114 of the [Clean Air] Act.

The EPA has further clarified situations in which the permit shield does *not* apply in its regulations by stating that "a part 70 permit that does not expressly state that a permit shield exists shall be presumed not to provide such a shield."[156] Likewise, the permit shield does not protect a source from liability "for any violation of applicable requirements prior to or at the time of permit issuance," or "the ability of EPA to obtain information from a source pursuant to section 114" of the CAA.[157]

General Permits

The state agency responsible for granting permits may issue a general permit covering numerous similar sources. States are given latitude in determining what are (or are not) similar sources qualified for a general permit.[158] The state agency must provide public notice and an opportunity for public participation as provided by EPA regulations.[159]

Any general permit must comply with all requirements applicable to permits for individual sources. In addition, general permits must identify the criteria by which the multiple sources may qualify for the general permit. To sources that qualify, the state agency that grants the permit must grant the conditions and terms of the general permit.

Notwithstanding the permit shield provisions discussed above, the source is subject to enforcement action for operation without a permit if the source is later determined not to qualify for the conditions and terms of the general permit. General permits are not be authorized for affected sources under the acid rain program unless otherwise provided in regulations promulgated under Title IV of the CAA.[160]

Sources electing to apply for a general permit must apply to the appropriate state agency. The permitting agency may, in the general permit, provide for applications which deviate from the requirements of individual permit applications. However, applications for general permits must meet the requirements of Title V of the CAA, and include all

---

[156] 40 CFR §70.6(f).
[157] Id.
[158] See Wooley, *Clean Air Act Handbook: aAPractical Guide to Compliance*, 8th ed. (West Group, 1998) at § 5.04[5]. See also Martineau and Novello (eds.), *Clean Air Act Handbook* (Amer. Bar Assn., 1997).
[159] 40 CFR § 70.7(h), discussed more fully below.
[160] 40 CFR § 70.6(d)(1).

information necessary to determine qualification for, and to assure compliance with, the general permit. The state agency may grant a source's request for authorization to operate under a general permit without repeating the public participation procedures required under EPA regulations,[161] but such the grant is then not a final permit action for purposes of judicial review.[162]

Permits for Temporary Sources

The permitting authority may issue a single permit authorizing emissions from similar operations by the same source owner or operator at multiple temporary locations. The operation must be temporary and involve at least one change of location during the term of the permit. No affected source shall be permitted as a temporary source.[163]

Permits for temporary sources must include the following:

(1) Conditions that will assure compliance with all applicable requirements at all authorized locations;

(2) Requirements that the owner or operator notify the permitting authority at least 10 days in advance of each change in location; and

(3) Conditions that assure compliance with all other provisions of this section.[164]

Public Participation

With the exception of modifications qualifying for minor permit modification procedures, all permit proceedings, including initial permit issuance, significant modifications, and renewals, must provide adequate procedures for public notice and comment, and a hearing on the draft permit if requested.[165]

The public notice must be given by publication in a newspaper of general circulation in the area where the source is located,[166] or by other means if necessary to assure adequate notice to the affected public. The notice must identify the affected facility; give the name and address of the permittee; the name and address of the permitting authority processing the permit; the activity or activities involved in the permit action; the emissions change involved in any permit modification; the name, address, and telephone number of a person from whom interested persons may obtain additional information, including copies of the permit draft, the

---

[161] 40 CFR § 70.7(h), discussed below.

[162] 40 CFR § 70.6(d)(2).

[163] 40 CFR § 70.6(e).

[164] Id.

[165] 40 CFR § 70.7(h).

[166] Public notice may also be in a state publication designed to give general public notice to persons on a mailing list developed by the state permitting authority, including any persons who request in writing to be on the list. Id.

application, all relevant supporting materials, and all other materials available to the permitting authority that are relevant to the permit decision; a brief description of the comment procedures; and the time and place of any hearing that may be held, including a statement of procedures to request a hearing (unless a hearing has already been scheduled).[167] The permitting authority must provide at least 30 days for public comment and give notice of any public hearing at least 30 days in advance of the hearing.

The permitting authority keeps a record of the commenters and also of the issues raised during the public participation process. This is to assure that the EPA will be able to fulfill its obligation under Section 505(b)(2) of the CAA[168] to determine whether a petition from a citizen objecting to a permit may be granted. These records are made available to the public.[169]

Permit Issuance

Once an application for a CAA permit under 40 CFR Section 70 has been completed, the permit is then issued by the state permitting authority if the permit application is complete; the permitting authority has complied with the requirements for public participation; the permitting authority has complied with the requirements for notifying and responding to affected states under EPA regulations;[170] the conditions of the permit provide for compliance with all applicable requirements; and the EPA has received a copy of the proposed permit and any notices required under EPA regulations,[171] and has not objected to issuance of the permit within the specified time period.[172]

Except for certain circumstances specified in EPA regulations,[173] or for permitting of affected sources under the acid rain program in CAA Title IV, the permitting authority must take final action on each permit application within 18 months after receiving a complete application.[174]

The state permitting authority shall provide a statement that sets forth the legal and factual basis for the draft permit conditions (including

---

[167] 40 CFR § 70.7(h)(2).
[168] 42 U.S.C. § 7661d(b)(2), which states "If the Administrator [of the EPA] does not object in writing to the issuance of a permit pursuant to paragraph (1) [describing the process for transmitting the permit application from the state agency to the EPA], any person may petition the Administrator within 60 days after the expiration of the 45-day review period specified in paragraph (1) to take such action."
[169] 40 CFR § 70.7(h)(3).
[170] 40 CFR § 70.8(b).
[171] 40 CFR §§ 70.8(a) and 70.8(b).
[172] 40 CFR § 70.8(c).
[173] 40 CFR § 70.4(b)(11).
[174] 40 CFR § 70.7(2).

references to the applicable statutory or regulatory provisions). This statement is sent to the EPA and to any other person who requests it.[175]

Permit Renewal, Revision, and Modification

All permits issued under Title V of the CAA have a built-in expiration date. Permit expiration terminates the source's right to operate unless a timely and complete renewal application has been submitted consistent with EPA regulations.[176] Ordinarily, Title V permits can be renewed. Permits being renewed follow the same procedures as when they were originally issued, and are subject to the same procedural requirements, including those for public participation, affected state and EPA review, that apply to initial permit issuance.[177] If the state permitting authority fails to act in a timely way on a permit renewal, then the EPA can invoke its authority under Section 505(e) of the CAA to terminate or revoke and reissue the permit.[178]

An "administrative permit amendment" is a simple change to a permit that may be issued by the state permitting authority for certain minor changes and errors in a permit, such as typographical errors; a change in the name, address, or phone number of any person identified in the permit; or a similar minor administrative change at the source.[179] Other changes that allow administrative permit amendments include sources that require more frequent monitoring or reporting, a change in ownership or operational control of a source where the permitting authority determines that no other change in the permit is necessary, preconstruction review permits authorized under an EPA-approved program, or any other similar type of change that the EPA has approved. An administrative permit amendment takes no more than 60 days from receipt of a request until final action, and does not require public notice and comment. Moreover, the source may implement the changes addressed in the request for an administrative amendment immediately upon submittal of the request.[180]

A "permit modification" is any revision to a 40 CFR Part 70 permit that cannot be accomplished under the provisions for administrative permit amendments. The state permitting authority programs provide adequate, streamlined, and reasonable procedures for expeditiously processing permit modifications. The state permitting authority can develop different procedures for different types of modifications, depending on the significance and complexity of the requested modification. Permit modifications may be "minor," or "significant."

---

[175] 40 CFR § 70.7(a)(5).
[176] 40 CFR §§ 70.5(a)(1)(iii) and 70.7(b).
[177] 40 CFR § 70.7(c)(1)(ii).
[178] 40 CFR § 70.7(c)(2).
[179] 40 CFR § 70.7(d)(1).
[180] Id.

Minor permit modifications are those that do not violate any applicable requirement; do not involve significant changes to existing monitoring, reporting, or recordkeeping requirements in the permit; do not require or change a case-by-case determination of an emission limitation or other standard, or a source-specific determination for temporary sources of ambient impacts, or a visibility or increment analysis; do not seek to establish or change a permit term or condition for which there is no corresponding underlying applicable requirement and that the source has assumed to avoid an applicable requirement to which the source would otherwise be subject; are not modifications under any provision of Title I of the CAA; and are not required by the state program to be processed as a significant modification.[181] Minor permit modification procedures may be used for permit modifications involving the use of economic incentives, marketable permits, emissions trading, and other similar approaches, as long as they are consistent with the SIP and EPA regulations.[182]

An application requesting the use of minor permit modification follows the same procedures as for a 40 CFR Section 70 permit application, and includes the following:

(A) A description of the change, the emissions resulting from the change, and any new applicable requirements that will apply if the change occurs;

(B) The source's suggested draft permit;

(C) Certification by a responsible official, consistent with EPA regulations,[183] that the proposed modification meets the criteria for use of minor permit modification procedures and a request that such procedures be used; and

(D) Completed forms for the state permitting authority to use to notify the EPA and affected states as required under EPA regulations.[184]

The state permitting authority may not issue a final permit modification until after EPA's 45-day review period,[185] or until EPA has notified the state permitting authority that EPA will not object to issuance of the permit modification, whichever is first, although the permitting authority

---

[181] 40 CFR § 70.7(e)(2). The regulations list two examples of minor permit modifications: (A) A federally enforceable emissions cap assumed to avoid classification as a modification under any provision of CAA Title I; and (B) an alternative emissions limit approved pursuant to regulations promulgated under CAA §112(i)(5) for hazardous air pollutants.

[182] 40 CFR § 70.7(e)(2)(1)(B).

[183] 40 CFR § 70.5(d).

[184] 40 CFR § 70.8.

[185] Under 40 CFR § 70.8(c).

can approve the permit modification prior to that time.[186] Within 90 days of the receipt of an application for a minor permit modification, or 15 days after the end of the EPA's 45-day review period, whichever is later, the permitting authority must either issue the permit for a minor modification, deny the request, determine that the request is for a significant permit modification, or revise the application.[187]

The state permitting authority may choose to consider applications for minor permit modifications in groups that collectively are below the threshold level approved by the EPA, but which meet the state's alternative threshold requirements.[188]

Permit modifications that are not administrative amendments or minor permit modifications are "significant permit modifications."[189] Criteria for determining whether a change is significant include, at a minimum, every significant change in existing monitoring permit terms or conditions and every relaxation of reporting or recordkeeping permit terms or conditions. An applicant for a significant permit modification must meet all requirements for a 40 CFR Part 70 permit, including the requirements for applications, public participation, review by affected states, and review by EPA, as they apply to permit issuance and permit renewal. The state permitting authority must complete review on the majority of significant permit modifications within 9 months after receipt of a complete application.

Every issued permit must include provisions specifying the conditions under which the permit will be reopened prior to expiration. A permit can be reopened and revised by the state permitting authority under any of the following circumstances:

● Additional applicable requirements under the CAA become applicable to a major source with a remaining permit term of 3 or more years. Such a reopening shall be completed not later than 18 months after promulgation of the applicable requirement. No such reopening is required if the effective date of the requirement is later than the date on which the permit is due to expire, unless the original permit or any of its terms and conditions has been extended pursuant to EPA regulations.[190]

---

[186] 40 CFR § 70.7(e)(2)(iii).
[187] Id.
[188] 40 CFR § 70.7(e)(2)(v). The state can set alternative limits based on an EPA-approved schedule, or on emissions that are 10 percent of the emissions allowed by the permit for the emissions unit for which the change is requested, 20 percent of the applicable definition of major source in 40 CFR § 70.2, or 5 tons per year, whichever is least.
[189] 49 CFR § 70.7(e)(4).
[190] 40 CFR §§ 70.4(b)(10)(i) or (ii).

- Additional requirements (including excess emissions requirements) become applicable to an affected source under the acid rain program. With EPA approval, excess emissions offset plans will be deemed to be incorporated into the permit.
- The state permitting authority or the EPA determines that the permit contains a material mistake or that inaccurate statements were made in establishing the emissions standards or other terms or conditions of the permit.
- The EPA or the state permitting authority determines that the permit must be revised or revoked to assure compliance with the applicable requirements.[191]

Proceedings to reopen and issue a permit follow the same procedures as apply to initial permit issuance, and affect only those parts of the permit for which cause to reopen exists. Before a permit can be reopened, the state permitting authority must provide notice to the affected source at least 30 days in advance of the date that the permit is to be reopened. In the case of an emergency, however, the state permitting authority can provide less than 30 days notice.[192]

Permits can be reopened for cause by the EPA. If the EPA determines that cause exists to terminate, modify, or revoke and reissue a permit, the EPA notifies the state permitting authority and the permittee of their finding in writing. The state permitting authority, within 90 days after receipt of the notification, then forwards a proposed determination of termination, modification, or revocation and reissuance, as appropriate, to the EPA.[193] The EPA can extend the 90-day period for an additional 90 days if it is determined that a new or revised permit application is necessary, or that the state permitting authority must require the permittee to submit additional information.

The EPA then reviews the proposed determination from the state permitting authority within 90 days of receipt. The state permitting authority then has 90 days from receipt of an EPA objection to resolve any objection that the EPA makes and to terminate, modify, or revoke and reissue the permit in accordance with the EPA's objection. If the state refused to submit a proposed determination or fails to resolve any objection, then the EPA will terminate, modify, or revoke and reissue the permit after providing at least 30 days' notice to the permittee in writing of the reasons for the action.[194]

---

[191] 40 CFR § 70.7(f)(2).
[192] 40 CFR § 70.7(f)(2).
[193] 40 CFR § 70.7(g).
[194] Id.

7.10 Enforcement

Under CAA Section 113(b), any time the EPA finds that "any person" has violated or is in violation of any requirement or prohibition of an applicable SIP or permit, the EPA must notify the person and the state in which the violation occurred. Thirty days after the notice of a violation is issued, the EPA can either issue an order requiring the person to comply with the requirements or prohibitions of such plan or permit, issue an administrative penalty order under CAA Section 113(d),[195] or bring a civil action against the person.[196] Similar penalties are specified for state permitting agencies. Civil penalties for violations can exceed $25,000 per day.[197]

Section 113(c)(2) determines criminal penalties:

(2) Any person who knowingly -

(A) makes any false material statement, representation, or certification in, or omits material information from, or knowingly alters, conceals, or fails to file or maintain any notice, application, record, report, plan, or other document required pursuant to this chapter to be either filed or maintained (whether with respect to the requirements imposed by the Administrator or by a State);

(B) fails to notify or report as required under this chapter; or

(C) falsifies, tampers with, renders inaccurate, or fails to install any monitoring device or method required to be maintained or followed under this chapter shall, upon conviction, be punished by a fine pursuant to title 18 or by imprisonment for not more than 2 years, or both. If a conviction of any person under this paragraph is for a violation committed after a first conviction of such person under this paragraph, the maximum punishment shall be doubled with respect to both the fine and imprisonment.[198]

Fines and/or imprisonment are also authorized for knowingly failing to pay any fees,[199] negligently releasing hazardous air pollutants into the

---

[195] 42 U.S.C. § 7413(d).
[196] 42 U.S.C. § 7413(a).
[197] 42 U.S.C. § 7413(b). For court decisions discussing civil penalties, see United States v. Hoechst Celanese Corp., 964 F. Supp. 967 (D.S.C. 1996); and United States v. SCM Corp., 667 F. Supp. 1110 (D. Md. 1987).
[198] 42 U.S.C. § 7413(c)(2). See United States v. Fern, 155 F.3d 1318 (11th Cir. 1998) [discussing the elements of a "knowing" offense].
[199] 42 U.S.C. § 7413(c)(3).

ambient air and placing other persons in imminent danger of death or serious injury.[200]

7.11 Appeals and Remedies Under the Clean Air Act

There are several mechanisms by which a CAA permit decision can be challenged. First, the CAA has a "citizens suit" provision under which "any person" can bring suit in federal court against a source to compel compliance with an emission limitation or EPA order, or against the EPA to compel a nondiscretionary duty under the CAA.[201] The plaintiff may seek only injunctive relief, but not damages or penalties. The U.S. Supreme Court has allowed such citizens suits, but requires a good faith allegation by the plaintiff that the violation had not been terminated.[202] However, amendments to CAA Section 304(a) in 1990 effectively prohibit citizen suits for violations that are wholly in the past.[203] Citizen suits have been used to force compliance with state SIPs,[204] and HAP standards.[205] The CAA citizens suit provision has been used to enforce *state* odor regulations under its SIP.[206]

There is a provision for judicial review by private parties in CAA Section 307(b)(1) to challenge national emission limitations and final actions by the EPA in federal courts of appeals.[207] Challenges to national standards and other actions with national importance must be brought in the Court of Appeals for the D.C. Circuit within 60 days of final action. The U.S. Supreme Court has given a broad interpretation to what decisions are reviewable under CAA Section 307(b)(1).[208]

While CAA citizen suits are not uncommon, the more usual process involves a disappointed applicant for a CAA permit who wishes either to challenge the denial of a permit, or to challenge limitations placed on a

---

[200] 42 U.S.C. § 7413(c)(4). Interestingly, it is a defense to such a charge that the person endangered had freely consented to the conduct.

[201] 42 U.S.C. § 7604.

[202] Gwaltney of Smithfield v. Chesapeake Bay Found., Inc., 484 U.S. 49 (1987).

[203] See Fried v. Sungard Recovery Services, Inc., 916 F. Supp. 465 (E.D. Penn. 1996).

[204] See Sierra Club v. Publ. Service Co. of Colorado, Inc., 894 F. Supp. 1455 (D. Colo. 1995).

[205] See Fried v. Sungard Recovery Services, Inc., 925 F. Supp. 364 (E.D. Penn. 1996).

[206] See Save our Health Organization v. Recomp of Minnesota, 829 F. Supp. 288 (D. Minn. 1993).

[207] 42 U.S.C. § 7607(b)(1).

[208] See Harrison v. PPG Indus. (446 U.S. 578 (1980) [informal decision by regional EPA administrator in letter was reviewable]).

permit by the permitting authority. It is possible for a source to pursue an appeal of a final Title V permit decision, although the mechanism generally depends on whether a state or federal agency issued the decision on the permit. If the state permitting authority makes the final permit decision, appeals ordinarily proceed in the state administrative or judicial system in accordance with the state's Title V program and relevant state law principles. If the EPA makes the final decision, appeals ordinarily proceed first to EPA's Environmental Appeals Board, and then to the federal courts of appeals. In addition, EPA's denial of a public petition requesting that it object to a permit can give rise to direct judicial review in the federal courts of appeals.[209]

When state permitting authorities grant or deny CAA Title V permits, judicial review must be sought in the state's judicial system. Section 502(b)(6) of the CAA requires that the state Title V program provide:

[A]n opportunity for judicial review in State court of the final permit action by the applicant, any person who participated in the public comment process, and any other person who could obtain review of that action under applicable law.[210]

If the state permitting authority fails to take timely action on a permit application, then CAA Section 502(b)(7) states that the state's failure will be treated as a final action for purposes of obtaining judicial review in state court.

EPA regulations contain requirements for judicial review which all state Title V programs must satisfy. A petition for judicial review of a final permit decision must be filed within 90 days of the decision, or such shorter time as the state may designate, and must be the exclusive means for judicial review of the Title V permit.[211] A petition can be filed later than the designated deadline for judicial review only if it is based on grounds arising after the deadline. A challenge to the state permitting authority's failure to take final action on a permit may be filed any time before such action is taken.[212]

The precise route for obtaining judicial review of permit actions by state permitting authorities will vary from state to state. In some states, appeals from final decisions of the permitting authority are filed directly in a state

---

[209] See Morgan, Lewis & Bockius LLP, *EPA's Operating Permit Regulations: Title V Permit Appeals and Strategic Considerations* (1996), found at http://envinfo.com/caain.

[210] 42 U.S.C. § 7661a(b)(6).

[211] 40 CFR § 70.4(b)(3)(xii).

[212] Id. See Morgan, Lewis & Bockius LLP, *EPA's Operating Permit Regulations: Title V Permit Appeals and Strategic Considerations* (1996), found at http://envinfo.com/caain.

court. In other states, initial appeals of final permit actions are filed before an environmental appeals board or similar administrative body. Appeals from the appeals boards' decisions in those states are subsequently filed in a state court.[213]

Most initial Title V permit decisions will be made by state permitting authorities, although EPA itself has the authority to issue or deny Title V permits in certain situations. Recall from above that EPA's exercise of this authority can arise in two different ways.[214] EPA regulations contain the procedures for obtaining review of EPA Title V permit decisions (or decisions of a state with delegated regulatory authority).[215] Any person who filed comments on the draft permit or participated in the public hearing may petition EPA's Environmental Appeals Board (EAB) to review any condition of the permit decision. An appeal of EPA's issuance or denial of a Title V permit must be filed with the EAB as a prerequisite to seeking judicial review,[216] although the EAB may review any condition of any permit issued by the EPA under EPA regulations on its own initiative.[217]

Petitions for review of EPA Title V permit decisions must be filed with the EAB within 30 days of the permit decision. The petition must be limited to specific issues as described in EPA regulations:

The petition shall include a statement of the reasons supporting that review, including a demonstration that any issues raised were raised during the public comment period (including any public hearing) to the extent required by these regulations unless the petitioner demonstrates that it was impracticable to raise such objections within such period or unless the grounds for such objection arose after such period, and, when appropriate, a showing that the condition in

---

[213] For more detailed information on state court challenges, see Morgan, Lewis & Bockius LLP, *EPA's Operating Permit Regulations: Title V Permit Appeals and Strategic Considerations* (1996), found at http://envinfo.com/caain; Nicewander, *Clean Air Act Permitting : A Guidance Manual* (Pennwell Pub., 1995); and Martineau and Novello (eds.), *Clean Air Act Handbook* (Amer. Bar Assn, 1997).

[214] If a state does not obtain Title V program approval, or if the EPA withdraws approval from a state, then the EPA becomes the permitting authority. 42 U.S.C. §§ 7661(d)(3) and (i)(4). If EPA objects to a proposed permit and the state permitting authority fails to revise and submit a proposed permit in response to the objection within 90 days, EPA is to issue or deny that permit. 40 CFR § 70.8(c)(4).

[215] 40 CFR Part 71.

[216] 40 CFR § 71.11(l)(4).

[217] 40 CFR § 71.11(l)(2). See Morgan, Lewis & Bockius LLP, *EPA's Operating Permit Regulations: Title V Permit Appeals and Strategic Considerations* (1996), found at http://envinfo.com/caain.

question is based on:

(i) A finding of fact or conclusion of law which is clearly erroneous; or

(ii) An exercise of discretion or an important policy consideration which the Environmental Appeals Board should, in its discretion, review.[218]

Within a reasonable time following the filing of the petition for review, the EAB must issue an order either granting or denying the petition for review. To the extent review is denied, the conditions of the final permit decision become final agency action for purposes of further appeal. The EAB must issue a public notice of a grant of review, which must set forth a briefing schedule for the appeal, and state that any interested person may file an amicus brief. However, a notice of denial of review is sent only to the permit applicant and to the person(s) requesting review.[219]

Under CAA Section 307(b), final decisions of the EAB may then be appealed to the federal courts of appeals.[220] Under CAA Section 307(b) guidelines, a petition for review of an EPA action must be filed within 60 days of the final decision, unless the petition is based "solely on grounds arising after such sixtieth day."[221] Exclusive jurisdiction over a final action which is "locally or regionally applicable," such as a permitting decision, rests with "the United States Court of Appeals for the appropriate circuit."[222] For purposes of Title V permit appeals, the "appropriate circuit" would presumably be the circuit in which the source is located.[223]

The appellate court's subsequent review of an EAB decision on a CAA Title V permit will usually consist only of the administrative record before the EAB, and will be conducted pursuant to the judicial review provisions of the Administrative Procedure Act.[224] This means that the appellate court will uphold the EAB's findings of fact unless they are "unsupported by substantial evidence," and will uphold the EAB's final decision unless the court concludes that the decision is "arbitrary, capricious, an abuse of

---

[218] 40 CFR § 71.11(l)(1). For an example of an EAB appeals decision (although for a Title IV permit), see Indianapolis Power & Light Company Petersburg Plant, 6 E.A.D. 23 (CAA Appeal No. 95-1) (5/15/95).
[219] 40 CFR § 71.11(l)(3).
[220] 42 U.S.C. § 7607(b).
[221] Id.
[222] Id.
[223] Morgan, Lewis & Bockius LLP, *EPA's Operating Permit Regulations: Title V Permit Appeals and Strategic Considerations* (1996), found at http://envinfo.com/caain; Nicewander, *Clean Air Act Permitting : A Guidance Manual* (Pennwell Pub., 1995); and Martineau and Novello (eds.), *Clean Air Act Handbook* (Amer. Bar Assn, 1997).
[224] 5 U.S.C. §§ 701-706.

discretion, or otherwise not in accordance with law."[225]

In addition, CAA Section 505(b)(2) allows "any person" (including the permit applicant) to file a petition requesting that EPA object to a proposed permit.[226] Section 505(b)(2) specifically provides that EPA's subsequent denial of a petition may be appealed directly to the federal courts of appeals pursuant under CAA Section 307(b). If EPA grants the public petition and objects to the permit, then, under CAA Section 505(c)(2), EPA's objection is not subject to judicial review until EPA takes final action on the permit itself.[227]

There is a particular problem where interstate pollution is concerned, because the CAA does little to address the problem. Attempts to force EPA to remedy interstate pollution by enforcing CAA Section 110(a)(2)(D)(I),[228] which requires every SIP to prohibit any source from interfering with another state's NAAQS or PSD requirements), have been unsuccessful.[229]

---

[225] 5 U.S.C. § 706(2). See In Re. Campo Landfill Project, Campo Band Indian Reservation, 6 E.A.D. (505 NSR Appeal No. 95-1) (6/19/96) [EAB review of offset permit denied because petitioners failed to meet burden].

[226] 42 U.S.C § 7661d(b)(2).

[227] 42 U.S.C. § 7661d(c).

[228] 42 U.S.C. § 7410(a)(2)(D)(I).

[229] For example, see New England Legal Found. v. Costle, 475 F. Supp. 425 (D. Conn. 1979), aff'd, 632 F.2d 936 (2d Cir. 1980) and 666 F.2d 30 (2d Cir. 1981); Connecticut v. EPA, 656 F.2d 902 (2d Cir. 1981); State of New York v. EPA, 710 F.2d 1200 (6th Cir. 1983); New York v. EPA, 716 F.2d 440 (7th Cir. 1983); Jefferson Co. v. EPA, 739 F.2d 1071 (6th Cir. 1984); and New York v. EPA, 852 F.2d 574 (D.C. Cir. 1988).

Hazardous and Toxic Substances Regulation

Resource Conservation and Recovery Act

8.1 Introduction

Congress enacted the Resource Conservation and Recovery Act (RCRA)[1] in 1976 out of concern that improper handling and disposal of solid and hazardous wastes posed a continuing threat to the environment and a danger to human health.[2] At the time of its original enactment, RCRA amended and completely revised its predecessor statute, the Solid Waste Disposal Act of 1965. RCRA sets forth a comprehensive "cradle-to-grave" framework for the management of solid and hazardous wastes from generation to final disposal.[3] The U.S. Environmental Protection Agency (EPA) is the primary agency vested with the authority to issue the regulations necessary to implement the goals and policies of the statute. The EPA has issued detailed regulations which set forth criteria for identifying solid and hazardous wastes, and which impose extensive storage, disposal, shipping, reporting, and recordkeeping requirements on generators and transporters of RCRA-regulated wastes.[4]

Delegation of RCRA Authority to States. RCRA empowers the EPA Administrator to delegate authority to the states to operate their own solid

---

[1] 42 U.S.C. §§ 6901 through 6991i, as amended by the Hazardous and Solid Waste Amendments of 1984, Pub. L. 98-616, 98 Stat. 3221 (Nov. 8, 1984).
[2] See 42 U.S.C. § 6901(b).
[3] Congress amended RCRA in 1984 with passage of the Hazardous and Solid Waste Amendments of 1984 (HSWA), Pub. L. 98-616, 98 Stat. 3221 (Nov. 8, 1984), which expanded RCRA's scope with additional comprehensive waste management requirements, most notably the underground storage tank program.
[4] 40 CFR pts. 260 through 265.

and hazardous waste management programs in lieu of the federal RCRA program.[5] Most states have received authorization from EPA to operate their own programs in lieu of the federal program. Program authorization indicates that a state has passed legislation and promulgated regulations to administer a solid and hazardous waste regulatory program that is fully equivalent to the federal RCRA program, with permitting, manifest, and operational requirements for generators, transporters, and treatment, storage, and disposal (TSD) facilities that are at least as stringent as the federal program. Where a state has received such EPA approval, generators, transporters, and TSDs doing business within that state must comply with the state program requirements, including obtaining applicable permits and forwarding all required reports to the state solid and hazardous waste regulatory agency.

In nonapproved states, the EPA administers the RCRA regulatory program. Thus, a generator, transporter, or TSD facility located within one of these states must obtain applicable RCRA permits directly from the EPA and comply with all federal regulations. It is important to note, however, that in these nonauthorized states, an EPA permit is not alone sufficient for RCRA compliance since these states still run their own solid and hazardous waste regulatory programs, which, until approved by the EPA, operate in tandem with the EPA-administered RCRA program. Thus, a generator, transporter, or TSD facility located in one of these states must obtain applicable EPA permits as well as state-issued permits, and must comply with both federal and state solid and hazardous waste regulations.

It must be kept in mind that although a state's solid and hazardous waste program must be equivalent to the federal program to receive EPA authorization, equivalent does not mean identical. RCRA does not prohibit states from establishing solid and hazardous waste regulatory regimes that are more stringent than those delineated in the EPA's regulations.[6] Thus, although the discussion that follows focuses on the requirements of RCRA and the federal implementing regulations, wastes generators, transporters, and TSD facilities must check state regulations for more stringent requirements.

RCRA Subtitles. RCRA consists of several subtitles, which comprise the framework for regulation of solid and hazardous wastes. A breakdown of the various subtitles is as follows:

Subtitle A - General Provisions
Subtitle B - Office of Solid Waste; Authorities of the Administrator

---

[5] 42 U.S.C. § 6926(b).
[6] 42 U.S.C. § 6929. See also Old Bridge Chemicals, Inc. v. New Jersey Dep't of Envt'l Protection, 965 F.2d 1287 (3d Cir. 1992); Hazardous Waste Treatment Council v. Reilly, 938 F.2d 1390 (D.C. Cir. 1991).

Subtitle C - Hazardous Waste Management
Subtitle D - State or Regional Solid Waste Plans
Subtitle E - Duties of the Secretary of Commerce in Resource and
 Recovery
Subtitle F - Federal Responsibilities
Subtitle G - Miscellaneous Provisions
Subtitle H - Research, Development, Demonstration, and Information
Subtitle I - Regulation of Underground Storage Tanks
Subtitle J - Medical Waste Tracking

Three of these subtitles are of greatest concern to companies and individuals who generate, store, or dispose of solid and hazardous wastes. Subtitle C deals with hazardous waste management; Subtitle D sets forth standards for state solid waste management programs; and Subtitle I covers underground storage tank regulation. An overview of these three subtitles is provided in this chapter.

8.2 Solid Waste Classification

Identifying the wastes that are regulated under RCRA is necessarily dependent on a clear understanding of the legal definitions of solid and hazardous waste. RCRA's regulatory framework requires proper identification of wastes that are defined as being "solid wastes" and "hazardous wastes." For purposes of Subtitle C, the hazardous waste management provisions, a two-step analysis must be performed. Waste material must be both a solid waste and a hazardous waste to be regulated under Subtitle C of RCRA. Solid wastes that are not classified as being hazardous, and hazardous substances that are not classified as being solid wastes, will not be regulated as hazardous wastes under Subtitle C. However, it should be noted that solid wastes that are not hazardous may nevertheless be regulated under RCRA Subtitle D, the solid waste management provisions, and hazardous substances that are not also solid wastes may still be regulated under another of the environmental laws discussed in this book.

Accordingly, the initial analysis for identifying a RCRA-regulated waste requires a determination of whether it is a "solid waste." RCRA generally defines the term solid waste to mean garbage, refuse, sludge, and any other discarded material.[7] The RCRA solid waste definition is further explained in EPA regulations, wherein the term solid waste is defined as "any discarded material."[8] The difficult issue under this definition is

---

[7] 42 U.S.C. § 6903(27).
[8] 40 CFR § 261.2(a)(1).

understanding what is meant by "discarded" material. Examination of the regulatory definition of "discarded" reveals which materials the EPA considers discarded and thereby subject to classification as solid wastes. According to the EPA regulations, "[a] discarded material is any material which is: (i) Abandoned ...; or (ii) Recycled ...; or (iii) Considered inherently waste-like ..."[9] unless some exclusion applies.[10] A discussion of each of these three types of "discarded materials" follows. If a material has been "discarded" and thus classified as a solid waste, the next step--for purposes of regulation under Subtitle C--would be to determine whether the solid waste is also a "hazardous waste.[11]

8.2.1 Discarded Materials That Are Abandoned

One of the three types of discarded materials set forth in the RCRA regulations is comprised of discarded materials that are "abandoned." Discarded materials are solid wastes if "abandoned" by being:
1. Disposed of; or
2. Burned or incinerated; or
3. Accumulated, stored, or treated (but not recycled) before, or in lieu of, being abandoned by being disposed of, burned or incinerated.[12]
This definition of the term "abandoned," as provided in the regulations, is hardly a model of clarity and requires further explanation.

"Disposed of." In order to understand what is meant by the term "disposed of" in this regulatory provision, one must read RCRA's statutory definition of disposal. "Disposal" is defined in RCRA as "the discharge, deposit, injection, dumping, spilling, leaking, or placing of any solid waste or hazardous waste into or on any land or water" so that it or any of its constituents may enter the environment or be emitted into the air or discharged into any waters, including groundwaters.[13] This definition of "disposal" is nearly identical to a previous definition of the term "disposed of," which was found in former RCRA regulations.[14] Thus, it is presumed

---

[9] 40 CFR § 261.2(a)(2).
[10] See section 8.2.4 for discussion of wastes specifically excluded from the definition of solid waste.
[11] See section 8.3 for guidance on making this determination. Remember, however, that even though the material is not a hazardous waste, it may still be subject to regulation as a solid waste under Subtitle D of RCRA. See section 8.5 for full discussion of regulation of solid wastes under Subtitle D.
[12] 40 CFR § 261.2(b).
[13] RCRA 1004(3), 42 U.S.C. § 6903(3). See also 40 CFR § 260.10, containing the identical definition of "disposal."
[14] 40 CFR § 261.2(d) (1984).

that proper interpretation of the term "disposed of" as it applies to an understanding of the term "abandoned" will depend on the statutory definition of "disposal."

Virtually any material that a facility is releasing or potentially releasing to the environment can be considered "abandoned" under the broad definition of the term disposal. The term disposal encompasses any method through which a solid waste can enter the land, water, or air, including nonintentional disposals such as through spills and leaks.[15] Further, a material need not be finally disposed of to constitute a solid waste since the term dispose does not mean finally and forever discarded.[16]

"Burned or Incinerated." The second way that waste material may be considered discarded (and thus a solid waste) by abandonment is when the material is abandoned by being burned or incinerated. To determine whether burning or incineration constitute abandonment depends on the device used to burn or incinerate the material, and the purpose for which the device is being used. If the burning or incineration takes place in any device other than a boiler[17] or industrial furnace,[18] the material is considered abandoned, regardless of whether energy or materials are recovered from the process. If the burning or incineration takes place in a boiler or industrial furnace, the material will not be considered discarded by abandonment unless the purpose of the burning is to destroy the material.[19]

"Accumulated, stored, or treated (but not recycled) before or in lieu of being abandoned by being disposed of burned, or incinerated." The third and final way for a material to be considered discarded (and thus a solid waste) by abandonment is when the material is abandoned by being "accumulated, stored, or treated (but not recycled) before or in lieu of being abandoned by being disposed of, burned, or incinerated."[20] This catch-all provision basically means that when a company or individual is storing or treating a material with the intention to subsequently abandon it, it will be classified as a discarded material and solid waste. Generators may clearly show an intent to abandon materials, such as when waste-containing drums are stored for shipment to an incinerator, landfill, or TSD facility. In other instances, the generator may show that the waste is

---

[15] 42 U.S.C. § 6903(3); 40 CFR § 261.2(b).
[16] See, for example, United States v. ILCO, 996 F.2d 1126 (11th Cir. 1993).
[17] See 40 CFR § 260.10 (definition of "boiler.").
[18] See 40 CFR § 260.10 (definition of "industrial furnace.").
[19] See 50 *Fed. Reg.* 614, 630-31 (1985).
[20] 40 CFR § 261.2(b)(3). The recycled parenthetical is necessary because recycling falls within the definition of "treatment." RCRA 1004(34), 42 U.S.C. § 6903(34); 40 CFR § 260.10 ("treatment").

being stored for later recycling.[21] However, material that has been used and is being accumulated with the intent that it will be recycled at some point in the future constitutes a solid waste until it is actually recycled.[22]

8.2.2 Discarded Materials That Are Recycled

Another category of discarded materials that falls within the scope of the solid waste definition is recycled materials.[23] The regulations governing recycled materials are especially complex.[24] Recycled materials might not ordinarily be thought of as discarded materials; however, the EPA purposely included recycled materials in the definition of discarded material to bring it within the scope of the solid waste definition. The determination of whether a recycled material is subject to regulation as a RCRA solid waste requires an evaluation of the type of recycling method used and the type of material produced from the recycling activity.

Materials that are recycled by one of the following methods are exempt from classification as solid wastes, and thus not be subject to regulation under Subtitle C of RCRA:[25]

1. Materials used or reused as ingredients in an industrial process to make a product, provided the materials are not being reclaimed;[26] or

2. Materials used or reused as effective substitutes for commercial products; or

3. Materials returned to the original process from which they are generated, without first being reclaimed. The material must be returned as a substitute for raw material feedstock, and the process must use raw materials as principal feedstocks.

On the other hand, seven specific types of materials will be considered solid wastes regardless of whether the recycling involves use, reuse, or return to the original process if produced by any of four types of recycling

---

[21] EPA severely limited the likelihood of a large percentage of "sham" recycling claims by including regulations that accord solid waste status to materials that are "accumulated speculatively." 40 CFR § 261.2(c)(4).

[22] See, for example, United States v. ILCO, 996 F.2d 1126 (11th Cir. 1993).

[23] 40 CFR § 261.2(c).

[24] See 40 CFR § 261.10 (definition of "recycled"); 40 CFR § 261.1 through 261.6, 260.30 through 260.33, 260.40, 260.41; 50 *Fed. Reg.* 614 (1985).

[25] 40 CFR § 261.2(e)(1).

[26] A reclaimed material is one that has been processed to recover a usable product of one that is regenerated. Examples include recovery of lead values from spent batteries and the regeneration of spent solvents. 40 CFR § 261.1(c)(4).

activities.[27] The seven types of materials are:
1. Spent materials[28]
2. Sludges[29] listed in 40 CFR Section 261.31 or 261.32
3. Sludges exhibiting a hazardous waste characteristic
4. By-products[30] listed in 40 CFR Section 261.31 or 261.32
5. By-products exhibiting a hazardous waste characteristic
6. Commercial chemical products listed in 40 CFR Section 261.33
7. Scrap metal.[31]

These materials are considered solid wastes if they are recycled, or accumulated, stored, or treated before recycling, and they are:

1. *Used in a manner constituting disposal.*[32]

Materials are used in a manner consituting disposal if applied to or placed on the land in a manner that constitutes disposal, or used to produce products that are applied to or placed on land or otherwise contained in products that are applied to or placed on the land (in which case the product itself remains a solid waste.) Commercial chemical products listed in 40 CFR Section 261.33, however, are not solid wastes if they are applied to the land and that is their ordinary manner of use.

2. *Burned for energy recovery.*[33]

The materials are solid wastes if recycled by being burned to recover energy, or used to produce a fuel or are otherwise contained in fuels (in which case the fuel itself remains a solid waste.) Again, commercial chemical products listed in 40 CFR Section 261.33, however, are not solid wastes if they are themselves meant to be fuels.

---

[27] 40 CFR § 261.2(e)(2). These seven materials are listed in in Table 1 to 40 CFR § 261.2 and the four types of recycling activities are porvided in the RCRA definition of solid wastes at 40 CFR § 261.2(c).

[28] Spent materials are materials that have been used and as a result of contamination can no longer serve the purpose for which they were produced without processing. 40 CFR § 261.1(c)(1).

[29] "Sludge means any solid, semi-solid, or liquid waste generated from a municipal, commercial, or industrial wastewater treatment plant, water supply treatment plant, or air pollution control facility exclusive of the treated effluent from a wastewater treatment plant." 40 CFR § 260.10 ("sludge").

[30] "A 'by-product' is a material that is not one of the primary products of a production process and is not solely or separately produced by the production process. Examples are process residues such as slags or distillation column bottoms. The term does not include a co-product that is produced for the general public's use and is ordinarily used in the form it is produced by the process." 40 CFR § 261.1(c)(3).

[31] Scrap metal consists of bits and pieces of metal parts (e.g., bars, turnings, rods, sheets, wire) or metal pieces that may be combined with bolts or soldering (e.g., radiators, scrap automobiles, railroad box cars), which when worn or superfluous can be recycled.

[32] 40 CFR § 261.2(c)(1)(i)–(ii).

[33] 40 CFR § 261.2(c)(2)(i)–(ii).

3. *Reclaimed.*[34]

Spent materials, listed sludges and by-products, and scrap metal are solid wastes if they are recycled by being reclaimed. Characteristic sludges and by-products and commercial chemical products are not. A material is reclaimed if it is processed to recover a usable product or if it is regenerated. Examples are recovery of lead values from spent batteries and regeneration of spent solvents.

4. *Accumulated speculatively.*[35]

All of the materials listed above, except commercial chemical products are classified as solid wastes when they are "accumulated speculatively." A material is accumulated speculatively if it is accumulated before being recycled. A material is not accumulated speculatively, however, if the person accumulating it can show that the material is potentially recyclable and has a feasible means of being recycled; and that during the calendar year the amount of material that is recycled, or transferred to a different site for recycling equals at least 75 percent by weight or volume of the amount of that material accumulated at the beginning of the period.[36]

It is important to note that even if a company examines a process it is using and determines that the particular recycling activity exempts certain materials from classification as a solid waste, it must be prepared to document that the approved recycling process was actually used to recycle the materials. The EPA included this provision in the recycling regulations to guard against "sham" recycling claims.[37]

8.2.3 Discarded Materials That Are Inherently Waste-Like

The third category of discarded materials that will be considered solid wastes are certain materials that are "inherently waste-like" materials when recycled in any manner.[38] These materials are:

1. Materials with hazardous waste Nos. F020, F021 (unless used as an ingredient to make a product at the site of generation), F022, F023, F026, and F028; or

2. Secondary materials fed to a halogen acid furnace that are characteristic hazardous waste or listed hazardous wastes, except brominated material

---

[34] 40 CFR § 261.2(c)(3).
[35] 40 CFR § 261.2(c)(4).
[36] 40 CFR § 261.1(c)(8). Refer to this section of the regulations for specific guidance on calculating the 75 percent requirement.
[37] 40 CFR § 261.2(f).
[38] 40 CFR § 261.2(d).

that meets several criteria.[39] This provision implies that materials listed as inherently waste-like may not be solid wastes if not recycled, however, even if not recycled such materials would likely be deemed solid wastes under the abandonment or recycled provisions discussed earlier.[40] Only if the material were being accumulated, stored or treated before recycling could it possibly be fall outside of the scope of the solid waste definition. However, even then, it would probably be a considered a solid waste under the speculative accumulation provision,[41] and would unquestionably become a solid waste at the time of recycling.[42] Thus, a listed inherently waste-like material would virtually always be considered a solid waste.[43] The EPA may add additional materials to the list of inherently waste-like materials. The criteria for listing materials as inherently waste-like are complicated,[44] and no material can be considered an inherently waste-like solid waste unless and until the EPA finds, in a rulemaking proceeding, that the material has at least one of the two waste-like attributes set forth in the regulations and it poses a potential hazard when recycled.

8.2.4 Materials Excluded from Classification as Solid Wastes

The regulations expressly exclude certain materials from the definition of solid wastes.[45] By checking this list of exclusions, a company or individual can determine that some materials will not be subject to RCRA regulation. These excluded materials are as follows:

Materials That Are Not Solid Wastes
1. Domestic sewage;
2. Any mixture of domestic sewage and other wastes that passes through a sewer system to a publicly-owned treatment works for treatment. "Domestic sewage" means untreated sanitary wastes that pass through a sewer system.

---

[39] The brominated material exception requires that (1) the material contains a bromine concentration of at least 45 percent; (2) the material contains less than a total of one percent of toxic organic compounds listed in Appendix VIII to 40 CFR Part 261; and (3) the material is processed continually onsite in the halogen acid furnace via direct conveyance (hard piping).
[40] See sections 8.2.1 and 8.2.2.
[41] 40 CFR § 261.1(c)(8), 261.2(c)(4).
[42] 40 CFR § 261.2(d).
[43] Furthermore, if a material listed as inherently waste-like is being recycled and the manner of recycling is not exempted in the language of the listing provision, the material is considered a solid and hazardous waste and a "recyclable material" for purposes of RCRA Subtitle C. 40 CFR § 261.6.
[44] 40 CFR § 261.2(d)(2).
[45] 40 CFR § 261.4(a).

3. Industrial wastewater discharges that are point source discharges subject to regulation under Section 402 of the Clean Water Act, as amended.[46]
4. Irrigation return flows.
5. Source, special nuclear or by-product material as defined by the Atomic Energy Act of 1954.[47]
6. Materials subjected to in-situ mining techniques which are not removed from the ground as part of the extraction process.
7. Pulping liquors (i.e., black liquor) that are reclaimed in a pulping liquor recovery furnace and then reused in the pulping process, unless it is accumulated speculatively as defined in 40 CFR Section 261.1(c).
8. Spent sulfuric acid used to produce virgin sulfuric acid, unless it is accumulated speculatively as defined in 40 CFR Section 261.1(c).
9. Secondary materials that are reclaimed and returned to the original process or processes in which they were generated when they are reused in the production process provided:
(a) Only tank storage is involved, and the entire process through completion of reclamation is closed by being entirely connected with pipes or other comparable enclosed means of conveyance;
(b) Reclamation does not involve controlled flame combustion (such as occurs in boilers, industrial furnaces, or incinerators);
(c) The secondary materials are never accumulated in such tanks for over 12 months without being reclaimed; and
(d) The reclaimed material is not used to produce a fuel, or used to produce products that are used in a manner constituting disposal.
10. Spent wood preserving solutions that have been reclaimed and are reused for their original intended purpose; and wastewaters from the wood preserving process that have been reclaimed and are reused to treat wood.
11. EPA Hazardous Waste No. K087, and any wastes from the coke by-products processes that are hazardous only because they exhibit the Toxicity Characteristic specified in 40 CFR Section 261.24 when, subsequent to generation, these materials are recycled to coke ovens, to the tar recovery process as a feedstock to produce coal tar or are mixed with coal tar prior to the tar's sale or refining. This exclusion is conditioned on there being no land disposal of the wastes from the point they are generated to the point they are recycled to coke ovens or the tar refining process.

---

[46] This exclusion applies only to the actual point source discharge. It does not exclude industrial wastewaters while they are being collected, stored, or treated before discharge, nor does it exclude sludges that are generated by industrial wastewater treatment. See, for example, United States v. Lean, 969 F.2d 187 (6th Cir. 1992) (manufacturing plant's discharges into an open lagoon were commonly categorized as solid waste even though the lagoon subsequently discharged into surface waters).
[47] 42 U.S.C. § 2011 et seq.

12. Nonwastewater splash condenser dross residue from the treatment of K061 in high temperature metals recovery units, provided it is shipped in drums (if shipped) and not land disposed before recovery.

8.3 Hazardous Waste Classification

This section explains the process for identifying whether a solid waste is considered a "hazardous waste" which would subject it to regulation under Subtitle C of RCRA.[48] The regulations identify three separate categories of solid wastes that are considered hazardous wastes:
1. "Listed" hazardous wastes;[49]
2. "Characteristic" hazardous wastes;[50] and
3. Solid wastes that have been mixed with listed or characteristic hazardous wastes.[51]
An additional category of hazardous wastes is comprised of solid wastes "derived from" treatment, storage or disposal of hazardous wastes.[52] A discussion of each of these categories of hazardous waste follows.

8.3.1 Listed Hazardous Wastes

Once it has been determined that a material is classified as a "solid waste,"[53] the initial method for determining whether the solid waste may be subject to regulation as a hazardous waste is to check the three EPA lists of RCRA hazardous wastes found at 40 CFR Sections 261.31, 261.32 and 261.33.[54] Each of the hazardous wastes on these lists has been assigned an EPA Hazardous Waste Number.

In order to make an accurate identification of any listed hazardous wastes, it helps to understand the EPA criteria for listing hazardous waste, particularly when using the "commercial chemicals" list provided at 40 CFR Section 261.33. The EPA can only list a solid waste as a hazardous waste upon determining that the solid waste meets one of the following

---

[48] It is the sole responsibility of the solid waste generator to determine whether a waste is a hazardous waste. 40 CFR § 262.11.
[49] See section 8.3.1.
[50] See section 8.3.2.
[51] See section 8.2.4.
[52] See section 8.2.5.
[53] See section 8.2 for discussion of the process for making this determination.
[54] The three lists are: 40 CFR § 261.31, Hazardous Wastes from Non-Specific Sources; 40 CFR § 261.32, Hazardous Wastes from Specific Sources; and 40 CFR § 261.33, Discarded Commercial Chemical Products, Off-Specification Species, Container Residues, and Spill Residues Thereof.

criteria:[55]

1. It exhibits any of the four characteristics of hazardous waste identified in 40 CFR Sections 261.20 through 261.24.

2. It has been found to be fatal to humans in low doses or, in the absence of data on human toxicity, it has been shown to be dangerous in animal studies or is otherwise capable of causing or significantly contributing to an increase in serious irreversible, or incapacitating reversible, illness. Waste listed in accordance with these criteria are designated "Acute" Hazardous Waste.

3. It contains any of the toxic constituents listed in 40 CFR Section Part 260, Appendix VIII and, after considering the following factors, the EPA concludes that the waste is capable of posing a substantial present or potential hazard to human health or the environment when improperly treated, stored, transported or disposed of, or otherwise managed:

(a) The nature of the toxicity presented by the constituent.

(b) The concentration of the constituent in the waste.

(c) The potential of the constituent or any toxic degradation product of the constituent to migrate from the waste into the environment under the types of improper management considered in (g) below.

(d) The persistence of the constituent or any toxic degradation product of the constituent.

(e) The potential for the constituent or any toxic degradation product of the constituent to degrade into non-harmful constituents and the rate of degradation.

(f) The degree to which the constituent or any degradation product of the constituent bioaccumulates in ecosystems.

(g) The plausible types of improper management to which the waste could be subjected.

(h) The quantities of the waste generated at individual generation sites or on a regional or national basis.

(i) The nature and severity of the human health and environmental damage that has occurred as a result of the improper management of wastes containing the constituent.

(j) Action taken by other governmental agencies or regulatory programs based on the health or environmental hazard posed by the waste or waste constituent.

(k) Such other factors as may be appropriate.

Substances are listed on Appendix VIII only if they have been shown in scientific studies to have toxic, carcinogenic, mutagenic or teratogenic effects on humans or other life forms. Wastes listed in accordance with these criteria are designated "Toxic" wastes.

---

[55] These criteria are found in 40 CFR § 261.11. See also 57 *Fed. Reg.* 28087, 28088 (June 24, 1992).

Appendix VIII to Part 261 contains a list of toxic constituents that, if found in a solid waste, the EPA may use to list a waste as hazardous. It is important not to confuse this list of toxic constituents with the three hazardous waste lists. Just because a solid waste is found to contain one of these constituents does not mean that the waste is automatically deemed hazardous.[56] Although the EPA may consider the waste to be a candidate for listing, it is not a listed hazardous waste unless the EPA adds it to one of the three hazardous waste lists. Nevertheless, by exhibiting one of these toxic constituents, the solid waste may be a characteristic hazardous waste, discussed in the next section.

8.3.2 Characteristic Hazardous Wastes

The second category of RCRA hazardous waste is known as "characteristic" hazardous waste. If upon determining that a particular solid waste is not a listed hazardous waste and is not one of the solid wastes excluded from the RCRA hazardous waste definition, the waste generator must determine whether the solid waste exhibits certain hazardous waste characteristics. A solid waste will be deemed a hazardous waste for purposes of Subtitle C of RCRA if it exhibits any of the following four types of characteristics: ignitability,[57] corrosivity,[58] reactivity,[59] or toxicity.[60] The EPA devised the characteristic hazardous waste regulations in recognition of the fact that it would be virtually impossible to list every possible waste that would pose a danger to human health and the environment if improperly handled.[61]

Characteristic Hazardous Wastes[62]

Ignitability (I)
Ignitable waste is a liquid with a flash point less than 60 degrees centigrade, a solid capable of causing fire that burns vigorously and continuously so that it creates a hazard, an ignitable compressed gas or an oxidizer.

---

[56] The EPA promulgated a revision to its hazardous waste listing criteria to clarify that a waste is not presumptively hazardous merely because it contains an Appendix VIII hazardous constituent. 57 *Fed. Reg.* 12 (Jan. 2, 1992).
[57] 40 CFR § 261.21.
[58] 40 CFR § 261.22.
[59] 40 CFR § 261.23.
[60] 40 CFR § 261.24.
[61] Section 3001 of RCRA establishes EPA authority to identify hazardous wastes by this method.
[62] 40 CFR §§ 261.20 through 261.24.

Wastes exhibiting the (I) characteristic have the EPA Hazardous Waste Number of D001.

Corrosivity (C)
Corrosive waste has a pH less than 2 or greater than 12.5 or corrodes steel at a rate greater than 6.35 mm/year.
Wastes exhibiting the (C) characteristic have the EPA Hazardous Waste Number of D002.

Reactivity (R)
Reactive waste exhibits one or more of the following characteristics: normally unstable and undergoes violent change without detonating; reacts violently with water; forms explosive mixtures with water; generates toxic gases, vapors, or fumes; is a cyanide or sulfide bearing waste that generates gases, vapors or fumes when exposed to a pH between 2 and 12.5; is capable of detonation or explosive reaction; or is a forbidden explosive.
Wastes exhibiting the (R) characteristic have the EPA Hazardous Waste Number of D003.

Toxicity (E)
Wastes fall into this class when tested under the Toxicity Characteristic Leachate Procedure (TCLP)[63] or equivalent method and are found to produce an extract that contains specific quantities of certain contaminants, including arsenic, barium, cadmium, chromium, lead, mercury, selenium, silver, endrin, lindane, methoxychlor, toxaphene, 2,4-D and 2,4,5-TP.
Wastes exhibiting the (E) characteristic have an EPA Hazardous Waste Number as specified in Table 1 to 40 CFR Section 261.24, which corresponds to the toxic contaminant causing it to be hazardous.

8.3.3 Solid Wastes Excluded from Classification as Hazardous Wastes

The regulations for Subtitle C of RCRA list several types of solid wastes that are excluded from the definition of hazardous waste. Waste generators should consult this list of excluded materials provided at 40 CFR Section 261.4(b). In addition to the excluded wastes found on this list, hazardous waste samples and other air, soil, or water samples collected solely for testing for characteristics or composition are also exempt from RCRA

---

[63] The TCLP test replaced the formerly used EP Toxicity test on September 9, 1990.

regulation as hazardous waste, subject to several detailed conditions, until the testing and any necessary storage or transporation has been completed. Waste generators should consult the specific regulations regarding samples to determine whether an exemption applies.[64]

Even upon determining that a particular solid waste is excluded from regulation as a hazardous waste under RCRA, it is important to note that such wastes may still be subject to regulation under one of the other environmental laws discussed in this book.

8.3.4 Solid Wastes Mixed with Hazardous Wastes

When solid wastes become mixed with listed or characteristic hazardous wastes they may be regulated as RCRA hazardous wastes depending on the particular circumstances. A mixture of a solid waste and a characteristic hazardous waste will only be regulated as a RCRA hazardous waste if the mixture exhibits hazardous waste characteristics when tested. Determining whether a mixture of solid waste and listed hazardous waste will be classified as a hazardous waste requires a more complex analysis. Any mixture of a solid waste and listed hazardous waste will be classified as a hazardous waste unless it qualifies for one of two exemptions. First, if the listed hazardous waste mixed with the solid waste was listed solely because it exhibits a hazardous waste characteristic, and the mixture itself does not exhibit that characteristic, the mixture will not be considered a hazardous waste.[65] Second, if the mixture consists of certain specified hazardous wastes and discharged wastewater subject to regulation under Sections 402 or 307(b) of the Clean Water Act, the mixture will be exempt from RCRA regulation as a hazardous waste as long as the concentrations of the specified hazardous waste in the wastewater do not exceed certain levels, and the mixture is not otherwise characteristic or listed.[66] The specified hazardous wastes that come within this second exemption are:

1. Certain spent solvents listed on the "non-specific source list," i.e., carbon tetrachloride, tetrachloroethylene, and trichloroethylene, provided the maximum total weekly usage of the solvents, other than amounts that can be demonstrated not to be discharged to wastewater, divided by the average weekly flow of wastewater into the headworks of the company's wastewater treatment or pretreatment system does not exceed 1 ppm;[67]

2. Certain other spent solvents listed on the non-specific source list, i.e.

---

[64] 40 CFR § 261.4(d)(1)–(3).
[65] 40 CFR § 261.3(a)(2)(iii).
[66] 40 CFR § 261.3(a)(2)(iv).
[67] 40 CFR § 261.3(a)(2)(iv)(A).

methylene chloride, 1,1,1-trichloroethane, chlorobenzene, o-dichlorobenzene, creosols, cresylic acid, nitrobenzene, toluene, methyl ethyl ketone (MEK), carbon disulfide, isobutanol, pyridine, and spent chloroflurocarbon solvents, provided the maximum total weekly usage of these solvents, other than the amounts that can be demonstrated not to be discharged to wastewater, divided by the average weekly flow of wastewater into the headworks of the company's wastewater treatment or pretreatment system does not exceed 25 ppm;[68]

3. Heat exchanger bundle cleaning sludge from the petroleum refining industry;[69]

4. Discarded commercial chemical products or chemical intermediates listed on the commercial chemical list arising from de minimis losses of these materials from manufacturing operations in which the materials are used as raw material or by which they are produced;[70] and

5. Wastewater resulting from laboratory operations containing toxic (T) listed hazardous wastes provided that the annual average flow of laboratory wastewater does not exceed 1.0 percent of the total wastewater flow from the company's facility, or that the annual average concentration does not exceed 1 ppm in the headworks of the company's treatment or pre-treatment facility.[71]

The mixture rule has been the subject of considerable criticism. The EPA's mixture and "derived-from"[72] rules were challenged and invalidated in an important decision by the United States Court of Appeals for the D.C. Circuit.[73] The court invalidated the mixture and derived-from rules entirely on procedural grounds, holding that the EPA violated the Administrative Procedure Act by failing to give the required public notice and comment before issuing these regulations in 1980.[74] However, the court permitted the EPA to reinstate the rules on an interim basis, subject to an April 1993 expiration date.[75] The EPA removed the expiration date in October 1992 and the rules have since remained in effect.[76]

---

[68] 40 CFR § 261.3(a)(2)(iv)(B).
[69] 40 CFR § 261.3(a)(2)(iv)(C).
[70] 40 CFR § 261.3(a)(2)(iv)(D).
[71] 40 CFR § 261.3(a)(2)(iv)(E).
[72] See section 8.3.5 for a discussion of the "derived-from" rule.
[73] Shell Oil Co. v. EPA, 950 F.2d 741 (D.C. Cir. 1991).
[74] The court noted that the mixture rule and the derived-from rule were never part of EPA's proposed hazardous waste regulations, and they appeared for the first time (without prior notice to the regulated community) only when EPA issued the final regulations.
[75] 57 *Fed. Reg.* 7628 (Mar. 3, 1992).
[76] 57 *Fed. Reg.* 49278 (Oct. 30, 1992).

8.3.5 Solid Wastes Derived from Treatment, Storage, and Disposal of Hazardous Wastes

Solid wastes generated from the "treatment, storage, or disposal of hazardous wastes" are commonly known as "derived-from" hazardous wastes and are regulated as RCRA hazardous wastes under 40 CFR Section 261.3(c)(2) unless exempted from regulation. Solid wastes are not regulated pursuant to the "derived-from" rule if:
1. Reclaimed and used beneficially (unless burned for energy recovery or used in a manner constituting disposal);[77] or
2. Specifically exempted;[78] or
3. Meet specific criteria.[79]
The specifically exempted solid wastes are not considered hazardous wastes even though generated from the treatment, storage, or disposal of hazardous wastes unless they exhibit a hazardous waste characteristic. The regulations list three types of solid wastes that are eligible for this exemption:
1. Waste pickle liquor sludge generated by lime stabilization of spent pickle liquor from the iron and steel industry (SIC Codes 331 and 332).
2. Waste from burning any of the materials exempted from regulation by 40 CFR Section 261.6(a)(3)(v) through (viii).
3. Nonwastewater residues, such as slag, resulting from high temperature metals recovery processing of K061 waste, in units identified as rotary kilns, flame reactors, electric furnaces, plasma arc furnaces, slag reactors, rotary hearth furnace/electric furnace combinations or industrial furnaces [as defined in 40 CFR Sections 260.10(6), (7), and (12)], that are disposed in Subtitle D units, provided that these residues meet generic exclusion levels identified for all constituents, and exhibit no characteristics of hazardous waste.
Other solid wastes will not be considered hazardous wastes even though generated from the treatment, storage, or disposal of hazardous wastes if they meet the following criteria set forth in 40 CFR Section 261.3(d):
1. In the case of any solid waste, it does not exhibit any of the characteristics of hazardous waste identified in 40 CFR Sections 261.20 through 261.24.[80]
2. In the case of a solid waste that contains a listed hazardous waste (listed under 40 CFR Sections 261.30 through 261.33), or is derived from a listed

---

[77] 40 CFR § 261.3(c)(2)(i).
[78] 40 CFR § 261.3(c)(2)(ii).
[79] 40 CFR § 261.3(d).
[80] However, wastes that exhibit a characteristic at the point of generation may still be subject to the requirements of 40 CFR part 268, even if they no longer exhibit a characteristic at the point of land disposal.

hazardous waste, it has been delisted as provided in 40 CFR Sections 260.20 and 260.22.

As explained in the preceding section, the "mixture"[81] and derived-from rules were challenged and invalidated in an important decision by the United States Court of Appeals for the D.C. Circuit.[82] However, at the advice of the court, the EPA reinstated the mixture and derived-from rules on an interim basis, subject to an April. 28, 1993 expiration date. In October 1992, the EPA removed this expiration date[83] and the rules have since remained in effect.

8.3.6 Recyclable Materials

The "recyclable material" provisions of the RCRA regulations are especially complex and difficult to understand.[84] A "recyclable material" is essentially the term used by the EPA to refer to a hazardous waste that is recycled.[85] These materials are regulated separately because somewhat different regulations apply to recycled hazardous wastes than to RCRA hazardous wastes in general.

Whether or not a recycled material constitutes a "recyclable material" starts with the same process used to identify all wastes subject to regulation under Subtitle C of RCRA. Initially, a determination must be as to whether the particular recycled material is classified as a "solid waste." Most recycling activities will remove materials from classification as solid wastes. However, materials produced from four specific types of recycling activities (listed at 40 CFR Section 261.2(c) and discussed earlier in section 8.2.2) will subject those materials to regulation as solid wastes. Upon determining that a recycled material is classified as a solid waste, if the material is also hazardous, it will constitute a "recyclable material" subject to regulation under Subtitle C of RCRA. In addition, certain materials known as "inherently waste-like materials" are considered solid wastes regardless of how they are recycled, even if the recycling process would otherwise render them exempt from regulation.[86] These materials, because of their hazardous attributes, are also considered hazardous wastes and are therefore always subject to hazardous waste regulation under Subtitle C.

Recycled materials that are classified as a solid waste will automatically

---

[81] See section 8.3.4 for a discussion of the "mixture" rule.
[82] Shell Oil Co. v. EPA, 950 F.2d 741 (D.C. Cir. 1991).
[83] 57 *Fed. Reg.* 49278 (Oct. 30, 1992).
[84] 40 CFR § 261.6.
[85] 40 CFR § 261.6(a).
[86] 40 CFR § 261.2(d). See Section 8.2.3 for discussion of inherently waste-like materials.

be considered recyclable materials if they are characteristic or listed hazardous wastes. However, in the case of spent materials, and scrap metal, which are not already hazardous by definition, an independent analysis will have to be performed to determine whether these recycled materials are listed or characteristic hazardous wastes. If they are identified as hazardous wastes, they will, of course, be classified as recyclable materials. If they are not listed or characteristic hazardous wastes, but only solid wastes, they will not be subject to RCRA Subtitle C regulation.

Once a material has been identified as a recyclable material, it is necessary to determine how it is regulated. This determination is not so straightforward because recyclable materials are not regulated in precisely the same manner as hazardous waste in general. The specific requirements for handling recyclable materials are set forth at 40 CFR Section 261.6. Importantly, certain recyclable materials are listed as altogether exempt from RCRA regulation and these are worth listing here:[87]

1. Industrial ethyl alcohol that is reclaimed

2. Used batteries (or used battery cells) returned to a battery manufacturer for regeneration;

3. Used oil that exhibits one or more of the characteristics of hazardous waste but is recycled in some other manner than being burned for energy recovery;[88]

4. Scrap metal;

5. Fuels produced from the refining of oil-bearing hazardous wastes along with normal process streams at a petroleum refining facility if such wastes result from normal petroleum refining, production, and transportation practices;

6. Oil reclaimed from hazardous waste resulting from normal petroleum refining, production, and transportation practices, which oil is to be refined along with normal process streams at a petroleum refining facility;

7. Hazardous waste fuel produced from oil-bearing hazardous wastes from petroleum refining, production, or transportation practices, or produced from oil reclaimed from such hazardous wastes, where such hazardous wastes are reintroduced into a process that does not use distillation or does not produce products from crude oil so long as the resulting fuel meets the used oil specification under 40 CFR Section 266.40(e) and so long as no other hazardous wastes are used to produce the hazardous waste fuel;

8. Hazardous waste fuel produced from oil-bearing hazardous waste from petroleum refining production, and transportation practices, where such hazardous wastes are reintroduced into a refining process after a point at which contaminants are removed, so long as the fuel meets the used oil

---

[87] 40 CFR § 261.6(a)(3).
[88] See section 8.3.8 for further discussion of used oil regulations.

fuel specification under 40 CFR Section 266.40(e);

9. Oil reclaimed from oil-bearing hazardous wastes from petroleum refining, production, and transportation practices, which reclaimed oil is burned as a fuel without reintroduction to a refining process, so long as the reclaimed oil meets the used oil fuel specification under 40 CFR Section 266.40(e);

11. Petroleum coke produced from petroleum refinery hazardous wastes containing oil at the same facility at which such wastes were generated, unless the resulting coke product exceeds one or more of the characteristics of hazardous waste in 40 CFR Section 261.21 through 261.24.

Classification Variances. On a case-by-case basis, subject to specific standards and criteria, the EPA may grant variances from classification of recyclable materials as solid wastes (in which case the material would be excluded from Subtitle C regulation). In order to obtain a variance, an application must be submitted to the EPA Regional Administrator for the Regional Office with jurisdiction where the recycling operation is located. The application must address certain relevant regulatory criteria.[89] The application will be evaluated and the Regional Administrator will issue a draft notice tentatively granting or denying the variance. An informal public notice and comment period will follow and public notice of a tenative decision will be provided by newspaper advertisement or radio announcement. The public comment period lasts for 30 days, and a public hearing may be held at the discretion of the Regional Administrator. Following the notice and comment period, the EPA issues a final decision on the application. The decision is final and not subject to administrative appeal to EPA.[90]

Variances may only be sought to exclude the following three types of materials from classification as a solid waste:

1. Materials that are accumulated speculatively without sufficient amounts being recycled[91];

2. Materials that are reclaimed and then reused within the original primary production process in which they were generated;

3. Materials that have been reclaimed but must be reclaimed further before the materials are completely recovered.

The criteria that need to be addressed in the variance application vary depending on the type of material for which the variance is being sought.

---

[89] 40 CFR § 260.33(a).
[90] 40 CFR § 260.33(b). The regulations are silent on whether a judicial appeal may be pursued.
[91] See 40 CFR § 261.1(c)(8) (definition of accumulated speculatively).

8.3.7 Hazardous Waste Residues in Empty Containers

Containers that have held RCRA hazardous wastes, or containers with inner-liners that have held hazardous waste, may be excluded from regulation as hazardous waste under Subtitle C of RCRA provided that they are "empty."[92] The key to gaining exclusion of these containers from regulation as hazardous waste is properly managing the containers in such a way that they fit the regulatory definition of empty.[93] The regulations set forth various requirements and procedures for rendering the containers empty.

The criteria for determining whether a container is empty depends on the type of RCRA hazardous waste in the container. Basically, three different methods can be used to render the container empty, depending on whether the containers have held: (1) compressed gas hazardous waste; (2) wastes listed as acute hazardous wastes (the "H" wastes); or (3) other RCRA listed and characteristic hazardous wastes. The procedures for making sure that containers are properly empty of each of these categories of waste are as follows:

1. Containers that Have Held Compressed Gas Hazardous Waste. A container that has held a hazardous waste that is a compressed gas is empty when the pressure in the container approaches ambient atmospheric air pressure.[94]

2. Containers that Have Held Acute Hazardous Waste. A container, or an inner liner removed from a container, that has held an acute hazardous waste listed in 40 CFR Section 261.31, 261.32, or 261.33(e) is empty if:

(a) the container or inner liner has been triple rinsed using a solvent capable of removing the commercial chemical product or manufacturing chemical intermediate;

(b) the container or inner liner has been cleaned by another method that has been shown in the scientific literature, or by tests conducted by the generator, to achieve equivalent removal; or

(c) in the case of a container, the inner liner that prevented contact of the commercial chemical product or manufacturing chemical intermediate with the container, has been removed.[95]

3. All Other RCRA Listed and Characteristic Hazardous Wastes. A container, or an inner liner removed from a container, that has held any hazardous waste (except a waste that is a compressed gas or that is identified as a listed acute hazardous waste) is empty if:

---

[92] 40 CFR § 261.7(a)(1).
[93] Containers not considered "empty" are subject to regulation as a RCRA hazardous waste. 40 CFR § 261.7(a)(2).
[94] 40 CFR § 261.7(b)(2).
[95] 40 CFR § 261.7(b)(3).

(a) all wastes have been removed that can be removed using the practices commonly employed to remove materials from that type of container, e.g., pouring, pumping, and aspirating, and
(b) No more than 2.5 centimeters (one inch) of residue remains on the bottom of the container or inner liner, or
(c) No more than 3 percent by weight of the total capacity of the container remains in the container or inner liner if the container is less than or equal to 110 gallons in size, or no more than 0.3 percent by weight of the total capacity of the container remains in the container or inner liner if the container is greater than 110 gallons in size.[96]

Obviously, the key to assuring that the container is empty depends on properly identifying what was in the container. Then the appropriate procedures can be followed to render it "empty."

Containers that have held the acute hazardous wastes are the most problematic because they must be triple-rinsed with an appropriate solvent. The rinsate from this process may be hazardous waste itself if it is a listed hazardous waste because of the type of solvent used (many spent solvents are listed hazardous wastes). Furthermore, if the rinsate contains a listed hazardous waste (which it probably will if used to rinse acute hazardous waste from the container), it too will be regulated as a hazardous waste unless an exemption applies.

It is also important to recall that if the containers are not "discarded material" they will not be solid wastes, and therefore excluded from regulation as a hazardous waste.[97] Thus, if a container is to be beneficially reused, such as by shipping a commercial chemical container back to a supplier to be refilled, it will be excluded from the definition of "solid waste," and cannot, therefore, be regulated as a "hazardous waste."[98]

8.3.8 Used Oil Management Standards

Used oil is defined under RCRA as "any oil which has been . . . refined from crude oil, used, and as a result of such use, contaminated by physical or chemical impurities."[99] Used oil may be in the form of spent industrial equipment oils from compressors, turbines, hydraulic and refrigeration equipment, as well as spent vehicle transmission fluids, brake fluid and crankcase oils. Used oil presents a special problem for waste generators due to its unusual status as a waste product under RCRA.

---

[96] 40 CFR § 261.7(b)(1).
[97] See section 8.2.1 for a full discussion of the "discarded material" component of the RCRA solid waste definition.
[98] See section 8.2.2 for discussion of recycled materials.
[99] RCRA 1004(36), 42 U.S.C. § 6903(36).

Even though the EPA chose not to list used oils as a hazardous waste, this does not mean that used oil is exempt from regulation under RCRA. In fact, used oils that are mixed with "listed" hazardous wastes are subject to regulation as hazardous wastes under Subtitle C of RCRA, unless otherwise exempted by the mixture rule.[100] By contrast, used oils that are mixed with "characteristic" hazardous wastes are regulated in accordance with special used oil management standards issued by the EPA in 1992.[101] Pursuant to these regulations, if, upon testing, the resultant mixture does not exhibit a hazardous waste characteristic, it may be managed as a solid waste. Importantly, even if the mixture of used oil and characteristic hazardous waste does exhibit a hazardous waste characteristic, such used oil will be exempt from otherwise applicable requirements of Subtitle C if it is recycled.[102] Used oil contaminated with polychlorinated biphenyls (PCBs) is subject to special requirements under RCRA and the Toxic Substances Control Act (TSCA), depending on the quantity of PCBs contained in the used oil.[103]

The standards for used oil generators are found in 40 CFR part 279, subpart C. A used oil generator is any person, by site, whose act or process produces used oil or whose act first causes used oil to become subject to regulation. The used oil management regulations do not exempt any class of generators based on a generation rate. However, household "do-it-yourself" (DIY) used oil generators or individuals who generate used oil through their maintenance of personal vehicles are not subject to regulation. Further, the EPA decided not to impose an accumulation limit on generator storage since some amount of used oil is almost always stored at generator sites. Also, since used oil is a marketable commodity, there is an incentive for generators to send used oil off-site for recycling rather than storing it on-site for prolonged periods.

Used oil generators are required to store used oil in tanks or containers and must maintain all tanks and containers in good operating condition.[104] In maintaining all tanks and containers in good condition, generators must

---

[100] 40 CFR § 261.3(a)(2)(ii)-(iii). See Section 8.3.4 for discussion of the mixture rule.
[101] 57 *Fed. Reg.* 41566 (Sept. 10, 1992), codified at 40 CFR pt. 279. See also 58 *Fed. Reg.* 26420 (May 3, 1993); 61 *Fed. Reg.* 33691 (June 28, 1996).
[102] 40 CFR § 279.10(b)(2).
[103] In 1998, the EPA issued a direct final rule to clarify when PCB-contaminated used oil is subject to the used oil management standards and when it is not. 63 *Fed. Reg.* 24963 (May 6, 1998). However, the parameters for determining whether PCB-contaminated used oil is subject to regulation under RCRA or TSCA, or both, remain unclear because the EPA removed the direct final rule amendments in response to public comments concerning the applicability of the used oil management standards to PCB-contaminated used oil. 63 *Fed. Reg.* 37780 (July 14, 1998).
[104] 40 CFR § 279.22(a), (b).

ensure that all tanks and containers are free of any visible spills or leaks, as well as structural damage or deterioration. Generators storing used oil in aboveground tanks and containers must clearly label all tanks and containers with the term "used oil." Generators who store used oil in underground tanks must label all fill pipes with the words "used oil."[105]

Whenever a release of used oil to the environment occurs from aboveground storage tanks and containers, response action must be taken to: (1) stop the release, (2) contain the released used oil, (3) clean up and properly manage released used oil and materials used for cleaning up/containing the release, and (4) remove the tank or container from service, and repair or replace the tank or container before returning it to service.[106] This requirement applies only when there is a release to the environment. Under the used oil management regulations, this would not include releases within contained areas such as concrete floors or impervious containment areas, unless the releases go beyond the contained areas. Releases of used oil from an underground storage tank (UST) are subject to the separate requirements of RCRA's UST regulations (Subtitle I), as applicable (see discussion in section 8.6).

Used oil generators are required to ensure that all shipments of used oil in quantities greater than 55 gallons are transported off-site only by transporters who have an EPA identification number. Generators may transport, in their own vehicles, up to 55 gallons of used oil that is either generated on-site or collected from household "do-it-yourself" (DIY) used oil generators, to a used oil collection center, or aggregation point (e.g., one that is licensed or recognized by a state or municipal government to manage used oil or solid waste).[107] A generator is not required to obtain an EPA identification number for this off-site transportation activity. A generator may also self-transport up to 55 gallons of used oil, in his own vehicle, to an aggregation point owned by the generator without obtaining an EPA identification number.

In addition to management of the used oil itself, used oil generators must separate used oils from other materials or solid wastes, and manage the remaining material or solid waste in accordance with all applicable RCRA requirements. The generator must determine whether or not the materials that previously contained used oil exhibit a characteristic of hazardous waste (with the exception of non-terne-plated used oil filters),[108] and if so, manage them in accordance with existing RCRA controls.[109] If the

---

[105] 40 CFR § 279.22(c).
[106] 40 CFR § 279.22(d).
[107] 40 CFR § 279.24(a).
[108] See 57 *Fed. Reg.* 21524 (May 20, 1992).
[109] 40 CFR § 279.81.

material does not exhibit a hazardous waste characteristic (and is not mixed with a listed hazardous waste) then the material can be managed as a solid waste.

8.3.9 Municipal Waste Combustion Ash

The issue of whether combustion ash, produced as a result of municipal waste incineration, is a hazardous waste under RCRA if it meets one of the characteristics used to identify hazardous waste, has been the subject of conflicting rulemaking. The EPA originally took the position that such ash waste was not excluded from regulation as a hazardous waste under RCRA's household waste exclusion.[110] However, the EPA later reversed itself and ruled that ash waste was excluded from regulation as a hazardous waste under RCRA's household waste exclusion.

In 1994, the U.S. Supreme Court resolved the issue by holding that the ash created by waste-to-energy incinerators burning municipal solid waste is subject to RCRA's hazardous waste requirements if the municipal combustion ash meets one of the characteristics used to identify hazardous waste.[111] The Supreme Court's ruling requires that the ash be tested for hazardous waste characteristics. Ash that fails the Toxicity Characteristic Leaching Procedure (TCLP) test will have to be disposed of in hazardous waste landfills instead of in sanitary landfills.

8.3.10 Delisting Procedures for Specific Wastes

In appropriate circumstances, a waste generator may file a delisting petition with the EPA to have certain wastes excluded from classification as listed hazardous wastes. Pursuant to Section 3001(f) of RCRA,[112] the EPA has established procedures[113] for petitioning the agency to delist a specific waste stream or substance which otherwise constitutes a characteristic or listed hazardous waste.[114] However, even if the petitioner is successful in obtaining the delisting of a particular waste, the delisting does not remove the waste from any of the hazardous waste lists. The

---

[110] 42 U.S.C. § 6921(i).
[111] City of Chicago v. Environmental Defense Fund, 511 U.S. 328 (U.S. 1994).
[112] 42 U.S.C. § 6921(f).
[113] See 40 CFR § 260.22(a) for delisting procedures and criteria.
[114] See, for example, Horsehead Resource Dev. Co. v. EPA, 130 F.3d 1090 (D.C. Cir. 1997); United States v. Bethlehem Steel Corp., 829 F. Supp. 1047 (N.D. Ind. 1993).

delisting is specific to the individual waste generated by the delisting applicant and the particular facility where the waste is generated.[115] If, following public comment, the EPA decides to grant the petition, the delisting goes into effect as a final regulation upon publication in the Federal Register.[116] The EPA has sometimes denied[117] and sometimes granted such petitions[118] and continually updates an inventory of delisted wastes denoting, where applicable, the specific facilities that generate these delisted wastes.[119]

8.4 RCRA Regulation of Hazardous Wastes (Subtitle C)

Concern over the dangers associated with the mishandling and improper disposal of hazardous waste lead Congress to establish, within Subtitle C of RCRA,[120] a comprehensive federal program for hazardous waste management. Generators of solid waste are required to determine whether the waste constitutes a hazardous waste as defined by RCRA and its regulations.[121] A hazardous waste generator is any person, by site, whose act or process produces characteristic or listed hazardous waste.[122]

Hazardous waste generators are subject to specific reporting, storage, treatment, disposal, and shipping requirements depending on the quantities of hazardous waste generated. A generator may not treat, store, dispose of, or ship hazardous waste without first having obtained an EPA identification number[123] and, if necessary, a state identification number.[124] All generators are also required to maintain accurate and up-to-date records of generated wastes, training programs, safety procedures, hazardous waste manifests, and other important waste management activities.

---

[115] 40 CFR § 260.22(k).
[116] See, for example, Horsehead Resource Dev. Co. v. EPA, 130 F.3d 1090 (D.C. Cir. 1997).
[117] See, for example, United States v. Bethlehem Steel Corp., 829 F. Supp. 1047, 1052 (N.D. Ind. 1993).
[118] See, for example, 60 *Fed. Reg.* 31107 (1995); 58 *Fed. Reg.* 40067 (1993); 58 *Fed. Reg.* 42238 (1993).
[119] 40 CFR § 261, Appendix IX.
[120] 42 U.S.C. § 6921 through 6939b.
[121] 40 CFR § 262.11. See Section 8.4 for discussion of procedures for determining whether a solid waste constitutes a hazardous waste under RCRA.
[122] 42 U.S.C. § 6903(6); 40 CFR § 260.10.
[123] To apply for this I.D. number, the generator must submit an EPA Notification of Hazardous Waste Activity (EPA Form 8700-12)
[124] 40 CFR § 262.12(a).

8.4.1 Waste Generator Requirements

The RCRA regulations impose different requirements on waste generators depending on the volume of hazardous wastes that the generator produces each month. The quantity of hazardous wastes produced will determine whether the generator is classified as (1) a conditionally exempt small quantity generator, (2) a small quantity generator, or (3) large quantity generator for purposes of the regulatory requirements.

Conditionally Exempt Small Quantity Generator (CESQG). In general, a Conditionally Exempt Small Quantity Generator (CESQG) must not generate in a calendar month (1) more than 100 kilograms of hazardous waste, (2) more than 1 kilogram of a "listed" acutely hazardous waste, and (3) more than 100 kilograms of any residue or contaminated soil, or other debris resulting from the cleanup of a spill of any "listed" acute hazardous waste.[125] These generators are exempted from the full set of RCRA generator requirements provided that they do not exceed these limits during any calendar month.[126] If the generator exceeds these prescribed limits, it is subject to the regulations governing Small Quantity Generators (SQGs) or Large Quantity Generators (LQGs).

CESQGs are exempt from the requirement that shipments leaving the site of generation be accompanied by a hazardous waste manifest.[127] These generators are also allowed to bring their hazardous wastes to sanitary solid waste landfills, rather than the more costly hazardous waste disposal facilities.

Other important regulations governing CESQGs include requirements concerning mixing of hazardous waste with nonhazardous solid waste. If the generator mixes a hazardous waste subject to the reduced CESQG requirements with a nonhazardous solid waste, the mixture is only subject to the reduced requirements even though the mixture may exceed the CESQG quantity limitations, unless the mixture would be classified as a characteristic hazardous waste.[128] If a solid waste is mixed with a hazardous waste that exceeds the quantity limitations for the reduced CESQG requirements, then the mixture will be subject to full regulation.[129]

---

[125] 40 CFR § 261.5 contains the special requirements for CESQGs.
[126] "Full regulation" means all regulations applicable to generators of more than 1,000 kilograms of non-acutely hazardous waste in a calendar month. See Comment to 40 CFR § 261.5(e)(2).
[127] 40 CFR § 261.5(a).
[128] 40 CFR § 261.5(h).
[129] 40 CFR § 261.5(i).

Small Quantity Generator (SQG). EPA regulations for Small Quantity Generators (SQGs)[130] are applicable to generators who produce between 100 kilograms (about 220 pounds) and 1,000 kilograms (about 2,200 pounds) of hazardous waste per calendar month, unless the state has more stringent regulations.[131] SQGs are permitted to accumulate hazardous waste onsite for up to 180 days without a permit provided that the total amount of accumulated waste is never greater than 6,000 kilograms and an employee is available onsite or on call to handle hazardous waste emergencies.[132] If the waste is to be transported more than 200 miles, the generator may accumulate waste onsite for 270 days.[133] If the generator exceeds the 180 (or 270) day limit or stores more than 6,000 kilograms of waste, the generator must notify the EPA and becomes subject to additional regulatory requirements applicable to treatment, storage and disposal (TSD) facilities.

SQGs must inspect all waste containers at least once a week. If a leak is detected, the waste must be transferred to another container. Special labeling and other requirements also apply to SQGs.[134]

Large Quantity Generator (LQG). If a generator produces hazardous waste in amounts exceeding 1,000 kilograms per month, it will become subject to the full set of regulations governing hazardous waste generators.[135] Large quantity generators must file a waste management report with the EPA Regional office on March 1st of each even numbered year and maintain an operating log that details how wastes are managed. LQGs must also dispose of their wastes within 90 days.

8.4.2 Hazardous Waste Manifest

One of the generator's most important duties regarding a specific load of hazardous waste is to fill out a hazardous waste manifest, which must accompany the shipment as it leaves the site of generation. Under RCRA, the term "manifest" refers to the form used for identifying the quantity, composition, and the origin, routing and destination of hazardous waste during its transportation from the point of generation to the point of

---

[130] 40 CFR pt. 262.
[131] Several states have more stringent requirements for SQGs, and California, Louisiana, and Rhode Island have no exemption at all for SQGs. In these states, the SQG is subject to the same requirements as Large Quantity Generators.
[132] 40 CFR § 262.34(d).
[133] 40 CFR § 262.34(e).
[134] See 40 CFR § 262.34.
[135] 40 CFR pt. 264.

disposal.[136] Manifests are multiple copy documents. The EPA designed a four-copy national Uniform Hazardous Waste Manifest (EPA Forms 8700-22 and 8700-22A);[137] however, some states with EPA-authorized programs utilize their own manifest forms with 6 to 8 copies.

Manifests are designed to facilitate the ability of regulators to ascertain that potentially dangerous wastes are produced only by permitted generators, handled and transported only by authorized conveyors, and treated, stored and disposed of only at licensed facilities. The generator, at a minimum, must complete those sections of the manifest concerning listing of the type of hazardous waste being shipped (utilizing the EPA hazardous waste number), the number and type of containers, the total quantity of the waste, and name and address of the disposal destination.[138] Any generator who exports hazardous waste overseas is subject to special rules,[139] including a requirement that a copy of the receiving country's written consent be attached to the manifest accompanying the shipment.[140]

The generator must sign and date the manifest and make sure that the initial transporter also signs it as he accepts the waste.[141] The generator retains one copy of the manifest and gives the remaining copies to the transporter. Although EPA does not require generators, transporters or TSD facilities to forward copies of their manifests to the EPA, some state programs require that the generator and/or the final disposal facility forward a copy of the manifest to the state's environmental regulatory agency.[142]

If the generator does not receive a signed copy of the manifest from the listed disposal facility within 35 days of the date the generator turns the waste over to the transporter, the generator must contact the transporter and inquire about the status of the shipment. If 45 days pass from the date the transporter acquired the shipment and the generator has still not received a signed copy of the manifest from the disposal facility listed on the manifest, the generator must forward a copy of the manifest to the EPA and file an Exception Report with the EPA detailing the efforts the generator has made to locate the shipment.[143]

Generators must maintain copies of all manifests, Exception Reports, and results of tests conducted to determine if a generated waste is

---

[136] 42 U.S.C. § 6903(12).
[137] 49 *Fed. Reg.* 10490 (Mar. 20, 1984).
[138] 42 U.S.C. § 6921(d)(3).
[139] 42 U.S.C. § 6938.
[140] 42 U.S.C. § 6938(a)(1)(C).
[141] 40 CFR § 262.23.
[142] See, for example, N.Y. Comp. Codes R. & Regs. tit. 6, 372.2 (b) and 372.4(b); N.J. Admin. Code 7:26-7.4 and 7:26-7.6; 25 Pa. Code 262.23(d) and 264.71(b)(6).
[143] 40 CFR § 262.42.

hazardous, for at least three years.[144] In addition, generators who ship hazardous wastes off-site to a TSD facility are required to submit to the EPA, by March 1st of each even numbered year, a report detailing hazardous waste activities during the previous calendar year.[145] This report must be submitted on EPA Form 8700-13A. The report must describe what hazardous wastes were shipped off-site, the generator's efforts to reduce the volume and toxicity of the waste sent to TSD facilities, and comparative changes in volume and toxicity of wastes that the generator actually achieved.

8.4.3 Waste Transporter Requirements

Pursuant to Section 3003 of RCRA,[146] the EPA has established regulations, in consultation with the U.S. Department of Transportation (DOT), applicable to those who transport hazardous waste.[147] A transporter must apply for an EPA identification number by completing and filing EPA Form 8700-12, Notification of Hazardous Waste Activity. On this form, the transporter must indicate which specific hazardous waste(s) will be transported and by what mode of transport (e.g., highway, rail, water, airplane, etc.). A transporter cannot legally haul a specific hazardous waste unless the waste is listed, by EPA hazardous waste number, on the transporter's permit.

Transporters are permitted to store manifested hazardous waste shipments, stored in proper containers, at a transfer facility for up to ten days. Storage for more than ten days is illegal unless the transfer facility has a storage (i.e., TSD) permit.

The driver of a vehicle transporting a hazardous waste must have a copy of the transport manifest in his possession and available for inspection at all times. A transporter can only deliver the hazardous waste being hauled to the TSD facility site listed by the generator on the manifest and the entire quantity of hazardous waste that was transported must be left at this TSD facility. Transporters must obtain the signature of the operator of the TSD facility (or the subsequent transporter) to whom the waste is delivered. Transporters must retain manifests, signed by the generator and the TSD operator, for at least three years.[148]

Where all or part of the hazardous waste being transported leaks or is otherwise discharged from the transport vehicle while en route to the TSD

---

[144] 40 CFR § 262.40.
[145] 40 CFR § 262.41(a).
[146] 42 U.S.C. § 6923.
[147] 40 CFR pt. 263.
[148] 40 CFR § 263.32.

facility, the transporter must clean up the waste or take other action to prevent a hazard to human health or the environment.[149] Where such a discharge has occurred, the transporter is also required to give immediate notice to the EPA's National Response Center[150] and to report in writing to the DOT.[151]

In addition to RCRA's requirements and regulations, transporters of hazardous waste are subject to the Hazardous Materials Transportation Act (HMTA)[152] and the DOT regulations promulgated pursuant to this Act. These regulations mandate that hazardous waste transporters utilize certain hazard communication labels, placards, and truck markings.[153]

8.4.4 Treatment, Storage and Disposal (TSD) Facility Requirements

RCRA lumps both interim and final destinations for hazardous waste under a single description: Treatment, Storage and Disposal (TSD) facilities. However, it is important to keep in mind that storage, treatment and disposal of hazardous wastes are separately permitted functions. Thus, a specific facility may only be permitted to store but not treat hazardous wastes, while another facility may be permitted to treat but not dispose of hazardous wastes, while yet another facility may be permitted to perform all three functions.

As with generators and transporters, the owner or operator of any facility which intends to either treat, store, or dispose of any hazardous waste must first obtain an EPA identification number by completing and filing EPA Form 8700-12, Notification of Hazardous Waste Activity.[154] In addition, as is the case with transporters, a TSD facility is authorized to accept for treatment, storage, and/or disposal only the specific types of hazardous waste delineated in its permit (by EPA identification number), not any and all types of hazardous waste.

Congress has authorized EPA to develop detailed standards for permitting TSD facilities covering, among other things: facility location, design and construction; operating methods and practices; personnel training; financial responsibility; facility maintenance; reporting and recordkeeping requirements; contingency plans to minimize damage

---

[149] 40 CFR § 263.30.
[150] 49 CFR § 171.51.
[151] 49 CFR § 171.61.
[152] 49 U.S.C. § 1801 et seq.
[153] 49 CFR pt. 171.
[154] 40 CFR § 264.11.

resulting from spills; and compliance with the transport manifest system.[155]

EPA has promulgated numerous, detailed regulations delineating operating standards for TSD facilities.[156] A facility cannot obtain a TSD permit until it has met these requirements. New TSD facilities can initially operate under an interim status before obtaining final permit approval.[157] However, a TSD facility whose interim status has expired cannot legally operate without a final permit.

8.4.5 Land Disposal Restrictions

Three classes of land disposal sites are in use in the United States. Class I, "Waste Management Units for Hazardous Waste," are used primarily for hazardous waste treatment residues.[158] Class II, "Waste Management Units for Designated Waste," are designated solely for certain solid hazardous wastes (e.g., asbestos). Class III, "Landfills for Nonhazardous Wastes," accept common household waste and construction debris.[159]

The Hazardous and Solid Waste Amendments of 1984 (HWSA)[160] instituted new requirements for land disposal of hazardous wastes. These land disposal restrictions (LDRs) are commonly referred to as the "Land Ban." Pursuant to the HSWA, the EPA established a schedule that was implemented over a six-year period to restrict land disposal of all listed and characteristic hazardous wastes. Land disposal of hazardous wastes is now prohibited unless EPA treatment standards have been met.[161]

Although the EPA did not eliminate the availability of land disposal for disposing of hazardous wastes, land disposal cannot be used unless wastes have first been properly treated according to the EPA standards.[162] In general terms, the EPA treatment standards usually require that, before land disposal, wastes must be treated to meet a certain concentration level or be treated by a specified technology. For each waste, the EPA is

---

[155] 42 U.S.C. § 6925.

[156] 40 CFR pt. 264.

[157] 40 CFR pt. 265.

[158] Since May 1992, only hazardous waste treatment residues are accepted at Class I sites.

[159] Class III sites are commonly known as sanitary landfills and are regulated under Subtitle D of RCRA.

[160] Hazardous and Solid Waste Amendments of 1984, Pub. L. 98-616, 98 Stat. 3221 (Nov. 8, 1984).

[161] "Land disposal means placement in or on the land and includes, but is not limited to placement in a land fill, surface impoundment, waste pile, injection well, land treatment facility, salt dome formation, salt bed formation, underground mine or cave, or placement in a concrete vault or bunker intended for disposal purposes." 40 CFR § 268.2(c).

[162] 40 CFR § 268.40-.44.

responsible for identifying a best demonstrated available technology (BDAT) for treatment prior to land disposal. RCRA prescribes a stringent standard for treatment. The treatment standards must be "levels or methods of treatment which substantially diminish the toxicity of the waste or substantially reduce the likelihood of migration of hazardous constituents from the waste so that short-term and long-term threats to human health and the environment are minimized."[163]

Dilution to meet the treatment standards, or to change the nature of the waste so it is no longer a restricted waste, is prohibited.[164] Intentional mixing to avoid a treatment standard is also prohibited. Dilution of characteristic hazardous wastes in a wastewater treatment system permitted under the Clean Water Act is not impermissible dilution unless a treatment technology has been specified in 40 CFR Section 268.42.[165]

8.4.6 RCRA Corrective Action

The original provisions of RCRA focused on ongoing solid and hazardous waste management issues and did not provide the EPA with the authority to require treatment, storage, and disposal (TSD) facilities to take corrective actions to cure past releases of waste.[166] However, when Congress passed the the Hazardous and Solid Waste Amendments of 1984 (HSWA),[167] corrective action provisions were added to RCRA to fill this regulatory gap.[168]

Under RCRA's corrective action provisions, the EPA may require a TSD facility to investigate and remedy a release of hazardous waste or hazardous waste constituents from solid waste management units (SWMUs) and other areas of concern (AOCs). SWMUs include any area at a facility at which solid wastes have been routinely and systematically released.[169] An AOC is any other area of known or suspected contamination, such as a spill area.

---

[163] RCRA 3004(m)(1), 42 U.S.C. § 6924.
[164] 40 CFR § 268.3(a).
[165] 40 CFR § 268.3(b).
[166] See, for example, Ciba-Geigy Corp. v. Sidamon-Eristoff, 3 F.3d 40 (2d Cir. 1993).
[167] Hazardous and Solid Waste Amendments of 1984, Pub. L. 98-616, 98 Stat. 3221 (Nov. 8, 1984).
[168] 42 U.S.C. § 6924(u)-(v); 6928(h). See also United States v. Bethlehem Steel Corp., 829 F. Supp. 1047 (N.D. Ind. 1993).
[169] The EPA defines a SWMU as "any discernible unit at which solid wastes have been placed at any time, irrespective of whether the unit was intended for the management of solid or hazardous waste. Such units include any area at a facility at which solid wastes have been routinely and systematically released." 55 *Fed. Reg.* 30798 (July 27, 1990).

Three principal provisions of RCRA provide the EPA with authority to require corrective action: (1) Section 3004(u) governs continuing releases; (2) Section 3004(v) governs off-site releases; and (3) Section 3008(h) governs interim status corrective action orders.[170] In 1993, the EPA issued final regulations to implement its corrective action authority.[171]

8.5 RCRA Regulation of Solid Wastes (Subtitle D)

That portion of the RCRA regulatory framework known as Subtitle D primarily deals with the management of nonhazardous and exempt hazardous solid wastes. The Subtitle D requirements mostly pertain to the design and monitoring of wastes that are disposed of in sanitary landfills.[172] These requirements, therefore, are mostly of concern to owners and operators of such landfills.

8.5.1 State Solid Waste Management Plans

Subtitle D of RCRA sets forth a system for development and implementation of state solid waste management plans to regulate landfills that accept for disposal nonhazardous and exempt hazardous wastes.[173] Section 4002(a) of RCRA requires that the EPA promulgate guidelines for the development of such plans.[174] Through 1988, federal financial assistance was available[175] when a state prepared a plan containing minimum statutory criteria specified in Section 4003[176] and promulgated regulations that complied with the EPA guidelines.[177]

Under Section 4004(b) of RCRA, each state plan was required to prohibit the establishment of new open dumps and require the disposal of nonhazardous solid waste in sanitary landfills.[178] Under Section 4005(a),

---

[170] 42 U.S.C. § 6924(u)-(v); 6928(h).
[171] 58 *Fed. Reg.* 8658 (Feb. 16, 1993).
[172] Sanitary landfills are referred to in the land disposal regulations as Class III landfills ("Landfills for Nonhazardous Wastes").
[173] 42 U.S.C. § 6941 through 6949a.
[174] 42 U.S.C. § 6942(a).
[175] 42 U.S.C. § 6948.
[176] Section 4003 of RCRA sets forth the minimum requirements for state plans, including: (1) use of resource conservation or recovery; (2) prohibiting the opening of new open dumps; (3) closure or upgrading of all existing open dumps; and (4) disposal of all solid waste in a sanitary landfill or in a manner that is environmentally sound. 42 U.S.C. § 6943.
[177] 42 U.S.C. § 6947.
[178] 42 U.S.C. § 6944(b).

RCRA directly prohibits any act of "open dumping."[179] All open dumps must be upgraded or closed.[180] In addition, under Sections 1008(a) and 4004(a) of RCRA, the EPA was required to issue minimum criteria establishing what solid waste management practices constituted "open dumping."[181]

Many factors contributed to an eventual overhaul of this system, which rendered these provisions wholly inadequate to address contamination from municipal solid waste landfills. First, with the elimination of federal financial aid under Section 4008 of RCRA, the major incentive for states to develop such plans disappeared. Second, the EPA's criteria for sanitary landfills lacked certain key provisions necessary to prevent future contamination and to remediate existing contamination. The criteria generally consisted of broadly worded performance standards. No monitoring of any kind was required. As a result, there was no assurance that contamination could be checked before significant degradation had already occurred. Another significant gap was the lack of any requirement to clean up contamination once detected. Finally, the criteria did not require protective measures after the landfill was closed.

Thus, when Congress amended RCRA in 1984,[182] it made Subtitle D a main focus of its revisions. Congress realized that because the exemption of some hazardous wastes from Subtitle C regulation, many municipal solid waste landfills were really pseudo-hazardous waste landfills.[183] Under the 1984 amendments to RCRA, Congress mandated that the EPA promulgate revised management standards for landfills that accept household hazardous waste and small quantity generator waste.[184]

The 1984 amendments required that the EPA revise its existing federal regulatory criteria for all solid waste facilities that receive hazardous household wastes or hazardous wastes from small quantity generators. The

---

[179] EPA has defined open dumps as land disposal sites at which solid wastes are exposed to the elements and scavengers and which are susceptible to open burning. 40 CFR § 240.101(s).

[180] 42 U.S.C. § 6945(a).

[181] 42 U.S.C. § 6907, 6944(a). The EPA satisfied the statutory mandates of both sections in 1990 by promulgating criteria for classification of solid waste disposal facilities and practices. 40 CFR pt. 257.

[182] Hazardous and Solid Waste Amendments of 1984, Pub. L. 98-616, 98 Stat. 3221 (Nov. 8, 1984).

[183] H.R. Conf. Rep. 1133, 98th Cong. 2d Sess. (1984), reprinted in 1984 U.S. Code Cong. & Admin. News 5649, 5688.

[184] These wastes are currently exempted from regulation as hazardous waste; thus, they may be disposed of in municipal landfills rather than in special hazardous waste landfills, which are subject to the EPA's hazardous waste regulations under RCRA Subtitle C. 42 U.S.C. § 6921.

revisions were to be "those necessary to protect human health and the environment," although they "may" consider the "practicable capability" of such facilities.[185] At a minimum, Congress specified that the criteria must require the same environmental controls found in the Subtitle C regulations, namely, groundwater monitoring, location requirements for new and existing facilities, and corrective action. In 1991, the EPA promulgated final revised Subtitle D regulations to institute these changes.[186]

8.5.2 State Solid Waste Disposal Regulations

A state's Solid Waste Management Plan must: establish regional solid waste management planning districts;[187] identify and articulate strategies for managing the various types of solid waste (e.g., residential, commercial, institutional, industrial, mining, agricultural, etc.); and address the issues of solid waste collection, transportation, storage, transfer, processing, treatment and, most importantly, disposal.[188]

RCRA requires that nonhazardous solid waste be channeled, wherever possible, to resource recovery facilities (e.g., recycling programs and waste-to-energy incinerators) and that where solid waste must be disposed of on land it can only be deposited at sanitary landfills.[189] RCRA also requires that every solid waste landfill conduct regular groundwater monitoring to detect possible contamination of groundwater resulting from the leaching of wastes dumped at the landfill.[190]

To enforce RCRA's mandated phase out of open dumps, the EPA requires that State Solid Waste Management Plans establish criteria for classifying solid waste disposal facilities.[191] The EPA further requires that "the State plan shall provide for the establishment of State regulatory

---

[185] 42 U.S.C. § 6949a.
[186] 56 *Fed.Reg.* 50978 (Sept. 6, 1991).
[187] See, for example, Hofer v. Mack Trucks, Inc., 981 F.2d 377 (8th Cir. 1992).
[188] 40 CFR § 256.02.
[189] 42 U.S.C. § 6944.
[190] 42 U.S.C. § 6949a(c). Although the EPA promulgated regulations creating a small municipal landfill groundwater monitoring exemption, (40 CFR § 258.1(f)(1)) the U.S. Court of Appeals for the D.C. Circuit vacated this exemption as contrary to RCRA's plain meaning, and Congressional intent, that all municipal landfills be required to conduct groundwater monitoring. Sierra Club v. EPA, 992 F.2d 337 (D.C. Cir. 1993).
[191] 40 CFR § 257.

powers"[192] including a mandatory permit program covering all disposal facilities.[193]

Thus, under state laws enacted pursuant to the EPA's guidelines for the land disposal of solid wastes, solid waste can no longer be legally discarded at an unlicensed dump site,[194] and licensed sanitary landfills are subject to detailed regulations concerning minimum levels of performance and recordkeeping requirements.[195] As an alternative to land disposal of solid wastes, the EPA has promulgated guidelines containing recommended procedures for the disposal of solid wastes via thermal processing (i.e. burning).[196]

To enforce the components of their State Plans that deal with solid waste collection, transportation, storage, and transfer, states have established permit requirements for regulated waste collectors, haulers, storage facilities, and transfer stations.[197] In addition, as with disposal facilities, states have promulgated regulations detailing how permittees must conduct their operations and what records they must keep.[198] A nonpermitted individual, or business entity, who engages in any type of solid waste handling activity for which a state permit is required, pursuant to a state regulations promulgated under an EPA-approved state plan, is in violation of RCRA. Similarly, a permit holder who violates any state solid waste regulation, or any condition of a State permit, is in violation of RCRA.

It is important to recognize that RCRA does not prohibit states from establishing, or retaining existing,[199] solid waste regulatory regimes which are more stringent[200] than those delineated by EPA in its guidelines.[201] Thus, where a solid waste has been mishandled, in order to determine what specific state regulatory requirement has been violated, it does not suffice to merely consult the EPA guidelines since it cannot be assumed that a state's solid waste regulations simply mirror those guidelines.

---

[192] 40 CFR § 256.21.

[193] 40 CFR § 256.21(c) and 256.22(d).

[194] See, for example, N.Y. Comp. Codes R. & Regs. tit. 6, 360-2.

[195] 40 CFR § 241.100(b).

[196] 40 CFR § 240.

[197] See, for example, N.Y. Comp. Codes R. & Regs. tit. 6, 364 (waste transporter permits) and N.Y. Comp. Codes R. & Regs. tit. 6, 360-11 (waste transfer stations).

[198] Id.

[199] 42 U.S.C. § 6947(c).

[200] See 40 CFR § 256.21(a).

[201] 42 U.S.C. § 6929.

8.6 RCRA Regulation of Underground Storage Tanks (Subtitle I)

In the Hazardous and Solid Waste Amendments of 1984 (HSWA),[202] Congress enacted RCRA Subtitle I[203] to address problems associated with leaking underground storage tanks (USTs) through which regulated substances can enter soil and groundwater.[204] The federal regulations cover the design, construction, and operation of USTs from installation to closure, require the cleanup of leaks and spills, and impose recordkeeping, reporting, and financial responsibility requirements on owners and operators of USTs.[205]

Subtitle I requires the owners of certain types of USTs to notify state and local officials as to the existence, location and size of the UST and what the UST is used to hold.[206] This information is to be used by states to create two statewide inventories. One inventory covers all USTs in the state which hold petroleum products and the other inventory lists all USTs which hold hazardous substances.[207]

Subtitle I of RCRA requires the EPA to promulgate regulations[208] requiring owners of regulated USTs to install leak detection systems and to take corrective actions to prevent potential leakages.[209] Under the EPA's regulations, owners of new, regulated USTs, installed after May 6, 1986, must notify the state within 30 days of the date the UST is brought into use.[210]

Subtitle I of RCRA also creates a federal response program and provides the EPA with the power to step in and take corrective actions where a UST leak endangers a community's water supply. Such corrective actions can include providing alternative household water supplies and even permanently relocating residents.[211]

To encourage states to enter into cooperative agreements with the EPA[212] and to establish their own UST programs which, after EPA

---

[202] Hazardous and Solid Waste Amendments of 1984, Pub. L. 98-616, 98 Stat. 3221 (Nov. 8, 1984).
[203] 42 U.S.C. § 6991–6991i.
[204] See, for example, Zands v. Nelson, 779 F. Supp. 1254 (S.D. Cal. 1991).
[205] 40 CFR pt. 280.
[206] 42 U.S.C. § 6991a(a)(1).
[207] 42 U.S.C. § 6991a(c).
[208] 40 CFR pt. 280.
[209] 42 U.S.C. § 6991b(a)–(c).
[210] 40 CFR § 280.3(c).
[211] 42 U.S.C. § 6991b(h)(5).
[212] 42 U.S.C. § 6991b(h)(7).

approval, would operate in lieu of the federal UST program,[213] Congress authorized the EPA to draw funds from a Leaking Underground Storage Tank Trust Fund to pay for federal and state response and corrective action costs.[214] Most states have received EPA approval to run their own UST programs in lieu of the federal program.[215] Where a state has established its own regulations regarding USTs but the state's UST program has not received the EPA's approval to operate in lieu of the federal UST program, the state's UST regulations are preempted by the federal regulations.[216]

8.6.1 Regulated Tanks

The EPA's UST regulations apply to all USTs containing petroleum or any of more than seven hundred chemicals designated as hazardous substances under RCRA.[217] USTs which are used to store hazardous wastes are not regulated under Subtitle I, but are instead subject to EPA regulations adopted pursuant to RCRA's provisions regarding hazardous waste management (Subtitle C).

A UST is defined as any tank (including its connected piping) holding an "accumulation of regulated substances" that has ten percent or more of its volume underground.[218] Federal regulations and most state regulations contain a list of tanks that are exempt from regulation.[219] Congress explicitly exempted: tanks used for storing heating oil which is consumed on the premises;[220] septic tanks; tanks which hold less than 1,100 gallons of motor fuel for noncommercial purposes; and various pipeline facilities which are regulated under other laws.[221]

[213] 42 U.S.C. § 6991c.

[214] 42 U.S.C. § 6991b(h)(7).

[215] In 1993, EPA established a new section in the UST regulations for codifying decisions to approve state UST programs. 58 *Fed. Reg.* 58624 (1993). State program approvals will appear at 40 CFR pt. 282, subpt. B.

[216] See, for example, G.J. Leasing v. Union Elec. Co., 825 F. Supp. 1363 (S.D. Ill. 1993).

[217] 42 U.S.C. § 6991(2). Hazardous substances are defined as those substances listed as hazardous pursuant to CERCLA. 42 U.S.C. § 9601(14).

[218] 42 U.S.C. § 6991(1); 40 CFR § 280.12.

[219] 42 U.S.C. § 6991(1); 40 CFR § 280.12. Note, however, that state programs may choose to regulate tanks that are excluded under the federal program.

[220] See, for example, Rockford Drop Forge Co. v. Pollution Control Bd., 221 Ill.App.3d 505, 164 Ill.Dec. 45, 582 N.E.2d 253 (1991).

[221] 42 U.S.C. § 6991(1).

8.6.2 New Tank System Requirements

Under the federal UST regulations bare steel tanks or piping are now prohibited. All new UST systems must be safeguarded against corrosion through the use of either cathodic protection, fiberglass coating, or other comparable means.[222] Cathodic protection systems must be properly maintained and periodically tested.[223] Tanks containing regulated substances other than petroleum require a double-wall or secondary containment system with continuous interstitial monitoring.[224]

All new tanks must be equipped with spill and overfill prevention equipment, including a spill catchment basin to prevent the inadvertent release of product when the transfer hose is detached from the fill pipe.[225] Overfill prevention equipment must automatically shut off the flow to the tank before the 95 percent full point is reached, or else alert the operator when the tank is no more than 90 percent full, by restricting flow or setting off an alarm.[226] Spill or overfill prevention is not required if the UST is filled by transfers of no more than 25 gallons at a time.[227]

New tanks must have leak detection equipment and procedures that can detect releases from any part of the system routinely containing product, with a leak rate of 0.2 gallon per hour, a probability of detection of 95 percent, and a false alarm rate of 5 percent. Acceptable methods include: (1) automatic tank gauging combined with monthly inventory control, (2) vapor, groundwater, or interstitial monitoring performed at least monthly, or (3) monthly inventory control combined with periodic tank tightness testing.[228] All pressurized piping must be equipped with automatic line leak detectors and have annual line tightness tests.[229] Suction piping meeting certain criteria is exempt from the release detection requirements and other suction piping is subject to more relaxed requirements.

Tanks that meet the corrosion protection standards may use monthly inventory controls plus tank tightness testing every five years until the later of December 22, 1998, or 10 years after installation.[230] Tanks that do not meet the corrosion protection standards must use automatic tank gauging, vapor, groundwater, or interstitial monitoring or monthly inventory control plus annual tightness testing.[231]

---

[222] 40 CFR § 280.20.
[223] 40 CFR § 280.31.
[224] 40 CFR § 280.42.
[225] 40 CFR § 280.20(c)(i).
[226] 40 CFR § 280.20(c)(ii)(A)-(B).
[227] 40 CFR § 280.20(c), 280.30.
[228] 40 CFR § 280.43(a)-(e).
[229] 40 CFR § 280.42(b)(4).
[230] 40 CFR § 280.41(a)(1).
[231] 40 CFR § 280.41(a)(2).

Inventory control consists of daily gauging of the tank with a dipstick calibrated to at least one-eighth inch, plus monthly reconciliation of inventory. Measurements of inputs, withdrawals, and amounts remaining in the tank must be recorded each operating day. The product dispenser must be metered to an accuracy of six cubic inches for every five gallons of product withdrawn. In general, monthly inventory controls must be able to detect a release of one percent of flow-through, plus 130 gallons.[232] After the later of December 22, 1998, or 10 years after installation, inventory control ceases to be a valid release detection device unless the owner can establish its validity under 40 CFR Section 280.43(h).

All tanks and piping must be installed according to the manufacturer's instructions and applicable industry standards. Certification of proper installation is required. The authorized agency must be notified, on a prescribed form, within 30 days after installation of a new UST.[233] Some states may require notification or approval prior to commencement of installation. The tank owner or operator should check with the authorized agency in its particular state.[234]

8.6.3 Upgrading Existing Tanks

All USTs installed after May 8, 1985, were previously subject to an interim prohibition against bare steel tanks and should have the required corrosion protection.[235] Unless closed or replaced, all other USTs had to be retrofitted with corrosion protection or fiberglass lining by December 22, 1998.[236]

Upgrading of corrosion protection may be accomplished by retrofitting existing UST systems with cathodic protection, installation of an interior lining, or both. In most cases, the tank must be internally inspected prior to retrofitting to ensure that the tank is structurally sound and free of corrosion. Alternatively, a tank less than 10 years old can be assessed for corrosion holes by conducting tightness tests both before and three to six months after installation of cathodic protection. All existing tanks were

---

[232] 40 CFR § 280.43(a). Less stringent requirements apply to tanks with a capacity of less than two thousand gallons. 40 CFR § 280.43(b)(5).
[233] 40 CFR § 280.20(d), (e), 280.22. A copy of the form is contained in 40 CFR pt. 280, Appendix I.
[234] 40 CFR pt. 280, Appendix II contains a listing of designated agencies.
[235] 42 U.S.C. § 6991b(g).
[236] 40 CFR § 280.21.

required to be retrofitted with leak detection systems.[237] Installation of line leak detectors was required by December 22, 1990.[238] Spill and overfill protections for upgraded tanks are identical to those required for new USTs.

8.6.4 Tank Closure

A tank owner or operator may temporarily close a UST and leave it in the ground, provided that the UST is emptied and all other applicable tank regulations are observed, with the exception of release detection requirements. If the tank is closed for more than three months, the owner or operator must open all vent lines and cap and secure all other lines and openings.[239] If the UST complies with the new tank standards or has been upgraded, it may remain indefinitely in a temporarily closed conditions. All other tanks must be permanently closed within one year, unless the authorized agency has granted an extension to the tank owner or operator.

If the owner or operator permanently closes a UST, it must be emptied, cleaned and removed from the ground or filled with inert material, such as sand or concrete.[240] If the owner or operator decides to continue use of a UST for storing nonregulated substances, such use is considered a change-in-service and requires compliance with all conditions applicable to permanent closure, except for removal.[241]

The owner or operator must notify the authorized agency at least 30 days before a tank is permanently closed or a change-in-service takes place.[242] Before the tank is permanently closed, the owner must perform a site assessment, check for contamination, and clean up any contamination to permissible levels.[243] A proper site assessment should include sampling and analyses of soil underneath and near the tank. Tank owners and operators must maintain closure records for at least three years following closure or removal of the tank.[244]

---

[237] Compliance deadlines were December 22, 1989 for tanks installed before 1965 and tanks of unknown age, and December 22, 1990, for tanks installed between 1965 and 1969. Other existing USTs had December of 1991, 1992, and 1993 deadlines.
[238] 40 CFR § 280.40(c).
[239] 40 CFR § 280.70(b).
[240] 40 CFR § 280.71(b).
[241] 40 CFR § 280.71(c).
[242] 40 CFR § 280.71(a).
[243] 40 CFR § 280.72.
[244] 40 CFR § 280.74.

8.7 RCRA Enforcement and Penalties

The EPA and Department of Justice are the two federal agencies primarily responsible for enforcement of RCRA. In addition, state environmental agencies possess enforcement powers in states where the EPA has delegated authority to the state to administer their own hazardous and solid waste management programs in lieu of the RCRA program. RCRA's enforcement and penalty provisions are found in Section 3008.[245] If the EPA or state agency determines that there has been a RCRA violation, the government may initiate administrative and civil judicial actions to compel conpliance[246] or, in the case of "knowing" violations, take criminal enforcement action[247] against the alleged violator. In addition, under appropriate circumstances, private parties may bring citizen suits to enforce RCRA in cases where governmental authorities have failed to act.[248]

Administrative and Civil Penalties. RCRA's penalty provisions authorize the EPA to issue an administrative compliance order assessing a penalty of up to $25,000 per day (without a cap) for any past or current violation, or the EPA may commence a civil action for injunctive relief and for civil penalties of up to $25,000 per day of violation.[249] The actual penalties assessed will vary according to the particular circumstances, and the EPA must take into account the seriousness of the violation and any good faith efforts of compliance. Often, the EPA will negotiate a settlement with the alleged violator, and the EPA's RCRA Civil Penalty Policy may be used to determine the amount of penalties assessed.

Criminal Penalties. The most controversial aspect of RCRA's penalty provisions concerns the culpability elements required for environmental criminal convictions. Section 3008(d) of RCRA provides criminal sanctions for:[250]
1. Knowingly transporting, or causing to be transported, hazardous waste to a facility that does not have a permit to receive the waste;
2. Knowingly treating, storing, or disposing of hazardous waste (a) without a permit, (b) in knowing violation of a permit condition, or (c) in knowing violation of interim permit status standards;
3. Knowingly omitting material information or making false statements in

---

[245] RCRA 3008, 42 U.S.C. § 6928.
[246] 42 U.S.C. § 6928(a)–(c), (g).
[247] 42 U.S.C. § 6928(d) and (e).
[248] 42 U.S.C. § 6972(a)(1).
[249] RCRA 3008(a)–(c), 42 U.S.C. § 6928(a)-(c).
[250] RCRA 3008(d), 42 U.S.C. § 6928(d).

documents required by RCRA;

4. Knowingly destroying, altering, concealing, or failing to file records required by RCRA;

5. Knowingly transporting hazardous waste without a manifest;

6. Knowingly exporting hazardous waste without compliance with export procedures; and

7. Knowingly storing, treating, transporting, or causing to be transported, disposed of, or otherwise handled, any used oil in knowing violation of a RCRA permit, condition, or regulation.

For a first conviction of any of the above crimes, criminal penalties of up to $50,000 for each day of violation and/or imprisonment of up to two years (up to five years for violations of 1 or 2 above) may be assessed, and these penalties may be double in the case of repeat offenders. The court will use sentencing guidelines to determine the appropriateness of the penalties sought by the government.[251]

In addition to the criminal violations set forth in Section 3008(d), RCRA was the first environmental statute to include a "knowing endangerment" provision. Under Section 3008(e), if any person commits any of the crimes listed in Section 3008(d) with knowledge that such action may place another person in imminent danger of death or serious bodily injury, that person is subject to a fine of up to $250,000 per count or imprisonment for up to 15 years or both.[252] Fines of up to $1 million may be imposed on a corporation.

The courts have focused their review of Section 3008(d) offenses on whether the alleged violator's "knowledge" of the lack of a valid permit or other violation is required before criminal liability can be imposed. For example, most courts have held that the government does not have to prove that the defendant knew that the material at issue was subject to regulation as a hazardous waste,[253] nor that the defendant had knowledge of the lack of a valid permit,[254] as a prerequisite to a finding of criminal liability under RCRA.

In one case, for example, the Sixth Circuit rejected the "knowledge of illegality" defense asserted by a paint manufacturing company and its vice

---

[251] See United States Sentencing Commission, Guidelines Manual 2Q1.2 (sentencing guidelines applicable to environmental crimes).
[252] RCRA 3008(e), 42 U.S.C. § 6928(e).
[253] See, for example, United States v. Kelly, 167 F.3d 1176 (7th Cir. 1999).
[254] See, for example, United States v. Laughlin, 10 F.3d 961 (2d Cir. 1993); United States v. Dee, 912 F.2d 741 (4th Cir. 1990); United States v. Kelley Technical Coatings, Inc., 157 F.3d 432 (6th Cir. 1998); United States v. Dean, 969 F.2d 187 (6th Cir. 1992); United States v. Wagner, 29 F.3d 264 (7th Cir. 1994); United States v. Hoflin, 880 F.2d 1033 (9th Cir. 1989), cert. denied, 493 U.S. 1083 (1990). But see United States v. Johnson & Towers, Inc., 741 F.2d 662 (3d Cir. 1984).

president convicted of knowingly storing and disposing of hazardous waste without a permit in violation of Section 3008(d)(2)(A) of RCRA.[255] In that case, an inspection by state environmental officials found between 600 and 1,000 rusting and leaking drums of hazardous waste at the defendants' paint manufacturing plant, which had been stored on-site for more than 90 days, and in some cases for many years, without a permit. The company was fined $225,000 and the vice president was sentenced to 21 months in prison and a fine of $5,000. The defendants challenged their convictions on the basis that the jury was improperly instructed that "the United States is not required to prove that the defendant knew that the material was listed or identified by law as hazardous waste or that he was required to obtain a permit before storing or disposing of [the] material." The defendants contended that the district court erred because it failed to instruct the jury that they could not convict unless they found that the defendants "knowingly" violated the law by determining that the defendants knew that the material in question was regulated hazardous waste and knew that a permit was required. The Sixth Circuit stated that the defendants' "knowledge of illegality" argument had been rejected by the Sixth Circuit and by every other circuit court that had considered the issue. Applying this precedent, the court stated that all the government had to prove was the defendant's knowledge of the storage or disposal, the defendant's knowledge that the material was waste, and the defendant's knowledge that it had the potential to be harmful to others or to the environment. The court concluded that the jury instructions adequately required that the defendants have knowledge of the facts that made the conduct a crime.

Citizen Suits. RCRA also authorizes private citizens, who can meet certain procedural preconditions, to file suits against "any person ... who is alleged to be in violation of any permit, standard, regulation, condition, requirement, prohibition, or order" effective under RCRA,[256] or any person whose handling "of any solid or hazardous waste ... may present an imminent and substantial endangerment to health or the environment."[257] Citizens seeking redress under Section 7002(a)(1)(A) of RCRA must, however, provide 60-day notice to the generator, the EPA, and the appropriate state environmental authority, and a 90-day notice is required

---

[255] United States v. Kelley Technical Coatings, Inc., 157 F.3d 432 (6th Cir. 1998).
[256] RCRA 7002(a)(1)(A), 42 U.S.C. § 6972(a)(1)(A). See, for example, White & Brewer Trucking, Inc. v. Donley, 952 F. Supp. 1306 (C.D. Ill. 1997).
[257] RCRA 7002(a)(1)(B), 42 U.S.C. § 6972(a)(1)(B). See, for example, Singer v. Bulk Petroleum Corp., 9 F.Supp.2d 916 (N.D. Ill. 1998); Raytheon Co. v. McGraw-Edison Co., 979 F. Supp. 858 (E.D. Wis. 1997).

in the case of a substantial and imminent endangerment action under Section 7002(a)(1)(B).[258] Such pre-suit notice is meant to provide the EPA or state environmental agency with the opportunity to initiate enforcement proceedings or take other actions concerning the RCRA violation which would preclude the need for a citizen suit.[259] A number of citizen suit plaintiffs have found their claims barred because they failed to comply with these notice provisions.[260]

In addition to the notice prerequisites, RCRA's citizen suit provisions contain other requirements that may preclude a private party from bringing a citizen suit action.[261] RCRA does not permit the filing of a Section 7002(a)(1)(A) citizen suit claim in situations where the EPA or state environmental agency is "diligently prosecuting" a civil or criminal action to require the defendant to comply with the permit, standard, regulation, condition, requirement, prohibition, or order that is the subject of the citizen suit claim.[262] The term "diligently prosecuting" is not defined in RCRA, but it has been interpreted to mean actual litigation or the entry of a court-approved consent order whose provisions encompass the goals of the citizen action.[263] In addition, a citizen suit claim asserted under Section 7002(a)(1)(A) cannot be brought to address "wholly past" violations of RCRA; the alleged violation must be a "continuing" one at the time suit is filed.[264]

A citizen suit based on an endangerment claim under Section 7002(a)(1)(B) of RCRA may not be commenced if the EPA is attempting to restrain or abate the conditions which have resulted in the endangerment situation through any one of the following specific activities:[265]
The filing and diligent prosecution of an imminent hazard action pursuant to Section 7003 of RCRA;[266]

---

[258] See Hallstrom v. Tillamook County, 493 U.S. 20 (1989).
[259] See, for example, Zands v. Nelson, 779 F. Supp. 1254 (S.D. Cal. 1991).
[260] See, for example, Agricultural Excess & Surplus Ins. Co. v. ABD Tank & Pump Co., 878 F. Supp. 1091 (N.D. Ill. 1995); Brandywine Indus. Paper, Inc. v. Chemical Leaman Tank Lines, Inc., 1998 WL 855502 (ED Pa. Dec. 10, 1998); Portmouth Redevelopment and Housing Auth. v. BMI Apartments Assocs., 847 F. Supp. 380 (E.D. Va. 1994).
[261] See 42 U.S.C. § 6972(b).
[262] 42 U.S.C. § 6972(b)(1)(B).
[263] See, for example, Supporters to Oppose Pollution v. Heritage Group 973 F.2d 1320 (7th Cir. 1992).
[264] See, for example, Chartrand v. Chrysler Corp., 785 F. Supp. 666 (E.D. Mich. 1992); Gache v. Town of Harrison, 813 F. Supp. 1037 (S.D.N.Y. 1993).
[265] 42 U.S.C. § 6972(b)(2)(B).
[266] 42 U.S.C. § 6973.

- The filing and diligent prosecution of an administrative order under Section 106(a) of CERCLA;[267]
- The pursuit of a removal action under Section 104 of CERCLA;[268]
- The incurrence of costs to initiate a Remedial Investigation and Feasibility Study (RI/FS) and diligent pursuit of a CERCLA remedial action;[269] or
- The EPA has obtained a court order or a consent decree, or has issued an administrative order (pursuant to Section 7003 of RCRA or Section 106 of CERCLA) which has resulted in a responsible party diligently pursuing a remedial action, a removal action or a RI/FS.

With regard to the final category, if the EPA has issued an administrative order pursuant to Section 7003 of RCRA or Section 106 of CERCLA, an endangerment claim brought under 7002(a)(1)(B) of RCRA is prohibited only as to the scope and duration of the administrative order.[270]

In addition to the above-mentioned EPA actions, a citizen suit based on an endangerment claim under Section 7002(a)(1)(B) may not be brought if state authorities are attempting to restrain or abate the conditions which have resulted in the endangerment situation by:[271]

- Diligently prosecuting a RCRA 7002(a)(1)(B) action brought by the state; or
- Engaging in a removal action under Section 104 of CERCLA; or
- Initiating a RI/FS and diligently proceeding with a remedial action under CERCLA.

If a state is not pursuing one of these three specific actions, a citizen suit claim based on Section 7002(a)(1)(B) of RCRA is not barred even though the state has initiated other administrative actions.[272]

Finally, unlike citizen suit claims asserted under Section 7002(a)(1)(A) of RCRA, which are subject to the "continuing" violation standard, a Section 7002(a)(1)(B) claim can be brought against a defendant whose

---

[267] 42 U.S.C. § 9606(a). See, for example, In re Tutu Wells Contamination Litig., 994 F. Supp. 638 (D.V.I. 1998).
[268] 42 U.S.C. § 9604.
[269] See, for example, McGregor v. Industrial Excess Landfill, Inc., 856 F.2d 39 (6th Cir. 1987).
[270] See, for example, Organic Chems. Site PRP Group v. Total Petroleum, Inc., 6 F.Supp.2d 660 (W.D. Mich. 1998); A-C Reorganization Trust v. E.I. DuPont de Nemours & Co., 968 F. Supp. 423 (E.D. Wis. 1997).
[271] 42 U.S.C. § 6972(b)(2)(C).
[272] See, for example, Gilroy Canning Co., Inc. v. California Canners & Growers, 15 F.Supp.2d 943 (N.D. Cal. 1998).

actions occurred wholly in the past if substantial endangerment presently exists.[273] Prospective injunctive relief is available for remaining contamination that may pose an imminent and substantial threat to persons or the environment;[274] however, it is important to note that a private party may not recover past cleanup costs under RCRA for contamination that has been removed.[275]

8.8 RCRA Inspections

Section 3007 of RCRA, provides EPA officials with authority to conduct inspections.[276] The premises subject to inspection are defined as any place where hazardous wastes are or have been generated, stored, treated, or disposed of, or any place from which hazardous wastes have been transported. Entry must be permitted, upon request, to an EPA officer, employee or representative. The entry must, however, be during a reasonable time and must be commenced and completed with reasonable promptness. EPA inspectors may obtain samples, and split samples must be provided if requested. Inspectors also have authority to access company records and make photocopies.[277]

Ordinarily, government officials will provide prior notification that they intend to conduct an inspection of company facilities and records, although they could just show up requesting entry for inspection purposes. In either case, the facility must be prepared to handle the inspection effectively. There are number of important practical considerations when it comes to dealing with environmental inspections, which are outlined here.

Deciding Whether to Grant or Deny Access. Certain company employees should be designated to greet government inspectors upon arrival on the premises. The company may choose to instruct these employees to request identification and credentials from inspectors, as well as have them sign a visitor's logbook. If the inspector does not have an inspection warrant, the company should consider whether to deny access. In cases where the inspector arrives without advance notice, it may make sense to deny access if the company is not properly prepared to handle the inspection at that time. If prior notification has been provided,

---

[273] See, for example, City of Toledo v. Beazer Materials & Servs., Inc., 833 F. Supp. 646 (N.D. Ohio 1993).
[274] See, for example, Raytheon Co. v. McGraw-Edison Co., 979 F. Supp. 858 (E.D. Wis. 1997).
[275] See Meghrig v. KFC Western, Inc., 516 U.S. 479 (1996).
[276] 42 U.S.C. § 6927.
[277] 42 U.S.C. § 6927(a).

but the inspector shows up without a warrant, the company must weigh out whether it is worth giving the inspector a difficult time by denying entry since the inspector is certain to come back soon with the necessary warrant, and will will surely be less congenial upon his or her return.

It is important that employees designated to deal with the inspector be aware of any operating permits that contain express authorization for the permitting agency to conduct inspections. Conditions of many environmental permits authorize inspections so relevant company personnel should be correctly advised and have properly reviewed the facility's permits for possible inspection authorizations.

From a legal standpoint, denial of access to inspectors based solely on the lack of a warrant will not result in assessment of any civil or criminal penalties as long as an emergency situation does not exist. Access may also be denied for other reasons, including:

1. The inspector lacks necessary safety equipment or has not undergone training required under OSHA or other federal laws; or
2. The inspector is seeking entry other than during the working hours of the facility.

If access is denied for one of these reasons, it is important that the inspector be told that access will be allowed upon compliance with the company's objection.

Access may not be denied for any of the following reasons:

1. The inspector's use of cameras or video recorders;
2. Strikes or plant shutdowns; or
3. The inspector's refusal to sign a waiver restricting liability or obligations of the facility owner or operator.

Valid Search Warrant. If an inspector presents a valid search warrant upon arrival, and access is denied, the company may be subject to criminal penalties. Therefore, it is crucial that designated company personnel ask for a copy of the warrant and read it. Determine the scope and limits of the warrant. To the extent that the warrant is limited to certain portions of the facility, the company may deny access to the remaining parts. Further, it is important to verify that (1) the warrant has been signed by a magistrate or judge; and (2) the warrant authorizes entry by the agents who have appeared at the facility to conduct the inspection. The company should also request copies of any affidavits that support issuance of the warrant, although the company does not have an absolute right to receive copies of this documentation.

Written Agreement About Scope of Inspection. The facility's policy concerning environmental inspections may include a request that inspectors provide designated employees with a summary or checklist of

records and facility components that the inspector wishes to review. If possible, it is best to have the inspector agree in writing to the exact scope of the inspection, including the following items:
1. Witness interview schedules
2. Documents to be produced
3. Sampling protocols
4. Confidentiality issues
5. Return of documents and
6. Inspector's use of photographic equipment while on the premises.

Oversight of Inspector's Activities. It is important to keep a watchful eye on the inspector. The inspector should not be given "carte blanche" to peruse the facility and company files. Company personnel should maintain a reasonable degree of control over the inspection process. The inspector should have an employee escort at all times so that the inspector does not go beyond the scope of the inspection, as agreed-upon in advance or provided by the terms of the inspection warrant. If possible, it is a good idea to take photographs or videotape the entire inspection. If the inspector takes photographs or uses a videocamera, ask for copies.

Participation in Employee Interviews. Inspectors should not be given free access to speak with company employees. If the inspector has a warrant, he or she is generally limited to seizure of documentary and tangible evidence. However, with or without a warrant, inspectors frequently will attempt to speak directly with company employees. Although the company cannot prohibit employees from speaking with inspectors, it can exercise some control over discussions between inspectors and employees by requesting that company management be present during any employee interviews. Further, personnel may be advised that they are not required to speak with inspectors and that they may refuse to answer any questions asked by government inspectors. It is important, however, to make sure that this advice is not given in such a way that it could be construed as forbidding employees from speaking with inspectors because, in certain circumstances, this might subject the company to obstruction of justice charges under 18 U.S.C. Section 1612.

Sampling and Split Sample Requests. The inspection authority of RCRA and most environmental laws permits the inspecting agency to perform sampling. If samples are collected, it is important to ensure that representative samples are taken and collected properly. The company should always request a split sample and perform an independent sampling analysis to verify the accuracy of the government's sampling analysis. Under RCRA and most environmental laws, inspectors are not required to

provide split samples unless a specific request is made. When requesting a split sample, first make sure that the split sample is equal in weight and volume. The following additional information should also be obtained:
1. Written receipts that describe the samples
2. Description of the tests to be performed on the samples and
3. Government test results.
The facility also should take special care to observe valid chain of custody procedures and appropriate analytical techniques when handling all split samples.

Inspection Conclusion. Following completion of the inspection, the facility should request copies of all photographs and videotapes taken by the inspector, as well as receipts for samples and a document inventory. If documents are seized pursuant to a valid search warrant, the facility is entitled to an inventory of the documents and the inspecting agency must file a copy of same with the court that issued the warrant. Although the inspector may be reluctant to do so, it doesn't hurt to ask if he or she has any preliminary findings from the inspection. Finally, as a follow-up, it may be worthwhile to make a written request for copies of photographs, videotapes, sample analyses, and the field report submitted by the inspector.

8.9 Municipal Solid Waste Flow Control

Many communities have constructed waste-to-energy incinerators, trash transfer stations, and recycling facilities as a cost-effective way to handle municipal solid waste (MSW). However, because such facilities need to process large quantities of waste to be financially viable (much more than is generated solely by the local community), municipalities often use MSW flow control ordinances to ensure the necessary steady stream of MSW. Flow control, as the term implies, consists of methods for directing the flow of MSW to designated local facilities. The higher the volume of trash sent to a facility, the greater the ability of the facility to recover its own costs, and the less burdensome the facility is for the locality that has sponsored its construction and operation.

There appears to be no reliable statistics as to the number of counties, municipalities, or other local government units that have actually adopted flow control ordinances. However, more than half of the states have enacted statutes that authorize local governments within the state to adopt such ordinances. These state laws are summarized in Table 8.1. Most of these statutes provide that a local authority may designate one or more facilities to receive all trash in the local jurisdiction. These local ordinances may require all trash to go to the designated facility and prohibit the export of trash to facilities in other localities and other states.

Table 8.1. State Authority for Local Flow Control Ordinances

State	*Statutory Authority*
Colorado	Colo. Rev. Stat. 30-20-107
Connecticut	Conn. Gen. Stat. 22A-220A
Delaware	Del. Code Ann. tit. 7, 6406(31)
Florida	Fla. Stat. ch. 403.713
Hawaii	Haw. Rev. Stat. 340A-3(a)
Illinois	Ill. Rev. Stat. ch. 34, para. 5-1047
Indiana	Ind. Code 36-9-31-3 and -4
Iowa	Iowa Code 28G.4
Louisiana	La. Rev. Stat. 30:2307(9)
Maine	Me. Rev. Stat. Ann. tit. 38, 1304-B(2)
Minnesota	Minn. Stat. 115A.80
Mississippi	Miss. Code Ann. 17-17-319
Missouri	Mo. Rev. Stat. 260.202
New Jersey	N.J. Stat. Ann. 13:1E-22, 48:13A-5
North Carolina	N.C. Gen. Stat. 130A-294
North Dakota	N.D. Cent. Code 23-29-06(6) & (8)
Ohio	Ohio Rev. Code Ann. 343.01(H)(2)
Oregon	Or. Rev. Stat. 268.317(3) & (4)
Pennsylvania	Pa. Stat. Ann. tit. 53, 4000.303(e)
Rhode Island	R.I. Gen. Laws 23-19-10(40)
Tennessee	Tenn. Code Ann. 68-211-814
Vermont	Vt. Stat. Ann. tit. 24, 2203a, 2203b
Virginia	Va. Code Ann. 15.1-28.01
Washington	Wash. Rev. Code 35.21.120, 36.58.040
West Virginia	W. Va. Code 240-2-1h
Wisconsin	Wis. Stat. 159.13(3), (11)

Although this authority is available for municipalities to implement flow control measures, local governments must be cautious about how they use this authority in light of a significant 1994 decision of the U.S. Supreme Court dealing with the issue of flow control. In *C & A Carbone, Inc. v. Town of Clarkstown*,[278] the Court invalidated a MSW flow control

---

[278] C & A Carbone, Inc. v. Town of Clarkstown, 511 U.S. 383 (1994).

ordinance, finding that it placed an improper restraint on interstate commerce. In that case, the Town of Clarkstown entered into a consent decree with the New York State Department of Environmental Conservation in 1989, agreeing to close its town landfill and build a new solid waste transfer station on the same site. The station was designed to receive bulk solid waste and separate recyclable from nonrecyclable items. Recyclable waste were to be shipped to a recycling facility and nonrecyclable waste to a suitable landfill or incinerator. The cost of building the transfer station was approximately $1.4 million. A local private contractor agreed to construct the facility and operate it for five years, after which time the town agreed to buy it for one dollar. For this five-year period, the town guaranteed a minimum waste flow of 120,000 tons per year, for which the contractor could charge the hauler a so-called tipping fee of $81 per ton. However, if the station received less than 120,000 tons in a year, the town was obligated to make up any revenue deficit. Thus, the town could amortize the cost of the transfer station with the income generated by the tipping fees. The town's main problem was how to meet the yearly guarantee, especially since the $81 per ton tipping fee exceeded the disposal cost of unsorted solid waste on the private market. The town decided to adopt a flow control ordinance.[279]

The ordinance required that all nonhazardous solid waste within the town be deposited at its transfer station. (Ordinance Section 3.C - waste generated within the town; Ordinance Section 5.A - waste generated outside and brought in). Noncompliance was punishable by as much as a $1,000 fine and up to 15 days in jail. (Ordinance Section 7.) C & A Carbone, Inc., and other parties challenged the constitutionality of the ordinance. Carbone operated a recycling center in Clarkstown, where it received bulk solid waste, sorted and baled it, and then shipped it to other processing facilities. The New York state courts upheld the ordinance and the U.S. Supreme Court reversed the state court rulings, invalidating it on Commerce Clause grounds.

The Supreme Court found that even though the flow control ordinance permitted local recyclers like Carbone to continue receiving solid waste, it required them to bring the nonrecyclable residue from that waste to the transfer station. It thus forbade Carbone to ship the nonrecyclable waste itself, and it required Carbone to pay a tipping fee on trash that Carbone had already sorted. Furthermore, the Court concluded that the ordinance squelched competition in the waste-processing industry, leaving no room for outside investment. The Court deemed these impacts discriminatory and of the type the Commerce Clause was intended to guard against.[280]

[279] Town of Clarkstown, Local Laws of 1990, No. 9.
[280] U.S. Const. art. I, sect. 8, cl. 3.

The Court stated that:

"While the immediate effect of the ordinance is to direct local transport of solid waste to a designated site within the local jurisdiction, its economic effects are interstate in reach. The Carbone facility in Clarkstown receives and processes waste from places other than Clarkstown, including from out of State. By requiring Carbone to send the nonrecyclable portion of this waste to the Route 303 transfer station at an additional cost, the flow control ordinance drives up the cost for out-of-state interests to dispose of their solid waste. Furthermore, even as to waste originant in Clarkstown, the ordinance prevents everyone except the favored local operator from performing the initial processing step. The ordinance thus deprives out-of-state businesses of access to a local market. These economic effects are more than enough to bring the Clarkstown ordinance within the purview of the Commerce Clause. It is well settled that actions are within the domain of the Commerce Clause if they burden interstate commerce or impede its free flow."

The majority, led by Justice Kennedy, relied on prior decisions that hold that discrimination against interstate commerce in favor of local interests is per se invalid, "save in a narrow class of cases in which the municipality can demonstrate, under rigorous scrutiny, that it has no other means to advance a legitimate local interest." The municipality's principal argument was that the flow control ordinance was necessary to sustain the financial viability of the facility and protect the public fisc. The Court rejected this justification for the ordinance, declaring that "revenue generation is not a local interest that can justify discrimination against interstate commerce." The Court stated that the town could subsidize the facility by other means, such as through general taxes or municipal bonds.

Although the Supreme Court found the Clarkstown ordinance overly restrictive of interstate commerce, the Court's ruling did not sound a deathknell to the use of MSW flow control. State authority still remains for local flow control measures. However, municipalities must exercise greater care in crafting flow control ordinances that do not unduly burden interstate commerce. Since the Supreme Court issued its ruling in the Clarkstown case, a number of flow control ordinances have been invalidated on commerce clause grounds,[281] but others have survived

---

[281] See, for example, Waste Management, Inc. of Tenn. v. Metropolitan Government of Nashville & Davidson County, 130 F.3d 731 (6th Cir. 1997); National Solid Wastes Management Ass'n v. Meyer, 63 F.3d 652 (7th Cir. 1995); Connecticut Carting Co. v. Town of East Lyme, 946 F. Supp. 152 (D. Conn. 1995).

commerce clause attack.[282] A few of these cases are summarized here for illustrative purposes.

In one case, for example, a local ordinance that required waste haulers to pay a per ton fee of $86 for all commercial waste collected within the Town of East Lyme, Connecticut was held to place an undue restraint on interstate commerce.[283] The $86 per ton fee exceeded tipping fees at nearby facilities which were as low as $40 to $45 per ton. Under the ordinance, the haulers could deliver the collected waste to a designated town facility for no further fee, or take it to a different facility and pay its tipping fee. Various haulers of commercial waste challenged the constititionality of the collection fee. They contended that the ordinance had the practical effect of directing all commercial waste to the town facility because the fee created an economic disincentive to take the waste elsewhere. The court ruled that the fee provision of the town ordinance violated the Commerce Clause. The court found that adding the town's fee to the cost of using other disposal facilities discouraged use of those facilities because it was cost-prohibitive. The added expense discriminated against other facilities by prohibiting them from competing with the town facility on an equal footing. Relying on the U.S. Supreme Court's decision in *C & A Carbone, Inc., v. Town of Clarkstown,* the court stated that because the ordinance "hoards solid waste, and the demand to get rid of it, for the benefit of the preferred processing facility," the effect was to remove the town's waste from the free flow of interstate commerce.

On the other hand, in a Second Circuit case, the court ruled that a local ordinance regulating the collection and disposal of solid wastes within a designated commercial garbage collection district passed constitutional muster.[284] The challenged ordinance prohibited local businesses within the district from hiring their own garbage haulers, required the businesses to use the services of a single garbage hauler hired by the town, and permitted the hauler to dump the garbage collected from the district free of charge at an incinerator owned by the town. In finding that the ordinance was not an impermissible restraint on interstate commerce, the court distinguished the challenged ordinance from the flow control ordinance struck down by the U.S. Supreme Court in *C & A Carbone, Inc., v. Town of Clarkstown,* finding that the town was not favoring local

[282] See, for example, Sal Tinnerello & Sons, Inc. v. Town of Stonington, 141 F.3d 46 (2d Cir. 1998); USA Recycling, Inc. v. Town of Babylon, 66 F.3d 1272 (2d Cir. 1995); SSC Corp. v. Town of Smithtown, 66 F.3d 502 (2d Cir. 1995); Waste Management of Michigan v. Ingham County, 941 F. Supp. 656 (W.D. Mich. 1996); National Solid Waste Management Ass'n v. Williams, 1997 WL 345667 (D. Minn. June 19, 1997).
[283] Connecticut Carting Co. v. Town of East Lyme, 946 F. Supp. 152 (D. Conn. 1995).
[284] USA Recycling, Inc v. Town of Babylon, 66 F.3d 1272 (2d Cir. 1995)

garbage service companies over out-of-state competitors. Instead, the town had chosen to exclude all garbage service companies from the market, both local and out-of-state, by itself becoming the sole provider of garbage services to businesses within the commercial garbage collection district. Thus, the town was not acting as a business selling solid waste collection and disposal services (as was found in *Carbone*); rather, it was merely carrying out the traditional local governmental function of providing municipal sanitation services to local businesses. The court stated that the case really boiled down to two simple propositions: one, that towns can assume exclusive responsibility for the collection and disposal of local garbage, and two, that towns can hire private contractors to provide municipal services to residents. The court said that in neither case does a town discriminate against, or impose any burden on, interstate commerce.

9

Comprehensive Environmental Response, Compensation, and Liability Act

9.1 Introduction

For decades, the nation's commercial, industrial, and government enterprises have generated, stored, and disposed of millions of tons of hazardous waste annually. Some of this waste has escaped from its intended on-site or off-site containment in storage drums, holding ponds, impoundments, and landfills (and in worse cases, has been recklessly or deliberately dumped); permeated underlying soils; polluted lakes, streams, and underground waters; and placed human health and the environment at risk from exposure to hazardous substances. As one court observed, "the vast carelessness that created the conundrum of hazardous waste, which has continued for decades, will not be quickly or easily remedied."[1]

In 1980, Congress enacted the Comprehensive Environmental Response, Compensation and Liability Act (CERCLA) to facilitate the cleanup of sites that have been contaminated by hazardous substances.[2] With thousands of contaminated sites located throughout the land, an enormous amount of litigation has ensued since the law's enactment, primarily focused on questions of who should be held accountable for paying the cleanup costs incurred at sites where releases of hazardous substances have occurred. This chapter examines CERCLA's liability framework, the threat of strict liability it poses to virtually anyone connected with a site

---

[1] Avondale Indus. Inc. v. Travelers Indem. Co., 887 F.2d 1200, 1201 (2d Cir. 1989).
[2] 42 U.S.C. §§ 9601 through 9675.

where hazardous substance contamination has occurred, the law's limited defenses, and the ongoing legal battles being fought between the government and private parties over the responsibility for CERCLA cleanup costs. Certain methods of minimizing liability are also discussed, including settlement agreements, insurance coverage for environmental cleanup costs, and liability protections and assurances given to private parties who undertake cleanup as part of various federal and state voluntary cleanup initiatives.

9.2 Nature of CERCLA Liability

Liability for the cleanup of sites contaminated by hazardous substances is established under Section 107(a) of CERCLA[3] if the government or a private party plaintiff can establish that:
1. The contaminated site in question is a "facility" as defined in Section 101(9) of CERCLA;[4]
2. The defendant is a responsible party under CERCLA Section 107(a);[5]
3. A release or threat of release[6] of a hazardous substance[7] has occurred at the facility; and
4. The release or the threatened release has caused the government or a private party to incur "necessary" response costs[8] which are consistent with the National Contingency Plan (NCP).[9]

In private cost recovery or contribution actions[10] brought against potentially responsible parties (PRPs),[11] a private party plaintiff bears the burden of proving each of these four elements by a preponderance of the

---

[3] 42 U.S.C. § 9607(a).
[4] 42 U.S.C. § 9601(9). See, for example, Clear Lake Properties v. Rockwell Int'l Corp., 959 F. Supp. 763 (S.D. Tex. 1997).
[5] See section 9.3 for discussion of potentially responsible parties.
[6] See, for example, Westfarm Assocs. v. Washington Suburban Sanitary Comm'n, 66 F.3d 669 (4th Cir. 1995); Companies for Fair Allocation v. Axil Corp., 853 F. Supp. 575 (D. Conn. 1994); Bunger v. Hartman, 851 F. Supp. 461 (S.D. Fla. 1994).
[7] Tex. Admin. Code tit. 30, 333.1 through 333.11 (Voluntary Cleanup Rules).
[8] In order to prove that response costs are "necessary," the plaintiff must demonstrate that the alleged contamination was serious enough to warrant a response action. Licciardi v. Murphy Oil U.S.A., Inc., 111 F.3d 396 (5th Cir. 1997); Amoco Oil Co. v. Borden, Inc., 889 F.2d 664 (5th Cir. 1989). See also Soo Line R.R. Co. v. Tang Indus., Inc., 998 F. Supp. 889 (N.D. Ill. 1998); Acushnet Co. v. Coaters, Inc., 937 F. Supp. 988 (D. Mass. 1996).
[9] See section 9.7.4 for discussion of consistency with the NCP.
[10] See section 9.7 for discussion of private cost recovery and contribution actions.
[11] See section 9.3 for discussion of categories of PRPs.

evidence before CERCLA liability can be imposed on the defendant.[12] In government actions to recover cleanup costs from PRPs,[13] the government bears the burden of proof with regard to each of the first three elements;[14] however, with regard to the fourth element, the government enjoys a rebuttable presumption that its costs were necessary and consistent with the NCP.[15] Private party plaintiffs, by contrast, must demonstrate affirmatively that their response costs were necessary and consistent with the NCP.[16]

9.2.1 Definition of "Hazardous Substance"

The term "hazardous substance" is defined broadly in Section 101(14) of CERCLA by reference to substances defined as hazardous in a number of other environmental statutes, including the Resource Conservation and Recovery Act,[17] the Clean Water Act,[18] and the Clean Air Act.[19] Petroleum is specifically exempted from CERCLA's definition of hazardous substances.[20]

On its face, CERCLA liability applies to the release of "any" hazardous substance, and it does not impose quantitative requirements. Although the EPA has listed certain substances as hazardous in 40 CFR Section 302.4 and the accompanying table, the quantitative limitations provided in the EPA regulations are only for purposes of CERCLA reporting requirements; they do not bear any relationship to a PRP's liability for CERCLA cleanup and response costs. The absence of any quantity requirements pertaining to CERCLA liability has inevitably lead to the

---

[12] See, for example, T & E Indus. v. Safety Light Corp., 680 F. Supp. 696 (D.N.J. 1988).

[13] See section 9.8 for discussion of government response and cost recovery actions.

[14] See, for example, United States v. Aceto Agric. Chem. Corp., 872 F.2d 1373 (8th Cir. 1989); United States v. Serafini, 706 F. Supp. 346 (M.D. Pa. 1988).

[15] See, for example, United States v. Chapman, 146 F.3d 116 (9th Cir. 1998); United States v. Amtreco, Inc., 846 F. Supp. 1578 (M.D. Ga. 1994).

[16] See, for example, United States v. Northeastern Pharmaceutical & Chem. Co., 579 F. Supp. 823 (W.D. Mo. 1984), aff'd on other grounds, 810 F.2d 726 (8th Cir. 1986), cert. denied 484 U.S. 848 (1987).

[17] 42 U.S.C. § 6921.

[18] 33 U.S.C. § 1317(a).

[19] 42 U.S.C. § 7412.

[20] 42 U.S.C. § 9601(14). See, for example, Textron, Inc. v. Barber-Colman Co., 903 F. Supp. 1546 (W.D.N.C. 1995) (adding nonhazardous additives to petroleum, such as kerosene, did not bring products outside of petroleum exclusion); Acme Printing Ink Co. v. Menard, Inc., 881 F. Supp. 1237 (E.D. Wis. 1995) (petroleum exclusion does not apply to waste oil contaminated with substances other than those that are constituents of petroleum products).

conclusion that Congress planned for the "hazardous substance" definition to include even minimal amounts of pollution. The Court of Appeals for the Second, Third, Fifth, Ninth, and D.C. Circuits have specifically addressed this issue and all agree that CERCLA's definition of hazardous substance has no minimum level requirement.[21]

9.2.2 Strict, Joint and Several, and Retroactive Liability

The liability imposed by Section 107(a) of CERCLA is both strict,[22] joint and several,[23] and retroactive.[24] Strict liability essentially means that it is unnecessary for the government or a private party to prove that the owner or operator of a facility was negligent or otherwise responsible for the release. "CERCLA section 107 requires only a minimal causal nexus between the defendant's hazardous waste and the harm caused ... CERCLA only requires that the plaintiff prove by a preponderance of the evidence that the defendant deposited his hazardous waste at the site and that the hazardous substances containing the defendant's waste are also found at the site."[25] There is no need to prove causation.[26] It is merely

---

[21] See, for example, B.F. Goodrich Co. v. Murtha, 958 F.2d 1192 (2d Cir. 1992); United States v. Alcan Aluminum Corp., 964 F.2d 252 (3d Cir. 1992); Amoco Oil Co. v. Borden, Inc., 889 F.2d 664 (5th Cir. 1989); A & W Smelter and Refiners, Inc. v. Clinton, 146 F.3d 1107 (9th Cir. 1998); Eagle-Picher Indus., Inc. v. EPA, 759 F.2d 922 (D.C.Cir. 1985).

[22] See, for example, Kerr-McGee Chemical Corp. v. Lefton Iron & Metal Co., 14 F.3d 321 (7th Cir. 1994); In re Chicago, Milwaukee, St. Paul & Pacific R.R., 974 F.2d 775 (7th Cir. 1992); Tanglewood East Homeowners v. Charles-Thomas, Inc., 849 F.2d 1568 (5th Cir. 1988).

[23] See, for example, United States v. Monsanto Co., 858 F.2d 160 (4th Cir. 1988), cert. denied, 490 U.S. 1106 (1989); New York v. Shore Realty Corp., 759 F.2d 1032 (2d Cir. 1985); United States v. New Castle County, 642 F. Supp. 1258 (D. Del. 1986); United States v. Conservation Chem. Co., 589 F. Supp. 59 (W.D. Mo. 1984); United States v. Chem-Dyne Corp., 572 F. Supp. 802 (S.D. Ohio 1983).

[24] See, for example, United States v. Olin Corp., 107 F.3d 1506 (11th Cir. 1997); Raytheon Co. v. McGraw-Edison Co., Inc, 979 F. Supp. 858 (E.D. Wis. 1997); Ninth Avenue Remedial Group v. Fiberbond Corp., 946 F. Supp. 651 (N.D. Ind. 1996); Nova Chems., Inc. v. GAF Corp., 945 F. Supp. 1098 (E.D. Tenn. 1996).

[25] Violet v. Picillo, 648 F. Supp. 1283 (D.R.I. 1986). See also Textron, Inc. v. Barber-Colman Co., 903 F. Supp. 1546 (W.D.N.C. 1995) (plaintiff failed to show that alleged wastes disposed of at the site by the defendant contained any hazardous substances).

[26] See, for example, United States v. Stringfellow, 661 F. Supp. 1053 (C.D. Cal. 1987) (court explained that the legislative history of CERCLA indicates that a showing of traditional causation is not a necessary element of a CERCLA claim.); United States v. Bliss, 667 F. Supp. 1298 (E.D. Mo. 1987) (court

necessary to prove a nexus between the defendant and the site.[27] Joint and several liability basically means that any single defendant can be held responsible for the entire cost of a cleanup or other response costs. Joint and several liability is traditionally imposed when the action of two or more defendants causes a single indivisible result.[28]

After CERCLA was passed in 1980, the courts unanimously held that Congress intended for CERCLA's strict liability scheme to be applied retroactively, encompassing liability for contamination that occurred before the statute was enacted. Although the issue was thought to have been firmly resolved, and had gone essentially unquestioned for more than a decade, in 1996, a highly controversial and potentially far-reaching decision by the U.S. District Court for the Southern District of Alabama caused a stir over the retroactive application of CERCLA.[29] In *United States v. Olin Corp.,*[30] the district court became the first to conclude that CERCLA's liability provisions did not apply retroactively to acts occurring before the law's passage in 1980. In what appeared to be a routine case, Olin Corporation and the U.S. EPA asked the district court to approve a consent decree between the parties which obligated Olin to spend more than $10 million to clean up a contaminated site in McIntosh, Alabama. The site was contaminated with wastes from two plants that together operated between 1955 and 1982. The district court refused to approve the consent decree insofar as it imposed liability on Olin for acts that occurred prior to CERCLA's passage in 1980. In holding that CERCLA liability does not apply retroactively, the court relied on a 1994 decision of the U.S. Supreme Court in *Landgraf v. USI Film Products,* a civil rights case.[31] The court acknowledged that previous rulings had

---

described the causation requirement in hazardous waste litigation cases as "released" due to the technical difficulties in tracing hazardous waste). See also United States v. Maryland Sand, Gravel and Stone Co., 39 Env't Rep. Cas. (BNA) 1761, 1994 WL 541069 (D. Md. Aug. 12, 1994); Northwestern Mutual Life Ins. Co. v. Atlantic Research Corp., 847 F. Supp. 389 (E.D. Va. 1994); Acme Printing Ink Co. v. Menard, Inc., 870 F. Supp. 1465 (E.D. Wis. 1994).

[27] See, for example, Environmental Transportation Systems v. Ensco, Inc., 969 F.2d 503 (7th Cir. 1992); New York v. Shore Realty Corp., 759 F.2d 1032 (2d Cir. 1985).

[28] See W. Page Keeton et al., Prosser and Keeton on The Law of Torts (5th ed. 1984).

[29] See Tucker, "Retroactive Liability Is Challenged," Nat'l L.J., Oct. 14, 1996, at C1.

[30] United States v. Olin Corp., 927 F. Supp. 1502 (S.D. Ala. 1996), *rev'd,* 107 F.3d 1506 (11th Cir. 1997).

[31] Landgraf v. USI Film Prods., 511 U.S. 244 (1994).

applied CERCLA retroactively,[32] but argued that those cases were decided before more stringent standards governing retroactivity were articulated by the Supreme Court in *Landgraf*.[33]

However, the Court of Appeals for the Eleventh Circuit reversed the district court's decision, concluding that Congress clearly intended retroactive application of CERCLA's cleanup liability provisions.[34] The court found that although Congress did not include explicit language in the statute to indicate that CERCLA's liability provisions are to be applied retroactively, the legislative history made such intent clear and unmistakable.[35] The court also noted that the district court's holding ran contrary to the rulings of all federal courts that had considered the retroactivity issue and that courts that had ruled on retroactivity challenges since the district court's decision had unanimously repudiated the district court's holding.[36] In addition, the court noted that Congress reauthorized CERCLA twice since 1980, once with substantive changes, without suggesting that the courts had misconstrued the statute regarding retroactivity.[37]

The only lingering significance of the district court case seems to be that it has caused some courts to revisit the issue of retroactivity; however, no court has agreed with the district court's view in *Olin*,[38] including those courts that considered the issue prior to the Eleventh Circuit's reversal.[39] Some commentators are still trying to revive the issue of retroactivity.[40]

---

[32] United States v. Olin Corp., 927 F. Supp. 1502 (S.D. Ala. 1996), *rev'd,* 107 F.3d 1506 (11th Cir. 1997), at 927 F. Supp.1507 & n. 25 (recognizing that of the 22 federal courts "which have directly addressed the issue of CERCLA's retroactivity, none have declined to apply CERCLA on retroactivity grounds.").

[33] In *Landgraf*, the Supreme Court ruled that a provision of the Civil Rights Act of 1991 creating new rights to compensatory and punitive damages for certain discriminatory acts did not apply to cases pending on appeal when the law was enacted. The Court, finding that the civil rights law included no explicit retroactivity provisions, pointed to the legislative history to show that such provisions had been introduced but not included in the final language.

[34] United States v. Olin Corp., 107 F.3d 1506 (11th Cir. 1997).

[35] Id. at 1514.

[36] Id. at 1512 n. 13.

[37] Id. at 1512 n. 12.

[38] See Raytheon Co. v. McGraw-Edison Co., Inc, 979 F. Supp. 858 (E.D. Wis. 1997).

[39] See, for example, Ninth Avenue Remedial Group v. Fiberbond Corp., 946 F. Supp. 651 (N.D. Ind. 1996); Nova Chems., Inc. v. GAF Corp., 945 F. Supp. 1098 (E.D. Tenn. 1996); Gould, Inc. v. A & M Battery & Tire Serv., 933 F. Supp. 431 (M.D. Pa. 1996).

[40] See, for example, Howard & Harr, Environmental Law--CERCLA Retroactivity, Nat'l L.J., Dec. 28, 1998, at B7; Howard, A New Justification for Retroactive Liability in CERCLA: An Appreciation of the Synergy Between Common and Statutory Law, 42 St. Louis U. L.J. 847 (1998).

9.3 Categories of Potentially Responsible Parties

Congress designated four broad categories of PRPs who, regardless of fault, may be held liable for CERCLA cleanups if they contributed any amount of a hazardous substance to the contaminated site.[41] The four classes of PRPs are:

1. Current owners or operators of a site.[42]

2. Past owners or operators of a site at the time hazardous substances were disposed of at the site.[43]

3. Anyone who arranged for the disposal, transport or treatment of hazardous substances found at the site (generators or "arrangers").[44]

4. Anyone who accepted hazardous substances for disposal and selected the site now slated for cleanup. (transporters).[45]

Applying CERCLA's strict liability scheme to these four broad categories of PRPs, courts have imposed liability under Section 107(a) on a wide array of individuals and companies, including:

- Parent corporations[46]
- Successor corporations[47]
- Corporate officers who were active in site operations[48]
- Active shareholders[49]
- Lessees of current and former landowners[50]

---

[41] CERCLA § 107(a), 42 U.S.C. § 9607(a).

[42] CERCLA § 107(a)(1), 42 U.S.C. § 9607(a)(1).

[43] CERCLA § 107(a)(2), 42 U.S.C. § 9607(a)(2).

[44] CERCLA § 107(a)(3), 42 U.S.C. § 9607(a)(3).

[45] CERCLA § 107(a)(4), 42 U.S.C. § 9607(a)(4).

[46] See, for example, United States v. TIC Investment Corp., 866 F. Supp. 1173 (N.D. Iowa 1994); Kelley v. Thomas Solvent Co., 727 F. Supp. 1554 (W.D. Mich. 1989). See also United States v. Bestfoods, 524 U.S. 51, 118 S.Ct. 1876 (1998), *vacating and remanding United States v. Cordova Chemical Co., 113 F.3d 572 (6th Cir. 1997).*

[47] See, for example, Aluminum Co. of Am. v. Beazer E., Inc., 124 F.3d 551 (3d Cir. 1997); Kleen Laundry & Dry Cleaning Serices v. Total Waste Management, 867 F. Supp. 1136 (D.N.H. 1994).

[48] See, for example, Sydney S. Arst Co. v. Pipefitters Welfare Education Fund, 25 F.3d 417 (7th Cir. 1994); FMC Corp. v. Aero Indus., Inc., 998 F.2d 2079 (10th Cir. 1993); Marriott Corp. v. Simkins Indus., Inc., 929 F. Supp. 396 (S.D. Fla. 1996).

[49] See, for example, John S. Boyd Co. v. Boston Gas Co., 992 F.2d 401 (1st Cir. 1993); Jacksonville Electric Auth. v. Bernuth Corp. 996 F.2d 1107 (11th Cir. 1993).

[50] See, for example, Caldwell v. Gurley Refining Co., 755 F.2d 645 (8th Cir. 1985); Clear Lake Properties v. Rockwell Int'l Corp., 959 F. Supp. 763 (S.D. Tex. 1997); Northwestern Mutual Life Ins. Co. v. Atlantic Research Corp., 847 F. Supp. 389 (E.D. Va. 1994); Folino v. Hampden Color and Chemical Co., 832

- Bankruptcy estates[51]
- Trustees[52]
- Foreclosing lenders[53]

9.3.1 Current Owners and Operators

The first category of liable parties--established under Section 107(a)(1)--consists of current owners and operators of a facility where a hazardous substance release has occurred.[54] Section 107(a)(1) imposes liability on current owners and operators regardless of whether they were owners or operators at the time hazardous substances were disposed of or released at the facility.[55] The current owner of a facility is within the scope of Section 107(a)(1) liability notwithstanding that the current owner never operated the facility as a hazardous waste disposal site and hazardous substances were never deposited at the facility during the current ownership.[56]

9.3.2 Past Owners and Operators

The second category of liable parties--established by Section 107(a)(2)--consists of past owners and operators of a facility where a hazardous substance release has occurred.[57] Former owners or operators of a site on which there has been a release of hazardous substances may be

---

F. Supp. 757 (D. Vt. 1993).

[51] See, for example, In re T.P. Long Chemical Inc., 45 B.R. 278 (Bankr. N.D. Ohio 1985).

[52] See, for example, North Carolina v. W.R. Peele, Sr. Trust, 1994 U.S. Dist. LEXIS 16335 (E.D.N.C. 1994); City of Phoenix v. Garbage Services Co., 816 F. Supp. 564 (D. Ariz. 1993).

[53] See, for example, United States v. Fleet Factors Corp., 821 F. Supp. 707 (S.D. Ga. 1993); United States v. Maryland Bank & Trust Co., 632 F. Supp. 573 (D.Md. 1986); United States v. Mirabile, 15 Envt'l L. Rep. (ELI) 20994 (E.D. Pa. 1985).

[54] See, for example, Kerr-McGee Chem. Corp. v. Lefton Iron & Metal Co., 14 F.3d 321 (7th Cir. 1994); New York v. Shore Realty Corp., 759 F.2d 1032 (2d Cir. 1985); G.J. Leasing Co. v. Union Electric Co., 854 F. Supp. 539 (S.D. Ill. 1994); Clear Lake Properties v. Rockwell Int'l Corp., 959 F. Supp. 763 (S.D. Tex. 1997).

[55] See, for example, Tanglewood East Homeowners v. Charles-Thomas, Inc., 849 F.2d 1568 (5th Cir. 1988).

[56] See, for example, United States v. Tyson, 25 Env't Rep. Cas. (BNA) 1897 (E.D. Pa. 1986).

[57] See, for example, ABB Industrial Sys, Inc. v. Prime Technology, Inc., 120 F.3d 351 (2d Cir. 1997).

held liable if they owned or operated the facility at "the time of disposal" of hazardous substances.[58]

9.3.3 Generators or "Arrangers" of Waste Disposal

The third category of liable parties--established under Section 107(a)(3)--is commonly referred to as the "generator" category and consists of persons who "arranged for" the treatment or disposal of hazardous substances at a facility from which there has been a release.[59] The law imposes liability upon "any person who by contract, agreement or otherwise, arranged for disposal or treatment of hazardous substances owned or possessed by such person by another party or entity and containing such hazardous substances."[60] Generator liability will attach regardless of whether the PRP knew the location of the disposal site where a waste transporter took its hazardous wastes.[61] Further, a generator's liability extends to any instance where he "arranged for" the disposal of a hazardous substance, even if not disposed of in accordance with the generator's instructions,[62] and even if the transporter disposes of the substances illegally.[63]

Since the "arranged for" language of Section 107(a)(3) is not defined in CERCLA, the courts have struggled to determine the precise parameters of arranger liability. Judicial interpretation of Section 107(a)(3) has

---

[58] See, for example, Smith Land & Improvement Corp v. Celotex Corp., 851 F.2d 86 (3d Cir. 1988), cert. denied 488 U.S. 1029 (1989); Nurad, Inc. v. William E. Hooper & Sons Co., 966 F.2d 837 (4th Cir.1992); United States v. Carolina Transformer Co., 739 F. Supp. 1030 (E.D.N.C. 1989), aff'd, 978 F.2d 832 (4th Cir. 1992); United States v. Mottolo, 695 F. Supp. 615 (D.N.H. 1988); United States v. Hooker Chems. & Plastics Corp., 680 F. Supp. 546 (W.D.N.Y. 1988).
[59] See, for example, United States v. Monsanto Co., 858 F.2d 160 (4th Cir. 1988), cert. denied, 490 U.S. 1106 (1989); Amcast Industrial Corp. v. Detrex Corp., 2 F.3d 746 (7th Cir. 1993); United States v. Aceto Agricultural Chems. Corp., 872 F.2d 1373 (8th Cir. 1989); South Florida Water Management Dist. v. Montalvo, 84 F.3d 402 (11th Cir. 1996); United States v. Vertac Chem. Co., 966 F. Supp. 1491 (E.D. Ark. 1997).
[60] CERCLA § 107(a)(3), 42 U.S.C. § 9607(a)(3).
[61] See, for example, Ekotek Site PRP Comm. v. Self, 932 F. Supp. 1328 (D. Utah 1996).
[62] See, for example, United States v. Northeastern Pharmaceutical & Chem. Co., 579 F. Supp. 823, 847 (W.D. Mo. 1984), aff'd on other grounds, 810 F.2d 726 (8th Cir. 1986), cert. denied, 484 U.S. 848 (1987).
[63] See, for example, United States v. Ward, 618 F. Supp. 884 (E.D.N.C. 1984).

essentially resulted in three different approaches to determining arranger liability.[64] The most liberal approach focuses on the defendant's ownership and control over the hazardous substances.[65] The most conservative approach evaluates whether the defendant specifically intended to dispose of hazardous waste.[66] The trend of more recent cases appears to be toward use of a third approach, whereby the court performs a case-by-case evaluation of all relevant factors, including ownership, control, and intent.[67]

9.3.4 Transporters

The fourth and final category of liable parties--established by Section 107(a)(4)--consists of persons who transport hazardous substances to treatment or disposal facilities. The courts have interpreted the language of Section 107(a)(4) as requiring plaintiffs to demonstrate that the transporter actually chose the treatment or disposal facility.[68]

---

[64] See, for example, Mathews v. Dow Chem. Co., 947 F. Supp. 1517 (D. Colo. 1996) (in which the court evaluated all three approaches).

[65] See, for example, General Elec. Co. v. AAMCO Transmissions, Inc., 962 F2d 281 (2d Cir. 1992); United States v. Aceto Agric. Chems. Corp., 872 F2d 1373 (8th Cir. 1989); New York v. SCA Servs., Inc., 844 F. Supp. 926 (S.D.N.Y. 1994); United States v. North Landing Line Constr. Co., 3 F.Supp.2d 694 (E.D. Va. 1998); Chesapeake & Potomac Tel. Co. v. Peck Iron & Metal Co., Inc., 814 F. Supp. 1293 (E.D. Va. 1993).

[66] See, for example, United States v. Cello-Foil Prods., Inc., 100 F.3d 1227 (6th Cir. 1996); AM Int'l, Inc. v. International Forging Equip. Corp., 982 F.2d 989 (6th Cir. 1993); Amcast Indus. Corp. v. Detrex Corp., 2 F.3d 746 (7th Cir. 1993); Struhar v. City of Cleveland, 7 F.Supp.2d 948 (N.D. Ohio 1998).

[67] See, for example, South Florida Water Management Dist. v. Montalvo, 84 F3d 402 (11th Cir. 1996); Mathews v. Dow Chem. Co., 947 F. Supp. 1517 (D. Colo. 1996); United States v. Gordon Stafford, Inc., 952 F. Supp. 337 (N.D. W.Va. 1997).

[68] See, for example, United States v. South Carolina Recycling and Disposal, Inc., 21 Env't Rep. Cas. (BNA) 1577 (D.S.C. 1984) (plaintiff must prove that transporter selected the facility to which wastes were delivered). See also Ascon Properties, Inc. v. Mobil Oil Co., 34 Env't Rep. Cas. (BNA) 1176 (C.D. Cal. 1991) (transporters of hazardous wastes not liable because they did not select the disposal site for such hazardous wastes); United States v. Parsons, 723 F. Supp. 757 (N.D. Ga. 1989) (imposing liability on parties who accepted hazardous wastes for transport and selected the disposal site); United States v. Bliss, 667 F. Supp. 1298 (E.D. Mo. 1987) (holding liable a defendant who accepted hazardous wastes for transportation and disposal at sites chosen by defendant).

9.4 Defenses to CERCLA Liability

Congress provided for certain limited defenses to CERCLA liability. A PRP may escape liability if able to establish entitlement to one of the following statutory defenses:
1. Act of God.[69]
2. Act of war.[70]
3. Act or omission of a third party.[71]
4. Innocent landowner defense.[72]
5. Security interest exemption.[73]

There is conflicting opinion among the courts on whether equitable defenses, such as waiver, unclean hands, laches, and estoppel, may also be available.(See Section 9.4.5 for discussion of equitable defenses.) Of the enumerated defenses, the act of God and act of war defenses have rarely been used or been successful.[74] PRPs have primarily relied on the third-party defense, innocent landowner defense, and security interest exemption to avoid liability.

9.4.1 Third-Party Defense

The third-party defense is the most heavily litigated of CERCLA's statutory defenses. Section 107(b)(3) of CERCLA provides an affirmative defense for a party who can establish that the offending "release ... of a hazardous substance and the damages resulting therefrom were caused solely by ... an act or omission of a third party," provided that:
1. The third party is not "one whose act or omission occurs in connection with a contractual relationship, existing directly or indirectly, with the defendant,"
2. The defendant "took precautions against foreseeable acts or omissions of any such third party and the consequences that could foreseeably result from such acts or omissions," and
3. The defendant "exercised due care with respect to the hazardous

---

[69] CERCLA § 107(b)(1), 42 U.S.C. § 9607(b)(1).
[70] CERCLA § 107(b)(2), 42 U.S.C. § 9607(b)(2).
[71] CERCLA § 107(b)(3), 42 U.S.C. § 9607(b)(3).
[72] CERCLA § 101(35)(A), 42 U.S.C. § 9601(35)(A).
[73] CERCLA § 101(20)(A), 42 U.S.C. § 9601(20)(A).
[74] See, for example United States v. Barrier Indus., Inc., 991 F. Supp. 678 (S.D.N.Y. 1998) (rejecting defendant's act of God defense); United States v. Alcan Aluminum Corp., 892 F. Supp. 648 (M.D. Pa. 1995) (rejecting act of God defense). But see Wagner Seed Co. v. Daggett, 800 F.2d 310 (2d Cir. 1986) (recognizing act of God defense); United States v. Shell Oil Co., 34 Env't Rep. Cas. (BNA) 1342 (C.D. Cal. 1992) (recognizing act of war defense).

substance concerned, taking into consideration the characteristics of such hazardous substance, in light of all relevant facts and circumstances."[75]

Under the first prong, the defense is only available if a defendant can demonstrate that a unrelated third party is the sole cause of the hazardous substance release at the site and the defendant does not have a direct or indirect contractual relationship with the third party whose act or omission caused the release.[76] A contractual relationship between the landowner and the third party will only bar the landowner from raising the third-party defense if the contract between the landowner and the third party either relates to the hazardous substances or allows the landowner to exert some element of control over the third party's activities.[77] Thus, a real estate contract is not the type of "contractual relationship" that would preclude assertion of CERCLA's third-party defense unless there is proof of a connection between the contract and the act or omission that resulted in the contamination of the site.[78] Similarly, several courts have made clear that a lease is a "contractual relationship," which precludes landlords and tenants from claiming third-party defenses with regard to each other's actions.[79] Similarly, courts have found that a generator's business relationship with a waste transporter prevents successful assertion of the third-party defense even when the generator was unaware of the location to which the transporter was taking its wastes.[80]

Under the second and third prongs of the third-party defense, the defendant must to show by a preponderance of the evidence that he "took precautions against foreseeable acts or omissions" of any third party and exercised "due care with respect to the hazardous substance concerned."[81]

---

[75] CERCLA § 107(b)(3), 42 U.S.C. § 9607(b)(3).
[76] See, for example, Chatham Steel Corp. v. Brown, 858 F. Supp. 1130 (N.D. Fla. 1994); G.J. Leasing Co. v. Union Electric Co., 854 F. Supp. 539 (S.D. Ill. 1994); United States v. Maryland Sand, Gravel and Stone Co., 39 Env't Rep. Cas. (BNA) 1761 (D.Md. 1994).
[77] See, for example, Westwood Pharmaceuticals, Inc. v. National Fuel Gas Distribution Corp., 964 F.2d 85, 91-92 (2d Cir. 1992).
[78] See, for example, American Nat'l Bank & Trust Co. v. Harcros Chems., Inc., 997 F. Supp. 994 (N.D. Ill. 1998).
[79] See, for example, United States v. Monsanto, 858 F.2d 160 (4th Cir. 1988), cert. denied, 490 U.S. 1106 (1989); International Clinical Laboratories, Inc. v. Stevens, 710 F. Supp. 466 (E.D.N.Y. 1989); United States v. Northernaire Plating Co., 670 F. Supp. 742 (W.D. Mich. 1987).
[80] See, for example, United States v. Mottolo, 695 F. Supp. 615 (D.N.H. 1988).
[81] Compare Idylwoods Associates v. Mader Capital, Inc., 915 F. Supp. 1290 (W.D.N.Y. 1996), motion for reconsideration denied, 956 F. Supp. 410 (W.D.N.Y. 1997) (defendants failed to exercise due care with regard to hazardous substances at the site) with State v. Lashins Arcade, 91 F.3d 353 (2d Cir. 1996) (defendant-purchaser of a contaminated site satisfied all requirements of the defense, including "due care" element). See also Kerr-McGee Chem. Corp. v. Lefton Iron & Metal Co., 14 F.3d 321 (7th Cir. 1994); United States v.

In one case, for example, *Idylwoods Associates v. Mader Capital, Inc.*,[82] the owners of a contaminated site were precluded from asserting CERCLA's third-party defense because they failed to prove that they exercised due care in regard to the hazardous substances at the site. In an earlier action, the federal district court ruled that the defendants were not entitled to the third-party defense;[83] however, the defendants filed a motion for reconsideration of the defense based on newly discovered evidence and the subsequent decision of the Second Circuit in *State v. Lashins Arcade*,[84] concerning the "due care" element of the third-party defense. The defendants contended that the *Lashins* decision required that summary judgment on the third-party defense be entered in their favor on the basis that the due care inquiry of the third-party defense focuses on all relevant facts and circumstances of the case at hand, and that whatever the facts and circumstances, the affirmative defense cannot be nullified by requiring the defendant to pay some of the response costs to prove that the defendant took affirmative steps necessary to establish the third-party defense. The court held that the evidence clearly showed that the defendants failed to exercise the requisite "due care" necessary to maintain the third-party defense. Noting that, under the facts present in *Lashins*, the Second Circuit stated that a party is not required to pay remediation costs in order to prove that it took affirmative steps to address a hazardous waste problem after receiving notice of the contamination, the court found the facts in *Lashins* to be distinguishable from the instant case. The court reasoned that in *Lashins*, the defendant, upon receiving notice that the groundwater around his property was contaminated, continued to maintain a water filter installed as part of the remediation of the land, and began to monitor his property, including inspecting the tenants' stores, and incorporating terms into the lease agreements with each tenant to ensure that no dumping of hazardous waste took place. Such actions represented affirmative action taken to prevent further contamination of the property. In contrast, in the instant case, although it was clear that the New York Central Railroad engaged in dumping activities at the site prior to the defendants' ownership of the property, upon the defendants learning of the hazardous waste problem, rather than taking affirmative steps to prevent continued contamination of the site, they instead attempted to distance themselves from the property, going so far as to cease paying property taxes on the site, in the hope that town and county officials would foreclose on the

---

A & N Cleaners & Launderers, Inc., 854 F. Supp. 229 (S.D.N.Y. 1994).
[82] Idylwoods Assocs. v. Mader Capital, Inc., 956 F. Supp. 410 (W.D.N.Y. 1997).
[83] Idylwoods Assocs. v. Mader Capital, Inc., 915 F. Supp. 1290 (W.D.N.Y. 1996).
[84] State v. Lashins Arcade, 91 F.3d 353 (2d Cir. 1996).

property and take it off their hands. As soon as preliminary investigative efforts by federal and state authorities began, rather than to take affirmative steps to assist the investigation and cleanup, the defendants took affirmative steps to take themselves out of the picture, and to distance themselves from any potential CERCLA liability. Moreover, during this period of uncooperation and inactivity, the contamination spread to a neighboring creek so that the remediation costs in 1993, when the defendants were brought into this action, were significantly greater than in 1984, when the investigation began. In addition, the court found that the new evidence set forth by the defendants was irrelevant to the issue of due care. That evidence, indicating that the defendants placed telegraph poles at the northwest corner of the property in an attempt to stop unauthorized dumping, did not address the defendants' actions with respect to the discovery of contamination at the property.

9.4.2 Innocent Landowner Defense

The "innocent landowner" defense was added to CERCLA in 1986 with passage of the Superfund Amendments and Reauthorization Act (SARA).[85] The defense was expected to moderate CERCLA liability by excluding from the group of potentially responsible parties those "innocent" landowners who:
1. Did not know that the property was contaminated at the time of purchase;
2. Reacted responsibly to the contamination when found; and
3. Made reasonable inquiries into the past uses of the property before purchase to determine whether the property was contaminated.

The innocent landowner defense provides that a defendant may avoid liability by establishing that property was acquired by the defendant after the disposal or placement of the hazardous substances on the property. To be entitled to the defense, the new property owner must, at the time of purchase, make appropriate inquiry into the previous ownership and uses of the property and take steps to minimize liability consistent with good commercial or customary practice.[86] In most cases, PRPs have had difficulty establishing their entitlement to the innocent purchaser defense.[87]

---

[85] CERCLA § 101(35)(A), 42 U.S.C. § 9601(35)(A).
[86] See Cross, Establishing Environmental Innocence, 23 Real Est. L.J. 332 (Spring 1995).
[87] See, for example, Kerr-McGee Chemical Corp. v. Lefton Iron & Metal Co., 14 F.3d 321 (7th Cir. 1994); Acme Printing Ink Co. v. Menard, Inc., 870 F. Supp. 1465 (E.D. Wis. 1994); In re Hemingway Transp. 993 F.2d 915 (1st Cir.), cert. denied, 114 S.Ct. 303 (U.S. 1993); Washington v. Time Oil Co., 687 F. Supp. 529 (W.D. Wash. 1988).

Performance of an environmental site assessment is generally required to satisfy the "appropriate inquiry" requirement of the innocent landowner defense. Paradoxically, however, a few courts have allowed the innocent landowner defense in cases where the purchaser failed to inspect the property prior to its acquisition.[88]

9.4.3 Security Interest Exemption

Section 101(20)(A) of CERCLA limits the scope of owner/operator liability under Section 107(a) by exempting from liability a person "who, without participating in the management of a vessel or facility, holds indicia of ownership primarily to protect his security interest."[89] This provision has generally been construed as shielding lenders from liability that might otherwise result from holding a security interest in contaminated property, provided they have not participated in the management of the day-to-day activities of the facility. A lender seeking to invoke the security interest exemption has the burden of establishing its entitlement to the exemption.[90] To sustain this burden, the lender must prove both that (1) it holds indicia of ownership primarily to protect its security interest in the subject property, and (2) it did not participate in the management of the property.[91]

Even with the security interest exemption, lenders have remained at risk for cleanup liability. The limited exemption has provided little assurance that a lender will not find itself party to a costly CERCLA litigation.[92] If the lender forecloses on contaminated property and takes title, it may become liable as an "owner" of the property.[93] Further, if the lender holds a mortgage on contaminated property and is actively involved in the

---

[88] See, for example, United States v. Pacific Hide & Fur Depot, Inc., 716 F. Supp. 1341 (D. Idaho 1989); United States v. Serafini, 706 F. Supp. 346 (M.D. Pa. 1988).
[89] CERCLA § 101(20)(A), 42 U.S.C. § 9601(20)(A).
[90] See, for example, United States v. Fleet Factors Corp., 901 F.2d 1550 (11th Cir. 1990); United States v. Fleet Factors Corp., 821 F. Supp. 707 (S.D. Ga. 1993); United States v. Maryland Bank & Trust Co., 632 F. Supp. 573 (D.Md. 1986); United States v. Wallace, 893 F. Supp. 627 (N.D.Tex. 1995).
[91] See, for example, Kemp Indus. v. Safety Light Corp., 857 F. Supp. 373 (D.N.J. 1994).
[92] See Mukatis & Nielsen, Real Estate Lending Activities of Commercial Banks Under Superfund, 24 Real Est. L.J. 358 (1996).
[93] See section 9.6.1 for full discussion of the potential liability of a lender as an "owner" under CERCLA.

operation or management of the property, it may be found liable as an "operator."[94]

9.4.4 Equitable Defenses

The EPA takes the position that the only defenses to CERCLA liability are those listed in the statute.[95] In CERCLA actions brought by the government, most courts have agreed with the EPA's view,[96] although a few courts have indicated a willingness to recognize such defenses under limited circumstances.[97] Still, even when the court recognizes the availability of equitable defenses, they have proven difficult to establish. For example, in *United States v. Mottolo,*[98] the government allegedly obtained the PRP's consent to enter the cleanup site by representing that it would not seek to recover its response costs. The court found that because the PRP could not otherwise have prevented the government's entry, it could not have detrimentally relied on the government's representations. Thus, the court held that the government was not equitably estopped from pursuing a cost recovery action.

Similarly, most courts have rejected equitable defenses in private cost recovery and contribution actions.[99] The courts have generally held that equitable defenses to CERCLA liability are precluded by the exclusive affirmative defenses set forth in Section 107(b).[100] Still, a few courts have ruled that equitable defenses may be available in private cost recovery

---

[94] See section 9.6.2 for full discussion of the potential liability of a lender as an "operator" under CERCLA.

[95] See, for example, United States v. Smuggler-Durant Mining Corp., 823 F. Supp. 873 (D. Colo. 1993); United States v. Hardage, 26 Env't Rep. Cas. (BNA) 1049 (W.D. Okla. 1987).

[96] See, for example, United States v. Davis, 794 F. Supp. 67 (D.R.I. 1992); United States v. Western Processing Co., 734 F. Supp. 930 (W.D.Wash. 1990).

[97] See, for example, United States v. Martell, 844 F. Supp. 454 (N.D. Ind. 1994) (the court denied the government's motion to strike equitable defenses of waiver, estoppel, laches, and unclean hands, finding that although equitable defenses are difficult to assert against the government when it acts in its sovereign capacity to protect the public welfare, such defenses are not precluded as a matter of law.)

[98] United States v. Mottolo, 695 F. Supp. 615 (D.N.H. 1988).

[99] See, for example, Atlantic Richfield Co. v. Blosenski, 847 F. Supp. 1261 (E.D. Pa. 1994) (court rejected defenses of waiver, estoppel, laches); Velsicol Chemical Corp. v. Enenco, Inc., 9 F.3d 524 (6th Cir. 1993) (laches not available as a defense in a Section 107 action); Amcast Industrial Corp. v. Detrex Corp., 779 F. Supp. 1519 (N.D. Ind. 1991) (court rejected a clean hands defense as a bar to CERCLA liability).

[100] See, for example, Versatile Metals, Inc. v. Union Corp., 693 F. Supp. 1563 (E.D. Pa. 1988).

actions under CERCLA Section 107(a) or in contribution actions under CERCLA Section 113(f).[101]

Although most courts hold that CERCLA does not allow equitable defenses in private cost recovery and contribution actions, they often explain that the statute does allow the courts to give consideration to equitable factors in apportioning costs between various responsible parties.[102] For example, the defense of unclean hands has been allowed when apportioning responsibility for CERCLA response costs.[103] Likewise, the defense of laches has been held available in CERCLA actions between private parties.[104]

9.5 Corporate Liability under CERCLA

A sizable body of case law has interpreted the scope of CERCLA liability to include corporations, including parent corporations, successor corporations, dissolved corporations, corporate officers who were active in site operations, and active shareholders.

9.5.1 Parent-Subsidiary Corporation Liability

Actions that support a finding of "owner" or "operator" status will be sufficient to impose CERCLA liability on corporations. CERCLA Section 107(a)(2), imposes liability on any "person" who owned or operated a facility at the time of disposal of hazardous waste. The term "person" is defined in the statute to include "corporations."[105]

Where a subsidiary corporation is deemed a potentially responsible party (PRP) under CERCLA Section 107(a), the parent corporation is likely to be sued as a defendant by the EPA or a private party on the basis that the

---

[101] See, for example, Thaler v. PRB Metal Prods., Inc., 815 F. Supp. 99 (E.D.N.Y. 1993); United States v. Conservation Chemical, 619 F. Supp. 162 (W.D. Mo. 1985).

[102] See, for example, Town of Munster v. Sherwin-Williams Co., 27 F.3d 1268 (7th Cir. 1994); Brookfield-North Riverside Water Comm'n v. Martin Oil Marketing, Ltd., 1992 WL 63273 (N.D. Ill. Mar. 12, 1992).

[103] See, for example, Thaler v. PRB Metal Products, Inc. 815 F. Supp. 99 (E.D.N.Y. 1993); Chesapeake & Potomac Tel. Co. v. Peck Iron & Metal Co., 814 F. Supp. 1285, opinion clarified, 822 F. Supp. 322 (E.D. Va. 1993).

[104] See, for example, Merry v. Westinghouse Electric Corp., 684 F. Supp. 852 (M.D. Pa. 1988) (but concluding that a "worst-case" of three years before filing suit was not barred by laches).

[105] CERCLA § 101(21), 42 U.S.C. § 9601(21). See also Kleen Laundry & Dry Cleaning Services v. Total Waste Management, 867 F. Supp. 1136 (D.N.H. 1994).

parent corporation controls the actions of its subsidiary. Ordinarily, in order for a corporate parent to be held liable for the acts of its subsidiary, the plaintiff must establish circumstances that require the court to pierce the parent's corporate veil.[106] However, in the context of CERCLA liability of parent corporations, many courts have dispensed with the corporate veil-piercing requirement altogether.[107] On the other hand, a few courts continue to apply the traditional veil piercing requirement from general corporation law to CERCLA litigation involving the parent-subsidiary relationship.[108]

Significantly, in 1997, the U.S. Supreme Court endorsed the minority view requiring corporate veil-piercing in the context of "operator" liability.[109] The U.S. Supreme Court granted *certiorari* in a Sixth Circuit case to resolve a split of opinion among the federal circuits on the issue of whether the corporate veil must be pierced before CERCLA liability may be imposed on parent corporations for acts of their subsidiaries. In the Sixth Circuit case, *United States v. Cordova Chemical Co.,*[110] the court held that when a parent corporation is sued for the cleanup of contamination resulting from the acts of a subsidiary that owns the site, the parent corporation can only be held liable as an "operator" under CERCLA, based on its control over the subsidiary, if state law requirements for piercing the corporate veil have been met. The court stated that it was not persuaded that Congress, in enacting CERCLA, intended to abandon the traditional concepts of limited liability associated with the corporate form. The court ruled that whether the parent will be liable as an operator depends upon whether the degree to which it controls its subsidiary and the extent and manner of its involvement with the

---

[106] Although corporate veil-piercing is usually done to impose liability on the parent for acts of a subsidiary, the corporate veil may likewise be pierced to hold a subsidiary liable for its parent's actions as well. See, for example, Chrysler Corp. v. Ford Motor Co., 972 F. Supp. 1097 (E.D. Mich. 1997) (evidence of alleged functional integration of subsidiary and parent was insufficient to pierce corporate veil and hold subsidiary liable for parent's pollution as alter-ego of parent).

[107] See, for example, John S. Boyd Co. v. Boston Gas Co., 992 F.2d 401 (1st Cir. 1993); United States v. Northeastern Pharmaceutical & Chem. Co., 810 F.2d 726 (8th Cir. 1986); Quadion Corp. v. Mache, 738 F. Supp. 270 (N.D. Ill. 1990); Kelley v. Thomas Solvent Co., 727 F. Supp. 1554 (W.D. Mich. 1989); Vermont v. Staco, Inc., 684 F. Supp. 822 (D. Vt. 1988).

[108] See, for example, Joslyn Corp. v. T.L. James & Co. 696 F. Supp. 222 (W.D. La. 1988), aff'd, 893 F.2d 80 (5th Cir. 1990); United States v. Cordova Chemical Co., 113 F.3d 572 (6th Cir. 1997).

[109] United States v. Bestfoods, 524 U.S. 51, 118 S.Ct. 1876 (1998), *vacating and remanding* United States v. Cordova Chemical Co., 113 F.3d 572 (6th Cir. 1997).

[110] United States v. Cordova Chemical Co., 113 F.3d 572 (6th Cir. 1997).

facility, amount to an abuse of the corporate form that will warrant piercing the corporate veil and disregarding the separate corporate entities of the parent and subsidiary. The court found that officials of the parent corporation participated on the subsidiary's board of directors, were involved in the subsidiary's decisionmaking and daily operations; actively participated in environmental matters; and exerted financial control over the subsidiary through approval of budgets and capital expenditures. However, the court concluded that "[w]hile these factors reveal a parent that took an active interest in the affairs of its subsidiary, they do not indicate such a degree of control that the separate personalities of the two corporations ceased to exist and that [the parent] utilized the corporate form to perpetrate the kind of fraud or other culpable conduct required before a court can pierce the veil."[111]

Notably, the Supreme Court essentially adopted the minority view espoused by the Sixth Circuit by ruling that a parent corporation may be charged with derivative liability for its subsidiary's pollution under CERCLA only when principles of common law permit piercing of the corporate veil.[112] The Court stated that CERCLA's failure to speak to the liability implications of corporate ownership demands application of the rule that, to abrogate common-law principles, a statute must speak directly to the issue. However, the Court continued, a corporate parent that actively participates in or exercises control over a polluting facility owned by its subsidiary may be held directly liable for its own actions as an "operator" under Section 107(a)(2) of CERCLA. Operator liability may attach if the parent manages, directs, or conducts operations specifically related to the leakage or disposal of hazardous waste, or makes decisions about compliance with environmental regulations. Thus, a parent's direct liability is not limited to those situations in which it actually operates the facility in the stead of its subsidiary or participates in its operation as a joint venturer.

In a subsequent decision of the Sixth Circuit, the court ruled that the Supreme Court's reasoning, although applied in a case of operator liability, must logically be extended to cases involving arranger liability since both are categories of PRPs under CERCLA. Thus, in a case involving the liability of a corporation and corporate officer as arrangers of the disposal of PCB-contaminated transformers at a site, the court held that the corporate officer could be held liable as an arranger for disposal due to his status as the sole shareholder of the corporation if Ohio law would allow the piercing of the corporate veil, and he could also be held liable in his own right due to his intimate participation in the arrangement

---

[111] Id., 113 F.3d at 581.
[112] United States v. Bestfoods, 524 U.S. 51 (1998), vacating and remanding United States v. Cordova Chemical Co., 113 F.3d 572 (6th Cir. 1997).

for disposal.[113] The court stated that "[h]e may not hide behind his officer or employee status in [the corporation] to claim that because he took all actions on behalf of the company he cannot be personally liable."[114]

9.5.2 Parent's Degree of Control over Subsidiary

In the context of the parent-subsidiary relationship, the courts have struggled in particular over the appropriate standard to apply in determining "operator" liability. The primary focus of this determination has been on the degree of control the parent exercises over the subsidiary's business affairs and waste-handling practices. There is considerable discrepancy among the courts as to how much actual authority must be shown before liability arises. Some courts require that the parent exert "actual" and pervasive control over the subsidiary to the extent that it actually involves itself in daily operations of the subsidiary.[115] Other courts employ a similar but less exacting standard, whereby a parent corporation could be held liable as an "operator" of its subsidiary under CERCLA if the parent corporation was "actively involved" in the subsidiary's activities.[116] A minority of courts apply the least stringent test dependent on whether the parent corporation had the "authority to control" the subsidiary's waste handling practices.[117] Whichever standard a court has chosen to apply, the determination whether the parent has exerted sufficient control over the subsidiary, such that liability should attach, is essentially performed on a case-by-case basis.

9.5.3 Successor Corporation Liability

Successor corporations have frequently found themselves entangled in CERCLA litigation over their liability as owners and operators of polluted

---

[113]   Carter-Jones Lumber Co. v. Dixie Distributing Co., 166 F.3d 840 (6th Cir. 1999).
[114]   Id., 166 F.3d at 846. See also Silecchia, Pinning the Blame and Piercing the Veil in the Mists of Metaphor: The Supreme Court's New Standards for CERCLA Liability of Parent Companies and a Proposal for Legislative Reform, 67 Fordham L. Rev. 115 (1998).
[115]   See, for example, Jacksonville Elec. Auth. v. Bernuth Corp., 996 F.2d 1107, 1110 (11th Cir. 1993).
[116]   See, for example, John S. Boyd Co. v. Boston Gas Co., 992 F.2d 401, 408 (1st Cir. 1993); CPC International, Inc. v. Aerojet-General, Inc., 777 F. Supp. 549, 575 (W.D. Mich. 1991).
[117]   See, for example, United States v. Northeastern Pharmaceutical & Chemical Co., 810 F.2d 726, 743 (8th Cir. 1986), cert. denied, 484 U.S. 848 (1987); Idaho v. Bunker Hill Co., 635 F. Supp. 665 (D. Idaho 1986).

sites they acquired from their corporate predecessors. For the most part, the courts have taken a very expansive approach to extending CERCLA liability to successor corporations. Although CERCLA is silent on the liability of successor corporations, most courts, in looking to federal common law to cure this vagueness in the statute, have concluded that successor corporations may be held liable for CERCLA cleanup and response costs.[118] Cases have reflected judicial concern that successor entities might be in the position to avoid CERCLA liability for contamination caused by predecessor companies.[119]

In the context of successor liability, courts have repeatedly examined the remedial purpose and public policy goals behind the enactment of CERCLA to find that Congress intended for successor corporations to be held liable for the improper waste handling and disposal practices of their predecessor companies. As stated by one federal district court:[120]

"Congressional intent supports the conclusion that, when choosing between the taxpayers or a successor corporation, the successor should bear the cost. Benefits from the use of the pollutant as well as savings resulting from the failure to use non-hazardous disposal methods inured to the original corporation, its successors and their respective stockholders and accrued only indirectly, if at all, to the general public. We believe it in line with the thrust of the legislation to permit - if not require - successor liability under traditional concepts."[121]

In line with the statute's public policy directives, some courts have reasoned that successor liability should be imposed under CERCLA to prevent the imposition of cleanup costs on the general public, by holding liable those responsible for creating or continuing the hazardous condition, since they benefited from such action. Offering further support for this view, one court declared that "[i]n the absence of successor liability, the government may find itself without any practical recourse against polluters where, as here, the predecessor corporation is long disbanded, its assets long disbursed, and its shareholders difficult if not impossible to locate

---

[118] See, for example, John S. Boyd Co. v. Boston Gas Co., 992 F.2d 401 (1st Cir. 1993); Anspec Co. v. Johnson Controls, 922 F.2d 1240 (6th Cir. 1991); HRW Systems, Inc. v. Washington Gas Light Co., 823 F. Supp. 318 (D. Md. 1993); Kleen Laundry & Dry Cleaning Services v. Total Waste Management, 867 F. Supp. 1136 (D.N.H. 1994).
[119] See, for example, United States v. Carolina Transformer Company, 978 F.2d 832 (4th Cir. 1992); United States v. Lang, 864 F. Supp. 610 (E.D. Tex. 1994).
[120] Chicago Cutlery, Inc. v. Hurlin, 1994 WL 605739, unreported (D.N.H. Oct. 31, 1994).
[121] Id. at *2-3, citing Smith Land and Improvement Corp. v. Celotex Corp., 851 F.2d 86, 91-92 (3d Cir. 1988).

should they be held personally liable in any way."[122]

The settled rule is that a corporation which acquires the assets of another corporation does not take the liabilities of the predecessor corporation from which the assets are acquired unless one of four generally recognized exceptions are met:[123]

1. The successor expressly or impliedly agrees to assume the liabilities of the predecessor.
2. The transaction may be considered a de facto merger.
3. The successor may be considered a "mere continuation" of the predecessor.
4. The transaction is fraudulent.

Most of the litigation concerning the CERCLA liability of successor corporations has focused on the "de facto merger" and "mere continuation" elements this general rule.

De facto merger exception. When a party alleges that a de facto merger has occurred, the court must focus on the substance of the agreement, not on the name the parties have attached to it.[124] A de facto merger exists where one corporation is absorbed by another but without compliance with the statutory requirements for a merger.[125] This type of merger makes the surviving corporation liable for the claims against the predecessor.[126] The court may hold the surviving corporation liable for the conduct of the transferor corporation if the parties have achieved "virtually all the results of a merger," even if they have not observed the statutory requirements of a de jure merger.[127]

The de facto merger doctrine is essentially a judge-made rule that rests on equitable principles. The courts have examined several factors when considering whether a de facto merger occurred, including whether:

1. There is a continuation of the enterprise of the predecessor in terms of continuity of management, personnel, physical location, assets and operations.

---

[122] In Re Acushnet River & New Bedford Harbor, 712 F. Supp. 1010, 1014 (D. Mass. 1989).

[123] See United States v. Carolina Transformer Co., 978 F.2d 832, 837 (4th Cir. 1992).

[124] See, for example, In re Acushnet River and New Bedford Harbor: Proceedings re Alleged PCB Pollution, 712 F. Supp. 1010, 1015 (D.Mass. 1989).

[125] See, for example, Allied Corp. v. Acme Solvents Reclaiming, Inc., 812 F. Supp. 124, 128 (N.D.Ill. 1993).

[126] See, for example, Kleen Laundry & Dry Cleaning Services, Inc. v. Total Waste Management, Inc., 867 F. Supp. 1136 (D.N.H. 1994).

[127] See In re Acushnet River and New Bedford Harbor: Proceedings re Alleged PCB Pollution, 712 F. Supp. 1010, 1015 (D.Mass. 1989).

2. There is a continuity of shareholders which results from the purchasing corporation paying for the acquired assets with its own shares of stock, this stock thereby coming to be held by the shareholders of the selling corporation so that they become a constituent part of the purchasing corporation.

3. The seller ceases operations, liquidates, and dissolves as soon as legally and practically possible.

4. The purchasing corporation assumes the obligations of the seller necessary for uninterrupted continuation of business operations.[128]

These factors are, however, only guiding principles; while all of these factors favor the finding of a de facto merger, "no one of these factors is either necessary or sufficient to establish a de facto merger."[129] One court limited the de facto merger exception to situations where the asset purchaser possessed knowledge of the potential liability and responsibility for such liability.[130]

"Mere continuation" exception. Under the traditional application of the "mere continuation" exception, the court should not find a corporation to be the continuation of a predecessor unless only one corporation remains after the transfer of assets and unless there is an identity of stock, stockholders, and directors between the two corporations.[131] Some courts have found such a continuation of the predecessor's enterprise as to impose successor liability.[132] Other courts have refused to hold successor corporations liable under the mere continuation exception where there was no overlap of stock ownership between the parent and successor corporations.[133]

---

[128] See, for example, Gould, Inc. v. Alter Metal Co., 39 Env't Rep. Cas (BNA) 1669 (N.D. Ill., Aug. 1, 1994) (evidence of de facto merger lacking).

[129] See, for example, Allied Corp. v. Acme Solvents Reclaiming, Inc., 812 F. Supp. 124, 127 (N.D.Ill. 1993); In re Acushnet River and New Bedford Harbor: Proceedings re Alleged PCB Pollution, 712 F. Supp. 1010, 1015 (D.Mass. 1989).

[130] Allied Corp. v. Acme Solvents Reclaiming, Inc., 812 F. Supp. 124 (ND Ill. 1993).

[131] See, for example, Gould, Inc. v. Alter Metal Co., 39 Env't Rep. Cas (BNA) 1669 (N.D. Ill., Aug. 1, 1994).

[132] See, for example, North Shore Gas Co. v. Salomon, Inc., 152 F.3d 642 (7th Cir. 1998) (utility company that purchased assets from sister company could be held liable as successor under mere continuation exception if sister company found liable on remand); HRW Systems, Inc. v. Washington Gas Light Co., 823 F. Supp. 318 (D. Md. 1993) (successor liable under mere continuation exception).

[133] See, for example, United States v. Carolina Transformer Company, 978 F.2d 832, 837 (4th Cir. 1992) (but employing the "substantial continuation" test to hold a successor liable).

"Continuity of enterprise" theory. When determining the CERCLA liability of successor corporations, some courts have applied an expanded version of the traditional "mere continuation" test, often called the "continuity of enterprise" or "substantial continuation" theory.[134] Under this exception to successor liability, a successor will be deemed to have assumed its predecessor's CERCLA liability if its activities constitute a substantial continuation of the predecessor's activities, whether as the result of a formal or merely de facto merger.[135] Courts have deemed use of this broadened test of successorship appropriate in situations where public policy dictates that traditional notions of successor liability should be overridden, such as in the context of environmental liability.[136] A number of courts, however, refuse to recognize the continuity of enterprise theory of successor liability.[137]

---

[134] See, for example, United States v. Carolina Transformer Co., 978 F.2d 832 (4th Cir. 1992); United States v. Mexico Feed & Seed Co., 980 F.2d 478 (8th Cir. 1992); Blackstone Valley Elec. Co. v. Stone & Webster, Inc., 867 F. Supp. 73 (D. Mass. 1994); Atlantic Richfield Co. v. Blosenski, 847 F. Supp. 1261 (E.D. Pa. 1994); Northwestern Mut. Life Ins. Co. v. Atlantic Research Corp., 847 F. Supp. 389 (E.D. Va. 1994); Hunt's Generator Comm. v. Babcock & Wilcox Co., 863 F. Supp. 879 (E.D. Wis. 1994).

[135] See United States v. Carolina Transformer Co., 739 F. Supp. 1030 (E.D.N.C. 1989) (calling for a uniform federal test for determining successor liability and employing the "substantial continuity" test to hold a successor liable), aff'd, 978 F.2d 832 (4th Cir. 1992).

[136] See Hunt's Generator Comm. v. Babcock & Wilcox Co., 863 F. Supp. 879 (E.D. Wis. 1994) (successor liability is justified by a showing that in substance, if not in form, the successor, not the public, is the one who should bear the burden of the cleanup). See also Schnapf, CERCLA and the Substantial Continuity Test: A Unifying Proposal for Imposing CERCLA Liability on Asset Purchasers, 4 Envtl. Law. 435 (1998).

[137] See, for example, City Management Corp. v. U.S. Chemical Co., Inc., 43 F.3d 244 (6th Cir. 1994) (expressly disapproving earlier district court decisions of the Sixth Circuit which had adopted the continuity of enterprise theory); Grand Labs., Inc. v. Midcon Labs of Iowa, 32 F.3d 1277, 1283 (8th Cir. 1994) (refusing to apply the "continuity of enterprise exception" and stating that in those jurisdictions where the exception is recognized, it "applies only in the products liability context."); Atchison, Topeka & Santa Fe Ry. Co. v. Brown & Bryant, Inc., 132 F.3d 1295, 1301-02 (9th Cir. 1997) (applying California law, the court concluded that the purchaser of assets from an agricultural chemical company could not be held liable under the continuing business enterprise exception to successor liability, since California, like most states, does not recognize the exception), overruling Louisiana-Pacific Corp. v. ASARCO, Inc., 909 F.2d 1260 (9th Cir. 1990); Sylvester Bros. Dev. Co. v. Burlington N. R.R., 772 F. Supp. 443, 449 (D. Minn. 1990) (declining to adopt the continuing enterprise theory in a CERCLA case).

Courts utilizing this approach have considered a series of factors in determining whether one corporation may be held liable as the successor to another:[138]

1. Retention of the same employees
2. Retention of the same supervisory personnel
3. Retention of the same production facilities in the same location
4. Production of the same product
5. Retention of the same name
6. Continuity of assets
7. Continuity of general business operations
8. Whether the successor holds itself out as the continuation of the previous enterprise.

Further, if the transfer to the new corporation was part of an effort to continue the business of the former corporation, yet avoid its existing or potential state or federal environmental liability, that also should be considered.[139]

Some courts have found successor corporations liable under CERCLA for the actions of their predecessor companies when employing the "continuity of enterprise" theory,[140] while others, under the particular facts, have found insufficient indicia of continuity of enterprise to impose successor liability.[141] Successor liability may be imposed under the continuity of business enterprise exception even where the successor acquired the assets of the former corporation prior to CERCLA's enactment. For example, a corporation that acquired all of a wood treatment facility operator's assets prior to enactment of CERCLA was held liable for environmental cleanup as the operator's successor.[142]

---

[138] See Atlantic Richfield Co. v. Blosenski, 847 F. Supp. 1261, 1284 (E.D. Pa. 1994).
[139] See, for example, United States v. Carolina Transformer Co., 978 F.2d 832 (4th Cir. 1992).
[140] See, for example, Blackstone Valley Elec. Co. v. Stone & Webster, Inc., 867 F. Supp. 73 (D. Mass. 1994); Northwestern Mut. Life Ins. Co. v. Atlantic Research Corp., 847 F. Supp. 389 (E.D. Va. 1994).
[141] See, for example, United States v. Mexico Feed & Seed Co., 980 F.2d 478 (8th Cir. 1992); United States v. Atlas Minerals and Chems., Inc., 824 F. Supp. 46 (E.D. Pa. 1993); Hunt's Generator Comm. v. Babcock & Wilcox Co., 863 F. Supp. 879 (E.D. Wis. 1994).
[142] Aluminum Co. of Am. v. Beazer E., Inc., 124 F.3d 551 (3d Cir. 1997).

9.5.4 Dissolved Corporations

Most courts have ruled that CERCLA preempts state corporation laws regarding a dissolved corporation's capacity to be sued.[143] Dissolved corporations cannot rely their dissolution status under state law to escape CERCLA liability.[144] Courts have reasoned that if corporations were allowed to escape CERCLA liability by simply dissolving before the government brought suit, the overall purpose of CERCLA--to effectuate cleanup of hazardous wastes by placing financial liability on those responsible for creating the harmful conditions--would be thwarted.[145]

Many federal courts have taken the view that CERCLA liability of a dissolved corporation depends on the distinction between "dead" and "dead and buried" corporations[146] One court has expressed this distinction as follows: "[a] 'dead' corporation is one that has dissolved but still holds assets that can be reached by CERCLA. A 'dead and buried' corporation has dissolved and has no assets remaining. It has ceased to exist as a 'person' that can be held liable under CERCLA."[147] Under this approach,

---

[143] See, for example, Barton Solvents, Inc. v. Southwest Petro-Chem, Inc., 836 F. Supp. 757 (D. Kan. 1993); Idylwoods Assocs. v. Mader Capital, Inc., 915 F. Supp. 1290 (W.D.N.Y. 1996); Burlington N. & Santa Fe Ry. Co. v. Consolidated Fibers, Inc., 7 F.Supp.2d 822 (N.D. Tex. 1998); United States v. Sharon Steel Corporation, 681 F. Supp. 1492 (D. Utah 1987). But see Global Landfill Agreement Group v. 280 Development Corp., 992 F. Supp. 692 (D.N.J. 1998) (stating that CERCLA mandates that state law must be utilized in order to determine a party's capacity to be sued and that state capacity statutes are not preempted under CERCLA).

[144] See, for example, State ex rel. Howes v. Peele, 876 F. Supp. 733 (E.D.N.C. 1995).

[145] See, for example, Barton Solvents, Inc. v. Southwest Petro-Chem, Inc., 836 F. Supp. 757 (D. Kan. 1993); Idylwoods Assocs. v. Mader Capital, Inc., 915 F. Supp. 1290 (W.D.N.Y. 1996).

[146] See, for example, Burlington N. & Santa Fe Ry. Co. v. Consolidated Fibers, Inc., 7 F.Supp.2d 822 (N.D. Tex. 1998) (CERCLA's preemption extends only to dead corporations--ones that have lawfully dissolved under state law--and not to dead and buried corporations--ones that have dissolved and distributed all of their assets); Chesapeake & Potomac Tel. Co. v. Peck Iron & Metal Co., 814 F. Supp. 1285, 1291 (E.D. Va. 1993) (stating that "while CERCLA liability can generally be imposed upon dissolved corporations [i.e. 'dead' corporations], dissolved corporations whose assets have been fully distributed [i.e. 'dead and buried' corporations] are beyond the reach of CERCLA."). See also Barton Solvents, Inc. v. Southwest Petro-Chem, Inc., 836 F. Supp. 757 (D. Kan. 1993); Burlington N. & Santa Fe Ry. Co. v. Consolidated Fibers, Inc., 7 F.Supp.2d 822 (N.D. Tex. 1998).

[147] Traverse Bay Area Intermediate School Dist. v. Hitco, Inc., 762 F. Supp. 1298, 1301 (W.D. Mich. 1991).

CERCLA liability can only be imposed on a "dead" corporation whose assets have not yet been fully distributed.[148]

9.5.5 Personal Liability of Corporate Officers, Directors, and Shareholders

Under traditional corporate law principles, the corporate shield may protect corporate officers from personal liability for the acts of the corporation. However, in the context of hazardous substance liability under CERCLA, there has been a general erosion of the protections traditionally afforded by the corporate veil.[149]

In regard to the personal liability of corporate officers, directors, and shareholders, the primary area of their concern is over "operator" liability under Section 107(a)(1)–(2) of CERCLA. Although the courts have generally rejected claims that corporate officers and shareholders should be held personally liable as "owners," they have ruled that CERCLA allows corporate officers and shareholders to be held personally liable if they constitute current or past "operators" within the meaning of the statute.[150]

9.5.6 Nature and Degree of Officer/Shareholder Involvement

Under CERCLA, in order for corporate officers, directors, and shareholders to be held personally accountable for the environmental transgressions of the corporation, there must be a showing that the individual personally participated in the conduct that violated CERCLA.[151] In determining whether personal liability is justified, courts generally consider such factors as:

---

[148] See, for example, Idylwoods Assocs. v. Mader Capital, Inc., 915 F. Supp. 1290 (W.D.N.Y. 1996); Burlington N. & Santa Fe Ry. Co. v. Consolidated Fibers, Inc., 7 F.Supp.2d 822 (N.D. Tex. 1998).
[149] See Oswald and Schipani, CERCLA and the "Erosion" of Traditional Corporate Law Doctrine, 86 Nw. U. L. Rev. 259 (Winter 1992).
[150] See, for example, United States v. Carolina Transformer Co., 978 F.2d 832 (4th Cir. 1992); Sydney S. Arst Co. v. Pipefitters Welfare Education Fund, 25 F.3d 417 (7th Cir. 1994); FMC Corp. v. Aero Indus., Inc., 998 F.2d 2079 (10th Cir. 1993); Truck Components, Inc. v. Beatrice Co., 1994 WL 520939, unreported (N.D. Ill. Sept. 21, 1994).
[151] See, for example, Armotek Indus., Inc. v. Freedman, 790 F. Supp. 383 (D. Conn. 1992) (officer/shareholder not liable); Commonwealth of Massachusetts v. Blackstone Valley Electric Co., 777 F. Supp. 1036 (D. Mass. 1991) (corporate officers and directors not personally liable).

1. The individual's stock ownership in the corporation.
2. The individual's active participation in the management of the corporation.
3. The individual's authority to control the corporation's waste handling practices.

The courts have struggled over the appropriate standard to apply in determining "operator" liability under CERCLA. A discrepancy exists among the courts over the degree of control or management a corporate officer, director, or shareholder must exert over the corporation before being held liable as an "operator."[152]

Authority to control test. Some courts, adopting a minority view, apply a CERCLA liability standard that hinges on whether the person in question had the mere "authority to control" the corporation's handling and disposal of hazardous substances.[153] When determining "operator" liability, courts using the "authority to control" standard have determined that actual control over daily operations is not necessary.[154]

Applying the less rigorous "authority to control" test, one district court held that the evidence was sufficient to withstand a motion to dismiss the government's claim of operator liability against the president of a corporation that polluted the site.[155] The federal government sued a corporation and various of its officers and directors under CERCLA, seeking to recover the costs of cleaning up a hazardous waste site. The president of the corporation filed a motion to dismiss the government's claim of operator liability against him, contending that he was but a "nominal" president of the corporation without sufficient participation in day-to-day operations to be deemed an "operator" under CERCLA. The court denied the motion, finding that the government's complaint set forth sufficient allegations to meet the "authority to control" standard of operator liability. The court stated that, under the governing Fourth Circuit precedent, the relevant inquiry must be whether the president possessed the "authority to control" environmental policy and activity at the

---

[152] See Fry, Liability of Shareholders and Corporate Directors, Officers, and Employees for CERCLA Response Costs, 1 Envtl. Law. 253 (Sept. 1994).
[153] See, for example, United States v. Carolina Transformer Co., 978 F.2d 832, 837 (4th Cir. 1992); United States v. TIC Investment Corp., 866 F. Supp. 1173, 1180 (N.D. Iowa 1994); Robertshaw Controls Co. v. Watts Regulator Co., 807 F. Supp. 144 (D. Me. 1992); Kelley v. Thomas Solvent Co., 727 F. Supp. 1554, 1561-1562 (W.D.Mich. 1989); Chicago Cutlery, Inc. v. Hurlin, 1994 WL 605739, unreported (D.N.H. Oct. 31, 1994); United States v. High Point Chem. Corp., 7 F.Supp.2d 770 (W.D. Va. 1998).
[154] See, for example, Donahey v. Bogle, 987 F.2d 1250 (6th Cir. 1993); Pierson Sand and Gravel, Inc. v. Pierson Twp., 851 F. Supp. 850 (W.D.Mich. 1994).
[155] United States v. High Point Chem. Corp., 7 F.Supp.2d 770 (W.D. Va. 1998).

contaminated site. The court interpreted the authority to control standard to be "rather less rigorous, particularly at the pleadings stage of litigation, than would be the 'actively participate' test" applied in other cases.[156] The court found that the facts, as pleaded, did not rule out operator liability because, during discovery, sufficient evidence could be unearthed that would enable the government to meet the authority to control standard. The court stated that although the defendant asserted that he was but the "nominal" president of the corporation, the complaint specifically alleged that he "participated in day-to-day production decisions at [the site]." Moreover, the court continued, the defendant's alleged power over the administrative and financial divisions of the corporation provided sufficient indicia of an authority to control the hazardous waste storage and disposal processes of the company, "at least indirectly (but no less definitely)."[157] The court further remarked that it "[did] not deem it a necessary predicate condition for the attachment of CERCLA liability that a defendant have 'dirt on his hands' to the point of having personally disposed of hazardous materials."[158]

Actual control or active participation in management test. On the other hand, the majority view has been for courts to ask whether the person in question had "actual control" over or "actively participated" in the company's business operations and/or waste disposal practices. These courts require that the corporate officer or shareholder have some level of personal involvement before CERCLA liability can be imposed.[159] In one case, for example, a federal district court found that a company president's exertion of actual control over a polluting paper mill's operation made him personally liable for cleanup costs as an operator under CERCLA.[160] The court noted that there are two lines of authority addressing the personal liability of corporate officers and directors under CERCLA: one holding that the officer or director need only have the authority to control a company's operations; the other that an individual must not only have the

---

[156] Id., 7 F.Supp.2d at 776.
[157] Id., 7 F.Supp.2d at 778.
[158] Id., 7 F.Supp.2d at 778 n. 7.
[159] See, for example, Lansford-Coaldale Joint Water Auth. v. Tonolli Corp., 4 F.3d 1209 (3d Cir. 1993); Riverside Market Development Corp. v. International Building Prods., Inc., 931 F.2d 327 (5th Cir. 1991); Jacksonville Electric Auth. v. Bernuth Corp. 996 F.2d 1107 (11th Cir. 1993); Levin Metals Corp. v. Parr-Richmond Terminal Co., 781 F. Supp. 1454 (N.D.Cal. 1991); Mathews v. Dow Chem. Co., 947 F. Supp. 1517 (D.Colo. 1996); Marriott Corp. v. Simkins Indus., Inc., 929 F. Supp. 396 (S.D. Fla. 1996); CBS, Inc. v. Henkin, 803 F. Supp. 1426 (N.D.Ind. 1992); Commonwealth of Massachusetts v. Blackstone Valley Elec. Co., 777 F. Supp. 1036 (D. Mass. 1991).
[160] Marriott Corp. v. Simkins Indus., Inc., 929 F. Supp. 396 (S.D. Fla. 1996).

authority to control a company's operations, but must be shown to have actually exercised that authority by participating in the company's operations. The court found that it was unnecessary to decide which line of cases was controlling because it was clear that the company president exercised actual operational control at the polluting facility.

9.6 Lender Liability under CERCLA

Lenders have been particularly concerned about owner/operator liability under CERCLA for cleanup of property held as loan collateral. Even though CERCLA's definition of "owner or operator" exempts persons who, without participating in management of a facility, hold indicia of ownership primarily to protect a security interest,[161] this so-called security interest exemption has not always provided the "safe harbor" that lenders were expecting. Even with the security interest exemption, lenders have remained at risk for cleanup liability, largely because of judicial disagreement over when a lender's actions are "primarily to protect a security interest" and what degree of "participating in management" of the property will forfeit the lender's eligibility for the exemption. After some important CERCLA cases were decided against lenders in the mid-80s, lenders learned that they still might be held liable for cleanup of hazardous substances located on property used as loan collateral if they were to become (1) an "owner" of the property (e.g., through foreclosure) or (2) an "operator" by participating in management of the property (e.g., during a loan workout).[162]

Further, in 1990, the Eleventh Circuit only complicated matters with its controversial decision in *United States v. Fleet Factors Corporation.*[163] In that case, the court stated in *dicta* that a bank could be held liable as an "operator" under CERCLA for costs incurred to cleanup hazardous substances on property held as loan collateral, if the lender participated in the management of the property "to a degree indicating a capacity to influence" the borrower's handling of hazardous substances. The court announced that the lender did not have to actually participate in the borrower's decisions concerning hazardous substances. The mere fact that the lender "could have influenced" such decisions was sufficient to impose liability.

---

[161] CERCLA § 101(20)(A), 42 U.S.C. § 9601(20)(A).

[162] See, for example, United States v. Maryland Bank & Trust Co., 632 F. Supp. 573 (D.Md. 1986); United States v. Mirabile, 15 Envt'l L. Rep. (ELI) 20994 (E.D. Pa. 1985).

[163] United States v. Fleet Factors Corp., 901 F.2d 1550 (11th Cir. 1990), cert. denied, 498 U.S. 1046 (1991).

The breadth of the Eleventh Circuit's "capacity to influence" language concerned all lenders, regardless of the extent of their activities in regard to borrowers' properties. However, no other court adopted this broad interpretation of the participation in management prong of the security interest exemption and the Eleventh Circuit's view was subsequently abrogated by EPA regulations[164] and clarifying amendments to CERCLA.[165]

9.6.1 Lender's Liability as "Owner"

The courts have not been in total agreement as to whether CERCLA's security interest exemption shields a lender from liability as an "owner" when the lender takes title to the borrower's contaminated property by foreclosure. Some courts, underscoring the fact that the language of the exemption is written in the present tense, have reasoned that the security interest must exist at the time of cleanup and, accordingly, have held that secured lenders that had foreclosed on property were not covered by the exemption for the period during which they held title, because the security interest no longer existed.[166] However, most courts have not adopted this narrow reading of the security interest exemption and, looking more to the overall purpose behind the exemption, have more generally noted that exemption is meant to shield from liability those owners who are, in essence, lenders holding title to the property as security for the debt.[167] Under this view, when determining whether a lender holds indicia of ownership primarily to protect a security interest, courts have stated that it is necessary to examine the intent of parties and the purpose of the particular transaction.[168] The mere fact that a lender holds title to the subject property does not, alone, make it an owner of the facility for purposes of CERCLA; rather, under the security interest exemption, it is

---

[164] The EPA expressly repudiated the Eleventh Circuit's view in *Fleet Factors* by establishing a general test for "participation in management" that specifically excludes the mere capacity to influence, or ability to influence, or the unexercised right to control facility operations. 40 CFR § 300.1100(c)(1). See section 9.6.3 for a detailed discussion of these regulations.

[165] See section 9.6.4 for discussion of these amendments.

[166] See, for example, Guidice v. BFG Electroplating & Manufacturing Co., 732 F. Supp. 556 (W.D. Pa. 1989); United States v. Maryland Bank & Trust Co., 632 F. Supp. 573 (D.Md. 1986).

[167] See, for example, Waterville Indus., Inc. v. Finance Auth. of Me., 984 F.2d 549 (1st Cir. 1993).

[168] See, for example, Kemp Indus. v. Safety Light Corp., 857 F. Supp. 373 (D.N.J. 1994).

necessary to determine why the lender holds indicia of ownership.[169] The nature of the title held by the lender is irrelevant to this determination; instead, it is the reason the lender took title that guides the inquiry--such as whether the lender took title "primarily" to protect its security interest or whether the lender was seeking to reap benefits from long-term ownership of the property.[170] Thus, a key factor in the decisions has been whether the lender has made a reasonably prompt effort to divest itself of unwelcome ownership after acquiring an unwanted title.[171] Although the determination of whether a lender's taking of title falls within the security interest exemption depends on the particular facts of each case, when applying the principles noted above, most courts have concluded that lenders were protected from owner liability by the security exemption.[172] Only in a few cases have courts found, under the circumstances, that lenders were not entitled to the security interest exemption from owner liability.[173]

9.6.2 Lender's Liability as "Operator"

If a lender becomes overly entangled in the affairs of the actual owner or operator of a facility, the lender may acquire the status of an "operator"

---

[169] See, for example, United States v. McLamb, 5 F.3d 69 (4th Cir. 1993); In re Bergsoe Metal Corp., 910 F.2d 668 (9th Cir. 1990).

[170] See, for example, United States v. McLamb, 5 F.3d 69 (4th Cir. 1993); Kemp Indus. v. Safety Light Corp., 857 F. Supp. 373 (D.N.J. 1994).

[171] See, for example, Northeast Doran v. Key Bank, 15 F.3d 1 (1st Cir. 1994) (lender protected by security interest exemption where it divested itself of property within six months following foreclosure); United States v. McLamb, 5 F.3d 69 (4th Cir. 1993) (lender protected by security interest exemption where it placed property on market within days of acquiring ownership even though property was sold to first able buyer over six months after lender took title by foreclosure); Ashland Oil, Inc. v. Sonford Prods. Corp., 810 F. Supp. 1057 (D.Minn. 1993) (lender protected by security interest exemption where it foreclosed on security interest and held title for no longer than a month before selling to another buyer).

[172] See, for example, Northeast Doran v. Key Bank, 15 F.3d 1 (1st Cir. 1994); Waterville Indus., Inc. v. Finance Auth. of Me., 984 F.2d 549 (1st Cir. 1993); United States v. McLamb, 5 F.3d 69 (4th Cir. 1993); Kemp Indus. v. Safety Light Corp., 857 F. Supp. 373 (D.N.J. 1994); Snediker Developers Ltd. Partnership v. Evans, 773 F. Supp. 984 (E.D. Mich. 1991).

[173] See, for example, United States v. Fleet Factors Corp., 821 F. Supp. 707 (S.D. Ga. 1993) (Fleet Factors IV); Guidice v. BFG Electroplating & Mfg. Co., 732 F. Supp. 556 (W.D. Pa. 1989); United States v. Maryland Bank & Trust Co., 632 F. Supp. 573 (D.Md. 1986).

under CERCLA and be held liable for cleanup costs.[174] Although the courts have struggled to define the precise parameters of control that a lender may exert over a borrower's affairs before it will be deemed "participating in management" and held liable for cleanup as an operator of the borrower's property, they have enunciated a number of guiding principles to be applied to the particular facts of each case. Generally, absent some actual participation in the management of the borrower's facility, the security interest exemption will shield a lender from liability as an operator.[175] The courts have further qualified that the participation that is critical to determining a lender's entitlement to the security interest exemption is participation in the borrower's operational, production, or waste disposal activities.[176] For example, a number of courts have made clear that participation in purely financial aspects of a borrower's operation is insufficient to bring a lender within the scope of CERCLA liability.[177] Still, although a lender must be permitted to monitor certain aspects of a debtor's business that relate to the protection of its security interest without incurring liability,[178] participation in the day-to-day operational aspects of a site, especially involvement in waste disposal activities and environmental decisionmaking, may cause a lender to lose the protection afforded by the security interest exemption.[179] However, a lender will not lose its exemption status merely by performance of an environmental site assessment, "even one that reveals the existence of

---

[174] See, for example, United States v. Mirabile, 15 Envt'l L. Rep. (ELI) 20994 (E.D. Pa. 1985). See also F.P. Woll & Co. v. Fifth & Mitchell Street Corp., 1997 WL 535936, unreported (E.D. Pa. July 31, 1997) (complaint stated sufficient allegations to raise triable issues of fact as to whether bank that foreclosed upon and took title to contaminated property was liable under CERCLA as an "operator" by participating in management of the property prior to foreclosure).

[175] See, for example, In re Bergsoe Metal Corp., 910 F.2d 668 (9th Cir. 1990).

[176] See, for example, United States v. Mirabile, 15 Envt'l L. Rep. (ELI) 20994 (E.D. Pa. 1985); United States v. Wallace, 893 F. Supp. 627 (N.D. Tex. 1995).

[177] See, for example, United States v. Fleet Factors Corp., 901 F.2d 1550 (11th Cir. 1990); In re Bergsoe Metal Corp., 910 F.2d 668 (9th Cir. 1990); United States v. Fleet Factors Corp., 821 F. Supp. 707 (S.D. Ga. 1993); Kelley ex rel. Michigan Natural Resources Comm'n v. Tiscornia, 810 F. Supp. 901 (W.D. Mich. 1993); Grantors of the Silresim Site Trust v. State Street Bank & Trust Co., 23 Envt'l L. Rep. (ELI) 20428, 1992 WL 494718 (D.Mass. 1992); United States v. Mirabile, 15 Envt'l L. Rep. (ELI) 20994 (E.D. Pa. 1985); United States v. Wallace, 893 F. Supp. 627 (N.D. Tex. 1995).

[178] See, for example, Z & Z Leasing, Inc. v. Graying Reel, Inc., 873 F. Supp. 51 (E.D. Mich. 1995).

[179] See, for example, United States v. Fleet Factors Corp., 821 F. Supp. 707 (S.D. Ga. 1993); United States v. Maryland Bank & Trust Co., 632 F. Supp. 573 (D.Md. 1986); United States v. Mirabile, 15 Envt'l L. Rep. (ELI) 20994 (E.D. Pa. 1985).

possible environmental contamination,"[180] or by requiring the borrower to abide by all applicable environmental laws.[181]

Applying these principles, courts have often found, under the particular facts, that a lender was protected by the security interest exemption and not liable as an operator based on a determination that the lender was not impermissibly involved in environmental decisionmaking at facilities in which they held a security interest,[182] or otherwise impermissibly involved in the operations of such facilities,[183] although in some cases, under the particular facts, courts have reached an opposite determination and concluded that the lender's involvement in facility operations fell outside the scope of the exemption.[184]

9.6.3 EPA's Lender Liability Rule

In direct response to the lending community's concerns about the Eleventh Circuit decision in *Fleet Factors*, the EPA promulgated regulations pursuant to CERCLA, commonly known as the "lender liability rule," to clarify the scope of lender liability under CERCLA and to identify a range of protected activities that lenders and other holders of security interests might take that would be considered consistent with holding indicia of ownership primarily to protect a security interest.[185] After the lender liability rule was issued in 1992, the Chemical

---

[180] Northeast Doran v. Key Bank, 15 F.3d 1, 3 (1st Cir. 1994). See also United States v. McLamb, 5 F.3d 69 (4th Cir. 1993); Waterville Indus., Inc. v. Finance Auth. of Me., 984 F.2d 549 (1st Cir. 1993).
[181] See, for example, Z & Z Leasing, Inc. v. Graying Reel, Inc., 873 F. Supp. 51 (E.D. Mich. 1995).
[182] See, for example, Z & Z Leasing, Inc. v. Graying Reel, Inc., 873 F. Supp. 51 (E.D. Mich. 1995); Grantors of the Silresim Site Trust v. State Street Bank & Trust Co., 23 Envt'l L. Rep. (ELI) 20428, 1992 WL 494718 (D.Mass. 1992); In re River Capital Corp., 155 B.R. 382 (Bankr. E.D. Va. 1991).
[183] See, for example, United States v. McLamb, 5 F.3d 69 (4th Cir. 1993); Z & Z Leasing, Inc. v. Graying Reel, Inc., 873 F. Supp. 51 (E.D. Mich. 1995); Kemp Indus. v. Safety Light Corp., 857 F. Supp. 373 (D.N.J. 1994); Kelley ex rel. Michigan Natural Resources Comm'n v. Tiscornia, 810 F. Supp. 901 (W.D. Mich. 1993); Ashland Oil, Inc. v. Sonford Prods. Corp., 810 F. Supp. 1057 (D.Minn. 1993); Grantors of the Silresim Site Trust v. State Street Bank & Trust Co., 23 Envt'l L. Rep. (ELI) 20428, 1992 WL 494718 (D.Mass. 1992); In re River Capital Corp., 155 B.R. 382 (Bankr. E.D. Va. 1991); Guidice v. BFG Electroplating & Mfg. Co., 732 F. Supp. 556 (W.D. Pa. 1989).
[184] See, for example, United States v. Fleet Factors Corp., 821 F. Supp. 707 (S.D. Ga. 1993); United States v. Maryland Bank & Trust Co., 632 F. Supp. 573 (D.Md. 1986); United States v. Mirabile, 15 Envt'l L. Rep. (ELI) 20994 (E.D. Pa. 1985).
[185] 57 *Fed. Reg.* 18344 (Apr. 29, 1992), codified at 40 CFR § 300.1100 *et seq.*

Manufacturers Association challenged the validity of the rule and, in 1994, the D.C. Circuit invalidated the EPA rule in *Kelley v. EPA*,[186] holding that the agency had exceeded its rulemaking authority under CERCLA. The result of this decision was to force back into the courts the debate concerning the scope of lender liability under CERCLA. However, notwithstanding the decision in *Kelley*, several courts still interpreted CERCLA in a way that was consistent with the invalidated rule, concluding that a lender exerting no unusual control over a borrower could not be held liable for contamination on that borrower's property, even if the lender foreclosed upon or took title to such property.[187] Furthermore, the decision in *Kelley* did not preclude the EPA from applying the rule as agency enforcement guidance.[188] As it turned out, any effects of the D.C. Circuit's invalidation of the lender liability rule were short-lasted because, in 1996, Congress reinstated the EPA's interpretations of the security interest exemption as set forth in the lender liability rule when it amended CERCLA with passage of the Asset Conservation, Lender Liability, and Deposit Insurance Protection Act of 1996.[189] Thus, the guidelines set forth in lender liability rule will be given careful consideration by the courts when called upon to determine whether a lender's actions fall within the scope of the security interest exemption to owner/operator liability under CERCLA.

The EPA rule provides a general standard for judging when a lender's "participation in management" would cause the lender to forfeit its exemption.[190] Under the general management standard, there must be actual participation in the management or operational affairs of a facility by the lender--not the mere capacity to influence, or ability to influence, or the unexercised right to control facility operations.[191] A lender is only considered to be participating in management when the borrower is still

[186] Kelley v. EPA, 15 F.3d 1100 (D.C. Cir. 1994), cert. denied sub. nom. American Bankers Ass'n v. Kelley, 115 S.Ct. 900 (1995).
[187] See, for example, Northeast Doran Inc. v. Key Bank of Me., 15 F.3d 1 (1st Cir. 1994); Z & Z Leasing, Inc. v. Graying Reel, Inc., 873 F. Supp. 51 (E.D. Mich. 1995); Kemp Indus., Inc. v. Safety Light Corp., 857 F. Supp. 373 (D.N.J. 1994); United States v. Pesses, No. 90-654, 1996 U.S. Dist. LEXIS 2597 (W.D. Pa., Feb. 14, 1996).
[188] See CERCLA Enforcement Against Lenders and Government Entities That Acquire Property Involuntarily, 60 *Fed. Reg.* 63517 (Dec. 11, 1995) (policy memorandum stating intention to follow the provisions of the rule as an enforcement policy).
[189] See Humphreys, Environmental Policy Alert: Congress Reinstates EPA's Lender Liability Rule, 44 Fed. Law. 34 (Mar./Apr. 1997). See also Section 9.6.4 for full discussion of the Asset Conservation, Lender Liability, and Deposit Insurance Protection Act of 1996.
[190] 40 CFR § 300.1100(c)(1).
[191] 40 CFR § 300.1100(c)(1).

in possession of the property encumbered by the security interest, if the lender either: (1) exercises decisionmaking control over the borrower's. environmental compliance; or (2) exercises control at a level comparable to that of a manager of the borrower's enterprise, such that the lender has assumed or manifested responsibility for the overall management of the enterprise encompassing the day-to-day decisionmaking of the enterprise with respect to environmental compliance, or all, or substantially all, of the operational aspects of the enterprise other than environmental compliance.[192] Operational aspects of the enterprise include functions such as that of facility or plant manager, operations manager, chief operating officer, or chief executive officer; whereas, financial or administrative aspects include functions such as that of credit manager, accounts payable or receivable manager, personnel manager, controller, chief financial officer, or similar position.[193]

The lender liability rule also specifies particular actions that are not considered participation in management. According to the regulations, a lender can--without incurring liability--undertake preloan investigations;[194] thereafter police the loan by monitoring or inspecting the property, and requiring that the borrower comply with all environmental standards;[195] and, when a loan nears default, engage in workout negotiations and activities, including ensuring that the collateral property does not violate environmental laws.[196] The rule also specifies that a lender does not participate in management by taking any response action under Section 107(d)(1) of CERCLA or under the direction of an on-scene coordinator.[197]

The rule also was designed to protect a secured lender that acquires full title to the collateral property through foreclosure, as long as the lender did not participate in the property's management prior to foreclosure and made certain diligent efforts to divest itself of the property following foreclosure.[198] Indicia of ownership that are held primarily to protect a security interest include legal or equitable title acquired through or incident to foreclosure or its equivalents, with "foreclosure or its equivalents" including: purchase at foreclosure sale; acquisition or assignment of title in lieu of foreclosure; termination of a lease or other repossession; acquisition of a right to title or possession; an agreement in satisfaction of the obligation; or any other formal or informal manner (whether pursuant to law or under warranties, covenants, conditions,

---

[192] 40 CFR § 300.1100(c)(1).
[193] 40 CFR § 300.1100(c)(1)(ii)(B).
[194] 40 CFR § 300.1100(c)(2)(i).
[195] 40 CFR § 300.1100(c)(2)(ii)(A).
[196] 40 CFR § 300.1100(c)(2)(ii)(B).
[197] 40 CFR § 300.1100(c)(2)(iii).
[198] 40 CFR § 300.1100(d).

representations, or promises from the borrower) by which the holder acquires title to or possession of the secured property.[199] However, protection of the security interest exemption is temporary, and the indicia of ownership held after foreclosure continue to be maintained primarily as protection for a security interest only where the holder undertakes to sell, re-lease property held pursuant to a lease financing transaction (whether by a new lease financing transaction or substitution of the lessee), or otherwise divest itself of the property in a reasonably expeditious manner, using whatever commercially reasonable means are relevant or appropriate with respect to the property.[200] The rule specifies post-foreclosure activities that may be undertaken to establish that ownership indicia continue to be held primarily to protect a security interest. A lender that did not participate in management prior to foreclosure or its equivalents, may liquidate, maintain business activities, wind up operations, and take measures to preserve, protect or prepare the secured asset prior to sale or other disposition.[201]

9.6.4 Asset Conservation, Lender Liability, and Deposit Insurance Protection Act of 1996

President Clinton signed the Asset Conservation, Lender Liability, and Deposit Insurance Protection Act of 1996 into law on September 30, 1996, commonly referred to as the Asset Conservation Act or ACA.[202] The ACA amended CERCLA and clarified the application of the security interest exemption to owner/operator liability under CERCLA.[203] Most significantly, the ACA reinstates and effectively codifies the EPA's lender liability rule.[204] The ACA amendments to CERCLA list several "safe harbor" activities for purposes of the security interest exemption, shield fiduciaries from CERCLA liability, extend lender liability protection to underground storage tanks, and provide protection for certain involuntary

---

[199] 40 CFR § 300.1100(d)(1).
[200] 40 CFR § 300.1100(d)(1).
[201] 40 CFR § 300.1100(d)(2).
[202] Pub.L. 104-208, 110 Stat. 3009.
[203] Pub.L. 104-208, Subtitle E, 2501-2505, 110 Stat. 3009, codified as amended at 42 U.S.C. § 9601(20)(A)-(F).
[204] See Kelley ex rel. Michigan Natural Resources Comm'n v. Tiscornia, 104 F.3d 361 (Table), 1996 WL 732323, at *2 (6th Cir. Dec 19, 1996) (unpublished disposition, text available on WESTLAW). The ACA prohibits further judicial review of that rule, but notes that any amendments to the rule would be subject to judicial challenge. See Humphreys, Environmental Policy Alert: Congress Reinstates EPA's Lender Liability Rule, 44 Fed. Law. 34 (Mar./Apr. 1997).

acquisitions of contaminated property by government entities.[205] It is important to note that the ACA applies only to liability under CERCLA, not to state laws or other federal laws, with the exception of RCRA's underground storage tank program. Lenders and fiduciaries--as owners and managers of real estate--remain subject to liability under environmental laws other than CERCLA. The ACA applies to any claim of lender liability that "has not been finally adjudicated" as of September 30, 1996. It specifically applies to the holder of title in lease financings and also applies to certain governmental mortgage lenders and "any other entity that in a bona fide manner buys or sells loans or interests in loans."

There are several clear statements of conditions under which liability will not be imposed. A lender will not incur CERCLA liability as long as it does not "participate in management" of the borrower's property. The ACA defines "participate in management" as "actually participating in the management or operational affairs" of the borrower's enterprise,[206] including exercising "decision-making control" with respect to hazardous substance management and disposal at a facility; exercising overall, day-to-day management of a facility; and exercising all or substantially all of the operational (as opposed to financial) control of the enterprise, other than environmental compliance functions. As to the latter, a lender cannot effectively take over business operations yet try to avoid environmental liability by insulating itself only from the environmental functions.

On the other hand, many types of activities that lenders might engage in are specifically not deemed participation in management, including:[207]

- Providing financial or administrative advice and assistance to the borrower;
- Restructuring a loan agreement;
- Having the mere capacity to influence facility management, or the unexercised right to control operations;
- Policing the terms of the loan contract, including environmental compliance terms; and
- Requiring environmental response actions to be undertaken by the borrower, or directly undertaking such actions.

In addition, actions taken prior to obtaining the security interest in the affected property are irrelevant with regard to the security interest exemption.

---

[205] See Note, Relief From CERCLA's "Rock and a Hard Place": The Asset Conservation, Lender Liability, and Deposit Insurance Protection Act, 3 Envtl. Law. 859 (1997).
[206] 42 U.S.C. § 9601(20)(E), (F)(i).
[207] 42 U.S.C. § 9601(20)(F)(iii).

The ACA also lists specific actions that lenders can take prior to and after foreclosure without losing protection of the security interest exemption, including:
- Selling, leasing or liquidating the facility in question;
- Maintaining business activities or winding up operations at a facility; or
- Undertaking response actions with respect to hazardous substance contamination.

However, these activities are only exempted if the lender is actively attempting to sell, re-lease (in the case of a lease finance transaction), or otherwise divest itself of the facility "at the earliest practicable, commercially reasonable time, on commercially reasonable terms, taking into account market conditions and legal and regulatory requirements."[208]

On June 30, 1997, EPA issued a "Policy on Interpreting CERCLA Provisions Addressing Lenders and Involuntary Acquisitions by Government Entities." The Policy clarifies the circumstances under which the EPA will apply the provisions of the lender liability rule in interpreting the CERCLA security interest exemption as amended by the ACA. The Policy states: "In light of the substantial similarities between CERCLA's amended secured creditor exemption and the CERCLA Lender Liability Rule, where the Rule and its [Federal Register] preamble provide additional clarification of the same or similar terms used in the secured creditor exemption, EPA intends to treat those portions of the Rule and preamble as guidance in intepreting the exemption."

9.7 Private Cleanup Cost Recovery Actions

CERCLA litigation typically commences when the U.S. EPA or state environmental agency sues PRPs to undertake cleanup or reimburse the government for cleanup costs incurred at sites contaminated by hazardous substances.[209] In addition, a private party may undertake cleanup--voluntarily or in response to a government order--and then seek to recover all or a portion of the cleanup and response costs from other responsible parties by bringing a private cost recovery action under Section 107(a) of CERCLA or a contribution action under Section 113(f).[210]

Section 107(a) permits a party to bring a private action against a PRP to recover cleanup and response costs. Section 107(a)(4)(B) specifically makes parties liable under CERCLA for "all costs of removal or remedial

---

[208] 42 U.S.C. § 9601(20)(E)(ii).
[209] See Section 9.8 for discussion of government response actions.
[210] CERCLA § 107(a), 42 U.S.C. § 9607(a); CERCLA § 113(f), 42 U.S.C. § 9613(f).

action incurred by [government entities and] any other necessary costs of response incurred by any other person consistent with the national contingency plan."[211] In addition, Section 113(f)(1) of CERCLA provides that "[a]ny person may seek contribution from any other person who is liable or potentially liable" for response costs under Section 107(a).[212] Together, these two sections permit private parties to recover from PRPs costs incurred in response to the release or threatened release of hazardous substances at a site.

9.7.1 Cost Recovery Versus Contribution Claims

Although CERCLA provides two mechanisms for parties to recover some or all of the costs associated with environmental response and cleanup of contamination at a site, a private party's success in recovering cleanup and response costs can be significantly impacted by the type of claim asserted--whether a cost recovery action under CERCLA Section 107(a) or a contribution action under CERCLA Section 113(f). Many courts have explained that cost recovery and contribution are two separate and distinct types of claims that do not overlap and that are governed by different liability standards.[213] Since liability is joint and several under Section 107(a), the private party may recoup all recoverable costs from any responsible party.[214] By contrast, because liability is merely several under Section 113(f), a PRP may only to recoup an equitable share of cleanup and response costs from other PRPs.[215] In addition, when a private party declines to settle its claims with the government and then sues for contribution from other PRPs, that party may find its equitable share of costs reduced because, under Section 113(f)(2), other parties that settle with the government are given "contribution protection" from claims by nonsettling PRPs.[216] Another disadvantage of having to resort to a contribution action, instead of a private cost recovery action, is that the statute of limitations for contribution claims is only three years, as

---

[211] CERCLA § 107(a)(4)(B), 42 U.S.C. § 9607(a)(4)(B). See also 40 CFR § 300.700(c)(2).

[212] CERCLA § 113(f), 42 U.S.C. § 9613(f).

[213] See, for example, United Technologies Corp. v. Browning-Ferris Indus., Inc., 33 F.3d 96 (1st Cir. 1994), cert. denied, 115 S.Ct. 1176 (1995).

[214] See section 9.2.2 for discussion of joint and several liability under Section 107(a) of CERCLA.

[215] See, for example, Town of New Windsor v. Tesa Truck, Inc. 919 F. Supp. 662 (S.D.N.Y. 1996).

[216] See, for example, Foamseal, Inc. v. Dow Chem. Co., 991 F. Supp. 883 (E.D. Mich. 1998).

opposed to a six-year period for cost recovery claims.[217] Finally, when a private party asserts a contribution claim, the claim is subject to the court's broad discretion to allocate costs among liable parties, using such equitable factors as the court should deem appropriate.[218]

9.7.2 Cost Recovery Claims Limited to "Innocent Parties"

A number of courts have considered the important question whether a PRP is restricted to bringing a contribution claim under Section 113(f) or whether it may also pursue a private cost recovery action under Section 107(a). A growing number of courts have clarified when the two different kinds of legal actions are appropriate. The general consensus is that a claim by one PRP against another PRP is a contribution claim, regardless of label, and is controlled by Section 113(f), not Section 107(a). In fact, the U.S. Court of Appeals for almost every one of the federal circuits has adopted this view.[219] Thus, a PRP cannot take advantage of the joint and several liability of Section 107 to recover all of its response costs from other PRPs, but instead is limited to asserting a contribution claim under Section 113 to recover an equitable share of the costs from the other PRPs.[220] Still, a few courts have adopted a minority view, holding that a

---

[217] See, for example, United Technologies Corp. v. Browning-Ferris Indus., Inc., 33 F.3d 96 (1st Cir. 1994) (PRP's action to recover cleanup costs from other PRPs was a contribution action barred by the three-year statute of limitations).

[218] Section 113(f)(1) of CERCLA states that "in resolving contribution claims, the court may allocate response costs among liable parties using such equitable factors as the court determines are appropriate." See, for example, Acushnet Co. v. Coaters, Inc., 937 F. Supp. 988 (D. Mass. 1996); United States v. Kramer, 19 F.Supp.2d 273 (D.N.J. 1998).

[219] See, for example, United Technologies Corp. v. Browning-Ferris Indus., Inc., 33 F.3d 96 (1st Cir. 1994), cert. denied, 115 S.Ct. 1176 (1995); Bedford Affiliates v. Sills, 156 F.3d 416 (2d Cir. 1998); New Castle County v. Halliburton NUS Corp., 111 F.3d 1116 (3d Cir. 1997); Amoco Oil Co. v. Borden, Inc., 889 F.2d 664 (5th Cir. 1989); Centerior Serv. Co. v. Acme Scrap Iron & Metal Corp., 153 F.3d 344 (6th Cir. 1998); Akzo Coatings. Inc. v. Aigner Corp., 30 F.3d 761 (7th Cir. 1994); Control Data Corp. v. S.C.S.C. Corp., 53 F.3d 930 (8th Cir. 1995); Pinal Creek Group v. Newmont Mining Corp., 118 F.3d 1298 (9th Cir. 1997); United States v. Colorado & Eastern R. Co., 50 F.3d 1530 (10th Cir. 1995); Redwing Carriers, Inc. v. Saraland Apartments, 94 F.3d 1489 (11th Cir. 1996).

[220] See, for example, Pinal Creek Group v. Newmont Mining Corp., 118 F.3d 1298 (9th Cir. 1997); In re Dant & Russell, Inc., 951 F.2d 246 (9th Cir. 1991); Sun Co., Inc. (R&M) v. Browning-Ferris, Inc., 124 F.3d 1187 (10th Cir. 1997); United States v. Vertac Chem. Co., 966 F. Supp. 1491 (E.D. Ark. 1997); Boyce v. Bumb, 944 F. Supp. 807 (N.D. Cal. 1996); Kaufman v. Unisys Corp., 868 F. Supp. 1212 (N.D. Cal. 1994); New Castle County, Rhone-Poulenc, Inc. v.

PRP is not limited to bringing a contribution action to recover CERCLA cleanup costs.[221]

Some courts, in adopting the majority view that a PRP is limited to assertion of a contribution claim, have reasoned that Section 107 and 113 work together--the first section creating the claim for contribution between PRPs, and the second qualifying the nature of that claim.[222] In one case,[223] for example, the Ninth Circuit explained that a claim of contribution is governed by the joint operation of Sections 107 and 113; Section 107(a) governs liability and Section 113(f) creates a mechanism for apportioning that liability among responsible parties. In arriving at this conclusion, the Ninth Circuit found that Section 107 implicitly incorporates a claim for contribution and that CERCLA's legislative history indicates that Congress enacted Section 113(f) to confirm and clarify the existing claim for contribution under Section 107. The court stressed that because a PRP's claim can only be for contribution, the liability of other PRPs cannot be joint and several; rather, their liability is only several and governed by the mechanisms of equitable apportionment that Congress provided in Section 113(f).

Some courts have held that when a PRP has entered into a consent decree to remedy future hazardous releases without formally admitting liability and without having been formally adjudicated liable, the PRP is not prohibited from bringing a Section 107 claim.[224] Likewise, one court allowed a plaintiff to pursue a Section 107 claim where the plaintiff, the current owner of the property, voluntarily cleaned up the property and none of the parties involved were subject to a judgment, consent decree,

---

Halliburton NUS Corp., 903 F. Supp. 771 (D. Del. 1995); Stearns & Foster Bedding Co. v. Franklin Holding Corp., 947 F. Supp. 790 (D.N.J. 1996); Seneca Meadows, Inc. v. ECI Liquidating, Inc., 16 F.Supp.2d 255 (W.D.N.Y. 1998); Hydro-Manufacturing, Inc. v. Kayser-Roth Corp., 903 F. Supp. 273 (D.R.I. 1995); Clear Lake Properties v. Rockwell Int'l Corp., 959 F. Supp. 763 (S.D. Tex. 1997).

[221] See, for example, Pneumo Abex Corp. v. Bessemer and Lake Erie Railroad Co., Inc., 921 F. Supp. 336 (E.D. Va. 1996); Chesapeake & Potomac Tel. Co. v. Peck Iron & Metal Co., 814 F. Supp. 1269 (E.D. Va. 1992).

[222] See, for example, United Technologies Corp. v. Browning-Ferris Industries, Inc., 33 F.3d 96, 102 n. 10 (1st Cir. 1994); New Castle County v. Halliburton NUS Corp., 111 F.3d 1116, 1122 (3d Cir. 1997); Pinal Creek Group v. Newmont Mining Corp., 118 F.3d 1298, 1302 (9th Cir. 1997); Sun Co., Inc. (R&M) v. Browning-Ferris, Inc., 124 F.3d 1187, 1191 (10th Cir. 1997); Boeing Co. v. Cascade Corp., 920 F. Supp. 1121, 1132 (D.Or. 1996); United States v. Bay Area Battery, 895 F. Supp. 1524, 1533 (N.D. Fla. 1995).

[223] Pinal Creek Group v. Newmont Mining Corp., 118 F.3d 1298 (9th Cir. 1997).

[224] See, for example, United States v. SCA Servs. of Ind., Inc., 865 F. Supp. 533, 543 (N.D. Ind. 1994); Laidlaw Waste Systems, Inc. v. Mallinckrodt, Inc., 925 F. Supp. 624, 631 & n. 5 (E.D. Mo. 1996).

or other agreement with the state or federal government regarding liability for site contamination or site cleanup.[225] On the other hand, in a Sixth Circuit case, the court explicitly rejected the "adjudged liable" distinction made by some courts and ruled that a PRP was limited to assertion of contribution claims against other PRPs.[226]

The Seventh Circuit has adopted the view that where a PRP alleges that it bears no responsibility for the contamination at a site where it has incurred response costs, the PRP may bring a direct cost recovery action under Section 107 with an alternative Section 113 claim should the facts later establish that the PRP actually was partially responsible for the contamination.[227] In that case, a landowner filed a cost recovery suit under Sections 107 and 113, alleging that he did not pollute the site in any way. Acting to ensure that volatile organic compounds on his property did not become a threat to health or the environment, the landowner incurred cleanup costs without being subject to an administrative cleanup order from any public authority, such as the state or EPA, and was not the subject of civil actions under either Sections 106 or 107 of CERCLA. Thus, the landowner was a PRP under CERCLA only because of his ownership of the site--ownership allegedly acquired without knowledge of the environmental hazards. Under the approach adopted by the Seventh Circuit, one of two outcomes would follow from the landowner's suit under Section 107(a): either the facts would establish that the landowner was truly blameless, in which case the other PRPs would be entitled to bring a suit under Section 113(f) within three years of the judgment to establish their liability among themselves, or the facts would show that the landowner was also partially responsible, in which case it would not be entitled to recover under its Section 107(a) theory and only the Section 113(f) claim would go forward. The court stated that neither one of those outcomes was inconsistent with the statutory scheme promoting allocation of liability. Other federal district courts have followed this reasoning, allowing a PRP to assert a Section 107 claim, with an alternative claim for contribution under Section 113(f) in the event that the PRP is later found liable.[228] However, in order to proceed in this fashion, some courts have ruled that the PRP's complaint must contain a specific allegation that it did

---

[225] Bethlehem Iron Works, Inc. v. Lewis Indus., Inc., 891 F. Supp. 221 (E.D. Pa. 1995).
[226] Centerior Serv. Co. v. Acme Scrap Iron & Metal Corp., 153 F.3d 344 (6th Cir. 1998).
[227] Rumpke of Indiana, Inc. v. Cummins Engine Co., Inc., 107 F.3d 1235 (7th Cir. 1997).
[228] See, for example, Wolf, Inc. v. L & W Service Center, Inc., 1997 WL 141685, unreported (D. Neb. Mar. 27, 1997).

not pollute the site in any way.[229] Thus, one district court held that a railroad company, as owner of a contaminated railroad yard, was limited to bringing a contribution action against a former lessee because the railroad company failed to allege specifically that it did not contribute to the site contamination.[230] Since such allegations were absent from the complaint, the court dismissed the railroad company's Section 107(a) claim without prejudice, granting it leave to amend the complaint to include the allegation that it did not pollute the site. The court stated that if the railroad company amended its complaint, it could proceed under Section 107; however, as the case progressed, if the facts revealed that it was partially responsible for the pollution, its Section 107 claim would be stricken, and it would have to proceed exclusively with a contribution claim under Section 113(f).

Other courts, however, have ruled that mere allegations by a PRP that it is not responsible for the contamination are insufficient to allow a PRP to proceed with a Section 107 cost recovery claim.[231] These courts hold that a PRP's complaint must set forth allegations sufficient to establish entitlement to one of CERCLA's statutory defenses to liability,[232] such as the third-party[233] or "innocent landowner" defense,[234] before a PRP can go forward with a direct cost recovery claim.

9.7.3 Meaning of "Necessary Costs of Response"

In cost recovery actions, private party plaintiffs must demonstrate that they have incurred "necessary costs of response" that are "consistent with

---

[229] See, for example, Rumpke of Indiana, Inc. v. Cummins Engine Co., Inc., 107 F.3d 1235, 1241 (7th Cir. 1997).

[230] Soo Line R.R. Co. v. Tang Indus., Inc. 998 F. Supp. 889 (N.D. Ill. 1998).

[231] See, for example, Bedford Affiliates v. Sills, 156 F.3d 416 (2d Cir. 1998); Sinclair Oil Corp. v. Dymon, Inc., 988 F. Supp. 1394 (D. Kan. 1997); Lefebvre v. Central Me. Power Co., 7 F.Supp.2d 64 (D. Me. 1998).

[232] See Section 9.4 for discussion of defenses to CERCLA liability.

[233] See, for example, Bedford Affiliates v. Sills, 156 F.3d 416 (2d Cir. 1998) (owner of contaminated property limited to bringing contribution claims against former lessees and a sublessee that operated a dry cleaning business at site because owner clearly had a contractual relationship with the sublessee, which precluded assertion of CERCLA's third-party defense).

[234] See, for example, Sinclair Oil Corp. v. Dymon, Inc., 988 F. Supp. 1394 (D. Kan. 1997) (owner of a contaminated oil refinery site limited to bringing a contribution action against a former lessee because the owner's complaint failed to plead the requisite elements of the innocent landowner defense); Lefebvre v. Central Me. Power Co., 7 F.Supp.2d 64 (D. Me. 1998) (current owner of site contaminated with coal gas waste set forth sufficient allegations for innocent landowner defense so as to permit him to proceed with cost recovery claim under Section 107).

the National Contingency Plan (NCP)" before recovery of such costs will be allowed.[235] There are two sources of limited guidance for determining whether a plaintiff's cleanup and response costs are recoverable. First, the EPA has promulgated regulations--commonly referred to as the National Contingency Plan (NCP)--which specify how response actions must be performed and provide the standard against which those actions are measured for consistency.[236] In addition, the statutory definitions of "removal" and "remedial action" list response actions for which private parties may recover the corresponding costs if the actions are "necessary" and "incurred consistent with the National Contingency Plan (NCP)."[237]

Although Congress and the EPA have given some guidance on the costs that are recoverable, private parties often must resort to the courts for a determination of what constitutes recoverable costs. As construed by the courts, allowable costs of response may include monitoring and investigation costs,[238] prejudgment interest,[239] security and fencing costs,[240] RCRA closure costs,[241] the costs of providing an alternate water supply,[242]

[235] CERCLA § 107(a)(1)-(4)(B), 42 U.S.C. § 9607(a)(1)-(4)(B). See also Licciardi v. Murphy Oil U.S.A., Inc., 111 F.3d 396 (5th Cir. 1997); Gregor v. Industrial Excess Landfill, Inc., 856 F.2d 39 (6th Cir. 1988); Ascon Properties, Inc. v. Mobil Oil Co., 866 F.2d 1149 (9th Cir. 1989); Marriott Corp. v. Simkins Indus., Inc., 929 F. Supp. 396 (S.D. Fla. 1996).
[236] See Section 9.7.4 for full discussion of the National Contingency Plan.
[237] See CERCLA § 101(23), 42 U.S.C. § 9601(23) (definition of "removal"); CERCLA § 101(24), 42 U.S.C. § 9601(24) (definition of "remedial action").
[238] See, for example, Amoco Oil Co. v. Borden, Inc., 889 F.2d 664 (5th Cir. 1989); Donahey v. Bogle, 987 F.2d 1250 (6th Cir. 1993); Soo Line R.R. Co. v. Tang Indus., Inc. 998 F. Supp. 889 (N.D. Ill. 1998); Hatco Corp. v. W.R. Grace & Co.-Conn., 849 F. Supp. 931 (D.N.J. 1994); Gache v. Town of Harrison, 813 F. Supp. 1037 (S.D.N.Y. 1993).
[239] See, for example, United States v. R.W. Meyer, Inc., 889 F.2d 1497 (6th Cir. 1989), cert. denied, 110 S.Ct. 1527 (1990); Bancamerica Commercial Corp. v. Mosher Steel of Kansas, Inc., 100 F.3d 692 (10th Cir. 1996); Colorado v. United States, 867 F. Supp. 948 (D. Colo. 1994). But see United States v. Ottati & Goss, Inc., 900 F.2d 429 (1st Cir. 1990).
[240] See, for example, Cadillac Fairview/California v. Dow Chemical Co., 840 F.2d 691 (9th Cir. 1988); Amland Properties Corp. v. Aluminum Co. of Am., 711 F. Supp. 784 (D.N.J. 1989).
[241] See, for example, Mardan Corp. v. C.G.C. Music, Ltd., 600 F. Supp. 1049 (D.Ariz. 1984), aff'd on other grounds, 804 F.2d 1454 (9th Cir. 1986); Chemical Waste Management, Inc. v. Armstrong World Indus., 669 F. Supp. 1285 (E.D. Pa. 1987).
[242] See, for example, Artesian Water Co. v. New Castle County, 659 F. Supp. 1269 (D.Del. 1987), aff'd, 851 F.2d 643 (3d Cir.1988); Lutz v. Chromatex, 718 F. Supp. 413 (M.D. Pa. 1989).

and temporary relocation costs.[243] A suit to recover such costs may be brought as soon as preliminary removal costs have been incurred.[244] One of the most contested potential expenditures in private cost recovery actions is medical monitoring. Although some courts have held that medical monitoring costs may be recoverable as response costs,[245] the clear trend is to deny the costs of medical monitoring.[246] Finally, contrary to earlier rulings by various federal courts, the U.S. Supreme Court ruled in *Key Tronic Corp. v. United States*, 511 U.S. 809 (1994), that private parties may not recover attorney's fees from other PRPs in CERCLA cost recovery actions. In a subsequent decision by the Ninth Circuit, noting that the Supreme Court had left open the question whether the government was also precluded from recovering attorney fees in a cost recovery action, the Ninth Circuit held that the EPA was entitled to reasonable attorney fees for litigation brought to recover its response costs.[247]

9.7.4 Consistency with National Contingency Plan

Under appropriate circumstances, private parties faced with the potential costs of cleanup or serious limitations on the use or transfer of contaminated property can file suit to hold responsible parties liable for payment of all or part of the cleanup costs.[248] The CERCLA regulations, known as the National Contingency Plan (NCP),[249] state that "[r]esponsible parties shall be liable for necessary costs of response actions to releases of hazardous substances incurred by any other person consistent with the NCP."[250] The courts have expressed different views on

---

[243] See, for example, Tanglewood East Homeowners v. Charles-Thomas, Inc., 849 F.2d 1568 (5th Cir. 1988); T & E Indus. v. Safety Light Corp., 680 F. Supp. 696 (D.N.J. 1988).

[244] See, for example, Artesian Water Co. v. New Castle County, 851 F.2d 643 (3d Cir. 1988); Southland Corp. v. Ashland Oil, Inc., 696 F. Supp. 994 (D.N.J. 1988).

[245] See, for example, Sinclair Oil Corp. v. Dymon, Inc., 988 F. Supp. 1394 (D. Kan. 1997); Williams v. Allied Automotive, 704 F. Supp. 782 (N.D. Ohio 1988); Brewer v. Ravan, 680 F. Supp. 1176 (M.D. Tenn. 1988); Pneumo Abex Corp. v. Bessemer and Lake Erie R.R. Co., 936 F. Supp. 1250 (E.D. Va. 1996).

[246] See, for example, Durfey v. E.I. DuPont De Nemours Co., 59 F.3d 121 (9th Cir. 1995); Price v. U.S. Navy, 39 F.3d 1011 (9th Cir. 1994); Daigle v. Shell Oil Co., 972 F.2d 1527 (10th Cir. 1992); Murray v. Bath Iron Works, 867 F. Supp. 33 (D. Me 1994).

[247] United States v. Chapman, 146 F.3d 116 (9th Cir. 1998).

[248] CERCLA § 107(a), 42 U.S.C. § 9607(a).

[249] National Oil and Hazardous Substances Pollution Contingency Plan, 40 CFR § pt. 300.

[250] 40 CFR § 300.700(c)(2).

the significance of the requirement that costs must be "consistent with the NCP" before they may be recoverable by private parties in CERCLA actions.

Applicable Version of NCP. The NCP provides a step-by-step process for investigating a site, identifying site risks, evaluating remedial alternatives, and selecting and implementing a CERCLA quality cleanup, referred to as the site remedy. The NCP has been revised several times with the major versions being the 1982, 1985 and 1990 versions. The 1982 version differs significantly from the 1985 and 1990 versions. For example, it allowed and even required that consideration be given to in-place closures and the effectiveness of natural barriers to migration. The 1985 version provided a more rigorous process for investigating a site and selecting a remedy. The 1990 version of the NCP, in response to the Superfund Amendments and Reauthorization Act (SARA), required a preference for using more expensive advanced technologies, such as incineration, which destroy or reduce the volume of waste at a site.

Since the NCP has undergone several major revisions, and of course is more or less continually under review, several courts have confronted the question of which NCP is controlling in determining cost recovery claims. The courts have concluded that consistency with the NCP should be determined by the NCP in effect when response costs are incurred, not when the response action commences or the claims are evaluated.[251]

Definition of "CERCLA-quality" Cleanup. While the original 1982 version of the NCP required strict compliance with the NCP, the current 1990 version of the NCP requires only that the response activities be in "substantial" compliance with the provisions of the NCP for the response costs to be recoverable. The substantial compliance standard of the 1990 NCP has been applied in several recent court cases.[252]

For the purpose of cost recovery claims, a private party response action will be considered consistent with the NCP "if the action, when evaluated as a whole, is in substantial compliance with the requirements of 40 CFR Section 300.700(c)(5) and (6) and results in a CERCLA-quality

---

[251] See, for example, Artesian Water Co. v. New Castle County, 659 F. Supp. 1269 (D.Del. 1987), aff'd, 851 F.2d 643 (3d Cir. 1988); Versatile Metals, Inc. v. Union Corp., 693 F. Supp. 1563 (E.D. Pa. 1988).

[252] See, for example, Louisiana Pacific Corp. v. ASARCO Inc., 6 F.3d 1332 (9th Cir. 1993); Marriott Corp. v. Simkins Indus., Inc., 929 F. Supp. 396 (S.D. Fla. 1996); A.S.I., Inc. v. Sanders, 42 Env't Rep. Cas. (BNA) 1272, 1996 WL 91626 (D. Kan., Feb. 9, 1996); Boeing Co. v. Cascade Corp., 920 F. Supp. 1121 (D. Or. 1996).

cleanup."[253] The accompanying comment to this provision states that a "CERCLA-quality cleanup" is one that (1) satisfies the three basic remedy selection requirements of CERCLA (i.e., is protective of human health and the environment, utilizes permanent and alternative treatment methods "to the extent possible," and is cost effective); (2) attains applicable and relevant and appropriate requirements (ARARs); and (3) provides for meaningful public participation.

The notion of a CERCLA-quality cleanup is an important concept that has evolved from the EPA's private party cleanup policy, because it has been promoted as the ultimate test for a cost recovery claim and was incorporated into the 1990 version of the NCP. Most courts suggest that the cleanup (or planned cleanup) must be of "CERCLA quality" in order for a private cost recovery claim to be successful. In one case, *Marriott Corp. v. Simkins Industries, Inc.*,[254] the court evaluated whether Marriott's cleanup of a site contaminated by paper pulp sludge from a former paper board manufacturing plant was in substantial compliance with the NCP. The court concluded that "the voluminous assessment and remedial action reports, along with the testimony of Marriott's consultants and experts, establish Marriott performed all the core elements of a CERCLA-quality response action consistent with the NCP: (1) project scoping; (2) a remedial investigation (RI); (3) a baseline risk assessment; (4) screening, development and analysis of remedial alternatives (also known as a feasibility study); (5) the equivalent of a record of decision (ROD); and (6) implementation of the selected remedy (also known as the remedial design/remedial action phase)."

In several cases, the courts have held that a private party does not need to prove consistency with the NCP to recoup investigation and monitoring costs but a party does need to prove consistency with the NCP to recoup actual remedial action costs.[255] However, in 1996, the Sixth Circuit Court of Appeals clarified this exception to the NCP requirement by stating that that "[o]nly 'initial' or 'preliminary' investigative costs may be recovered despite failure to comply with the NCP. Any costs incurred after a cleanup has begun are recoverable only if incurred consistent with the NCP."[256]

---

[253]   40 CFR § 300.700(c)(3)(i).
[254]   Marriott Corp. v. Simkins Indus., Inc., 929 F. Supp. 396 (S.D. Fla. 1996).
[255]   See, for example, Donahey v. Bogle, 987 F.2d 1250 (6th Cir. 1993); Artesian Water Co. v. New Castle County, 659 F. Supp. 1269 (D.Del. 1987), aff'd, 851 F.2d 643 (3rd Cir. 1988); Gache v. Town of Harrison, 813 F. Supp. 1037 (S.D.N.Y. 1993); Versatile Metals, Inc. v. Union Corp., 693 F. Supp 1563 (E.D.Pa. 1988).
[256]   Pierson Sand & Gravel, Inc. v. Pierson Township, 1996 WL 338624, at *6 (6th Cir. 1996).

Necessity Test. The decision to implement a removal or remedial action must be shown to be a "necessary" action.[257] Consequently, to establish necessity, one must substantiate that there has been a release or is such a threat of release that the site poses a significant risk to human health and the environment if no remediation were conducted. In one case, for example, the Fifth Circuit dismissed a cost recovery claim brought by plaintiff-landowners against a neighboring oil refinery due to a lack of evidence that the release of lead contamination on the landowners' property necessitated a response action.[258] Although courts have ruled that an initial removal must be necessary,[259] some courts have indicated that removal actions are excused from the more exacting requirements of the NCP because they involve a rapid response to an environmental emergency."[260] Thus, a determination of whether the response action qualifies as a "removal" or "remedial" action can have a significant bearing on NCP consistency since different NCP requirements apply to each type of response. In determining which NCP requirements apply to a particular response action, many courts have adopted the view that when a response action is undertaken to provide a permanent site remedy it constitutes a remedial--not a removal--action.[261] Some courts have suggested that excavation of contaminated soil is a removal action even though excavation is listed in the NCP as a possible remedial action.[262]

Cost-Effectiveness Test. CERCLA and the NCP require that a remedy should not be unnecessarily costly. This cost-effectiveness test is applied after protective alternatives are identified that comply with applicable or relevant and appropriate regulatory requirements (ARARs). Relevant and appropriate requirements are those that are intended for similar circumstances but are not applicable. The application of the ARARs concept to remedy selection requires a knowledge of the broad range of environmental regulations and is often completed only after receiving community input. The least costly of the alternatives that satisfy ARARs

---

[257] See, for example, Amoco Oil Co. v. Borden, Inc., 889 F.2d 664 (5th Cir. 1989).
[258] Licciardi v. Murphy Oil U.S.A., Inc., 111 F.3d 396 (5th Cir. 1997).
[259] See, for example, United States v. Amtreco, Inc., 846 F. Supp. 1578 (M.D. Ga. 1994).
[260] See, for example, A.S.I., Inc. v. Sanders, 42 Env't Rep. Cas. (BNA) 1272, 1996 WL 91626 (D. Kan., Feb. 9, 1996); Versatile Metals, Inc. v. Union Corp., 693 F. Supp. 1563 (E.D.Pa. 1988).
[261] See, for example, Reynolds Metals Co. v. Arkansas Power & Light Co., 1997 WL 580361, unreported (E.D. Ark. July 29, 1997).
[262] See, for example, General Elec. Co. v. Litton Indus. Automation Sys., 920 F.2d 1415 (8th Cir. 1990); Analytical Measeurements, Inc. v. Keuffel & Esser Co., 843 F. Supp. 920 (D.N.J. 1993).

would be considered the cost-effective option(s). Alternatives of similar costs and effectiveness may be considered equally cost-effective. Some courts have held that the NCP only requires cost-effectiveness for remedial actions not removal actions.[263]

Public Participation Requirement. Another requirement of the NCP is that "[p]rivate parties undertaking response actions should provide an opportunity for public comment concerning the selection of the response action...."[264] Specific community relations requirements are set forth in 40 CFR Sections 300.415, 300.430, and 300.435, and apply to removal, remedial, and enforcement actions.[265] In some cases, noncompliance with the NCP's public participation requirements has precluded successful assertion of private cost recovery claims.[266] In one case,[267] for example, after a private party plaintiff undertook cleanup of hazardous substances at a contaminated site, it brought a CERCLA cost recovery action to recoup some of those costs from the defendant, Town of Greeneville, Tennessee. The town argued that the plaintiff was barred from recovering most or all of its response costs incurred with respect to the site because the plaintiff did not incur those costs in a manner consistent with the NCP. In particular, the town pointed to the plaintiff's failure to provide an opportunity for public comment concerning the selection of a remedy for cleaning up the site. The plaintiff never held any public hearings nor provided any opportunity for public comment regarding the remedy selection for the site. However, the plaintiff maintained that its failure to act in a manner consistent with the NCP with respect to community relations did not bar it from recovering response costs incurred in a "removal," as opposed to costs incurred in a "remedial" action. The court characterized "removal" action within the meaning of Section 101(23) of CERCLA as actions taken in response to an immediate threat to public welfare. Although the court acknowledged that the community relations requirements were diminished for removal actions, the court said that they were not abolished. Further, the court stated that the plaintiff bore the burden of proving that the costs it incurred were for removal instead of remedial action. However, because the plaintiff failed to produce sufficient evidence to make this showing, the court did not need to determine to what

---

[263] See United States v. Amtreco, Inc., 846 F. Supp. 1578 (M.D. Ga. 1994).
[264] 40 CFR § 300.700(c)(6). See also Reynolds Metals Co. v. Arkansas Power & Light Co., 1997 WL 580361, unreported (E.D. Ark. July 29, 1997).
[265] 40 CFR § 300.115(c).
[266] See, for example, Estes v. Scotsman Group, Inc., 16 F.Supp.2d 983 (C.D. Ill. 1998); C & C Millwright Maintenance Co., Inc. v. Town of Greeneville, Tenn., 946 F. Supp. 555 (E.D. Tenn. 1996).
[267] C & C Millwright Maintenance Co., Inc. v. Town of Greeneville, Tenn., 946 F. Supp. 555 (E.D. Tenn. 1996).

extent the NCP's community relations requirements applied to removal actions. Rather, the court ruled that, to the extent that the plaintiff's response action was remedial, the failure to provide an opportunity for public comment rendered its remedial action inconsistent with the NCP and barred recovery of costs.

9.8 Government Response Actions

After a contaminated site becomes targeted for environmental response and cleanup, the EPA will identify PRPs considered responsible for the site contamination and send "PRP letters" to those parties, notifying them of potential liability for the cleanup. The EPA then will attempt settlement negotiations with the PRPs for either cleanup costs incurred by the government or to induce the PRPs to commence voluntary cleanup of the site. If settlement negotiations fail with one or all of the PRPs, CERCLA provides the EPA with various mechanisms to compel cleanup of the site. One alternative is for the EPA to issue an administrative order for abatement of the hazardous substance release under Section 106 of CERCLA.[268] In addition, Section 104 authorizes the EPA to remove life-threatening toxic materials,[269] and then sue PRPs under Section 107 to recover funds spent on cleanup.[270] Although a single party is rarely responsible for the entire amount of hazardous substances released at a site, the EPA is not required to sue all PRPs. Since CERCLA liability is joint and several, the EPA can file suit against a single PRP and seek recovery of all cleanup costs from that one PRP.

9.8.1 PRP Notification Letters

With regard to parties identified as PRPs under CERCLA, the EPA initially encourages voluntary participation in cleanup efforts through its issuance of PRP notification letters.[271] PRP letters serve to inform the parties about their potential liability for CERCLA response costs, define the scope of potential liability, explain why they have been identified as PRPs, begin the exchange of information, and facilitate negotiation of settlement agreements. Settling with the EPA primarily involves

---

[268] CERCLA § 106, 42 U.S.C. § 9606.
[269] CERCLA § 104, 42 U.S.C. § 9604.
[270] CERCLA § 107, 42 U.S.C. § 9607.
[271] In addition, a state environmental agency may issue an equivalent PRP notice advising the recipient of its potential liability under a parallel state hazardous waste cleanup law.

formulating an acceptable proposal for cleaning up the pollution under the assumption of PRP liability.[272] Although the EPA designates recipients as PRPs, it is not the equivalent of a conventional demand letter or a simple accusation of fault. First, PRP notifications are sent after the EPA has established that there is sufficient evidence to make a preliminary determination of potential CERCLA liability. Second, parties who are simply identified as PRPs under Section 107(a) of CERCLA are strictly liable, regardless of fault.[273]

The EPA's initial PRP correspondence typically requests information from the PRP for the purpose of assisting the EPA in determining the need for response action. The EPA further requests that the PRP inform the government of its willingness to "voluntarily" participate in cleanup plans by submitting a "good faith" proposal for implementing and conducting remedial action.[274] If the PRP chooses not to respond to the initial PRP letter, the EPA will take one of several steps: (1) seek an injunction in federal district court forcing the PRP to act; (2) issue an administrative order pursuant to Section 104(e) or 106(a) of CERCLA, either demanding information or forcing the PRP to perform the cleanup;[275] or (3) send additional notice letters, known colloquially as "drop dead" letters, informing the PRPs that they must follow the EPA's suggested cleanup "voluntarily"—otherwise, the government will remove the contamination itself, and thereafter demand reimbursement through a cost recovery action.[276] Violations of these orders could subject the PRP to civil penalties of up to $25,000 for each day of noncompliance[277] and punitive damages up to three times the amount of costs incurred by the EPA as a result of the violation.[278] Whether the EPA attempts to compel cleanup or seeks reimbursement, once the agency notifies a party of its potential liability, the PRP is essentially faced with three alternatives: (1) engage in a voluntary settlement; (2) force the government to order cleanup; or (3) have the government unilaterally implement cleanup and litigate for reimbursement later.[279] A PRP may not seek judicial review of the EPA's

---

[272] See Johnson, Whether Insurers Must Defend PRP Notifications: An Expensive Issue Complicated by Conflicting Court Decisions, 10 N. Ill. U. L.Rev. 579 (1990).
[273] 42 U.S.C. § 9607(a).
[274] 42 U.S.C. § 9622.
[275] 42 U.S.C. § 9604(e); 9606(a).
[276] See, for example, Professional Rental, Inc. v. Shelby Ins. Co., 75 Ohio.App.3d 365, 599 N.E.2d 423, 429-430 (1991) (detailing these options).
[277] 42 U.S.C. § 9604(e)(5)(B); 9606(b)(1).
[278] 42 U.S.C. § 9607(c)(3).
[279] See, for example, City of Edgerton v. General Casualty Co., 184 Wis.2d 750, 517 N.W.2d 463, 467 n.4 (1994) (stating that the recipient of a PRP letter essentially has three options: "(1) do nothing and wait for the government to recover the costs of the cleanup; (2) clean up the affected site or join with other

actions until the EPA sues for cost recovery pursuant to Section 107(a) of CERCLA,[280] and even after a PRP gets to court, judicial review is limited to an administrative record prepared by the EPA itself[281] pursuant to Section 113(k).[282]

9.8.2 Government Oversight Costs

Most courts have ruled that the EPA may recover oversight costs incurred to monitor cleanup activities at a site, regardless of whether the EPA is itself implementing the cleanup or whether the cleanup is being conducted by private parties.[283] In one case, for example, the Fifth Circuit upheld a district court order requiring parties responsible for chromium contamination at a manufacturing facility to reimburse the EPA for costs incurred in overseeing the cleanup of the site.[284] Some courts have, however, limited the EPA's recovery of oversight costs to situations where the agency is overseeing the cleanup activities of its own contractors. Thus, these courts hold that the government is not entitled to reimbursement of costs associated with the monitoring of private party cleanup activities.[285] In addition, only those costs related to oversight of a CERCLA cleanup are recoverable. Accordingly, the Third Circuit ruled that the EPA's costs of overseeing a RCRA cleanup could not be recovered in a CERCLA action.[286]

Some courts have permitted state agencies to recover oversight costs,[287] with one federal district court holding that such costs were recoverable as a part of a consent decree.[288] In some instances, private parties may also be entitled to recovery of government oversight costs associated with remedial actions. According to the Tenth Circuit, government monitoring

---

PRPs to effect a cleanup; or (3) litigate with the government so as to possibly secure a more favorable future result.").

[280] 42 U.S.C. § 9607(a).

[281] 42 U.S.C. § 9613(j)(1).

[282] 42 U.S.C. § 9613(k).

[283] See, for example, United States v. Chromalloy American Corp., 158 F.3d 345 (5th Cir. 1998); United States v. Lowe, 118 F.3d 399 (5th Cir. 1997).

[284] United States v. Chromalloy American Corp., 158 F.3d 345 (5th Cir. 1998).

[285] See, for example, United States v. Rohm & Haas Co., 2 F.3d 1265 (3d Cir. 1993); United States v. Witco Corp., 853 F. Supp. 139 (E.D. Pa. 1994).

[286] United States v. Rohm & Haas Co., 2 F.3d 1265 (3d Cir. 1993).

[287] See, for example, New York v. Shore Realty, 759 F.2d 1032 (2d Cir. 1985); California v. SnyderGeneral Corp., 28 Chem. Waste Litig. Rep. 367 (E.D. Cal. 1994); California v. Louisiana-Pacific Corp., 28 Chem. Waste Litig. Rep. 360 (E.D. Cal. 1994).

[288] United States v. Atlas Minerals & Chems., Inc., 851 F. Supp. 639 (E.D. Pa. 1994).

is a necessary part of a private party cleanup, and the EPA oversight is included as part of the monitoring activities under the CERCLA definition of "remedial action."[289] In one case, however, a private party who had already compensated the EPA for oversight costs was denied the right to recover a portion of those costs in a contribution action.[290]

9.8.3 CERCLA Liens for Cleanup Costs

CERCLA provides for the imposition of a lien on property when the government has expended funds from the Superfund in connection with the cleanup of hazardous substances.[291] A CERCLA lien may be asserted against the property interest of any party who is potentially liable for CERCLA cleanup costs. The lien provides a mechanism for the EPA to secure payment when it undertakes cleanup and a responsible party refuses to pay its share of the costs. The lien takes effect when costs are first incurred by the federal government or when a PRP receives written notification of potential liability, whichever occurs later. Upon filing, a CERCLA lien takes priority over all subsequently filed liens, but liens in existence prior to filing of the CERCLA lien are not affected. Unlike the liens authorized by a number of states,[292] the CERCLA lien is not a "superlien," i.e., the lien does not take precedence over creditors who have a prior perfected security interest in the property. Thus, if a lender files a mortgage prior to the filing of a CERCLA lien, for example, the mortgage will have priority over the CERCLA lien. The federal lien is treated like a judgment lien for an unsecured debt under applicable state law. A CERCLA lien is superior to rights of all other creditors except a holder of a security interest or a judgment lien creditor whose interest was perfected under a state law before notice of the federal lien was filed in the appropriate state office.

The federal government will file notices of liens against individuals and corporations liable for the costs of CERCLA cleanup in several instances, including:
- The property is the chief or substantial asset of the potentially responsible party.
- The property has substantial monetary value.

---

[289] Atlantic Richfield Co. v. American Airlines, Inc., 98 F.3d 564 (10th Cir. 1996).
[290] Central Me. Power Co. v. FJ O'Connor Co., 838 F. Supp. 641 (D. Me. 1993).
[291] CERCLA § 107(e), 42 U.S.C. § 9607(e).
[292] See Section 9.8.4 for discussion of state environmental lien laws.

- There is a likelihood that the owner of the property may file for bankruptcy.
- The value of the property will increase significantly as a result of the cleanup.
- The potentially responsible party intends to sell the property.

Notice of the CERCLA lien is filed in accordance with the laws of the state in which the lien is filed and the property is located.[293] If the state in which the lien must be filed has not adopted legislation designating a place for filing, the lien may be filed in the office of the Clerk of the United States District Court in the district in which the property is located.

9.8.4 State Environmental Liens

Many states have adopted specific legislation providing for the filing of liens to secure environmental cleanup obligations. The nature and scope of state environmental lien laws vary, depending on a number of distinguishing features, including whether the lien:

1. Covers only the contaminated property or all personal and real property of the responsible party.
2. Covers both nonresidential and residential property.
3. Applies to all responsible parties or only the owner of the contaminated property.
4. Has priority over all existing liens and interests in the property or only those perfected after the state lien is filed.

There are two basic types of state environmental liens. The first type of state lien mirrors CERCLA's lien provisions and operates prospectively only. In other words, the lien does not have priority over mortgages and liens that already exist at the time of filing. The majority of states with environmental lien laws use this type of lien.[294] The lien may be perfected either when the state spends money on cleanup activities or notifies the property owner of potential liability.

The second type of environmental lien, adopted in only a few states, is a known as a "superlien." Unlike the first type, a superlien takes priority over all other liens or security interests, including those in existence prior

---

[293] CERCLA § 107(e), 42 U.S.C. § 9607(e).
[294] See Alaska Stat. 46.08.075; Ark. Code Ann. 8-7-417, 8-7-516; Ariz. Rev. Stat. Ann. 49-295; Cal. Health & Safety Code 25365; Fla. Stat. Ann. 403.709; 415 Ill. Comp. Stat. 5/21.3(a); Ind. Code Ann. 13-25-4-11; Iowa Code Ann. 424.11; Ky. Rev. Stat. Ann. 224.01-400(23); La. Rev. Stat. Ann. 30:2281; Md. Code Ann., Nat. Res. 3-109(d)(3); Mont. Code Ann. 75-10-720(2)-(3); N.Y. Nav. Law 181-a; Ohio Rev. Code. Ann. 3734.20, 3734.122; Or. Rev. Stat. 466.205; S.D. Codified Laws Ann. 34A-12-13; Tenn. Stat. Ann. 68-212-209; Tex. Health & Safety Code Ann. 361.194; Va. Code Ann. 10.1-1406(c).

to filing of the superlien. Superlien legislation has been enacted in Connecticut, Maine, Massachusetts, New Hampshire, New Jersey, and Wisconsin, and also to a limited extent in Michigan and Minnesota.[295] Somewhat surprisingly, the superlien statutes have generated only a handful of lawsuits and these statutes have withstood those challenges.[296]

Although a comprehensive review of state environmental lien laws is beyond the scope of this book, a summary of some of these state lien provisions is provided here for illustrative purposes.

California. The California lien law provides that cleanup costs or damages incurred and payable from the Hazardous Substance Cleanup Fund constitute a lien on real property owned by the responsible party.[297] The lien acts prospectively and takes effect upon filing in the county in which the property subject to the lien is located.

Illinois. The Illinois environmental lien statute provides that all costs of removal, remedial, preventive, corrective, or enforcement action incurred by the state pursuant to state environmental law constitutes a lien on all real property owned by a party who is liable for such costs.[298] The lien only takes effect following notice to the owners of property subject to the lien and filing of the lien in county where the property is located. It operates prospectively and takes priority over purchasers, mortgagees, and other lienholders as of the date of filing. However, the lien is subordinate to liens for general taxes, special assessments, and special taxes levied by any political subdivision of the state even if filed after the environmental reclamation lien is recorded.

Massachusetts. The Massachusetts superlien statute creates a lien on all real and personal property owned by persons who are liable for past and future response costs incurred by the state under the Massachusetts Oil and Hazardous Material Release Prevention and Response Act. The law imposes a superlien on contaminated property cleaned up by the state and an ordinary lien on all other property owned by a responsible party.[299] To

---

[295] Conn. Gen. Stat. Ann. 22a-452a; Me. Rev. Stat. Ann. tit. 38, 1371(2); Mass. Gen. Laws. Ann. ch 21E, 13; Mich. Stat. Ann. 13A.20138; Minn. Stat. Ann. 514.672; N.H. Rev. Stat. Ann. 147-B:10-b; N.J. Stat. Ann. 58:10-23.11f(f); Wis. Stat. 144.76(13)(c).
[296] See, for example, Kessler v. Tarrats, 194 N.J. Super. 136 (App. Div. 1984) (upholding constitutionality of New Jersey superlien statute); Chicago Title Ins. Co. v. Kumar, 506 N.E.2d 154 (Mass. App. 1987) (dismissing challenge to Massachusetts superlien statute).
[297] Cal. Health & Safety Code 25365.6.
[298] 415 Ill. Comp. Stat. 5/21.3(a).
[299] Mass. Gen. Laws. Ann. ch. 21E, 13.

be valid against real property, the state must file a statement of claim with the registry of deeds and, in the case of personal property, the statement must be filed in the office where U.C.C. Article 9 financing statements are filed. The law is unclear whether there is no lien at all on residential property or whether there is a lien but no superpriority.

New Jersey. The New Jersey lien law provides that any cleanup expenditure made by the state creates a lien on all real and personal property owned by the discharger when the notice of lien is filed.[300] Upon filing of the notice of lien with the clerk of the Superior Court, the state has a superlien on all contaminated real and personal property owned by a responsible party, except for residential real property, and a general lien on all other property owned by the responsible party. Thus, the lien takes priority over all existing liens or claims with respect to property subject to state cleanup action, and priority from the date of filing with respect to all other property owned by the responsible party.

Texas. Under the Texas lien law, all remediation costs for which a person is liable to the state constitute a lien on the property subject to or affected by the state's remedial action.[301] The lien attaches to the property when an affidavit is recorded in the county where the property is located. The lien operates prospectively and does not affect preexisting mortgages, liens, and other encumbrances against the property. However, the statute provides for an exception where a mortgagee or other lienholder had, or reasonably should have had, actual notice or knowledge that the property was subject to or affected by a cleanup action, or that the state had incurred cleanup costs, at the time it took the mortgage or acquired the lien.

9.9 CERCLA Settlements and Contribution Protection

CERCLA liability may be apportioned through administrative or judicially approved settlement agreements.[302] Settlement agreements with the United States or a state are governed by procedures set forth in Section 122 of CERCLA.[303] Parties settling their liability for CERCLA cleanup and response costs with the United States or a state are shielded from

---

[300] N.J. Stat. Ann. 58:10-23.11f(f).
[301] Tex. Health & Safety Code Ann. 361.194.
[302] See, for example, Matter of Bell Petroleum Servs., Inc., 3 F.3d 889 (5th Cir. 1993); General Time Corp. v. Bulk Materials, Inc., 826 F. Supp. 471 (M.D. Ga. 1993).
[303] 42 U.S.C. § 9622.

contribution actions brought by non-settling parties "regarding matters addressed in the settlement."[304] Section 113(f)(2) of CERCLA--the contribution action provision--specifically states that "[a] person who has resolved his liability to the United States or a State in an administrative or judicially approved settlement shall not be liable for claims for contribution regarding matters addressed in the settlement."[305] This provision, known as "contribution protection," provides an important incentive for settlement of CERCLA claims.[306]

9.9.1 Effect of Settlements on Non-Settling Parties

Section 113(f)(2) of CERCLA affects non-settling parties by reducing their potential liability "by the amount of the settlement."[307] This provision has been interpreted to mean that a settlement with the government will result in a "dollar-for-dollar reduction of the aggregate liability" which then must be fully apportioned among the non-settling parties.[308]

In determining how the liability of non-settling defendants to a CERCLA contribution action should be calculated, a court may chose to apply either the Uniform Contribution Among Tortfeasors Act (UCATA) or the Uniform Comparative Fault Act (UCFA). Under Section 4 of the UCATA, a non-settling defendant's liability is reduced by the amount specified in the release or covenant.[309] By contrast, under Section 6 of the UCFA, a non-settlor's liability is reduced by the amount of the settlor's equitable share of the obligation.[310] Both the UCATA and the UCFA expressly provide for contribution protection to all settling parties. Understandably, nonsettlors prefer the UCFA approach because it protects them from paying an equitably disproportionate amount of response costs, and settlors prefer the UCATA because it allows them to settle the amount of their liability with the certainty that it will not increase.

Courts deciding contribution suits between potentially responsible

---

[304] See United States v. Cannons Eng'g Corp., 899 F.2d 79 (1st Cir. 1990) (holding that "Congress plainly intended nonsettlors to have no contribution rights against settlors regarding matters addressed in settlement.").
[305] 42 U.S.C. § 9613(f)(2).
[306] See, for example, Foamseal, Inc. v. Dow Chemical Co., 991 F. Supp. 883 (E.D. Mich. 1998); United States v. Seymour Recycling Corp., 686 F. Supp. 696 (S.D. Ind. 1988); Allied Corp. v. Frola, 730 F. Supp. 626 (D.N.J. 1990); City of New York v. Exxon Corp., 697 F. Supp. 677 (S.D.N.Y. 1988).
[307] 42 U.S.C. § 9613(f)(2) ("Such settlement does not discharge any of the [nonsettling] parties unless its terms so provide, but it reduces the potential liability of the [nonsettling parties] by the amount of the settlement.").
[308] United States v. Cannons Eng'g Corp., 899 F.2d 79 (1st Cir. 1990).
[309] 12 U.L.A. 57, 98 (1975).
[310] 12 U.L.A. 37, 50 (Supp. 1988).

parties (PRPs) have almost unanimously adopted the UCFA approach.[311] One court acknowledged that disproportionate liability would invariably result under the UCFA approach but concluded that Congress intended "the disparities that inevitably arise ... to act as a catalyst for early and inexpensive settlements" and stated that a nonsettling party bears the risk that the ultimate liability of the settling parties may exceed the settlement amount.[312] On the other hand, some courts have applied the UCATA in private party settlements.[313] This different result is largely attributable to the contrasting language found in Sections 113(f)(1) and 113(f)(2) of CERCLA. Section 113(f)(2) states that when someone settles with the government, the liability of other nonsettling parties' will be reduced by the amount of the settlement, whereas Section 113(f)(1) says that "in resolving contribution claims, the court may allocate response costs among liable parties using such equitable factors as the court determines are appropriate." One court has indicated that the specific directive in Section 113(f)(1) of CERCLA to consider equitable factors, coupled with the omission of the same language in Section 113(f)(2) appears to dictate application of the UCFA in actions between PRPs and application of the UCATA where the EPA or a state agency brings the action.[314]

9.9.2 De Minimis Settlements

Under Section 122(g) of CERCLA, the EPA may settle with persons who contributed hazardous substances to a site which are minimal, both in terms of volume and toxicity or other hazardous effects relative to other hazardous substances at the site.[315] Settlements of CERCLA liability are usually very advantageous to de minimis contributors. These PRPs have the opportunity to settle CERCLA claims early with the EPA, agreeing to a payment commensurate with the small volume of waste they contributed to the site. Further, by agreeing to payment of a premium, the de minimis

---

[311] See, for example, Hillsborough County v. A & E Road Oiling Serv., Inc., 853 F. Supp. 1402 (M.D. Fla. 1994); Foamseal, Inc. v. Dow Chem. Co., 991 F. Supp. 883 (E.D. Mich. 1998); New York v. Solvent Chem. Co., 984 F. Supp. 160 (W.D.N.Y. 1997).

[312] Foamseal, Inc. v. Dow Chemical Co., 991 F. Supp. 883 (E.D. Mich. 1998).

[313] See, for example, City and County of Denver v. Adolph Coors Co., 829 F. Supp. 340 (D.Colo. 1993); Allied Corp. v. Frola, 730 F. Supp. 626 (D.N.J. 1990).

[314] Hillsborough County v. A & E Road Oiling Serv., Inc., 853 F. Supp. 1402 (M.D. Fla. 1994). See also United States v. SCA Servs. of Ind., Inc., 827 F. Supp. 526 (N.D. Ind. 1993); United States v. Gencorp, Inc., 935 F. Supp. 928 (N.D. Ohio 1996).

[315] 42 U.S.C. § 9622(g).

contributor can usually be protected from government claims for future cleanup and response costs associated with the site. Perhaps most importantly, the de minimis contributor may obtain "contribution protection" against future cost recovery claims of other PRPs under Section 122(h)(4) of CERCLA.[316]

An important EPA Directive contains guidelines used by the agency for encouraging and expediting settlements with de minimis waste contributors.[317] The Directive is designed to streamline the EPA's approach to de minimis settlements through implementation of the following key provisions:

● *Timing.* The EPA is no longer required to prepare a waste-in list. Available documentary information may be used. The guidance suggests that the EPA now need only assess the individual PRP's waste relative to the total volume of waste at the site.

● *Volume.* Volume cutoff for settlement eligibility is site-specific, although one percent is suggested as a cutoff.

● *Toxicity.* Focus on relative toxicity of the settlor's waste in relation to other substances at the site.

● *Payment Amounts.* Guidance suggests preparation and use of a payment matrix, as well as the settlor's payment of a premium of an additional 50 to 100 percent in exchange for the EPA's covenant not to sue, with or without a cost remedy re-opener.

● *Settlement Implementation.* Guidelines for implementing the settlement are similar to those provided in earlier directives, including encouraging settlement after a de minimis group forms, discouraging negotiation of settlement terms, and the use of model settlement documents.

To determine whether a PRP is eligible for de minimis settlement, the EPA assesses the individual PRP's waste contribution relative to the volume of waste at the site. Comparing these two pieces of information allows the EPA to determine whether that party's contribution was minor in comparison to other hazardous substances at the site. Generally, the agency will then divide the individual contribution by the volume of waste at the site to establish the PRP's volumetric percentage of waste contribution.

Pursuant to the Directive, the EPA will use available documentary evidence to identify the individual amount of hazardous substances contributed. The EPA may estimate the volume of waste present at the site using several methods, including review of site volumetric records, process engineering information, or site sampling results. The volumetric

---

[316] See, for example, Dravo Corp. v. Zuber, 13 F.3d 1222 (8th Cir. 1994).
[317] EPA, Streamlined Approach for Settlement with De Minimis Waste Contributors under CERCLA Section 122(g)(1)(A), OSWER Directive 9834.7-1D (July 30, 1993).

estimate should reflect the EPA's understanding of the waste present at the site; the amount does not need to be a precise figure. In circumstances where it is particularly difficult to quantify the waste amount (especially early in the response process), the EPA may identify the volumetric estimate as a range. When identifying the volume of waste at the site as a range, the EPA will use the lower estimate of the range for calculating the contributor's eligibility for a de minimis settlement. For example, if a PRP contributed 500 batteries to a site where the EPA estimates that between 50,000 and 100,000 batteries are present at the site, the PRP's assigned volumetric percentage would be calculated as one percent (500/50,000).

In addition to a volumetric determination, the EPA must also evaluate the relative toxicity of the wastes contributed to the site. In earlier guidance, the EPA stated that the toxicity finding is met when the hazardous substances are not "significantly more toxic and not of significantly greater hazardous effect" than other hazardous substances at the site.[318] For example, if the hazardous substances at a site are of similar toxicity and hazardous nature, the EPA does not have to engage in further evaluation to make the toxicity determination.

After evaluating the volumetric and toxicity information, the EPA needs to determine the appropriate cutoff for de minimis contributors at the site. Although the guidance does not establish a set percentage for eligibility for a de minimis waste contributor settlement, the guidance does outline an acceptable range for permitting de minimis settlements. The guidance provides a de minimis payment matrix that starts at .001 percent and has a eligibility cutoff at one percent.

Consistent with earlier guidance, the EPA establishes a baseline payment amount by applying several factors: the individual's percentage of waste contributed to the site, the total past costs expended, and an estimate of future costs. To determine the future cost estimate, the EPA generally uses its *Methodology for Early De Minimis Waste Contributor Settlements under CERCLA Section 122(g)(1)(A)*.[319] To identify the past and future cost baseline payment, the EPA first multiplies the individual volumetric precentage by the total past cost amount, which provides a PRP's pro-rata share of past costs. A similar multiplication is made to establish the pro-rata share of future costs. The pro-rata shares are added together to form the baseline payment amount.

---

[318] EPA, Interim Guidance on Settlements with De Minimis Waste Contributors under Section 122(g) of SARA, OSWER Directive 9834.7 (June 19, 1987); EPA, Methodologies for Implementation of CERCLA Section 122(g)(1)(A) De Minimis Waste Contributor Settlements, OSWER Directive 9834.7-1B (Dec. 20, 1989).
[319] OSWER Directive 9834.7-1C (June 2, 1992).

If the EPA can establish an individual's percentage, identify past costs, and estimate future costs with relative ease, that is the preferred approach for establishing the baseline amount. However, there may be situations where there is uncertainty in the overall volume of waste at a site (used to establish the individual percentage) or where the future estimate of site costs is particularly difficult to establish other than to estimate the amount within a range. In such situations, the EPA may construct a payment matrix to assist in establishing the PRP's baseline payment amount.

9.9.3 De Micromis Settlements

In certain situations, companies or individuals may have contributed such miniscule amounts of hazardous substances to a site that the EPA did not even know of their existence. Alternatively, the EPA may have known about these parties but did not actively pursue enforcement actions against them, preferring to focus its limited resources on more significant hazardous waste contributors. Under the settlement authorities provided in Section 122(g) of CERCLA, the EPA has also developed guidance for settling with those parties, known as "de micromis" contributors, who contributed minuscule amounts of hazardous substances to a site.[320]

De micromis settlements are a subset of de minimis settlements. Therefore, in considering parties for de micromis settlements, the EPA must first be able to make Section 122(g) findings required for a de minimis settlement. The next step in determining eligibility is establishing a de micromis volumetric cutoff, above which no party could qualify for a de micromis settlement (although they may still qualify for other settlements).

The EPA may consider several factors in determining the eligibility cutoff for PRPs who would qualify as de micromis, including the settlor's contribution of hazardous substances in relation to the overall volume of waste at the site, and the toxic or hazardous effects of such hazardous substances. The EPA can enter into a de micromis settlement as soon as it reasonably determines that a party meets its eligibility requirements and the EPA can calculate the appropriate payment of site costs by the de micromis party. The EPA will ordinarily evaluate several information sources before offering a de micromis settlement, which may include:

● Information about hazardous substances sent to the site by the de micromis contributor.

---

[320] See U.S.E.P.A, *Guidance on CERCLA Settlements With De Micromis Waste Contributors, OSWER Directive # 9834.17* (July 30, 1993).

- Total estimate of wastes at the site.
- State records.
- Manifests.
- Site records.
- Waste-in lists, if available.
- CERCLA Section 104(e) information request responses.

The agency will need to make a reasonable estimate of past and future response costs at the site. To estimate costs, the EPA may use the procedures outlined in the Directive for de minimis settlements. The EPA may choose to use a payment matrix to calculate the appropriate payment amount for a de micromis party. Another alternative is for the EPA to establish a standardized payment for everyone in the de micromis settlement class at a particular site.

De micromis settlors will not be required to pay a premium whenever they seek settlement with the EPA. This approach departs from the typical de minimis settlement where the EPA often charges a separate premium payment for parties who were eligible to settle earlier, but who entered the de minimis settlement later in the process. The EPA has decided that a premium payment for persons who enter the settlement late would be inappropriate for de micromis settlors because the agency, in exercising its enforcement discretion, generally would not pursue these parties, and because the de micromis party's share represents such a minuscule amount of the site's total cleanup costs.

De micromis settlements will address a party's potential liability under Sections 106 and 107 of CERCLA and provide the settlor with an immediately effective covenant not to sue for past and future liability. The EPA intends for de micromis settlements to be a final resolution of the de micromis party's potential liability unless new information shows that the settlor does not qualify as a de micromis settlor or that the settlor falsified data in its certification statement. Otherwise, the payment of the party's de micromis settlement amount should satisfy the government's potential CERCLA claims against it. The EPA can reopen the settlement if it discovers that the party is not eligible for the de micromis settlement. Finally, the de micromis settlement should contain language that the settlor receives protection against contribution actions by other PRPs to the full extent provided in Section 113(f) of CERCLA and as provided in Section 122(g)(5). Contribution protection is, however, generally only applicable to "matters addressed in the settlement" with the EPA.

9.9.4 Settlement Guidance for Municipalities and Municipal Solid Waste Contributors

In 1998, the EPA issued agency guidance for negotiating settlements with generators and transporters of municipal solid waste, and with municipal owners and operators of co-disposal facilities.[321] Styled as a policy, the document sets out certain numerical figures for settlements with such parties. For example, with respect to generators and transporters, the policy identifies $5.30 per ton of waste contributed as the desired rate to charge; for owners and operators, the policy sets 20 to 35 percent of the site's estimated total cleanup costs as the desired rate.[322] However, because the policy only serves as a guide to future settlements,[323] the EPA has discretion to deviate from the baseline figures, or chose not to apply the policy's principles to particular sites and parties, such as when "the resulting settlement would not be fair, reasonable, or in the public interest."[324]

In a federal district court case,[325] the Chemical Manufacturers Association filed suit to enjoin the EPA from future use of the policy. The Association was primarily concerned that the policy would diminish its bargaining power to broker settlements with owners, operators, generators, and transporters before having to assert contribution claims. The court granted the EPA's motion to dismiss for lack of subject matter jurisdiction, ruling that because the settlement policy was merely agency guidance, not a binding regulation, it did not constitute final agency action subject to judicial review.

9.10 Insurance Coverage Against Environmental Liability

Under CERCLA's strict liability scheme, any individual or company having even the slightest involvement with contaminated property is faced with the risk of being held responsible for cleanup costs, regardless of whether that party actually caused or contributed to the contamination.[326] Once contamination is identified at a site, the question becomes who is to

---

[321] EPA, Policy for Municipality and Municipal Solid Waste; CERCLA Settlements at NPL Co-Disposal Sites, 63 *Fed. Reg.* 8197 (1998).

[322] Id., 63 *Fed. Reg.* at 8199.

[323] Id., 63 *Fed. Reg.* at 8201 (policy is "intended exclusively as guidance for employees of the U.S. Government").

[324] Id., 63 *Fed. Reg.* at 8200.

[325] Chemical Manufacturers Ass'n v. Environmental Protection Agency, 26 F.Supp.2d 180 (D.D.C. 1998).

[326] See section 9.2 for discussion of the nature of CERCLA liability and Section 9.3 discussion of categories of PRPs.

clean it up and who is to pay for the damages it caused. One facet of the problem is evaluated here: whether an insured may rely upon its insurer to defend and indemnify it against claims that the insured is liable for environmental contamination.

9.10.1 Insurer's Duty to Defend and Indemnify the Insured

Historically, insurance carriers offered Comprehensive General Liability (CGL) insurance with the expectation that they would cover the insured's liability for any claim for damages not expressly excluded in the policy.[327] All CGL insurance policies have an insuring clause which contains the insurer's promise to defend and indemnify the insured for damages occurring during the policy period. The standard language of the insuring clause states that:

"The Company will pay on behalf of the insured all sums which the insured shall become legally obligated to pay as damages because of

A. bodily injury or

B. property damage

to which this insurance applies, caused by an occurrence, and the company shall have the right and duty to defend any suit against the insured seeking damages on account of such bodily injury or property damage, even if any of the allegations of the suit are groundless, false or fraudulent, and may make such investigation and settlement of any claim or suit as it deems expedient, but the company shall not be obligated to pay any claim or judgment or to defend any suit after the applicable limit of the company's liability has been exhausted by payment of judgments or settlements."

Pursuant to this language, insurers must fulfill two obligations under standard form CGL insurance policies: (1) the duty to indemnify the insured in the event of a loss; and (2) the duty to defend the insured against "suits" seeking "damages" within the terms of the policy.[328] The two duties--to defend and to indemnify--impose different obligations on the insurer under a CGL policy and must be examined independently of one another.[329] The duty to defend is broader than the duty to indemnify and does not depend on whether a third party will ultimately prevail against the

---

[327] See American Home Prods. Corp. v. Liberty Mut. Ins. Co., 565 F. Supp. 1485 (S.D.N.Y. 1983) (where the court discusses the drafting history of CGL insurance policies).

[328] See M. Lathrop, *Insurance Coverage for Environmental Claims*, § 8.03[1][a] (1994).

[329] See, for example, Weyerhaeuser Co. v. Aetna Casualty & Sur. Co., 123 Wash.2d 891, 874 P.2d 142 (1994).

insured.[330] An insurer's duty to defend is triggered immediately upon the filing of a claim based on the potential for coverage of that claim, rather than upon actual proof that coverage exists.[331] The insurer owes a broad duty to defend its insured against any "suit" that creates a potential of liability under the policy.[332] In contrast to the duty to defend, the insuring clause does not require a "suit" in order to trigger the duty to indemnify; it requires coverage for all sums the insured shall be obligated to pay by reason of the liability imposed upon the insured by law.[333]

The duty to defend has been broadly interpreted by the courts to cover defense of "suits" by private parties alleging bodily injury or property damage resulting from the insured's release of hazardous substances into the environment, as well as government "suits" brought against the insured pursuant to federal and state environmental laws. The insurer's duty to defend is not invoked, however, in instances where no "suit" has been initiated against insured within the meaning of the policy,[334] where no covered "occurrence" exists during the relevant policy period,[335] where no "damages" are found within the meaning of the policy,[336] or where a policy exclusion is found to preclude coverage.[337]

9.10.2. Meaning of "Suit" Invoking Insurer's Duty to Defend Environmental Claims

Since the term "suit" in the insuring clause of standard form CGL insurance policies is not defined, numerous courts have been asked to the determine its meaning with respect to the scope of the insurer's duty to defend third party claims alleging the insured's liability for environmental contamination. A major controversy has developed over whether the term "suit" should be construed to mean only a lawsuit in the traditional sense or whether the term should also include administrative actions initiated

---

[330] See, for example, Seymour Mfg. Co., Inc. v. Commercial Union Ins. Co., 665 N.E.2d 891 (Ind. 1996); American Bumper & Mfg. Co. v. Hartford Fire Ins. Co., 452 Mich. 440, 550 N.W.2d 475 (1996).

[331] See, for example, Technicon Electronics Corp. v. American Home Assurance Co., 74 N.Y.2d 66, 544 N.Y.S.2d 531, 542 N.E.2d 1048 (1989).

[332] See, for example, Harford County v. Harford Mut. Ins. Co., 327 Md. 418, 610 A.2d 286 (1992); St. Paul Fire & Marine Ins. Co. v. McCormick & Baxter Creosoting Co., 126 Or.App 689, 870 P.2d 260 (1994).

[333] See, for example, Bausch & Lomb, Inc. v. Utica Mut. Ins. Co., 330 Md. 758, 625 A.2d 1021 (1993).

[334] See section 9.10.2 for discussion of the meaning of the term "suit."

[335] See sections 9.10.3 and 9.10.4 for discussion of the meaning of the term "occurrence" and the trigger of coverage for pollution damage claims.

[336] See section 9.10.5 for discussion of the meaning of the term "damages."

[337] See section 9.10.6 for discussion of pollution exclusion clauses.

against the insured which may create a potential for liability, such as the government's issuance of potentially responsible party (PRP) letters[338] and regulatory compliance orders, or commencement of administrative enforcement proceedings.

In construing the term "suit" as used in the insuring clause of CGL insurance policies, one group of courts has adopted the view that the term is unambiguous and only meant to refer to the filing of a traditional lawsuit against the insured by means of summons and complaint. Accordingly, an insurer's duty to defend can only be triggered when an action is commenced against the insured in a court of law, and not by such administrative mechanisms as PRP letters, government compliance orders, or agency enforcement proceedings.[339] On the other hand, another group of courts has adopted the view that the term "suit" as used in CGL insurance policies is ambiguous and capable of more than one reasonable interpretation, so that its meaning is not limited solely to a traditional court proceeding, but may also include administrative actions that assert claims of potential environmental liability against the insured. Thus, construing the term broadly, these courts hold that an insurer's duty to defend may also be invoked by the government's issuance of PRP letters[340] and regulatory compliance orders,[341] or initiation of administrative enforcement proceedings against the insured.[342]

9.10.3 Policy Definition of "Occurrence"

Most CGL insurance policies providing coverage for pollution damage claims are "occurrence," as distinguished from "claims made" policies.

---

[338] PRP letters serve to inform recipients about their potential liability for environmental cleanup costs, begin the exchange of information, and facilitate negotiation of settlement agreements. See section 9.8.1 for further discussion of PRP letters.

[339] See, for example, Forest Preserve Dist. v. Pacific Indem. Co., 279 Ill.App.3d 728, 665 N.E.2d 305 (1996); A.Y. McDonald Indus., Inc. v. Insurance Co. of N. Am., 475 N.W.2d 607 (Iowa 1991); Patrons Oxford Mut. Ins. Co. v. Marois, 573 A.2d 16 (Me. 1990); City of Edgerton v. General Casualty Co., 184 Wis.2d 750, 517 N.W.2d 463 (1994).

[340] See, for example, Hazen Paper v. U.S. Fidelity and Guar., 407 Mass. 689, 555 N.E.2d 576 (1990); American Bumper & Mfg. Co. v. Hartford Fire Ins. Co., 452 Mich. 440, 550 N.W.2d 475 (1996).

[341] See, for example, Coakley v. Maine Bonding & Casualty Co., 136 N.H. 402, 618 A.2d 777 (1992); C.D. Spangler Constr. Co. v. Industrial Crankshaft & Eng'g Co., 326 N.C. 133, 388 S.E.2d 557 (1990).

[342] See, for example, A.Y. McDonald Indus. v. Insurance Co. of N. Am., 475 N.W.2d 607 (Iowa 1991); St. Paul Fire & Marine Ins. Co. v. McCormick & Baxter Creosoting Co., 126 Or.App. 689, 870 P.2d 260 (1994).

Occurrence policies generally cover the insured for losses arising from "occurrences" during the policy period.[343] Assuming that proper notice requirements are met, an insured may file a claim for damages years after the pollution-causing event occurred.[344] Claims-made coverage, on the other hand, requires that the insured file all claims during the policy period.[345] Claims filed after the policy has expired are not covered.

The standard CGL policy definition of "occurrence" states that:

"'Occurrence' means an accident, including continuous or repeated exposure to conditions, which results in personal injury or property damage neither expected nor intended from the standpoint of the insured."

In order for coverage to be triggered, it is necessary for an "occurrence" to take place during the policy period.[346] Two areas of contention arise with respect to "occurrences" within the meaning of such policies: (1) whether the "occurrence" was "neither expected or intended from the standpoint of the insured" and (2) whether an occurrence happened within the particular policy period.

In determining whether there has been an occurrence, there must be an accident (which courts have construed to mean some fortuitous event) that was "neither expected nor intended" by the insured. Many courts apply a subjective standard and have stated that this "neither expected nor intended" language means that the insured must not have willfully intended to damage property.[347] Other courts apply an objective standard and construe this language to mean that a reasonable person in the position of the insured could not have anticipated the resulting property damage.[348]

---

[343] See, for example, Hoppy's Oil Serv., Inc. v. Insurance Co. of N. Am., 783 F. Supp. 1505 (D. Mass. 1992).

[344] See, for example, Harford County v. Harford Mut. Ins. Co., 327 Md. 418, 610 A.2d 286, 294 (1992) ("occurrence policies cover liability inducing events occurring during the policy term, irrespective of when an actual claim is presented.")

[345] See, for example, Mutual Fire, Marine & Inland Ins. v. Vollmer, 306 Md. 243, 508 A.2d 130 (1986) (noting that a "claims made" or "discovery" policy covers liability inducing events if and when a claim is made during the policy term, irrespective of when the events occurred).

[346] See, for example, Armotek Indus. v. Employers Ins. of Wausau, 952 F.2d 756 (3d Cir. 1991) (no "occurrence" caused property damage while the policies were in effect).

[347] See, for example, New Castle County v. Hartford Accident & Indem. Co., 933 F.2d 1162 (3d Cir. 1991); Shell Oil Co. v. Accident & Casualty Ins. Co., 12 Cal.App.4th 715, 15 Cal.Rptr.2d 815 (1993).

[348] See, for example, City of Farragut v. Hartford Accident & Indem. Co., 837 F.2d 480 (8th Cir. 1987); County of Broome v. Aetna Casualty & Sur. Co., 146 A.D.2d 337, 540 N.Y.S.2d 620 (1989).

Which standard should be used to judge expectation and intent remains hotly contested, with insurers championing the objective test, and insureds advocating the subjective standard.

9.10.4 Trigger of Coverage for Pollution Damage Claims

In CGL insurance coverage disputes involving environmental claims, insurers may acknowledge that an "occurrence" took place but will often refuse coverage by contending that the occurrence did not take place during the particular period covered by the policies that they sold to the insured. Since an occurrence policy only covers the insured for damages caused by an "occurrence" that takes place during the policy period, coverage can only be triggered under those policies in effect at the time of the occurrence. Pinpointing the precise time when the pollution-causing event occurred is often a difficult task in the case of environmental contamination, especially where pollutants have gradually been released into the environment for many years before they cause injury or the harm is discovered.[349] The question of which policies are triggered by an occurrence--the so-called "trigger of coverage" issue--is of obvious importance in determining the scope of liability coverage available to an insured for environmental claims. In cases where the insured changed insurance carriers several times over a period of years when the pollution-related damage began, continued, and was finally discovered, a court must determine the time of occurrence in order to sort out which of the insurers' policies afford coverage for the insured's environmental claims. Even where the insured has purchased all of its CGL policies from one carrier over a number of years, the time of the occurrence may not implicate coverage under all of the policies.

The difficulties inherent in determining the exact time of an occurrence has resulted in divergent judicial approaches to the trigger of coverage issue. Four disparate trigger of coverage theories have emerged--commonly referred to as the "exposure," "manifestation," "injury in fact," and "triple or continuous" trigger theories of coverage.[350]

---

[349] See, for example, Zuckerman v. National Union Fire Ins. Co., 100 N.J. 304, 495 A.2d 395, 399 (1985) (observing that in the use of "occurrence" policies for perils that can cause latent damage, as in environmental litigation, there is a difficulty in determining precisely when the essential causal event occurred).

[350] These coverage triggers were largely developed by analogy to asbestos injury cases. See Eagle-Picher Indus., Inc. v. Liberty Mut. Ins. Co., 682 F.2d 12 (1st Cir. 1982), cert. denied, 460 U.S. 1028 (1983) (adopting manifestation trigger); American Home Prods. v. Liberty Mut. Ins. Co., 565 F. Supp. 1485 (S.D.N.Y. 1983), aff'd as modified, 748 F.2d 760 (2d Cir. 1984) (adopting injury in fact trigger).

Although no single methodology has gained general acceptance, the courts have most frequently applied the injury in fact trigger, which requires that there be actual damage during the policy period in order to trigger coverage.

Courts applying the exposure trigger hold that coverage is triggered when the first exposure to injury-causing conditions occurred. Thus, the insurer whose policy is in effect at the time of the initial release of hazardous substances is obligated to provide coverage for the insured's pollution damage claims.[351] Under the manifestation trigger, coverage is triggered at the time when bodily injury or property damage caused by a hazardous substance first manifests itself.[352] Under the injury in fact trigger, the court looks to when damage actually occurred and not to the time of initial exposure or when injury first manifested itself. This trigger falls somewhere between manifestation and exposure. The policy in effect at the time that actual damage resulted from exposure to hazardous substances determines which insurer is liable to defend and indemnify the insured.[353] Finally, under the "triple or continuous" trigger, courts hold that there is no temporal limitation on the term "injury" in the CGL policy and reason that an insurer's risk should cover the period from the time of initial exposure to the time that injury manifests itself. Thus, all insurers whose policies were in effect during the time that covered persons or property were exposed, injured in fact, or when the injury was manifested, are each obligated to defend and indemnify the insured.[354]

9.10.5 Environmental Cleanup Costs as "Damages"

The insuring clause of CGL insurance policies requires the insurer to defend and indemnify the insured against "damages" resulting from injury

---

[351] See, for example, Insurance Co. of N. Am. v. Forty-Eight Insulations, Inc., 633 F.2d 1212 (6th Cir. 1980); Hancock Laboratories, Inc. v. Admiral Ins. Co., 777 F.2d 520 (9th Cir. 1985).

[352] See, for example, Safeco Ins. Co. of Am. v. Federated Mut. Ins. Co., 915 F.2d 1565 (4th Cir. 1991); Peerless Ins. Co. v. Strother, 765 F. Supp. 866 (E.D.N.C. 1990).

[353] See, for example, Spartan Petroleum Co. v. Federated Mut. Ins. Co., 1999 WL 182512 (4th Cir. Jan. 13, 1999); Ray Indus., Inc. v. Liberty Mut. Ins. Co., 974 F.2d 754 (6th Cir. 1992); County of San Bernardino v. Pacific Indem. Co., 56 Cal.App.4th 666, 65 Cal.Rptr.2d 657 (1997); Jenoff, Inc. v. New Hampshire Ins. Co., 558 N.W.2d 260 (Minn. 1997).

[354] See, for example, Keene Corp. v. Insurance Co. of N. Am., 667 F.2d 1034 (D.C. Cir. 1981), cert. denied, 455 U.S. 1007 (1982); New Castle County v. Continental Casualty Co., 725 F. Supp. 800 (D. Del. 1989), aff'd in relevant part, 933 F.2d 1162 (3d Cir. 1991); Gottlieb v. Newark Ins. Co., 238 N.J. Super. 531, 570 A.2d 443 (App. Div. 1990).

to persons or property that occurs during the policy period. With regard to the latter, a pivotal coverage issue concerns whether CGL policies were meant to cover the insured's liability for environmental cleanup costs. Although most courts have recognized that cleanup costs are "damages" covered by CGL policies, others do not, so the issue remains unsettled.

In support of the position that cleanup costs are not covered as "damages" under CGL policies, insurers argue that the term "damages" is only intended to mean "legal damages" and not environmental cleanup costs, which are generally considered to be a form of equitable relief. In several cases, the First, Fourth, Seventh, and Eighth Circuit Courts of Appeals have agreed with the insurers' position.[355] Insureds, on the other hand, contend that the term "damages" should be given its everyday ordinary meaning to include any damage to property. This interpretation makes no distinction between whether losses are incurred on a legal or equitable basis. The Second, Third, Ninth, and D.C. Circuit Courts of Appeals have agreed with the insureds' position.[356] Of course it is important to note that the federal courts apply state law when deciding these environmental insurance disputes. Thus, the split among the federal circuits on the "damages" issue is largely attributable to particular federal court interpretations of state law.

State courts have more consistently taken the position that cleanup costs are "damages" within the meaning of CGL insurance policies, with the highest courts of the states of California, Iowa, Massachusetts, Minnesota, Missouri, New Hampshire, North Carolina, and Washington having ruled that cleanup costs are recoverable as "damages."[357] On the other hand, the

[355] See, for example, A. Johnson & Co. v. Aetna Casualty and Sur. Co., 933 F.2d 66 (1st Cir. 1991); Cincinnati Ins. Co. v. Milliken & Co., 857 F.2d 979 (4th Cir. 1988); Wisconsin Power & Light Co. v. Century Indem. Co., 130 F.3d 787 (7th Cir. 1997); Continental Ins. Cos. v. Northeastern Pharmaceutical & Chem. Co., 842 F.2d 977 (8th Cir. 1988).
[356] See, for example, Gerrish Corp. v. Universal Underwriters Ins. Co., 947 F.2d 1023 (2d Cir. 1991); New Castle County v. Hartford Accident & Indem. Co., 933 F.2d 1162 (3d Cir. 1991); Intel Corp. v. Hartford Accident & Indem. Co., 952 F.2d 1551 (9th Cir. 1991); Independent Petrochemical Corp. v. Aetna Casualty and Sur. Co., 944 F.2d 940 (D.C. Cir. 1991).
[357] See AIU Ins. Co. v. Superior Court, 51 Cal.3d 807, 799 P.2d 1253, 274 Cal. Rptr. 820 (1990); A.Y. McDonald Indus., Inc. v. Insurance Co. of N. Am., 475 N.W.2d 607 (Iowa 1991); Hazen Paper Co. v. United States Fidelity & Guar. Co., 407 Mass. 689, 555 N.E.2d 576 (1990); Minnesota Mining & Mfg. Co. v. Travelers Indem. Co., 457 N.W.2d 175 (Minn. 1990); Farmland Indus., Inc. v. Republic Ins. Co., 941 S.W.2d 505 (Mo. 1997) (en banc); Coakley v. Maine Bonding & Casualty Co., 136 N.H. 402, 618 A.2d 777 (1992); C.D. Spangler Constr. Co. v. Industrial Crankshaft & Eng'g Co., 326 N.C. 133, 388 S.E.2d 557 (1990); Boeing Co. v. Aetna Casualty & Sur. Co., 113 Wash.2d 869, 784 P.2d 507 (1990).

highest courts of the states of Maine and Wisconsin have adopted a minority view by holding that cleanup costs are not recoverable "damages."[358]

Another facet of the "damages" issue that has been presented in environmental insurance coverage disputes is whether the insured is covered for expenses incurred for performance of "voluntary" cleanup actions that have been undertaken in advance of a government order or third party lawsuit. A minority of courts hold that an insurer has no duty to provide indemnification for such voluntary cleanup actions on grounds that these cleanup costs are not sums which the insured is "legally obligated" to pay as damages within the meaning of the insuring clause of standard form CGL policies.[359] Most courts hold to the contrary, however, and require the insurer to indemnify the insured for the costs of voluntary cleanup measures.[360] These courts recognize the unfavorable public policy implications of a rule that would discourage insureds from initiating voluntary cleanup efforts.[361] According to this view, if the insured had to wait for initiation of formal government action or a third-party claim to be certain of coverage under a CGL insurance policy, the insured would have a strong disincentive to initiate voluntary cleanup measures.[362] By contrast, courts also have addressed the issue of whether preventive measures taken before pollution has occurred are costs incurred because of property damage. In most cases, courts have failed to consider such costs "damages" within the meaning of CGL policies.[363]

---

[358] See Patrons Oxford Mut. Ins. Co. v. Marois, 573 A.2d 16 (Me. 1990); City of Edgerton v. General Casualty Co., 184 Wis.2d 750, 517 N.W.2d 463 (1994). Cf. General Casualty Ins. Co. v. Hills, 209 Wis.2d 167, 561 N.W.2d 718 (1997) (distinguishing *Edgerton* to hold that cleanup costs may constitute covered "damages" under CGL insurance policies where an independent third party--not a governmental agency--files a complaint against the insured seeking "substitutionary, monetary relief" to compensate for damage to property, as opposed "preventive relief for future conduct" as in the case of a federal or state agency order directing the insured to develop a remediation plan or incur remediation and response costs).

[359] See, for example, Curran Composites, Inc. v. Liberty Mut. Ins. Co., 874 F. Supp. 261 (W.D. Mo. 1994).

[360] See, for example, Aetna Cas. & Sur. Co. v. Pintlar Corp., 948 F.2d 1507 (9th Cir. 1991); Metex Corp. v. Federal Ins. Co., 290 N.J. Super. 95, 675 A.2d 220 (App. Div. 1996).

[361] See, for example, Weyerhaeuser Co. v. Aetna Casualty & Sur. Co., 123 Wash.2d 891, 874 P.2d 142 (1994).

[362] See, for example, Broadwell Realty Servs., Inc. v. Fidelity & Casualty Co., 218 N.J. Super. 516, 528 A.2d 76 (App. Div. 1987) ("the policy does not require parties to calmly await further catastrophe.").

[363] See, for example, AIU Ins. Co. v. Superior Court, 51 Cal.3d 807, 274 Cal.Rptr. 820, 799 P.2d 1253 (1990); Hazen Paper Co. v. United States Fidelity & Guar. Co., 407 Mass. 689, 555 N.E.2d 576 (1990).

Whether or not a court considers cleanup costs to be "damages" may be the deciding factor in an action seeking a determination with regard to the insurer's duty to defend and indemnify the insured against a suit alleging liability for environmental contamination. Should the court follow the minority view that cleanup costs do not constitute "damages," the insurer's duty to defend or indemnify would not be triggered since there would be no "suit seeking damages" against the insured within the meaning of the CGL policy's insuring clause. In other words, in those jurisdictions where this view is followed, in the absence of any potential for "damages," the insured would have no grounds for asserting the insurer's duties of defense and indemnification because, simply put, there would be nothing to defend or indemnify.[364]

9.10.6 Pollution Exclusions

Beginning in the 1970s, insurers sought to limit their liability for coverage of pollution-related damage claims by including pollution exclusion clauses in the standard CGL policy.

Sudden and Accidental Pollution Exclusion. In 1973, the insurance industry made its first attempt at carving pollution out of the standard CGL insurance policy when the Insurance Service Office added the following provision, commonly known as the "sudden and accidental" pollution exclusion, stating that the policy did not apply:
"to bodily injury or property damage arising out of the discharge, dispersal, release or escape of smoke, vapors, soot, fumes, acids, alkalis, toxic chemicals, liquids or gases, waste materials or other irritants, contaminants or pollutants into or upon land, the atmosphere or any water course or body of water; but this exclusion does not apply if such discharge, dispersal, release or escape is sudden and accidental."
    The sudden and accidental pollution exclusion has spawned a great deal of litigation. Judicial interpretation of the last phrase of this pollution exclusion, the "sudden and accidental" language, has often been pivotal in determining whether coverage is available for pollution damage claims. Federal and state courts have been almost evenly divided over the meaning of the phrase.[365] One line of cases holds that the word "sudden" is unambiguous, always has a temporal quality, and means "abrupt" or

---

[364] See, for example, City of Edgerton v. General Casualty Co., 184 Wis.2d 750, 517 N.W.2d 463 (1994).
[365] See, for example, New Castle County v. Hartford Accident & Indem. Co., 933 F.2d 1162, 1195, ns. 60 & 61 (3d Cir. 1991) (listing 24 cases holding that the pollution clause bars coverage and 26 cases holding the opposite).

"brief." Under this interpretation, coverage is usually denied for damage caused by gradual or ongoing pollution.[366] A second line of cases holds that the word "sudden" is ambiguous and, construing the term in favor of the insured, means "unexpected." Most courts in this latter group essentially consider the word "sudden" to be a restatement of the definition of the term "occurrence" and therefore coverage should not be excluded where the damage to the environment was "neither expected nor intended from the standpoint of the insured."[367]

The outcome of an insured's challenge to application of the sudden and accidental pollution exclusion often hinges on whether the court finds a temporal component to the meaning of the term "sudden." The highest courts of California, Florida, Massachusetts, Michigan, New Jersey, North Carolina, and Ohio have held that the word "sudden," as used in the pollution exclusion, contains an inherent temporal element.[368] These courts, holding that the term "sudden" has a temporal meaning, find that the exclusion bars coverage for gradual discharges. Because most cases involve some kind of gradual release of pollutants into the environment over an extended period of time, courts finding a bar to coverage under the exclusion have construed "sudden" as unambiguously meaning "abrupt," "brief," or "immediate."[369] These courts hold that long-term hazardous

---

[366] See, for example, Stamford Wallpaper Co., Inc. v. TIG Ins., 138 F.3d 75 (2d Cir. 1998); FL Aerospace v. Aetna Casualty & Sur. Co., 897 F.2d 214 (6th Cir. 1990); Smith v. Hughes Aircraft Co., 22 F.3d 1432 (9th Cir. 1993); A-H Plating, Inc. v. American Nat'l Fire Ins. Co., 57 Cal.App.4th 427, 67 Cal.Rptr.2d 113 (1997).

[367] See, for example, CPC Int'l, Inc. v. Northbrook Excess & Surplus Ins. Co., 962 F.2d 77 (1st Cir. 1992); Avondale Indus., Inc. v. Travelers Indem. Co., 887 F.2d 1200 (2d Cir. 1989); Patz v. St. Paul Fire & Marine Ins. Co., 15 F.3d 699 (7th Cir. 1994); Outboard Marine Corp. v. Liberty Mut. Ins. Co., 154 Ill.2d 90, 180 Ill.Dec. 691, 607 N.E.2d 1204 (1992).

[368] See Montrose Chem. Corp. v. Superior Court, 6 Cal.4th 287, 24 Cal.Rptr.2d 467, 861 P.2d 1153 (1993) (en banc); Dimmitt Chevrolet v. Southeastern Fidelity Ins., No. 78293, 1993 WL 241520 (Fla. July 1, 1993); Lumbermens Mut. Casualty Co. v. Belleville Indus. Inc., 407 Mass. 675, 555 N.E.2d 568 (1990); Upjohn v. New Hampshire Ins., 476 N.W.2d 392 (Mich. 1991); Morton Int'l, Inc. v. General Accident Ins. Co. of Am., 629 A.2d 831 (N.J. 1993); Waste Management of Carolinas, Inc. v. Peerless Ins. Co., 315 N.C. 688, 340 S.E.2d 1096 (1986); Hybud Equip. v. Sphere Drake Ins., 597 N.E.2d 1096 (Ohio 1992).

[369] See, for example, ACL Technologies, Inc. v. Northbrook Property & Casualty Ins. Co., 17 Cal.App.4th 1773, 22 Cal.Rptr.2d 206 (1993); Shell Oil Co. v. Winterthur Swiss Ins. Co., 12 Cal.App.4th 715, 15 Cal.Rptr.2d 815 (1993).

waste disposal cannot meet the definition of "sudden and accidental."[370]

On the other hand, the highest courts of Colorado, Georgia, Illinois, Indiana, West Virginia, and Wisconsin construe the word "sudden" narrowly, finding that it does not have a temporal component, and holding that coverage is available for gradual pollution as long as the discharge was "unintended and unexpected."[371] Some of these courts make a distinction between an intentional act not intended to discharge wastes into the environment and an intentional discharge.[372] Nevertheless, insureds have generally argued to no avail that their entrustment of hazardous wastes to a waste hauler who then disposed of the wastes at a licensed landfill or disposal facility could not be considered a "discharge" or "release" within the meaning of the pollution exclusion.[373] Such an argument, premised on the fact the insured never "intended" for the hazardous wastes to be discharged into the environment (but rather contained at the landfill), has met with rare success.[374]

Absolute Pollution Exclusion. In response to court decisions finding the "sudden and accidental" pollution exclusion ambiguous, and construed in favor of insureds, the insurance industry modified the wording of the exclusion in an effort to bar recovery for all pollution-related damages. This exclusion, known as the "absolute pollution exclusion," excludes coverage whether or not the discharge or release is sudden and accidental. Although the insurance industry began to include an absolute pollution

---

[370] See, for example, Stamford Wallpaper Co., Inc. v. TIG Ins., 138 F.3d 75 (2d Cir. 1998); Bituminous Casualty Corp. v. Tonka Corp., 9 F.3d 51 (8th Cir. 1993); Smith v. Hughes Aircraft Co., 22 F.3d 1432 (9th Cir. 1993); Bell Lumber and Pole Co. v. U.S. Fire Ins. Co., 847 F. Supp. 738 (D. Minn. 1994).

[371] See Hecla Mining v. New Hampshire Ins., 811 P.2d 1083 (Colo. 1991); Claussen v. Aetna Casualty & Sur., 380 S.E.2d 686 (Ga. 1989); Outboard Marine Corp. v. Liberty Mut. Ins. Co., 154 Ill.2d 90, 180 Ill.Dec. 691, 607 N.E.2d 1204 (1992); American States Ins. Co. v. Kiger, 662 N.E.2d 945 (Ind. 1996); Joy Technologies v. Liberty Mut. Ins., 421 S.E.2d 493 (W.Va. 1992); Just v. Land Reclamation Ltd., 155 Wis.2d 737, 456 N.W.2d 570 (1990).

[372] See Patz v. St. Paul Fire & Marine Ins. Co., 15 F.3d 699 (7th Cir. 1994); Outboard Marine Corp. v. Liberty Mut. Ins. Co., 607 N.E.2d 1204 (Ill. 1992). But see St. Paul Fire and Marine Ins. Co. v. Warwick Dyeing Corp., 26 F.3d 1195 (1st Cir. 1994); Broderick Inv. Co. v. Hartford Accident & Indem. Co., 954 F.2d 601 (10th Cir. 1992).

[373] See, for example, St. Paul Fire and Marine Ins. Co. v. Warwick Dyeing Corp., 26 F.3d 1195 (1st Cir. 1994); Stamford Wallpaper Co., Inc. v. TIG Ins., 138 F.3d 75 (2d Cir. 1998). But see Nestle Foods Corp. v. Aetna Casualty & Sur., 842 F. Supp. 125 (D.N.J. 1993) (holding that transfer of waste to an independent hauler who disposed of the waste at a landfill did not amount to "discharge, dispersal, release or escape" under a policy's pollution exclusion).

[374] See, for example, St. Paul Fire and Marine Ins. Co. v. Warwick Dyeing Corp., 26 F.3d 1195 (1st Cir. 1994).

exclusion clause in CGL insurance policies as early as the late 1970s, the exclusion was not in common usage until the mid-1980s. Of those courts to have considered the application of the absolute pollution exclusion, most have ruled that it is unambiguous and, accordingly, have found coverage barred for pollution damage claims.[375] Very few courts have refused to apply an absolute pollution exclusion, and have done so only where factual issues remained concerning policy language.[376]

Owned Property Exclusion. Most CGL policies also exclude coverage for damage to property owned, occupied or rented by the insured. The standard CGL policy excludes coverage for damage to "(1) property owned or occupied by or rented to the insured, (2) property used by the insured; or (3) property in the care, custody or control of the insured or as to which the insured is for any purpose exercising physical control." Insurers usually take the position that this language precludes coverage for the costs of cleanup on property owned by the insured.[377] However, notwithstanding the exclusion, some courts have found coverage where the evidence showed that the insured's failure to clean up on-site contamination could lead to extensive contamination off-site.[378] Where contaminated groundwater has migrated to another's property, the insured may be entitled to recover costs associated with removing the source of the contamination.[379] Other courts, however, hold that coverage should be excluded if the insured cannot show actual damage to a third party interest. These courts refuse to recognize coverage for future damages.[380] In this vein, some courts have refused coverage when the insured has made capital improvements on its own property that are essentially measures

---

[375] See, for example, Ascon Properties v. Illinois Union Ins. Co., 908 F.2d 976 (9th Cir. 1990); Legarra v. Federated Mut. Ins. Co., 35 Cal.App.4th 1472, 42 Cal.Rptr.2d 101 (1995); Titan Corp. v. Aetna Casualty & Sur. Co., 22 Cal.App.4th 457, 27 Cal.Rptr.2d 476 (1994); Heyman Assocs. No. 1 v. Insurance Co. of State of Pa., 231 Conn. 756 (Conn. 1995).

[376] See, for example, Titan Holdings Syndicate, Inc. v. City of Keene, 898 F.2d 265 (1st Cir. 1990); Norfolk Southern Ry. Co. v. Roberts, 1996 WL 931575 (N.D. Ala. Oct. 29, 1996); American States Ins. Co. v. Kiger, 662 N.E.2d 945 (Ind. 1996).

[377] See, for example, Diamond Shamrock Chem. Co. v. Aetna Casualty and Sur. Co., 231 N.J. Super. 1, 554 A.2d 1342 (1989).

[378] See, for example, Reese v. Travelers Ins. Co., 129 F.3d 1056 (9th Cir. 1997); Claussen v. Aetna Casualty & Sur. Co., 754 F. Supp. 1576 (S.D. Ga. 1990).

[379] See, for example, CPS Chem. v. Continental Ins., 222 N.J.Super. 175, 536 A.2d 311 (App.Div. 1988); Broadwell Realty Inc. v. Fidelity & Cas. Co., 218 N.J.Super. 516, 528 A.2d 76 (App.Div. 1987).

[380] See, for example, State v. Signo Trading Int'l, Inc., 130 N.J. 51, 612 A.2d 932 (1992).

aimed at preventing future contamination.[381] A critical inquiry appears to be the imminency of any threat of contamination to third-party property.[382]

9.11 Voluntary Cleanup programs

Various programs have been introduced at the federal and state government levels to encourage the cleanup and rehabilitation of contaminated sites. Of particular note, in 1995, the U.S. EPA unveiled several initiatives to assist cities and private businesses in the cleanup and redevelopment of brownfield sites.[383] "Brownfield" is the term commonly used to describe property that has been abandoned or taken out of productive use as a result of actual or perceived risks from environmental contamination.[384] The EPA's initiatives include removal of sites from the Comprehensive Environmental Response, Compensation and Liability Information System (CERCLIS) database, execution of prospective purchaser agreements, funding of brownfield redevelopment pilot projects, the EPA's issuance of comfort/status letters for brownfield properties, and the establishment of a Brownfields Internet Homepage. In addition to the EPA initiatives, many states have implemented voluntary cleanup programs which offer financial incentives and liability protections in exchange for voluntary investigation and cleanup of contaminated properties, including brownfields. This section describes the EPA and state programs designed to encourage the voluntary cleanup of contaminated property.

9.11.1 Removal of Sites From EPA's CERCLIS Database

On March 29, 1995, the EPA adopted new procedures for maintaining its Comprehensive Environmental Response, Compensation and Liability Information System (CERCLIS).[385] CERCLIS is the database and data management system used by the EPA to track activities at sites considered

---

[381] See, for example., AIU Ins. Co. v. Superior Court, 51 Cal.3d 807, 799 P.2d 1253, 274 Cal. Rptr. 820 (1990).

[382] See, for example, Summit Assocs., Inc. v. Liberty Mut. Fire Ins. Co., 229 N.J. Super. 56, 550 A.2d 1235 (App. Div. 1988).

[383] U.S.E.P.A., *The Working Draft of the Brownfields Action Agenda* (Jan. 25, 1995).

[384] See Dennison, *Brownfields Redevelopment: Programs and Strategies for Rehabilitating Contaminated Real Estate* (Government Institutes 1998).

[385] "Amendment to the National Oil and Hazardous Substances Pollution Contingency Plan (NCP); CERCLIS Definition Change," 60 *Fed. Reg.* 16053 (Mar. 29, 1995).

for cleanup under CERCLA. The EPA rule announced the agency's decision to remove from CERCLIS those sites that the agency found to warrant no further evaluation under the Superfund program. The EPA specifically included sites that the agency has given a designation of "No Further Response Action Planned" (NFRAP), to eliminate any possible disincentive to purchase, improve, redevelop, and revitalize sites, related to inclusion on CERCLIS. Many of these NFRAP sites are not contaminated and others are currently being cleaned up by the states. The EPA has removed more than 25,000 NFRAP sites from the list of 38,000 sites included in CERCLIS.

9.11.2 Prospective Purchaser Agreements

On June 21, 1995, the EPA issued long-awaited supplemental guidance on prospective purchaser agreements.[386] This guidance document supersedes the agency's 1989 policy concerning agreements with prospective purchasers of contaminated property.[387] The new guidance is designed to facilitate greater use of prospective purchaser agreements by expanding the universe of eligible sites and the circumstances under which the EPA will consider entering into such agreements.

During the past few years, numerous prospective purchasers of contaminated property have requested that the EPA limit their CERCLA liability by offering covenants not to sue. Although Section 122 of CERCLA[388] empowers the EPA to enter into settlement agreements concerning CERCLA liability, including covenants not sue and contribution protection,[389] this authority only extends to agreements with potentially responsible parties (PRPs), as defined in Section 107(a) of CERCLA.[390] Since prospective purchasers are not yet owners or operators of contaminated property, they fall outside the contemplated reach of the statutory covenant not to sue. Thus, the basis for EPA's authority to enter into settlement agreements with prospective purchasers, is derived from the U.S. Department of Justice (DOJ)'s inherent authority to settle matters for the United States. A prospective purchaser must have a mandatory

---

[386] Announcement and Publication of Guidance on Agreements With Prospective Purchasers of Contaminated Property and Model Prospective Purchaser Agreement, 60 *Fed. Reg.* 34792 (July 3, 1995).

[387] Guidance on Landowner Liability under Section 107(a) of CERCLA, De Minimis Settlements under Section 122(g)(1)(B) of CERCLA, and Settlements with Prospective Purchasers of Contaminated Property, OSWER Directive No. 9835.9, 54 *Fed. Reg.* 34235 (Aug. 18, 1989).

[388] 42 U.S.C. § 9622.

[389] See section 9.9 concerning settlements and contribution protection.

[390] See section 9.3 for discussion of categories of PRPs.

consultation with the Director of the EPA Regional Support Division, Office of Site Remediation Enforcement. Any agreement negotiated between the EPA and a prospective purchaser requires the express approval of the Department of Justice.

The EPA has determined that prospective purchaser agreements might be both appropriate and beneficial in more circumstances than contemplated by its 1989 guidance. The 1989 guidance limited the use of these agreements to situations where the EPA planned to take an enforcement action, and where the agency received a substantial benefit, not otherwise available, from cleanup of the site by the purchaser.

The EPA now believes that it may be appropriate to enter into agreements resulting in somewhat reduced benefits to the agency. The new guidance authorizes use of prospective purchaser agreements, if the agreement results in either (1) a substantial direct benefit to the agency in terms of cleanup or funds for cleanup or (2) a substantial indirect benefit to the community, coupled with a lesser direct benefit to the EPA. The new guidance is also applicable to persons seeking prospectively to operate or lease contaminated property.

A significant component of the 1995 guidance, not contained in the earlier guidance concerning agreements with prospective purchasers of contaminated property, is a model prospective purchaser agreement. The model agreement functions as a starting point for negotiations between the EPA and prospective purchasers.

9.11.3 Criteria for Prospective Purchaser Agreements

The 1995 guidance outlines several criteria that must be met before the EPA will consider entering into prospective purchaser agreements. These criteria are intended to reflect the EPA's commitment to removing the barriers imposed by potential CERCLA liability while ensuring protection of human health and the environment. The EPA will consider five criteria when evaluating prospective purchaser agreements, each of which is summarized below.

1. EPA Response Action Undertaken, Ongoing, or Anticipated. This criterion is meant to ensure that the EPA does not become unnecessarily involved in purely private real estate transactions or expend its limited resources in negotiations which are unlikely to produce a sufficient benefit to the public. The EPA, however, recognizes the potential gains in terms of cleanup and public benefit that may be realized with broader application of prospective purchaser agreements. Therefore, this criterion has been expanded beyond the limitation in the 1989 guidance to sites where enforcement action is anticipated, to now include sites where federal involvement has occurred or is expected to occur.

When requested, the EPA may consider entering into prospective purchaser agreements at sites listed or proposed for listing on the National Priorities List (NPL), or sites where the EPA has undertaken, is undertaking, or plans to conduct a response action. If the agency receives a request for a prospective purchaser agreement at a site where the EPA has not yet become involved, the EPA will first evaluate the realistic possibility that a prospective purchaser may incur Superfund liability when determining the appropriateness of entering into a prospective purchaser agreement. This evaluation should clearly show that the EPA's covenant not to sue is essential to remove Superfund liability barriers and allow the private party to cleanup and undertake productive use, reuse, or redevelopment of the site.

2. Substantial Benefit to Agency. A cornerstone of the Agency's evaluation process under the 1995 guidance is the measurement of environmental benefit, in the form of direct funding, or cleanup, or a combination of reduced direct funding or cleanup and an indirect public benefit. The EPA believes that its past practice of limiting prospective purchaser agreements to those situations where substantial benefit was measured only in terms of cost reimbursement or work performed may have decreased the effectiveness of this tool.

Thus, the new guidance encourages a more balanced evaluation of both the direct and indirect benefits of a prospective purchaser agreement to the government and the public. The EPA recognizes that indirect benefits to a community is an important consideration and may justify the commitment of the agency's resources necessary to negotiate a prospective purchaser agreement, even where there are reduced direct benefits to the agency in terms of cleanup and cost reimbursement. The EPA may now consider negotiating prospective purchaser agreements that will result in substantial indirect benefits to the community as long as there is still some direct benefit to the agency. Examples of indirect benefits to the community include measures that serve to reduce substantially the risk posed by the site, creation or retention of jobs, development of abandoned or blighted property, creation of conservation or recreation areas, or provision of community services (such as improved public transportation and infrastructure.) Examples of reduced but measurable benefits to EPA include partial cleanup or compensation.

3. Site Operation Will Not Aggravate Existing Contamination or Interfere with EPA's Response Action. The EPA will not enter into an agreement if available information is insufficient for purposes of evaluating the impact of continued operation or new site development activities. Information that should be considered by the agency to evaluate the effect of these activities could include site assessment data and the Engineering Evaluation Cost Analysis (EE/CA) or remedial investigation/feasibility study (RI/FS), if available, and all other information relevant to the condition of the site. If the prospective purchaser intends to continue the operations of an existing facility, the prospective purchaser should submit information sufficient to allow the agency to determine whether the continued operations are likely to aggravate or contribute to the existing contamination or interfere with the remedy. If the prospective purchaser plans to undertake new operations or development of the property, comprehensive information regarding these plans should be provided to the EPA. If the planned activities of the prospective purchaser are likely to aggravate or contribute to the existing contamination or generate new contamination, the EPA generally will not enter into an agreement, or will include restrictions in the agreement which prohibit those operations.

4. Site Operation Will Not Pose Health Risks to Community. The EPA believes it is important to consider the environmental implications of site operations on the surrounding community and to those likely to be present or have access to the site.

In addition, due to the fact that prospective purchaser agreements will provide contribution protection to the purchaser, the surrounding community and other members of the public should be afforded opportunity to comment on the settlement, whenever feasible. Because settlements with prospective purchasers are not expressly governed by Section 122 of CERCLA, there is no legal requirement for public notice and comment. Whenever practicable, however, the EPA intends to publish notices in the Federal Register and undertake other appropriate action to ensure that adequate notification of the agreement is given to all interested parties.

5. Prospective Purchaser Is Financially Viable. A settling party, including a prospective purchaser of contaminated property, should demonstrate that it is financially viable and capable of fulfilling any obligation under the agreement. In appropriate circumstances, the EPA may structure payment or work to be performed so as to avoid or minimize an undue financial burden on the purchaser.

9.11.4 Brownfields Pilot Projects

During 1995 and 1996, the EPA issued grants of up to $200,000 each for 50 brownfields pilot projects as part of a two-year demonstration of redevelopment solutions. These grants were issued to an number of State, county and municipal governments, including Boston, MA; Buffalo, NY; Dallas, TX; Duwamish Coalition, Seattle, WA; Philadelphia, PA; Pittsburgh, PA; Sand Creek Corridor, CO; West Jordan, UT; and the states of Illinois, Indiana and Minnesota. The EPA expects that the pilot program will be instrumental in initiating nationwide redevelopment projects. The objectives of the brownfield pilot program include:
- Increase the participation of interested parties in shaping the cleanup and productive reuse of contaminated sites.
- Stimulate a national search for innovative ways to overcome the current obstacles to the reuse of contaminated properties.
- Coalesce federal, state, and municipal efforts to examine new approaches to achieving cleanup and reuse.
- Explore the potential for combining economic stimuli and prompt environmental cleanup to contribute to the achievement of environmental justice.

9.11.5 "Comfort/Status" Letters for Brownfield Properties

On January 30, 1997, the EPA issued a new policy statement, primarily designed to assist parties who seek to clean up and reuse brownfield properties.[391] EPA headquarters and regional offices often receive requests from parties for some level of comfort that if they purchase, develop, or operate on brownfield property, the EPA will not pursue them for the costs to clean up any contamination resulting from the previous use. The EPA expects to provide a measure of "comfort" by helping an interested party to better understand the agency's potential or actual involvement at a brownfield site. The new policy contains four sample comfort/status letters which address the most common inquiries for information that the EPA receives regarding contaminated or potentially contaminated properties. While the sample comfort/status letters do not account for every possible situation, the EPA believes that the letters contained in this policy will address the most common requests for comfort. The policy is not a rule, and does not create any legal obligations. The extent to which the EPA applies the policy will depend on the facts of each case.

---

[391] 62 *Fed. Reg.* 4624 (Jan. 30, 1997).

Purpose of Policy Statement. Uncertainty about potential contamination and/or CERCLA liability may prevent otherwise interested parties from purchasing or redeveloping brownfields. To allay the fear of potential federal pursuit of parties for cleanup of brownfields, the EPA may provide varying degrees of comfort by communicating the agency's intentions toward a particular piece of property. Comfort may range from a formal legal agreement containing a covenant not to sue, which releases a party from liability for cleanup of existing contamination, to agency policy statements regarding the exercise of the EPA's enforcement discretion as it relates to specific site circumstances or activities of a party.

Upon receiving a request from an interested party for information about a particular property, EPA regional offices may issue comfort/status letters, at their discretion, when there is a realistic perception or probability of incurring CERCLA liability and such comfort will facilitate the cleanup and redevelopment of a brownfield property, and there is no other mechanism available to adequately address the party's concerns. With the information provided by the EPA, the party inquiring about the property can decide whether the risk of EPA action is enough to forego involvement, whether to proceed as planned, whether additional investigation into site conditions is necessary, or whether further information from the EPA or other agencies is needed.

Sample Comfort/Status Letters. The EPA has developed four sample comfort/status letters to address the most common inquiries received regarding brownfield properties. Each of the sample comfort letters is intended to address a particular set of circumstances and provide whatever information is contained within the EPA's databases. The sample letters are structured with opening and closing paragraphs applicable to all scenarios falling under that category of letter. EPA regional offices may then choose and combine the applicable substantive paragraphs to tailor the sample letter to address a party's particular request. The following is a brief summary of the sample letters:

1. *No Previous Federal Superfund Interest Letter*: This letter may be provided to parties when there is no historical evidence of federal Superfund program involvement with the property/site in question [i.e., site is not found in the Comprehensive Environmental Response, Compensation, and Liability Information System (CERCLIS)].
2. *No Current Federal Superfund Interest Letter:* This letter may be provided when the property/site either has been archived and is no longer part of the CERCLIS inventory of sites, has been deleted from the National Priorities List (NPL), or is situated near, but not within, the defined boundaries of a CERCLIS site.

3. *Federal Superfund Interest Letter:* This letter may be provided at sites where the EPA either plans to respond in some manner or already is responding at the site. This letter is intended to inform the recipient of the status of the EPA's involvement at the property. Additionally, language is included to respond to requests regarding the applicability of a CERCLA policy, regulation, or statutory provision to a party or particular set of circumstances.

4. *State Action Letter:* This letter may be provided when the state has the lead for day-to-day activities and oversight of a response action at the site.

9.11.6 EPA's Brownfields Internet Homepage

The EPA has developed a Brownfields Internet homepage to maximize distribution of brownfields information, increase the timeliness of the information, and reduce document distribution costs. The EPA's Brownfields Internet homepage went on-line in January 1996. The homepage is an effective vehicle for providing local governments, businesses, affected community members, and other brownfields stakeholders with access to the brownfields information and tools that they need to understand, address, and solve brownfields problems.

The homepage provides the user with access to a range of brownfields information, including several key brownfields documents. The brownfields homepage is updated frequently to ensure that users have access to the most current brownfields information and redevelopment tools. Future versions of the homepage may allow users to query information pertaining to their specific information needs; access more in-depth pilot information, including maps, photographs, and updates of ongoing activities; and access cleanup and redevelopment tools, such as the CERCLIS archive list and the LandView database.

The homepage information is organized into the following four categories: General Information; Tools; Other Brownfields-Related EPA Sites; and Brownfields Initiative Information.

1. General Information--provides general information about the Brownfields Initiative and includes:
- Mission Statement
- Brownfields Action Agenda
- Major Milestones/Accomplishments
- Frequently Asked Questions
- Announcements/What's New

2. Tools--provides information on available brownfields tools and includes:
- Index of Brownfields Publications

- Brownfields in the News
- Endorsements
- Starting a Brownfields Effort
- Contacts
- Tools

3. Other Brownfields-Related EPA Sites--provides information on and links to related EPA Web sites, including:

- Superfund
- Office of Enforcement and Compliance Assurance
- Office of Environmental Justice
- Environmental Finance Advisory Board
- Common Sense Initiative Iron and Steel Sector Subcommittee

Additional state and non-EPA sites will be linked to the Brownfields homepage in the future.

4. Brownfields Initiative--provides detailed information on:

- Regional Brownfields Initiatives
- Liability and Cleanup Issues
- Partnerships and Outreach
- Job Development and Training
- Brownfields Pilots

The Brownfields Pilot page gives the user additional information on pilot announcements, application information, pilot fact sheets, and tools for new pilot participants. These pilot tools include organizations and publications of interest, advice from other pilot participants, available funding mechanisms, and a discussion of various brownfields stakeholders.

The Brownfields homepage can be accessed in any of the following ways:

1. Type the URL address: http://www.epa.gov/swerosps/bf/
2. From EPA's homepage (http://www.epa.gov/), choose the highlighted link Offices, then Solid Waste and Emergency Response, then Outreach Programs, Special Projects, and Initiatives, then Brownfields.
3. From EPA's homepage (http://www.epa.gov/), choose the highlighted link Initiatives, then Brownfields Home Page.

9.11.7 State Voluntary Cleanup Programs

Over the past several years, more than 30 states have unveiled programs to encourage productive reuse of abandoned, idle, or underutilized sites that are hampered by actual or suspected contamination. These state programs--called voluntary cleanup programs, brownfields programs, land recycling programs, and similar names--basically offer financial incentives and liability protections in exchange for voluntary investigation and cleanup of contaminated properties. Many states have created voluntary cleanup funds, grant programs, low-interest loan programs, tax breaks, and other economic incentives to assist project sponsors in their cleanup and redevelopment efforts. Importantly, under most programs, once cleanup is completed to the satisfaction of the state environmental agency, a participant in the voluntary cleanup program may be issued various liability protections and assurances, including a No Further Action letter, Certificate of Completion, or formal covenant not to sue.

Voluntary cleanup programs are particularly popular because they allow private parties to initiate cleanups and avoid some of the costs and delays associated with other enforcement-driven programs. Most of the voluntary programs provide technical guidance and oversight, in some cases assisting with site assessment and cleanup. Many programs attempt to provide clearer standards on permissible levels of various types of contamination. Some programs apply special cleanup standards to participants in the program. Others incorporate land use controls that anticipate future use that usually involves less public exposure to the site (e.g., 500 employees at an industrial site as opposed to thousands of consumers at a mall). The land use controls are not meant to eliminate all risks to human health or the environment but provide assurance of an appropriate public exposure/use of a site.

Not all contaminated sites are eligible for participation in state voluntary cleanup programs. For instance, the existence of groundwater contamination may bar participation in voluntary cleanups. In addition, most programs apply only to parties not responsible for existing site contamination. It is important to note at the outset that these voluntary cleanup programs generally only apply to sites that are not listed on the CERCLA National Priorities List (NPL), the EPA's CERCLIS database, or state hazardous waste remediation priority lists. Thus, the most severely contaminated sites fall outside the scope of most voluntary cleanup programs. As such, the primary focus is on accelerated cleanup and redevelopment of those sites that can be cleaned up and returned to productive, sustainable use in a relatively short period of time and that pose lesser degrees of financial and environmental risk to potential developers and investors.

Unfortunately, space limitations preclude a full discussion of these state programs.[392] However, a brief description of the programs for Minnesota,

---

[392] For a full discussion of state voluntary cleanup programs, see Dennison, *Brownfields Redevelopment: Programs and Strategies for Rehabilitating Contaminated Real Estate* (Government Institutes 1998). See also Breggin & Pendergrass, Voluntary and Brownfields Remediation Programs: An Overview of the Environmental Law Institute's 1998 Research, 29 Envtl. L. Rep. (ELI)

New Hampshire and Texas is provided here as an illustration of how these programs work. For other programs, the reader is directed to refer to the state laws listed in Table 9.1.

TABLE 9-1. State voluntary cleanup Legislation

State	Statutory Reference
Arizona	Ariz. Rev. Stat. 33-4-3; 49-104A.17, 49-282.05, 49-285B
Arkansas	Ark. Stat. 8-7-1101--8-7-1104
California	Cal. Health & Safety Code 512; 25260--25268; 25395--26300
Colorado	Colo. Rev. Stat. Ann. 25-16-301
Connecticut	Conn. Gen. Stat. Ann. 22a-133, 22a-454, 22a-471
Delaware	Del. Code Ann. tit. 7, 9102--9116; tit. 30, 2010--2011
Florida	Fla. Stat. 376.77--376.85
Georgia	Ga. Code Ann. 12-8-200
Hawaii	Haw. Rev. Stat. 128D-31-41
Idaho	Idaho Code 39-7201--39-7210
Illinois	Ill. Rev. Stat. ch. 415, para. 5/4(y), 5/58--5/58.12; Ill. Admin. Code tit. 35, 740.100--740.625
Indiana	Ind. Code Ann. 6-1-1-42-10; 13-25-5-1
Iowa	Iowa Code Ann. 455H
Kansas	Kan. Stat. Ann. 65-34.161--34.174
Louisiana	La. Rev. Stat. Ann. 30:2285
Maine	Me. Rev. Stat. Ann. tit. 38, 342-343
Maryland	Md. Code Ann., Envir. 4-401(1), 7-201(n-1), 7-501--7-516; Md. Code Ann. art. 83A, 3-901--3-905; Md. Code Ann., Tax-Prop. 9-229 & 14-902
Massachusetts	Mass. Gen. Laws ch. 21E
Michigan	Mich. Stat. Ann. 299.601--299.618; 324.20101--324.20142
Minnesota	Minn. Stat. 115B.17(14), 115B.175
Missouri	Mo. Rev. Stat. 260.565--260.575; 447.714
Mississippi	Miss. Code Ann. 17-17-54; 49-35-1

10339 (June 1999).

TABLE 9-1. State voluntary cleanup Legislation (*Continued*)

State	Statutory Reference
Montana	Mont. Code Ann. 75-10-730--75-10-738
Nebraska	Neb. Rev. Stat. 81-15.181--15.188
New Hampshire	N.H. Rev. Stat. Ann. 147-A, 147-B, 147-F
New Jersey	N.J. Stat. Ann. 58:10-23.11, 58:10B-1; N.J. Admin. Code tit. 7, 26
New Mexico	N.M. Stat. Ann. 74-4G-1
New York	N.Y. Envtl. Conserv. Law 3-0301.2, 56-0101, 56-0502, 56-0503; 6 N.Y. Comp. Codes R. & Regs. tit. 6, pt. 375-4
North Carolina	N.C. Gen. Stat. 130A-310
Ohio	Ohio Rev. Code Ann. ch. 3746
Oklahoma	Okla. Stat. tit. 27A, 2-15-101--2-15-110; Okla. Reg. tit. 252, 220-1-1--220-7-3
Oregon	Or. Rev. Stat. 465.260, 465.327; Or. Admin R. 340-122-010--340-122-140
Pennsylvania	35 Pa. Stat. 6026.101--6026.908; 25 Pa. Code ch. 250
Rhode Island	R.I. Gen. Laws 23-19.14
Tennessee	Tenn. Code Ann. 68-212-104
Texas	Tex. Health & Safety Code Ann. 361.601--361.613; Tex. Admin. Code tit. 30, 333.1--333.11
Utah	Utah Code Ann. 19-8-101--19-8-118
Vermont	Vt. Stat. tit. 10, 6615a
Virginia	Va. Code Ann. 10.1-1429.1--10.1-1429.4; Va. Admin. Code 20-160-10--20-160-130
Washington	Wash. Rev. Code ch. 70.105D; Wash. Admin. Code 173-340-300
West Virginia	W. Va. Code 22-22-1
Wisconsin	Wis. Stat. 144.765; Wis. Stat. Ann. 292.15

Minnesota

Minnesota was the first state to implement a voluntary cleanup program. Minnesota's Voluntary Investigation and Cleanup (VIC) Program addresses the liability and technical issues associated with buying, selling, and developing property contaminated with hazardous substances.

Because of the potential for liability as an owner of contaminated property, property owners, buyers, developers, financial institutions, and other participants in property transactions frequently need to determine the nature and extent of possible contamination on the subject property.

In response to a growing need for agency review and oversight of voluntary investigations and response actions, primarily involving property transactions, a Property Transfer Program was established in 1988 under the Minnesota Environmental Response and Liability Act (MERLA).[393] The Property Transfer Program consists of two distinct components. Under the first component, referred to as File Evaluation Program, parties interested in information about potentially contaminated property can request information assistance from the Minnesota Pollution Control Agency (MPCA). The MCPA provides MPCA file and database information that might be used to determine whether the property of interest, or surrounding properties within a one-mile radius, have been the site of a release or threatened release of hazardous substances. The second component, originally referred to as the Property Transfer/Technical Assistance Program, is the Voluntary Investigation and Cleanup (VIC) Program. The key functions of the VIC Program are to set standards for a site investigation, to provide MPCA review of the adequacy and completeness of such investigation, and to approve cleanup plans (response action plans) to address identified contamination. By obtaining MPCA approval of investigation and response action plans, landowners, lenders, and potential developers can determine the extent of environmental contamination on the property, can devise the most appropriate cleanup action, and can calculate the cost of cleanup measures needed to satisfy statutory requirements.

The VIC program is designed to provide information needed to make sensible financial decisions about developing or transferring contaminated or potentially contaminated property. Implicit in the voluntary nature of the program is the recognition that voluntary parties have a choice to participate or not participate in the VIC Program. A voluntary party can terminate participation in the program at any point by written notification to appropriate VIC Program staff. If a voluntary party decides to terminate participation in the VIC Program and the voluntary party is not otherwise a responsible party, as defined by the MERLA, the MCPA staff would not take further administrative action to mandate future investigation or cleanup by the voluntary party. However, if the voluntary party is the owner of the property, it will be required to cooperate with the MCPA or other responsible parties so that the MPCA or the responsible parties can complete additional investigation and response actions.

---

[393] Minn. Stat. 115B.17, subd. 14.

Various improvements have been made to the VIC Program since it was originally established in 1988. Most significantly, the Minnesota legislature amended MERLA with enactment of the Land Recycling Act of 1992,[394] which clarifies the application of cleanup liability to specific parties and provides statutory mechanisms to obtain liability protections. The Land Recycling Act offers incentives to promote voluntary investigation and cleanup activities under oversight and approval of the VIC Program. Future liability protection is available to eligible parties when MCPA-approved response actions are conducted and completed by VIC Program participants. Liability protection applies to the party who undertakes and completes response actions and to the owner of the identified property (if those parties are not responsible for the release or threatened release), as well as financing parties, and successors and assigns of the person to whom liability protection applies.[395]

The Land Recycling Act allows the MPCA to approve partial response action plans--plans that do not address all identified releases or threatened releases--but additional conditions and requirements must be met.[396] Voluntary response actions may also be undertaken by responsible parties; however, the response action of a responsible party must address all releases and threatened releases. A partial cleanup is not allowed and a responsible party is not eligible for statutory liability protection.[397] VIC Program participants can obtain written assurances from the MCPA in the form of a technical approval letter, a "no action" letter, an "off-site source determination" letter, a "no association determination" letter, or a Certificate of Completion.

New Hampshire
In July 1996, New Hampshire joined a host of other states by enacting brownfields legislation designed to encourage voluntary cleanup and redevelopment of contaminated properties.[398] New Hampshire's Brownfields Program is designed to provide incentives for both environmental cleanup and redevelopment of contaminated properties by parties who did not cause the contamination. This is accomplished under a process by which eligible parties can obtain a "Covenant Not to Sue" from the New Hampshire Department of Justice (DOJ) and a "Certificate of Completion" from the New Hampshire Department of Environmental Services (DES) when investigation and cleanups are performed in accordance with DES cleanup requirements.

---

[394] Minn. Stat. 115B.175 (Land Recycling Act).
[395] Minn. Stat. 115B.175, subd. 6.
[396] Minn. Stat. 115B.175, subd. 2.
[397] Minn. Stat. 115B.175, subd. 6(a).
[398] N.H. Rev. Stat. Ann. 147-F, effective July 1, 1996.

Eligibility Criteria. Essentially, any person who did not cause the existing contamination of the property is eligible for participation in the program. This may include:

1. Prospective purchasers
2. Current property owners if they did not cause or contribute to the contamination
3. Secured creditors or mortgage holders
4. Municipalities owed real estate taxes on the property.

Any property contaminated with hazardous waste, hazardous materials, or oil is eligible for the program, *unless*:

1. There is noncompliance with an environmental or corrective action order and DES determines that the property will not be brought into substantial compliance as a result of participation in the Brownfields Program or
2. The property is eligible for substantial reimbursement from one of the state petroleum discharge reimbursement funds (the Oil Discharge and Disposal Cleanup Fund, the Fuel Oil Discharge Cleanup Fund, or the Motor Oil Discharge Cleanup Fund) toward the total costs of cleanup. If, however, cleanup costs for a petroleum-contaminated site exceed petroleum reimbursement fund coverage limits, the site may then be eligible for participation in the Brownfields Program.

Eligibility Determination. To apply for an eligibility determination, the applicant must submit the following information:
- A signed, complete application form (provided on request by DES).
- All supporting information required as part of the application package.
- An environmental site assessment report. This may also include submittal of an initial characterization report or site investigation and/or remedial action plan for sites that are further along on the investigation and cleanup process.
- A non-refundable application fee of $500.00.

After receipt of an application package, DES will provide a completeness determination within 10 days and, if the application package is complete, a written notice of eligibility determination within 30 days.

Liability Protection. The New Hampshire program contains the following specific liability protections of benefit to eligible parties:
- An eligible person is not liable for the remediation of additional contamination or increased environmental harm caused by pre-remedial or site investigation activities, unless attributable to negligence or reckless conduct by the eligible person.
- If the eligible person cannot complete the site cleanup, the "Covenant Not to Sue" provides protection from liability as long as the site is

stabilized to the satisfaction of DES and the site is not left in worse condition than it was before the cleanup was started.

● The "Covenant Not to Sue" is transferable to other eligible parties. The conditions for transfer to new persons may vary depending on the status of site cleanup at the time of transfer.

● Both the "Covenant Not to Sue" and the "Certificate of Completion" are recorded in the county registry of deeds to permanently document the extent of these protections.

It is important to note that these liability protections extend only to actual or potential liabilities arising under state law. The New Hampshire Brownfields Program does not relieve parties from compliance with other applicable state, federal, and local laws and regulations. Still, the EPA has recently implemented several policies which provide liability relief under certain circumstances and guidelines under which the EPA will issue "comfort letters" for EPA-listed sites. If federal liability issues are a concern for a brownfield site, DES will provide guidance and assistance upon request.

Remedial Action and Certificate of Completion. The eligible party may submit a workplan for additional site investigation with an initial non-refundable program participation fee of $3,000. The total fees paid to the State will vary depending on the complexity of the site and the amount of DES time required to review and approve reports; however, in most cases, the fees are not expected to exceed the initial program participation fee.

After a work plan is approved by DES, the eligible person will perform the necessary investigations and data analysis. After review of the investigation reports or at any other stage in this process, if DES concludes that cleanup goals have been fully attained, DES will issue a "Certificate of No Further Action" and close the site. If the reports confirm site contamination, the eligible person must develop a remedial action plan (RAP), which describes the proposed actions to clean up the site and submit the RAP to DES for approval. Upon RAP approval, DES will issue a "Notice of Approved Remedial Action Plan" and the DOJ will issue a "Covenant Not to Sue" to the eligible person, each of which may contain conditions relative to the required actions at the site. The "Notice of Approved RAP" must be recorded in the registry of deeds by the eligible party. Upon completion of active site cleanup and DES approval of a completion report prepared by the eligible party, DES will issue a "Certificate if Completion." Depending on the site, the "Certificate of Completion" may include conditions such as use restrictions, environmental monitoring requirements, and routine site maintenance requirements. When received by the eligible party, the "Certificate of

Completion" and the related "Covenant Not to Sue" will also be recorded in the county registry of deeds.

Texas

In 1995, the Texas legislature enacted the Texas Voluntary Cleanup Law[399] to provide incentives for cleanup of thousands of contaminated properties necessary to complete real estate transactions by removing liability of future landowners and lenders and by providing a process to facilitate completion of voluntary response actions in a timely and efficient manner. Pursuant to the new law, the Voluntary Cleanup Program (VCP) was established as part of the Pollution Cleanup Division of the Texas Natural Resource Conservation Commission (TNRCC). The TNRCC adopted final rules for implementation of the VCP on March 27, 1996.[400]

The VCP addresses sites that represent a real or perceived threat to public health and the environment through contaminated soil, groundwater or surface water, and air. By entering into the VCP and successfully cleaning up the property, landowners, lenders, and potential developers can be reasonably confident that they know the nature and extent of any environmental problems on the property. Once the property is successfully remediated, as necessary, the TNRCC issues a Certificate of Completion which is recorded in the county deed registry where the property is located.

The Certificate of Completion releases lenders and future landowners from liability to the state with regard to existing contamination at the site, allowing the sale or transfer of the property where the previous contamination might have otherwise posed a barrier to the sale or transfer of the property. They are also protected in the event that more stringent regulations are passed, which, without the Certificate of Completion, might have required additional cleanup. Also, the TNRCC will not initiate enforcement action against persons fulfilling the terms of the VCP agreement regarding performance of response actions. By becoming an applicant, any person (not just owners and lenders) who is not already a responsible party, may obtain a release of liability upon issuance of the Certificate of Completion.

Virtually any site is eligible for the VCP, provided that it is not subject to a TNRCC order or permit, or under the jurisdiction of the Texas Railroad Commission. Additionally, a site may be rejected from participation in the VCP if the site is subject to any other administrative, state, or federal enforcement action, or where a federal grant requires an enforcement action be taken.

---

[399] Tex. Health & Safety Code Ann. 361.601 through 361.613.
[400] Tex. Admin. Code tit. 30, 333.1 through 333.11 (Voluntary Cleanup Rules).

The first step toward participation in the VCP is to obtain and complete the VCP application package. A complete application includes an application form, an environmental site assessment, and an application fee of $1,000. Once the application is received and the site is accepted into the program, the voluntary party then enters into a VCP Agreement with the TNRCC.

The VCP Agreement is a nonbinding agreement between the applicant and the TNRCC that sets forth the terms and conditions of evaluation of the workplans and reports, and commits the applicant to pay the TNRCC's costs. After acceptance into the VCP, a TNRCC project manager contacts the applicant and the agreement negotiations begin. To ensure a complete agreement, the applicant should designate applicable rules and regulations and provide a schedule of activities that will be necessary to achieve a Certificate of Completion for the site. Once all of the terms of the agreement are agreed upon, both the applicant and the TNRCC signs it. Once the agreement is signed, the assigned VCP project manager may begin reviewing and commenting on any work plan and report submittals. Either party may terminate the agreement at any time by giving 15 days written notice. However, without a Certificate of Completion, there is no release of liability.

10

Pesticide Regulation

10.1 The Problems of Pesticide Pollution

For over a century, farmers and ranchers have applied various chemical compounds to agricultural fields in an attempt to restrict the damage caused by various pests, such as insects, rodents, and fungi. Unfortunately, many pesticides have a powerful deleterious effect on nontarget species in the environment. Chlorinated hydrocarbons insecticides (e.g., DDT, dieldrin, toxaphene, chlordane, and heptachlor) are known to have caused significant damage to nontarget populations of fish, birds, and beneficial insects.[1] Some species perceived as "pests" in agricultural situations may be critical parts of an ecosystem in other situations. To make matters worse, many of the most potent pesticides are persistent, remaining in the environment for decades.[2]

The first commercial pesticides became available in the United States in 1902. Applications of pesticides during the twentieth century have been largely responsible for the tremendous increase in agricultural productivity enjoyed by the United States (the dollar value of U.S. agricultural products rose from $440 *million* in 1964 to $12 *billion* by 1969).[3]

The problems associated with pesticide pollution are virtually unique. In few other circumstances is it the specific intent that a toxic chemical be introduced into the environment for the express purpose of destroying biological organisms.

---

[1] Office of Science and Technology, Ecological Effects of Pesticides on Non-Target Species (1971).
[2] See Rodgers, The Persistent Problem of the Persistent Pesticides: A Lesson in Environmental Law, 70 Colum. L. Rev. 567 (1970).
[3] HEW, *Report of the Secretary's Commission on Pesticides and Their Relationship to Environmental Health* (1969). See Rodgers, *Environmental Law*, 2d ed. (West, 1994) at § 5.1.

Unfortunately, pesticide use has been on the increase for many years. One of the problems is that pest species tend eventually to develop immunities to most pesticides.[4] This means that larger and larger quantities of pesticides with increasing toxicity must be applied, which greatly compounds the problems associated with pesticide pollution. One potential solution is the use of Integrated Pest Management (IPM), which utilizes procedures such as crop and pesticide rotations, the use of low-strength pesticides, and biological controls in the context of the ecological integrity of the ecosystem. Unfortunately, IPM has not received much support from the agricultural community.

An additional problem is that the technologies for pesticide application have improved to the point that small-scale individual farmers can now apply their own pesticides. For the reasons noted above, farmers tend to over-apply pesticides in an attempt to improve crop productivity. The large numbers of small farms has made it very difficult for governmental agencies to regulate pesticide use in a broad and equitable manner.

The magnitude of pesticide use is staggering. The Environmental Protection Agency (EPA) has estimated that pesticides were used on 50 percent of U.S. agricultural land in 1971, but had increased to 70 percent by 1976. From 1950 to 1978, pesticide usage increased fivefold. Approximately one billion pounds of pesticides have been used every year in the United States since 1980.[5]

It is easy to criticize the United States for our over-use of pesticides, but many (most?) agricultural countries around the world have even greater problems. The lack of regulatory controls, the use of dangerous pesticides (many of which have been banned in the United States, like DDT), and pressures to increase crop productivity at any cost have made pesticide pollution a critical worldwide problem, especially in so-called third world countries. The World Commission on Environment and Development estimated in 1987 that there are 10,000 deaths and 400,000 acute poisonings from pesticides in "developing" countries every year!

10.2 History of Pesticide Regulation

The United States recognized the dangers of unregulated pesticide use quite early. The first regulatory law was the federal Insecticide Act of 1910, which generally provided for grants and educational programs, but

---

[4] National Academy of Science, *Study on Pesticide Resistance* (Nat'l Academy Press, 1986).
[5] Rodgers, *Environmental Law*, 2d ed. (West, 1994) at § 5.1. See U.S.E.P.A., Office of Pesticides, *Pesticide Industry Sales and Usage* (1985); and Ware, *The Pesticide Book*, 5th ed.(Thomson, 1999).

had little regulatory force.

In response to concerns about the increased use of synthetic compounds as pesticides, Congress passed the Federal Insecticide, Fungicide, and Rodenticide Act (FIFRA) in 1947.[6] Under the 1947 version of FIFRA, the U.S. Department of Agriculture was required to register all pesticides before they could be put to commercial uses. The major restriction on registration was that a pesticide must be properly labeled, including instructions on how to use the pesticide safely and effectively. Congress apparently felt that health and environmental problems were the result of misuse, and that the problem could be solved by proper labeling.

In 1962, Rachel Carson published her famous book *Silent Spring*, which contemplated a world in which most species of birds (and other animals as well) had been driven to extinction by pesticides such as DDT. Her book enjoyed remarkable private and public support for its focus on the dangers of DDT, and was responsible for a new emphasis on pesticide regulation in general.

In 1964, FIFRA was amended to require the Secretary of Agriculture to suspend the registration of a pesticide if necessary to prevent an "imminent hazard." The Secretary refused to suspend the registration for DDT as it was applied to cotton (its major use at the time) pending "further study."

By 1970, the administration of FIFRA was given to the newly created EPA. Soon after, it was held that the EPA had a duty to suspend any pesticide registration when there is a "substantial question of an imminent hazard."[7] EPA Administrator William Ruckelshaus then issued an extremely controversial opinion canceling virtually all uses of DDT, which was sustained on appeal.[8] Agricultural interests had argued that a pesticide could only be withdrawn based on evidence of carcinogenicity (cancer) or mutagenicity (tumors) in humans, a very difficult burden for EPA to meet, but EPA then and now relies on animal studies.

In 1972, Congress amended FIFRA with the Federal Environmental Pesticide Control Act (FEPCA). FEPCA continued the consumer protection philosophy of the original FIFRA, but supplemented it with health-based regulations. The FEPCA amendments require that risks and benefits of pesticides be considered at four stages: registration, restricted registration, cancellation, and suspension. It should be noted that the definitions of risk and benefit are different at the different stages.

Further amendments in 1975 and 1978 served primarily to extend deadlines that could not possibly have been met, and provide rollbacks on several procedures that certain powerful lobbies had supported.

---

[6] 7 U.S.C. § 136.
[7] Environmental Defense Fund, Inc. v. Ruckelshaus, 439 F.2d 584 (D.C. Cir. 1971).
[8] Environmental Defense Fund, Inc. v. EPA, 489 F.2d 1247 (D.C. Cir., 1973).

The most sweeping amendments were those in 1988. The 1988 amendments changed several critical aspects respecting the registration process (specifically reregistration, storage and disposal of pesticides, and enforcement). The 1988 amendments will be discussed below.

10.3 The Federal Insecticide, Fungicide, and Rodenticide Act

10.3.1 What Is a "Pest," and What Is a "Pesticide"?

Under FIFRA, a "pesticide" is:

(1) any substance or mixture of substances intended for preventing, destroying, repelling, or mitigating any pest, (2) any substance or mixture of substances intended for use as a plant regulator, defoliant, or desiccant, and (3) any nitrogen stabilizer, except that the term "pesticide" shall not include any article that is a "new animal drug" . . .[9]

Other sections define plant regulators, defoliants, and desiccants more precisely, and the regulations contain further classifications of pesticides by classes.[10]

The term "pest" is defined in FIFRA Section 2(t) as:

The term "pest" means (1) any insect, rodent, nematode, fungus, weed, or (2) any other form of terrestrial or aquatic plant or animal life or virus, bacteria, or other micro-organism (except viruses, bacteria, or other micro-organisms on or in living man or other living animals) which the Administrator declares to be a pest.[11]

Section 25(c)(1) of FIFRA authorizes the EPA, after notice and opportunity for public hearing, to declare as a "pest any form of plant or animal life [except for humans and human micro-organisms] which is injurious to health or the environment."[12]

Rather than compiling a list of known pests, EPA has decided to declare virtually every living thing a potential pest. The regulations state that any "organism is declared to be a pest under circumstances that make it deleterious to man or the environment, if it is: (a) any vertebrate animal

---

[9] 7 U.S.C. § 136(u).
[10] 7 U.S.C. §§ 136(v), (f), and (g) 40 CFR § 152.3(s)
[11] 7 U.S.C. § 136(t).
[12] 7 U.S.C. § 136w(c)(1).

other than man; (b) any invertebrate animal . . . [except internal animal parasites]; (c) any plant growing where not wanted . . .; or (d) any fungus, bacterium, virus or other microorganisms [except for those growing on or in living humans or other animals, or those in processed food, beverages, drugs, cosmetics, etc.]."[13]

Under these inclusive definitions, most attempts to exclude materials from FIFRA regulation have been unsuccessful.[14] In an interesting case, a federal district court held that an "oral larvicide" fed to cattle to control fly larvae in manure is a pesticide regulated under FIFRA, not a "chemical substance" regulated under TSCA.[15] Pesticides are exempt from TSCA (see chapter 11) because they are regulated by FIFRA. Does this seem to be a reasonable tradeoff? Note that TSCA's pesticide exemption applies only to the pesticide in useable form.[16] The exemption does *not* apply to raw chemicals and ingredients, nor does it apply to pesticides during disposal.

Does FIFRA seem to be "taking sides" in the environmental arena by pitting humans against all other species? Could there be any problems with the "environmental ethics" of such a system?

10.3.2 Registration of Pesticides

The number of registered pesticides is high. The U.S. Government Accounting Office in 1986 stated that about 50,000 pesticides were registered at that time, containing about 600 active chemical ingredients.

What must be registered prior to commercial use is a pesticide that is a mixture of substances: (a) "intended for preventing, destroying, repelling, or mitigating any pest," or (b) to defoliate or desiccate plants.[17]

The process for registration of pesticides is found generally at FIFRA Section 3, which prohibits the distribution or sale of any unregistered pesticide to any person.[18]

While the statute does not say so explicitly, the burden of proving that a pesticide meets the regulatory standards is on "the proponent of initial or continued registration."[19]

FIFRA Section 3(c)(5) recognizes four criteria, which, if met, require

---

[13] 40 CFR § 152.15.

[14] For example, see Mariner Water Renaturalizer of Washington, Inc. v. Aqua Purification Systems, Inc., 665 F.2d 1066 (D.C. Cir. 1981) [water regulators characterized as eliminating disease-carrying bacteria are pesticides].

[15] Koch v. Shell Oil Co., 820 F.Supp. 1336 (D. Kan. 1993).

[16] 15 U.S.C. § 2602(B)(ii).

[17] 7 U.S.C. § 136(u).

[18] 7 U.S.C. § 136a.

[19] 50 *Fed. Reg.* 1119 (Jan. 9, 1985).

EPA to register the pesticide. The four criteria are that: (a) "its composition is such as to warrant the proposed claims for it"; (b) labeling and supporting paperwork meet the requirements of FIFRA; (c) "it will perform its intended function without unreasonable adverse effects on the environment"; *and* (d) when used "in accordance with widespread and commonly recognize practice it will not generally cause unreasonable adverse effects on the environment."[20]

"Unreasonable adverse effects on the environment" is further defined in FIFRA Section 2(bb) as "any unreasonable risk to man and the environment, taking into account the economic, social, and environmental costs and benefits of the use of any pesticide."[21] This cost-benefit requirement was the result of strong political pressures from agricultural interests when the 1972 amendments were passed.

The references in the four Section 3(c)(5) criteria to "unreasonable adverse effects on the environment" was a 1972 response to a much-litigated question of whether FIFRA imposed a substantive standard with respect to safety for humans and the environment.[22] There seems to be little question today that such substantive standards exist.

10.3.3 Classification of Pesticides

Under FIFRA Section 3(d), pesticides are classified as either "general use" or "restricted use." A "general use" pesticide is one that is determined by EPA "will not generally cause unreasonable adverse effects on the environment."[23] General use pesticides still must list any restrictions in their labeling.[24] An example of general use would be a pesticide authorized for use as a general herbicide on rangeland.

A "restricted use" pesticide is one that is not a "general use" pesticide.[25] An example of a restricted use would be a pesticide for specific use in killing rodents by certified exterminators. Restricted use pesticides require "additional regulatory restrictions" (such as a requirement that it be applied only by a certified applicator) to prevent unreasonable environmental effects. If the additional restrictions cannot prevent unreasonable effects, then the pesticide cannot be registered.

---

[20] 7 U.S.C. § 136a(c)(5).
[21] 7 U.S.C. § 136(bb).
[22] See Stearns Elec. Paste Co. v. EPA, 461 F.2d 293 (7th Cir. 1972); and Nor-Am Agricultural Products, Inc. v. Hardin, 435 F.2d 1133 (7th Cir. 1970), *rev'd on procedural grounds*, 435 F.2d 1151 (7th Cir. 1970), *cert. dismissed*, 91 S.Ct. 1399 (1971).
[23] 7 U.S.C. § 136a(d)(1)(B).
[24] 7 U.S.C. § 136a(ee).
[25] 7 U.S.C. § 136a(d)(1)(C).

Restricted pesticides are those that are subject to "acute dermal or inhalation toxicity," or that require additional regulatory restrictions. Restricted pesticides must be registered only for application under the direct supervision of a "certified applicator," or with other restrictions.[26]

Pesticides may be reclassified by EPA from general to restricted use with 45 days notice (and publication in the *Federal Register*) if warranted.[27] A registrant may seek reclassification from restricted to general use by following the procedures in Section 3(d)(3).[28]

EPA may delegate to states the operation of certification and training plans for restricted use pesticides.[29] EPA sets standards for state certification plans, but for private applicators the only qualification is the ability to fill out a form.[30] This process is often called "self certification."

State certification and training plans have been extremely popular. EPA estimated in 1985 that over 2 million applicators had been trained, and 1.2 million certified under state programs. EPA may de-authorize an outdated or inadequate state plan, but it must do so only with careful consideration of the plan.[31]

FIFRA Section 14 permits EPA to assess civil penalties (after a hearing) of up to $5,000 per offense against any registrant, commercial applicator, wholesaler, dealer, or distributor who violates any part of FIFRA.[32] Any registrant, applicant for a registration, or producer who *knowingly* violates any part of FIFRA may be subject to criminal penalties, including fines up to $50,000, and imprisonment up to one year (commercial applicators and distributors may be fined up to $25,000 and imprisoned for up to one year).

10.4 FIFRA and the Endangered Species Act

Section Section 7(a)(2) of the federal Endangered Species Act[33] requires every federal agency to ensure that its actions do not "jeopardize" a protected species, or result in "destruction or adverse modification" of

---

[26] 7 U.S.C. §§ 136a(d)(1)(C)(I) and (ii).
[27] 7 U.S.C. § 136a(d)(2).
[28] 7 U.S.C. § 136a(d)(3).
[29] 7 U.S.C. § 136i(a)(2).
[30] See 7 U.S.C. § 136i(a)(1).
[31] See Nat'l Cattlemen's Ass'n v. USEPA, 773 F.2d 268 (10th Cir. 1985).
[32] 7 U.S.C. § 136(l).
[33] The Endangered Species Act is discussed in more detail in chapter 15 of this book.

critical habitat.[34] Under inter-agency cooperative regulations, the EPA must consult with the United States Fish and Wildlife Service (or the National Marine Fisheries Service) if a pesticide is suspected of having an impact on an endangered or threatened species.[35] If a pesticide is found to "jeopardize" a protected species, then special regulatory controls are triggered. However, a 1986 study by Council on Environmental Quality demonstrated that the EPA failed to take any action for nearly half of the jeopardy objections they received.[36]

Section 9 of the ESA prohibits the "taking" of an endangered species by any person, where a taking is defined as practically any action that would harm individuals of the species or their habitat.[37] In *Defenders of Wildlife v. Administrator, E.P.A.*,[38] an environmental group sued the EPA over the continued registration and use of strychnine, which was used to kill prairie dogs but also killed endangered black-footed ferrets. The Circuit Court held that the EPA's continued registration of strychnine "constituted takings under [Section 9 of the Endangered Species] Act."[39] It is likely that an individual who harmed an endangered species by misapplying a pesticide, even one that is properly labeled, would be guilty of a violation of ESA Section 9.

10.5 Problems in Labeling and Branding

FIFRA depends heavily on proper labeling of pesticides as a technique to reduce risk of damage to health and the environment. Reliable and accurate labels are a condition for registration[40] and for lawful sale.[41]

Courts have dealt with labeling issues by assuming that approved labeling properly describes the risks,[42] by assuming that users will follow

---

[34] 16 U.S.C. § 1536(a)(2). See Roosevelt Campobello Park Comm'n v. U.S.E.P.A., 684 F.2d 1041 (1st Cir. 1982); and Palila v. Hawaii Dept. of Land and Natural Resources, 639 F.2d 495 (9th Cir. 1981). See also Houck, The Institutionalization of Caution Under Section 7 of the Endangered Species Act: What Do You Do When You Don't Know, 12 Envtl. L. Rept. 15001 (1982); and Rosenberg, Federal Protection of Unique Environmental Interests: Endangered and Threatened Species, 58 N. Car. L. Rev. 491 (1980).
[35] 40 CFR Part 402.
[36] See Rodgers, *Environmental Law*, 2d ed. (West, 1994) at § 5.5(B).
[37] 16 U.S.C. § 1538.
[38] 688 F.Supp. 1334 (D. Minn. 1988), *aff'd in part*, 882 F.2d 1294 (8th Cir. 1989).
[39] Id. at 1301.
[40] 7 U.S.C. § 136a(c)(5)(B).
[41] 7 U.S.C. §§ 136(q) and 136j(a)(1)(E).
[42] See First Nat'l Bank of Albuquerque v. U.S., 552 F.2d 370 (10th Cir. 1977), *cert. den'd*, 98 S.Ct. 122 (1977).

the labels,[43] and by limiting EPA's involvement in pesticide quality to that of regulating labels.[44]

However, many regulators and environmentalists have criticized FIFRA's heavy reliance on labeling to enforce pesticide regulations. Although EPA has clarified its labeling requirements,[45] there are still many instances where products are either intentionally or negligently mislabeled. Several court cases have emphasized the inability of labeling to offset the effects of misuse.[46]

10.6 Enforcement and Judicial Review

FIFRA enforcement operates at three levels: (1) FIFRA Section 12 describes unlawful acts; (2) FIFRA Section 13 contains "stop use" provisions, which prevent further use of a pesticide but do not penalize the user; and FIFRA Section 14 allows the Administrator of the EPA to assess penalties against violators.[47]

FIFRA Section 12 The distinction between private and commercial applicators is significant when penalties are at issue. Private applicators who use a pesticide unlawfully, such as "in a manner inconsistent with its labeling," are subject to written warning or citation from the EPA.[48] Subsequent violations are punishable by a civil penalty of not more than $1,000 for each offense. Private applicators who apply pesticides for others, but who do not come within the definition of a commercial applicator, may be assessed a civil penalty of not more than $500 for the first offense instead of a written warning or citation.[49] Subsequent violations are punishable by civil penalties of not more than $1,000 for each offense.

A person charged with a violation of FIFRA is given notice and an opportunity for a hearing before any civil penalty is assessed. In determining the amount of the penalty, the EPA considers the appropriateness of the penalty to: (1) the gravity of the violation; (2) the effect on the person's ability to continue in business, and (3) the size of the business of the person charged. If the agency finds that the violation occurred despite the exercise of due care or did not cause significant harm

---

[43] See Stearns Elec. Paste Co. v. EPA, 461 F.2d 293 (7th Cir. 1972).
[44] See S.L. Cowley & Sons Mfg. Co. v. EPA, 615 F.2d 1312 (10th Cir. 1980).
[45] See 40 CFR § 156
[46] For example, see Hercules, Inc. v. EPA, 598 F.2d 91 (D.C. Cir. 1978) [pesticide "endrin" implicated in 52 fish and wildlife kills].
[47] 7 U.S.C. §§ 136j-136l.
[48] 7 U.S.C. § 136j(a)(2).
[49] Id.

to the health or the environment, the EPA may issue a warning instead of assessing a penalty.[50]

Under FIFRA Section 14(b), private applicators may be subject to civil penalties for violations committed by persons acting for or employed by them.[51] Private applicators are also subject to criminal penalties for knowingly violating any provisions of the statute.[52] A knowing violation of the statute is a misdemeanor, punishable by a fine of not more than $1,000, or imprisonment for not more than 30 days, or both. Private applicators are also subject to criminal penalties for knowing violations committed by persons acting for or employed by them.[53]

A commercial applicator, wholesaler, dealer, retailer, or other distributor who uses, stores or disposes of a registered pesticide in violation of FIFRA may be assessed a civil penalty of not more than $5,000 for each offense.[54] A person charged with a violation must be given notice and an opportunity for a hearing before assessment of the penalty.[55] In determining the amount of the penalty, the EPA will consider: (1) the gravity of the violation; (2) the effect on the person's ability to continue in business; and the appropriateness of the penalty to the size of the business of the person charged.[56]

Upon conviction, any registrant, applicant for registration, or producer who knowingly violates a provision of FIFRA will be fined not more than $50,000, or imprisoned for not more than one year, or both.[57] Upon conviction, any commercial applicator of a restricted use pesticide, or any other person who is not a registrant, applicant for a registration, or producer, but who distributes or sells pesticides and knowingly violates any provision of FIFRA will be fined not more than $25,000, or imprisoned for one year, or both.[58]

FIFRA does not contain a "citizen's suit" provision allowing a private right of action as do several environmental laws contain provisions. As a result, suits by private citizens for improper pesticide application, storage, or disposal must be brought under common law theories of liability.

Most judicial review under FIFRA results from cancellations or suspensions of registration for particular pesticides. Only about one-third

---

[50] 7 U.S.C. § 136l(a)(4).
[51] 7 U.S.C. § 136l(b)(4).
[52] 7 U.S.C. § 136l(b)(2).
[53] 7 U.S.C. § 136l(b)(4).
[54] 7 U.S.C. § 136j.
[55] 7 U.S.C. § 136l(a)(3).
[56] 7 U.S.C. § 136l(a)(4).
[57] 7 U.S.C. § 136l(b)(1).
[58] 7 U.S.C. § 136l(b)(2).

of the pesticide cancellations or suspensions have actually resulted in litigation, largely because the number of challenges dwindled during the Reagan administration.

Many pesticide cancellations or restrictions are covered by other environmental laws, such as "takings" under the Endangered Species Act,[59] and the Clean Water Act.[60]

Although reviewing courts have tended to give deference to EPA's judgement on questions of cancellation and suspension, it is clear that reviewing courts have not hesitated to reverse an EPA decision where EPA's reasoning seemed superficial,[61] impulsive,[62] or hasty.[63] Where the EPA has followed its procedural requirements carefully, its decisions on cancellations and suspensions have been supported generally by the courts.[64] Interestingly, judicial review in the district courts uses the "arbitrary and capricious" standard,[65] while in the appellate courts it is the "substantial evidence" test.[66]

In 1992, the U.S. Supreme Court held that FIFRA preempts labeling-based state tort law cases, meaning that a state cannot permit a state suit against a pesticide producer or user who is in compliance with FIFRA.[67] On the other hand, such state claims are *not* preempted by FIFRA if advertisements for the pesticide differ substantially from claims made during the FIFRA registration process.[68]

---

[59] See Defenders of Wildlife v. EPA, 882 F.2d 1294 (8th Cir. 1989) [strychnine].

[60] See Environmental Defense Fund, Inc. v. EPA, 598 F.2d 62 (D.C. Cir. 1978 [PCBs]; and Hercules, Inc. v. EPA, 598 F.2d 91 (D.C. Cir. 1978) [chlordane and heptachlor].

[61] See Environmental Defense Fund, Inc. v. EPA, 465 F.2d 528 (D.C. Cir. 1972).

[62] See Environmental Defense Fund, Inc. v. Blum, 458 F. Supp. 650 (D.D.C. 1972).

[63] See Love v. Thomas, 858 F.2d 1347 (9th Cir. 1988), *cert. den'd*, 109 S.Ct. 1932 (1989).

[64] For example, see Northwest Food Processors Ass'n v. Reilly, 886 F.2d 1075 (9th Cir. 1989), *cert. den'd*, 110 S.Ct. 3239 (1990); and Nat'l Cattlemen's Ass'n v. USEPA, 773 F.2d 268 (10th Cir. 1985).

[65] 7 U.S.C. § 136d(c)(4); see Nat'l Coalition Against Misuse of Pesticides v. USEPA, 670 F.Supp. 55 (D.D.C. 1988), *judgment rev'd*, 867 F.2d 636 (D.C. Cir. 1989).

[66] 7 U.S.C. § 136n(b).

[67] See Cipollone v. Liggett Group, Inc., 112 S.Ct. 2608 (1992); Taylor AG Indus. v. Pure-Gro, 54 F.3d 555 (9th Cir. 1995); and Welchert v. Amer. Cyanamid, Inc., 59 F.3d 69 (8th Cir. 1995).

[68] Lowe v. Sporicidin Int'l., 47 F.3d 124 (4th Cir. 1995).

10.7 The Food Quality Protection Act

The effects of pesticides on animals are relatively well documented, but there is great uncertainty as to the effects of pesticides on humans.[69] Several studies in the 1990s focused on the effects of certain "estrogen disruptor" pesticides, which have been shown to interfere with normal reproductive patterns in animals.[70] Alarmed by the possibility that these same pesticides could have a deleterious effect on human reproduction, Congress passed the Food Quality Protection Act (FQPA) in 1996. The FQPA amended portions of FIFRA Section 25,[71] and the federal Food, Drug and Cosmetics Act (FDCA).[72]

10.7.1 The FQPA and Endocrine Disruptors

Among the mandates of the FQPA, the EPA was directed to develop an Estrogenic Substances Screening Program:

Not later than 2 years after August 3, 1996, the Administrator [of the EPA] shall in consultation with the Secretary of Health and Human Services develop a screening program, using appropriate validated test systems and other scientifically relevant information, to determine whether certain substances may have an effect in humans that is similar to an effect produced by a naturally occurring estrogen, or such other endocrine effect as the Administrator may designate.[73]

This screening program was to be conducted under the guidance of the Scientific Advisory Panel established under FIFRA Section 25(d).[74]

The final report of the Endocrine Disruptor Screening and Testing Advisory Committee (EDSTAC) was released in August, 1998. EPA is directed to implement the screening program, and:

---

[69] The Mrak Commission report in 1969 found that "200 million Americans are undergoing lifelong exposure, yet our knowledge of what is happening to them is at best fragmentary and for the most part indirect and inferential." HEW, Report of the Secretary's Commission on Pesticides and Their Relationship to Environmental Health (1969) at 37.

[70] For reviews of the literature regarding estrogen disruptor pesticides, see U.S.E.P.A., Endocrine Disruptor Screening Program (EDSP): Priority Setting Workshop, 63 *Fed. Reg.* 71,541–71,570.

[71] 7 U.S.C. § 136w.

[72] 21 U.S.C. § 346a.

[73] 21 U.S.C. § 346a(p)(1).

[74] 7 U.S.C. § 136w(d).

(A) shall provide for the testing of all pesticide chemicals; and
(B) may provide for the testing of any other substance that may have
an effect that is cumulative to an effect of a pesticide chemical if the
Administrator [of the EPA] determines that a substantial population
may be exposed to such substance.[75]

Pesticides that are determined not to have an effect on humans will be
exempted from further regulation.[76] If a manufacturer fails to offer a
pesticide for testing, or if the testing indicates that the substance may have
an endocrine effect on humans, then the EPA can take appropriate action
to reduce the likelihood of exposure by humans.[77]

10.7.2 FIFRA, the FQPA, and Organophosphates

In order to protect the safety of the food supply of the United States, the
FQPA also mandates that the EPA set a tolerance, or maximum residue
limit, for each registered pesticide. The tolerance limits are the amount of
pesticide residue that may lawfully remain in each food commodity that
has been treated with a pesticide. In establishing tolerances, the EPA
considers the toxicity of the pesticide, how much of the pesticide is
applied and how often, and how much of the pesticide (i.e., the residue)
typically remains in food and ensures that this level will be safe. The
pesticide tolerances set by the EPA are enforced by the United States Food
and Drug Administration and the Department of Agriculture, which
monitor food produced in the United States and food imported from other
countries to the United States[78]

The EPA is required to complete all tolerance reassessments by August
2006. There are 469 pesticide active ingredients or high-hazard inert
ingredients with food use tolerances; approximately 9,700 tolerances were
in effect at the passage of FQPA. EPA divided registered pesticides into
three groups, which provide the framework for the scheduling of the
pesticides for reassessment.[79]

---

[75] 21 U.S.C. § 346a(p)(3).
[76] 21 U.S.C. § 346a(p)(4).
[77] 21 U.S.C. § 346a(p)(5)(D).
[78] See U.S.E.P.A., Organophosphate Pesticides in Food. A Primer on
Reassessment of Residue Limits (Publ. 735-F-99-014, May 1999).
[79] The three groups are Group 1 (228 pesticides)—Organophosphates,
Carbamates, Probable carcinogens—Reference dose exceeders (tolerances that
are at levels above the amount that is believed to be safe for life-long, daily
consumption)—High-hazard inerts; Group 2 (93 pesticides)—Possible
carcinogens —All remaining reregistration chemicals (those that were first
registered before 1984); and Group 3 (148 pesticides)—Remaining pre-FQPA

The FQPA requires EPA to complete one-third of the tolerance reassessments by August 1999. Organophosphates were the first group to be reassessed, largely because they are used on many food crops, as well as in residential and commercial buildings and for ornamental plants and lawn care, which means that people may be exposed to them on a regular basis. In addition, they cause known effects (both acute and chronic) to humans as well as to wildlife.[80]

Organophosphates account for about half (by amount sold) of all insecticides used in the United States. In addition to major crops such as cotton, corn, and wheat, they are used on many important minor crops. Some also are used for mosquito control to protect public health against diseases such as malaria, dengue fever, and encephalitis. Approximately 60 million pounds of organophosphates are applied to approximately 60 million acres of U.S. agricultural crops annually. Nonagricultural uses account for about 17 million pounds per year.

The wide use of organophosphates is based on several factors: (1) They are relatively inexpensive; (2) They are broad spectrum (most organo-phosphates can be used on several crops to control a variety of insect pests); (3) Because of this broad spectrum of activity, one organophosphate might control the insects that would require three or four non-organo-phosphate insecticides; and (4) In general, insects have not developed resistance to organophosphates as they have to some other pesticides.

Most organophosphates are insecticides. Because they have a wide variety of uses, there are many opportunities for exposure. Organophosphates were developed during the early nineteenth century, but their effects on insects, which are similar to their effects on humans, were discovered in 1932. Some are very poisonous (they were used in World War II as nerve agents). However, they usually are not persistent in the environment. Organophosphates affect the nervous system by reducing the ability of cholinesterase, an enzyme, to function properly in regulating a neurotransmitter called acetylcholine. Acetylcholine helps transfer nerve impulses from nerve cells to muscle cells or other nerve cells. If acetylcholine is not properly controlled by cholinesterase, the nerve impulses or neurons remain active longer than they should,

---

pesticides with reregistration eligibility decisions—Remaining post-1984 pesticides—Biological pesticides—Remaining inerts. U.S.E.P.A., Organophosphate Pesticides in Food. A Primer on Reassessment of Residue Limits (Publ. 735-F-99-014, May 1999).

[80] Id.

overstimulating the nerves and muscles and causing symptoms such as weakness or paralysis of the muscles.[81]

While the acute effects of organophosphates are well documented and generally understood to cause acute cholinesterase inhibition, the chronic effects are less certain. Some studies suggest that at certain dose levels there may be long-term consequences of repeated acute exposures to these pesticides. Chronic toxicity must be evaluated on a case-by-case basis; these effects are difficult to generalize. An important aspect of the human health risk assessment is whether workers may be exposed to harmful effects of pesticides. The EPA includes provisions for protection of workers in its registration or reregistration of a pesticide.

The EPA has also determined that organophosphates might cause contamination of water and injury to plants or animals that were not the targets of the pesticide application. The EPA looks at the potential for contamination of water through runoff or seepage into groundwater, as well as the effects on other plants and animals when registering or reviewing pesticides. Problems in these areas might be addressed by restrictions on where and how a pesticide is applied.[82]

EPA completed its reassessment of organophosphates in July, 1998 (although it was not released to the public until October, 1999). In its report, the Hazard Identification Review Committee determined that one particular group of organophosphates known as "chlorpyrifos" pose a safety risk for people who use it in their gardens, fields and homes.[83] Specifically, the EPA report states that extensive exposure to "Dursban," an insecticide manufactured by Dow Chemical Company, is linked to blurred vision, muscle weakness, headaches and problems with memory, depression and irritability. Dursban is a popular insecticide, and is applied to home gardens, large-scale agricultural fields, and to many commercial and private buildings for termite and cockroach control.[84] Dow Chemical Company, in a letter included in the EPA report, argued that the EPA risk analysis was misleading and based on fundamental scientific errors.[85]

---

[81] Id.

[82] Id.

[83] Hazard Identification Assessment Review Committee (U.S.E.P.A.), *Hazards of the Organophosphates* (USEPA, July 7, 1998); and U.S.E.P.A., *Preregistration Eligibility Sceince Chapter for Chlorpyrifos. Fate and Environmental Risk Assessment* (USEPA, October 1999).

[84] In 1997 Dow Chemical Co. voluntarily stopped selling Dursban for use in pet shampoos and dips and household foggers.

[85] Hazard Identification Assessment Review Committee (U.S.E.P.A.), *Hazards of the Organophosphates* (USEPA, July 7, 1998).

10.8 Integrated Pest Management (IPM)

FIFRA Section 20-1 describes Integrated Pest Management (IPM) as "a sustainable approach to managing pests by combining biological, cultural, physical, and chemical tools in a way that minimizes economic, health, and environmental risks."[86] More specifically, IPM combines various pest management techniques like crop rotation, timed crop planting and biological controls.[87] In many situations, IPM techniques result in an equally effective program of pest control compared to reliance on pesticides alone, but with less expensive through reduced consumption of expensive chemicals.[88]

FIFRA Section 11 requires that the Administrator of the EPA make IPM information available to those who request it through Cooperative State Extension Services and applicator certification programs.[89] Many state extension services provide additional training and assistance in IPM techniques. Such services might include "scouting" fields for pest population levels and planning appropriate IPM programs. The use of IPM techniques, where practical, are recommended as a means of reducing pesticide use, and adventitiously reducing the chance of accidents and lawsuits resulting from injury to health or the environment.

10.9 Biopesticides

Biological pesticides ("biopesticides") are various types of pesticides derived from natural materials such as animals, plants, bacteria, and certain minerals. For example, garlic, mint, and baking soda all have pesticidal applications and are considered biopesticides.

At the end of 1998, there were approximately 175 registered biopesticide active ingredients and 700 products. Biopesticides fall into three major categories.

● Microbial pesticides contain a microbial and microorganism (bacterium, fungus, virus, antimicrobial protozoan or alga) as the active ingredient. Pesticides: The most widely known microbial pesticides are varieties of

---

[86] 7 U.S.C. § 136r-1.
[87] For more information on IPM nethods, see Committee on Pest and Pathogen Control, *Ecologically Based Pest Management: New Solutions for a New Century* (Nat'l Acad. Press, 1996); and Van Emden, *Beyond Silent Spring: Integrated Pest Management and Chemical Safety* (Chapman & Hall, 1996).
[88] See Bottrell, *Integrated Pest Management* (Council on Environmental Quality, 1979).
[89] 7 U.S.C. § 136i(c).

the bacterium Bacillus. Certain other microbial pesticides act by out-competing pest organisms. Microbial pesticides need to be continuously monitored to ensure they do not become capable of harming non-target organisms, including humans.

- Plant-pesticides are pesticidal substances that plants produce from genetic material that has been added to the plant. For example, scientists can take the gene for antimicrobial pesticides for the Bt pesticidal protein, and introduce the gene into the plants' own genetic control material. Then the plant manufactures the substance that destroys the pest. Both the protein and its genetic material are regulated by the EPA (although the plant itself is not regulated).
- Biochemical pesticides are naturally occurring substances that control pests by non-toxic mechanisms. Conventional pesticides, by contrast, are synthetic materials that usually kill or inactivate the pest. Biochemical pesticides include substances that interfere with growth or mating, such as plant growth regulators, or substances that repel or attract pests such as pheromones. Because it is sometimes difficult to determine whether a pesticide controls the pest by a non-toxic mode of action, EPA has established a committee to determine whether a pesticide meets the criteria for a biochemical pesticide.[90]

Some of the advantages of biopesticides are:

- Biopesticides are inherently less harmful than conventional pesticides.
- Biopesticides are designed to affect only one specific pest or, in some cases, a few target organisms, in contrast to broad spectrum, conventional pesticides that may affect organisms as different as birds, insects, and mammals.
- Biopesticides often are effective in very small quantities and often decompose quickly, thereby resulting in lower exposures and largely avoiding the pollution problems caused by conventional pesticides.
- When used as a component of Integrated Pest Management (IPM) programs, biopesticides can greatly decrease the use of conventional pesticides, while crop yields remain high.

Since biopesticides tend to pose fewer risks than conventional pesticides, EPA generally requires much less data to register a biopesticide than to register a conventional pesticide.

New biopesticides are often registered in less than a year, compared with an average of more than three years for conventional pesticides. While biopesticides require less data and are registered in less time than

---

[90] See U.S.E.P.A., *Biopesticides* (January 6, 1999).

conventional pesticides, the EPA must always conduct rigorous reviews to ensure that pesticides will not have adverse effects on human health or the environment. For the EPA to be sure that a pesticide is safe, it requires that registrants submit a variety of data about the composition, toxicity, degradation, and other characteristics of the pesticide.[91]

---

[91] Id.

11

Toxic Substance Regulation

11.1 The Problems with Toxic Chemicals

11.1.1 Early History

Prior to the industrial revolution, few deadly chemicals were available for human use. However, substances such as cyanide (used to extract gold and other metals from ore), as well as some poisons, were known. The Industrial Revolution introduced a variety of toxic chemicals into the environment, many of which were byproducts of the manufacture of metals, fabrics, and dyes. In addition, toxic chemicals became common in airborne wastes.

During World War II, many toxic chemicals were introduced into the environment either accidentally as byproducts of wartime manufacturing techniques, or intentionally as various poisons intended to kill or incapacitate humans.

After World War II, the synthetic chemical industry proliferated, largely as a result of the expanded uses of synthetic plastics. New uses for "polymer chemistry" resulted in the production of thousands of new chemicals, very few of which were submitted to testing for possible toxicity. For example, polyvinyl plastics were commonly used in the 1950s for everything from automobile seat covers to milk containers. In the early 1960s, researchers discovered that toxins released by polyvinyl plastic containers had leached out of the plastic and into stored blood, and was probably responsible for a disease in hospitalized soldiers in Vietnam called "shock lung."

Several federal laws, such as the federal Food, Drug, and Cosmetics Act[1]

---

[1] 21 U.S.C. § 301

were designed to address some of these problems, but no comprehensive statute dealt with potentially dangerous chemicals until the federal Toxic Substances Control Act (TSCA).

11.1.2 What Are Toxic Substances?

There are millions of chemical compounds currently in existence, and thousands of new ones are developed every year. Many of these existing and new chemical substances are lethal or injurious to human health or the environment (e.g., polychlorinated biphenyls (PCBs), vinyl chloride, and chlorofluorocarbons).

The medical and legal interpretations of "toxic substances" differ. The medical definition of a toxic substance is any substance that interferes with normal physiology (function) when taken into the body by ingestion, inhalation, injection, or absorption. Virtually any substance *can* be toxic if consumed in sufficient quantity (e.g., aspirin is not usually considered a toxin, but accidental aspirin overdoses kill more children annually than do any traditional poisons).

There is no specific legal definition of a "toxic substance," although TSCA implies one in discussing: "chemical substances and mixtures . . . whose manufacture, processing, distribution in commerce, use, or disposal may present an *unreasonable risk of injury to health or the environment*."[2] As we will see, the TSCA definition of "toxic substances" seems to overlap definitions for "hazardous" substances as they appear in other environmental statutes, such as RCRA, the Clean Air Act, Clean Water Act (CWA), CERCLA, etc. In fact, there is no consistent way to distinguish "hazardous" and "toxic" substances other than the way they are listed by the various statutes (this will be discussed further).

TSCA is not the only statute to use the term "toxic" substance. Some other statutes refer to toxic substances. For example, the federal CWA requires the Environmental Protection Agency (EPA) to prepare a list of toxic water pollutants based on "toxicity of the pollutant, its persistence, degradability . . . and the extent and effect of the toxic pollutant on [aquatic organisms]."[3] Unfortunately, the CWA never specifically defines what it means by "toxic" or "toxicity."

---

[2] 15 U.S.C. § 2601(a)(2) (emphasis added).
[3] 33 U.S.C. § 1317(a)(1).

11.2 The Toxic Substances Control Act of 1976

TSCA was passed in 1976, and its basic purpose is to regulate various chemical substances and mixtures. It does so by regulating both the distribution of existing chemicals and the manufacture of new chemicals based on their risks to health and the environment.[4] The burden under TSCA is placed on manufacturers to supply information on environmental and health effects of chemical substances and mixtures to the EPA. In turn, the EPA has broad power to regulate the manufacture, use, distribution in commerce, and disposal of chemical substances and mixtures, but it must do so only after balancing the economic and social benefits of a chemical against the risks.

Examination of Congress' findings in TSCA Section 2(a) indicates that Congress recognized the problem:

The Congress finds that —
  (1) human beings and the environment are being exposed each year to a large number of chemical substances and mixtures;
  (2) among the many chemical substances and mixtures which are constantly being developed and produced, there are some whose manufacture, processing, distribution in commerce, use, or disposal may present an unreasonable risk of injury to health or the environment; and
  (3) the effective regulation of interstate commerce in such chemical substances and mixtures also necessitates the regulation of intrastate commerce in such chemical substances and mixtures.[5]

In the TSCA Section 2(b) statement of policy, Congress set out the basic strategy to be employed by TSCA:

It is the policy of the United States that —
  (1) adequate data should be developed with respect to the effect of chemical substances and mixtures on health and the environment and that the development of such data should be the responsibility of those who manufacture and those who process such chemical substances and mixtures;
  (2) adequate authority should exist to regulate chemical substances and mixtures which present an unreasonable risk of injury to health or the environment, and to take action with respect to chemical substances and mixtures which are imminent hazards; and

---

[4] For an in-depth review of all TSCA programs, see McKenna & Cuneo, *Tsca Handbook*, 3rd ed. (Government Institutes, 1997).
[5] 15 U.S.C. § 2601(a).

(3) authority over chemical substances and mixtures should be exercised in such a manner as not to impede unduly or create unnecessary economic barriers to technological innovation while fulfilling the primary purpose of this chapter to assure that such innovation and commerce in such chemical substances and mixtures do not present an unreasonable risk of injury to health or the environment.[6]

Finally, Congress makes clear in TSCA Section 2(c) that all factors, economic as well as environmental, are to be considered in regulating potentially chemicals:

It is the intent of Congress that the Administrator [of the EPA] shall carry out this chapter in a reasonable and prudent manner, and that the Administrator shall consider the environmental, economic, and social impact of any action the Administrator takes or proposes to take under this chapter.[7]

As an environmental statute, TSCA is unusual in several ways. For one, few other statutes have the immediate and direct impact that TSCA did. TSCA imposes sudden, complete regulation on a major industry, which employs 1.5 million people at 12,500 establishments, and produces chemical products valued at over $275 billion in 1990.

Interestingly, TSCA has remained remarkably unchanged through the years. The only major amendments occurred in 1986. TSCA's regulatory scheme is lengthy (although less so than many other federal environmental laws), but it is relatively easy to understand.

TSCA's mandates apply to "chemical substances and mixtures." "Chemical substance" is defined in TSCA Section 3(2) as:

[A]ny organic or inorganic substance of a particular molecular identity, including —
    (i) any combination of such substances occurring in whole or in part as a result of a chemical reaction or occurring in nature and
    (ii) any element or uncombined radical.[8]

However, the term "chemical substance" specifically excludes:

    (i) any mixture [defined below],
    (ii) any pesticide (as defined in the Federal Insecticide, Fungicide,

---

[6] 15 U.S.C. § 2601(b).
[7] 15 U.S.C. § 2601(c).
[8] 15 U.S.C. § 2602(2)(A).

and Rodenticide Act [7 U.S.C. Sections 136 et seq.]) when manufactured, processed, or distributed in commerce for use as a pesticide,

    (iii) tobacco or any tobacco product,

    (iv) any source material, special nuclear material, or byproduct material (as such terms are defined in the Atomic Energy Act of 1954 [42 U.S.C. Sections 2011 et seq.] and regulations issued under such Act),

    (v) any article the sale of which is subject to the tax imposed by section 4181 of the Internal Revenue Code of 1986 [liquor], and

    (vi) any food, food additive, drug, cosmetic, or device (as such terms are defined in section 201 of the Federal Food, Drug, and Cosmetic Act [21 U.S.C. Section 321]) when manufactured, processed, or distributed in commerce for use as a food, food additive, drug, cosmetic, or device.[9]

A "mixture" is defined as:

[A]ny combination of two or more chemical substances if the combination does not occur in nature and is not, in whole or in part, the result of a chemical reaction; except that such term does include any combination which occurs, in whole or in part, as a result of a chemical reaction if none of the chemical substances comprising the combination is a new chemical substance and if the combination could have been manufactured for commercial purposes without a chemical reaction at the time the chemical substances comprising the combination were combined.[10]

TSCA specifically regulates "new" chemical substances and mixtures, which are defined as: "any chemical substance which is not included in the chemical substance list compiled and published under section 2607(b) of this title."[11]

11.3 TSCA Section 5 and Premanufacture Notice

Under TSCA Section 5's mandates, the EPA is required to prepare a list of toxic substances that are new or have new uses. In preparing the list the EPA must consider "all relevant factors," including the volume of the substance, the extent of changes in the type of exposure to humans and the

---

[9] 15 U.S.C. § 2602(2)(B).
[10] 15 U.S.C. § 2602(8).
[11] 15 U.S.C. § 2602(9).

environment, increases in the magnitude and duration of exposure, and anticipated methods of manufacture, distribution, and disposal.[12]

Section 5 of TSCA requires manufacturers or those involved in commerce or chemical disposal to give to the EPA premanufacture notice (PMN) regarding any "new chemical substance."[13]

The key to the TSCA Section 5 program is the requirement that any person who manufactures or processes a new chemical for commercial purposes must submit a "notice of intent" (the PMN) to the EPA at least 90 days before they begin manufacturing or processing.[14] The person must submit test data for the substance, which must be performed under carefully controlled circumstances spelled out in the TSCA Section 4 and the regulations.[15] These will be discussed in more detail below.

The TSCA Section 5 PMN requirement is, however, limited to persons who "manufacture for commercial purposes" the chemical in question. TSCA Section 3 defines "manufacture" as "to import into the customs territory of the United States . . . , produce, or manufacture."[16] In its TSCA regulations, however, the EPA makes it very clear that the term "Manufacture for commercial purposes" is interpreted very broadly:

(1) Manufacture for commercial purposes means to import, produce, or manufacture with the purpose of obtaining an immediate or eventual commercial advantage for the manufacturer, and includes, among other things, such ``manufacture" of any amount of a chemical substance or mixture:

(i) For distribution in commerce, including for test marketing.

(ii) For use by the manufacturer, including use for product research and development, or as an intermediate.

(2) Manufacture for commercial purposes also applies to substances that are produced coincidentally during the manufacture, processing, use, or disposal of another substance or mixture, including both byproducts that are separated from that other substances or mixture and impurities that remain in that substance or mixture. Such byproducts and impurities may, or may not, in themselves have commercial value. They are nonetheless produced for the purpose of obtaining a commercial advantage since they are part of the manufacture of a chemical product for a commercial purpose.[17]

---

[12]   15 U.S.C. § 2604(a)(2).
[13]   15 U.S.C. § 2604(a)(1)(A).
[14]   15 U.S.C. § 2604(a)(1).
[15]   40 CFR §§ 790-792 and 796 et seq., and in § 4 of TSCA, 15 U.S.C. § 2603, discussed below.
[16]   15 U.S.C. § 2602(7).
[17]   40 CFR § 717.3(e).

The TSCA Section 5 PMN requirement applies to "process" chemicals for commercial purposes, which is defined in EPA regulations as:

> Process for commercial purposes means the preparation of a chemical substance or mixture, after its manufacture, for distribution in commerce with the purpose of obtaining an immediate or eventual commercial advantage for the processor. Processing of any amount of a chemical substance or mixture is included. If a chemical substance or mixture containing impurities is processed for commercial purposes, then those impurities are also processed for commercial purposes.[18]

If the information in a PMN is insufficient to allow a reasoned conclusion on health and the environment, or if the chemical might present unreasonable risk, then TSCA Section 5(e) allows the EPA to issue an administrative order or seek a court injunction to stop or limit its use.[19]

The EPA's record in implementing PMN requirements has not been good. It was years late in promulgating final regulations, and only a few chemicals have been subjected to Section 5(e) sanctions.[20]

11.3.1 Contents of the PMN

Under TSCA Section 5(d), the PMN:

> [S]hall (1) identify the chemical substance or mixture for which data have been received; (2) list the uses or intended uses of such substance or mixture and the information required by the applicable standards for the development of test data; and (3) describe the nature of the test data developed.[21]

A PMN form is available from local U.S. EPA offices. EPA regulations provide additional information on the contents of the PMN:

> The [PMN] must contain the following information:
> (i) The specific chemical identity of the PMN substance.
> (ii) A generic chemical name (if the chemical identity is claimed as

---

[18] 40 CFR § 717.3(g).
[19] 15 U.S.C. § 2604(e).
[20] See Rodgers, *Environmental Law*, 2d ed. (West, 1994) at § 6.3; and Applegate, The Perils of Unreasonable Risk: Information, Regulatory Policy, and Toxic Substances Control, 91 Colum. L. Rev. 261 (1981).
[21] 15 U.S.C. § 2603(d).

confidential by the submitter).

(iii) The premanufacture notice (PMN) number assigned by EPA.

(iv) The date of commencement for the submitter's manufacture or import for a non-exempt commercial purpose (indicating whether the substance was initially manufactured in the United States or imported). The date of commencement is the date of completion of non-exempt manufacture of the first amount (batch, drum, etc.) of new chemical substance identified in the submitter's PMN. For importers, the date of commencement is the date the new chemical substance clears United States customs.

(v) The name and address of the submitter.

(vi) The name of the authorized official.

(vii) The name and telephone number of a technical contact in the United States.

(viii) The address of the site where commencement of manufacture occurred.

(ix) Clear indications of whether the chemical identity, submitter identity, and/or other information are claimed as confidential by the submitter.[22]

Some submitters (manufacturers) may be concerned that divulging confidential information about their chemical substances and mixtures will provide an unfair advantage to competitors. To this end, a manufacturer may claim that the identity of the chemical substance is confidential, and request that the identity to be listed on the confidential portion of the Chemical Use Inventory.[23] The confidentiality claim must be reasserted and substantiated in accordance with EPA regulations,[24] or the EPA will list the specific chemical identity on the public inventory. However, submitters who do not claim the chemical identity, submitter identity, or other information to be confidential in the PMN cannot claim this information as confidential in the notice of commencement.[25]

11.3.2 Exclusions and Exemptions from the PMN Requirement

Under some circumstances, a manufacturer of a "new chemical substance or mixture" might be excused from the TSCA Section 5 PMN requirement. For example, in section 11.2 it was noted that the definition

---

[22] 40 CFR § 720.102(c).
[23] Id.
[24] 40 CFR § 720.85(b).
[25] 40 CFR § 720.102(c).

of "chemical substance" specifically excludes any mixture, any pesticide (when manufactured, processed, or distributed in commerce for use as a pesticide), tobacco or any tobacco product, nuclear material, and any food, food additive, drug, or cosmetic.[26] In short, these materials are all regulated under other statutes, such that TSCA regulation would be redundant to a degree.

The EPA may exempt manufacturers, processors, and persons involved in disposal of chemicals from the PMN requirement if the EPA determines that the material will not present unreasonable risk of health or environmental effects or if the substance is the chemical equivalent of a substance that has already been submitted, such that the another PMN would be duplicative.[27] Interestingly, however, an exempted person may have to reimburse any other person who previously submitted a PMN for a portion of the costs of preparing the PMN, unless the original PMN was based solely on uses in scientific research.[28] Because testing is time consuming and costly, Congress decided that a person who takes advantage of the original PMN should share in the expense.

TSCA Section 5(h)(1) allows a Test Market Exemption (TME) for a manufacturer who produces a chemical solely to test its marketability "upon a showing by such person satisfactory to the Administrator [of the EPA] that the manufacture, processing, distribution in commerce, use, and disposal of such substance, and that any combination of such activities, for such purposes will not present any unreasonable risk of injury to health or the environment, and . . . under such restrictions as the Administrator considers appropriate."[29] An additional exemption is available when the new chemical substance or mixture is produced "solely for purposes of . . . scientific experimentation or analysis."[30]

Perhaps the most familiar TSCA Section 5(h) exemption is for certain low-volume or low-release chemicals. The Low-Volume Exemption (LVE) and the Low-Exposure Exemptions (LoREX) regulations were substantially amended in 1995. The amended LVE exempts certain new, low-volume substances from the requirement to submit a "full" PMN. Under the new LVE, there is no $2,500 filing fee (which is required for a PMN), and the review period is only 30 days as compared to the 90-day review period for a full PMN. To be eligible for the LVE, the manufacturer or importer must manufacture or import no more than 10,000 kg/yr of the substance and must file a low-volume exemption application (LVEA) at least 30 days prior to manufacturing or importing the substance

---

[26] 15 U.S.C. § 2602(2)(B).
[27] 15 U.S.C. §§ 2604(h)(1) and (2).
[28] 15 U.S.C. § 2604(h)(2)(B).
[29] 15 U.S.C. § 2604(h)(1).
[30] 15 U.S.C. § 2604(h)(3)(A).

for a nonexempt commercial purpose.[31] Benefits of the amended LVE include: (1) the predecessor LVE was limited to manufacture or import volumes of no greater than 1000 kg/yr, while the new limit is 10,000 kg/yr; (2) under the original LVE, no more than one person could be granted an LVE for a substance, while the amended LVE permits multiple parties to hold LVE's for the same substance. However, when reviewing subsequent LVEA's, EPA will consider whether the potential human exposure to, and environmental release of, the substance at the higher aggregate production volume will present an unreasonable risk of injury to human health or the environment. Thus, there is a clear advantage to being the first holder of an LVE for a particular substance. In addition, the original LVE rule did not allow manufacturers to change their manufacturing site from that described in their LVEA unless they submitted a new LVEA at least 21 days prior to beginning manufacture of the substance at the new site. (3) The new LVE rule permits a change in manufacturing site without a new LVEA filing if, through records kept at the new manufacturing site, the manufacturer can show that exposure to individual workers at the new site is equal to or less than that reported in the original filing, and certain environmental release and exposure criteria are satisfied.

Limitations of the new LVE rules include are worth noting as well. First, substances that are manufactured or imported under the LVE are not added to the TSCA inventory, which can be problematic from a customer assurance standpoint. Second, a holder of an LVE is bound to the conditions described in the low volume exemption application (LVEA) including: (1) use; (2) site of manufacture; (3) exposure and release controls (including physical form, if applicable); (4) importation only, if so specified; and (5) a production volume of less than 10,000 kg/yr, if so specified. In addition, EPA requires manufacturers or importers of LVE substances to notify processors and industrial users of the substance that it is a new chemical substance whose use is restricted to the uses specified in the LVEA. Customers must also be notified if the substance is subject to any exposure and environmental release controls specified in the LVEA (including physical form). Also, while a new LVEA need not be submitted, a manufacturer that changes the manufacturing site listed in the LVEA must inform EPA within thirty days of commencement of manufacturing at the new site.[32] Despite a few minor shortcomings, with its 10,000 kg/yr production limit and 30-day review period, the new LVE is proving to be an attractive alternative to the PMN.

---

[31] 40 CFR § 723.50.
[32] U.S.E.P.A., *The TSCA Low Volume Exemption -- A Practical Alternative to a PMN* (10/9/97).

11.4 TSCA Section 4 Testing Requirements

Section 4 of TSCA requires the EPA to require testing for any chemical or mixture of chemicals that may "present an unreasonable risk of injury to health or the environment."[33] Although it is not explicitly required to do so by the statute, the EPA has undertaken a list of all existing and new chemicals and mixtures that the EPA determines present an unreasonable risk of injury to health or the environment or lack sufficient information for an informed preliminary assessment.

The EPA's record on developing a list of toxic chemicals has not been without problems. Over 55,000 chemicals and mixtures were listed when the "Inventory of Existing Chemicals" was closed in 1980. Since then, the EPA has been forced to spend more time on TSCA mandates than on assessing risks and controlling potentially dangerous chemicals. Over 1,000 inventoried chemicals have been screened so far, but only a handful have received more than cursory attention.[34]

The EPA is changing its inventory update rules. It is implementing the Chemical Use Inventory (CUI), which will compile data related to uses of chemicals in commerce resulting in human and environmental exposures.

TSCA Section 4(b)(1)(B) requires EPA to develop by rule "standards for the development of test data" for chemical substances and mixtures.[35] These standards must be reviewed and amended if necessary every twelve months.

Standards created under TSCA Section 4 may contain information on:

The health and environmental effects for which standards for the development of test data may be prescribed include carcinogenesis, mutagenesis, teratogenesis, behavioral disorders, cumulative or synergistic effects, and any other effect which may present an unreasonable risk of injury to health or the environment. The characteristics of chemical substances and mixtures for which such standards may be prescribed include persistence, acute toxicity, subacute toxicity, chronic toxicity, and any other characteristic which may present such a risk. The methodologies that may be prescribed in such standards include epidemiologic studies, serial or hierarchical tests, in vitro tests, and whole animal tests, except that before prescribing epidemiologic studies of employees, the Administrator

---

[33] 15 U.S.C. § 2603(a).
[34] For specific EPA actions, see 43 *Fed. Reg.* 11,318 (1979) [chlorofluorocarbon propellants in aerosol containers]; and 45 *Fed. Reg.* 61,966 (1980) [asbestos building materials in schools].
[35] 15 U.S.C § 2603(b)(1)(B).

shall consult with the Director of the National Institute for Occupational Safety and Health.[36]

As the terms are used here, "carcinogenesis" means that the chemical might cause cancer, "mutagenesis" means that it might cause mutations, and "teratogenesis" referes to developmental (birth) defects. Cumulative or synergistic effects); (b) chemical characteristics (e.g., persistence, and acute, subacute, and chronic toxicity); and (c) methodologies (epidemiologic studies, serial or hierarchical tests, in vitro and whole animal tests).[37]

TSCA Section 4(e) creates an interagency committee whose job it is to recommend to the EPA which chemicals should receive priority.[38] This committee (now called the Interagency Testing Committee, or ITC) makes recommendations to the EPA regarding rules for specific chemical substances and mixtures "to which the Administrator [of the EPA] should give priority consideration" in developing regulations. In making its recommendation with respect to any chemical substance or mixture, the committee considers "all relevant factors," including:

(i) the quantities in which the substance or mixture is or will be manufactured,

(ii) the quantities in which the substance or mixture enters or will enter the environment,

(iii) the number of individuals who are or will be exposed to the substance or mixture in their places of employment and the duration of such exposure,

(iv) the extent to which human beings are or will be exposed to the substance or mixture,

(v) the extent to which the substance or mixture is closely related to a chemical substance or mixture which is known to present an unreasonable risk of injury to health or the environment,

(vi) the existence of data concerning the effects of the substance or mixture on health or the environment,

(vii) the extent to which testing of the substance or mixture may result in the development of data upon which the effects of the substance or mixture on health or the environment can reasonably be determined or predicted, and

(viii) the reasonably foreseeable availability of facilities and personnel for performing testing on the substance or mixture.[39]

---

[36] Id.
[37] 15 U.S.C. § 2603(b)(2)(A).
[38] 15 U.S.C. § 2603(e)(1)(A).
[39] Id.

In addition, the ITC is to:

[G]ive priority attention to those chemical substances and mixtures which are known to cause or contribute to or which are suspected of causing or contributing to cancer, gene mutations, or birth defects.[40]

The recommendations of the ITC are in the form of a list of chemical substances and mixtures in the order in which the ITC feels the EPA should take action under TSCA Section 4(a).

Unfortunately, the EPA has been slow in implementing TSCA Section 4. No final test rules at all were promulgated until late in 1983. The EPA was forced to promulgate the first 18 test rules after in lost the decision in *NRDC v. Costle*.[41]

Because it was so slow in developing test rules, the EPA attempted to rely on negotiated test rules with manufacturers, rather than on promulgating its own test rules. Under this plan, the EPA would allow manufacturers of new chemical substances and mixtures to use their expertise in devising test rules that the EPA would then apply under TSCA Section 4. Initially, courts did not look on negotiated test rules with favor.[42] However, the leading court decision on TSCA Section 4 testing authority is *Chemical Manufacturers Association v. USEPA*,[43] which generally upholds the EPA's testing rule authority.[44] Nevertheless, there have been challenges to the EPA's findings on specific chemicals.[45]

11.5 TSCA Section 6 Regulation

TSCA Section 6 deals with TSCA regulation of "hazardous substances and mixtures." However, TSCA Section 6 refers consistently to any "chemical substance or mixture" whose "manufacture, processing, distribution in commerce, use, or disposal" presents "an unreasonable risk of injury to health or the environment."[46] This characterization of a

---

[40] Id.

[41] NRDC v. Costle, 10 Envtl. Law. Rep. (ELI) 20,274 (S.D.N.Y. 1980).

[42] See NRDC v. USEPA, 595 F. Supp. 1255 (S.D.N.Y. 1984); Citizens for a Better Environment v. Thomas, 704 F. Supp. 149 (N.D. Ill. 1989); and Citizens for a Better Environment v. Reilly, 33 Env. Rep. (Cas.) 1460 (N.D. Ill. 1991).

[43] Chemical Manufacturers Association v. USEPA, 859 F.2d 977 (D.C. Cir. 1988).

[44] See also Ausimont USA, Inc. v. EPA, 838 F.2d 93 (3d Cir. 1988); and Shell Chem. Co. v. EPA, 826 F.2d 295 (5th Cir. 1987).

[45] See, for example, Chemical Mfrs. Ass'n v. EPA, 899 F.2d 344 (5th Cir. 1990) ["cumene," or isopropyl benzene].

[46] See, for example, 15 U.S.C. § 2605(a).

"hazardous" chemical differs substantially from that in most other environmental laws, such as the Clean Air Act (CAA, Chapter 7), the Solid Waste Disposal Act and RCRA (Chapter 8), and the Comprehensive Environmental Response, Compensation, and Liability Act (CERCLA). However, the TSCA Section 6 characterization of a "hazardous" chemical is similar to the description of "toxic" chemicals found in TSCA Section 4 and Section 5.

TSCA Section 6 sets out a series of regulatory tools that create a continuum roughly from the most strict to the more lenient, although the least burdensome is prescribed for a particular situation.[47] These range from outright bans on manufacturing, production, and distribution, to limits on amounts, to mere warnings, instructions, public notices, and monitoring and testing obligations.[48]

The EPA can inquire into the quality control procedures of manufacturers, which may lead to notice and recall actions.[49] TSCA Section 6(a) also allows the EPA to confine certain requirements to "specified geographic areas," which has led to a series of regional compromises (e.g., phosphorous in detergents are limited in areas with eutrophic waters).

TSCA Section 6(a) also imposes a "least to most" burdensome requirement on manufacturers, so long as the requirements will "protect adequately" against the risk.[50] A federal court held that for the EPA to impose a complete ban on asbestos, it must demonstrate not only that its action reduces the risk to an adequate level, but also that less burdensome actions would be inadequate.[51] While this version of cost-benefit analysis has appeal to the regulated community, some environmentalists worry that the least burdensome requirements may also be the least dependable.

The various regulatory options in TSCA often seem to be in competition with each other. For example, those who manufacturer, transport, or dispose of toxic materials must know when it is sufficient to report under TSCA Section 8, to test under Section 4, develop information under Section 5(e)(2), or examine processes under Section 6(b)(2). They must also be concerned whether EPA will act quickly under TSCA Sections 5(f) or 7, or hold off for 180 days under Section 4(f).

TSCA ultimately regulates toxic and hazardous chemicals, but so do many other environmental laws. The EPA often must choose among several laws (including TSCA) to determine which applies. For example, is it TSCA or another statute that applies to hazardous air pollutants

---

[47] 15 U.S.C. § 2605(a).
[48] 15 U.S.C. §§ 2605(a)(1)(A) and (B).
[49] 15 U.S.C. § 2605(b).
[50] 15 U.S.C. § 2605(a).
[51] Corrosion Proof Fittings v. EPA, 947 F.2d 1201 (5th Cir. 1991).

(CAA?), used oil, toxins in water discharges (CWA?), solid wastes (RCRA, CERCLA?) or radioactive-contaminated sites (Atomic Energy Act?). These conflicts have created yet more headaches for the agencies, and for the regulated community as well. TSCA Section 6(c) states that the EPA must favor risk-reduction actions under other laws, unless "it is in the public interest" to proceed under TSCA.[52]

Section 6(a)(1) of TSCA provides limitations on the amount of a chemical that may be manufactured or distributed.[53] These production limits have caused controversy because they establish what amounts to quotas, which leads to allocations of the right to produce a certain chemical. Allocations have often led to painful choices of who gets to produce a particular chemical, and who does not.

TSCA rulemaking under Section 6 provides for notice and comment procedures, subject to an informal hearing.[54] At this hearing, an interested person is entitled to present a position either orally or in writing.[55] If the EPA determines that there are "disputed issues of material fact," then the parties must be allowed to submit "rebuttal submissions" and conduct appropriate cross examination.[56] Citizen groups may be awarded attorney's fees and expert witness fees.[57]

11.5.1 Section 6 and Polychlorinated Biphenyls (PCBs)

Only six chemical substances are specifically regulated under Section 6 of TSCA. These are asbestos, chlorofluorocarbons (CFCs), dioxins, hexavalent chromium, certain metal-working fluids, and polychlorinated biphenyls.[58] EPA regulation of CFCs will be discussed in section 11.5.2.

Probably the best-known regulation of chemicals under Section 6 of TSCA is that of PCBs. PCBs are actually over 200 flame-resistant compounds that were used for many years as adhesives, textile coatings, and in transformers and other electrical equipment. Although its use has

---

[52] 15 U.S.C. § 2605(c).
[53] 15 U.S.C. § 2605(a)(1).
[54] 15 U.S.C. §§ 2605(c)(2) and (3).
[55] 15 U.S.C. § 2605(c)(3)(A)(I).
[56] 15 U.S.C. § 2605(c)(3)(A)(ii)). See Corrosion Proof Fittings v. EPA, 947 F.2d 1202 (5th Cir. 1991) [EPA failed to provide adequate cross examination on rule banning asbestos].
[57] 15 U.S.C. § 2605(c)(4).
[58] See generally the discussion in McKenna & Cuneo, L.L.P., *TSCA Handbook*, 3d ed. (Government Institutes, 1997). The EPA's regulations regarding asbestos were eventually overturned in Corrosion Proof Fittings v. EPA, 987 F.2d 1201 (5th Cir. 1991) [EPA failed to justify ban on asbestos because it did not demonstrate that an alternative action would not be adequate].

been prohibited since 1978, it has been estimated that over 150 million pounds are dispersed in the air and water.[59]

EPA regulations specifically state the dangers of PCB exposure in a fashion seen for no other substance regulated by TSCA. The regulations state that:

[T]he manufacture, processing, and distribution in commerce of PCBs at concentrations of 50 ppm or greater and PCB Items with PCB concentrations of 50 ppm or greater present an unreasonable risk of injury to health within the United States. This finding is based upon the well-documented human health and environmental hazard of PCB exposure, the high probability of human and environmental exposure to PCBs and PCB Items from manufacturing, processing, or distribution activities; the potential hazard of PCB exposure posed by the transportation of PCBs or PCB Items within the United States; and the evidence that contamination of the environment by PCBs is spread far beyond the areas where they are used. In addition, the Administrator [of the EPA] hereby finds, for purposes of section 6(e)(2)(C) of TSCA, that any exposure of human beings or the environment to PCBs, as measured or detected by any scientifically acceptable analytical method, may be significant, depending on such factors as the quantity of PCBs involved in the exposure, the likelihood of exposure to humans and the environment, and the effect of exposure.[60]

Section 6(e) of TSCA requires the EPA to:

(A) prescribe methods for the disposal of polychlorinated biphenyls, and
(B) require polychlorinated biphenyls to be marked with clear and adequate warnings, and instructions with respect to their processing, distribution in commerce, use, or disposal or with respect to any combination of such activities.[61]

Moreover, TSCA Section 6(e)(2) attempts to minimize PCB exposure by humans or the environment by mandating that "no person may manufacture, process, or distribute in commerce or use any polychlorinated biphenyl in any manner other than in a totally enclosed

---

[59] Over 300 million pounds are in landfills, and over 400 million pounds are used in industry. See Anderson, Mandelker, and Tarlock, *Environmental Protection: Law and Policy*, 2d ed. (Little, Brown & Co., 1990).
[60] 40 CFR § 761.20.
[61] 15 U.S.C. § 2605(e)(1).

manner," unless the EPA "finds that such manufacture, processing, distribution in commerce, or use (or combination of such activities) will not present an unreasonable risk of injury to health or the environment."[62]

In June, 1998 the EPA published new rules which significantly amended the regulations affecting the use, manufacture, processing, distribution in commerce, and disposal of PCBs.[63] There are special requirements for proper labeling and marking of PCB-containing materials.[64]

Under the new regulations, virtually any release of PCBs into the environment in concentrations greater than 50 ppm is considered "disposal." Disposal includes spills, leaks, and other uncontrolled discharges of PCBs as well as actions related to containing, transporting, destroying, degrading, decontaminating, or confining PCBs and PCB Items."[65] There are special requirements for reporting spills.[66]

The regulations contain special rules for dealing with the cleanup of spills.[67] PCBs may be stored for reuse for periods up to 5 years,[68] and for up to one year before disposal. There are special disposal requirements for PCBs depending on the type of material involved.[69]

11.5.2 Section 6 and Chlorofluorocarbons (CFCs)

Studies in the early 1970s convinced many scientists that chlorofluorocarbons (CFCs), used as propellants for aerosol containers among other uses, were entering the atmosphere, interacting with other chemicals, and destroying the layer of ozone in the stratosphere. It is thought that this ozone layer keeps much harmful ultraviolet radiation from entering the atmosphere where it increases skin cancers, and causes other dangerous health problems.

Using its authority under TSCA Section 6(a), the EPA (with cooperation from other agencies) initiated a ban on CFCs as aerosol propellants in 1978. While the idea seemed good at the time, and enjoyed much support from the legislature and citizens, it was gradually chipped away by a series of "special" and "essential use" exemptions.[70] The number and nature of exemptions soon became ludicrous, and quickly swallowed the rule (e.g., there are exemptions for over-the-counter asthma bronchiodilators,

---

[62] 15 U.S.C. § 2605(e)(2).
[63] 63 *Fed. Reg.* 35,383 (1995).
[64] 40 CFR §§ 761.40 and 761.45.
[65] 40 CFR § 761.3.
[66] 40 CFR §§ 761.20-761.135.
[67] 40 CFR § 761.125.
[68] 40 CFR § 761.35.
[69] 40 CFR § 761.60.
[70] See 40 CFR §§ 762.59 and 762.58.

hand-held tear-gas devices, pharmaceutical rotary tablet press punch lubricants, artificial smoke and smog machines used in the entertainment industry). The EPA withdrew its CFC regulations from the CFR in 1995 because the Clean Air Act had made regulation of CFCs under TSCA essentially obsolete.[71]

The failure of the CFC ban stands as one of the low points in environmental regulation. Since the ban was imposed, a considerable body of scientific evidence has been developed that suggests that ozone depletion is responsible for global warming, worldwide extinction of thousands of species of plants and animals, and health effects that go far beyond skin cancer. There is even evidence to suggest that the amount of CFCs already present in the air, which is gradually working its way higher into the atmosphere, is sufficient to destroy most of the ozone layer. Nevertheless, CFCs remain the major aerosol propellant used in the United States. While the current use of CFCs is only 1/10 of the volume in 1977, it is estimated that over 10 million tons of CFCs have entered the atmosphere since 1975.

11.5.3 What Is "Unreasonable Risk"?

The term "unreasonable risk" permeates the language of TSCA, but it is never defined. This ambiguity was recognized by Congress when the Act was first considered, but it was felt that the ambiguity would work itself out as more information became available. It has not!

TSCA Section 6(c) requires the EPA to consider and publish a "succinct and precise" statement in promulgating its rules with respect to toxic substances or mixtures. The statement must include consideration of:

(A) the effects of such substance or mixture on health and the magnitude of the exposure of human beings to such substance or mixture,

(B) the effects of such substance or mixture on the environment and the magnitude of the exposure of the environment to such substance or mixture,

(C) the benefits of such substance or mixture for various uses and the availability of substitutes for such uses, and

(D) the reasonably ascertainable economic consequences of the rule, after consideration of the effect on the national economy, small

---

[71] 60 *Fed. Reg.* 31,919 (1995).

business, technological innovation, the environment, and public health.[72]

Moreover, TSCA Section 6(c)(3) states that TSCA may only supercede another federal environmental law when "it is in the public interest" to protect against risk under TSCA. But in making such a decision, the Administrator of the EPA must consider:

(i) all relevant aspects of the risk, as determined by the Administrator in the Administrator's discretion,
(ii) a comparison of the estimated costs of complying with actions taken under this chapter and under such law (or laws), and
(iii) the relative efficiency of actions under this chapter and under such law (or laws) to protect against such risk of injury.

The "unreasonable risk" terminology of TSCA clearly contains a cost-benefit element. The language of the statute, congressional history, and subsequent court decisions demonstrate a willingness to consider the health and environmental consequences of new chemicals in the context of relative costs to both large and small businesses.[73]

11.6 TSCA Enforcement

The penalties for violations of TSCA are discussed in TSCA Section 16.[74] TSCA Section 16(a)[75] specifies civil penalties of up to $25,000 per violation for persons violating TSCA Section 15 or Section 409 (prohibited acts). The Administrator of the EPA must take into account the circumstances, extent, and gravity of the violation in assessing civil penalties, and has the authority to compromise, modify, or remit the penalties.[76]

In some cases, the EPA's interpretation of its TSCA regulations is unclear (or the EPA has struggled with a reasonable interpretation). One court recently held that the EPA must provide "fair warning" of its interpretation before civil penalties can be assessed.[77]

TSCA Section 16(b) specifies criminal penalties (in addition to any civil

---

[72] 15 U.S.C. § 2605(c)(3).
[73] See Corrosion Proof Fittings v. EPA, 947 F.2d 1201 (5th Cir. 1991) [a view of "reasonableness" that accommodates economic considerations].
[74] 15 U.S.C. § 2615.
[75] 15 U.S.C. § 2615(a).
[76] 15 U.S.C. § 2615(a)(2)(B-C).
[77] General Electric Co. v. EPA, 53 F.3d 1324 (D.C. Cir. 1995).

penalties) of up to $25,000 per day, and up to one year in prison for persons who knowingly or willingly violate TSCA Section 15 or Section 409.[78]

11.7 Toxic Torts

11.7.1 What Are "Toxic Torts"?

The concept of "torts" in general, and "toxic torts" specifically, was introduced in chapter 2. Recall that "torts" are private or civil wrongs or injury resulting from the breach of a duty owed to one party by another. We have discussed several such torts, such as private and public nuisances.

Although toxic tort cases go back to 17th-century England, they gained prominence in the United States following incidents in the 1950s and 1960s resulting from conditions in which people were intentionally or negligently exposed to dangerous conditions and materials by representatives of the government.

Some of the more prominent situations involving "toxic torts" include the following:
- The exposure of soldiers in Vietnam to Agent Orange;[79]
- Exposure to uranium by miners and nuclear reactor workers;[80]
- Casualties of the Army's simulated biological warfare attack on San Francisco;[81]
- Unsuspecting victims of the CIA's brainwashing research who were given hallucinogenic drugs;[82] and
- Victims of asbestos exposure at military bases.[83]

As in many tort cases, toxic tort trials are often resolved based on the testimony of expert witnesses. Both sides in these cases frequently supply conflicting and contradictory expert testimony on the effects of pollutants (the so-called "battle of the experts"). Obviously, it can be very difficult for a judge or a jury to know what testimony is most credible. In *Daubert v. Merrell Dow Pharmceuticals, Inc.*, the U.S. Supreme Court held that expert testimony is admissable only if it reflects actual "scientific

---

[78] 15 U.S.C. § 2615(b).

[79] See In re. Agent Orange Product Liability Litigation, 818 F.2d 210 (2d Cir. 1987).

[80] See Fried v. United States, 674 F. Supp. 636 (N.D. Ill. 1987); and Barnson v. U.S., 816 F.2d 549 (10th Cir. 1987).

[81] Nevin v. United States, 696 F.2d 1229 (9th Cir. 1983), *cert. denied*, 104 S.Ct. 70 (1983).

[82] Orlikow v. United States, 682 F. Supp. 77 (D.D.C. 1988).

[83] Shuman v. United States,765 F.2d 283 (1st Cir. 1985).

knowledge," and if it is "helpful" to the court or the jury.[84] In other words, the scientific evidence must not only prove the proposition for which it is proffered, but also be relevant. Other courts were quick to adopts the "Daubert Standard."[85] The Supreme Court extended the Daubert standard to nonscientific evidence in *Kumho Tire Co. v. Carmichael.*[86]

11.7.2 The Federal Tort Claims Act

The Federal Tort Claims Act (FTCA) was passed in 1946, immediately after World War II.[87] Many of the early cases involved automobile accidents resulting from faulty design and construction, but it has since become the major mechanism by which citizens seek relief from the federal government for intentional or negligent exposure to toxic chemicals.

With the passage of the FTCA, the federal government agreed to abrogate its "sovereign immunity" where certain kinds of tort actions were concerned. The basic structure of the FTCA allows a plaintiff to sue the U.S. government for wrongful or negligent acts by employees of the federal government (within the scope of their employment). The local law of the place where the tort occurs applies, and makes the government (or the employee) liable "in the same manner and to the same extent as a private individual under like circumstances."[88] Under the FTCA, the government may be liable for damages to persons and property, but it is *not* liable for punitive damages.[89]

There are several controversial exceptions under the FTCA that give the federal government immunity from tort claims. An important example is misrepresentation by the federal employee. In *Rey v. United States*, the government received immunity even though a government doctor erroneously told a farmer that his hogs were infected with cholera, prompting the farmer to kill the hogs to protect against the spread of the disease.[90] In *Wells v. United States*, the government received immunity

---

[84] Daubert v. Merrell Dow Pharmaceuticals, Inc., 113 S.Ct. 2786 (1993).

[85] See Joiner v. General Electric Co., 118 S.Ct. 512 (1997); Schmaltz v. Norfolk & Western Ry. Co., 878 F. Supp. 1119 (N.D. Ill. 1995); and McCullock v. H.B. Fuller Co., 61 F.3d 1038 (2d Cir. 1995).

[86] Kumho Tire Co. v. Carmichael, 526 U.S. 137 (1999). See Targ and Feldman, Courting Science: Expert Testimony after *Daubert* and *Carmichael*, 13(4) Natural Resources & Env't 507 (1999).

[87] 28 U.S.C. §§ 1291 ff..

[88] 28 U.S.C. § 1346(b).

[89] See United States v. Hooker Chemicals & Plastics Corp., 850 F. Supp. 993 (W.D.N.Y. 1994), one of the famous Love Canal cases.

[90] Rey v. United States, 484 F.2d 45 (5th Cir. 1973).

when a government employee mede false statements and failed to warn of the hazards of lead pollution.[91]

Another important exception to the FTCA (where the government is immune from tort claims) is independent contractors In *Dickerson, Inc. v. Holloway*, the government received immunity from errors involving the selection of a hazardous waste operator.[92] A third exception to the FTCA is for certain uniquely governmental activities that have no private counterparts. In *C.P. Chemical Co. v. United States*, the government was held immune from tort claims for business losses resulting from a ban on urea-formaldehyde foam use because there is no comparable private rulemaking activity.[93]

By far the most contentious exception to the FTCA is the immunity given the government for "discretionary" functions or duties of a federal employee.[94] The leading decision on the discretionary exemption is *United States v. Varig Airlines*, a Supreme Court decision allowing immunity for the government failure by government employees to check airlines and prevent midflight fires.[95] While the Varig Airlines case is not actually a toxic tort case, there have been many other discretionary exemption cases that are toxic tort cases. In *United States Fidelity & Guaranty Co. v. United States*, there was government immunity from tort claims for property damage from a cloud of toxic gas released from a superfund cleanup site.[96] Nevertheless, the courts have carved out several exceptions to the exception, where the FTCA applies and immunity is denied. In *Dickerson, Inc. v. Holloway*, the U.S. Navy was not immune from liability for damages under the Solid Waste Disposal Act's (RCRA, see Chapter 8) cradle to grave requirements.[97] In *Starrett v. United States*, there was no government immunity for damage resulting from failure to follow Clean Water Act requirements for secondary treatment of sewage.[98]

Some plaintiffs have sued the individual governmental employee under the FTCA. For example, a civilian warehouseman sued an Army supervisor for burns he received when he was exposed to toxic soda ash at a weapons depot.[99]

---

[91] Wells v. United States, 655 F. Supp. 715 (D.D.C. 1987), aff'd on other grounds, 851 F.2d 1471 (D.C. Cir. 1988), *cert. denied*, 109 S.Ct. 836 (1989).
[92] Dickerson, Inc. v. Holloway, 685 F. Supp 1555 (M.D. Fla. 1987).
[93] C.P. Chemical Co. v. United States, 810 F.2d 34 (2d Cir. 1987).
[94] 28 U.S.C. § 2680(a).
[95] United States v. Varig Airlines, 104 S.Ct. 2755 (1984).
[96] United States Fidelity & Guaranty Co. v. United States, 837 F.2d 116 (3d Cir. 1988), cert. denied, 108 S.Ct. 2902 (1988).
[97] Dickerson, Inc. v. Holloway, 685 F. Supp. 1555 (M.D. Fla. 1987).
[98] Starrett v. United States, 847 F.2d 539 (9th Cir. 1988). See chapter 4.
[99] Westfall v. Erwin, 108 S.Ct. 580 (1988) [immunity for the Army, but not for the supervisor].

VI

Pollution Prevention Law and Policy

12

Pollution Prevention

12.1 Introduction

Since passage of the National Environmental Policy Act (NEPA) in 1969, and establishment of the U.S. Environmental Protection Agency (EPA) in 1970, public environmental advocacy, good corporate citizenry, and committed governmental action have helped to improve the quality of the environment. Emissions of air pollutants from cars and industrial facilities has been reduced, over 5,000 wastewater treatment facilities have been constructed, ocean-dumping of wastes has been prohibited, production and use of hazardous substances, such as asbestos, polychlorinated biphenyls (PCBs), and chlorofluorocarbons (CFCs), have been banned or are being phased out.[1] Before the 1990s, the predominant waste management practice had been "end-of-pipe" treatment or land disposal of hazardous and nonhazardous wastes. While this approach provided substantial progress in improving the quality of the environment, obvious limits exist on how much environmental improvement can be achieved using methods that manage pollutants after they have been generated. Regulators learned that traditional end-of-pipe approaches were not only expensive and less than fully effective, but sometimes transferred pollution from one medium to another. It became clear that additional improvements in environmental quality could only be realized if measures were implemented to prevent pollution from occurring in the first place. Acknowledging the slow progress and the limited success of various "command-and-control" measures to regulate pollution, Congress, the EPA, and the states have been incorporating pollution prevention requirements into the environmental regulatory framework. The purpose

---

[1] U.S.E.P.A., *The New Generation of Environmental Protection: EPA's Five-Year Strategic Plan* (USEPA 200-B-94-002, 1994), p.1.

511

of this chapter is to explain the pollution prevention regulatory requirements and offer practical pollution prevention strategies that can be applied to business operations.

12.2 Types of Pollution Prevention

Pollution prevention, in its broadest sense, refers to the reduction in volume and/or toxicity of waste prior to discharge or disposal. Pollution prevention techniques generally consist of source reduction and recycling activities. Although treatment may be used to reduce the toxicity of some waste streams, it is not generally thought of as pollution prevention in its truest sense.

12.2.1 Source Reduction

Source reduction means the reduction or elimination of waste at its source. Facilities seeking to implement source reduction techniques need to evaluate their manufacturing, production, and general waste generating operations for opportunities to reduce wastes before they are generated. Each process or manufacturing operation must be closely examined to determine material inputs, transformations that occur as part of production processes, and material outputs. The impact of quality control parameters, product specifications, and production goals must also be considered.

Good operating practices are key in achieving source reduction goals. These practices typically have been used to improve efficiency and reduce production costs. Improving yields by reducing production losses has long been a common practice by industries where raw materials account for a significant portion of operating costs. Good operating practices generally require little or no capital investment, are easily implemented, and result in significant savings. Good operating practices might include the following measures:
- Waste reduction programs
- Management and personnel practices
- Material handling and inventory practices
- Loss prevention
- Waste segregation
- Production scheduling

A waste reduction program should be formalized to indicate management support for the program. Management gives strong recognition to company progress in meeting program goals by identifying employees or department that make significant strides toward achieving the goals of the company's waste reduction program. Certain personnel practices may be used to encourage employee participation on pollution prevention. Incentives may be used as an effective means of rewarding employees that make contributions to company's pollution prevention efforts.

Materials handling and inventory practices include programs to reduce loss of input materials caused by mishandling, expired shelflife, and poor storage conditions. The proper control over materials as they are handled or transferred from one location to another reduces the chances of spills. Simple procedures may be implemented to ensure proper materials handling, such as properly training employees in the operation of each type of transfer equipment, allowing adequate spacing of containers, stacking containers so as to avoid the chances of punctures and breaks, and proper labeling of containers to indicate the name and type of substance inside. Computerized systems are the most efficient method of inventory control and materials tracking. Poor inventory control can result in overstocking or disposal of expired materials.

Loss prevention minimizes wastes by avoiding spills and leaks from production equipment and storage areas. The most effective ways to minimize wastes that are needlessly generated by spills and leaks is to take precautionary measures to ensure that spills and leaks don't occur in the first place.

Waste segregation reduces the volume of wastes by preventing the mixing of hazardous and nonhazardous components of the company's waste streams. Mixing of hazardous and nonhazardous waste may cause the entire mixture to become classified as a hazardous waste. Thus, by separating nonhazardous from hazardous waste, the overall volume of hazardous waste requiring disposal will be reduced, resulting in significantly reduced hazardous waste management and disposal costs.

Production scheduling changes, especially where batch processes are used, can also be an effective waste reduction technique. Schedule changes in batch production runs can reduce the frequency of equipment and tank cleaning that results in large amounts of solvent waste. To reduce cleaning frequency, batch sizes should be maximized or batches of one material followed with a similar product, which may not necessitate cleaning between batches.

12.2.2 Recycling

If a waste stream or component of the waste stream cannot be reduced or eliminated through source reduction, recycling presents the next best option for pollution prevention. Recycling can take two basic forms: preconsumer and postconsumer recycling. As the term implies, preconsumer recycling involves raw materials, products, and by-products that have not yet reached consumers for intended end-use, but are typically reused within the original production process. On the hand, postconsumer recycled materials are those that have served their intended end-use, and have then been separated from the municipal solid waste stream. Although postconsumer recycling is beneficial toward recovery of some discarded materials for reuse, it is clearly not as effective as preconsumer recycling in achieving pollution prevention because of the high waste management costs and regulatory burdens associated with recovery of postconsumer wastes. Preconsumer recycling, by contrast, enables facilities to reclaim or reuse certain components of process waste streams for a beneficial purpose in a different process.

Facilities can implement recycling to help eliminate waste disposal costs, reduce raw material costs, and provide income from saleable waste. A material is recycled if used, reused, or reclaimed. Recycling through use and/or reuse involves returning the waste material either to the original process as a substitute for an input material, or to another process as an input material. Recycling through reclamation is the processing of waste for recovery of valuable material or for regeneration.

12.2.3 Treatment

Unfortunately, not all wastes can be entirely eliminated at the source, nor recovered for recycling. When these pollution prevention opportunities are unavailable, effective treatment methods should be used to reduce the toxicity of remaining wastes. According to a an EPA definition, treatment is "any practice, other than recycling, designed to alter the physical, chemical, or biological character or composition of a hazardous substance, pollutant, or contaminant, so as to neutralize said substance, pollutant, or contaminant or to render it non-hazardous through a process or activity separate from the production of a product or the providing of a service." Treatment can generally be divided into three categories: (1) chemical treatment, (2) biological treatment, or (3) physical treatment. These methods are basically available for treating wastes after they have been generated and, as such, are not true methods of pollution prevention. Accordingly, this chapters does not discuss the use of treatment technologies as a pollution prevention alternative.

12.3 Regulatory Initiatives For Pollution Prevention

The EPA and state environmental agencies have primarily attempted to encourage business and industry to make voluntary commitments to pollution prevention, largely through participation in various EPA and state regulatory programs. Still, although most pollution prevention measures are considered voluntary, there are a number of mandatory pollution prevention requirements at the federal and state levels. These pollution prevention regulatory requirements are reviewed in this section.

12.3.1 Pollution Prevention Act of 1990

In 1990, Congress passed the Pollution Prevention Act of 1990 (PPA)[2] to focus more attention on reducing the volume and toxicity of wastes at the source. Thus, source reduction, recycling, and other waste minimization strategies are fast becoming a significant environmental regulatory compliance issue. In Section 6602(b) of this law, Congress declared it a national policy to prevent or reduce pollution at the source whenever feasible.[3]

This statute is basically an enabling act which states congressional commitment to waste reduction and recycling activities and which mandates that the EPA implement pollution prevention strategies and regulations. The PPA requires that the EPA provide grants to the states to implement their own pollution prevention programs, and requires that the EPA set up an information clearinghouse and conduct pollution prevention research. The PPA also requires that companies report their pollution prevention practices under SARA Title III, also known as the Emergency Planning and Community Right-to-Know Act (EPCRA).[4]

The PPA sets forth a hierarchy of waste management options in descending order of preference: prevention/source reduction, environmentally sound recycling, environmentally sound treatment, and environmentally sound disposal. Pollution should be prevented or reduced at the source whenever feasible, if it cannot be prevented it should be recycled in an environmentally safe manner. In the absence of feasible prevention or recycling opportunities, pollution should be treated. Disposal or other release into the environment should be used as a last resort.

---

[2] Pub. L. 101-508, 42 U.S.C. §§ 13101 through 13109
[3] PPA § 6602(b), 42 U.S.C. § 13101(b).
[4] 42 U.S.C. §§ 11001–11050. EPCRA was originally enacted as part of the Superfund Amendments and Reauthorization Act (SARA) of 1986, which amended CERCLA.

Pollution prevention is explained in the PPA to mean source reduction and other practices that reduce or eliminate the generation of pollution. Section 6602 notes that:

> There are significant opportunities for industry to reduce or prevent pollution at the source through cost-effective changes in production, operation, and raw materials use.... The opportunities for source reduction are often not realized because existing regulations, and the industrial resources they require for compliance, focus upon treatment and disposal, rather than source reduction.... Source reduction is fundamentally different and more desirable than waste management and pollution control.[5]

Source reduction is defined in the law as any practice which reduces the amount of any hazardous substance, pollutant, or contaminant entering any waste stream or otherwise released into the environment (including fugitive emissions) prior to recycling, treatment, or disposal, and which reduces the hazards to public health and the environment associated with the release of such substances, pollutants, or contaminants.[6]

The PPA mandates establishment of an EPA office to carry out various functions required by the Act.[7] The EPA office must function independently of the Agency's other "single-medium" programs; it must promote a multi-media approach to source reduction. The law contains a listing of more than a dozen specific measures that the EPA must develop and implement, including:[8]

- Facilitate the adoption of source reduction techniques by businesses and by other federal agencies,
- Establish standard methods of measurement for source reduction,
- Review regulations to determine their effect on source reduction,
- Investigate opportunities to use federal procurement to encourage source reduction,
- Develop improved methods for providing public access to data collected under federal environmental statutes,
- Develop a training program on source reduction opportunities, model source reduction auditing procedures, a source reduction clearinghouse, and an annual award program.

---

[5] PPA § 6602(a), 42 U.S.C. § 13101(a).
[6] PPA § 6603(5)(A), 42 U.S.C. § 13102(5)(A).
[7] PPA § 6604(a), 42 U.S.C. § 13103(a). Although not specifically referred to as such in the PPA, this independent office is the Office of Pollution Prevention and Toxics, headquartered in Washington, D.C.
[8] PPA § 6604(b), 42 U.S.C. § 13103(b).

The PPA authorizes a grant program for state technical assistance programs to promote source reduction techniques by businesses. Approved state programs may receive up to 50 percent in federal matching funds for each year that a state participates in the program.[9]

The PPA also requires that the EPA report to Congress within 18 months (and biennially afterwards) on actions needed to implement a strategy to promote source reduction, and an assessment of the clearinghouse and the grant program.[10]

12.3.2 EPA's Pollution Prevention Strategy

Under the PPA, the EPA was required to develop a pollution prevention strategy that reduces pollution at the source. In February 1991, the EPA published its pollution prevention strategy, commonly referred to as the Industrial Toxics Project or the "33/50" Initiative.[11] Under the plan, the EPA's goal is to reduce releases and off-site transfers of 17 high-volume EPCRA Section 313 toxic chemicals. Each of the 17 chemicals was selected from the EPA's Toxic Release Inventory (TRI), based on a number of factors, including high production volume, high releases and off-site transfers of the chemical relative to total production, opportunities for pollution prevention, and potential for causing detrimental health and environmental effects. The EPA initially set a goal to reduce the releases of these chemicals by 33 percent by the end of 1992, and 50 percent by the end of 1995. The TRI will be used to track these reductions using 1988 data as a baseline. In numeric terms, the goal was to reduce the amount of releases and off-site transfers from the 1.4 billion pounds reported in 1988 to 700 million pounds by 1995.

The 17 target chemicals are:
- Benzene
- Cadmium and Cadmium Compounds
- Carbon Tetrachloride
- Chloroform (Trichloromethane)
- Chromium and Chromium Compounds
- Cyanide and Cyanide Compounds
- Lead and Lead Compounds
- Mercury and Mercury Compounds
- Methyl Ethyl Ketone

---

[9] PPA § 6605, 42 U.S.C. § 13104.
[10] PPA § 6608, 42 U.S.C. § 13107.
[11] 56 *Fed. Reg.* 7849 (Feb. 26, 1991). Copies of the 33/50 Initiative are available free of charge by calling the EPCRA hotline at (800) 535-0202. Ask for Pub. No. EPA/741/R-92/001.

- Methyl Isobutyl Ketone
- Methylene Chloride (Dichloromethane)
- Nickel and Nickel Compounds
- Tetrachloroethylene (Perchloroethylene)
- Toluene
- 1,1,1-Trichloroethane (Methyl Chloroform)
- Trichloroethylene
- Xylene (all xylenes)

These industrial chemicals include known and potential carcinogens, developmental toxins, chemicals that bioaccumulate, ozone-depleting chemicals, and chemicals contributing to ozone pollution at ground level. All of the 33/50 Program chemicals are regulated under one or more existing environmental statutes, and the 33/50 Program is intended to complement, not replace, ongoing EPA programs. All 17 targeted chemicals will be subject to the Maximum Achievable Control Technology (MACT) standards of the Clean Air Act Amendments of 1990. The EPA believes that the incentive for early reductions offered by the MACT provisions will further the progress of the 33/50 Program.

Under the program, the EPA encourages companies to reach the 33 percent and 50 percent goals by using the pollution prevention hierarchy outlined in the PPA. Since the PPA accords prevention/source reduction the highest value, the EPA correspondingly makes source reduction the preferred method of pollution prevention. Recycling is considered the next best method, and treatment is the least preferred method of pollution prevention. Although participation in the 33/50 Program is completely voluntary and the program goals are not enforceable, companies should take steps to implement pollution prevention techniques because such measures are destined to become mandatory components of pollution control laws.

The 33/50 Program is part of the EPA's overall Pollution Prevention Strategy and the first of the agency's pollution prevention initiatives. It is also a major component of the Office of Pollution Prevention and Toxics' Existing Chemicals Revitalization Program. The EPA is seeking reductions primarily through pollution prevention practices that go beyond regulatory requirements. The EPA is also encouraging industry to develop a preventive approach, seeking continuous environmental improvements beyond these reductions and the 17 priority chemicals. Success in the program will be measured by nationwide reductions, rather than results at each company or facility. This approach provides flexibility and allows participating companies to develop reduction strategies that are the most cost-effective for their facilities.

12.3.3 EPCRA Toxic Chemical Release Inventory

In 1986, the Emergency Planning and Community Right-to-Know Act (EPCRA),[12] also known as SARA Title III, was added as a freestanding part of the Superfund Amendments and Reauthorization Act (SARA), which amended CERCLA. The primary goal of EPCRA is to facilitate public awareness and emergency response planning for chemical hazards. To fulfill this purpose, EPCRA requires that certain companies that manufacture, process, and use chemicals in specified quantities must file written reports, provide notification of spills/releases, and maintain toxic chemical inventories.

Certain companies must submit an annual report of releases of listed "toxic chemicals" pursuant to EPCRA Section 313, known as the Toxic Chemical Release Inventory (Form R).[13] On Form R, the company reports any releases made during the preceding twelve months. Form R must be filed if the business has ten or more full-time employees, has a Standard Industrial Classification Code (SIC) 20-39, and the business manufactures, stores, imports, or otherwise uses designated toxic chemicals at or above threshold levels. Generally, a company must file an annual Form R if it manufactures, imports, or processes at least 25,000 pounds of a listed toxic chemical, or if it uses at least 10,000 pounds of a listed toxic chemical during the previous calendar year. Toxic chemicals subject to the Form R reporting requirements and their respective threshold quantities are listed at 40 CFR Section 372.65. Certain exemptions from the Form R reporting of toxic chemical releases may apply.[14]

With the passage of the Pollution Prevention Act of 1990 (PPA),[15] new requirements were added to the Form R. Section 6607 of the PPA expands and makes mandatory source reduction and recycling information on the EPCRA list of toxic chemicals. The information includes: the quantities of each toxic chemical entering the waste stream and the percentage change from the previous year, the quantities recycled and percentage change from the previous year, source reduction practices, and changes in production from the previous year. For more information about the Form R, call the EPCRA hotline at (800) 535-0202.

---

[12] 42 U.S.C. §§ 11001 through 11050.
[13] Violations of EPCRA Section 313 reporting are punishable by fines of up to $25,000 per day. 40 CFR § 372.18.
[14] See 40 CFR § 372.38.
[15] 42 U.S.C. § 13101 through 13109.

12.3.4 RCRA Waste Minimization Regulatory Guidance

Before the PPA became law, some early consideration was given to waste reduction activities. In fact, with the passage of the Hazardous and Solid Waste Amendments (HSWA) to RCRA in 1984, Congress established a significant new policy concerning hazardous waste management. Specifically, Congress declared that the reduction or elimination of hazardous waste generation at the source should take priority over the management of hazardous wastes after they are generated. In Section 1003(b) of RCRA, Congress declared it to be national policy that, whenever feasible, the generation of hazardous waste is to be reduced or eliminated as expeditiously as possible. Waste that is nevertheless generated should be treated, stored, or disposed of so as to minimize the present and future threat to human health and the environment.[16]

In furtherance of this national policy toward pollution prevention, the 1984 amendments to RCRA added a significant new waste minimization requirement. Under Section 3002(b) of RCRA, hazardous waste generators who transport their wastes off-site are required to certify on their hazardous waste manifests that they have programs in place to reduce the volume or quantity and toxicity of hazardous waste generated to the extent economically practicable. Certification of a waste minimization "program in place" is also required as a condition of any permit issued under section 3005(h) for the treatment, storage, or disposal of hazardous waste at facilities that generate and manage hazardous wastes on-site.

In May 1993, the EPA issued interim final guidance to assist hazardous waste generators and owners and operators of hazardous waste treatment, storage, or disposal (TSD) facilities to comply with the waste minimization certification requirements of Sections 3002(b) and 3005(h) of RCRA, as amended by the HSWA.[17] The guidance document fulfills a commitment made by the EPA in its 1986 report to Congress entitled *The Minimization of Hazardous Waste*[18] to provide additional information to generators on the meaning of the certification requirements added by the HSWA.[19] In addition, the interim guidance set forth a detailed plan for waste generators to develop and implement hazardous and solid waste minimization programs. The waste minimization program guidance is discussed in detail in section 12.5.

---

[16] 42 U.S.C. § 6902(b).
[17] Guidance to Hazardous Waste Generators on the Elements of a Waste Minimization Program, 58 *Fed. Reg.* 31114 (May 28, 1993).
[18] U.S.E.P.A., *The Minimization of Hazardous Waste* (EPA/530-SW-86-033, Oct. 1986).
[19] See 51 *Fed. Reg.* 44683 (Dec. 11, 1986).

12.3.5 CWA Best Management Practices

The primary objective of the Clean Water Act (CWA) is the restoration and maintenance of the chemical, physical, and biological integrity of our Nation's waters.[20] To achieve its objective, the CWA sets forth a series of goals, including attaining fishable and swimmable designations and eliminating the discharge of pollutants into navigable waters. As part of the CWA strategy to eliminate discharges of pollutants to receiving waters, National Pollutant Discharge Elimination System (NPDES) permit limitations have become more stringent.

The principal mechanism for reducing the discharge of pollutants from point sources is through implementation of the NPDES program, established by Section 402 of the CWA. All facilities with point source discharges must apply for and obtain a NPDES permit. The EPA has delegated authority to most states to issue NPDES permits. Where state NPDES authorization has not yet occurred, EPA Regions issue NPDES permits.

A NPDES permit is essentially a license that allows a facility to discharge contaminated water. On the NPDES permit application, the facility provides information about the type of facility and type of discharge being requested. If the facility's application is approved, the permitting authority will issue a permit that contains various conditions related to the facility's pollutant discharges. The permit will generally contain specific limitations on contamination levels or specific actions that the facility must take, such as sampling or inspections.

Four minimum elements are typically included in each permit issued:

1. Effluent discharge limitations,
2. Monitoring and reporting requirements,
3. Standard conditions,
4. Special conditions.

The numeric effluent discharge limits contained in a NPDES permit are based on the most stringent value among technology-based effluent guidelines limitations, water quality-based limitations, and limitations derived on a case-by-case basis. Permits also contain standard conditions that prescribe administrative and legal requirements to which all facilities are subject. Finally, permits may contain any supplemental controls, referred to as special conditions, that may be needed in order to ensure that the regulations driving the NPDES program and, ultimately, the goals of

---

[20] CWA § 101(a), 33 U.S.C. § 1251(a).

the CWA are met. Best management practices (BMPs) are one such type of supplemental control.

Section 304(e) of the CWA authorizes the EPA Administrator to publish regulations to control discharges of significant amounts of toxic pollutants listed under Section 307 or hazardous substances listed under Section 311 from industrial activities that the Administrator determines are associated with or ancillary to industrial manufacturing or treatment processes.

In 1978, the EPA proposed regulations addressing the use of procedures and practices to control discharges from activities associated with or ancillary to industrial manufacturing or treatment processes. The proposed rule indicated how BMPs would be imposed in NPDES permits to prevent the release of toxic and hazardous pollutants to surface waters.[21] While this Subpart (40 CFR Part 125, Subpart K) never became effective, it remains in the Code of Federal Regulations and can be used as guidance by permit writers.

Although these regulations were never finalized, the EPA and states continue to incorporate BMPs into permits based on the authority contained in Section 304(e) of the CWA and the regulations set forth in 40 CFR Section 122.44(k). While Section 304(e) of the CWA restricts the application of BMPs to ancillary sources and certain chemicals, the regulations contained in 40 CFR Section 122.44(k) authorize the use of BMPs to abate the discharge of pollutants under the following circumstances:

1. They are developed in accordance with Section 304(e) of the CWA,
2. Numeric limitations are infeasible,
3. The practices are necessary to achieve limitations/standards or meet the intent of the CWA.

Thus, permit writers are afforded considerable latitude in employing BMPs as pollution control mechanisms.

As defined by CWA Section 304(e), the discharges to be controlled by BMPs are plant site runoff, spillage or leaks, sludge or waste disposal, and drainage from raw material storage. These activities have historically been found to be amenable to control by BMPs. Some examples include the following:

●*Material storage areas* for toxic, hazardous, and other chemicals including raw materials, intermediates, final products, or byproducts. Storage areas may be piles of materials or containerized substances. Typical storage containers could include liquid storage vessels ranging in size from large tanks to 55-gallon drums; dry storage in bags, bins,

---

[21] 43 *Fed. Reg.* 37078 (Aug. 21, 1978) (40 CFR pt. 125, subpt. K, Criteria and Standards for Best Management Practices Authorized Under Section 304(e) of the CWA).

silos and boxes; and gas storage in tanks and vessels. The storage areas can be open to the environment, partially enclosed, or fully contained.

● *Loading and unloading operations* involving the transfer of materials to and from trucks or rail cars, including in-plant transfers. These operations include pumping of liquids or gases from truck or rail car to a storage facility or vice versa, pneumatic transfer of dry chemicals during vehicle loading or unloading, transfer by mechanical conveyor systems, and transfer of bags, boxes, drums, or other containers from vehicles by forklift, hand, or other materials handling methods.

● *Facility runoff* generated principally from rainfall on a plant site. Runoff can become contaminated with harmful substances when it comes in contact with material storage areas, loading and unloading areas, in-plant transfers areas, and sludge and other waste storage/disposal sites. Fallout, resulting from plant air emissions that settle on the plant site, may also contribute to contaminated runoff. In addition to BMPs, facility runoff from industrial sites may also be directly regulated under the NPDES storm water permitting program.

● *Sludge and waste storage and disposal areas* including landfills, pits, ponds, lagoons, and deep-well injection sites. Depending on the construction and operation of these sites, there may be a potential for leaching of toxic pollutants or hazardous substances to groundwater, which can eventually reach surface waters. In addition, liquids may overflow to surface waters from these disposal operations.

Many facilities currently implement successful measures to reduce and control environmental releases of all types of pollutants. These measures have been successfully implemented both formally as part of BMP plans and informally as part of unwritten standard operating procedures. In the context of the NPDES permit program, permittees are required to develop BMP plans to address specific areas of concern. The BMP plan developed by the permittee becomes an enforceable condition of the permit.

BMPs may apply to an entire site or be appropriate for discrete areas of an industrial facility. Many of the same environmental controls promoted as part of a BMP plan may currently be used by industry in storm water pollution prevention plans, spill prevention control and countermeasure (SPCC) plans, Occupational Safety and Health Administration (OSHA) safety programs, fire protection programs, insurance policy requirements, or standard operating procedures. Additionally, where facilities have developed pollution prevention programs, controls such as source reduction and recycling/reuse may be similar to those promoted as part of a BMP plan.

With the increasing awareness of pollution prevention opportunities, as well as the increase in legislation and regulatory policies directing efforts toward pollution prevention, much of the traditional focus of BMP activities is being redirected from ancillary activities to industrial manufacturing processes. This redirection is resulting in the integrated application of traditional BMPs and pollution prevention practices into cohesive and encompassing plans that cover all aspects of industrial facilities. Specific guidance for implementing a BMP plan to control water pollution discharges is provided in section 12.6.

12.3.6 Stormwater Pollution Prevention Plans

In addition to the general BMPs commonly incorporated into NPDES permits, the EPA specifically mandates that all permits for storm water pollution discharges contain storm water pollution prevention plans. For each type of storm water permit (i.e., general or individual), a storm water pollution prevention plan must be developed for each facility covered by the permit. Storm water pollution prevention plans must be prepared in accordance with good engineering practices and in accordance with factors outlined in 40 CFR Section 125.3(d) (2) or (3), as appropriate. The plan must identify potential sources of pollution which may reasonably be expected to affect the quality of storm water discharges associated with activities at the facility. In addition, the plan must describe and ensure the implementation of practices which are to be used to reduce the pollutants in storm water discharges associated with activities at the facility and to ensure compliance with the terms and conditions of the permit. Facilities must implement the provisions of the storm water pollution prevention plan as a condition of permit issuance. The pollution prevention requirements of storm water permits are discussed in detail in chapter 5.

12.3.7 State Pollution Preventions Requirements

In response to the new regulatory trend toward pollution prevention, in contrast to the traditional "end-of-pipe" pollution controls, numerous states have been incorporating source reduction and recycling provisions into their regulatory regimes. Many states have made these requirements mandatory through enactment of various types of pollution prevention laws.[22] Others have initiated pollution prevention activities through voluntary programs. Although the voluntary state provisions set forth

---

[22] See Rabe, From Pollution Control to Pollution Prevention: The Gradual Transformation of American Environmental Policy, 8 Envtl & Plan. L.J. 226 (1991) (examining the mandatory pollution prevention programs developed in Massachusetts, Minnesota, and New Jersey).

goals to encourage source reduction, waste minimization, and recycling activities, these state programs attempt to achieve these goals through voluntary practices. The voluntary programs have generally adopted educational outreach and technical assistance mechanisms to promote pollution prevention technologies and techniques. Table 12.1 provides a listing of state pollution prevention programs.

Table 12.1. State Pollution Prevention Programs

Mandatory State Programs:

Arizona: Ariz. Rev. Stat. Ann. § 49-961 to -73.
California: Cal. Health & Safety Code § 25244.12 to .24.
Georgia: Ga. Code Ann. § 12-8-60 to -83.
Louisiana: La. Rev. Stat. Ann. § 30.2291 to .2295.
Maine: Me. Rev. Stat. Ann., tit. 38, § 2301 to 2312.
Massachusetts: Mass. Ann. Laws ch. 211, § 1 to 23.
Minnesota: Minn. Stat. Ann. § 115D.01 to .12.
Mississippi: Miss. Code Ann. § 49-31-1 to -27.
New Jersey: N.J. Stat. Ann. § 13:1D-35 to -50.
New York: N.Y. Envtl Conserv. Law § 27-0900 to -0925.
Oregon: Or. Rev. Stat. § 465.003 to .037.
Tennessee: Tenn. Code Ann. § 68-212-301 to -312.
Texas: Tex. Health & Safety Code Ann. § 361.501 to .510.
Washington: Wash. Rev. Code § 70.95C.010 to .240

Voluntary State Programs:

Alaska: Alaska Stat. § 46.06.021 to .041.
Colorado: Colo. Rev. Stat. Ann. § 25-16.5-101 to -110.
Connecticut: Conn. Gen. Stat. Ann. Appendix Pamphlet, P.A. 91-376.
Delaware: 7 Del. Code Ann. § 7801 to 7805.
Florida: Fla. Stat. Ann. § 403.072 to .074.
Illinois: Ill. Ann. Stat. ch. 111 1/2, § 7951 to 7957.
Indiana: Ind. Code Ann. § 13-9-1 to -7.
Iowa: Iowa Code Ann. § 455B.516 to .518.
Kentucky: Ky. Rev. Stat. Ann. § 224.46-310 to -325.
Rhode Island: R.I. Gen. Laws § 37-15.1-1 to .11
South Carolina: S.C. Code Ann. § 68-46-301 to -312.
Wisconsin: Wis. Stat. Ann. § 144.955.

Unfortunately, space limitations preclude a full discussion of these state programs.[23] However, a brief description of the programs for Arizona, California, and Minnesota is provided here to illustrate how some states are making pollution prevention a mandatory component of environmental regulatory compliance. For other state programs, the reader is directed to refer to the laws outlined in Table 12.1.

Arizona

Arizona's pollution prevention law applies to facilities that (1) must file the annual Toxic Chemical Release Inventory Form R required by EPCRA § 313 or (2) during the preceding 12 months, generated an average of one kilogram per month of an acutely hazardous waste (as defined in 40 CFR Part 261) or an average of 1,000 kilograms per month of hazardous waste.[24] These facilities must file an annual toxic data report with the state environmental protection agency and implement a pollution prevention plan designed to reduce the use of toxic substances and the generation of hazardous wastes. Facilities subject to the reporting requirements of the state's pollution prevention act must file the toxic data report annually until (1) the facility ceases operation or (2) it did not have to file a Form R for the preceding calendar year or (3) for two consecutive years it did not generate enough hazardous wastes to meet the prescribed threshold quantities.[25] The toxic data report must contain a copy of the Form R filed pursuant to EPCRA § 313,[26] and an annual progress report concerning the facility's pollution prevention plan. The pollution prevention plan must, at a minimum, cover a two-year time period.[27]

The pollution prevention plan must include the following components:[28]
1. The name and location of and principal business activities at the facility.
2. The name, address and telephone number of the owner or operator of the facility and of the senior official with management responsibility at the facility.
3. A certification by the senior official with management responsibility at the facility that he or she has read the plan and that it is to the best of her or his knowledge true, accurate, and complete.
4. Specific performance goals for the prevention of pollution, including an explanation of the rationale for each performance goal. The plan must

---

[23] For a full discussion of state pollution prevention programs, see Dennison, *Pollution Prevention Strategies and Technologies* (Government Institutes, 1995).
[24] Ariz. Rev. Stat. Ann. § 49-962(A).
[25] Ariz. Rev. Stat. Ann. § 49-962(B).
[26] 42 U.S.C. § 13106.
[27] Ariz. Rev. Stat. Ann. § 49-963(K).
[28] Ariz. Rev. Stat. Ann. § 49-963(J).

include a goal for the facility and may include goals for individual production processes.

5. A written policy setting forth management and corporate support for the pollution prevention plan and a commitment to implement the plan to achieve the plan goals.

6. A statement of the plan's scope and objectives.

7. An analysis identifying pollution prevention opportunities to reduce or eliminate toxic substance releases and hazardous waste generation.

8. An analysis of pollution prevention activities that are already in place and that are consistent with the requirements of this article.

9. Employee awareness and training programs to involve employees in pollution prevention planning and implementation to the maximum extent feasible.

10. Provisions to incorporate the plan into management practices and procedures in order to ensure its institutionalization.

11. A description of the options considered and an explanation of why the options considered were not implemented.

In addition to preparing and implementing this pollution prevention plan, each facility must also file an annual progress report. The annual progress report must "analyze the progress made, if any, in pollution prevention including toxics use reduction, source reduction and hazardous waste minimization relative to each performance goal established and relative to the prescribed plan contents. Pollution prevention achieved under previously implemented activities may also be included. In the progress report, the facility also must set forth amendments to the pollution prevention plan and explain the need for any amendments.[29]

If a facility that is required to submit a pollution prevention or an annual progress fails to do so, the state environmental agency must order the facility to submit an adequate plan or report within a reasonable time period of at least 90 days. If the facility fails to comply with this order, the agency may take action against the facility, including inspection of the facility, gathering necessary information and preparing a plan or progress report at the facility's expense, or entering an administrative compliance order that is enforceable in a proceeding.[30]

In addition to these mandatory requirements, Arizona's law also directs the state Department of Environmental Quality to establish a pollution prevention technical assistance program, which includes a hazardous waste reduction clearinghouse, hazardous waste minimization workshops and training, on-site technical assistance to hazardous waste generators, and incentives for innovative hazardous waste management.[31]

---

[29] Ariz. Rev. Stat. Ann. § 49-963(L).
[30] Ariz. Rev. Stat. Ann. § 49-964(F).
[31] Ariz. Rev. Stat. Ann. § 49-965.

Arizona has also developed a parallel pollution prevention program for state agencies.[32] State agencies that produce hazardous wastes or use toxic substances in excess of the threshold quantities and time limits applicable to other facilities must also file pollution prevention plans. The pollution prevention plan must have a goal of 20 percent reduction in hazardous waste within two years, 50 percent reduction in hazardous waste within five years and a 70 percent reduction in hazardous waste in 10 years. The pollution prevention plan must address a reduction in the use of toxic substances and the generation of hazardous wastes. The plan must initially be filed on or before January 1, 1993, and every five years thereafter.[33] Just like other regulated facilities, these state agencies must also submit annual progress reports concerning their pollution prevention plans and file annual toxic data reports.[34]

California

In 1989, California passed the Hazardous Waste Source Reduction and Management Review Act.[35] The primary purpose of the Act is to:[36]
1. Reduce the generation of hazardous waste.
2. Reduce the release into the environment of chemical contaminants which have adverse and serious health or environmental effects.
3. Document hazardous waste management information and make that information available to state and local government.

The California Department of Toxic Substances Control is required to establish a program, in coordination with other state agencies, that promotes hazardous waste source reduction. The Act promotes the reduction of hazardous waste at its source, and wherever source reduction is not feasible or practicable, encourages recycling. The goal of the Act is to reduce the generation of hazardous wastes in the state by 5 percent per year from the year 1993 through the year 2000.[37]

The Act only applies to hazardous waste generators who produce more than 12,000 kilograms of hazardous waste in a calendar year, or more than 12 kilograms of extremely hazardous waste in a calendar year. Hazardous waste refers to the those wastes considered hazardous under the state's Hazardous Waste Control Law.[38] Until December 31, 1997, however, generators of more than 5,000 kilograms in a calendar year of certain

---

[32] Ariz. Rev. Stat. Ann. §§ 49-972, 49-973.
[33] Ariz. Rev. Stat. Ann. § 49-972.
[34] Ariz. Rev. Stat. Ann. §§ 49-972(I), 49-973.
[35] Cal. Health & Safety Code § 25244.12 to .24.
[36] Cal. Health & Safety Code § 25244.13(b).
[37] Cal. Health & Safety Code § 25244.15(e).
[38] See Cal. Health & Safety Code § 25110.

listed hazardous wastes[39] are also subject to the requirements of the Hazardous Waste Source Reduction and Management Review Act.[40]

Each generator regulated under the Act must conduct a source reduction evaluation review and plan every four years, commencing on or before September 1, 1991.[41] The source reduction evaluation review and plan must be conducted and completed for each site according to a specified format and include information concerning each of the following components:

1 The name and location of the site.

2. The SIC Code of the site.

3. Identification of all routinely generated hazardous waste streams which result from ongoing processes or operations that have a yearly volume exceeding 5 percent of the total yearly volume of hazardous waste generated at the site, or, for extremely hazardous waste, 5 percent of the total yearly volume generated at the site.

4. For each hazardous waste stream, the review and plan must include the following information:

● An estimate of the quantity of hazardous waste generated.

● An evaluation of source reduction approaches available to the generator which are potentially viable, including consideration of input change, operational improvement, production process change, and product reformulation.

5. A specification of, and a rationale for, the technically feasible and economically practicable source reduction measures which will be taken by the generator with respect to each hazardous waste stream. The review and plan shall fully document any statement explaining the generator's rationale for rejecting any available source reduction approach.

6. An evaluation, and, to the extent practicable, a quantification, of the effects of the chosen source reduction method on emissions and discharges to air, water, or land.

7. A timetable for making reasonable and measurable progress toward implementation of the selected source reduction measures.

8. Certification by a registered professional engineer.

9. Four-year numerical goals for reducing the generation of hazardous waste streams, based upon its best estimate of what is achievable in that four-year period.

10. A progress report as part of its submittal of a biennial report on March 1 of each even-numbered year pursuant to Section 66262.41 of Title 22 of the California Code of Regulations. If the generator is not required to submit a biennial report pursuant to that regulation, it shall prepare a

---

[39] See Cal. Health & Safety Code § 25179.7(a)(1)-(3).

[40] Cal. Health & Safety Code § 25244.15(d)(3)(A).

[41] Cal. Health & Safety Code § 25244.19.

separate progress report on the same time schedule required for the biennial report. Generators not required to submit a biennial report shall not be required to submit their prepared progress report. Progress reports must address plan implementation activities undertaken by the generator during the two years preceding the year in which the biennial report is required to be submitted.

11. The progress report must briefly summarize and, to the extent practicable, quantify, in a manner which is understandable to the general public, the results of implementing the source reduction methods identified in the generator's source reduction evaluation, review, and plan for each waste stream addressed by that plan. The progress report due on March 1, 1994, and every other progress report thereafter, shall also include an estimate of the amount of reduction the generator anticipates will be achieved by the implementation of source reduction methods during the period between the preparation of the progress report and the preparation of its next progress report.

For generators who generate less than 12,000 kilograms per year, the Act requires that the California Department of Toxic Substances Control modify the review and plan requirements of Section 25244.19 by substituting a compliance check list approach for source reduction evaluation reviews and plans.[42] The purpose of the compliance checklist is to provide a simple, understandable method for small generators to comply with the waste reduction requirements of the Act in an inexpensive, convenient manner.

The Act also directs the California Department of Toxic Substances Control to establish a technical and research assistance program to assist generators in identifying and applying source reduction approaches.[43]

Minnesota

The state of Minnesota also enacted mandatory pollution prevention legislation.[44] The Minnesota Toxic Pollution Prevention Act identifies the "preferred means of preventing toxic pollution as techniques and processes that are implemented at the source and that minimize the transfer of toxic pollutants from one environmental medium to another."[45]

Minnesota, much like the Arizona and California programs, seeks to achieve its goals through mandatory toxic pollution prevention plans.[46] Facilities that are required to file toxic chemical release reporting forms pursuant to EPCRA or Minnesota's emergency planning and community

---

[42] Cal. Health & Safety Code § 25244.15(d)(3)(B).
[43] Cal. Health & Safety Code § 25244.17.
[44] Minn. Rev. Stat. Ann. § 115D.01 to .12.
[45] Minn. Rev. Stat. Ann. § 115D.02(a).
[46] Minn. Rev. Stat. Ann. § 115D.07.

right-to-know law[47] must prepare a toxic pollution prevention plan for the facility. The Minnesota Toxic Pollution Prevention Act contains different deadlines for plan completion, depending on the type of facility.[48] Each toxic pollution prevention plan must establish a program identifying the specific technically and economically practicable steps that could be taken during at least the three years following the date the plan is due, to eliminate or reduce the generation or release of toxic pollutants reported by the facility. Toxic pollutants resulting solely from research and development activities need not be included in the plan.

The plan must be updated every two years and contain the following information:[49]

1. A policy statement articulating upper management support for eliminating or reducing the generation or release of toxic pollutants at the facility.

2. A description of the current processes generating or releasing toxic pollutants that specifically describes the types, sources, and quantities of toxic pollutants currently being generated or released by the facility.

3. A description of the current and past practices used to eliminate or reduce the generation or release of toxic pollutants at the facility and an evaluation of the effectiveness of these practices.

4. An assessment of technically and economically practicable options available to eliminate or reduce the generation or release of toxic pollutants at the facility, including options such as changing the raw materials, operating techniques, equipment and technology, personnel training, and other practices used at the facility. The assessment may include a cost-benefit analysis of the available options.

5. A statement of objectives based on the assessment and a schedule for achieving those objectives. Wherever technically and economically practicable, the objectives for eliminating or reducing the generation or release of each toxic pollutant at the facility must be expressed in numeric terms. Otherwise, the objectives must include a clearly stated list of actions designed to lead to the establishment of numeric objectives as soon as practicable.

6. An explanation of the rationale for each objective established for the facility.

7. A listing of options that were considered not to be economically and technically practicable.

---

[47] Minn. Rev. Stat. Ann. § 299K.08.
[48] Minn. Rev. Stat. Ann. § 115D.07, subd.1(b)-(e).
[49] Minn. Rev. Stat. Ann. § 115D.07, subd. 2.

8. A certification, signed and dated by the facility manager and an officer of the company attesting to the accuracy of the information in the plan.

The Act requires that regulated facilities prepare and submit annual progress reports on October 1 of each year.[50] Minnesota's Toxic Pollution Prevention Act also requires that the state's pollution control agency establish a pollution prevention technical assistance program and provide grants to study or demonstrate the feasibility of applying specific pollution prevention technologies.[51]

12.4 Solid Waste Reduction Programs

Companies have seen a dramatic increase in the complexity and costs of managing solid wastes. In response, innovative companies have been incorporating waste reduction strategies into daily business operations. Waste reduction refers to actions taken to reduce the amount and/or toxicity of wastes that require disposal. It includes waste prevention, recycling, and the purchase and manufacture of goods that have recycled content or produce less waste. Some companies are adopting simple waste reduction options, such as reducing paper consumption through the use of electronic mail. Other businesses are reviewing their entire operations to identify and implement as many opportunities for reducing waste as possible. Whether simple alterations or large-scale initiatives, companies are finding that waste reduction can offer impressive dividends.

In addition to saving money through lower waste removal costs-sometimes thousands of dollars annually-waste reduction also makes good business sense in other ways. Waste reduction can help reduce expenditures on raw materials, office supplies, equipment, and other purchases. Streamlining operations to reduce waste often can enhance overall efficiency and productivity. Furthermore, waste reduction measures can help demonstrate concern for the environment, thus enhancing a company's image in the eyes of existing and potential customers. For many companies, waste reduction is rapidly becoming an important component of long-term business planning. Several different approaches can be taken to reduce the amount and toxicity of solid wastes, including waste prevention (or source reduction), recycling, and purchasing. This section provides a methodology that companies can follow to implement an effective solid waste reduction program.

---

[50] Minn. Rev. Stat. Ann. § 115D.08, subd. 1.
[51] Minn. Rev. Stat. Ann. § 115D.04.-.05

12.4.1 Developing a Solid Waste Reduction Program

Successful waste reduction programs hinge on careful planning and organization. The key steps for establishing an effective program are:

- Obtaining management support and involvement.
- Establishing a waste reduction team and team leader.
- Setting preliminary program objectives.
- Getting the whole company on board by announcing the program and its goals to all employees.

Management Support. The support of company management is essential for developing a lasting and successful waste reduction program. At the outset of a program, an endorsement from company management is needed to help establish a waste reduction team. Throughout the program, company management can support the team by endorsing program goals and implementation, communicating the importance of reducing waste within the company, guiding and sustaining the program, and encouraging and rewarding employee commitment and participation in the effort. Stressing the range of benefits that can come from waste reduction, such as cost savings and enhanced company image, will help sell the program to management.

Waste Reduction Team. The waste reduction team is a group of employees who are responsible for many of the tasks involved in planning, designing, implementing, and maintaining the program. A team approach allows these tasks to be distributed among several employees and enables employees from all over the company to directly contribute to reducing waste.
   Typically, members of a waste reduction team are responsible for:

- Working with company management to set the preliminary and long-term goals of the waste reduction program.
- Gathering and analyzing information relevant to the design and implementation of the program.
- Promoting the program to employees and educating them about how they can participate in the effort.
- Monitoring the progress of the program.
- Periodically reporting to management about the status of the program.

The size of the team should relate to the size of the company and be representative of as many departments or operations as possible. For a modest waste reduction program, an effective team might consist of just one or two people. Larger businesses might opt to create a team of employees from different departments to encourage widespread input and support.

Company management or the team should appoint a knowledgeable and motivated team leader. Depending on the size of the company and the type of program being implemented, the position can require a significant amount of time and energy. The leader must be capable of directing team efforts; administering the planning, implementation, and operation of the waste reduction program; and acting as a liaison between management and the team. Likely candidates include a facilities manager, an environmental manager, or an employee who has championed waste reduction in the company. If possible, the task should be incorporated into the person's job description.

Once the team has been established, members should meet regularly to develop a plan and begin program implementation. The time needed to design and implement a waste reduction program will vary. Generally, large facilities incorporating many different options will need several months to start up a program. Department-specific or more modest programs might be implemented in less than a month. Some businesses might even be able to implement simple options within a matter of days. In any case, the investment of time and resources at this stage will likely be returned by the savings realized through a successful waste reduction program.

Program Objectives. While the general objective of any waste reduction program is to reduce the amount and/or toxicity of solid waste being generated, the team's first task will be to work with management to establish and record specific, preliminary goals for the program. These goals might include enhancing the company's corporate image or increasing operational efficiency. The goals should be based primarily on how much waste reduction is possible given the level of effort that the company is willing to dedicate to the task. The goals set by the team will provide a framework for specific waste reduction efforts to follow.

Employee Involvement. Once the general direction of the waste reduction program has been established, the program should be presented to the rest of the company. This is a good opportunity to get employees involved and generate some momentum behind the team's efforts. An announcement should be made by the company president or representative of the upper management, demonstrating that the program has full management support

and is a high priority for the company. The announcement should:
- Introduce employees to waste reduction.
- Explain how waste reduction can benefit both the company and the environment.
- Outline the design and implementation stages of the program.
- Offer the team leader's name and number and encourage employees to contact him or her with any ideas or suggestions.

The program is more likely to succeed if suggestions for reducing solid wastes are solicited from employees. To reduce paper, the announcement should be posted in a prominent place, circulated, or distributed through electronic or voice mail, if available.

Throughout the duration of the program, periodic communications (e.g., centrally posted memos or announcements) can help maintain employee support. Employees are likely to appreciate being asked to join in the company's waste reduction efforts, and such offers will encourage consistent participation.

12.4.2 Conducting Waste Assessments

Having established the framework of the company's waste reduction program, the next step for the waste reduction team is to consider conducting a waste assessment. Some teams, especially those planning very limited programs or in companies where the waste stream is well understood, might opt to forego a waste assessment. In fact, many effective waste reduction measures can be adopted without the help of an assessment. The data generated in an assessment can, however, provide the team with a much greater understanding of the types and amounts of waste the company generates. These data can be invaluable in the design and implementation of a waste reduction program.

If the company does not have the time or resources to conduct a waste assessment, it might consider using industry averages of the amount of waste generated by companies in the same field to approximate the amounts and types of waste the company generates. Often, waste generation estimates by general waste category can be obtained for a company's specific type of business and used as the basis for designing a waste reduction program. While this may be the easiest way to approximate a waste generation rate, these estimates are unable to account for specific conditions and may, therefore, result in inaccuracies. In addition, these potentially inaccurate data can hinder the evaluation process, since measuring waste reduction progress depends on comparing current waste generation data with information regarding the amounts and types of waste produced before program implementation.

12.4.3 Waste Assessment Approaches

Planning and executing an appropriate waste assessment involves determining its scope, scheduling the different assessment activities, communicating the necessary information to employees, and performing the actual assessment. Depending on the objective of the company's waste reduction program, a waste assessment can involve:
- Examining facility records
- Conducting a facility walk-through
- Performing a waste sort

The assessment may require just one of these activities, or a combination of approaches.

The team should determine what type of assessment is best for the company based on such factors as the type and size of the facility, the complexity of the waste stream, the resources (money, time, labor, equipment) available to implement the waste reduction program, and the goals of the program. For example, if the company generates only a few types of waste materials, the team might only need to review company records and briefly inspect facility operations. On the other hand, if the company generates diverse types of waste and has established a goal to cut waste disposal by 50 percent, the team will need to thoroughly examine and quantify the wastes generated in most company operations by performing a waste sort.

12.4.4 Evaluating and Selecting Waste Reduction Options

Using the findings from the waste assessment, the team should list all the possible waste reduction measures that it feels might be effective based on the goals of the waste reduction program. It is important that the waste reduction team throughly review the potential effects of each waste prevention, recycling, and purchasing option. While a strong consideration is likely to be whether the option's costs are justified by potential savings, the waste reduction team also should consider:
- Effects on product or service quality and product marketing
- Compatibility with existing operations
- Equipment requirements
- Space and storage requirements

- Operation and maintenance requirements
- Staffing, training, and education requirements
- Implementation time
- Effects on employee morale, environmental awareness, and community relations

Based on these criteria, the team should screen options to identify a subset of options that deserve further analysis and possible inclusion in the waste reduction program.

Once a short list of waste reduction options has been identified, the team should begin the process of deciding which options are the most appropriate for the program. During this evaluation process, the team should be clear on the relative importance of the different criteria against which the options are being measured. Depending on the company's waste reduction goals, for example, cost-effectiveness may not always be the overriding criteria for selected options. Other criteria, such as improved environmental awareness, employee morale, and community relations, may be equally important. In addition, teams whose companies feel cost-effectiveness must be a key criteria should be sure to consider the long-term economic feasibility of an option. While the team may be inclined to disregard a particular option with large start-up costs, the measure may end up yielding impressive savings over several years.

Some options might not require extensive analysis. For example, if the company already has a copy machine with the ability to make two-sided copies efficiently, then a policy mandating double-sided copying usually can be implemented easily. On the other hand, complex options that require a significant change in operations or large capital investments should be analyzed carefully. For complex options, the team will want to contact suppliers, product refurbishers, packaging designers, and any other individuals who could help determine if the option is feasible. These individuals also can help pinpoint any unforeseen obstacles or complications that could hinder implementation.

12.4.5 Waste Prevention Options

When analyzing and selecting specific options, team members should focus first on waste prevention. The most effective way to reduce wastes is for the company to generate less wastes in the first place. Companies can adopt a wide range of waste prevention strategies, including:

- *Using or manufacturing minimal or reusable packaging.* Encourage suppliers to minimize the amount of packaging used to protect their products or seek new suppliers who offer products with minimal packaging. Work with suppliers to make arrangements for returning shipping materials, such as crates, cartons, and pallets for reuse. In addition, examine the packaging that the company uses for its own products to determine if it is possible to use fewer layers of materials or to ship merchandise in returnable or reusable containers.

- *Using and maintaining durable equipment and supplies.* Purchase quality, long-lasting supplies and equipment that can be repaired easily, and establish regular maintenance schedules for them. These items will stay out of the waste stream longer, and the higher initial costs are often justified by lower maintenance, disposal, and replacement costs. In addition, these items are replaced far less frequently, offering further cost savings.
- *Reusing products and supplies.* Using durable, reusable products rather than single-use materials is one of the most effective waste prevention strategies. Consider adopting simple, cost-effective measures, such as reusing common items like file folders and interoffice envelopes.
- *Reducing the use of hazardous constituents.* Often, substitutes for the standard cleaning solvents, inks, paints, glues, and other materials used by graphics and maintenance departments are available which are free of the hazardous ingredients that otherwise could end up being disposed of with the rest of the company's solid waste. Ask suppliers to suggest reformulated products, such as toners with no heavy metals and water-based paints and cleaning solutions.
- *Using supplies and materials more efficiently.* There are many strategies that a company can adopt to reduce waste and conserve materials. In addition, purchasing and inventory practices that generate waste unnecessarily can be eliminated. For example, some companies might order large quantities of an item to receive a discounted unit price, only to have a portion of the order end up unused and discarded. Be cautious about overordering products with a limited shelf life.
- *Eliminating unnecessary items.* When reviewing the company's operations for opportunities to reduce waste, don't overlook the obvious. The company may routinely use items that contribute little or nothing to its products or services. A number of effective waste reduction measures may involve simply eliminating the use of unnecessary materials and supplies.

After studying the company's waste generation and management practices, the team will likely have compiled a number of waste prevention options. Determine the capital and operating costs of these options and compare them against potential savings and revenues. Be sure to examine the potential operational effects, as well. For example, while modifying packaging can significantly reduce waste, the team will want to consider carefully how these changes will affect storage, operations, and labor costs. One waste prevention option may result in savings in several different areas, including avoided purchasing, storage, materials handling, and removal costs.

12.4.6 Recycling Options

Recycling options should be evaluated next. Recycling offers businesses a way to avoid disposing of wastes that cannot be prevented. Many businesses are collecting bottles, cans, paper, corrugated cardboard, and other materials for recycling. Before implementing any recycling option, the team needs to consider the marketability of the materials to be collected. To locate potential buyers, contact local recycling companies. Consult the Yellow Pages (under "recycling"), trade associations, chambers of commerce, and state or local government recycling offices for assistance. When conducting preliminary contract discussions with local buyers and haulers, there are a number of questions you should ask including:

1. What types of recylables will the company accept and how must they be prepared?
Recycling companies might request that the material be baled, compacted, shredded, granulated, or loose. Generally, recyclers will offer a better price for compacted or baled material. Compacting or densifying materials before transporting also can be a cost-effective method of lowering hauling costs for the buyer.

2. What contract terms will the buyer require?
Discuss the length of the potential contract with the buyer. Shorter contracts provide greater flexibility to take advantage of rising prices, while longer contracts provide more security in an unsteady market. Often, buyers favor long-term contracts to help ensure a consistent supply of materials. The terms of payment should be discussed as well, since some buyers pay after delivery of each load, while others set up a periodic schedule. Also, ask whether the buyer would be willing to allow changes to the contract over time.

3. Who provides transportation?
If transportation services are not provided by the buyer, locate a hauler to transport materials to the buyer. The Yellow Pages, local waste haulers, and state or local waste management authorities can help provide this information.

4. What is the schedule of collections?

If the recycling company offers to provide transportation, check on the frequency of collections. Some businesses might prefer to have the hauler be on call, picking up recyclables when a certain weight or volume has been reached. Larger companies might generate enough recyclable material to warrant a set schedule of collections.

5. What are the maximum allowable contaminant levels, and what is the procedure for dealing with rejected loads?

Inquire about what the buyer has established as maximum allowable contaminant levels for food, chemicals, or other contaminants. If these requirements are not met, the buyer might reject a contaminated load and send it back to the company. The buyer also might dispose of a contaminated load in a landfill or combustor, which can result in the company incurring additional costs and/or liability.

6. Are there minimum quantity requirements?

Find out whether the buyer requires a minimum weight or volume before accepting delivery. If a buyer's minimum quantity requirements are difficult to meet, consider working with neighboring offices or retail spaces. By working together, it might be possible to collect recyclables in central storage containers and thereby meet the buyer's requirements.

7. Where will the waste be weighed?

Ask where the material will be weighed, and at what point copies of the weight slips will be available. Weighing the material before it is transported will eliminate the problem of lost weight slips and confirm the accuracy of the weight recorded by the buyer.

8. Who will provide containers for recyclables?

Buyers should be asked whether they will provide containers in which to collect, store, and transport the material, and whether there is a fee for this service.

9. Can "escape clauses" be included in the contract?

Such clauses establish the right of a company to be released from the terms of the contract under conditions of noncompliance by the buyer.

10. Be sure to check references.

Obtain and thoroughly check the buyer's references with existing contract holders, asking these companies specifically whether their buyer is fulfilling all contract specifications.

Be sure to weigh carefully the cost-effectiveness and potential operational effects of recycling options. Recycling programs, especially more ambitious efforts, often require purchases of equipment like containers, compactors, and balers. Additional labor also might be required. Moreover, steps might be necessary to ensure that contamination of collected materials is minimized. Some companies also may have to pay a fee to have their collected recyclable material removed. In many cases, however, the savings and revenues (such as reduced removal costs and revenues from selling collected materials) will offset these costs. In addition, consider whether the new recycling program will affect current purchasing practices. For instance, the company might want to begin buying exclusively white legal pads instead of yellow ones to take advantage of the strong market for white office paper. Also examine the extent to which internal collection, transfer, and storage systems are needed and whether these new systems will be compatible with existing operations.

12.4.7 Purchasing Options

During the waste assessment, the team may have noted puchasing changes that could help reduce waste, from buying supplies with reduced packaging to careful inventory control to avoid over-ordering and possibly throwing away perishable items. In addition, during the team's exploration of local recycling markets, the need for favoring products made with recycled content also may have become evident. In any business, many opportunities exist to use the company's buying power to reduce waste and encourage the growth of recycling markets. To identify specific changes in purchasing that the company could adopt, the team might contact its suppliers and discuss alternative products that would meet the new purchasing criteria. Check with other suppliers, as well, to see what they may be able to offer. In addition, various industry groups, state solid waste agencies, and federal information services, such as the EPA's RCRA Hotline, can help identify ways to reduce waste through product purchasing and sources of products made from recycled materials.

After having identified opportunities to purchase recycled products and products that can help to reduce waste, each item should be evaluated in terms of availability and cost. Reduced waste and recycled products do not necessarily cost more than other products. For example, while paper made from recycled fibers was once considerably more expensive than virgin paper, the price of paper with recovered content is now competitive with traditional paper. In addition, be sure to compare recycled or reduced-waste products to other products on the basis of long-term costs, rather than purchasing costs alone. Similarly, while reusable products may cost more to purchase initially, they often save money over time by avoiding frequent purchases of single-use items.

12.4.8 Implementing the Waste Reduction Program

Having determined the initial waste reduction measures to adopt, the team should now begin to implement the measures. Consider building the program slowly, implementing a few options at a time, so employees are not overwhelmed by changes in procedure. This is particularly important for more complex waste reduction programs. Building slowly also provides an opportunity to identify, assess, and solve any operational problems in the early stages. If, however, a program involves only a few simple measures, it might be possible to implement all options at once.

12.4.9 Program Evaluation

Waste reduction is a dynamic process. Once the program is under way, the team will need to evaluate its effectiveness to see if preliminary goals are being met. In addition, once the potential for reducing waste in the company becomes better understood, consider establishing long-term goals for the program. It is important to evaluate the program periodically to:
- Keep track of program success and to build on that success (e.g., waste reduced, recycling rates achieved, money saved).
- Identify new ideas for waste reduction.
- Identify areas needing improvement.
- Document compliance with state or local regulations.
- Determine the effect of any new additions to the program.
- Keep employees informed and motivated.

The best way to assess and monitor program operations is through continued documentation. Perform the first evaluation after the program has been in place long enough to have an effect on the company's waste generation rate. In addition, it might be worthwhile to conduct additional periodic waste assessments to determine further changes in the company's waste. Also consider reviewing the company's waste removal receipts and purchasing records, or preparing a summary of recycling receipts and waste assessment worksheets.

Many companies are finding that waste reduction makes economic and environmental sense. Not only can such a program look good on the bottom line, but it also can reflect well on the company. Table 12.2 outlines a number of common solid waste reduction practices.

Table 12.2. Common Solid Waste Reduction Practices

1. Writing/Printing Paper

- Establish a company-wide double-sided copying policy, and be sure future copiers purchased by the company have double-sided capability.
- Reuse envelopes or use two-way ("send-and-return") envelopes.
- Keep mailing lists current to avoid duplication.
- Make scratch pads from used paper.
- Circulate (rather than copy) memos, documents, periodicals, and reports.
- Reduce the amount of advertising mail received by writing to the Direct Marketing Association Mail Preference Service, P.O. Box 9008, Farmingdale, NY 11735-9008, and ask that the company be eliminated from mail lists.
- Use outdated letterhead for in-house memos.
- Put company bulletins on voice or electronic mail or post on a central bulletin board.
- Save documents on hard drives or floppy disks instead of making paper copies.
- Use central files to reduce the number of hard copies the company retains.
- Proof documents on the computer screen before printing.
- Eliminate unnecessary reports.
- Donate old magazines and journals to hospitals, clinics, or libraries.

2. Packaging

- Order merchandise in bulk.
- Purchase products with minimal packaging and/or in concentrated form.
- Work with suppliers to minimize the packaging used to protect their products.
- Establish a system for returning cardboard boxes and foam peanuts to suppliers for reuse.
- Request that deliveries be shipped in returnable and/or recyclable containers.
- Minimize the packaging used for the company's products.
- Use reusable and/or recyclable containers for shipping products.

Table 12.2. Common Solid Waste Reduction Practices (*Continued*)

- Repair and reuse pallets or return them to suppliers.
- Reuse newspaper and shredded paper for packaging.Reuse foam packing peanuts, "bubble wrap," and cardboard boxes, or donate to another organization.

3. Equipment

- Rent equipment that is used only occasionally.
- Purchase remanufactured office equipment.
- Establish a regular maintenance routine to prolong the life of equipment like copiers, computers, and heavy tools.
- Use rechargeable batteries where practical.
- Install reusable furnace and air conditioner filters.
- Reclaim usable parts from old equipment.
- Recharge fax and printer cartridges or return them to the supplier for remanufacture.
- Sell or give old furniture and equipment to other businesses, local charitable organizations, or employees.

4. Inventory/Purchasing

- Implement an improved inventory system (such as systems based on optical scanners) to provide more precise control over supplies.
- Avoid ordering excess supplies that may never be used.
- Advertise surplus and reusable waste items through a materials exchange.
- Set up an area for employees to exchange used items.
- Substitute less toxic or nontoxic inks, paints, and cleaning solvents.
- Use products that promote waste reduction (products that are more durable, of higher quality, recyclable, reusable).
- Where appropriate, order supplies in bulk to reduce excess packaging.

12.5 Hazardous Waste Minimization Programs

This section explains the meaning of the term waste minimization, discusses the benefits of waste minimization, and outlines the necessary components of an effective hazardous waste minimization program.

12.5.1 Waste Minimization Defined

The EPA considers waste minimization, the term employed by Congress in the RCRA statute, to include (1) source reduction, and (2) environmentally sound recycling.

The first category, source reduction, is defined in Section 6603(5)(A) of the Pollution Prevention Act,[52] as any practice which:

1. Reduces the amount of any hazardous substance, pollutant, or contaminant entering any waste stream or otherwise released into the environment (including fugitive emissions) prior to recycling, treatment, or disposal, and

2. Reduces the hazards to public health and the environment associated with the release of such substances, pollutants, or contaminants.

The term includes equipment or technology modifications, process or procedure modifications, reformulation or redesign of products, substitution of raw materials, and improvements in housekeeping, maintenance, training, or inventory control. The EPA relies on this definition for use in identifying opportunities for RCRA source reduction.

The second category, environmentally sound recycling, is the next preferred alternative for managing those pollutants that cannot be reduced at the source. In the context of hazardous waste management, there are certain practices or activities that the RCRA regulations define as "recycling." The definitions for materials that are "recycled" are found in 40 CFR Section 261.1(c).

The EPA considers recycling activities that closely resemble conventional waste management activities not to constitute waste minimization. Unfortunately, it is not always easy to distinguish recycling from conventional treatment.[53] Treatment for the purposes of destruction or disposal is not part of waste minimization, but is, rather, an activity that occurs after the opportunities for waste minimization have been pursued.

Transfer of hazardous constituents from one environmental medium to another also does not constitute waste minimization. For example, the use of an air stripper to evaporate volatile organic constituents from an aqueous waste only shifts the contaminant from water to air. Furthermore, concentration activities conducted solely for reducing volume does not constitute waste minimization unless, for example, concentration of the waste is an integral setup in the recovery of useful constituents prior to treatment and disposal. Similarly, dilution as a means of toxicity reduction would not be considered waste minimization, unless dilution is a necessary step in a recovery or a recycling operation.

---

[52] 42 U.S.C. § 13102(5)(a).
[53] See 56 *Fed. Reg.* 7143 (Feb. 21, 1991); 53 *Fed. Reg.* 522 (Jan. 8, 1988).

12.5.2 Benefits of Hazardous Waste Minimization

Waste minimization provides additional environmental improvements over "end of pipe" control practices, often with the added benefit of cost savings to generators of hazardous waste and reduced levels of treatment, storage, and disposal. Waste minimization has already been shown to result in significant benefits for industry, including:

1. Minimizing quantities of hazardous waste generated, thereby reducing waste management and compliance costs and improving the protection of human health and the environment.
2. Reducing or eliminating inventories and possible releases of "hazardous chemicals."
3. Possible decrease in future CERCLA and RCRA liabilities, as well as future toxic tort liabilities.
4. Improving facility mass/energy efficiency and product yields.
5. Reducing worker exposure.
6. Enhancing organizational reputation and image.

Waste minimization programs are being implemented by a wide array of organizations. Numerous state governments have also enacted legislation requiring facility specific waste minimization programs, and other states have legislation pending that may mandate some type of facility-specific waste minimization program.

12.5.3 Components of Hazardous Waste Minimization Program

In 1993, the EPA issued guidance to hazardous waste generators on how to develop and implement a RCRA hazardous waste minimization plan.[54] The EPA's guidance on the elements of a waste minimization program is intended to assist companies and individuals to properly certify that they have implemented a program to reduce the volume and toxicity of hazardous waste to the extent "economically practicable." The guidance is directly applicable to generators who generate 1000 or more kilograms per month of hazardous waste ("large quantity" generators) or to owners and operators of hazardous waste treatment, storage, or disposal facilities who manage their own hazardous waste on-site.

Small quantity generators who generate greater than 100 kilograms but less than 1000 kilograms of hazardous waste per month are not subject to the same "program in place" certification requirement as large quantity

---

[54] Guidance to Hazardous Waste Generators on the Elements of a Waste Minimization Program, 58 *Fed. Reg.* 31114 (May 28, 1993).

generators. Instead, they must certify on their hazardous waste manifests that they have "made a good faith effort to minimize" their waste generation. Nevertheless, the EPA encourages small quantity generators to develop their own waste minimization programs to show good faith efforts.

According to the EPA guidance (which is not a formal regulation, and therefore, not enforceable) the following basic elements should be part of most waste minimization programs:

1. Top management support
2. Characterization of waste generation and waste management costs
3. Periodic waste minimization assessments
4. Appropriate cost allocation
5. Encouragement of technology transfer
6. Program implementation and evaluation

Thus, generators should consider these elements when designing multimedia pollution prevention programs directed at preventing or reducing wastes, substances, discharges, and/or emissions to all environmental media — air, land, surface water, and groundwater. Each of these elements is discussed below.

12.5.4 Management Support

Top management should support a company-wide effort to minimize hazardous wastes. There are many ways to accomplish this goal. Some of the methods described below may be suitable for some companies, while not for others. However, some combination of these techniques or similar ones will demonstrate top management support:

1. Make waste minimization a part of the company policy. Put this policy in writing and distribute it to all departments and individuals. Each individual, regardless of status or rank, should be encouraged to identify opportunities to reduce waste generation. Encourage workers to adopt the policy in day-to-day operations and encourage new ideas at meetings and other organizational functions. Waste minimization, especially when incorporated into company policy, should be a process of continuous improvement. Ideally, a waste minimization program should become an integral part of the company's strategic plan to increase productivity and quality.
2. Set explicit goals for reducing the volume and toxicity of waste streams that are achievable within a reasonable time frame. These goals may be quantitative or qualitative. Both can be successful.
3. Commit to implementing recommendations identified through assessments, evaluations, and waste minimization teams.

4. Designate a waste minimization coordinator who is responsible for facilitating effective implementation, monitoring, and evaluation of the program. In some cases (particularly in large multifacility organizations), an organizational waste minimization coordinator may be needed in addition to facility coordinators. In other cases, a single coordinator may have responsibility for more than one facility. In these cases, the coordinator should be involved or be aware of operations and should be capable of facilitating new ideas at each facility. It is also useful to set up self-managing waste minimization teams chosen from a broad spectrum of operations: engineering, management, research and development, sales and marketing, accounting, purchasing, maintenance, and environmental staff personnel. These teams can be used to identify, evaluate, and implement waste minimization opportunities.

5. Publicize success stories. Set up an environment and select a forum where creative ideas can be heard and tried. These techniques can inspire additional ideas.

6. Recognize individual and collective accomplishments. Reward employees that identify cost-effective waste minimization opportunities. These rewards can take the form of collective and/or individual monetary or other incentives for improved productivity/waste minimization.

7. Train employees on the waste-generating impacts that result from the way they conduct their work procedures. For example, purchasing and operations departments could develop a plan to purchase raw materials with less toxic impurities or return leftover materials to vendors. This approach can include all departments, such as those in research and development, capital planning, purchasing, production operations, process engineering, sales and marketing, and maintenance.

12.5.5 Waste Generation and Management Costs

Maintain a waste accounting system to track the types and amounts of wastes as well as the types and amounts of the hazardous constituents in wastes, including the rates and dates they are generated. Each organization must decide the best method to obtain the necessary information to characterize waste generation. Many organizations track their waste production by a variety of means and then normalize the results to account for variations in production rates.

In addition, a waste generator should determine the true costs associated with waste management and cleanup, including the costs of regulatory oversight compliance, paperwork and reporting requirements, loss of production potential, costs of materials found in the waste stream (perhaps based on the purchase price of those materials), transportation/treatment/

storage/disposal costs, employee exposure and health care, liability insurance, and possible future RCRA or CERCLA corrective action costs. Both volume and toxicities of generated hazardous waste should be taken into account. Substantial uncertainty in calculating many of these costs, especially future liability, may exist. Therefore, each organization should find the best method to account for the true costs of waste management and cleanup.

12.5.6 Waste Minimization Assessments

Different and equally valid methods exist by which a waste minimization assessment can be performed. Some organizations identify sources of waste by tracking materials that eventually wind up as waste, from point of receipt to the point at which they become a waste. Other organizations perform mass balance calculations to determine inputs and outputs from processes and/or facilities. Larger organizations may find it useful to establish a team of independent experts outside the organization structure, while some organizations may choose teams comprised of in-house experts. Most successful waste minimization assessments have common elements that identify sources of waste and calculate the true costs of waste generation and management. Each organization should decide the best method to use in performing a waste minimization assessment that addresses these two general elements:

1. Identify opportunities at all points in a process where materials can be prevented from becoming a waste (for example, by using less material, recycling materials in the process, finding substitutes that are less toxic and/or more easily biodegraded, or making equipment/process changes). Individual processes or facilities should be reviewed periodically. In some cases, performing complete facility material balances can be helpful.

2. Analyze waste minimization opportunities based on the true costs associated with waste management and cleanup. Analyzing the cost effectiveness of each option is an important factor to consider, especially when the true costs of treatment, storage, and disposal are considered.

12.5.7 Cost Allocation

If practical and implementable, organizations should appropriately allocate the true costs of waste management to the activities responsible for generating the waste in the first place (e.g., identifying specific operations that generate the waste, rather than charging the waste management costs to "overhead"). Cost allocation can properly highlight the parts of the organization where the greatest opportunities for waste minimization exist; without allocating costs, waste minimization opportunities can be obscured by accounting practices that do not clearly identify the activities generating the hazardous wastes.

12.5.8 Technology Transfer

Many useful and equally valid techniques have been evaluated and documented that are useful in a waste minimization program. It is important to seek or exchange technical information on waste minimization from other parts of the organization, from other companies, trade associations, professional consultants, and university or government technical assistance programs. EPA- and/or state-funded technical assistance programs (e.g., California Waste Minimization Clearinghouse, Minnesota Technical Assistance Program (MnTAP), EPA Pollution Prevention Information Clearinghouse) are becoming increasingly available to assist in finding waste minimization options and technologies.

12.5.9 Program Implementation and Evaluation

Implement recommendations identified by the assessment process, evaluations, and waste minimization teams. Conduct a periodic review of program effectiveness. Use these reviews to provide feedback and identify potential areas for improvement.

General documents to assist organizations with more detailed guidance on conducting waste minimization assessments and developing pollution prevention programs include:

1. *Waste Minimization Opportunity Assessment Manual*, EPA 625/7-88/003, July 1988 (Pub. No. PB 92-216 985), available by calling NTIS at (703) 487-4650.

2. *Facility Pollution Prevention Guide*, EPA/600/R-92/088, available by calling the CERI Publications Unit at the EPA's Cincinnati office at (513) 569-7562.

3. *Waste Minimization: Environmental Quality with Economic Benefits*, EPA/530-SW-90-044, April 1990, available by calling the RCRA Information Center at (202) 260-9327.

The EPA has also developed numerous waste minimization and pollution prevention documents that are tailored to specific manufacturing and other types of processes, and periodically sponsors pollution prevention workshops and conferences.

12.6 BMP Plans For Water Pollution Control

Pollution prevention has become an important part of the Clean Water Act's National Pollutant Discharge Elimination System (NPDES) program, working in conjunction with best management practices (BMPs) to prevent the release of toxic and hazardous pollutants to receiving waters. BMPs are inherently pollution prevention practices. Traditionally, BMPs have focused on good housekeeping measures and good management techniques intending to avoid contact between pollutants and water media as a result of leaks, spills, and improper waste disposal. However, based on the authority granted under the Clean Water Act regulations, BMPs may include the universe of pollution prevention encompassing production modifications, operational changes, materials substitution, materials and water conservation, and other such measures. The EPA believes that the intent of pollution prevention practices and BMPs are similar and that they can be concurrently developed in a technologically sound and cost-effective manner. Thus, although the purpose of this section is to provide guidance for NPDES permittees in the development of BMPs to control and reduce water pollution discharges, permittees should consider pollution prevention for all environmental media.

12.6.1 Types of BMPs

BMPs may be divided into general BMPs, applicable to a wide range of industrial operations, and facility-specific (or process-specific) BMPs, tailored to the requirements of an individual site. General BMPs are widely practiced measures that are independent of chemical compound, source of pollutant, or industrial category. General BMPs are also referred to as baseline practices, and are typically low in cost and easily implemented. General BMPs are practiced to some extent at almost all facilities. Common general BMPs include good housekeeping, preventive maintenance, inspections, security, employee training, and recordkeeping and reporting.

Facility-specific BMPs are measures used to control releases associated with individually identified toxic and hazardous substances and/or one or more particular ancillary source. Facility-specific BMPs will vary from site to site depending upon site characteristics, industrial processes, and pollutants. The following general factors should be considered when selecting specific BMPs for the facility:

- *Chemical nature.* The need to control materials based on toxicity and fate and transport.
- *Proximity to water bodies.* The need to control liquid spills prior to their

release to media, such as water, from which materials may not later be separated.

- *Receiving waters.* The need to protect sensitive receiving waters which are more severely impacted by releases of toxic or hazardous materials. The need to protect the water uses including recreational waters, drinking water supplies, and fragile aquatic and biota communities.
- *Proximity to populace.* The need to control hazardous materials with potential to be released near populated areas.
- *Climate.* The need to prevent volatilization and ignitability in warmer climates. The need to reduce wear on moving parts in freezing climates. The need to avoid spills in climates and under circumstances where mitigation cannot occur.
- *Age of the facility/equipment.* The need to prevent releases caused by older equipment with greater capacity for failure. The need to address obsolete and outdated instruments and processes which are not environmentally protective.
- *Process complexity.* The need to address problems of materials incompatibility.
- *Engineering design.* The need to address design flaws and deficiencies.
- *Employee safety.* The need to prevent unnecessary exposure between employees and chemicals.
- *Environmental release record.* The need to control releases from specific areas demonstrating previous problems.

Facility-specific BMPs are often developed when a facility notes a history of problem releases of toxic or hazardous chemicals, or when facility personnel believe that actual or potential pollutant discharge problems should be addressed. Facility-specific BMPs may include many different practices such as source reduction and on-site recycle/reuse.

12.6.2 Components of BMP Plans

The suggested elements of a baseline BMP plan can be separated into three phases:

1. Planning phase
- BMP committee
- BMP policy statement
- Release identification and assessment

2. Development and implementation phase
- Good housekeeping

- Preventive maintenance
- Inspections
- Security
- Employee training
- Recordkeeping and reporting

3. Evaluation/reevaluation phase
- Evaluate plan implementation benefits

Generally, the planning phase, includes demonstrating management support for the BMP plan and identifying and evaluating areas of the facility to be addressed by BMPs. The goal of plan development should be to ensure that its implementation will prevent or minimize the generation and the potential for release of pollutants from the facility to U.S. waters. The development phase consists of determining, developing, and implementing general and facility-specific BMPs. The evaluation/reevaluation phase consists of an assessment of the components of a BMP plan and reevaluation of plan components periodically, or as a result of factors such as environmental releases and/or changes at the facility.

12.6.3 Planning Phase

In the planning phase, a facility must decide who will take the responsibility for establishing and carrying out the BMP plan. The plan should be initiated with clear support and input from facility management and employees. The facility must also identify and evaluate areas of the facility that, because of the substances involved and their management, will be addressed in the BMP plan. The elements of the planning phase generally consist of establishing a BMP committee, developing a BMP policy statement for the facility, and conducting a release identification and assessment. Each of these planning phase components are discussed in turn.

12.6.4 BMP Committee

The first step in the planning phase involves setting up a BMP committee. A BMP committee is comprised of interested staff within the facility's organization. The committee represents the company's interests in all phases of BMP plan development, implementation, oversight, and plan evaluation. It should be noted that a BMP committee may function similarly to other committees that might already exist at an industrial facility (e.g., pollution prevention committee) and may include the same employees.

The BMP committee is established to assist the facility in managing all aspects of the BMP plan. The committee's functions include responsibility for the following activities:

- Developing the scope of the BMP plan
- Making recommendations to management in support of company BMP policy
- Reviewing any existing accidental spill control plans to evaluate existing BMPs
- Identifying toxic and hazardous substances
- Identifying areas with potential for release to the environment
- Conducting assessments to prioritize substances and areas of concern
- Determining and selecting appropriate BMPs
- Establishing standard operating procedures for implementation of BMPs
- Overseeing the implementation of the BMPs
- Establishing procedures for recordkeeping and reporting
- Coordinating facility environmental release response, cleanup, and regulatory agency notification procedures
- Establishing BMP training for plant and contractor personnel
- Evaluating the effectiveness of the BMP plan in prevention and mitigation of releases of pollutants
- Periodically reviewing the BMP plan to evaluate the need to update and/or modify the BMP plan

The BMP committee is responsible for developing the BMP plan and assisting the facility management in its implementation, periodic evaluation, and updating. While the BMP committee is responsible for developing the plan and overseeing its implementation, all activities need not be limited to committee members. Rather, appropriate company personnel who are knowledgeable in the areas of concern can carry out certain activities associated with BMP plan development. With this in mind, the selection of the committee members can be limited to a select set of individuals, while the resources of interested and knowledgeable employees can still be utilized.

In order to ensure a properly run organization, one person should be designated as the lead committee member. The determination of a single leader will assist in the smooth conduct of meetings and the designation of tasks, and will aid in the decision-making process. Generally, the designated chairperson should be highly motivated to develop and implement the BMP plan, be familiar with all committee members and their areas of expertise, and be experienced in managing tasks of this magnitude. The chairperson will be responsible for ensuring that all tasks are assigned to appropriate personnel, keeping facility management and employees informed, and cohesively developing the BMP plan. Potential candidates for this role are plant managers, environmental coordinators, or other knowledgeable technical and management personnel.

All affected facility areas should be represented when selecting the appropriate individuals to serve on the BMP committee. Members might also be selected based on their areas of expertise (e.g., industrial processes). Personnel might be selected who have a full understanding of the manufacture processes from raw materials to final products, as well as of the recycling, treatment, and disposal of wastes. Possible candidates include plant supervisors in manufacturing, production, or waste treatment and disposal; maintenance engineers; environmental and safety coordinators; and materials storage and transfer managers. Not only must BMP committee members understand the activities conducted throughout the entire facility, members of the committee must also include individuals who are in the decision-making positions within the company structure. Some committee members must represent company management and have the authority to implement measures adopted by the committee.

While the BMP committee should reflect the lines of authority within the company, it should also be sensitive to general employee interests. It is crucial to ensure that employees are aware of and in support of the BMP plan and the responsible committee, as it is primarily the employees who will implement the changes resulting from committee decisions. Forming a committee comprised solely of upper-level management and administrative personnel would exclude general personnel whose input is critical for the development and implementation of the plan. Selecting employee-chosen representatives, such as union stewards, may be an appropriate means to ensure employee involvement.

The size of a BMP committee should reflect the size and complexity of the facility, as well as the quantity and toxicity of the materials at the facility. The committee must be small enough to communicate in a open and interactive manner, yet large enough to allow for input from all necessary parties.

Where needed, committee members should call upon the expertise of others through the establishment of project-specific task forces. For example, personnel involved in research and development may be asked to research the effectiveness of product substitution and process changes that are being considered as part of BMP plan development. This method of calling upon specialists, when the need arises, should allow the committee to remain a manageable size.

12.6.5 BMP Policy Statement

The next step in the planning phase involves the development of a written BMP policy statement. A BMP policy statement describes the objectives of the BMP program in clear, concise language and establishes the company policies related to BMPs. The policy statement provides two major functions: (1) it demonstrates and reinforces management's support of the BMP plan; and (2) it describes the intent and goals of the BMP plan. It is very important that the BMP policy represent both the company's goals and general employee concerns. The policy statement may include references to the company's commitment to being a good environmental citizen, expected improvements in plant safety, and potential cost savings.

The author of the BMP policy statement should be a person who performs policy- or decision-making functions for the facility. The length and level of detail of the policy statement will vary depending on the author's personal style. The following variations may be included in a BMP policy statement:

- An outline of steps that will be taken
- A discussion of the time frames for development and implementation
- An indication of the areas and pollutants of focus
- A projection of the end result of the BMP plan

The tone of the BMP policy statement is also important. The projected positive impacts of BMP implementation should be discussed in general terms. If specific goals are outlined, the level of information and the expectations presented should be reasonable to encourage support by all employees. Ultimately, the policy should provide an upbeat message of the improved working environment that will result from BMP implementation. Since gaining employee support is so important, it may be appropriate to solicit employee concerns prior to the development of the BMP policy. These concerns can be highlighted as areas which will be evaluated during BMP plan development.

Finally, to ensure that all employees are aware of the impending BMP plan, the policy statement should be printed on company letterhead and distributed to all employees. Complete distribution can be best ensured if the statement is both delivered to each employee and posted in common areas. To indicate management's commitment, the policy statement should be signed by a responsible corporate officer.

12.6.6 Release Identification and Assessment

The final step in the planning phase involves an identification and assessment of potential releases at the facility. Release identification is the systematic cataloging of areas at a facility with ongoing or potential releases to the environment. A release assessment is used to determine the impacts on human health and the environment of any on-going or potential releases identified. The identification and assessment process involves the evaluation of both current discharges and potential discharges.

The release identification and assessment process can provide a focus for the range of BMPs being considered on those activities and areas of a facility where the risks (considering the potential for release and the hazard posed) are the greatest. In some cases, the assessment may be performed based on experience and knowledge of the substances and circumstances involved. In other cases, more detailed analyses may be necessary to provide the correct focus, and release assessments may then rely on some of the techniques of risk assessment (e.g., pathway analysis, toxicity, relative risk). Understanding the dangers of releases involves both an understanding of the hazards each potential pollutant poses to human health and the environment, as well as the probability of release due to the facility's methods of storage, handling, and/or disposal of hazardous materials and wastes.

Some facilities may identify a number of situations or circumstances representing actual or potential hazards that should all be addressed in some detail through the BMP plan. However, in other instances prioritizing potential hazards is the most sensible and cost effective approach. Identifying and assessing the risk of pollutant releases for purposes of a BMP plan can best be accomplished in accordance with a five-step procedure:

1. Reviewing existing materials and plans, as a source of information, to ensure consistency, and to eliminate duplication.
2. Characterizing actual and potential pollutant sources that might be subject to release.
3. Evaluating potential pollutants based on the hazards they present to human health and the environment.
4. Identifying pathways through which pollutants identified at the site might reach environmental and human receptors.
5. Prioritizing potential releases.

Once established, these criteria may be used in developing a BMP plan that places the greatest emphasis on the sources with the greatest overall risk to human health and the environment, considering the likelihood of release and the potential hazards if a release should occur, while still implementing low cost BMPs that might contribute to safety or other employee-driven needs.

The first step in the conduct of a release identification and assessment involves the review of existing materials and plans to gather needed information. Many industrial facilities are already subject to regulatory requirements to collect and provide information that may be useful in the identification and assessment of releases. In some cases, these plans may have been developed by persons in plant safety or process engineering who do not normally consider themselves part of the environmental staff. In particular, the following plans should be identified and reviewed:

Preparedness, prevention, and contingency plans (see 40 CFR Parts 264 and 265) require the identification of hazardous wastes handled at a facility.

Spill control and countermeasures (SPCC) plans (see 40 CFR Part 112) require the prediction of direction, rate of flow, and total quantity of oil that could be discharged.

* Storm water pollution prevention plans (see 40 CFR Section 122.44) require the identification of potential pollutant sources which may reasonably be expected to affect the quality of storm water discharges.[55]

Toxic organic management plans (see 40 CFR Parts 413, 433, and 469) may require the identification of toxic organic compounds.

* Occupational Safety and Health Administration (OSHA) emergency action plans (see 29 CFR Part 1910) require the development of a list of major workplace fire and emergency hazards.

Other sources of information that might be pertinent to the release identification and assessment process include the facility's NPDES permit application and, where applicable, information collected for the Toxic Chemical Release Inventory (Form R) which certain facilities must file annually to satisfy the reporting requirements of Section 313 of the Emergency Planning and Community Right-to-Know Act (EPCRA),[56] also known as SARA Title III.[57]

The second step of conducting a release identification and assessment

---

[55] See chapter 5, section 5.9 for full discussion of Storm Water Pollution Prevention Plans.
[56] 42 U.S.C. § 11001 through 11050.
[57] See section 12.3.3 for discussion of the EPCRA Toxic Chemical Release Inventory.

is to characterize current and potential pollutant sources. This step may be conducted through assembling a description of facility operations and chemical usage and then verifying information through inspections. This process allows facility personnel to confirm the accuracy of information on hand (e.g., the amount of chemicals used in a specific location) while also tracking changes that might have evolved over time (e.g., changing the staging of lubricants in a particular part of the plant).

Generally, the preparation of a site map or maps covering the entire facility is very useful in this evaluation. Maps should cover the entire property and illustrate plant features including material storage areas for raw materials, by-products, and products; loading and unloading areas; manufacturing areas; and waste/wastewater management areas. The map should also indicate site topography, including facility drainage patterns. Any existing structural control measures already used to reduce pollutant releases should be highlighted, and conveyance mechanisms or pathways to surface water bodies should be noted. The facility site map should also indicate property boundaries, buildings, and operation or process areas. Any neighboring properties that have potential sources of contaminants that might migrate onto the facility (because of drainage patterns) should also be noted on the map.

Following preparation of a site map, a materials inventory should be prepared. Generally, purchasing records should be helpful in determining the raw materials that are part of the inventory. However, the products manufactured and the byproducts resulting during the manufacturing process should also be considered. The materials inventory should include descriptions of the amounts of pollutants released or with the potential to be released based on methods of storage or on-site disposal, loading and access methods, and management and control practices (including structural measures or treatment). The inventory should refer to the location of the material keyed to the site map. Materials inventories will vary with the size and the complexity of the facility. It may be helpful to conduct separate inventories for different areas (e.g., manufacturing areas 1, 2, and 3; water drainage areas 1 and 2).

The site map and materials inventory developed to this point have been created solely in reliance on plant records. The next part of the process requires a field evaluation/inspection that verifies the facts compiled to this point, and determines the reasons for any discrepancies. Determining the cause of discrepancies is an important step as it may result in the identification of new locations of concern (e.g., storage areas or process lines have been moved to a different part of the plant). This process may also add/delete chemicals or other materials to/from the list being evaluated (e.g., where a chemical is no longer in use, or where a chemical substitution has been made).

The field evaluation also provides an opportunity to look for evidence of past releases or situations that represent potential releases to the environment. Notes should be assembled indicating the substances that might be released and the migration pathway that would be followed by any such release. This information should be correlated with the facility map. Where evidence of past leaks is found, further study should be undertaken to determine if the evidence correlates with the release information already obtained.

The third step in the release identification and assessment process involves evaluating potential pollutants based on the hazards they present to human health and the environment. No single measure of toxicity or hazardous characteristics exists because chemicals may have a variety of effects (both direct and indirect) that are characterized by a range of physical/chemical properties and associated effects. Some chemicals, for example, may be hazardous because of flammability and therefore represent fire hazards. Other products may be toxic and represent a threat to waterways and their associated flora and fauna, contaminate groundwaters, and/or threaten workers who are cleaning up spills and have not been provided with the proper protective equipment (e.g., respirators). Potential releases of pollutants to the environment might be subject to regulation under environmental permits, and represent threats to the facility in the form of noncompliance.

Detailed information on material properties should be available from plant safety personnel. When evaluating the threats posed by chemicals, facility personnel should consult available technical literature, manufacturer's representatives, and technical experts, such as safety coordinators within the plant. A variety of technical resources can provide information of chemical properties, including the following:

- Material safety data sheets.
- American Council of Government and Industrial Hygienist publications.
- N. Sax, *Dangerous Properties of Industrial Materials,* Eighth Edition, Volumes 1–3, Van Nostrand Reinhold, New York, New York (1994).
- *National Institute of Occupational Safety and Health (NIOSH) Pocket Guide to Chemical Hazards, U.S.* Department of Health and Human Services (1990).
- M. Dennison, *Understanding Solid and Hazardous Waste Identification and Classification,* John Wiley & Sons, Inc, New York, New York (1993).
- EPA guidance documents. (Call the EPA Public Information Center at (202) 260-7751.)

These references can provide information on specific physical/chemical properties that should be considered in evaluating hazards, including toxicity, ignitability, explosivity, reactivity, and corrosivity. Careful evaluation of these data will provide a basis for determining the intrinsic threat posed by materials at the facility. Armed with such understanding and subsequent identification of exposure pathways and potential receptors (the next step in the process), the need for developing BMPs comes into focus.

The fourth step in the release identification and assessment process involves identifying pathways by which pollutants identified at the site might reach environmental and human receptors. Identifying the pathways of current releases can usually be accomplished based on visual observations. However, identifying the pathways of potential releases requires the use of sound engineering judgement in determining the point of release, estimating the direction and rate of flow of potential releases toward receptors of concern, and identification and technical evaluation of any existing means of controlling chemical releases or discharges (such as dikes or diversion ditches). Information from the site map and observations made during the visual inspection (e.g., location of materials, potential release points, drainage patterns) should prove useful in this analysis.

The final step in the release identification and assessment process requires the application of best professional judgment in prioritizing potential releases. Priorities should be established for both known and potential releases. A combination of information identified in the previous steps about releases (the probability of release, the toxicity or hazards associated with each pollutant, and descriptions of the potential pathways for releases) should be evaluated. Using this information, a facility can rank actual and potential sources as high, medium, or low priority. These priorities can then be used in developing a BMP plan that places the greatest emphasis on BMPs for the sources that present the greatest risk to human health and environment.

12.6.7 Development Phase

After the BMP committee and policy statement have been established and the potential release identification and assessment has defined those areas of the facility that will be targeted for BMPs, the committee can begin the development phase to determine the most appropriate BMPs for controlling environmental releases. The BMP plan should consist of both general and facility-specific BMPs. General BMPs are relatively simple to evaluate and adopt. All BMP plans should include the following general BMPs:

1. *Good housekeeping.* A practice designed to maintain the facility in a clean and orderly fashion.
2. *Preventive maintenance.* A practice focused on preventing releases caused by equipment problems, rather than repair of equipment after problems occur.
3. *Inspections.* A practice established to oversee facility operations and identify actual or potential problems.
4. *Security.* A practice designed to avoid releases due to accidental or intentional entry.
5. *Employee training.* A practice developed to instill an understanding of the BMP plan in employees.
6. *Recordkeeping and reporting.* A practice designed to maintain relevant information and foster communication.

Each of these six general BMPs is discussed in the sections that follow.

12.6.8 Good Housekeeping

Good housekeeping is essentially the maintenance of a clean, orderly facility. Maintaining good housekeeping is at the heart of a facility's overall pollution control effort. Good housekeeping cultivates a positive employee attitude and contributes to the appearance of sound management principles at a facility. Some of the benefits that may result from good housekeeping practices include ease in locating materials and equipment; improved employee morale; improved manufacturing and production efficiency; lessened raw, intermediate, and final product losses due to spills, waste or releases; fewer health and safety problems arising from poor materials and equipment management; environmental benefits resulting from reduced releases of pollution; and overall cost savings.

Good housekeeping measures can be easily and simply implemented. Some examples of commonly implemented good housekeeping measures include the orderly storage of bags, drums, and piles of chemicals; prompt cleanup of spilled liquids to prevent significant runoff to receiving waters; expeditious sweeping, vacuuming, or other cleanup of accumulations of dry chemicals to prevent them from reaching receiving waters; and proper disposal of toxic and hazardous wastes to prevent contact with and contamination of storm water runoff.

The primary impediment to a good housekeeping program is a lack of thorough organization. To overcome this obstacle, a three-step process can be used, as follows:
1. Determine and designate an appropriate storage area for every material and every piece of equipment.
2. Establish procedures requiring that materials and equipment be placed

in or returned to their designated areas.

3. Establish a schedule to check areas to detect releases and ensure that any releases are being mitigated.

The first two steps act to prevent releases that would be caused by poor housekeeping. The third step acts to detect releases that have occurred as a result of poor housekeeping.

As with any new or modified program, the initial stages will be the largest hurdle; ultimately, though, good housekeeping should result in savings that far outweigh the efforts associated with initiation and implementation. Generally, a good housekeeping plan should be developed in a manner that creates employee enthusiasm and thus ensures its continuing implementation.

In most cases, a thorough release identification and assessment has already generated the needed inventory of materials and equipment and has determined their current storage, handling, and use locations. This information together with that from further assessments can then be used to determine if the existing location of materials and equipment are adequate in terms of space and arrangement. Cramped spaces and those with poorly placed materials increase the potential for accidental releases due to constricted and awkward movement in these areas. A determination should be made as to whether materials can be stored in a more organized and safer manner (e.g., stacked, stored in bulk as opposed to individual containers). The proximity of materials to their place of use should also be evaluated. Equipment and materials used in a particular area should be stored nearby for convenience, but should not hinder the movement of workers or equipment. This is especially important for waste products. Where waste conveyance is not automatic (e.g., through chutes or pipes) waste receptacles should be located as close as possible to the waste generation areas, thereby preventing inappropriate disposal which can lead to environmental releases.

Appropriately designated areas (e.g., equipment corridors, worker passageways, dry chemical storage areas) should be established throughout the facility. Signs and adhesive labels are the primary methods used to assign areas. Many facilities have developed innovative labeling approaches, such as color coding the equipment and materials used in each particular process. Other facilities have stenciled outlines to assist in the proper positioning of equipment and materials. Once a facility site has been organized in this manner, it is important that employees maintain this organization. This can be accomplished through explaining organizational procedures to employees during training sessions, distributing written instructions, and most importantly, demonstrating by example.

Despite good housekeeping measures, the potential for environmental releases remains. Thus, good housekeeping requires the prompt identification and mitigation of actual or potential releases. Where potential releases are noted, measures designed to prevent release can be implemented. Where actual releases are occurring, mitigation measures may be required. Mitigation practices are simple in theory: the immediate cleanup of an environmental release lessens chances of spreading contamination and lessens impacts due to contamination. When considering choices for mitigation methods, a facility must consider the physical state of the material released and the media to which the release occurs. Generally, the ease of implementing mitigation actions should also be considered. For example, crushed stone, asphalt, concrete, or other covering may top a particular area. Consideration as to which substance would be easier to clean in the event of a release should be evaluated.

12.6.9 Preventive Maintenance

Preventive maintenance (PM) is a method of periodically inspecting, maintaining, and testing plant equipment and systems to uncover conditions which could cause breakdowns or failures. As part of a BMP plan, PM focuses on preventing environmental releases. Most facilities have existing PM programs. It is not the intent of the BMP plan to require development of a redundant PM program. Instead, the objective is to expand the current PM program to address concerns raised as part of the potential release identification and assessment.

A PM program accomplishes its goals by shifting the emphasis from a repair maintenance system to a preventive maintenance system. It should be noted that in some cases, existing PM programs are limited to machinery and other moving equipment. The PM program prescribed to meet the goals of the BMP plan includes all other items (man-made and natural) used to contain and prevent releases of toxic and hazardous materials. Ultimately, the well-operated PM program devised to support the BMP plan should produce environmental benefits of decreased releases to the environment, as well as reducing total maintenance costs and increasing the efficiency and longevity of equipment, systems, and structures.

In terms of BMP plans, the PM program should prevent breakdowns and failures of equipment, containers, systems, structures, or other devices used to handle the toxic or hazardous chemicals or wastes. To meet this goal, a PM program should include a suitable system for evaluating equipment, systems, and structures; recording results; and facilitating corrective actions. A PM program should, at a minimum, include the following activities:

- Identification of equipment, systems, and structures to which the PM program should apply.
- Determination of appropriate PM activities and the schedule for such maintenance.
- Performance of PM activities in accordance with the established schedule.
- Maintenance of complete PM records on the applicable equipment and systems, and structures.

Generally, all good PM programs will consist of the four components noted above. However, it is of particular importance that the PM program address those areas and pollutants identified during the release identification and assessment step of the planning phase.

At the outset of a PM program, an inventory should be devised. This inventory should provide a central record of all equipment and structures including: location; identifying information such as serial numbers and facility equipment numbers/names; size, type, and model; age; electrical and mechanical data; the condition of the equipment/structure; and the manufacturer's address, phone number, and person to contact. In addition to the equipment inventory, an inventory of the structures and other non-moving parts to which the PM program is to apply should also be determined.

Inventories can be developed through inspections and/or reviews of facility specifications and operations and maintenance manuals. In some cases, it is effective to label equipment and structure with assigned numbers/names and some of the identifying information. This information may be useful to maintenance personnel in the event of emergency situation or unscheduled maintenance where maintenance information is not readily available. Several different methods are effective for recording inventory information including the use of index cards, prepared forms and checklists, or a computer database.

Since the PM program involves the use of maintenance materials (i.e., spare parts, lubricants, etc.), some additional considerations may apply. First, good housekeeping measures are particularly important for organizing maintenance materials and keeping areas clean. A tracking system may also be necessary for organizing maintenance materials. The inventory should include information such as materials/parts description, number, item specifications, ordering information, vendor addresses and phone numbers, storage locations, order quantities, order schedules and costs. A large facility may require a parts catalog to coordinate such information. Large facilities may also need to develop a purchase order system which maintains the stock in adequate number and in the proper order by keeping track of the minimum and maximum number of items required to make timely repairs, parts that are vulnerable to breakdown, and parts that have a long delivery time or are difficult to obtain.

Once the inventory is completed, the facility should determine the PM requirements including schedules and specifications for lubrication, parts replacement, equipment and structural testing, maintenance of spare parts, and general observations. The selected PM activities should be based on the facility-specific conditions but should be at least as stringent as the manufacturer's recommendations. Manufacturer's specifications can generally be found in brochures and pamphlets accompanying equipment. An operations and maintenance manual also may contain this information. If these sources are not available, the suggested manufacturer's recommendation can be obtained directly from the manufacturer. In cases of structures or non-moving parts, the facility will need to determine an appropriate maintenance activities (e.g., integrity testing). As with inventory information, PM information should be recorded in an easily accessible format.

After establishment of the materials inventory and the development of PM requirements, a facility should schedule and carry out PM on a regular basis. Personnel with expertise in maintenance should be available to conduct maintenance activities. In a small facility where one person may conduct regular maintenance activities, specialized contractors may supplement the maintenance program for more complex activities. An up-to-date list of outside firms available for contract work beyond the capability of the facility staff should be readily available. Additionally, procedures explaining how to obtain such support should be provided in the pollution prevention plan. Larger facilities should have sufficient PM expertise within the staff, including a PM manager, an electrical supervisor, a mechanical supervisor, electricians, technicians, specialists, and clerks to order and acquire parts and maintain records. Ongoing training and continuing education programs may be used to establish expertise in deficient areas.

Maintenance activities should be coordinated with normal plant operations so that any shutdowns do not interfere with production schedules or environmental protection. The maintenance supervisory staff should also consider other timing constraints such as the availability of the PM staff for both regularly scheduled PM and unanticipated corrective repairs.

The final step in the development of a PM program involves the organization and maintenance of complete records. A PM tracking system which includes detailed upkeep, cost, and staffing information should be utilized. A PM tracking system assists facilities in identifying potential equipment or structural problems resulting from defects, general old age, inappropriate maintenance, or poor engineering design; preparation of a maintenance department budget; and deciding whether a piece of equipment or a structure should continue to be repaired or replaced.

There are many commercial software systems that enable facilities to track maintenance. Computer systems allow for input of inventory and PM information and generate daily, weekly, monthly, and/or yearly maintenance sheets which include the required item to be maintained, the maintenance duties, and materials to be used (e.g., oil, spare parts, etc.). The system can be continually updated to add information gathered during maintenance activities. Some of the maintenance information that proves useful includes the work hours spent, materials used, frequency of downtime for repairs, and costs involved with maintenance activities. This information in turn can generate budgets and determinations of the cost effectiveness of repair versus replacement and so forth. Computerized systems for maintenance tracking are usually most effective at larger facilities.

Maintenance logs should also be developed for each piece of equipment and each structure, and should contain information such as the maintenance specifications, and data associated with the completion of maintenance activities. Maintenance personnel should complete relevant information including the date maintenance was conducted, hours spent on duties, materials used, worker identification, and the nature of the problem.

12.6.10 Inspections

Inspections provide an ongoing method to detect and identify sources of actual or potential environmental releases. Inspections also act as oversight mechanisms to ensure that selected BMPs are being implemented. Inspections are particularly effective in evaluating the good housekeeping and PM programs previously discussed.

Many facilities may be currently conducting inspections, but in a less formalized manner. Security scans, site reviews, and facility walk-throughs conducted by plant managers and other such personnel qualify as inspections. These types of reviews, however, are often limited in scope and detail. To ensure the objectives of the BMP plan are met, these types of reviews should be conducted concurrently with periodic, in-depth inspections as part of a comprehensive inspection program.

Inspections implemented as part of the BMP plan should cover those equipment and facility areas identified during the release identification and assessment as having the highest potential for environmental releases. Since inspections may vary in scope and detail, an inspection program should be developed to prevent redundancy while still ensuring adequate oversight and evaluation.

A BMP inspection program should set out guidelines for each of the following:

- Scope of each inspection
- Personnel assigned to conduct each inspection
- Inspection frequency
- Format for reporting inspection findings
- Remedial actions to be taken as a result of inspection findings

Despite the different requirements of each type of inspection, the focus of inspections conducted as part of the BMP plan should not vary. Some of the areas within the facility that may be the focus of the BMP plan include solid and liquid materials storage areas, in-plant transfer and materials handling areas, activities with potential to contaminate storm water runoff, and sludge and hazardous waste disposal sites.

An inspection program's goal will be to ensure thoroughness, while preventing redundancy. Ultimately, this will ensure that the use of resources is optimized. In addition, it should be clear that the inspection team's efforts are directed to support the operating groups in carrying out their responsibilities for equipment and personnel safety, and work quality, and to ensure that all standards are met. In achieving these goals, written procedures discussing the scope, frequency and scheduling, personnel, format, and remediation procedures should be provided.

The scope of each inspection type should be discussed in the written procedures. Many different types of inspections are conducted as part of the inspection program. Guidelines for the scope of these inspections include:

- *Security scan.* Search for leaks and spills which may be occurring. Specifically examine problems areas which have been identified by the plant manager or equivalent persons.
- *Walk-through.* Conduct oversight of the duties associated with a security scan. In addition, ensure that equipment and materials are located in their appropriate positions.
- *Site review.* Conduct oversight of duties associated with a walk-through. Additionally, evaluate the effectiveness of the PM, good housekeeping, and security programs by visual oversight of their implementation.
- *BMP plan oversight inspection.* Conduct oversight of duties associated with a site review. Evaluate the implementation of all aspects of the written BMP plan including the review of the records generated as part of these programs (e.g., inspection reports, PM activity logs).

- *BMP plan evaluation/reevaluation inspection:* Conduct an evaluation/ reevaluation of the facility and determine the most appropriate BMPs to control environmental releases.

An appropriate mix of these types of inspections should be developed based on facility-specific considerations. The proper frequency for conducting inspections will vary based on the type of the inspection and other facility-specific factors. Some general guidelines for establishing frequency follow:

- Security scans can be conducted various times daily.
- Walk-through inspections can be conducted once per shift to once per week.
- Site reviews can be conducted once per week to once per six months.
- BMP plan oversight inspections can be conducted once per month to once per year.
- BMP plan reevaluation inspections can be conducted once per year to once every five years.

There are no hard and fast rules for conducting inspections as part of the BMP plan. Inspection frequencies should be based on a facility's needs. Two points should be considered when establishing an inspection program: (1) As would be expected, more frequent inspections should be conducted in the areas of highest concern; and (2) inspections must be conducted more frequently during the initial BMP implementation until the BMP plan procedures become part of standard operating procedures.

It may be useful to set up a schedule to ensure a comprehensive inspection program. Varying the dates and times of inspection conduct is also good practice in that it ensures all stages of production and all situations are reviewed.

Individuals qualified to assess the potential for environmental releases should be assigned to conduct formal inspections. Members of the BMP committee can generally fulfill this requirement, but they may not be available to conduct all inspections. Thus, it may be appropriate to identify and train personnel to conduct specific types of inspections. For example, shift supervisors and other equivalent personnel may appropriately conduct walk-throughs and site reviews as a result of their position of authority and ability to require prompt correction if problems are observed. Personnel with immediate responsibility for an area should not be asked to conduct inspections of that area as they may be tempted to overlook problems. Additionally, plant security and other personnel who routinely conduct walk-throughs should not be assigned to conduct BMP plan inspections since their familiarity with the facility may result in their not being suited to best identify opportunities for improvement.

An inspection checklist of areas to inspect with space for a narrative report is a helpful tool when conducting inspections. A standard form helps ensure inspection consistency and comprehensiveness. Checklists may, however, not be necessary for each inspection performed. This may be particularly true for facilities conducting frequent inspections (once per hour, once per shift, etc.); procedures for using inspection checklists should be reasonable to prevent excessive paperwork.

The findings of inspections will be useless unless they are brought to the attention of appropriate personnel and subsequently acted upon. To ensure that reports are acted upon in an expeditious and appropriate manner, procedures for routing and review of reports should be developed and followed. Despite the usefulness of written reports, in no way should a written report replace verbal communication. Where a problem is noted, particularly environmental releases currently occurring or about to occur, it should be verbally communicated by the inspector to the responsible personnel as soon as possible.

12.6.11 Security

A security plan describes the system installed to prevent accidental or intentional entry to a facility that might result in vandalism, theft, sabotage, or other improper or illegal use of the facility. In relation to a BMP plan, a security system should prevent environmental releases caused by any of these improper or illegal acts.

Most facilities already have a program for security in place; this security program can be integrated into the BMP plan with minor modifications. Facilities developing a program for security as part of the BMP plan may be hesitant to describe their security measures in detail due to concerns of compromising the facility. The intent of including a security program as part of the BMP plan is not to divulge facility or company secrets; the specific security practices for the facility may be kept as part of a separate confidential system. The security program as part of the BMP plan should cover security in a general fashion, and discuss in detail only the practices which focus on preventing environmental releases.

The security program as part of the BMP plan should be designed to meet two goals. First, the security plan should prevent security breaches that result in the release of hazardous or toxic chemicals to the environment. The second goal is to effectively utilize the observation capabilities of the security plan to identify actual or potential releases to the environment. Some typical components of a security plan include the following:

- Routine patrol of the facility property by security guards in vehicles or on foot
- Fencing to prevent intruders from entering the facility site
- Good lighting to facilitate visual inspections at night, and of confined spaces
- Vehicular traffic control (i.e., signs)
- Access control using guardhouse or main entrance gate, where all visitors and vehicles are required to sign in and obtain a visitor's pass
- Secure or locked entrances to the facility
- Locks on certain valves or pump starters
- Camera surveillance of appropriate sites, such as facility entrance, and loading/unloading areas
- Electronic sensing devices supplemented with audible or covert alarms
- Telephone or other forms of communication

Security systems focus typically on the areas with the greatest potential for damage as a result of security breaches. As part of the BMP plan, the security program will focus on the areas that result in environmental releases. Typically, these areas have been identified in the release identification and assessment step of the planning phase. In many cases, the findings of this step may indicate a need to change the focus or broaden the scope of the security program to include areas of the facility addressed by the BMP plan. Since the security program may not be common knowledge, general BMP committee members may not be able to recommend changes. As a result, security personnel should be involved in the decisions made by the committee, with one person possibly serving as a member.

While performing their duties, security personnel can actively participate in the BMP plan by checking the facility site for indications of releases to the environment. This may be accomplished by checking that equipment is operating properly; ensuring no leaks or spills are occurring at materials storage areas; and checking on problem areas (i.e., leaky valves, etc).

The advantages of integrating security measures into the BMP plan are considerable. Security personnel are in positions that enable them to conduct periodic walk-throughs and scans of the facility, as well as covertly view facility operations. They are in an excellent position to identify and prevent actual or potential releases to the environment.

Where security personnel are utilized as part of the oversight program, two obstacles generally must be overcome: (1) support must be gained from the security staff; and (2) security personnel must be knowledgeable about what may and may not be a problem, and to whom to report when there is a problem. Involving the security staff in the BMP plan development at an early stage should assist in gaining their support.

Integration of the security staff into the employee training, and recordkeeping and reporting programs, respectively, can also be used to overcome these barriers.

12.6.12 Employee Training

Employee training conducted as part of the BMP plan is a method used to instill in personnel, at all levels of responsibility, a complete understanding of the BMP plan, including the reasons for developing the plan, the positive impacts of the plan, and employee and managerial responsibilities under the BMP plan. The employee training program should also educate employees about the general importance of preventing the release of pollutants to water, air, and land.

Training programs are a routine part of facility operations. Most facilities conduct regular employee training in areas, including fire drills, safety, and miscellaneous technical subject areas. Thus, the training program developed as a result of the BMP plan should be easily integrated into the existing training program. Employee training conducted as part of the BMP plan should focus on those employees with direct impact on plan implementation. This may include personnel involved with manufacturing, production, waste treatment and disposal, shipping/receiving, or materials storage; areas where processes and materials have been identified as being of concern; and PM, security, and inspection programs. Training programs, which include all appropriate personnel, should include instruction on spill response, containment, and cleanup. Generally, the employee training program should serve to improve and update technical, managerial, or administrative skills; increase motivation; and introduce incentives for BMP plan implementation.

Employee training programs function through the following four step process:
1. Analyzing training needs
2. Developing appropriate training materials
3. Conducting training
4. Repeating training at appropriate intervals in accordance with steps 1 through 3

The first stage in developing a training program is analyzing training needs. Generally, training should be conducted during the planning and development phases of the BMP plan, and as follow-up to BMP implementation for selected areas of concern. In all three situations, it is important to analyze training needs and develop appropriate training tools to use during conduct of the training.

The initial BMP development session educates employees of the need for, objectives of, and projected impact of the BMP plan. As would be expected, this initial training should be conducted at the outset of the BMP development. The message portrayed at this session should be the positive impacts of the BMP plan, including ease in locating materials and equipment; improved employee morale; improved manufacturing and production efficiency; lessened raw, intermediate and final product losses due to releases; fewer health and safety problems arising from unmitigated releases and/or poor placement of materials and equipment; environmental benefits resulting from reduced releases of pollution; and overall cost savings. When providing this message, it is essential that the benefits for employees, as well as the company itself be stressed. While it is important to point out the reasons that led to the decision to implement a BMP plan, it is also important to provide a realistic picture of the changes and impacts that will result. These modifications should be discussed in terms of their positive impact to help maintain a high level of enthusiasm.

After the BMP plan is developed, the BMP implementation training sessions should be developed. The training sessions should review the BMP plan and associated procedures, such as the following:

- The good housekeeping practices, including the use of labeling to assign areas and procedures to return materials to assigned areas
- The PM program, including new PM schedules and procedures
- Integration of the security plan with the BMP plan
- Inspection program
- Responsibilities under the recordkeeping and reporting system
-

In some cases, it may be appropriate to provide a general session explaining BMP plan implementation followed by specialized training for each area. For example, since all employees should be aware of the good housekeeping program, this program should be discussed at the general session. Training for selected facility-specific BMPs may be necessary only for employees in the production and manufacturing areas. PM information could be presented only to the personnel conducting maintenance, while security personnel need only be briefed of security-related responsibilities under the BMP plan.

Training sessions are only as effective as the level of preparation. It is vital that workshop materials are technically accurate, easily read, and well-organized. More importantly, training materials must leave a strong impression, such that their message is remembered and any distributed training materials are consulted in the future. The use of audiovisual aids supplemented with informational handouts is one of the best methods of conveying information. Including copies of any slide or overhead helps avoid distractions during presentation caused by employees' writing notes on the contents of overheads. Other techniques which assist in effectively

conveying information include the following:
- Providing aesthetically pleasing covers and professional looking handouts
- Developing detailed tables of contents with well numbered pages
- Frequently assimilating graphics into presentations
- Integrating break-out sections and exercises
- Incorporating team play during exercises
- Allowing for liberal question/answer sessions and discussions during or after presentations
- Providing frequent breaks
- Integrating field activities with class room training

The use of qualified personnel to conduct training presentations also supports the facility's commitment to BMP plan implementation. Speakers should be identified in the initial training preparation stages based on their expertise in the topics to be presented. However, expertise is not the only consideration. Expertise must be supplemented with a well-executed, interesting, enthusiastic presentation. Preparation prior to the training event will allow speakers to organize presentations, establish timing, and develop tone and content.

Proper planning should ensure the execution of an effective training event. Once the training event has been conducted, some follow-up activities should be conducted. For example, evaluation forms requesting feedback on the training should be distributed to employees. These evaluation forms can be used to identify presentation areas needing improvement, ideas needing clarification, and future training activities. Ultimately, information gathered from these forms can help direct the employee training program in the future.

Once BMP plan implementation is under way, training should be conducted both routinely and on an as-needed basis. Special training sessions may also be prompted when new employees are hired, environmental release incidents occur, recurring problems are noted during inspections, or changes in the BMP plan are necessary.

12.6.13 Recordkeeping and Reporting

As part of a BMP plan, recordkeeping focuses on maintaining records that are pertinent to actual or potential environmental releases. These records may include the background information gathered as part of the BMP plan, the BMP plan itself, inspection reports, PM records, employee training materials, and other pertinent information.

Maintenance of records is ineffective unless a program for the review of records is set forth. In particular, a system of reporting actual or potential problems to appropriate personnel must be included. Reporting, as it relates to the BMP plan, is a method by which appropriate personnel are kept informed of BMP plan implementation, such that appropriate actions may be determined and expeditiously taken. Reporting may be verbal or follow a more formal notification procedure. Some examples of reporting include the following:

- Informational memos distributed to upper management or employees to keep them updated on the BMP plan.
- Verbal notification by BMP inspectors to supervisors concerning areas of concern noted during inspections.
- Corrective action reports from the BMP committee to the plant manager which cite deficiencies with BMP plan implementation.
- Verbal and written notification to regulatory agencies of releases to the environment.

An effective recordkeeping and reporting program functions through the following three step procedure:

1. Developing records in a useful format.
2. Routing records to appropriate personnel for review and determination of actions to address deficiencies.
3. Maintaining records for use in future decision-making processes.

Recordkeeping and reporting play an overlapping role with the programs previously discussed. In general, these programs will involve the development, review, maintenance, and reporting of information to some degree. For example, an inspection program may include the development and use of an inspection checklist, submittal of the completed checklist to relevant personnel, evaluation of the inspection information, and determination of appropriate corrective actions. This may, in some cases, involve the development of a corrective action report to submit to appropriate persons (which may include regulatory agencies where necessary/required). The checklist and the corrective action reports should be maintained in organized files.

As part of the BMP plan, a recordkeeping and reporting program will primarily be developed for the PM and inspection programs. However, effective communication methods can also be useful in the development of the release identification and assessment portion of the BMP plan.

The first step to ensuring an effective recordkeeping and reporting program is the development of records in a useful format. The use of standard formats (i.e., checklists) can help to ensure the completion of necessary information, thoroughness in reviews, and understanding of the supplied data. For example, a standard inspection format may specify a summary of findings, recommendations, and requirements on the first page; then, detailed information by geographical area (e.g., materials storage area A, materials storage area B, the north loading and unloading zone) may be discussed. With a standard format, an inspection report reviewer may quickly review the findings summary to determine where problems exist, then refer to the detailed discussion of areas of concern. Ultimately, the use of a standard format minimizes the review time, expedites decisionmaking concerning corrective actions, and simplifies reporting.

Despite the recommended use of standard formats, inspectors should not feel constrained by the format. Sufficient detail must be provided in order for the report to be useful. Narratives should accompany checklists where necessary to provide detailed information on materials that have been released or have the potential to be released; nature of the materials involved; duration of the release or potential release; potential or actual volume; cause; environmental results of potential or actual releases; recommended countermeasures; people and agencies notified; and possible modifications to the BMP plan, operating procedures, and/or equipment.

The second step to ensuring an effective recordkeeping and reporting system involves routing information to appropriate personnel for review and determination of actions to address deficiencies. Regardless of whether the system for recordkeeping and reporting is structured or informal, the BMP plan should clearly indicate: (1) how information is to be transferred (i.e., by checklist, report, or simply by verbal notification); and (2) to whom the information is to be transferred (i.e., the plant manager, the supervisor in charge, or the BMP committee leader). Customarily, formal means to transfer information would be more appropriate in larger more structured companies. For example, reviews of findings and conclusions as part of inspection reports may be conducted by supervisory personnel and the information may be routed through the chain of command to the responsible personnel such as shift supervisors. Less formal communication methods such as verbal notification may be appropriate for smaller facilities. It should be noted that verbal communications of impending or actual releases should be made regardless of whether a formal communications process has been set forth.

The key to ensuring a useful communication system is identifying one person (or, at larger facilities, several persons) to receive and dispense records and information. This person will be responsible for ensuring that designated individuals review records where appropriate, that corrective actions are identified, and that appropriate personnel are notified of the need to make corrections. Additionally, this person will ensure that information is maintained on file for use in later evaluations of the BMP plan effectiveness.

A communications system for notification of potential or actual release should be designated. Such a system could include telephone or radio contact between transfer operations, and alarm systems that would signal the location of a chemical release. Provisions to maintain communication in the event of a power failure should be addressed. Reliable communications are essential to expedite immediate action and countermeasures to prevent incidents or to contain and mitigate chemicals released.

A reporting system should include procedures for notifying regulatory agencies. A number of federal and state agencies may require reporting of environmental releases. It is outside the scope of this chapter to provide a summary of all necessary reporting requirements.[58] However, reporting requirements specified under the NPDES permitting program include, at a minimum, the following:

- Releases in excess of reportable quantities which are not authorized by an NPDES permit.
- Planned changes which:
  -subject the facility to new source requirements
  -significantly change the nature or quantity of pollutants discharged
  -change a facility's sludge use or disposal practices
  -may result in noncompliance
- Notification within 24 hours of any unanticipated discharges (including bypasses and upsets) which may endanger human health or the environment, and the submission of a written report within five days.
- The discharge of any toxic or hazardous pollutant above notification levels.
- Any other special notification procedure or reporting requirement specified in the NPDES permit.

Reports maintained in the recordkeeping system can be used in evaluating the effectiveness of the BMP plans, as well as when revising the BMP plan. Additionally, these records provide an oversight mechanism which allows the BMP committee to ensure that any detected problem has been adequately resolved. As such, the final step in developing a recordkeeping and reporting program involves the development and maintenance of an organized recordkeeping system.

In general, an organized filing system involves selecting an area for maintaining files, labeling files appropriately, and filing information in an organized manner. A single location should be designated for receiving the data generated for and related to the BMP plan. At larger facilities, several locations may be appropriate (e.g., maintenance records in one location,

---

[58] For a complete discussion of environmental reporting requirements, see Dennison, *Environmental Reporting, Recordkeeping, and Inspections: A Compliance Guide for Business and Industry* (Van Nostrand Reinhold, 1995).

other BMP related documentation in another). A centralized location will help to consolidate materials for later review and consideration. Without a designated location, materials may become dispersed throughout a facility and subsequently lost.

Filing information by subject and date is a practice followed by most facilities. The most effective filing system usually includes hard copies of the information on file. Additionally, keeping inventory lists of documents maintained in file folders assists in quick reviews of file contents. Small facilities may be able to file all BMP-related information in the same folder in chronological order; larger facilities may have to file information by subject. In some cases, larger facilities may find it convenient to develop an automated tracking system (e.g., a database system) for efficiently maintaining records.

12.6.14 Plan Evaluation and Reevaluation Phase

Planning, development, and implementation of the BMP plan require the dedication of important resources by company management. The benefits derived, however, serve to justify the costs and commitments made to the BMP plan. To illustrate the plan's benefits, it may be appropriate and even necessary in some cases to measure the plan's effectiveness. An evaluation can be performed by considering a number of variables, including benefits to employees, environmental benefits, and reduced expenditures. Benefits to the employees can be assessed in terms of health and safety, productivity, and other factors such as morale. Comparisons before and after plan implementation can be made to determine trends that show BMP plan effectiveness.

Environmental benefits can be measured by several factors. First, pollutant monitoring prior to the inception of the BMP plan may show significant quantities of pollutants and or wastes that are minimized or eliminated after plan implementation. Discharge monitoring report records may show reductions in the quantity or variability of pollutants in the discharges. In addition, the reductions in volumes of and/or hazards posed by solid waste generation and air emissions may demonstrate the success of the BMP plan. Other derived environmental benefits may include reduced releases to the environment resulting from spills, volatilization, and losses to storm water runoff. These benefits may be measured through reductions in the number and severity of releases and of lessened losses of materials.

Reduced expenditures are the "bottom line" in substantiating the need for the BMP plan. Cost considerations can be easily tracked through expense records including chemicals usage, energy usage, water usage, and employee records. The development of production records on product per unit cost before and after BMP plan implementation may show a significant drop, thereby demonstrating the effectiveness of the plan.

The operations at an industrial facility are expected to be dynamic and therefore subject to periodic change. As such, the BMP plan cannot remain effective without modifications to reflect facility changes. At a minimum, the BMP plan should be revisited annually to ensure that it fulfills its stated objectives and remains applicable. This time-dated approach allows for the consideration of new perspectives gained through the implementation of the BMP plan, as well as the reflection of new directives, emerging technologies, and other such factors. However, plan revisions should not be limited to periodic alterations. In some cases, it may be appropriate to evaluate the plan due to changed conditions such as the following:

- Restructuring of facility management
- Substantial growth
- Significant changes in the nature or quantity of pollutants discharged
- Process or treatment modifications
- New permit requirements
- New legislation related to BMPs
- Releases to the environment

Many changes at a facility may warrant modifications to the BMP plan. Growth may require more frequent employee training or a redesign of the good housekeeping practices to ensure the site is maintained in a clean and orderly fashion. The evaluation or modification of existing process, treatment, and chemical handling methods may substantiate the need for additional facility-specific BMPs.

Where new permit requirements or legislation focus on a specific pollutant, process, or industrial technology, it may be appropriate to consider establishing additional controls. These permit requirements or legislative changes do not necessarily have to be directly related to environmental issues. For example, new OSHA standards may result in modification of the BMP plan to include procedures that address the protection of worker health and safety.

If there has been a spill or other unexpected chemical release, the reasons for the release and corrective actions taken should be investigated. This investigation should include evaluation of all general BMPs, including good housekeeping, PM, inspections, security, employee training, and recordkeeping and reporting. Additionally, facility-specific BMPs should be evaluated at that time to determine their effectiveness.

Ultimately, the BMP plan reevaluation may pinpoint areas of the facility not addressed by the plan, or activities that would benefit from further development of facility-specific BMPs or revision of the general practices contained in the BMP plan. It is useful to bear in mind that as the BMP plan improves, costs can continue to be minimized as a result of reduced waste generation, less hazardous or toxic materials use, and prevented environmental releases.

VII

Environmental Control of Land Use

Wetlands Regulation

13.1 Introduction to Wetlands

The term "wetland" refers to a variety of different habitats that often seem to have little in common other than that they contain water during at least part of the year. Terms such as "swamp," "marsh," or "bog" all refer to wetlands.[1] Historically, most Americans have used these terms in a negative context. The term "wetland" often induces images of mosquito-ridden, snake-infested, smelly areas where disease is rampant and danger lurks. While they are often not founded in fact, these attitudes have resulted in the destruction or degradation of nearly half of the nation's wetlands resources as wetland areas were converted to agricultural uses, housing, industry, or for a variety of "reclamation" projects.[2]

Despite the often negative view of wetlands, some positive values of wetlands were recognized early in our nation's history. Some wetlands were preserved for fishing, as habitat for migratory waterfowl, or for recreation as early as the nineteenth century. However, it has only been within the past half-century that wetland scientists and other interested people began studying the many positive attributes of wetlands in such a way that a full appreciation of the many values of wetlands were recognized. As a consequence, strong emphasis has been placed on the preservation and study of wetlands of all kinds.[3]

---

[1] See Tiner, *In Search of Swampland : A Wetland Sourcebook and Field Guide* (Rutgers Univ. Press, 1998).
[2] See and Giblett, Gilbert, and Giblett, Postmodern Wetlands: Culture, History, Ecology (Edinburgh Univ. Press, 1997); Dennison and Berry, Overview, Chap. 1, in Dennison and Berry, *Wetlands: Guide to Science, Law, and Technology* (Noyes, 1993); and Mitsch and Gosselink., *Wetlands*, 2d ed. (Van Nostrand Reinhold,1993).
[3] Giblett, Gilbert, and Giblett, *Postmodern Wetlands: Culture, History, Ecology* (Edinburgh Univ. Press, 1997).

13.1.1 Wetland Definitions

Wetlands are difficult to define and to classify. There is no single, formal definition of wetlands among wetland ecologists and managers, or even among government regulators. Wetland definitions often reflect the purposes for which they were created (e.g., regulation, scientific investigation, or conservation).[4] Federal laws and regulations often contain different and contradictory definitions, while state and local laws and regulations often differ from each other and from federal definitions.[5] Over fifty federal and state wetland definitions are in use, with varying requirements for their identification and delineation.[6]

Formal definitions of wetlands are of particular interest to the regulated community, and to those persons, organizations and agencies involved in wetland protection. Inconsistent wetland definitions place a severe burden on any private property owner who wishes to develop a parcel of land that may (or may not) contain wetlands which are potentially subject to regulation by a governmental agency. This burden is particularly harsh when the governmental agencies with jurisdiction have conflicting definitions.

Problems in defining wetlands stem from the nature of wetlands themselves. Wetlands vary enormously in their characteristics and functions.[7] Many wetlands are transient, and may seem to disappear for long periods of time during droughts or when water levels are lowered. In addition, many wetlands are variously degraded by dredging and filling such that many of their natural functions are depleted.[8]

The regulatory definition of wetlands used by the U.S. Army Corps of Engineers and the U.S. Environmental Protection Agency in administering

---

[4] See Tiner, *Wetland Indicators : A Guide to Wetland Identification, Delineation, Classification, and Mapping* (CRC Press–Lewis Publ., 1999); and Environmental Defense Fund/World Wildlife Fund, *How Wet Is a Wetland? The Impacts of the Proposed Revisions to the Federal Delineation Manual* (EDF/WWF Jan. 16, 1992).

[5] See 56 *Fed. Reg.* 40,446 (Aug. 14, 1991). See also National Academy of Sciences, *Wetlands: Characteristics and Boundaries* (Nat'l Academy Press, 1995).

[6] Environmental Defense Fund/World Wildlife Fund, *How Wet Is a Wetland? The Impacts of the Proposed Revisions to the Federal Delineation Manual* (EDF/WWF Jan. 16, 1992); and Willard, Leslie, and Reed, Defining and Delineating Wetlands in Bingham, Clark, Haygood, and Leslie, *Issues in Wetlands Protection: Background Papers Prepared for the National Wetlands Policy Forum* (The Conservation Foundation, 1990).

[7] See Berry, Ecological Principles of Wetland Ecosystems, Chap. 2 in Dennison and Berry, *Wetlands: Guide to Science, Law, and Technology* (Noyes, 1993).

[8] National Academy of Sciences, *Wetlands: Characteristics and Boundaries* (Nat'l Academy Press, 1995).

dredge and fill permitting under CWA Section 404 contains a useful definition of wetlands:

> The term "wetlands" means those areas that are inundated or saturated by surface or groundwater at a frequency and duration sufficient to support, and that under normal circumstances do support, a prevalence of vegetation typically adapted for life in saturated soil conditions. Wetlands generally include swamps, marshes, bogs and similar areas.[9]

Under this definition, an area is a wetland if it contains three characteristics: wetland hydrology (the presence of water at or near the surface for a period of time), hydrophytic vegetation (wetland plants), and hydric soils (periodically anaerobic soils resulting from prolonged saturation or inundation). Implementation of this definition depends on the wetlands "delineation manual" in place at the time.[10] Under most circumstances, an area must have *all three* attributes to qualify as a jurisdictional wetland (i.e., one subject to the permitting requirements of Section 404 of the Clean Water Act). Many states have adopted versions of this definition.[11]

By contrast, the U.S. Fish and Wildlife Service uses a more expansive definition which defines a wetland as an area with *any one* of the attributes of wetland hydrology, hydrophytic vegetation, or hydric soils. The result of application of the USFWS definition is that more areas would be considered wetlands than under the definition used by the Corps/EPA definition. The more inclusive USFWS definition reflects that agency's mandate to serve as a resource conservation agency, and also to facilitate its responsibility for a National Wetlands Inventory pursuant to Section 208(i) of the Clean Water Act.[12]

Both the USFWS and Corps/EPA definitions indicate the degree of scientific sophistication required to recognize and classify the various types of wetlands. For example, recognizing the various kinds of wetland plants requires an advanced knowledge of botany.[13] Similarly,

---

[9] 33 CFR § 323.2, and 40 CFR § 230.3.

[10] See discussion in United States v. Banks, 115 F.2d 916 (11th Cir. 1997).

[11] Willard, Leslie, and Reed, Defining and Delineating Wetlands, in Bingham, Clark, Haygood, and Leslie, *Issues in Wetlands Protection: Background Papers Prepared for the National Wetlands Policy Forum* (The Conservation Foundation, 1990).

[12] 33 U.S.C. § 1288(i).

[13] See National Academy of Sciences, *Wetlands: Characteristics and Boundaries* (Nat'l Academy Press, 1995); and Tiner, The Concept of A Hydrophyte for Wetland Identification, 41 BioScience 236 (1991). The USFWS has prepared plant lists for each of the regions of the country to facilitate wetland identification (see Reed, *National List of Plant Species That Occur in*

identification of the various kinds of hydric soils requires advanced knowledge of soil science, and wetland hydrology often requires training in geology and hydrology.[14]

A problem in identifying as well as classifying wetlands is that wetlands often do not form discrete tracts of easily identified habitat. Wetland types may overlap or grade into each other, creating a substantial transitional zone. For example, an estuarine embayment may gradually become a tidal marsh, then a brackish marsh, and finally freshwater marsh as one moves inland where freshwater inputs are higher. Moreover, one wetland type may develop from another through time in a process called "succession."[15]

The most significant wetland classifications have been those of the federal regulatory agencies, primarily the U.S. Fish and Wildlife Service. The earliest agency classification was the often-cited "Circular 39" classification, which organized wetlands into 20 categories of particular value for waterfowl management purposes.[16] The Circular 39 classification was subsequently replaced by the comprehensive classification of Cowardin, which includes wetland types that may not exhibit all three wetland attributes as required by the Corps definition (for example, areas lacking wetland vegetation like mudflats).[17] The Cowardin classification has provided a standard classification which can be used to compare wetlands from different parts of the country.

13.1.2 Wetland Functions and Values

The public perception of wetlands as dangers to health and welfare, or as obstacles to progress, has changed dramatically in the past 20 years. Much of this change has resulted from an increase in scientific information regarding various wetland functions and values.

There is general consensus that the primary values of wetlands (at least

---

Wetlands: National Summary (Biol. Rept. 88(24), U.S. Fish & Wildl. Serv., 1988)).

[14] See Tiner, Problem Wetlands for Delineation, Ch. 6 in Dennison and Berry, *Wetlands: Guide to Science, Law, and Technology* (Noyes, 1993).

[15] See National Academy of Sciences, *Wetlands: Characteristics and Boundaries* (Nat'l Academy Press, 1995); and Berry, Ecological Principles of Wetland Ecosystems, Chap. 2 in Dennison and Berry, *Wetlands: Guide to Science, Law, and Technology* (Noyes, 1993).

[16] Shaw and Fredine, *Wetlands of the United States. Their Extent, and Their Value for Waterfowl and Other Wildlife* (Circular 39, U.S. Fish & Wildl. Serv., 1956).

[17] Cowardin, Carter, Golet, and LaRoe, *Classification of Wetlands and Deepwater Habitats of the United States* (FWS/OBS 79-31, U.S. Fish & Wildl. Serv., 1979).

from the perspective of humans) are: (1) their ability to cleanse both surface and groundwater, either by filtering surface water as it percolates through wetland soils or by removing particulate material and pollutants before returning the water to flowing surface waters; (2) reducing the effects of flooding by storing storm water and gradually returning it to surface flow, and reducing the effects of erosion by stabilizing soils and dampening the effects of wave action; and (3) serving as critical feeding grounds and nurseries for a variety of fish, waterfowl, and other wildlife. Other values may be of equal importance, such as recreation and esthetics.[18]

The problem of determining the most important functions and values of wetlands has become critical, since proper management often means concentrating management efforts on specific wetland functions. Furthermore, successful efforts to restore degraded wetlands or to create new wetlands often depend on successful identification and duplication of important functions and values. Wetland science will continue to mature as a scientific discipline, and one of its primary responsibilities will be continuing study of wetland functions and values, and creating the technology necessary to replicate them as precisely as possible.

13.2 The Clean Water Act Section 404 Program

The primary federal authority for protecting the Nation's wetlands is Clean Water Act (CWA) Section 404.[19] The Army Corps of Engineers is primarily charged with oversight of the CWA Section 404 program, with guidance from the U.S. EPA.[20]

The Corps was first given authority to regulate construction activities involving dredging, filling, or obstructing "navigable waters" under the Rivers and Harbors Act of 1899, although that authority did not expressly extend to wetlands.[21] Because wetlands are usually outside the mean high

---

[18] See Dennison and Berry, Overview, Chap. 1 in Dennison and Berry, *Wetlands: Guide to Science, Law, and Technology* (Noyes, 1993). See also Mitsch and Gosselink, *Wetlands*, 2d ed. (Van Nostrand Reinhold,1993). The monetary valuation of wetlands is discussed by Keating, *The Valuation of Wetlands: An Appraisal Institute Handbook* (Appraisal Institute, 1995).

[19] 33 U.S.C. § 1344.

[20] For a detailed discussion of CWA § 404 and the wetlands regulation process, see Dennison and Berry, The Regulatory Framework, Chap. 7 in Dennison and Berry, *Wetlands: Guide to Science, Law, and Technology* (Noyes, 1993).

[21] The original purpose of the Rivers and Harbors Act regulation of dredge-and-fill activities was to protect and promote navigation. 33 U.S.C. § 403.

water mark, the Rivers and Harbors Act had very limited impact on the protection of wetlands.[22] For this reason, the Corps operated its permit program for almost 70 years while paying little attention to wetlands protection.

Following the passage of the National Environmental Policy Act of 1969 (NEPA),[23] the Corps' power to consider environmental factors in its permitting process was strengthened. Under NEPA, all federal agencies are required to consider the possible environmental impact of their proposed actions and projects.[24] The first test of Corps environmental protection powers came in *Zabel v. Tabb*,[25] in which two developers attempted to build a mobile home park on eleven acres of wetlands in Boca Ciega Bay, Florida. The developers applied to the Corps for a permit to fill the proposed site. Even though the Corps concluded that the development would not impede navigation, it denied the permit because the proposed construction would have had a detrimental impact on marine life in the bay. The U.S. Court of Appeals for the Fifth Circuit upheld the Corps decision and concluded that the Corps could refuse dredge and fill permits on the basis of environmental considerations.

Congress passed the Federal Water Pollution Control Act Amendments of 1972, which created the modern Section 404 program. The language of CWA Section 404 is actually quite limited:

(a) The Secretary may issue permits, after notice and opportunity for public hearings for the discharge of dredged or fill material into the navigable waters at specified disposal sites.[26]

Under Corps regulations, dredged material is defined as "material that is excavated or dredged from waters of the United States."[27] The regulations define fill material as "any material used for the primary purpose of replacing an aquatic area with dry land or of changing the bottom elevation of a waterbody."[28] The draining of wetlands, which is a major source of wetland losses, is not expressly regulated or prohibited by Section 404.

---

[22] See Borax Consolidated, Ltd. v. City of Los Angeles, 296 U.S. 10 (1935) [discussion of mean high water boundaries].
[23] 42 U.S.C. § 4321 et seq.
[24] 42 U.S.C. § 4332.
[25] Zabel v. Tabb, 430 F.2d 199 (5th Cir.1970), cert. denied, 401 U.S. 910 (1971).
[26] 33 U.S.C. § 1344.
[27] 33 CFR § 323.2(c).
[28] 33 CFR § 323.2(e)

13.2.1 Corps of Engineers/EPA Authority

Under the CWA Section 404 program, the Corps and EPA have concurrent jurisdictional authority over the dredging and filling of waters of the United States, including wetlands.[29] The Secretary of the Army, acting through the Chief of Engineers, is authorized to issue individual permits for the discharge of dredged or fill material into the waters of the United States, which includes wetlands.[30]

In some circumstances, the Corps may issue "nationwide permits" for certain activities in jurisdictional wetlands that are deemed to have minimal environmental impacts.[31] Although the Corps' field personnel are responsible for making the initial decision to grant or deny permits, the EPA is responsible for formulating the Section 404(b)(1) Guidelines used by the Corps to make the permit decisions.[32] The EPA is also empowered to veto or overrule the granting of permits by the Corps.[33] However, EPA has rarely overruled a Corps decision to issue a permit.[34]

Corps regulations describe the comprehensive procedures for the permit process.[35] In evaluating a permit application, the Corps is required to consider the recommendations of the USFWS and the National Marine Fishery Service (NMFS).[36] Under authority of the Fish and Wildlife

---

[29] Although the Clean Water Act is essentially silent on which agency has authority to make jurisdictional determinations under the § 404 Program, the EPA and Corps have formulated agreements detailing their respective jurisdictional responsibilities. EPA/Department of Defense, Memorandum of Understanding on "Geographical Jurisdiction of the Section 404 Program (MOU)," 45 *Fed. Reg.* 45,018 (July 2, 1980); Department of the Army/EPA Memorandum of Agreement Concerning the Geographic Jurisdiction of the Section 404 Program and the Application of the Exemptions under Section 404(f) of the Clean Water Act (MOA)" (January 19, 1989).

[30] 33 U.S.C. § 1344(a). Corps regulations governing individual permits are found at 33 CFR § 323.

[31] 33 U.S.C. § 1344(e). Corps regulations governing the Nationwide permit program are found at 33 CFR § 330. Nationwide permits will be discussed more fully in section 13.3.4.

[32] Department of the Army/EPA Memorandum of Agreement Concerning the Geographic Jurisdiction of the Section 404 Program and the Application of the Exemptions under Section 404(f) of the Clean Water Act (MOA) (January 19. 1989). Clean Water Act § 404(b)(1) Guidelines; Correction, 55 *Fed. Reg.*, 9210, 9211 (Feb. 7, 1990).

[33] 33 U.S.C. § 1344(b) and (c).

[34] U.S. General Accounting Office, *Wetlands - The Corps of Engineers' Administration of the Section 404 Program* (U.S. General Accounting Office, 1988).

[35] 33 CFR Parts 320, 323, and 325.

[36] 33 CFR § 320.4(c).

Coordination Act,[37] the USFWS and the NMFS review applications for these federal permits and provide comments to the Corps on the environmental impacts of proposed work. In addition, the USFWS has conducted an inventory of the nation's wetlands, and has produced a series of National Wetlands Inventory (NWI) maps for the entire country.[38]

Comments and objections from certain state agencies must also be considered. For instance, the Coastal Zone Management Act permits a state to object to a proposed permit if the state has an approved Coastal Zone Management Program (CMP) and the state determines that issuance of the permit will be inconsistent with the goals of the state CMP.[39] Although various state and federal agencies may object to a permit application, the Corps may decide to issue a permit over the objections of other agencies.[40]

13.2.2 Wetlands as "Waters of the United States"

Section 404(a) of the CWA states that a permit is necessary only for the discharge of dredged or fill material into the "navigable waters." The "navigable waters" language was construed to mean all "waters of the United States" after the EPA and other public interest groups had sought an expanded definition of the term. This expanded definition was found to be consistent with the definition of navigable waters found in the Clean Water Act.[41]

The Corps initially refused to expand its Section 404 wetlands jurisdiction, relying on prior judicial decisions under the Rivers and Harbors Act, that construed "navigable waters" as limited to the mean high water mark. The Corps now defines "waters of the United States" to mean:

> [a]ll waters which are currently used, or were used in the past, or may be susceptible to use in interstate or foreign commerce, including all waters which are subject to the ebb and flow of the tide. . . .[42]

---

[37] 16 U.S.C. § 661.
[38] Copies of NWI maps may be obtained be calling 800-USA-MAPS. Many of these maps are also available in digitized form. NWI digital databases can be purchased at cost from the NWI office in St. Petersburg, Florida.
[39] 16 U.S.C. § 1456(c)(3)(A). See chapter 14 for a full discussion of the Coastal Zone Management Act.
[40] 33 CFR § 325.
[41] 33 U.S.C. § 1362(7); see Natural Resources Defense Council, Inc. v. Calloway, 392 F. Supp. 685 (D.D.C. 1975).
[42] 33 CFR § 328.3(a)(1).

This definition also includes:

[a]ll other waters such as intrastate lakes, rivers, streams (including intermittent streams), mudflats, sandflats, wetlands, sloughs, prairie potholes, wet meadows, playa lakes, or natural ponds, the use, degradation or destruction of which could affect interstate or foreign commerce. . .[43]

However, these definitions do not anticipate whether Corps jurisdiction extends to waters that are "adjacent" to navigable waters as opposed to those waters that are "isolated."

13.2.3 Regulation of "Adjacent" and "Isolated" Wetlands

Corps' regulations also encompass wetlands "adjacent" to waters associated with interstate commerce, and have been interpreted to include jurisdiction over certain "isolated wetlands."[44] Several courts have upheld various aspects of the Corps' expansive interpretation of its wetland jurisdiction.[45] However, the Corps does not have regulatory authority over discharges into wetlands for which it did not issue a permit.[46]

In 1985, the U.S. Supreme Court issued a significant decision in the case of *United States v. Riverside Bayview Homes*, finding that the Corps' jurisdiction over wetlands extended to areas that are adjacent to navigable waters.[47] In its decision, the Court observed: "Congress evidently intended to repudiate limits that had been placed on federal regulation by earlier water pollution control statutes and to exercise its powers under the Commerce Clause to regulate at least some waters that would not be deemed 'navigable' under the classical understanding of that term."[48] The Court upheld the Corps' regulation of wetlands adjacent to navigable

---

[43] 33 CFR § 328.3(a)(3).
[44] 33 CFR 328.3(a)(5) and (7).
[45] United States v. Riverside Bayview Homes, 474 U.S. 121, 123 (1985). See also Mills v. United States, 36 F.3d 1052 (11th Cir. 1994) [delegation of authority to Corps to define "waters of the United States" was constitutional]; but see United States v. Wilson, 133 F.3d 251 (4th Cir. 1997) [holding that the Corps exceeded its authority under the CWA by expanding the statutory definition to include intrastate waters].
[46] United States v. Hallmark Construction Co., 14 F. Supp.2d 1069 (N.D. Ill. 1998).
[47] United States v. Riverside Bayview Homes, 474 U.S. 121 (1985).
[48] See also United States v. Hobbs, 21 Envt'l L. Rep. (ELI) 20,830 (E.D. Va. Aug. 24, 1990); and see 40 CFR § 230(s) and 33 CFR § 328.2).

waters because these are "waters that together form the entire aquatic system."[49] The Court reasoned that adjacent wetlands would "affect the water quality of the other waters within that aquatic system."[50]

Other federal courts have ruled on the adjacency issue in several other cases. For example, in *United States v. Lee Wood Contracting, Inc.*, it was held that a wetland area met the statutory definition of an adjacent wetland because it was connected to a river by a slough and, therefore, was contiguous.[51]

Wetlands that are not contiguous but that affect the water quality and aquatic ecosystems of navigable waters may also be considered "adjacent wetlands." For example, a wetland that traps undesirable pollutants and sediments before they reach a navigable river has been held to be "adjacent."[52]

In addition to jurisdiction over adjacent wetlands, Corps regulation includes CWA Section 404 jurisdiction over certain "isolated" waters. The CWA's jurisdiction also is extended by regulation over certain intrastate waters not part of a surface tributary system, that is "isolated" waters, if their "use, degradation or destruction ... could affect interstate or foreign commerce."[53]

Although the "isolated" wetlands issue seldom arose during the first fifteen years of the Section 404 Program, when it did, the Corps or EPA generally determined jurisdiction according to the effect the proposed wetland activity might have on interstate commerce. With this analysis, they generally considered whether the site in question served an interstate market or was visited by out-of-state residents (either for recreation or study) and whether the proposed work in the wetlands area would be likely to affect such visits.[54]

Other examples of the kinds of wetland activities that could affect interstate or foreign commerce include wetlands from which fish or shellfish are taken and sold in interstate or foreign commerce and waters

---

[49] United States v. Riverside Bayview Homes, 474 U.S. at 133, quoting from 42 *Fed. Reg.* 37,122, 37,128 (1977).

[50] Id. at 134. The regulations define "adjacent" as meaning "bordering, contiguous, or neighboring..." but do not establish a distance for "adjacency." 33 CFR § 323.2(d). See Dennison and Berry, The Regulatory Framework, Chap. 7 in Dennison and Berry, *Wetlands: Guide to Science, Law, and Technology* (Noyes, 1993).

[51] United States v. Lee Wood Contracting, Inc., 529 F. Supp. 119, 120 (E.D. Mich. 1981).

[52] Conant v. United States, 786 F.2d 1008 (11th Cir. 1986).

[53] 33 CFR § 328.3(a)(3); see also 33 CFR § 330.2(e).

[54] See United States v. Byrd, 609 F.2d 1204 (7th Cir. 1979). See also Memorandum of the Acting General Counsel, "Clean Water Act Jurisdiction Over Springs in Ash Meadows, Nevada" (July 5, 1983).

that are used or could be used for industrial purposes by industries in interstate commerce.[55]

Judicial analysis has been slow in interpreting which types of isolated waters are included under the Corps' jurisdiction.[56] In *United States v. Riverside Bayview Homes*, the Supreme Court specifically left open the question whether "isolated wetlands" (i.e., wetlands that do not have a hydrological connection to "waters of the United States") are within the scope of jurisdiction under the Section 404 Program.[57]

In *Leslie Salt Co. v. United States*, the court found that artificially created wetlands, which formed in crystallization basins and calcium chloride pits that had previously been used for salt production could be subject to Corps jurisdiction even though isolated from the tidal arm of San Francisco Bay by a quarter of a mile, as long as those waters had sufficient ties to interstate commerce.[58] The Ninth Circuit rejected the district court's assertion that Corps jurisdiction extends only to natural formations, holding that "the Corps intends to exempt from its own jurisdiction only those artificially created waters, which are currently being used for commercial purposes, and that even those waters are subject to such jurisdiction on a 'case-by-case' basis of review." The court noted that "the seasonal nature of the ponding presented no obstacle to Corps jurisdiction," because intermittent streams and playa lakes are specifically enumerated as types of isolated waters over which the Corps may have jurisdiction.[59]

The Fourth Circuit issued a surprising ruling in 1997 that invalidated the Corps' regulatory definition of "waters of the United States."[60] In *United States v. Wilson*, a developer who had been convicted of discharging fill material into a wetland without a permit challenged the validity of the Corps regulatory definition. The court held that the Corps had impermissibly exceeded its congressional authorization under the CWA, as limited by the Commerce Clause, by expanding the statutory definition of "waters of the United States" to include intrastate waters.[61] Whether

---

[55] See Dennison and Berry, The Regulatory Framework, Chap. 7 in Dennison and Berry, *Wetlands: Guide to Science, Law, and Technology* (Noyes, 1993).

[56] See Geltman, Regulation of Non-adjacent Wetlands under Section 404 of the Clean Water Act, 23 New England L. R. 615 (1989).

[57] United States v. Riverside Bayview Homes, 474 U.S. 121, 131 (1985).

[58] Leslie Salt Co. v. United States, 896 F.2d 354 (9th Cir. 1990), *cert. denied*, 111 S. Ct. 1089 (1991).

[59] Leslie Salt Co. v. United States, 896 F.2d at 355, 357 (9th Cir. 1990).

[60] 33 CFR § 328.3(a)(3).

[61] United States v. Wilson, 133 F.3d 251 (4th Cir. 1997).

other courts will follow the lead of the Fourth Circuit remains to be seen.[62]

In other cases, the Corps and EPA have used migratory birds to establish an effect on commerce for purposes of jurisdiction over isolated wetlands. The basis of the migratory bird rule is that if a bird might use an isolated wetland as a stopover or nesting grounds while on its migratory flight. Any activity disturbing the bird's wetland habitat could affect interstate commerce since bird watchers would be unable to observe or hunt the birds during interstate travels.[63]

The migratory bird rule has met with mixed reviews in the courts. In *Tabb Lakes, Ltd. v. United States*, the court held that the migratory bird rule was invalid because it was a substantive rather than an interpretive rule, and that the provision could not be effective without a prior opportunity for notice and comment.[64] In *Hoffman Homes, Inc. v. Administrator, United States Environmental Protection Agency*,[65] the Seventh Circuit held that it was reasonable for the EPA to interpret the regulatory definition[66] as allowing migratory birds to form the connection between "other waters" and interstate commerce. The court noted that millions of dollars are spent on hunting, trapping, and observing migratory birds, and that the destruction of wetlands impinges on that commerce.[67] By holding that the "other waters" regulation could be read to encompass waters used by migratory birds, the court implicitly treated this as a reasonable interpretation of the statutory term "waters of the United States."[68]

Other courts have upheld the validity of the migratory bird rule as a basis for Corps jurisdiction over isolated wetlands. In *United States v. Hallmark Construction Co.* the district court relied heavily on the reasoning from the *Hoffman Homes* case and ruled that Corps reliance on the migratory bird rule did not exceed its authority under the CWA.[69] The Seventh Circuit in *Solid Waste Agency of N. Cook County v. U.S. Army*

---

[62] See United States v. Sartori, 62 F. Supp. 1362 (S.D. Fla. 1999) [declining to address the "thorny issue" of determining the constitutionality of the Army Corps' "other waters" definition].

[63] See Leslie Salt Co. v. United States, 55 F.3d 1388 (9th Cir. 1995). See also Dennison and Berry, The Regulatory Framework, Chap. 7 in Dennison and Berry, *Wetlands: Guide to Science, Law, and Technology* (Noyes, 1993).

[64] Tabb Lakes, Ltd. v. United States, 715 F. Supp. 726, 729 (E.D.Va. 1988), aff'd, 885 F.2d 866 (4th Cir. 1989).

[65] Hoffman Homes, Inc. v. Administrator, United States Environmental Protection Agency, 999 F.2d 256 (7th Cir. 1993).

[66] 40 C.F.R. § 230.3(s)(3)

[67] Hoffman Homes, Inc. v. Administrator, United States Environmental Protection Agency, 999 F.2d 256, 261 (7th Cir. 1993).

[68] Leslie Salt Co. v. United States, 55 F.3d 1388 (9th Cir. 1995).

[69] United States v. Hallmark Construction Co., 14 F. Supp. 2d 1069 (N.D. Ill. 1998).

Corps of Engineers held that 17.6 acres of a 180 acre proposed balefill repository for non-hazardous waste contained navigable waters subject to Corps jurisdiction based on the use of the waters by migratory birds. The court held that the aggregate effects of wetland losses together with the billions of dollars spent each year on hunting, trapping, and observing migratory waterfowl established authority under the Commerce Clause for Corps jurisdiction.[70]

13.2.4 Discharges into Waters of the United States

Once there has been a determination of jurisdiction over a wetland, the next inquiry involves determining whether the activity requires a permit. Until recently, EPA, the Corps, and courts have held that CWA Section 404 regulates only physical discharges of dredged or fill material into navigable waters.[71] On its face, this language indicates that many activities that destroy wetlands are not regulated under the Clean Water Act. This is consistent with the plain language of CWA Section 404, which regulates only "discharges" of pollutants.[72] However, this narrow "discharge rule" has been eroded by regulatory and judicial decisions.

In *Avoyelles Sportsmen's League, Inc. v. Marsh*, the court held that certain land-clearing activities, which resulted in redeposits of fill material taken from the wetland, constituted a discharge of a pollutant and, therefore, required a dredge-and-fill permit.[73] Corps regulations define "fill material" as "any material used for the primary purpose of replacing an aquatic area with dry land or of changing the bottom elevation of a waterbody."[74]

In 1990, a Texas federal district court held for the first time that an activity need not involve a discharge to be regulated under CWA Section 404. In *Save Our Community v. EPA* an environmental group challenged an EPA/Corps determination that they did not have legal authority to require a permit when the only activity conducted on a jurisdictional wetland was drainage through mechanical means. No discharges were made during the draining process. Significantly, the court found that draining a "legally designated" wetland property is a regulated activity

---

[70] Solid Waste Agency of N. Cook County v. U.S. Army Corps of Engineers, 191 F.3d 845 (7th Cir. 1999).
[71] See United States v. Lambert, 589 F. Supp. 366 (M.D. Fla. 1984).
[72] 33 U.S.C. § 1311(a).
[73] Avoyelles Sportsmen's League, Inc. v. Marsh, 715 F.2d 897 (5th Cir. 1983).
[74] See 33 CFR § 323.2(e)).

under CWA Section 404(b), requiring a permit when the draining would alter or destroy the wetland.[75]

13.2.5 "Fallback," and the "Tulloch Rule"

An interesting issue has developed over the status of "fallback," or the return of material to essentially the same spot it originated during construction activities. In 1993, The Corps and EPA changed their regulations at the time[76] in response to a court decision in *North Carolina Wildlife Federation v. Tulloch*,[77] concerning a developer who sought to drain and clear 700 acres of wetlands in North Carolina. Because the developer's efforts involved only minimal incidental releases of soil and other dredged material, the Corps's field office personnel determined that, under the terms of the regulations at the time, CWA Section 404's permit requirements did not apply. Concerned by what they viewed as the adverse effects of the developer's activities on the wetland, environmental groups filed an action seeking enforcement of the Section 404 permit requirement. As part of the settlement of the Tulloch case (a settlement to which the developer was not a party), the two administering agencies agreed to propose stiffer rules governing the permit requirements for landclearing and excavation activities. The resulting regulation, which has come to be called the "Tulloch Rule," alters the preexisting regulatory framework primarily by removing the de minimis exception and by adding coverage of incidental fallback. Specifically, the rule defines "discharge of dredged material" to include "[a]ny addition, including any redeposit, of dredged material, including excavated material, into waters of the United States which is incidental to any activity, including mechanized landclearing, ditching, channelization, or other excavation."[78] "In effect the new rule subject[ed] to federal regulation virtually all excavation and dredging performed in wetlands."[79]

However, a group of trade groups successfully challenged the 1993 regulations in *National Mining Association v. United States Army Corps of Engineers*.[80] The court held that the 1993 regulation exceeded the Corps' authority under Section 404 of the CWA because it impermissibly

---

[75] Save Our Community v. EPA, 741 F. Supp. 605 (N.D. Texas 1990).
[76] See 58 *Fed. Reg.* at 45,016 (1993).
[77] North Carolina Wildlife Federation v. Tulloch, Civ. No. C90-713-CIV-5-BO (E.D. N.C. 1992).
[78] 33 CFR § 323.2(d)(1)(iii), as it appeared at the time.
[79] American Mining Congress v. United States Army Corps of Engineers, 951 F. Supp. 267 (D.D.C. 1997); aff'd sub nom, National Mining Association v. United States Army Corps of Engineers, 145 F.3d 1339 (D.C. Cir. 1998).
[80] Id.

regulated "incidental fallback" of dredged material, which is not subject to CWA Section 404 regulation as an "addition" of pollutants.[81]

In 1999 the Corps and EPA issued new, revised regulations which, once again, modified the definition of "discharge of dredged material" in order to comply with the court's holding in the *National Mining Association* case. The Corps made two changes to the regulations.[82] First, the rule deletes use of the word "any" as a modifier of the term "redeposit." Second, the new rule expressly excludes "incidental fallback" from the definition of "discharge of dredged material." The new regulation defines "Discharge of dredged material" as:

[A]ny addition of dredged material into, including redeposit of dredged material other than incidental fallback within, the waters of the United States. The term includes, but is not limited to, the following:

(i) The addition of dredged material to a specified discharge site located in waters of the United States;

(ii) The runoff or overflow, associated with a dredging operation, from a contained land or water disposal area; and

(iii) Any addition, including redeposit other than incidental fallback, of dredged material, including excavated material, into waters of the United States which is incidental to any activity, including mechanized landclearing, ditching, channelization, or other excavation.[83]

Deciding when a particular redeposit is subject to CWA Section 404 jurisdiction will require a case-by-case evaluation, based on the particular facts of each case. Based on the decision in *National Mining Association v. United States Army Corps of Engineers*,[84] redeposits associated with the following are subject to CWA jurisdiction: mechanized landclearing, redeposits at various distances from the point of removal (e.g.,

---

[81] See also United States v. Sartori, 62 F. Supp. 2d 1362 (S.D. Fla. 1999) [use of earth-moving equipment to excavate, move, and deposit dirt and other materials was discharge of a pollutant, not "incidental fallback," subject to Corps jurisdiction].

[82] 33 CFR § 323.2(f); 40 CFR § 232.2.

[83] 40 CFR § 232.2.

[84] American Mining Congress v. United States Army Corps of Engineers, 951 F. Supp. 267 (D.D.C. 1997); aff'd sub nom, National Mining Association v. United States Army Corps of Engineers, 145 F.3d 1339 (D.C. Cir. 1998).

sidecasting), and removal of dirt and gravel from a streambed and its subsequent redeposit in the waterway after segregation of minerals.[85]

13.3 Dredge and Fill Permits

Under Section 404(a) of the CWA, the Secretary of the Army may issue permits for the discharge of dredged or fill material into the navigable waters.[86] The process by which a person, called the "applicant" in Corps regulations, applies for a permit is described in Corps regulations.[87] The process itself is complex, and replete with snares for the unprepared. The cautious applicant may avoid unnecessary difficulties and delays by preparing carefully the necessary materials for the application. Many applications, particularly those that are large or controversial, require the assistance of scientific and legal consultants.

13.3.1 The Application Process

The requirement for dredge and fill permits under CWA Section 404 and Rivers and Harbors Act Section 10 applies to any private or governmental individual, entity, organization, or agency.[88] Certain federal projects which are specifically approved by Congress are exempted under CWA Section 404(r), if pertinent information is supplied in an environmental impact statement.[89] The Corps need not apply for a CWA Section 404 permit from itself, but it is required nevertheless to comply with the same laws and

---

[85] National Mining Association v. United States Army Corps of Engineers, 145 F.3d 1339, 1407 (D.C. Cir. 1998). See also, Avoyelles Sportsmen's League v. Marsh, 715 F.2d 897 (5th Cir. 1983) (mechanized landclearing requires section 404 permit); United States v. M.C.C. of Florida, 772 F.2d 1501 (11th Cir. 1985), vacated on other grounds, 481 U.S. 1034 (1987), readopted in relevant part on remand, 848 F.2d 1133 (11th Cir. 1988) (redeposit of river bottom sediments on adjacent sea grass beds is an "addition"); Rybachek v. EPA, 904 F.2d 1276 (9th Cir. 1990) (resuspension of materials by placer miners as part of gold extraction operations is an "addition of a pollutant" under the CWA subject to EPA's regulatory authority); NMA, 951 F. Supp. at 270 ("Sidecasting, which involves placing removed soil alongside a ditch, and sloppy disposal practices involving significant discharges into waters, have always been subject to section 404").
[86] 33 U.S.C. § 1344(a). The Secretary of the Army acts through the Chief of Engineers, 33 U.S.C. § 1344(d).
[87] 33 CFR §§ 320-344. See Want, *Law of Wetlands Regulation* (West, 1998).
[88] 33 CFR § 323.2(a) and (b).
[89] 33 U.S.C. § 1344(r).

follow the same procedures as any other applicant.[90]

The next step in the process involves obtaining the permit application itself. An applicant may obtain the proper Department of the Army permit form from the office of the local district engineer to whom the application will be submitted.[91] Some District offices require the use of forms with some slight variations in order to facilitate local coordination with other federal, state, or local agencies. For example, applicants in Illinois complete a joint form from the U.S. Army Corps of Engineers and the Illinois Department of Transportation; applicants in Florida fill out a joint form from the Corps, the Florida Department of Environmental Regulation, and the Florida Department of Natural Resources.

Regulatory guidance on the contents of the application is found in Corps regulations.[92] The general rule is that an application will be considered complete "when sufficient information is received to issue a public notice."[93] These requirements are met when the following information is provided:

(1) A complete description of the proposed activity, including necessary drawings, sketches, or plans sufficient for a public notice.

(2) The location, purpose and need for the proposed activity.

(3) Scheduling of the activity.

(4) The names and addresses of adjacent property owners.

(5) The location and dimensions of adjacent structures.

(6) A list of authorizations required by other federal, interstate, state or local agencies for the work.

(7) Preliminary jurisdictional determination.

(8) Signature.[94]

If an activity will involve dredging in navigable waters, the application must include a description of the type, composition, and quantity of the material to be dredged, as well as the method of dredging, and the site and plans for its disposal.[95] If there will be a discharge of dredged or fill materials (or transportation of these materials for ocean discharge), then the application must describe the source of material; the purpose of the

---

[90] 33 CFR §§ 209.145, 322.3(c)(1), 323.3(b), 336(1)(a), and 40 CFR 230.2(a)(2); see Minnesota v. Hoffman, 543 F.2d 1198 (8th Cir. 1976).

[91] ENG Form 4345, OMB Approval No. OMB 49-R0420; see 33 CFR § 325.1(c). A permit form is also found as Appendix A to 33 CFR § 325. See Want, *Law of Wetlands Regulation* (West, 1998) for an example of a completed application form).

[92] 33 CFR § 325.1(d).

[93] 33 CFR § 325.1(d)(9).

[94] 33 CFR § 325.1(d)(1). See Want, *Law of Wetlands Regulation* (West, 1998).

[95] 33 CFR § 325.1(c)(3).

discharge; the type, composition, and quantity of material; the method of transportation and disposal; and the site of disposal.[96] Certification is also required from the EPA under CWA Section 401, the National Pollutant Discharge Elimination System (NPDES).[97] If the project includes the construction of an impoundment structure, the application may be required to demonstrate compliance with state dam safety criteria.[98]

The most controversial application requirement is contained in Corps regulations:

> All activities which the applicant plans to undertake which are *reasonably related* to the same project and for which a DA [Corps] permit would be required should be included in the same permit application. District Engineers may reject, as incomplete, any permit application that fails to comply with this requirement.[99]

There has been considerable disagreement as to what constitutes "reasonably related" projects. For utilitarian reasons, applicants may wish to submit separate applications for two related projects and escape close Corps scrutiny of the cumulative impacts of the two projects taken together. In *Russo Development Corp v. Thomas*, a federal district court rejected the Corps' attempt to require a single application for two separate properties on the basis that they were contiguous (on one property the developer sought a permit to construct buildings, and on the other he sought an after-the-fact permit for a building already constructed).[100]

A related question occasionally arises when an applicant wishes to change a project after permit approval, but without filing a new permit application. Such changes are not uncommon, and are generally approved by the Corps if the scope of wetland fill and environmental impacts is the same. In the famous (if somewhat bizarre) case *Missouri Coalition for the Environment v. Army Corps of Engineers*, a federal appellate court held that a developer's change from an industrial park to a football stadium did *not* require a new permit, because the scope of wetland fill and potential impacts were similar.[101]

---

[96] 33 CFR § 325.1(d)(4).
[97] 33 U.S.C. § 1341.
[98] 33 CFR § 325.1(d)(6).
[99] 33 CFR § 325.1(d)(2), emphasis added.
[100] Russo Development Corp v. Thomas, 735 F. Supp. 631 (D.N.J. 1989).
[101] Missouri Coalition for the Environment v. Army Corps of Engineers, 866 F.2d 1025 (8th Cir. 1989).

13.3.2 Notice, Comment, and Conflict Resolution

Within 15 days of receipt of an application, the district engineer must review the application and respond in one of two ways: (1) either issue a public notice, or (2) advise the applicant that additional information is necessary for a complete application.[102] In practice, Corps offices are nearly always late despite the clear mandate of the regulations. Occasional contact with the Regulatory Division of the local Corps office may serve to satisfy an applicant that his application is receiving timely treatment.[103]

In general, the Corps attempts to make knowledge of the application available as generally as possible. Under Corps regulations, public notices are to be distributed to the public in general by posting in post offices "or other appropriate public places in the vicinity of the site of the proposed work."[104] Notices are sent specifically to the applicant, adjacent property owners, appropriate local officials and state agencies, Native American tribes, and to concerned federal agencies. Notices are also sent to concerned business interests, environmental organizations, state and regional clearing houses for such information, and local news media. In addition, the Corps will send a notice to "all parties who have specifically requested copies of public notices."[105] The public notice *may* specify that a public hearing will be held on the application,[106] but it is far more common for the Corps to decide if a public hearing will be held after comments are received.

Any interested person or organization may make comments on a CWA Section 404 permit application. The district engineer acknowledges receipt of comments "if appropriate," and will make them a part of the administrative record of the application.[107] In cases where the issuance of a CWA Section 404 permit (or failure to issue one) is challenged in court, the administrative record may be of critical importance to the litigants.

The district engineer must "consider all comments received in response to the public notice in his subsequent actions on the permit application.[108]

---

[102] 33 CFR § 325.2(a)(2).

[103] For more information on the application process, see Dennison and Berry, The Regulatory Framework, Chap. 7 in Dennison and Berry, *Wetlands: Guide to Science, Law, and Technology* (Noyes, 1993); and Want, *Law of Wetlands Regulation* (West, 1998).

[104] 33 CFR § 325.3(3)(d)(1).

[105] Id. In the authors' experience, any person (or organization) with an interest in dredge and fill issues in a particular area is well advised to request from the local Corps office that they be sent any public notices for CWA § 404 permit applications received by the office.

[106] 33 CFR § 327.4(b).

[107] 33 CFR § 325.2(a)(3).

[108] 33 CFR § 325.2(a)(3).

If the comments received by the Corps indicate that the views of the applicant are necessary in order to conduct a public interest determination on a particular issue, then the applicant will be given the opportunity to furnish his views on the issue.

At the earliest practicable time, the Corps will supply copies of substantive comments to the applicant. These comments may be supplied in the form of summaries, the actual letters (or portions of them), or representative comments. The applicant must be given the opportunity to contact the objectors in an attempt to resolve conflicts regarding the application, although the final decision on issuance of the permit rests with the Corps. In *Mall Properties, Inc. v. Marsh*, the court held that Corps regulations require that an applicant be given an opportunity to rebut negative comments made of the application.[109] Our experience has been that the conflict resolution phase of a permit application is often of critical importance. It is frequently the first time that parties on both sides of the issue have the opportunity to sit at the same table and attempt to resolve differences through proper discussion and negotiation. Timely, thoughtful, and responsive conflict resolution often leads to equitable agreements without the need for unpleasant confrontation and expensive litigation.

13.3.3 Issuing the Permit

After the district engineer has considered all the information on the permit application (including all public comments and hearings), a statement of findings (SOF), or a record of decision (ROD) must be issued if an environmental impact statement has been prepared pursuant to the requirements of NEPA.[110] The SOF or ROD must include the district engineer's views on the effects of the proposed project on "the public interest," including compliance with all aspects of the CWA Section 404(b)(1) guidelines.[111] In the case of zoning and land use issues, the district engineer must either accept the decisions of local and state governing bodies,[112] or include in the SOF or ROD an explanation of why national issues should override local or state decisions.[113]

Corps regulations specify a careful "public interest review" of a permit,

---

[109] Mall Properties, Inc. v. Marsh, 672 F. Supp. 561 (D.Mass. 1987), *rejecting remand as appealable order,* 841 F.2d 440 (1st Cir. 1988). Here, the governor of Connecticut had recommended rejection of the application at a meeting not attended by the applicant.

[110] 33 CFR § 325.2(a)(6). See Chapter 3.

[111] 40 CFR Part 230.

[112] 33 CFR § 320.4(j)(2).

[113] 33 CFR § 325.2(a)(6).

which involves a cost-benefit balance that is nearly unique among environmental regulations:

> The decision whether to issue a permit will be based on an evaluation of the probable impacts, including cumulative impacts, of the proposed activity and its intended use on the public interest. Evaluation of the probable impact ... on the public interest requires a careful weighing of all those factors which become relevant in each particular case. The benefits which reasonably may be expected to accrue from the proposal must be balanced against its reasonably foreseeable detriments.[114]

This general balancing process determines whether a permit will be issued, and what kinds of conditions it will require. Among the factors to be considered are conservation, economics, esthetics, general environmental concerns, fish and wildlife values, water quality, and public welfare.[115]

If a CWA Section 404 permit is warranted, the district engineer determines special conditions that will be required, and what the duration of the permit should be. Once the district engineer's name and signature are affixed to the permit, it normally becomes a final action unless he must forward it to his superiors in the Corps or other agencies for final action.[116]

Corps regulations require the district engineer to consult with the USFWS, NMFS, and state agencies responsible for fish and wildlife, and give full consideration to their views.[117] However, courts have held that the Corps need not defer to recommendations of these agencies.[118]

The duration of a CWA Section 404 permit generally depends on the nature of the permitted activity. Permits for permanent structures are usually indefinite with no expiration date, but the permit will be of limited duration with a definite expiration date where a structure or project is temporary in nature.[119]

13.3.4 Nationwide Permits and Exemptions

Section 404(e) of the Clean Water Act authorizes the Corps to issue "general" permits on a state, regional, and nationwide basis for any

---

[114] 33 CFR § 320.4(a).
[115] See 33 CFR § 320.4(a) for additional guidance on the balancing process
[116] 33 CFR § 325.8(b)
[117] 33 § CFR 320.4(c).
[118] See Sierra Club v. Alexander, 484 F. Supp. 455 (N.D.N.Y. 1980), aff'd, 633 F.2d 206 (2d Cir. 1981).
[119] 33 CFR § 325.6(b).

category of activities which are similar in nature and will have only minimal individual and cumulative environmental impacts.[120] The apparent intention of general permits is that unnecessary time and paperwork can be avoided for those activities that will have similar impacts in each case, such that complete, individual CWA Section 404 permit applications would be redundant.

In general, an activity that is covered by a general permit does not require a CWA Section 404 permit application, so long as the person complies with the conditions incorporated in the general permit.[121]

The most significant general permits are called "nationwide" permits. These are issued by Corps headquarters in Washington, D.C., and apply equally in all parts of the nation. Because of their broad reach, nationwide permits must undergo a review process prior to implementation separate from that for more localized general permits.[122]

There have been over 40 nationwide permits issued to date, and these are listed and described in Corps regulations.[123] Nationwide permits cover such activities as construction of aids to navigation and regulatory markers (Permit no. 1); scientific measurement devices (Permit no. 5); utility line backfill and bedding (Permit no. 12); minor discharges that do not exceed 25 cubic yards (Permit no. 18); oil spill cleanup activities (Permit no. 20); and maintenance dredging of existing marinas, canals, and boat slips (Permit no. 35).

Nationwide permits incorporate a series of general conditions that apply to all nationwide permits.[124] Many of these pertain to environmental controls and maintenance associated with the permitted activities. It should be noted that nationwide permits still require compliance with the federal Endangered Species Act,[125] as well as coastal zone management consistency requirements.[126]

District engineers have override authority over nationwide permits, which allows them to suspend, modify, or revoke nationwide permits if individual or cumulative impacts are more than minimal, or when the activity is contrary to the public interest.[127] It has been held that the Corps may suspend, modify, or revoke a nationwide permit without providing a hearing.[128]

---

[120] 33 U.S.C. § 1344(e)(1).

[121] 33 CFR § 320.1(c).

[122] See 33 CFR § 330.1.

[123] Appendix A to 33 CFR Part 330.

[124] Appendix A(C) to 33 CFR Part 330.

[125] 16 U.S.C. §§ 1531-1534. See Chapter 15.

[126] See 33 CFR § 330.4(d). See chapter 14 for a full discussion of the Coastal Zone Management Act.

[127] 33 CFR §§ 330.1(d), 330.4(e), and 330.5

[128] See O.Connor v. Corps of Engineers, 801 F. Supp. 185 (N.D.Ind. 1992).

Without question the most controversial nationwide permit is Nationwide Permit 26, which exempted from the permit process any discharges of dredged or fill materials into "headwaters and isolated waters," provided: (1) the discharge did not cause the loss of more than ten acres of waters; (2) the Corps district was notified of the discharge and loss; and (3) the discharge was part of a single and complete project. Nationwide Permit 26 applied to wetlands above the headwaters, as well as the water bodies to which they are adjacent. Nationwide permit 26 was the result of a 1984 settlement agreement between the Corps and 16 environmental groups which had challenged a predecessor of the permit.[129] Much of the controversy surrounding the Nationwide Permit 26 was that the potential for destruction of wetlands was enormous.[130]

In 1996, the Corps announced plans to modify and ultimately eliminate Nationwide Permit 26. Nationwide permit 26 expired in September, 1999.[131] It was replaced with a series of smaller Nationwide Permits that have about the same ultimate coverage as Permit 26.

CWA Section 404(f)(1) of the Clean Water Act lists several kinds of activities that are exempt from the requirement for a Section 404 dredge and fill permit.[132] No permit is required for discharges:

(A) from normal farming, silviculture, and ranching activities such as plowing, seeding, cultivating, minor drainage, harvesting for the production of food, fiber, and forest products, or upland soil and water conservation practices;

(B) for the purpose of maintenance ... of currently serviceable structures such as dikes, dams, levees, groins, riprap, breakwaters, causeways, ...

(C) for the purpose of construction or maintenance of farm or stock ponds or irrigation ditches, or the maintenance of drainage ditches;

(D) for the purpose of construction of temporary sedimentation basins on a construction site ...;

(E) for the purpose of construction or maintenance of farm roads or forest roads, or temporary roads for moving mining equipment;

---

[129] See National Wildlife Federation v. Marsh, 14 Envtl. L. Rep. (Envtl. L. Inst.) 29262 (D.D.C. 1984).

[130] See Blumm and Zaleha, Federal Wetlands Protection under the Clean Water Act: Regulatory Ambivalence, Intergovernmental Tension, and a Call for Reform. 60 U. Col. L.Rev. 695 (1989); and Thomas and Burns, Comment: The Army Corps of Engineers and Nationwide Permit 26: Wetlands Protection or Swamp Relocation? 18 Ecology L.Q. 619 (1991).

[131] 61 *Fed. Reg.* 65,874 (Dec. 13, 1996).

[132] 33 U.S.C. § 1344(f)(1).

(F) resulting from any activity with respect to which a State has an approved [statewide water quality plan under 33 U.S.C. Section 1288(b)(4)].

Several of these exemptions have been tested in court. For example, a federal district court in *United States v. Zanger* refused to allow a landowner to claim several different exemptions for an alleged stream bank erosion repair which was really a redirected stream channel and fill of the former streambed.[133] The same court held in *Leslie Salt Co. v. United States* that blockage of water flow through a culvert was not maintenance of a drainage ditch or existing structure.[134] In another case, the court ruled that the dredging of wetlands to construct a farm pond fell within the exemption of CWA Section 404(f)(1)(C).[135]

The CWA Section 404(f)(1) exemptions are limited by the CWA Section 404(f)(2) "recapture clause, which makes the exemptions inapplicable if the purpose of the activity is to bring navigable waters into a use to which they were not previously subject, or where the flow of water is impaired.[136] Courts have construed narrowly the reach of CWA Section 404(f)(1) exemptions based on the Section 404(f)(2) recapture clause.[137]

In addition to the agricultural exemption for normal farming, silviculture, and ranching activities, the Corps promulgated a rule in 1993 that excludes prior converted cropland from the definition of "waters of the United States," and from Corps jurisdiction under CWA Section 404.[138] Such prior converted cropland can be inundated for no more than fourteen consecutive days during the growing season. If prior converted cropland is abandoned for agricultural uses for five years or more, then it may revert to wetland status.[139]

13.3.5 The EPA Veto

Section 404(c) of the Clean Water Act authorizes the Administrator of the U.S. EPA to veto a CWA Section 404 dredge and fill permit:

---

[133] United States v. Zanger, 767 F. Supp. 1030 (N.D. Cal. 1991).
[134] Leslie Salt Co. v. United States, 22 Envtl. L. Rep. (ELI) 20361 (N.D. Cal. 1992).
[135] 33 U.S.C. § 1344(f)(1)(C). In re. Carsten, 211 B.R. 719 (D.Mont. 1997).
[136] 33 U.S.C. § 1344(f)(2).
[137] See review of cases in Want, *Law of Wetlands Regulation* (West, 1998).
[138] 33 CFR § 401.11(1). Prior converted cropland is defined as "areas that, prior to December 23, 1985, were drained or otherwise manipulated for the purpose of or having the effect of making production of a commodity crop possible." 58 *Fed. Reg.* 45,008, 45,031 (1993).
[139] 58 *Fed. Reg.* 45,008, 45,031 (1993).

[W]henever [the EPA Administrator] determines, after notice and opportunity for public hearings, that the discharge of such materials into such area will have an unacceptable adverse effect on municipal water supplies, shellfish beds and fishery areas (including spawning and breeding areas), wildlife, or recreational areas. Before making such determination, the Administrator shall consult with the Secretary [of the Army]. The Administrator shall set forth in writing and make public his findings and his reasons for making any determination under this subsection.[140]

While EPA has actually used its veto only rarely, the *threat* of a possible veto has led the Corps to place modifications and conditions on many CWA Section 404 permits.[141]

The scope of the EPA veto has been clarified by several major court cases. The famous *Sweedens Swamp* litigation addressed the issue in several separate decisions.[142] In *Newport Galleria v. Deland*, the court held that the EPA veto was not reviewable in court because it was not a final agency action.[143] Later, a different court in *Bersani v. Deland* rejected the developer's argument that EPA veto authority had expired because no decision regarding a veto was reached within the 30-day limit for a public hearing.[144] In another decision, yet a third court held in *Bersani v. U.S. EPA* that EPA could use the same criteria in its veto decision (i.e., availability of alternative sites) that the Corps had used in issuing the permit.[145]

The EPA veto has been upheld by most courts.[146] However, in *James City County, Va. v. U.S. Environmental Protection Agency*, the EPA had exercised its Section 404(c) veto over a Corps permit for fill to construct a dam for a proposed reservoir, but the court held that the evidence did not

---

[140] 33 U.S.C. § 1344(c).
[141] See Dennison and Berry, The Regulatory Framework, Chap. 7 in Dennison and Berry, *Wetlands: Guide to Science, Law, and Technology* (Noyes, 1993); and Want, *Law of Wetlands Regulation* (West, 1998).
[142] See review in Bosselman, Sweeden's Swamp: the Morass of Wetland Regulation, 1989 Land Use L. 3 (Mar. 1989).
[143] Newport Galleria v. Deland, 618 F. Supp. 1179 (D.D.C. 1985). In an unrelated case, Reid v. Marsh, 20 Env't Rep. Cas. (BNA) 1337 (N.D. Ohio 1984), the court held that the EPA veto is discretionary, and not subject to judicial review.
[144] Bersani v. Deland, 640 F. Supp. 716 (D. Mass. 1986).
[145] Bersani v. U.S. EPA, 674 F. Supp. 405 (N.D.N.Y. 1987), *aff'd*, 850 F.2d 36 (2d Cir. 1988).
[146] For example, see Creppel v. United States, 19 Envtl. L. Rep. (ELI) 20134 (E.D. La. 1988); Russo Development Corp. v. Reilly, 20 Envtl. L. Rep. (ELI) 20938 (D.N.J. 1990); and City of Alma v. United States, 744 F. Supp. 1546 (S.D. Ga. 1990).

support EPA's conclusion that the county had practicable water supply alternatives.[147] EPA vetoed the permit a second time on the basis that it would have unacceptable adverse impacts, and the court again reversed the decision, this time because EPA considered environmental concerns but not human needs.[148]

Despite the adverse decision in *James City County*, the EPA veto of dredge and fill permits under Section 404(c) remains a powerful force demanding environmental sensitivity in Section 404 permit decisions.

13.4 Remedies for CWA Section 404 Permit Denials

13.4.1 Lack of Administrative Appeals Mechanism

No mechanism exists for administrative appeal to the Corps, nor is there any right to an adjudicatory hearing for CWA Section 404 permit denials like there is for denial of other types of environmental permits. Still, a permittee or other party who is adversely impacted by a Corps wetland permit decision is not without legal remedy. After the Corps rules on a federal wetland permit application, a final decision may be subject to judicial review under the Administrative Procedure Act.[149] In some cases, parties opposing the approval of a permit may have standing to institute a "citizen suit" under the Clean Water Act, National Environmental Policy Act or other environmental law.

However, judicial review is usually difficult for a challenger because the court is bound by a "substantial evidence" standard in reviewing the Corps' decision. Under this standard, the court may set aside the permit decision only if it finds that the Corps' action is arbitrary, capricious or not supported by substantial evidence in the record.[150] The courts are bound by the administrative record and may not admit new evidence unless the record is so scant that it makes judicial review virtually impossible.[151] In other cases, applicants who have been denied a permit may challenge the Corps' decision in federal court on constitutional grounds. For example, regulatory takings actions have frequently been brought in federal court to challenge permit denials.

[147] James City County, Va. v. U.S. Environmental Protection Agency, 758 F. Supp. 348 (E.D. Va. 1990), aff'd in part and remanded, 955 F.2d 254 (4th Cir. 1992).

[148] James City County v. U.S. Environmental Protection Agency, 23 Envtl. L. Rep. (ELI) 20228 (E.D. Va. 1992).

[149] 5 U.S.C. §§ 701-706.

[150] See Avoyelles Sportsmen's League, Inc. v. Marsh, 715 F.2d 897, 904 (5th Cir. 1983).

[151] See Friends of the Earth v. Hintz, 800 F.2d 822, 828 (9th Cir. 1986).

13.4.2 Regulatory Takings Challenges

If a governmental regulation places such a burdensome restriction on a landowner's use of property that the government has for all intents and purposes "taken" the landowner's property, then a regulatory taking has probably occurred.   Like physical governmental takings such as condemnation, regulatory takings are governed by either the due process clause of the Fourteenth Amendment or the "taking" clause of the Fifth Amendment to the United States Constitution.[152] The Fifth Amendment is held applicable to the states by incorporation into the "due process" clause of the Fourteenth Amendment.[153] Whether or not the taking is caused by physical interference or regulatory interference with a landowner's property rights, just compensation is required under the Constitution whenever a taking occurs. State constitutions also contain due process and taking clauses, and form the basis for state claims for just compensation.

The courts have grappled with the regulatory takings issue for many years, but no clear test has emerged.[154] Although the U.S. Supreme Court has decided several land use takings cases since 1978, there does not appear to be a bright-line test that can be followed in the regulatory takings context.[155] Several factors have been utilized to determine when a

---

[152] The Fourteenth Amendment to the Constitution states: "Nor shall any state deprive any person of life, liberty, or property without due process of law; nor deny to any person within its jurisdiction the equal protection of the laws." U.S. Const. Amendment XIV.

The Fifth Amendment states: "No person shall be ... deprived of life, liberty or property without due process of law; nor shall private property be taken for public use, without just compensation." U.S. Const. Amendment V.

[153] Chicago, Burlington & Quincy R.R. v. Chicago, 166 U.S. 226 (1897).

[154] A great deal has been written on regulatory takings. A few of the many reviews are Meltz, Merriam, Frank, Banta, and Callies, *The Takings Issue: Constitutional Limits on Land-Use Control and Environmental Regulation* 1995 ed. (Island Press, 1999); Skouras, *Takings Law and the Supreme Court: Judicial Oversight of the Regulatory State's Acquisition, Use, and Control of Private Property* (Peter Lang Publ., 1998); Marzulla and Narzulla, *Property Rights : Understanding Government Takings and Environmental Regulation* (Government Inst., 1997); and Fischel, *Regulatory Takings : Law, Economics, and Politics* (Harvard Univ. Press, 1995).

[155] Penn Central Transportation Co. v. New York City, 438 U.S. 104, 124 (1978); Agins v. City of Tiburon, 447 U.S. 255 (1980); Williamson County Regional Planning Comm'n v. Hamilton Bank of Johnson City, 473 U.S. 172 (1985); MacDonald, Sommer & Frates v. County of Yolo, 477 U.S. 340 (1986); Keystone Bituminous Coal Ass'n v. De Benedictis, 480 U.S. 470 (1987); Nollan v. California Coastal Commission, 483 U.S. 825 (1987); First English Evangelical Lutheran Church v. County of Los Angeles, 482 U.S. 324 (1987); Lucas v. South Carolina Coastal Council, 505 U.S. 1003 (1992); Dolan v. City of Tigard, 512 U.S. 374 (1994).

land use regulation results in a "taking" of property, "the Supreme Court has characterized the analytic process as one relying instead on ad hoc, factual inquiries into the circumstances of each particular case."[156]

Government denial of a wetlands permit to promote environmental goals may serve a legitimate public purpose while depriving a landowner of beneficial use of his property. If this is the case, courts have employed a balancing of interests test focusing on the nature and extent of the benefit derived for the public and the nature and extent of the loss occasioned on the landowner. This balancing test is the most frequently relied on method for determining whether a regulatory taking has occurred. Under this type analysis, the courts look to three factors in making a taking determination: (1) the character of the government action; (2) the economic impact of the regulation; and (3) interference with the landowner's reasonable investment-backed expectations.[157] These factors are used as guidelines for courts to determine whether a taking has occurred.[158]

The United States Claims Court has narrowed the scope of inquiry, however, and increased the economic impact of the permit denial.[159] The court has used a three-part test to determine whether the property owner's land has been devalued to the point that a regulatory taking has occurred. First, the court asks whether the alleged taking was physical or regulatory. Second, if the alleged taking was regulatory, what was the relevant parcel for determining the economic impact of the regulation.[160] Third, the court will ask whether the regulatory action actually constituted a taking.[161]

In examining the second part of the test in wetlands cases, the court will often ask whether the relevant parcel was the entire parcel of land involved in the case, or just the wetlands portion. This relevant parcel inquiry is critical because the "test for regulatory taking requires us to compare the value that has been taken from the property with the value that remains in

---

[156]   Loveladies Harbor, Inc. v. United States, 21 Cl. Ct. 153, 155, 20 Envt'l Rep. 21,207 (1990).

[157]   This test was originally established by the landmark case of Pennsylvania Coal Co. v. Mahon, 260 U.S. 393 (1922).

[158]   See Loveladies Harbor, Inc. v. U.S., 21 Cl. Ct. 153, 160 n. 9, 20 Envt'l Rep. 21,207 (1990).

[159]   A disappointed applicant can bring suit against the government claiming that their property has been "taken" in the conservative U.S. Court of Claims in Washington D.C. Moreover, appeals of U.S. Court of Claims decisions go to the equally conservative Court of Appeals for the Federal Circuit (CAFC) in Washington, D.C. Some applicants have found this a more expedient approach to a suit in the geographical district courts, which are often less sympathetic to the predicaments of property owners.

[160]   See Loveladies Harbor, Inc. v. United States, 28 F.3d 1171, 1180 (Fed. Cir. 1994).

[161]   See id. at 1181-82.

the property.[162] In *Loveladies Harbor v. United States*, a developer argued that his property's value on 11.5 acres of wetlands and one acre of uplands was reduced by 99 percent because the Corps refused to grant a CWA Section 404 dredge and fill permit.[163] Interpreting the U.S. Supreme Court's decision in *Lucas v. South Carolina Coastal Council*,[164] the court held that the appropriate parcel for takings analysis was the full 11.5 acres of wetlands and one acre of uplands.[165]

With respect to the third part of the test for regulatory takings, the court will ask whether the regulatory taking may "den[y] all economically beneficial or productive use of land" (known as a "categorical" taking),[166] or have "crossed the line from a noncompensable 'mere diminution' to a compensable 'partial taking,'"[167] In *Florida Rock Industries, Inc. v. United States*,[168] the court found that denial of a Section 404 permit was a "partial taking" since it did not deprive the owner of all economically beneficial use of the land in question. The court again relied on the U.S. Supreme Court in *Lucas v. South Carolina Coastal Council*,[169] and held that the appropriate analysis in the partial taking context is a balance of the competing interests of the applicant's investment-backed expectations for the property versus the government's interest in denying the permit. It should be noted that a mere denial of a landowner's preferred use of the property does not establish a taking.[170]

Under some circumstances, a regulatory taking that is temporary (rather than permanent) will constitute a compensable taking.[171] However, delays attributable to normal and acceptable administrative review procedures do

---

[162] Keystone Bituminous Coal Ass'n v. DeBenedictis, 480 U.S. 470, 497 (1987). See also Penn Central Transp. Co. v. New York City, 438 U.S. 104, 130-31 (1978) ("In deciding whether a particular governmental action has effected a taking, this Court focuses . . . on the nature and extent of the interference with rights in the parcel as a whole.").

[163] Loveladies Harbor, Inc. v. United States, 28 F.3d 1171 (1994).

[164] Lucas v. South Carolina Coastal Council, 505 U.S. 1003 (1992).

[165] Loveladies Harbor, Inc. v. United States, 28 F.3d 1171 (1994). But see Forest Properties, Inc. v. United States, 27 Envtl. L. Rep. (ELI) 20185 (Fed. Cir. 1997) [the relevant parcel for takings analysis is the entire 62 acres, including uplands as well as submerged lands].

[166] Lucas v. South Carolina Coastal Council, 505 U.S. 1003 (1992) at 1015.

[167] Florida Rock Indus. v. United States, 18 F.3d 1560, 1570 (Fed. Cir. 1994).

[168] Id.

[169] Lucas v. South Carolina Coastal Council, 505 U.S. 1003 (1992).

[170] Florida Rock Indus., Inc. v. United States, 791 F.2d 893, 901 (Fed. Cir. 1986) [mere denial of highest and best use does not constitute a taking]; Ciampitti v. United States, 22 Cl. Ct. 310, 1991 U.S. Cl. Ct. LEXIS 21 (1991) [no taking where development prohibition resulted in property worth only 70 percent of its potential value].

[171] See First English Evangelical Lutheran Church v. County of Los Angeles, 482 U.S. 324 (1987).

not constitute a temporary taking.[172] Moreover, the U.S. Supreme Court in *Agins v. City of Tiburon* stated that mere fluctuations in [property] value during the process of governmental decision-making, absent extraordinary delay, are "incidents of ownership. They cannot be considered a 'taking' in the constitutional sense.'"[173] In *Norman v. United States*, development of a landowner's property was delayed for over six years due, in part, to a change in delineation procedures.[174] The landowners argued that a temporary taking of their property had taken place, and both parties moved for summary judgment. The court first asked whether "substantially all economic use of its property was denied during the period in question."[175] The court next asked whether whether the government is responsible for "extraordinary delay" in the regulatory process.[176] Finding sufficient evidence for both inquiries, the court denied the motions to dismiss.[177]

Where the property includes accessible and developable uplands that are beyond the jurisdiction of the Corps and may therefore be developed without a permit, courts have noted that all economically viable use may

---

[172] Smereka v. Haid, 1999 U.S. Dist. LEXIS 13356 (E.D. Mich., August 17, 1999).

[173] Agins v. City of Tiburon, 447 U.S. 255, 263 n.9 (1980) (quoting Danforth v. United States, 308 U.S. 271 (1939)).

[174] Norman v. United States, 38 Fed. Cl. 417, 1997 U.S. Claims LEXIS 168 (1997). The parties were in dispute as to whom, exactly, was the cause of the delay.

[175] Id., citing Anaheim Gardens v. United States, 33 Fed. Cl. 24, 36 (1995). See also Tabb Lakes, Inc. v. United States, 26 Cl. Ct. 1334 (1992) (rejecting temporary takings claim where landowner had "substantial economically beneficial use" of property during period of alleged temporary taking), aff'd, 10 F.3d 796 (Fed. Cir. 1993); 1902 Atlantic Limited v. United States, 26 Cl. Ct. 575 (1992) ("Just as in a permanent taking claim, plaintiff must show that substantially all economic use of its property was denied during the time in question."); and Dufau v. United States, 22 Cl. Ct. 156 (1990) (property owners asserting temporary takings claim "correctly recognize that they must prove substantially all economically viable use of their property has been denied").

[176] See, e.g., Anaheim Gardens v. United States, 33 Fed. Cl. 24, 36 (1995) [temporary taking requires showing of "unreasonable" or "extraordinary" delay]; Tabb Lakes, Inc. v. United States, 26 Cl. Ct. 1334, 1352-54 (1992) [rejecting plaintiff's temporary takings claim where plaintiff was unable to demonstrate "extraordinary delay" by the government), aff'd, 10 F.3d 796 (Fed. Cir. 1993); 1902 Atlantic Limited v. United States, 26 Cl. Ct. 575 (1992) ["The only way plaintiff can recover [for a temporary taking] is to prove that this period constituted extraordinary delay"]; Dufau v. United States, 22 Cl. Ct. 156 (1990) ["The Supreme Court has suggested that 'extraordinary delays' in the process of government decision-making may give rise to a temporary taking claim"].

[177] Norman v. United States, 38 Fed. Cl. 417, 1997 U.S. Claims LEXIS 168 (1997).

not have been eliminated by the regulation.[178] However, where uplands are completely surrounded by wetlands, and thus permit denial effectively precludes use of the uplands, the existence of such uplands may support a takings claim. In fact, such uplands surrounded by wetlands may be considered taken as a result of the wetland permit denial.[179]

In *Ciampitti v. United States*, the Claims Court ruled that the denial of a CWA Section 404 permit did not constitute a compensable taking where plaintiffs purchased their property in 1983 aware of state and federal restrictions on development in wetlands.[180] The Claims Court reasoned that because plaintiffs had ample warning of the likelihood that the wetlands could not be developed, the permit denials did not interfere with reasonable, distinct investment-backed expectations. In *Good v. United States*, a developer purchased a 54-acre lot that included over 10 acres of wetlands. The Federal Circuit held that the developer had actual knowledge that state and/or federal regulations could prevent him from developing the parcel, and that he could not have had a reasonable expectation that he would obtain approval to fill ten acres of wetlands in order to develop the land. [181]

In the context of wetland regulation, the courts often take an additional factor into account in ruling on whether a particular land use regulation constitutes a regulatory taking. Many courts will employ a harm-prevention analysis and uphold a regulation even when all economically viable use of the landowner's property has been taken so long as the regulation is aimed at preventing a serious public harm.[182] Although this harm prevention analysis is grounded in longstanding precedent that economic loss to a private property owner is irrelevant

---

[178] See Deltona Corp. v. United States, 657 F.2d 1184 (Cl. Ct. 1981), cert. denied, 455 U.S. 1017 (1982).

[179] Loveladies Harbor, 21 Cl. Ct. 153, 31 Env't Rep. (BNA) 1847, 1853 (1990).

[180] 22 Cl.Ct. 310, 32 Env't Rep. Cas.(BNA) 1608, 21 Envt'l L. Rep. (ELI) 20,866 (Cl. Ct. 1991).

[181] Good v. United States, 189 F.3d 1355 (Fed. Cir., 1999).

[182] See, for example, McNulty v. Town of Indialantic, 727 F. Supp. 604 (M.D. Fla. 1989), rejecting a landowner's contention that a setback requirement on oceanfront property worked a taking because "the government can destroy all economic use if necessary to avoid a public nuisance or nuisance-like use." 727 F. Supp. at 609; Presbytery v. King County, 114 Wash.2d 320, 787 P.2d 907 (1990), cert. denied, 111 S.Ct 284 (1990), holding that if a regulation protects the public from harm, and does not infringe on the landowner's right to possess, exclude others and dispose of his property, no taking has occurred; Claridge v. New Hamphire Wetlands Board, 125 N.H. 745, 485 A.2d 287 (1984), where the court upheld the denial of permit to fill tidal wetlands, thereby rendering the property of negligible economic value, based on a harm prevention rationale that the regulation prevented destruction the coastal habitat that would posing a further risk to the public welfare.

when government regulation furthers the legitimate purpose of preventing harm to the public,[183] it is questionable how valid this line of cases may be in light of a 1992 U.S. Supreme Court decision in *Lucas v. South Carolina Coastal Council*, where the court held that if a government regulation deprives a landowner of all beneficial and productive use of his property, just compensation is due even if the regulation is aimed at preventing a serious public harm.[184] The only exception to this categorical rule is when in the context of a total regulatory taking, the government can show that the landowner's expectations regarding use of the property are unreasonable in light of state common law nuisance and property law principles. Courts have generally found that dredging or filling a wetland does not constitute a nuisance.

13.5 Challenging Issuance of a CWA Section 404 Permit

It is not uncommon for a person other than the applicant or recipient of a CWA Section 404 permit to wish to challenge the issuance of the permit. Many such challenges are initiated by environmental organizations or other groups of concerned citizens, but challenges may also come from local or state governments, or from federal, state, or local agencies. Because there is no direct appeal mechanism to the Corps (see Section 13.5.1), the person wishing to challenge issuance of the permit must file a lawsuit in federal district court, although there are some shortcuts available under CWA Section 505, the "citizen suit" provision.[185]

A party wishing to challenge issuance of a CWA Section 404 permit must follow carefully the procedures for initiating and pursuing a lawsuit in federal court. The following sections discuss some of the major issues involved in such challenges, but it should be cautioned that the process in more complex than is represented here, and the prudent litigant will seek professional legal advice before proceeding.

Litigants in a wetlands lawsuit are subject to the same constraints as in other cases. For example, courts will dismiss a lawsuit which has not "ripened" into a case or controversy suitable for adjudication. In the context of CWA Section 404 permit challenges, this usually means that there has been final agency action on the permit. Most lawsuits on the ripeness issue have been brought by disappointed landowners who challenge negative permit decisions, but it is clear that courts will not hear a case on wetland determination issues until the Corps and EPA have

---

[183] Mugler v. Kansas, 123 U.S. 623, 668-69 (1887).
[184] Lucas v. South Carolina Coastal Council, 505 U.S. 1003 (1992).
[185] 33 U.S.C. § 1365.

made final decisions on the wetlands,[186] or on permit issuance issues until a permit has actually been issued or denied.[187] An EPA decision to veto a permit under CWA Section 404(c) does not become ripe until the veto is finalized.[188]

For most kinds of lawsuits, there exist various legal limitations on the time in which the lawsuit must be brought, known as "statutes of limitations." Interestingly, there are no specific statutes of limitations in the Clean Water Act or elsewhere that serve to limit the time in which an action against the Corps, EPA or other agencies may be brought in a wetland determination or permit case. In *North Carolina Wildlife Federation v. Woodbury*,[189] the court held that the general five-year limit for civil penalty actions applied to a citizens suit challenge.[190] In *United States v. Banks*,[191] the court held that time limits apply to civil penalties, but the government's equitable claims (injunctive relief) against Banks were not barred.

Right to Jury Trial. In a significant wetlands decision, the United States Supreme Court in *Tull v. United States* held that there is a right to a jury trial when the government seeks civil penalties, but *not* when it seeks an injunction or other equitable relief.[192] Even so, the jury decides only the substantive issue of whether the property owner violated the provisions of the Clean Water Act, while the judge decides the amount of the penalty.[193]

Attorney's Fees. An important aspect of wetlands litigation is the circumstances under which a prevailing party may be awarded attorney's fees and expert witness' fees. Under CWA Section 505(d), such fees are awarded at the court's discretion to a "prevailing or substantially prevailing party."[194] In *National Wildlife Federation v. Hanson*, the court

---

[186] See Avella v. U.S. Army Corps of Engineers, 20 Envtl. L. Rep. (Envtl. L. Inst.) 20920 (S.D. Fla. 1990), aff'd per curiam, 916 F.2d 721 (11th Cir. 1991).

[187] See Norman v. United States, 38 Fed. Cl. 417, 1997 U.S. Claims LEXIS 168 (August 12, 1997); and Route 26 Land Dev. Ass'n v. United States, 753 F. Supp. 532 (D.Del. 1990).

[188] Newport Galleria Group v. Deland (618 F. Supp. 1179 (D.D.C. 1985), the first Sweedens Swamp case.

[189] North Carolina Wildlife Federation v. Woodbury, 19 Envtl. L. Rep. (ELI) 21308 (E.D.N.C. 1989).

[190] 28 U.S.C. § 2462.

[191] United States v. Banks, 115 F.2d 916 (11th Cir. 1997). See also United States v. Telluride Co., 146 F.3d 1241 (10th Cir. 1998).

[192] Tull v. United States, 481 U.S. 412 (1987).

[193] See Want, *Law of Wetlands Regulation* (West, 1998) See also United States v. M.C.C. of Florida, 863 F.2d 802 (11th Cir. 1989) [jury trial is not required under § 10 of the Rivers and Harbors Act].

[194] Water Quality Act of 1987 § 505(c), amending 33 U.S.C. § 1365(d).

upheld an award of over $398,000 despite the government's claim that wetland determinations are discretionary actions, and only non-discretionary actions are subject to citizen suits.[195]

It is not uncommon, however, for a defendant to reach a settlement with the government after commencement of a CWA Section 505 citizen suit. The question then arises whether or not the plaintiff is considered a "prevailing party" entitled to attorneys' fees and expert witness' fees. The Second Circuit concluded that is was proper for a district court to award such costs to a plaintiff if it could be inferred that it was the citizen suit that motivated the defendant to reach the settlement.[196] The Eighth Circuit allowed such costs to a plaintiff where the defendant entered into a consent decree with the government that was less strict than the penalties requested by the plaintiff.[197] However, a federal district court held that environmental groups were not prevailing parties entitled to such costs where the defendant entered into a consent decree after the environmental groups sent a 60-day notice of intent to sue under CWA Section 505, but before the lawsuit actually commenced.[198]

13.6 Wetland Mitigation

A requirement for many wetlands development projects is some form of "mitigation" to restore or replace lost wetland values. Generally speaking, mitigation is the attempted replacement of the functions and values of wetlands proposed for filling through creation of new wetlands or enhancement of existing wetlands; that is, "compensating" for lost functions.[199] The logic behind this requirement is that some forms of damage to wetlands may be unavoidable, even though a particular project is environmentally sound. If mitigation can create a situation where the

---

[195] 18 Envtl. L. Rep. (Envtl. L. Inst.) 20008 (E.D.N.C. 1987, *aff'd*, 859 F.2d 313 (4th Cir. 1988). But see Golden Gate Audubon Society v. U.S. Army Corps of Engineers (738 F. Supp. 339 (N.D. Cal. 1988) [court refused to grant attorney's fees because wetland jurisdictional determination is a discretionary matter].

[196] Atlantic States Legal Found. v. Eastman Kodak Co., 933 F.2d 124 (2d Cir. 1991).

[197] Armstrong v. ASARCO, Inc., 138 F.3d 382 (8th Cir. 1998).

[198] United States v. Maine Dept. of Transp., 980 F. Supp. 546 (D. Me. 1997).

[199] See Kusler and Opheim, *Our National Wetland Heritage: A Protection Guide,* 2nd ed. (Environmental Law Institute, 1996). See also Kruczynski, Mitigation and the Section 404 Program: a Perspective in Kusler and Kentula (eds.). *Wetland Creation and Restoration: the Status of the Science* (Island Press, 1990). The U.S. Army Corps of Engineers and the USEPA specifically refer to "compensatory mitigation" as restoration of existing degraded wetlands, or creation of man-made wetlands.

overall effect is a net *gain* in wetland values (or, at least, no net loss), then the project should be allowed.

For many people mitigation represents a valuable compromise between a desire to protect wetland resources, while at the same time allowing property development for a variety of human uses. Developers are often required to readjust lot lines, redirect stormwater or other runoff, or completely relocate a development project in order to protect important wetland values. Many developers have argued that there is often a net *improvement* of the affected wetlands as a result.[200] Nevertheless, the concept of mitigation as applied to wetlands has been controversial since its inception.[201]

13.6.1 The Mitigation Regulatory Framework

The requirement for mitigation as compensation for damages sustained by wetlands is an important part of the federal government's "no net loss" policy.[202] However, mitigation is not specifically required under CWA Section 404. The mitigation requirement is found in other federal statutes, notably the National Environmental Policy Act (NEPA)[203] and the Fish and Wildlife Coordination Act,[204] which require mitigation for *all* federal actions which adversely affect the environment. The issuance of a CWA Section 404 dredge and fill permit is the kind of federal agency action which triggers NEPA and its mitigation requirement.

Current mitigation policies administered by the Corps under its CWA Section 404 regulations state that:

Mitigation is an important aspect of the review and balancing process . . .

Consideration of mitigation will occur throughout the permit application review process and includes avoiding, minimizing, rectifying, reducing, or compensating for resource losses. Losses will

---

[200] Salvesen, *Wetlands. Mitigating and Regulating Development Impacts* (Urban Land Institute, 1990).

[201] See Kusler, The mitigation banking debate, 14(1) Nat'l Wetlands Newsletter 4 (1992); and Redmond, How successful is mitigation? 14(1) Nat'l Wetlands Newsletter 5 (1992).

[202] See U.S. Fish and Wildlife Service, *Wetlands: Meeting the President's Challenge (Wetlands Action Plan)* (U.S. Fish & Wildl. Serv., 1990); and White House Office of Environmental Policy, *Protecting America's Wetlands: A Fair, Flexible, and Effective Approach* (Aug. 24, 1993).

[203] 42 U.S.C. § 4321 et seq.

[204] 16 U.S.C. § 661.

be avoided to the extent practicable. Compensation may occur on-site or at an off-site location.[205]

Despite the relatively straightforward language of the regulations, the Corps is currently modifying its methods of applying its wetland mitigation policies.[206] Many specific applications of Corps policy are regulated by the provisions of the 1990 Memorandum of Agreement ("MOA") between the Corps and EPA.

Current Corps policy permits mitigation to occur on-site (i.e., restoring those wetlands located on the project site which are degraded by the project or by previous actions), or off-site (either the purchase of wetlands at another site, or creation of new wetlands at a site where none exist), although there is a strong preference for on-site mitigation. The "on-site vs. off-site" question has received considerable attention, particularly from environmentalists, scientists, and some federal agencies who argue that off-site mitigation allows "the sacrifice of valuable natural wetlands for the ill-considered promise of some future, potentially less desirable, wetland replication."[207] Opponents of mitigation plans have argued successfully that neither a rock quarry[208] nor a marsh fed by urban runoff[209] was sufficient mitigation for destruction of valuable wetlands.[210]

13.6.2 Kinds of Mitigation Measures

Corps' regulations place mitigation measures into three categories.[211] First, minor project modifications (those considered feasible by the

---

[205] 33 CFR 320.4(r)(1).

[206] See 60 *Fed. Reg.* 58, 605 (Nov. 28, 1995) (wetland mitigation banking guidance).

[207] U.S. Environmental Protection Agency, *Wetlands: Region 4 Implementation and Management of the Section 404 Wetlands Program* (Rept. of Audit E1h7F8-04-0331-0100208, U.S.E.P. A., Atlanta, GA., 1990). See also Berry, Wetlands: a Way out of the Morass? 1992 Environ. & Dev. 1-3 (Feb. 1992).

[208] Bersani v. U.S. Envtl. Protection Agency 674 F. Supp. 405 (N.D.N.Y. 1987), aff'd sub nom. Robichaud v. U.S. Envtl. Protection Agency, 850 F.2d 36 (2d Cir. 1988), cert. denied, 109 S.Ct. 1556 (1989) [the famous "Sweeden's Swamp" case].

[209] National Audubon Soc'y v. Hartz Mountain Development Corp., 14 Envtl. L. Rep. (ELI) 20,724 (D.N.J. 1983).

[210] See Houck, Hard Choices: The Analysis of Alternatives Under Section 404 of the Clean Water Act and Similar Environmental Laws. 60 U. Colo. L. Rev. 773 (1989).

[211] 33 CFR § 320.4(r). See also Dennison, *Wetland Mitigation: Mitigation Banking and Other Strategies for Development and Compliance* (Government Institutes, 1997) [discussing mitigation options].

applicant) can include reductions in scope and size; changes in construction methods, materials, or timing; and operation and maintenance practices that reflect a sensitivity to environmental quality. For example, erosion control features could be required on a fill project to reduce sedimentation impacts.[212] Second, additional mitigation measures may be required to satisfy legal requirements under CWA Section 404(b)(1), which provides for the EPA review process of potential environmental impacts under Section 404 dredge and fill permits.[213] EPA's regulatory guidelines under CWA Section 404(b)(1) contain a list of some such measures.[214] Third, still other mitigation measures may be required as a result of the Corps' public interest review process if found to be reasonable and justified by the District Engineer, but only if the mitigation measures are required to ensure that "the project is not contrary to the public interest."[215]

The third requirement mandates that the Corps need not require mitigation measures unless it can be shown that failure to require the measures would be "contrary to the public interest."

The Corps' Section 404 regulations are reasonably explicit with respect to the kinds of wetlands losses for which mitigation is required:

> All compensatory mitigation will be for significant resource losses which are specifically identifiable, reasonably likely to occur, and of importance to the human or aquatic environment. Also, all mitigation will be directly related to the impacts of the proposal, appropriate to the scope and degree of those impacts, and reasonably enforceable.[216]

In other words, wetland degradation judged by the District Engineer to be minor, insignificant, unimportant, or speculative does *not* trigger the requirement for compensatory mitigation. Likewise, the Corps cannot require mitigation that is unrelated to the impacts that will result from the project itself (i.e., a developer *cannot* be forced to provide mitigation for a project unrelated to the one for which the CWA Section 404 permit is granted).[217]

---

[212] 33 CFR § 320.4(r)(i).
[213] 33 CFR § 320.4(r)(1)(ii).
[214] 40 CFR §§ 230.70-230.77.
[215] 33 CFR § 320(r)(1)(iii).
[216] 33 CFR 320(r)(2).
[217] See Berry and Dennison, Wetland Mitigation. Chap. 8 in Dennison and Berry, *Wetlands: Guide to Science, Law, and Technology* (Noyes 1993).

13.6.3 Corps/EPA Section 404(b)(1) Mitigation Guidelines and Joint MOA

In reviewing an application for dredge and fill activities in wetlands areas, the Corps is required to consider CWA Section 404(b)(1) Guidelines issued by EPA.[218] The guidelines protect wetlands by prohibiting discharges that have significant adverse effects on human health or welfare, recreation, aesthetics, economics, aquatic ecosystems, and wildlife dependent on aquatic ecosystems.[219] The CWA Section 404(b)(1) guidelines require applicants to take all appropriate and practicable steps to minimize the adverse impacts of proposed filling activities. The regulations do not define what will be considered appropriate and practicable. In the case of non-water-dependent projects, "no discharge of dredged or fill material shall be permitted if there is a practicable alternative to the proposed discharge which would have less adverse impact on the aquatic ecosystem ."[220]

Several controversial wetlands cases have considered the meaning of this "practicable alternatives" language. The leading case concerning the practicable alternatives provision is *Bersani v. EPA*, commonly referred to as the "Sweedens Swamp" decision.[221] In the *Bersani* case, the Second Circuit upheld the EPA's CWA Section 404(c) veto of a permit to fill wetlands to construct a shopping mall because "available" and "feasible" alternative sites were available to the developer, as specified in Corps regulations.[222] In a subsequent case, a district court held that the Corps correctly analyzed the practicable alternatives in permitting the filling of 7.4 acres of wetlands for construction of a baseball stadium complex.[223]

Because various conflicts arose between EPA and Corps over interpretation of the CWA Section 404(b)(1) guidelines, including the "practicable alternatives," the two agencies issued a joint Memorandum of Agreement (MOA) concerning the determination of mitigation measures under Section 404(b)(1). Under the joint MOA, before a permit may be issued, the applicant must first attempt to avoid wetlands impacts,

---

[218] Section 404(b)(1), 33 U.S.C. § 1344(b) authorizes the Secretary of the Army to specify disposal sites "through the application of guidelines developed by the Administrator" of the EPA "in conjunction with" the Corps. The guidelines are found at 40 CFR Part 230.

[219] 40 CFR § 230.10(c).

[220] 40 CFR § 230.10(a).

[221] Bersani v. EPA, 674 F. Supp. 405 (N.D.N.Y. 1987), aff'd, 850 F.2d 36 (2nd Cir. 1988), cert. denied, 489 U.S. 1089 (1989).

[222] 40 CFR § 230.10(a)(2).

[223] Sierra Club v. U.S. Army Corps of Engineers, 935 F. Supp. 1556 (S.D. Ala. 1996).

then minimize impacts, and finally as a last resort, compensate for unavoidable impacts.[224]

The EPA/Corps Joint MOA adopts the goal of no overall net loss of wetland values and functions. By emphasizing wetland "values and functions," the MOA authorizes a less than one to one replacement of wetland acreage where the wetlands lost are significantly degraded.

Although the MOA recognizes that the "no net loss" goal may not be achieved with respect to every permit application, it does not address whether such losses may be "made up" on other permit applications, thus meeting the "no overall net loss" requirement. The MOA sets forth a strict sequence of regulatory considerations: avoidance, minimization, and compensatory mitigation. Wetland impacts must be avoided to the maximum extent practicable. Cost, existing technology, and logistics in light of overall project purposes may be considered in determining what is "practicable." If unavoidable impacts remain, they must be minimized to the extent appropriate and practicable. Only as a last resort, where unavoidable adverse impacts remain after minimization procedures, may the Corps consider compensatory mitigation proposals. Thus, the Corps may not consider a compensatory mitigation proposal if it determines that adverse impacts may be avoided.

The MOA states a preference for in-kind, on-site compensatory mitigation over out-of-kind, off-site mitigation. Where off-site measures are necessary, they should be in close physical proximity to the affected site and preferably in the same watershed. Finally, restoration of degraded wetlands is preferred over creation of new wetlands, because of scientific uncertainty with respect to the success of wetland creation. The MOA requires at least a "one for one functional replacement (i.e., no net loss of values), with an adequate margin of safety to reflect the expected degree of success associated with the mitigation plan."[225]

13.6.4 Forms of Mitigation

A wide range of mitigation measures have been allowed by the Corps, such as: (1) increased public access to the area; (2) acquisition of other wetlands to provide enhanced protection, or acquisition with a management commitment; (3) restoration or creation of wetlands, either

---

[224] Section 404(b)(1) Guidelines Mitigation MOA, 55 *Fed. Reg.* 9210 (Feb. 7, 1990).

[225] For discussions of the legal details of mitigation, see Berry and Dennison, Wetland Mitigation. Chap. 8 in Dennison and Berry, *Wetlands: Guide to Science, Law, and Technology* (Noyes 1993); and Want, *Law of Wetlands Regulation* (West, 1998).

as general compensation or as replacement for a specific habitat type; (4) indemnification or direct monetary payment for lost wetland values; and (5) mitigation banking (compensatory off-site wetlands restoration or creation).[226] Most federal agencies and most states no longer permit approaches (1) or (2) unless the goal of increased public access is compensation for lost public recreational opportunities, or the acquisition includes enhancement or assurance of proper management to compensate for lost wetland values.[227]

Alternatives for compensatory mitigation fall generally into five categories, listed in relative order of their invasiveness:[228]

1. *Preservation.* Purchase of a parcel of land containing a valuable wetland, which is then placed in public ownership with provisions for long-term protection and/or management

2. *Exchange.* The exchange for a wetland area which will be damaged by a project for another wetland (typically of larger size and higher wetland values), which is placed in long-term protection as in (1).

3. *Enhancement.* A wetland in which some functions have been degraded (or lost) is "repaired," such that the degraded (or lost) wetland functions are again available.

4. *Restoration.* A former wetland with few (or no) remaining wetland functions is restored to a form in which specific functions (perhaps all) are available.

5. *Creation.* A wetland is created where none previously existed. The goal is usually to create specific wetland functions.

---

[226] See Dennison, *Wetland Mitigation: Mitigation Banking and Other Strategies for Development and Compliance* (Government Institutes, 1997); and Dial and Deis, *Mitigation Options for Fish and Wildlife Resources Affected by Port and Other Water-Dependent Developments in Tampa Bay, Florida* (Biol. Rept. 86(6), U.S. Fish & Wildl. Serv., 1986).

[227] See Berry and Dennison, Wetland Mitigation. Chap. 8 in Dennison and Berry, *Wetlands: Guide to Science, Law, and Technology* (Noyes 1993); and Want, *Law of Wetlands Regulation* (West, 1998).

[228] See Kruczynski, Options to Be Considered in Preparation and Evaluation of Mitigation Plans in Kusler and Kentula (eds.), *Wetland Creation and Restoration: the Status of the Science* (Island Press, 1990). See also Schmid, Wetland Mitigation Case Studies. Chap. 10 in Dennison, *Wetland Mitigation: Mitigation Banking and Other Strategies for Development and Compliance* (Government Institutes, 1997).

13.6.5 Wetland Mitigation Banking

Mitigation "banking" is a concept developed originally in the early 1980s by the U.S. Fish and Wildlife Service (USFWS) in an attempt to increase the effectiveness of wetlands mitigation while reducing the costs to the regulated community.[229] The practice remains controversial, but it is increasing in popularity and has been described as "the most promising solution to the loss of wetlands during development."[230]

In its simplest form, wetland banking consists of either a deteriorated (or deteriorating) wetland that is enhanced or restored, or a completely new wetland is created where none existed before.[231] The restored or created wetlands are owned by private or public entities (including federal, state, and local agencies) who receive wetland "credits" which can be applied at a later time for "debits" resulting from unavoidable impacts to natural wetlands. This approach has the appeal of apparently conforming with the federal governments "no net loss of wetlands" policies.

In 1995, the Army Corps of Engineers, the EPA, the Natural Resources Conservation Service, the USFWS, and the National Marine Fisheries Service issued joint guidance on the establishment, use, and generation of mitigation banks for the purpose of providing compensatory mitigation of adverse impacts to wetlands and other aquatic resources.[232] The purpose of the guidance is to clarify the manner in which mitigation banks might be used to satisfy the mitigation requirements of Section 404 of the CWA, and the wetland conservation provisions of the Food Security Act.[233]

---

[229] See Fish and Wildlife Service Mitigation Policy, 46 *Fed. Reg.* 7644 (Jan. 23, 1981).

[230] Sokolove and Huang, Privitization of Wetland Mitigation Banking, 7(1) Nat'l Resources & Environment 36 (1992).

[231] See Dennison, Wetland Mitigation Banking, Chap. 7 *in Wetland Mitigation: Mitigation Banking and Other Strategies for Development and Compliance* (Government Institutes, 1997); and Brooks, Restoration and Creation of Wetlands, Chap. 10 in Dennison and Berry, *Wetlands: Guide to Science, Law, and Technology* (Noyes, 1993).

[232] 60 *Fed. Reg.* 58,605 (Nov. 28, 1995).

[233] See Dennison, Wetland Mitigation Banking Guidance, Chap. 8 in *Wetland Mitigation: Mitigation Banking and Other Strategies for Development and Compliance* (Government Institutes, 1997).

13.7 Planning and Wetland Protection

13.7.1 Special Area Management Plans (SAMPS)

The Special Area Management Plan (SAMP) process is a comprehensive plan providing for natural resource protection and reasonable economic growth, which contains detailed statement of policies and criteria to guide land and water uses in specific geographic areas. The Coastal Zone Management Act defines "special area master plan" as:

> [A] comprehensive plan providing for natural resource protection and reasonable coastal-dependent economic growth containing a detailed and comprehensive statement of policies; standards and criteria to guide public and private land uses of lands and waters; and mechanisms for timely implementaion in specific geographic areas within the coastal zone.[234]

A SAMP provides predictability to developmental interests by establishing an area-wide basis for regulatory actions founded on cumulative effects of changes in the environment. The SAMP requires extensive study an planning by federal, state and local environmental and land use planning authorities. The nature of the geographic area targeted for SAMP development will determine the degree of involvement of various government and private interests in the process.

Because development of a SAMP would most likely be considered a major federal action under the National Environmental Policy Act, a SAMP would usuallly be developed in conjunction with an Environmental Impact Statement (EIS; see Chapter 3). The function of the EIS for the SAMP is to develop management plan alternatives, assess potential environmental, social, and economic consequences of each alternative, and identify the preferred alternative. A benefit of the EIS process is that it provides a forum for the informed identification and evaluation of management plan alternatives, while allowing opportunity for interested individuals and groups to participate in the development of the SAMP.

A SAMP can be especially useful as a wetlands mitigation plan for an area. The Corps/EPA joint MOA on mitigation determinations provides that mitigation consistent with an EPA and Corps approved comprehensive plan, such as a SAMP, would satisfy the avoidance, minimization, and compensatory mitigation requirements.[235]

---

[234] 16 U.S.C. § 1453(17).
[235] Section 404(b)(1) Guidelines Mitigation MOA, 55 *Fed. Reg.* 9210 (Feb. 7, 1990). The Army Corps Regulatory Guidance Letter No. 86-10 discusses the development of SAMPs.

13.7.2 Advance Identification of Wetlands (ADIDs)

A promising mechanism available for governmental protection of wetlands is the "Advance Identification of Wetlands" (ADID or AVID) process which is authorized by the EPA's CWA Section 404(b)(1) regulations. Under the regulations, EPA and the Corps, on their own initiative or at the request of any other party (and after consultation with any affected state) "may identify sites which will be considered as: (1) possible future disposal sites, including existing disposal sites and non-sensitive areas; or (2) areas generally unsuitable for disposal site specification."[236]

The designation of an area as acceptable for CWA Section 404 dredge and fill disposal activities does not, however, constitute a Section 404 permit, nor does identification of an area as not available necessary preclude a Section 404 permit in the future.[237] On the other hand, it operates as a form of notice to property owners as well as local, state, and federal agencies of the likely acceptability of an individual or general Section 404 permit in an affected area, and may facilitate permit approval. As such, the ADID is a valuable and powerful planning tool that is increasing in popularity across the nation.

As a general matter, the purpose of an ADID is to help protect and manage the nation's remaining wetlands by determining which wetlands are of high ecological value and should be protected from dredge and fill activities. By making these determinations *before* an application for a permit takes place, the intent is to prevent inadvertent (or intentional) unpermitted wetland losses.

The ADID process begins when the EPA initiates ADID procedures. Under CWA Section 404(b)(1) guidelines, EPA (and the Corps) must consider the likelihood that future dredge and fill activities within the ADID boundaries will result in compliance with the guidelines. To facilitate this analysis, EPA and the Corps review available water resources management data, including data from the public, other federal and state agencies, and information from approved coastal zone management plans and river basin plans.[238]

As stated by the regulations, however, the process can also be initiated by "any person."[239] Although initiation by a person other than the EPA can take place in a variety of ways, it is common for the person to apply

---

[236] 40 CFR § 230.80(a)). See U.S.E.P.A., Office of Wetlands, Oceans, and Watersheds, *Summary of ADID Projects under Section 230.80 of the 404(b)(1) Guidelines* (June 10, 1991).
[237] 40 CFR § 230.80(b).
[238] 40 CFR § 230.80(d).
[239] 40 CFR § 230.80(a).

directly to the nearest EPA Wetlands Division office. There is no standard form which the application must follow, but it is prudent for the following information to be included: (1) the purpose and scope of the proposed ADID; (2) the proposed ADID study area; (3) objectives of the ADID (including effects on ecological resources, public interest, fish and wildlife values, recreation, storm and flooding protection, health and welfare, and opportunities for interagency cooperation); (4) other potential uses of ADID study data and findings to other laws and regulations; and (5) identification of existing data resources.

Once the ADID process commences, an appropriate public notice is issued.[240] The Corps (or other permitting authority) maintains a public record of the identified areas, and a written statement of the basis for identification.[241]

13.8 State Wetland Regulation

Landowners and developers who wish to carry out activities in wetland areas must find their way through a maze of federal, state and local laws and regulations before finally securing the necessary approvals for a specific activity or project. Federal wetland approvals can not be obtained unless necessary state wetland permit approvals have been secured.[242]

A certification of water quality may be necessary under CWA Section 401, as well as state law.[243] Section 303(a) of the CWA requires states to adopt water quality standards to protect uses of waters within the state.[244] In addition, CWA Section 404 permits require as a prerequisite that there be certification from the state that the discharge will comply with state water quality standards.[245] If a state fails to respond within a "reasonable time" to a request for certification from the Corps, then the state certification requirement is waived.[246] If a state fails to assert its right to reject water quality certification during the Corps' permitting process, it does not waive that right.[247] Federal courts cannot review a state's

---

[240] 40 CFR § 230.80(c).

[241] 40 CFR § 230.80(e).

[242] 33 CFR § 320.4(j).

[243] 33 U.S.C. § 1341. See discussion in section 4.4.

[244] 33 U.S.C § 1313(a). See U.S.E.P.A., *Wetlands and 401 Certification* (April, 1989).

[245] 33 U.S.C. § 1341. See also 33 CFR § 320.4(d).

[246] Id. A "reasonable time" is 60 days under Corps regulations, although the time can be extended up to one year. 33 CFR § 325.2(b)(I)(ii).

[247] North Carolina v. Federal Energy Regulatory Comm'n, 112 F.3d 1175 (D.C. Cir. 1997).

certification decision on substantive issues,[248] although a federal court can review whether a state's attempted revocation of its certification was consistent with the CWA.[249]

For some projects, an environmental impact assessment may be required pursuant to a state's environmental quality review act.[250] Local zoning regulations, such as wetland ordinances, may impose additional requirements. In addition to all these laws, the landowner or developer may need to certify that coastal wetland activities will be carried out in a manner consistent with a state's coastal management program.[251]

13.8.1 State Assumption of Section 404 Program

Section 404(g)(1) of the CWA authorizes states to apply to the EPA for approval to administer their own individual and general CWA Section 404 permits for the discharge of dredged and fill material into jurisdictional waters shoreward of the ordinary high water mark and mean high water mark within the state.[252] Under Corps regulations, the state's application must include an identification of the state authority which authorizes funding for the program; assure record keeping, inspection, monitoring, and enforcement; and avoid interference with Corps' function.[253] To date, only Michigan and New Jersey have assumed permitting authority under CWA Section 404(g)(1).[254] The EPA is authorized to withdraw its approval to a state if the state does not comply with the statutory and regulatory requirements.

Once a state has assumed the CWA Section 404 program, then the appropriate state agency must transmit to EPA a copy of the public notice for every Section 404 permit application, along with any action taken by the state on the application.[255] The EPA can submit written objections to an application, at which time the state agency must request a hearing or resubmit a revised application within 90 days.[256] If the state agency does

---

[248] Roosevelt Campobello Int'l Park Comm'n v. U.S. EPA, 684 F.2d 1041 (1st Cir. 1982) [judicial review is properly in the state courts].
[249] Keating v. Federal Energy Regulatory Comm'n, 927 F.2d 616 (D.C. Cir. 1991).
[250] See chapter 3.
[251] See chapter 14.
[252] 33 U.S.C. § 1344(g)(1).
[253] 40 CFR § 233.
[254] See Want, *Law of Wetland Regulation* (West, 1997) at § 3.04.
[255] 33 U.S.C. § 1344(j).
[256] Id.

not satisfy the EPA's objections, then the authority to issue the permit shifts from the state agency to the Army Corps.[257] However, in *Friends of the Crystal River v. EPA*, the Sixth Circuit held that the EPA could not transfer permitting authority back to the state once that authority had been transferred to the Corps.[258]

13.8.2 New York State Tidal Wetland Regulation

Under the CWA Section 404 wetland program, individual states may adopt and administer their own wetland protection programs once approved by the Army Corps.[259] New York has separate regulatory programs for tidal and freshwater wetlands.[260] New York State statutory authority regarding tidal wetlands is found in Article 25 of the New York Environmental Conservation Law.[261] The New York Department of Environmental Conservation is responsible of administering the statute. The stated policy of Article 25 is "to preserve and protect tidal wetlands, and to prevent their despoliation and destruction, giving due consideration to the reasonable economic and social development of this state."[262] "Tidal Wetlands" are defined as "those areas which border on or lie beneath tidal waters, such as, but not limited to, banks, bogs, salt marsh, swamps, meadows, flats or other low lands subject to tidal action, including those areas now or formerly connected to tidal waters."[263]

Under the statute, the commissioner of environmental conservation is required to inventory all tidal wetlands in New York state, setting forth boundaries using photographic and cartographic techniques to clearly and accurately map the state's tidal wetlands.[264] The commissioner is given

---

[257] Id.

[258] Friends of the Crystal River v. EPA, 35 F.3d 1073 (6th Cir. 1994).

[259] 33 U.S.C. §§ 1344(g) and (h).

[260] N.Y. Envtl. Conserv. Law art. 24 (freshwater wetlands); N.Y. Envtl. Conserv. Law art. 25 (tidal wetlands).

[261] N.Y. Envtl. Conserv. Law art. 25 (McKinney).

[262] N.Y. Envtl. Conserv. Law § 25-0102).

[263] N.Y. Envtl. Conserv. Law § 25-0103(1)(a). The definition also contains a listing of vegetation indicative of tidal wetlands. N.Y. Envtl. Conserv. Law § 25-0103(1)(b). See also O'Brien v. Barnes Building Co., Inc., 85 Misc.2d 424, 380 N.Y.S.2d 405, aff'd mem. sub. nom. O'Brien v. Biggane, 372 N.Y.S.2d 992 (App. Div. 1975).

[264] N.Y. Envtl. Conserv. Law § 25-0201(1).

broad discretion in amending the wetlands inventory map.[265] The commissioner must file a detailed description of the technical criteria used to delineate the tidal wetlands with the secretary of state.[266]

Under the tidal wetlands program the Department of Environmental Conservation works with local governments to protect tidal wetlands found in the locality.[267] In addition to any permits required by the municipality in which the tidal wetland is located, the statute requires application to the commissioner for additional permits before any form of draining, dredging, excavation, dumping, filling, or erection of any structure can be undertaken.[268] The statute requires that any land use activity in tidal wetlands be compatible with land use regulations promulgated pursuant to the statute.[269]

13.8.3 New Jersey State Coastal Wetlands Regulation

The New Jersey coastal wetlands statute has provisions not much different from the New York tidal wetlands scheme. The New Jersey coastal wetlands law is part of Title 13, "Conservation and Development," codified as Chapter 9A, "Coastal Wetlands."[270] Under the statute, the state legislature declares tidal wetlands one of the most vital and protective areas of that state, and "that in order to promote the public safety, health and welfare, and to protect public and private property, wildlife, marine fisheries and the natural environment, it is necessary to perserve the ecological balance of this area and to prevent its further deterioration."[271] Like the New York statute, the Commissioner of Environmental Protection is required to inventory all tidal wetlands in the state and map them with boundaries showing the areas that are at or below high water.[272] "Coastal wetlands" are defined as "any bank, marsh, swamp, meadow, flat or other low land subject to tidal action in the State of New Jersey . . . , including

---

[265] See Thompson v. Department of Environmental Conservation, 130 Misc.2d 123, 495 N.Y.S.2d 107 (Sup. Ct. 1985), aff'd, 132 A.D.2d 665, 518 N.Y.S.2d 36 (App. Div. 1987), appeal denied, 71 N.Y.2d 803, 527 N.Y.S.2d 769, 522 N.E.2d 1067 (1988); Jack Coletta, Inc. v. New York State Dept. of Environmental Conservation, 128 A.D.2d 755, 513 N.Y.S.2d 465 (1987), appeal denied, 70 N.Y.2d 602, 518 N.Y.S.2d 1025, 512 N.E.2d 551 (1987).

[266] N.Y. Envtl. Conserv. Law § 25-0201(2); 6 N.Y. Comp. Codes R. & Regs.tit. 6, § 661.27 contains the regulations pertaining to maintenance and amendments to the inventory map).

[267] N.Y. Envtl. Conserv. Law § 25-0301.

[268] N.Y. Envtl. Conserv. Law § 25-0401

[269] N.Y. Envtl. Conserv. Law § 25-0302.

[270] N.J. Stat. Ann. § 13:9A.

[271] N.J. Stat. Ann. § 13:9A-1(a).

[272] N.J. Stat. Ann. § 13:9A-1(b).

those areas now or formerly connected to tidal waters whose surface is at or below an elevation of 1 foot above local extreme high water, and upon which may grow or is capable of growing some, but not necessarily all, of the following: [plant species]."[273]

The Commissioner is authorized to adopt, modify, or repeal orders regulating dredging, filling, removing, altering or polluting coastal wetlands. The New Jersey statute requires that a permit be secured to conduct any of these activities.[274] The Department of Environmental Protection oversees determinations on all permit applications.

13.8.4 Illinois

Illinois' Interagency Wetland Policy Act of 1989 sets out the goals of the state's wetlands protection, including a state goal of "no net loss of existing wetlands."[275] More specifically, the Act authorizes the Illinois Department of Energy and Natural Resources(IDENR) to develop regulations to implement and enforce the Act.

State agencies are directed to "preserve, enhance, and create wetlands where possible and avoid adverse impacts to wetlands" during state activities.[276] If adverse impacts to wetlands are unavoidable, then they must be offset through a wetlands Compensation Plan that includes constructing new wetlands. If the affected wetlands are on the Illinois Natural Areas Inventory, are under public ownership, or provide habitat for federal or state endangered species, then the agency must consult with IDENR and comply with the Illinois Endangered Species Act.[277]

Wetlands are defined such that they require hydric soils, wetland hydrology, and hydrophytic vegetation. Restored and created wetlands are included in the definition "even when all three parameters are not present."[278]

The Illinois Act establishes an Interagency Wetlands Commission to develop rules and regulations, adopt policies regarding wetlands, and regulate Agency Action Plans which guide each state agency in implementing the Act. Dredge and fill permits are issued by the Illinois Department of Transportation (IDOT). IDOT uses a joint form with the U.S. Army Corps of Engineers.

---

[273] N.J. Stat. Ann. § 13:9A-2.
[274] N.J. Stat. Ann. § 13:9A-4.
[275] 20 ILCS § 830/1-4.
[276] 20 ILCS § 830/1-3.
[277] 520 ILCS § 10.
[278] 20 ILCS § 830/1-6.

14

Coastal Zone Management

14.1 Introduction

Increasingly, the diverse values of coastal environments have been recognized, including flood control; barriers to waves and erosion; sedimentation control; habitat for fish, shellfish, and wildlife; recreation; filtration of water pollutants; and food and timber production. Coastal wetlands ecosystems are known to perform valuable environmental functions, such as prevention of erosion or saltwater intrusion into coastal areas, conveyance and storage of floodwaters, and formation of critical habitats for endangered species of fish and wildlife. Degradation of coastal water quality can cause sediment contamination, loss of submerged aquatic plants, adverse effects on fish and shellfish productivity, and closure of beach and shellfish beds.[1]

Although the coastal zone accounts for less than 10 percent of the nation's landmass, more than 75 percent of the population lives within 50 miles of coastal areas.[2] "[I]ncreasing coastal populations are creating new pressures for coastal amenities and taxing the productive but fragile coastal environment. By the year 2010, coastal population will have grown from 80 million to more than 127 million, an increase of almost 60 percent nationwide."[3]

As development tries to keep pace with the growing coastal population, the government has made efforts to protect the coastal environment. In the wake of an increased awareness of the extent of environmental pollution,

---

[1] Office of Technology Assessment Report, *Wastes in Marine Environments* (1987).
[2] See Rychlak, Coastal Zone Management and the Search for Integration, 40 DePaul L.Rev. 981 (Summer 1991).
[3] NOAA, *Biennial Report to Congress on Coastal Zone Management* (April 1990).

Congress ushered in a new era of environmental protection commencing in 1969 with enactment of the National Environmental Policy Act.[4] In 1972, Congress passed the Coastal Zone Management Act (CZMA) to address coastal environmental problems.[5] When Congress passed the CZMA, it recognized "a national interest in the effective management, beneficial use, protection and development of the coastal zone."[6] This language reflects the need identified by Congress to balance development in the coastal zone with protection of vital coastal resources. Congress appreciated the values inherent in coastal areas in its findings on the CZMA, stressing that coastal areas contain "a variety of natural, commercial, recreational, ecological, industrial, and esthetic resources of immediate and potential value to the present and future well-being of the Nation."[7]

Despite these strong policies to protect the coast, the 1980s saw a steady decline in the quality of coastal habitats. Pollution from medical wastes, domestic sewage, and plastics caused the closure of public beaches. Productive fishing areas and shellfishing beds were closed. Pollution of coastal waters only increased. The filling, dredging, and alteration of coastal wetlands caused further detriment to fish and wildlife habitats.

Therefore, with the Coastal Zone Act Reauthorization Amendments of 1990 ("Reauthorization Amendments"),[8] Congress reiterated the need for a balancing of interests, but underscored the importance of placing greater emphasis on environmental protection values. Congress concluded in its findings on the Reauthorization Amendments that there were clearly inadequacies in the management of the coastal zone that would only be compounded by population growth in the coastal zone.[9] With passage of the Reauthorization Amendments, Congress directed the states to place greater controls on land use activities in or affecting the coastal zone.

---

[4] 42 U.S.C. §§ 4321 through 4370a. Other environmental legislation was soon to follow, including: the Federal Water Pollution Control Act (Clean Water Act), 33 U.S.C. § 1251 through 1376, and the Clean Air Act, 42 U.S.C. § 7401 through 7642, in 1972; the Toxic Substances Control Act, 15 U.S.C. § 2601 through 2629, and the Resource Conservation and Recovery Act, 42 U.S.C. § 6901 through 6987, in 1976; and the Comprehensive Environmental Response, Compensation and Liability Act, 42 U.S.C. § 9601 through 9657, in 1980.
[5] Coastal Zone Management Act of 1972, 16 U.S.C. § 1451 through 1464.
[6] 16 U.S.C. § 1451(a).
[7] CZMA § 302(b), 16 U.S.C. § 1451(b).
[8] Pub. L. No. 101-508, 104 Stat. 1388 (1990). See also P.L. 104-150, 110 Stat. 1380 (1996) (reauthorizing the CZMA through 1999).
[9] 136 Cong. Rec. H8068-01, 101st Cong. 2nd Sess. (Sept. 26, 1990) (H.R. 4450).

14.2 State Authority Over Coastal Land Use

The primary goal of the CZMA is to preserve, protect, develop, and wherever possible, to restore or enhance, the resources in the coastal zone. Congress elected to make the states, instead of local governments or the federal government, "the focal point for developing comprehensive plans and implementing management programs for the coastal zone."[10] Thus, Section 1451(i) of the CZMA states that "The key to more effective protection and use of the land and water resources of the coastal zone is to encourage the states to exercise their full authority over the lands and waters in the coastal zone...."[11]

Under the CZMA, coastal states may apply for grants to develop coastal management plans (CMPs),[12] to complete the development and assist in the initial implementation of the plans,[13] and to administer the plans.[14] After development of its CMP, the coastal state submits the program to the Secretary of Commerce for review and approval.[15]

14.2.1 Federal Approval of State CMPs

Despite its delegation of authority to the states to oversee coastal zone management, Congress retained a strong safeguard for the federal interest. To ensure that the state coastal management plans conformed with the national goals of conserving the coastal zone and of promoting orderly development, Congress specifically empowered the Secretary of Commerce to approve each state's plan before the state could receive federal funding.[16]

Prior to approval of a state's management program, the Secretary[17] must make several findings. The Secretary must determine that the substantive requirements of Section 1454(b) are met,[18] and that the plan is consistent with Congress's policy declarations.[19] The CZMA requires the Secretary, in the course of approving a state's CMP, to find that the plan "provides

---

[10] Id.
[11] 16 U.S.C. § 1451(i).
[12] 16 U.S.C. § 1454(a)(1).
[13] 16 U.S.C. § 1454(a)(2).
[14] 16 U.S.C. § 1455.
[15] 16 U.S.C. § 1454(h).
[16] See 16 U.S.C. § 1455.
[17] The Secretary's authority under the CZMA to approve state plans has been delegated to the Assistant Administrator for Coastal Zone Management in the National Oceanic and Atmospheric Administration (NOAA). 15 CFR § 923.2(b).
[18] 16 U.S.C. § 1455(a).
[19] 16 U.S.C. § 1455(c).

for adequate consideration of the national interest involved in planning for, and in the siting of, facilities that are necessary to meet requirements which are other than local in nature."[20]

The regulations promulgated by the Secretary explain in detail how the states are to meet this requirement to consider the national interest. States are required to:

1. Describe the national interest in the planning for and siting of facilities considered during program development.

2. Indicate the sources relied upon for a description of the national interest in the planning for and siting of the facilities.

3. Indicate how and where the consideration of the national interest is reflected in the substance of the management program. In the case of energy facilities in which there is a national interest, the program must indicate the consideration given any interstate energy plans or programs, developed pursuant to section 309 of the Act, which are applicable to or affect a State's coastal zone.

4. Describe the process for continued consideration of the national interest in the planning for a siting of facilities during program implementation, including a clear and detailed description of the administrative procedures and decisions points where such interest will be considered.[21]

Once the Secretary approves a state CMP, the Secretary is authorized to make annual grants to the state for administration of the program.[22] The Secretary's approval continues indefinitely, until the Secretary next reviews the plan.[23] The Secretary must "conduct a continuing review of the performance of coastal states with respect to coastal management,"[24] which must be a written evaluation assessing and detailing the state's implementation and enforcement of its coastal management program, and the state's manner of addressing all Congressional policy declarations.[25] If the Secretary finds that a "coastal state is failing to make significant improvement in achieving the coastal management objectives," the Secretary must reduce any financial assistance received by the state for administration of its program.[26] If the Secretary finds that the state is failing to adhere to its approved management plan without justification, the Secretary must withdraw federal approval of a state's entire coastal management program and all financial assistance available to the state

---

[20] 16 U.S.C. § 1455(c)(8).
[21] 15 CFR § 923.52(c).
[22] 16 U.S.C. § 1455(a).
[23] 16 U.S.C. § 1458. See also Norfolk S. Corp. v. Oberly, 632 F. Supp. 1225 (D.Del. 1986), aff'd, 822 F.2d 388 (3d Cir. 1987).
[24] 16 U.S.C. § 1458(a).
[25] The congressional policy directives are specified in 16 U.S.C. § 1452(2)(A)-(I).
[26] 16 U.S.C. § 1458(c).

under the CZMA. This ongoing review mechanism allows the Secretary to ensure that state administrative and judicial interpretations of the state program remain consistent with national policy goals.

The Secretary is also required to submit a biennial report to Congress on the administration of the CZMA.[27] The report must include a description of each participating state's program and accomplishments; a statement of reasons for the disapproval of any state program reviewed; a summary of the continuing review over state programs and of any sanctions applied; a listing of all activities and projects found inconsistent with an approved state coastal management program; a summary of a coordinated national strategy and program for the nation's coastal zone; and a description of the economic, environmental, and social consequences of energy activity affecting the coastal zone.[28]

14.2.2 Federally Required Content

Congress gave the states some flexibility in deciding how best to implement their individual CMPs. Some states have used a policy format, which outlines specific goals to be achieved by the CMP in accordance with the federal Act. This approach utilizes existing state laws and regulations to implement the policies outlined in the CMP.[29] This method is popular because it does not require the adoption of a new statutory scheme aimed solely at coastal zone management. Other states have instead chosen to enact specific coastal zone management laws, which act as their coastal zone management programs.[30]

Whichever format (policy or statutory) the coastal state chooses for implementation of its CMP, the state must still ensure that certain mandatory elements are contained in its coastal zone management program in order to receive federal approval. The CZMA sets forth specific content requirements that must be met.[31] For example, a single state agency must be designated to administer the management program.[32] Other prerequisites to federal approval of a state's CMP include:

1. An identification of the boundaries of the coastal zone subject to the management program.[33]

2. A definition of what shall constitute permissible land uses and water uses within the coastal zone which have a direct and significant

---

[27] 16 U.S.C. § 1462(a).
[28] Id.
[29] See, for example, New York, Massachusetts, New Jersey, and Florida.
[30] See, for example, Connecticut, North Carolina, and South Carolina
[31] 16 U.S.C. § 1455(d).
[32] 16 U.S.C. § 1455(d)(6); See 15 CFR § 923.47.
[33] 16 U.S.C. § 1455(d)(2)(A).

impact on the coastal waters.[34]

3. An inventory and designation of areas of particular concern within the coastal zone.[35]

4. An identification of the means by which the State proposes to exert control over the land uses and water uses, including a list of relevant State constitutional provisions, laws, regulations, and judicial decisions.[36]

5. Broad guidelines on priorities of uses in particular areas, including specifically those uses of lowest priority.[37]

6. A description of the organizational structure proposed to implement such management program, including the responsibilities and interrelationships of local, areawide, State, regional, and interstate agencies in the management process.[38]

7. A definition of the term "beach" and a planning process for the protection of, and access to, public beaches and other public coastal areas of environmental, recreational, historical, esthetic, ecological, or cultural value.[39]

8. A planning process for energy facilities likely to be located in, or which may significantly affect, the coastal zone, including a process for anticipating the management of the impacts resulting from such facilities.[40]

9. A planning process for assessing the effects of, and studying and evaluating ways to control, or lessen the impact of, shoreline erosion, and to restore areas adversely affected by such erosion.[41]

The state CMP must also provide mechanisms to control land uses and water uses within the coastal zone,[42] including:

1. State establishment of criteria and standards for local implementation, subject to administrative review and enforcement.

2. Direct State land and water use planning and regulation.

3. State administrative review for consistency with the management program of all development plans, projects, or land and water use regulations, including exceptions and variances thereto, proposed by any State or local authority or private developer, with power to approve or disapprove after public notice and an opportunity for hearings.

Additionally, the CZMA requires that state CMPs provide for public participation in permitting processes,[43] consistency determinations, and

---

[34] 16 U.S.C. § 1455(d)(2)(B).
[35] 16 U.S.C. §§ 1455(d)(2)(C), 1455(d)(9).
[36] 16 U.S.C. § 1455(d)(2)(D). See 15 CFR §§ 923.40, 923.41, and 923.43.
[37] 16 U.S.C. § 1455(d)(2)(E).
[38] 16 U.S.C. § 1455(d)(2)(F). See 15 CFR § 923.46.
[39] 16 U.S.C. § 1455(d)(2)(G).
[40] 16 U.S.C. § 1455(d)(2)(H).
[41] 16 U.S.C. § 1455(d)(2)(I).
[42] 16 U.S.C. § 1455(d)(11); See 15 CFR §§ 923.3., 923.40-.43.
[43] 16 U.S.C. § 1455(d)(14).

other such decisions. Further, the CMP must provide a mechanism for ensuring that all state agencies will adhere to the program.[44] Finally, state CMPs must contain enforceable policies and mechanisms for implementation of mandatory Coastal Nonpoint Source Pollution Control Programs as required by section 6217 of the Coastal Zone Act Reauthorization Amendments of 1990.[45]

14.2.3 Consistency with Approved CMPs

Once approved, a state's CMP can be a powerful tool for regulating land use activities within its coastal zone. CZMA Section 307 requires that all federal activities and projects affecting the state's coastal zone, as well as activities carried out by private parties that require a federal permit or license, must be consistent with the state's approved CMP.[46]

Applicants for a federal permit must demonstrate to the state coastal zone authority that the proposed activity complies with the policies of the state's CMP.[47] If the state authority finds that the proposed activity is inconsistent with its CMP, the federal agency cannot issue the necessary permits unless the Secretary of Commerce overrides the state's objection on appeal or by his own initiative.[48]

14.2.4 Definition of "Coastal Zone"

In each approved CMP, the CZMA requires identification of the boundaries of the "coastal zone" subject to the state's management program.[49] Each state is permitted to determine the boundaries of its coastal zone. The varying geography of each state has resulted in different ways of mapping the state coastal zone. Some states have employed setback boundaries from the mean high water mark. Other states have decided to use the boundaries of counties located on the coast. In Florida,

---

[44] 16 U.S.C. § 1455(d)(15).
[45] 16 U.S.C. § 1455(d)(16).
[46] 16 U.S.C. § 1456(c), as amended by The Coastal Zone Act Reauthorization Amendments of 1990, Pub. L. 101-508, § 6208.
[47] 16 U.S.C. § 1456(c)(3)(A), as amended by The Coastal Zone Act Reauthorization Amendments of 1990, Pub. L. 101-508, § 6208(b)(2).
[48] Id. See also 15 CFR § 930.120-122.
[49] 16 U.S.C. § 1454(b)(1). Public Law Section 6204 of The Coastal Zone Act Reauthorization Amendments of 1990, Pub. L. 101-508, 104 Stat. 1388 (Nov. 5, 1990), amended the definition of the term "coastal zone" to expressly limit the seaward coastal zone boundary to the extent of state ownership and title (in most cases three nautical miles). Pub. L. 101-508, § 6204, 104 Stat. 1388 (Nov. 5, 1990).

the entire state is considered the coastal zone for purposes of the state's coastal management program. By allowing the states to determine their own coastal zone boundaries, the federal CZMA recognizes the unique physical features of each state and acknowledges the need for each state to establish boundaries that work well in administration of the coastal zone management program.

14.3 Local Authority Over Coastal Land Use

Although the states maintain primary authority for management of land use in the coastal zone via their federally approved CMPs, coastal municipalities also play a key role in controlling coastal land use activities. Congress expected local governments to assist the states in carrying out the individual state CMPs and in fulfilling the goals of the CZMA. Under the CZMA, each state must coordinate its CMP with local, areawide, and interstate plans applicable to areas within the coastal zone and establish a means for ongoing consultation and coordination between the designated state agency and local governments, interstate agencies, regional agencies, and areawide agencies within the jurisdiction of the state coastal zone.[50]

Although the CZMA requires that the state agency in charge of overseeing implementation of the CMP coordinate and ensure local government participation in the management process, the Congress was concerned that vesting too much power in coastal municipalities would result in inconsistent regulation of activities taking place within the coastal zone. In fact, a requirement for federal approval of a state's CMP is that the CMP contain a method of ensuring that local land use and water use regulations within the coastal zone do not unreasonably restrict or exclude land uses and water uses of regional benefit.[51]

Land use regulation has traditionally been a matter of local concern, largely because individual local governments are most closely involved and best equipped to understand the needs of the local community. For the most part, this local land use planning and zoning authority remains intact in the coastal zone despite the state's lead management role under the federally approved CMP. State CMPs are not meant to supersede regulation of land use activities at the local level. Instead, the state programs must coordinate regulatory efforts in the coastal zone between

---

[50] 16 U.S.C. § 1455(d)(3)(A),(B). The CMP must contain a description of the organizational structure proposed to implement the CMP, which includes the responsibilities and interrelationships of the various local, areawide, state, regional and interstate agencies involved in the management process. See 15 CFR § 923.46.

[51] 16 U.S.C. § 1455(d)(12).

state and local authorities. Generally, local coastal communities retain substantial regulatory authority within their jurisdictions. The primary mandate of the CZMA is that local regulation be consistent with the goals and policies of the state program in order to fully implement the purposes of the federal Act.[52] If a local regulation is inconsistent with the federally-approved CMP, then the local regulation must be changed to further the state scheme.[53]

14.4 State Coastal Zone Management Plans

The nature and structure of CMPs vary widely from state to state. Some states, like North Carolina, have passed comprehensive legislation as a framework for coastal management. Other states, like Oregon, have decided to use existing land use legislation as the foundation for their programs. Finally, some states, like Florida and Massachusetts, have linked existing, single-purpose laws into a comprehensive umbrella for coastal management. "The national program, therefore, is founded on the authorities and powers of the coastal states and local governments. Through the CZMA, these collective authorities are structured to serve the national interest in effective management of the coastal zone."[54]

14.4.1 Connecticut

The Connecticut coastal management program was approved and adopted in the form of the Connecticut Coastal Management Act.[55] The Act sets forth (1) general goals and policies[56] for implementation, (2) policies for federal, state and municipal agencies to carry out,[57] and (3) policies for federal and state agencies to carry out.[58] Some of these goals and policies are as follows:

---

[52] See, for example, Issuance of a CAMA Minor Development Permit No. 82-0010 to Ford S. Worthy v. Town of Bath, 82 N.C. App. 32, 345 S.E.2d 699 (1986) (zoning ordinance aimed at achieving aesthetic qualities and water quality objectives consistent with state CMP).
[53] See, for example, Lusardi v. Curtis Point Property Owners Association, 86 N.J. 217, 430 A.2d 881 (1981) (goal of preserving residential character of neighborhood must yield to state policy that public recreational use consistent with environmental concerns is uniquely appropriate for oceanfront beaches).
[54] 136 Cong. Rec. H8068-01, 101st Cong. 2nd Sess. (Sept. 26, 1990) (H.R. 4450).
[55] Conn. Gen. Stat. § 22a-90 through 22a-112.
[56] Conn. Gen. Stat. § 22a-92(a)(1)-(10).
[57] Conn. Gen. Stat. § 22a-92(b).
[58] Conn. Gen. Stat. § 22a-92(c).

1. Coordination of planning and regulatory activities of public agencies to insure maximum protection of coastal resources while minimizing conflicts with economic development.[59]

2. Assurance that state and municipal government provide adequate planning for facilities and resources that are in the national interest.[60]

3. Use existing municipal planning, zoning and other local regulatory authorities to manage uses in the coastal zone.[61]

4. Locate sewer and water lines so as to encourage concentrated development in suitable areas and disapproval of sewer and water extensions to undeveloped beaches, barrier beaches and tidal wetlands, except when necessary to abate existing sources of pollution.[62]

5. Require that structures located in tidal wetlands and coastal waters be designed, constructed and maintained to limit adverse impact on coastal resources.[63]

6. Require mitigation measures where development would adversely impact historical, archaeological or paleontological resources.[64]

7. Discourage uses that do not permit natural rates of erosion.[65]

8. Preserve natural beach systems,[66] tidal wetlands,[67] and undeveloped islands.[68]

9. Preserve public beach access.[69]

The state's coastal area includes land and water delineated by a line "on the landward side by the interior contour elevation of the one hundred year frequency coastal flood zone, ... or a one thousand foot linear setback measured from the mean high water mark in coastal waters, or a one thousand foot linear setback measured from the inland boundary of tidal wetlands mapped ... whichever is farthest inland."[70] This coastal zone boundary is shown on maps prepared by the Commissioner of Environmental Protection.[71] A municipal planning commission may adopt a municipal coastal boundary provided the boundary does not diminish the

---

[59] Conn. Gen. Stat. § 22a-92(a)(9).
[60] Conn. Gen. Stat. §22a-92(a)(10). See Conn. Gen. Stat. § 22a-93 for a definition of the national interest.
[61] Conn. Gen. Stat. § 22a-92(b)(1)(A).
[62] Conn. Gen. Stat. § 22a-92(b)(1)(B).
[63] Conn. Gen. Stat. § 22a-92(b)(1)(D).
[64] Conn. Gen. Stat. § 22a-92(b)(1)(J).
[65] Conn. Gen. Stat. § 22a-92(b)(2)(A).
[66] Conn. Gen. Stat. § 22a-92(b)(2)(C).
[67] Conn. Gen. Stat. § 22a-92(b)(2)(E); § 22a-92(c)(1)(E) (disallowing new dredging except where no feasible alternative exists).
[68] Conn. Gen. Stat. § 22a-92(b)(2)(H).
[69] Conn. Gen. Stat. § 22a-92(c)(1)(K).
[70] Conn. Gen. Stat. § 22a-94(b).
[71] Conn. Gen. Stat. § 22a-94(c).

area as mapped by the Commissioner.[72]

Under the Connecticut scheme, the Commissioner must develop a model municipal coastal program for use by the coastal municipalities, which includes model municipal coastal plans and regulations, regulatory planning methodologies for revising municipal coastal plans, and criteria and procedures for undertaking municipal coastal site plan reviews.[73]

The Commissioner is the designated state representative and must coordinate all activities of all regulatory programs under his jurisdiction with permitting authority in the coastal zone.[74] "Any person seeking a license, permit or other approval of an activity under the requirements of such regulatory programs shall demonstrate that such activity is consistent with all applicable goals and policies in section 22a-92 and that such activity incorporates all reasonable measures mitigating any adverse impacts of such actions on coastal resources and future water-dependent development activities."[75] Each coastal municipality may adopt its own coastal program to provide guidance to property owners and developers for municipal territory within the coastal zone.[76] The municipal coastal program must include a revision of the municipal plan of development, revisions to the municipal zoning regulations[77] and other listed ordinances.[78] Procedures for revising the municipal plan of development and municipal zoning regulations are specified in the Act.[79] The Connecticut CMP also provides for criteria and procedures for coastal site plan reviews.[80]

The CMP provides for substantial local authority to monitor land use activities in the coastal zone. In fact, the state Act provides for extensive use of existing municipal planning, zoning and other local regulatory authorities to manage uses in the coastal zone.[81] A key provision of the statute is a review mechanism for coastal site plans.[82] A developer must

---

[72] Conn. Gen. Stat. § 22a-94(f).
[73] Conn. Gen. Stat. § 22a-95(e).
[74] Conn. Gen. Stat. § 22a-98.
[75] Id.
[76] Conn. Gen. Stat. § 22a-101.
[77] Municipal zoning regulations are adopted pursuant to Conn. Gen. Stat. § 8-2. The municipal plan of development is adopted pursuant to Conn. Gen. Stat. § 8-23.
[78] Conn. Gen. Stat. § 22a-101(b)(2)(A)-(H).
[79] Conn. Gen. Stat. §§ 22a-102, 22a-103.
[80] Conn. Gen. Stat. § 22a-105.
[81] Conn. Gen. Stat. § 22a-92(b)(1)(A).
[82] "Coastal site plans" are defined as (1) site plans submitted to a zoning commission; (2) plans submitted to a planning commission for subdivision or resubdivision; (3) applications for a special exception or special permit submitted to a planning commission, zoning commission or zoning board of appeals; (4) applications for a variance submitted to a zoning board of appeals;

submit the coastal site plan for his or her proposed project to the appropriate municipal board or commission and the governing body must review the proposed site plan on the basis of criteria set out in section 22a-106 of the Connecticut Coastal Management Act to "ensure that the potential adverse impacts of the proposed activity on both coastal resources and future water-dependent development activities are acceptable."[83] If the coastal site plan is submitted to a local zoning commission, the commission must review the site plan "... to aid in determining the conformity of the proposed building, use, structure, or shoreline flood and erosion control structure ... with the specific provisions of the zoning regulations of the municipality ...."[84] The review of any coastal site plan is "... not deemed complete and valid unless the board or commission having jurisdiction over such plan has rendered a final decision.[85] Thus, the Connecticut CMP provides a mechanism whereby coastal municipalities not only review development projects for conformity with local regulation but also ensure that the goals of the state CMP are fully implemented when ruling on site plan applications.

When the appropriate local authority rules on a variance application for property in the coastal zone, the CMP specifically requires that the local authority apply not only the requirements for a variance specified by local regulation, but also the coastal area management criteria specified in the Act.[86]

The Connecticut scheme delegates the administration of state-wide policy for planned coastal development to local governmental agencies charged with the responsibility for zoning and planning decisions. The CMP sets up a single review process for development proposals, which are simultaneously reviewed for compliance with local zoning requirements and for consistency with the policies of the CMP.[87]

14.4.2 Massachusetts

Unlike Connecticut, the Massachusetts CMP relies solely on existing statutory authority to implement a series of regulatory and nonregulatory policies. The regulatory policies form a basis for administrative decisions

---

and (5) a referral of proposed project to a planning commission. Conn. Gen. Stat. § 22a-105(b).

[83] Conn. Gen. Stat. § 22a-105(e).

[84] Conn. Gen. Stat. § 22a-109(a).

[85] Conn. Gen. Stat. § 22a-105(f). If the board or commission fails to render a final decision within the statutory time period, the site plan is deemed rejected. Id.

[86] Conn. Gen. Stat. §§ 22a-105(b)(1),(4), 22a-105(c).

[87] See Vartuli v. Sotire, 192 Conn. 353, 358-363, 472 A.2d 336 (1984).

to approve or disapprove activities in the coastal zone to the extent that the policies are contained in the text of the state laws or regulations. The non-regulatory policies are provided as guidance to administrative decisions, however, have no legal binding effect on private parties.

Some of the regulatory policies of the Massachusetts CMP are:

Policy 1: Wetlands and buffers protection.
Policy 2: Preservation and restoration areas
Policy 3: Water quality goals support
Policy 4: Construction in water bodies; erosion control structures
Policy 5: Dredging and dredged material disposal
Policy 6: Offshore mining
Policy 7: Licensing harbor and port development
Policy 8: Energy facility siting
Policy 9: OCS and alternative source management
Policy 10: Conformance to existing air and water permit requirements
Policy 11: Scenic rivers, outdoor advertising
Policy 12: Impacts on historic districts and sites
Policy 13: Impacts on public recreation beaches

Under each of the policies listed, the CMP provides a policy statement and description, a statement on how the policy is to be implemented, and a listing of the applicable state and federal laws that govern the policy. This method sets out the goals to be achieved by the CMP and how existing law will be used to implement the goals. The program seeks to improve the administration of existing laws to provide optimal use of state coastal resources. Development activities are not subject to any additional requirements because of the CMP. Development is regulated solely by the various existing laws and regulations, which are guided to some extent by the policies contained in the CMP.

The coastal zone is defined as the area extending landward 100 feet beyond specified major roads, rail lines or other visible rights-of-way and seaward to the edge of the territorial sea and includes all of Cape Cod, Martha's Vineyard, Nantucket, and Gosnold. The Office of the Secretary within the Executive Office of Environmental Affairs is the designated lead agency for implementation of the Massachusetts CMP.

14.4.3 New Jersey

New Jersey's CMP is also implemented through existing laws and agencies. The principal legal authority for administering the program is the coordinated use of the state Coastal Area Facility Review Act (CAFRA),[88]

---

[88] N.J. Stat. Ann. § 13:19-4.

Wetlands Act,[89] Waterfront Development Act,[90] shore protection program, tidelands management program, the Pinelands Protection Act[91] and the activities governed by the Department of Energy Act and the Hackensack Meadowlands Reclamation and Development Act. A designated portion of the state's coastal zone will fall under the jurisdiction of one of these laws. Whenever a proposed development site is within the coastal zone boundary a permit must be obtained from the authority responsible for administering the pertinent law. Department of Environmental Protection CAFRA permits and Pinelands Commission development permits are most commonly sought because the largest percentage of coastal lands and waters are under the jurisdiction of Department of Environnmental Protection (DEP) and the Pinelands Commission.[92] The CMP consists of three basic elements: (1) a boundary defining the geographic scope of the program; (2) policies defining the standards for making decisions on what activities may take place within the coastal zone boundary; and (3) a management system defining the types of decisions subject to the program and the process by which those decisions are made.

The coastal boundary extends to areas governed by the CAFRA, the Waterfront Development Act, the Wetlands Act, and the Pinelands Protection Act or the landward boundary of state-owned wetlands, whichever extends farthest inland. The New Jersey DEP is the lead agency for purposes of the CZMA. An applicant for a permit submits the application to the agency responsible for administering the particular law and the agency determines whether the project would contravene any of the policies of the state CMP.

For example, the CAFRA governs all development in the coastal area covered by that Act. A person proposing to construct any facility[93] in the CAFRA coastal zone must file an application and environmental impact statement with the DEP.[94] Once the application is deemed complete by the DEP, DEP must hold a hearing to afford interested parties an opportunity to comment on the proposed project.[95] The granting of a permit for the proposed project is contingent upon meeting specific standards set out in

---

[89] N.J. Stat. Ann. § 13:9A-1.

[90] N.J. Stat. Ann. § 12:5-3.

[91] N.J. Stat. Ann. § 13:18A-1 et seq.

[92] Of the 37.1 percent of the land and water in New Jersey that is within the coastal zone, 19.4 percent is under the jurisdiction of the Pinelands Commission and 18.3 percent is under the jurisdiction of the DEP. Matter of Egg Harbor Associates, 464 A.2d 1115, 1119, 94 N.J. 358, 369 (1983).

[93] A "facility" is broadly defined to include almost all land uses. N.J. Stat. Ann. § 13:19-3(c).

[94] N.J. Stat. Ann. § 13:19-6, -7.

[95] N.J. Stat. Ann. § 13:19-9.

the DEP's Division of Coastal Resources regulations.[96] These regulations are the substantive policies regarding the use and development of coastal resources, and are used primarily by DEP to review permit applications pursuant to CAFRA, Coastal Wetlands, and Waterfront Development laws.

The New Jersey Administrative Code contains detailed rules for the state coastal zone management program.[97] It sets out the various policies of the CMP and provides the framework for coastal decisionmaking. The regulations group policies as location, use and resource policies. The application of the location policies, in conjunction with the use and resource policies, is known as the "Coastal Location Acceptability Method (CLAM)" for evaluating the suitability of development in coastal areas. The location policies classify all land and water locations into a General Area and some into Special Areas.[98]

The CLAM is a nine-step process which is used to determine DEP policy for any proposed coastal use in any coastal location. The first six steps of the CLAM are the mapping and policy determination process used to assess Location Acceptability of development.[99] Next, steps 7 and 8 refine the Location Acceptability of the development by reviewing the proposed use in terms of Uses and Resources Policies.[100] Finally, step 9 is used to determine the final acceptability of a proposed use.[101] Approval is granted only if a proposal satisfies all three sets of policies, location, use, and resource.[102]

---

[96] N.J. Admin. Code § 7:7E-1.1 et seq.
[97] N.J. Admin. Code § 7:7E-1.1 et seq.
[98] N.J. Admin. Code § 7:7E-2.2.
[99] CLAM Location Policy Analysis is as follows:
Step 1 - Identify and map site and surrounding region
Step 2 - Identify and map Special Areas
Step 3 - Determine the applicable Special Area Policies
Step 4 - Identify and map general areas
Step 5 - Determine the applicable General Area Policies
Step 6 - Map Final Location Acceptability and list Location Policy conditions.
[100] CLAM Use Policy Analysis is as follows:
Step 7 - Identify applicable Use Policies, evaluate the proposed use, and if necessary, modify the Location Acceptability Determination and list Use Policy conditions.
CLAM Resource Analysis is as follows:
Step 8 - Identify applicable Resource Policies, evaluate the proposed use, and, if necessary, modify the Location Acceptability Determination and list Resource Policy Conditions.
[101] CLAM Synthesis is as follows:
Step 9 - Determine the final acceptability of the proposed use, summarizing and synthesizing the final acceptability of a proposed use at the proposed location in terms of the applicable Location, Use and Resource Policies.
[102] N.J. Admin. Code § 7:7E-2.3.

CAFRA cannot be construed so as to impair local zoning authority.[103] Still, municipalities are required to refrain from exercising their zoning powers in a manner that would conflict with the CAFRA regulations.[104] Planning decisions "must be consistent with statewide policies concerning land use and resource allocation."[105]

14.4.4 New York

Within New York state, for purposes of its coastal management program, coastal areas are found in several regions of the state. The coastal zone includes areas that one might not regard as "coastal," including the state border on the St. Lawrence seaway and Great Lakes area, as well as the Hudson River Valley, which extends from the Atlantic Ocean at New York City north 150 miles into upstate New York. In addition to these sectors, New York City and Long Island make up the remaining areas governed by the state's CMP. Because of the unique geography of the state and the variety of areas comprising the coastal zone, New York needed to employ a rather complex process for determining the boundary lines for the coastal zone. The landward boundary varies from region to region. Generally, the inland boundary is located 1,000 feet from the shoreline of the mainland area, however, several special boundary criteria were used in delineation and mapping of the state's coastal zone.[106]

The New York management program relies on existing laws and regulations to implement the program's objectives. New York's CMP provides a system for coordinating activities among the various state agencies responsible for administering the state laws and regulations aimed at protecting coastal resources. The CMP contains 44 coastal policies with which the decisions of all agencies must be consistent.[107]

The Department of State is the designated agency for administering the state program, and conducting federal consistency reviews. The Department of Environmental Conservation has the major responsibility for protecting coastal erosion hazard areas, as well as carrying out its

---

[103] N.J. Stat. Ann. § 13:19-19.

[104] N.J. Admin. Code § 7:7E-1.1 to 9.23.

[105] See Lusardi v. Curtis Point Property Owners Ass'n, 86 N.J. 217, 227, 430 A.2d 881 (1981).

[106] For example, for urbanized and developed areas along the coast, the landward boundary is about 500 feet from the mainland's shoreline, or less than 500 feet where a major roadway or railway line runs parallel to the shoreline.

[107] Generally, the policies fall into three categories: (1) promotion of the beneficial use of coastal resources; (2) prevention of coastal resource impairment; and (3) management of activities substantially affecting coastal resources.

existing permit authority for tidal and freshwater wetlands, and air and water quality. Activities with the potential for significantly impacting coastal resources are reviewed in connection with the State Environmental Quality Review Act (SEQRA), which requires preparation of an Environmental Impact Statement (EIS) for actions that will have significant adverse environmental effects. Numerous other state agencies carry out their regulatory authority pursuant to existing laws, taking into account the policies contained in the state's CMP.

When development is proposed in the coastal zone, the developer must secure the necessary permits pursuant to the various state and federal laws, such as water quality, air quality, and wetlands permits. In addition, the developer must complete a federal consistency assessment form and certification attesting that the proposed activity will be consistent with the policies of the state CMP. When a developer applies for state permits, the state agency responsible for issuing the permit must determine whether the proposed activity is consistent with the coastal policies if the development will be situated in or impact upon the state coastal zone.[108] The developer must provide sufficient information in the permit application to enable the state agency to complete a coastal assessment form.[109] If the project is to be approved, the information must demonstrate that the proposed project is consistent with the policies of the coastal management program. If the proposed project is one that also requires a federal permit and, thus, submission of a federal consistency certification to the New York Department of State (DOS), the state agency will take the DOS's findings on the federal consistency determination into account when determining state consistency.[110]

In conjunction with the regulatory authority of the various state agencies, local government's are encouraged to participate in the CMP under the Waterfront Revitalization Act and Coastal Resources Act, which provide a means and incentive for municipalities to develop and implement local waterfront revitalization programs.[111] A participating municipality submits a local government waterfront revitalization program (LGWRP) to the New York Department of State for approval. In order to obtain approval, the local government must ensure that the program is

---

[108] The coastal policies are found at N.Y. Comp. Codes R. & Regs. tit. 19, pt. 600.

[109] N.Y. Comp. Codes R. & Regs. tit. 19, § 600.4; See also N.Y. Comp. Codes R. & Regs. tit. 6, § 621.3(a)(9). The coastal assessment forms contain virtually the same information as the federal certification forms that are submitted to the New York Department of State.

[110] See, for example, Matter of the Applications of Xanadu Properties Associates, DEC No. 3-5510-161-1-0, 1990 WL 263916 (N.Y. Dept. of Env. Conserv. Oct. 15, 1990).

[111] N.Y. Comp. Codes R. & Regs. tit. 19, pt. 601.

consistent with the policies of the CMP.[112] Once approved, the municipality will have authority to regulate land use practices within its coastal jurisdiction pursuant to the approved LGWRP. State agencies are expected to adhere to the details of the program and defer to the judgment of the local government when it makes land use decisions pursuant to the approved program.[113] Local government participation is optional, however, the LGWRP is a powerful tool for the municipality to exercise control over activities within its coastal area.

14.4.5 North Carolina

North Carolina's CMP is codified in the form of a state statute,[114] and employs a two pronged regulatory scheme for development activities taking place in its coastal zone. In North Carolina the Coastal Resources Commission[115] and the Division of Coastal Management are responsible for managing the state CMP. The state CMP exists in the form of the Coastal Area Management Act of 1974 (CAMA).[116] Under the state program, areas of environmental concern (AECs) are designated, policies and guidelines for coastal development are adopted, and local land use plans are certified. The rules and policies apply to the state's twenty coastal counties.

Any development in an AEC requires a CAMA permit.[117] The Division of Coastal Management issues permits for major development projects[118] and the local governments issue permits for minor development projects.[119] When a major development permit is sought, the project proponent applies to the Division of Coastal Management (DCM). A major permit is required if the proposed project involves any of the following:
1. Alteration of more than 20 acres of land and/or water within an AEC;
2. Construction of one or more buildings covering a ground area greater than 60,000 square feet on a single parcel of land;

---

[112] N.Y. Comp. Codes R. & Regs. tit. 19, § 601.3.

[113] The Secretary of State will, however, periodically review the local government's actions pursuant to the program to assure that it is adhering to the policies and goals of the approved LGWRP. N.Y. Comp. Codes R. & Regs. tit. 19, § 601.6.

[114] The state CMP exists in the form of the Coastal Area Management Act of 1974 (CAMA). N.C. Gen. Stat. art. 7, § 113A-100 et seq.

[115] N.C. Gen. Stat. art. 7, § 113A-104.

[116] N.C. Gen. Stat. art. 7, § 113A-100 et seq.

[117] N.C. Gen. Stat. art. 7, § 113A-118.

[118] A major development permit is required for projects that require another state or federal permit, cover more than 20 acres, or have a structure larger than 60,000 square feet. N.C. Gen. Stat. art. 7, § 113A-118(d)(1).

[119] N.C. Gen. Stat. art. 7, § 113A-118(d)(2).

3. Excavation or drilling for natural resources on land in an AEC or under water; or

4. Another state or federal permit, license, or authorization.

Certain projects are exempt from the major development permit.[120] A CAMA major development permit is most commonly required for dredge and fill activities.[121] Projects within an AEC that require a federal permit automatically require a major development permit. The developer must submit an application for a CAMA major development permit at the same time that the federal permit application is submitted.[122] When the applicant fills out the CAMA major development permit application, he must sign a statement certifying that the project is consistent with the state CMP.

If a project is proposed in an AEC and does not meet the criteria for a major development permit, a minor development permit is required. Certain projects are exempt from the minor development permit requirement.[123] The local governmental agencies are in charge of the minor development permit program. The state Division of Coastal Management trains local permit officers (LPOs) to review applications for CAMA minor development permits. The local permit officer is a local government employee, such as a building inspector, zoning administrator, or land use planner who reviews the application for consistency with the standards of the state CMP and is authorized to issue or deny the permit. The LPO visits the project site and determines whether the proposed activity complies with the local CAMA land use plan and local development ordinances.

The local land use plan is prepared pursuant to the CAMA and all twenty coastal counties must participate in and update their plan every five years.[124] Grants and technical assistance are provided to local governments for the development of land use plans.[125] Permits can be denied if the proposed project is inconsistent with county or municipal land use plans.[126] The local land use plan requirements are in addition to local zoning and

---

[120] The exemptions can be found in Title 7K, Section .0202 of the N.C. Administrative Code.

[121] As required under the N.C. Dredge and Fill Act, N.C. Gen. Stat. art. 7, § 113-229.

[122] North Carolina has developed a joint permit application with the Army Corps of Engineers for applicants seeking a federal Corps permit.

[123] The minor permit exemptions are found in Title 15, Subchapter 7K, Sections .0302, .0303, and .0304 of the N.C. Administrative Code.

[124] In addition to the twenty counties, about 65 municipalities also participate in the planning program.

[125] N.C. Gen. Stat. art. 7, § 113A-112.

[126] N.C. Gen. Stat. art. 7, § 113A-111.

subdivision regulations. A developer's proposed project must meet the requirements of each.[127]

14.4.6 South Carolina

The coastal management program for the state of South Carolina is implemented through the South Carolina Coastal Management Act.[128] The Coastal Council is responsible for administering the state's CMP. The South Carolina CMP is similar in concept to the North Carolina management program. It uses a two tier approach like North Carolina, however, it's emphasis is less on the type of activity (major or minor development permit) and more focused on the particular environment where a development may be situated. The South Carolina CMP designates certain areas within the state coastal zone as "critical areas" where the state Coastal Council maintains direct permit authority for all activities. Critical areas are defined as coastal waters, tidelands, beaches, and primary oceanfront sand dunes.[129] The coastal zone is defined as coastal waters and submerged lands seaward to the state's jurisdictional limits, as well as the lands and waters of the eight coastal counties.

Outside of the critical areas, but areas still within the coastal zone, the Council maintains review and certification authority while local and other state agencies exercise direct permitting authority. Thus, a developer must secure a permit from the Coastal Council for any activity proposed to take place in a critical area, and apply for the necessary permits from other state and local authorities if the proposed site is outside a critical area. Still, no permit can be issued unless the project conforms to the policies of the state CMP.

The management program lists types of uses, such as residential development, commercial development, dredging, and transportation, and provides review and certification policies for the Council in determining whether state and federal permits are consistent with the CMP. Rules and regulations are followed by the Council in determining whether it should offer a permit when the proposed use is located in a critical area.

---

[127] The Division of Coastal Management has developed a helpful guide to the permit process for projects governed by CAMA: A Guide to Protecting Coastal Resources Through the CAMA Permit Program (Div. of Coastal Management, N.C. Dept. of Natural Resources and Community Development).
[128] S.C. Code § 48-39-10 through 48-39-360.
[129] Rules and regulations for permitting in critical areas are found at R. 30-1 through 30-11 of the S.C. Code of Laws of 1976. R. 30-12 through 30-20 policies can be found in S.C. Coastal Council Permitting Rules and Regulations.

Local authority is restrained in the coastal zone because local governments with jurisdiction fronting the Atlantic Ocean are required to submit a local beach management plan to the Coastal Council for approval. If a local government fails to adopt an approved plan, the Coastal Council adopts one instead and the local government is bound by the provisions of the local plan adopted by the Coastal Council. Furthermore, the authority and responsibility of implementing the local plan then vests in the Coastal Council unless it agrees to delegate that authority to the local government.

14.5 Permit Applications

Under various environmental laws, applications for licenses or permits may need to be filed before conducting different land use activities in or affecting coastal areas. The process of securing all the necessary federal, state, and local permits is complex and time-consuming to say the least. However, proceeding without the necessary government approvals and regulatory compliance exposes landowners to the risk of severe penalties and other enforcement measures, such as removal of existing structures.

Most permits and approvals for activities in or affecting coastal areas are sought at the state and local level. However, at the federal level, various permits and approvals may be necessary depending on the type of activity and its impact on the coastal zone. The three most common federal permits include: (1) U.S. Army Corps of Engineers Section 10 permits for building structures in navigable waters; (2) U.S. Army Corps of Engineers Section 404 permits for placing fill material in wetlands; and (3) U.S. Coast Guard permits for the construction or modification of bridges and causeways over navigable waters. Although the state may review a application for virtually any federal permit sought by an applicant who wishes to carry out an activity in the coastal zone, the Section 404 permit is by far the most commonly reviewed by the state for consistency with its coastal management program.

After federal permits are applied for, the developer must determine which state and local permits are necessary to complete the project. The most common types of state permits are dredge and fill permits for wetlands, water quality permits pursuant to a state's clean water act, permits issued pursuant to a state's environmental quality review act (little NEPA permits), and construction permits pursuant to state and local regulation.

14.5.1 Federal Consistency with State CMP

Activities undertaken in or affecting the coastal zone must be consistent with approved state CMPs. This requirement applies to both federal and private development and land use activities. Any federal activity, including federal development projects must be consistent "the maximum extent practicable" with approved state management programs.[130] Thus, the federal government must conduct its land use activities in a manner that ensures adequate protection of the coastal zone.[131] Developers wishing to undertake activities in the coastal zone for which a federal license or permit is required are also subject to the federal consistency requirement.[132]

Applicants for a federal permit or license should first consult with the designated state agency to "obtain the views and assistance of that agency regarding the means for ensuring that the proposed activity will be conducted in a manner consistent with the State's management program.[133] This is the most common sense method of ensuring that the project will not later be contested for failure to comply with the state CMP. Consultation with the state agency gives the developer the best information concerning what the state would expect the developer to do to mitigate adverse coastal impacts.

The developer should also consult the management program document to understand the policies and goals of the CMP. Included in the CMP is a list of federal license and permit activities which the state deems likely to affect the coastal zone and which the state will expect to review for consistency with its program.[134] This list is mandatory as part of the approved program and must describe the types of federal licenses and permits involved.[135] The state is still authorized to review other unlisted activities that affect the coastal zone. If the state wishes to review the license or permit for an unlisted activity it does so by informing the federal agency and the applicant within 30 days from the notice of the license or permit application. The state must also notify the Assistant Administrator for Coastal Zone Management at NOAA of its wish to review the activity. The Assistant Administrator then determines whether the state is justified

---

[130] 15 CFR § 930.32.
[131] The consistency determination must include a statement indicating consistency to the maximum extent practicable with the state CMP based on the relevant provisions of the state management program. It must also contain a detailed description of the activity, its associated facilities, and the impact of the activity on the coastal zone. 15 CFR § 930.39.
[132] 15 CFR § 930.50 to 930.66.
[133] 15 CFR § 930.56(a).
[134] 15 CFR § 930.53(a).
[135] 15 CFR § 930.53(b).

in seeking a consistency review. If unjustified, the applicant need not comply with the consistency certification requirement for the federal license or permit. If the Assistant Administrator approves of the state's wish to review the activity, then the applicant must amend his application with a consistency certification.[136]

14.5.2 Federal Consistency Certification

After the federal permit applicant determines that his activity will be consistent with the state CMP, he provides a certification to that effect in the federal permit application. The applicant must simultaneously submit a copy of the certification to the designated state agency.[137] Along with the certification, the applicant must provide the designated state agency with "necessary data and information," including:
1. A detailed description of the proposed activity and its associated facilities along with a copy of the federal application and all supporting material.
2. Information requirements that the state may choose to list in its CMP pursuant to 15 CFR Section 930.56(b).
3. A brief assessment of the probable coastal zone effects of the proposed activity to the relevant elements of the CMP.
4. A brief set of findings indicating that the proposed activity are consistent with the enforceable and mandatory provisions of the management program. Only "adequate consideration" need be given to CMP policies that are only in the nature of recommendations.[138]

State agency review of a applicant's consistency certification commences as soon as the state agency receives a copy of the certification and supporting information.[139] The state is required to notify the applicant and the relevant federal agency "at the earliest practicable time" whether the it concurs with or objects to the consistency certification. If the state has not determined whether the proposed activity is consistent with the CMP within three months, it must advise the applicant and federal agency of the status of the matter and the basis for any further delay.[140] If no decision has been rendered by the state after six months, consistency is

---

[136] 15 CFR § 930.54(b)-(e). See also 15 CFR § 930.63 and 930.64.
[137] 15 CFR § 930.57(a). The consistency certification must contain the following language: "The proposed activity complies with (name of State) approved coastal management program and will be conducted in a manner consistent with such program." 15 CFR § 930.57(b).
[138] 15 CFR § 930.58(a)(1)-(4).
[139] 15 CFR § 930.60.
[140] 15 CFR § 930.63(b).

conclusively presumed.[141]

Should the state object to the consistency certification, it must notify the applicant, federal agency and the Assistant Administrator.[142] The state's objection must describe how the proposed activity will be inconsistent with the CMP and provide possible means for the applicant to carry out the activity in a way that would be consistent with the CMP if any such alternative measures exist.[143] In the event that the state objects to a consistency determination, various remedies are available to the applicant.[144]

14.5.3 State Consistency Certification

The issuance of state and local permits also depends upon whether the activity for which the permit is sought will be carried out in a manner that is consistent with the state CMP. The procedures for certifying consistency vary from state to state. Generally, the type of CMP will determine how the state consistency requirement is fulfilled. If the state CMP is in the form of a state statute, the statute will contain procedures governing how the consistency determination is to be made. If the CMP is in the form of a policy document relying on existing laws and regulations to carry out the policies, then the various state and local agencies will consider the policies of the state CMP when deciding whether to grant or deny the requested permit. In any case, when the proposed activity will be carried out in the coastal zone, or will impact upon the coastal zone, the state certification of consistency is a requirement in addition to the specific requirements that must be met for individual state and local permits.

14.5.4 Permit Denials

A leading cause for the denial of a permit to carry out a land use activity in the coastal zone is inconsistency with the state's CMP.[145] Technical

---

[141] 15 CFR § 930.63(a).
[142] 15 CFR § 930.64(a).
[143] 15 CFR § 930.64(b).
[144] See section 14.6
[145] For example, see:
California: Acme Fill Corp. v. San Francisco Bay Conservation and Development Commission, 187 Cal.App.3d 1056, 232 Cal. Rptr. 348 (1986).
Connecticut: Vom Saal v. Zoning Commission of the Town of Stratford, 1991 WL 288132 (No. CV91-27-90-21 Conn. Sup. Ct. Dec. 30, 1991);
New Jersey: Colonial Care Convalescent Center v. Division of Coastal Resources, 1989 WL 266476 (N.J. Dept. of Envt'l Prot. OAL Dkt. No. ESA

flaws in a permit applicant's development plan may lead to permit denial due to insufficient information to render a decision.[146] The failure to have a sedimentation control plan in place to further a CMP goal of protecting degradation of water quality has been a cause of a permit denial to construct a single-family residence in the coastal zone.[147] Likewise, evidence that a proposed sewage disposal system would present potential harm to the environment has been a justification for a permit denial in the coastal zone.[148] Building permit denials have also been supported by the adverse impact that the construction would have on the coastal environment.[149] Promotion of aesthetic values pursuant to a local ordinance has been deemed consistent with the goals a state's CMP and, consequently, grounds for a permit denial for construction of a marina.[150] A developer's misunderstanding of the application of a permit exception for reconstruction contained in the state CMP has lead to imposition of fines for violation of the CMP.[151]

14.6 State CMP Administrative Appeals Procedures

Procedures relating to administrative appeals of permit denials for activities proposed in or affecting a state's coastal zone vary from jurisdiction to jurisdiction. Federal permit denials may be taken to the federal agency responsible for issuing the permit. Administrative appeals

---

2519-88, Agency Dkt. No. 88-77 Jan. 12, 1989); Atrium Developers, Inc. v. Dept. of Environmental Protection, 1988 WL 179820 (N.J. Dept. of Env. Prot. OAL Dkt. No. ESA 1515-87, Agency Dkt No. 97-36 Sept. 26, 1988).
New York: Matter of Quogue Associates, Inc., 1983 WL 25872 (N.Y. Dept. Env. Conserv. Tidal Wetlands Permit Application No. 10-82-0688 Oct. 7, 1983).
South Carolina: Beard v. South Carolina Coastal Council, 403 S.E.2d 620, 304 S.C. 205, 22 Envt'l L. Rep.(ELI) 20,036 (1991); Lucas v. South Carolina Coastal Council, 404 S.E.2d 895, 21 Envt'l L.Rep. (ELI) 20,837 (1991).
[146] See, for example, Noank Shipyard, Inc. Noank Fire District Zoning Commission, No. 50-95-42, 1990 WL 265307 (Conn. Sup. Ct. Dec. 7, 1990) (failure to show contour lines, storm drainage, landscaping and traffic flow make it impossible to determine compliance with Coastal Management Act regulations).
[147] Lunghino v. Planning & Zoning Commission of the Town of Westport, No. CV89-0099958-S, 1990 WL 275834 (Conn. Sup. Ct. Oct. 24, 1990).
[148] Milardo v. Coastal Resources Management Council, 434 A.2d 266, 12 Envt'l L.Rep.(ELI) 20,133 (R.I. 1981).
[149] Santini v. Lyons, 448 A.2d 124, 13 Envt'l L.Rep.(ELI) 20,079 (R.I. 1982).
[150] Issuance of a CAMA Minor Development Permit No. 82-0010 to Ford S. Worthy v. Town of Bath, 345 S.E.2d 699, 82 N.C. App. 32 (1986).
[151] Pamlico Marine Co., Inc. v. North Carolina Dept. of Natural Resources and Community Dev., 341 S.E.2d 108, 80 N.C. App. 201 (1986).

regarding a state's objection to the permit applicant's federal consistency certification may be taken directly to the U.S. Secretary of Commerce.[152]

The forum and process of administrative appeals of state and local permit decisions varies from jurisdiction to jurisdiction.[153] The type of permit, the level of government responsible for issuing the permit, and the type of appeals mechanism set up for the permit issuing agency will determine the appropriate procedures for appealing a permit decision. The permit applicant will need to consult the individual state laws to determine the proper method of filing an appeal. Administrative appeals may go to the agency responsible for administering the state CMP (commonly the state department of environmental protection), to a review board set up by the state, or to the agency responsible for issuing the particular permit in question. For some permits, there is no administrative appeals mechanism in place, thereby making judicial review the only available remedy for someone aggrieved by the permit decision. Where local zoning and planning permits are denied, the appeal will ordinarily go to a Zoning Board of Appeals or similar local administrative body.

Under certain circumstances, and depending on the jurisdiction, the administrative appeals process may not need to be exhausted before seeking relief in state or federal courts. Where the administrative appeals process can be bypassed, it may be good strategy to proceed directly to court since the decision of an administrative appeals body is very difficult to overturn on judicial review. Generally speaking, courts will accord great deference to the judgment of the administrative agency, will only review the record of the administrative decision, and will not disturb the agency's ruling unless the decision is not supported by substantial evidence.[154] This standard makes it very difficult to overturn the agency's decision, however, occasionally an appellant will succeed.[155]

---

[152] See section 14.6.1.

[153] See section 14.6.2.

[154] See, for example, Milardo v. Coastal Resources Management Council, 434 A.2d 266, 12 Envt'l L. Rep. (ELI) 20,133 (R.I. 1981); Webb v. North Carolina Dept. of Environment, Health and Natural Resources, 102 N.C. App. 767, 404 S.E.2d 29 (1991).

[155] See, for example, Sakonnet Rogers, Inc. v. Coastal Resources Management Council, 536 A.2d 893 (R.I. 1988) (because agency failed to base its decision on any findings of fact addressing criteria specified in the CMP, decision could not be upheld on appeal); South Carolina Wildlife Fed. v. South Carolina Coastal Council, 296 S.C. 187, 371 S.E.2d 521 (1988) (environmental group successfully challenged agency's decision certifying development project where evidence failed to support decision).

14.6.1 Review of Federal Permit Consistency

The applicant for a federal permit, whose application cannot be approved because of a consistency objection by the state agency in charge of administering the state CMP, may appeal the state's decision directly to the Secretary of Commerce.[156] The applicant has 30 days to file an appeal with the Secretary and send a copy of the notice of appeal to the federal and state agencies involved.[157] The Secretary may dismiss the action for "good cause," which includes the applicant's failure to base the appeal on grounds that the proposed activity either (1) is consistent with the objectives or purposes of the Act or (2) is necessary in the interest of national security.[158] These two grounds are the only ones that the Secretary is empowered to consider in reviewing the proposed activity. The Secretary then issues a written opinion regarding the appeal.[159] The Secretary's decision is considered final agency action for purposes of the Administrative Procedures Act.[160]

14.6.2 Review of State/Local Permit Consistency

Although it is not possible to describe the administrative appeals process under every state's CMP, a general framework can be provided here. Procedural rules and filing periods are unique to each jurisdiction, therefore, it is crucial to determine the specific rules that apply to the particular appeal and forum. As previously shown, state CMPs generally can be categorized as one of two types: (1) where the state CMP has been enacted as its own statutory scheme; or (2) where the state CMP is a set of policies that are implemented through existing laws and regulations.

Generally, where a state's CMP is part of an existing statutory scheme, administrative appeals of permit decisions may be taken to various forums depending on the type of permit and statutory appeals mechanism in place. The state statute that governs the particular permit must be consulted to determine whether administrative appeals are available for the permit in question and, if so, whether the administrative appeals procedure must be

---

[156] 15 CFR § 930.120 to 930.134. Some authority holds that this administrative remedy must first be exhausted before seeking judicial review of a state's consistency determination. See, for example, Acme Fill Corp. v. San Francisco Bay Conservation and Development Comm'n, 187 Cal. App.3d 1056, 232 Cal. Rptr. 348 (1986) (corporation required to exhaust administrative remedy of appeal to Secretary of Commerce).
[157] 15 CFR § 930.125(a).
[158] 15 CFR § 930.128(a)-(d).
[159] 15 CFR § 930.130(c).
[160] 15 CFR § 930.130(d).

used before pursuing judicial review. Administrative appeals may be governed by a state administrative procedures act; appeals may go directly to the permit issuing agency; a specialized board may hear appeals for certain types of permits, such as wetland permits; or the state department of environmental protection may hear administrative appeals for all environmental permits, regardless of which agency issued the permit.

Where a state's CMP has been enacted as its own statutory scheme, the CMP should be consulted to determine the availability of administrative appeals of permit decisions for activities in or affecting the state's coastal zone. In some sense, it is easier to determine the appropriate procedures for filing an appeal because the CMP provides a single review process for permits.

It must be cautioned that where administrative remedies are pursued, careful examination of the appropriate state appeals mechanism is essential to determine the correct filing deadlines and hearing procedures for the permit at issue.

14.7 Coastal Nonpoint Source Pollution Control Program

Effective coastal zone management programs can halt the ongoing pollution that threatens coastal destruction while taking into account the need for various commercial activities that provide employment and economic growth in coastal areas. A difficult environmental problem to control is nonpoint source pollution. Congress determined that a large percentage of the pollutants found in coastal environments were coming from nonpoint sources. Nonpoint source pollution is increasingly recognized as a significant factor in coastal water degradation. Further, in urban areas, storm water and combined sewer overflow are linked to major coastal problems and, in rural areas, runoff from agricultural activities may add to coastal pollution. Compounding this pollution problem is the impact of inadequate sewage treatment on coastal resources, which is at least as significant as pollution from agricultural or industrial sources. According to a study by the Office of Technology Assessment, 1,300 major industries and 600 municipal treatment plants discharge into rivers that flow into coastal waters.[161]

To combat this pollution problem, the Coastal Zone Act Reauthorization Amendments of 1990[162] requires each state with an approved CMP to develop a "Coastal Nonpoint Source Pollution Control Protection Program" to implement coastal land use management measures for

---

[161] Office of Technology Assessment Report, Wastes in Marine Environments (1987), at 66-72.
[162] Pub.L. No. 101-508, 104 Stat. 1388 (1990).

controlling nonpoint source pollution.[163] This provision is certain to be the most difficult for states to comply with, given the difficulty in controlling nonpoint source pollution.

EPA has published national guidelines on "management measures" to control coastal nonpoint sources.[164] Management measures are defined as "economically achievable measures" for the control of pollutants from new and existing nonpoint sources that reflect the "greatest degree of pollutant reduction achievable" through application of the best available nonpoint source control measures and methods.[165] States were required to submit their proposed programs to the Secretary of Commerce and the EPA Administrator by July 1995. Following program approval, the state must implement the program through changes to the state plan for the control of nonpoint source pollution approved under Section 319 of the Clean Water Act,[166] and through changes to the state's CMP.[167] If the state fails to submit an approvable program, the Secretary may withhold a percentage of any CZMA Section 306 grant money, and EPA may withhold portions of any section 319 grant under the Clean Water Act. Grants are available for up to 50 percent of the costs of developing coastal nonpoint pollution control programs.

14.8 Other Coastal Zone Protection Laws

In addition to the CZMA, other federal environmental laws may regulate coastal land use activities. Some of these laws, which specifically control ocean dumping and coastal erosion, are discussed in this section.

14.8.1 Clean Water Act Section 403

Section 403 of the Clean Water Act provides that point source discharges (i.e., industrial and municipal facilities) to the territorial seas (3 nautical miles from the baseline), contiguous zone (to 12 nautical miles), and oceans are subject to regulatory requirements in addition to the technology or water quality-based requirements applicable to typical discharges.[168] The Section 403 requirements are intended to ensure that no unreasonable degradation of the marine environment will occur as a result

---

[163] 16 U.S.C. § 1455b(a)(1).
[164] Guidance Specifying Management Measures for Sources of Nonpoint Pollution of Coastal Waters, EPA/840-B-92-002 (Jan. 1993).
[165] 16 U.S.C. § 1455b(g)(5).
[166] 33 U.S.C. § 1329.
[167] 16 U.S.C. § 1455b(c)(2).
[168] 33 U.S.C. § 1343.

of the discharge and to ensure that sensitive ecological communities are protected. These requirements can include ambient monitoring programs designed to determine degradation of marine waters, alternative assessments designed to further evaluate the consequences of various disposal options, and pollution prevention techniques designed to further reduce the quantities of pollutants requiring disposal and thereby reduce the potential harm to the marine environment. If Section 403 requirements for protection of the ecological health of marine waters are not met, a National Pollutant Discharge Elimination System (NPDES) permit will not be issued.

In assessing the potential effects of a marine discharge during permit application review, the permitting authority evaluates the impact of a marine discharge on the biological community based on ecological, social, and economic factors. Under the provisions of Section 403, the permitting authority can require the permit applicant to provide the information necessary to conduct such an evaluation.

To implement Section 403, the EPA developed Ocean Discharge Guidelines which specify the ecological, social, and economic factors to be used by permit writers when they evaluate the impact of a discharge on the marine environment.[169] The 10 factors to be considered in determining whether unreasonable degradation of the marine environment will occur are as follows:

1. Quantities, composition, and potential bioaccumulation or persistence of the pollutants to be discharges.

2. Potential transport of the pollutants by biological, physical, or chemical processes.

3. Composition and vulnerability of potentially exposed biological communities, including unique species or communities, endangered or threatened species, and species critical to the structure or function of the ecosystem.

4. Importance of the receiving water area to the surrounding biological community, such as spawning sites, nursery/forage areas, migratory pathways, and areas necessary for critical life stages/functions of a marine organism.

5. The existence of special aquatic sites, including (but not limited to) marine sanctuaries/refuges, parks, monuments, national seashores, wilderness areas, and coral reefs/seagrass beds.

6. Potential direct or indirect impacts on human health.

7. Existing or potential recreational and commercial fishing.

8. Any applicable requirements of an approved Coastal Zone Management Plan.

---

[169] 40 CFR Part 125, Subpart M.

9. Such other factors relating to the effects of the discharge as may be appropriate.

10. Marine water quality criteria.

Much of the information necessary to make these evaluations is usually already available to the permitting authority from previous scientific studies, permit evaluations, or other data collection activities. Additional information may be requested from the applicant when necessary to help the permit writer make decisions regarding the permit.

In those cases where there is insufficient information to support a finding of "no unreasonable degradation," applicants must demonstrate that the discharge will not cause "irreparable harm." When the permitting authority makes a determination of no irreparable harm, a permit may be issued while confirmatory data on ecosystem health are gathered for evaluation prior to permit reissuance. These data are collected as part of a monitoring program to assess the impact of the discharge on water, sediment, and biological quality, as well as an assessment of alternative sites for the discharge or disposal of wastewater. Data are also gathered through monitoring compliance with all other conditions of the permit.

14.8.2 Coastal Barrier Resources Act

Much in the way the CZMA is intended to protect, preserve and restore resources in the coastal zone, the Coastal Barriers Resources Act of 1982 (CBRA)[170] was enacted "to minimize the loss of human life, wasteful expenditure of federal revenues, and the damage to fish, wildlife and other natural resources associated with the coastal barriers. . . ."[171] Coastal barriers ecosystems may be degraded or destroyed by development activities. The CBRA seeks to discourage development on coastal barriers through a system of federal subsidy denials.

Although the CZMA and CBRA fulfill similar environmental protectionist roles, the CBRA is quite different in its scope and in the manner by which it is implemented. Both statutes are aimed at protecting coastal resources, however, CZMA authority broadly covers activities in or affecting the "coastal zone," whereas CBRA authority is aimed solely at protection of coastal barriers. CBRA is implemented at the federal level, whereas CZMA implementation is delegated to state and local governments through their federally-approved CMPs.

The CBRA promotes coastal barrier protection through the denial of federal subsidies for new construction in hazard-prone and ecologically

---

[170] 16 U.S.C. § 3501 through 3510.
[171] 16 U.S.C. § 3501(b).

significant coastal areas.[172] These areas are designated for protection by Congress as the "Coastal Barrier Resources System (CBRS)."[173] Federal agencies that administer the various subsidy programs are responsible for ensuring that no federal assistance is provided to units within the CBRS. Without federal assistance, developers may decide that the prohibitive cost and risk of coastal construction far outweigh the profit derived from new development.

In addition to denying federal assistance to build in protected areas, the CBRA also amended the National Flood Insurance Program (NFIP) so that federal flood insurance coverage is prohibited for any new or substantially improved structures within the CBRS.[174] Building permits might still be secured to develop on a designated coastal barrier since the CBRA does not necessarily affect permit approvals under a state's CMP. However, the lack of federal assistance, coupled with the lack of federal flood insurance protection, surely provides a strong disincentive to development on coastal barriers.[175]

14.8.3 Marine Protection, Research and Sanctuaries Act

In 1972, Congress passed the Marine Protection, Research and Sanctuaries Act,[176] commonly referred to as the "Ocean Dumping Act," to "prevent or strictly limit the dumping in ocean waters of any material which would adversely affect human health, welfare, or amenities, or the marine environment, ecological systems, or economic potentialities."[177] To accomplish these goals, Congress prohibited most forms of ocean

---

[172] 16 U.S.C. § 3501(b).

[173] Pursuant to the CBRA, a Coastal Barrier Resource System (CBRS) lists the areas protected under the Act. Only Congress can amend the list. 16 U.S.C. §§ 3502(1)(A)-(B).

[174] A substantially improved structure is one that has received improvements that increase its value by 50 percent or more. U.S. Dep't of the Interior, Coastal Barriers Study Group, Report to Congress: Coastal Barrier Resources System With Recommendations as Required by Section 10 of the Public Law 97-348, The Coastal Barrier Resources Act of 1982, Vol. 1, 80 (1988).

[175] Certain projects are exempt from the restrictions of the Act as long as the projects are consistent with CBRA's goals, such as projects relating to wildlife management, recreation, navigation aids, and Land and Water Conservation Fund property acquisition. 16 U.S.C. § 4601-4 to 4604-11. Also exempt are projects approved pursuant to the the Coastal Zone Management Act, 16 U.S.C. § 1451 through 1464.

[176] 33 U.S.C. § 1401 through 1444.

[177] 33 U.S.C. § 1401(b).

dumping, unless authorized by a permit issued by EPA.[178] Pursuant to the Act, EPA may issue two types of permits: (1) short-term "research" permits or (2) "special" (commercial operating) permits. The Ocean Dumping Act requires that EPA promulgate regulations for "reviewing and evaluating" applications for ocean dumping permits, including consideration of specific environmental factors,[179] as well as alternatives to ocean dumping and incineration. EPA may issue a permit only after it determines that: (1) there are no practicable technological improvements that will reduce adverse impacts, and (2) there are no practicable alternatives available that have less adverse environmental impact or potential risk.[180] Depending on the type of coastal zone development proposed, an EPA Ocean Dumping Act permit may be an additional requirement for the developer or property owner.

Further, a portion of the Act known as the Marine Sanctuaries Act,[181] provides for regulation and monitoring of the use of selected areas of the marine environment valued for their uniqueness, beauty, and historical significance.[182] The U.S. Department of Interior designates certain marine areas as sanctuaries wherein land use activities are prohibited unless a special use permit is issued by the Secretary.

Section 1441(b) provides that a special use permit:

(1) shall authorize the conduct of an activity only if that activity is compatible with the purposes for which the sanctuary is designated and with protection of sanctuary resources;

(2) shall not authorize the conduct of any activity for a period of more than 5 years unless renewed by the Secretary;

(3) shall require that activities carried out under the permit be conducted in a manner that does not destroy, cause the loss of, or injure sanctuary resources; and

(4) shall require the permittee to purchase and maintain comprehensive general liability insurance against claims arising out of activities conducted under the permit and to agree to hold the United States harmless against such claims.[183]

---

[178] See 33 U.S.C. § 1402(b), (c), (e).
[179] The EPA is required to consider the effects of the proposed dumping or incineration on human health and welfare, economic, aesthetic and recreational values, fisheries resources, wildlife, shorelines and beaches, marine ecosystems, the persistence and permanence of such effects, the effect of dumping on alternate uses of the oceans, and whether feasible locations exist beyond the Continental Shelf. See 33 U.S.C. § 1412(a)(B), (C), (D), (E), (F), (H), (I).
[180] 40 CFR § 227.16(a)(1), (2).
[181] 16 U.S.C. § 1431 through 1444.
[182] 16 U.S.C. § 1431(b).
[183] 16 U.S.C. § 1441(b).

If a developer or owner has property within a coastal zone sanctuary, a special use permit would become an additional hurdle in the development process. These permits are only good for five years, unless renewed upon expiration.[184]

14.9 Public Trust Doctrine

The public trust doctrine has played an increasingly important role in protection of public natural resources. Congress enacted laws that implicitly delegated to various federal administrative agenices the power to protect public trust property.[185] The CZMA also contains explicit trust language, declaring as its purpose to "preserve, protect, develop, and where possible, to restore or enhance, the resources of the Nation's coastal zone for this and succeeding generations."[186] Thus, this common law doctrine may serve as an additional means of protecting the coastal zone when regulatory measures prove ineffective.[187]

The public trust doctrine is a longstanding common law doctrine that has its roots in Roman law.[188] The doctrine was later followed in English law[189] and then adopted as a part of the American Common Law.[190] Traditionally, the doctrine was used to establish public rights to navigable waters, such as oceans and rivers, and as a basis for states to assert their public trust ownership of tidelands.[191] The doctrine developed largely as a matter of state law.[192] In 1892, the landmark case on the public trust

---

[184] 16 U.S.C. § 1441(b)(2).

[185] See, for example, The Wild Free-Roaming Horses and Burros Act, 16 U.S.C. § 1331 et seq.; the Federal Land Policy and Management Act, 43 U.S.C. § 1701 et seq.; the National Park Service Act, 16 U.S.C. § 1701 et seq.; and the Comprehensive Environmental Response, Compensation, and Liability Act, 42 U.S.C. § 9601 et seq.

[186] 16 U.S.C. § 1452(1).

[187] See, for example, State v. South Carolina Coastal Council, 289 S.C. 445, 346 S.E.2d 716 (1986).

[188] See R. Sohm, The Institutes: A Textbook of the History and System of Roman Private Law 302-09 (J. Ledlie trans. 3d ed. 1970).

[189] 2 H. Bracton, *On the Laws and Customs of England* 39-40 (S. Thorne trans. 1977).

[190] See, for example, Shively v. Bowlby, 152 U.S. 1 (1894); Barney v. Keokuk, 94 U.S. 324 (1876); Pollard's Lessee v. Hagan, 44 U.S. (3 How.) 212 (1845).

[191] See Arnold v. Mundy, 6 N.J.L. 1 (1821), later overruled by Gough v. Bell, 22 N.J.L. 441, 458-61 (1850), aff'd, Bell v. Gough, 23 N.J.L.. 624 (1852).

[192] Under the common law rule, the boundary of land bordered by a tidal navigable stream extended only to the high water mark. See, for example, State v. Pinckney, 22 S.C. 484, 507-09 (1885).

doctrine, *Illinois Central Railroad v. Illinois,*[193] held that the navigable waters and the lands beneath them were held in trust for the public.[194] The state's control for purposes of this trust could never be alienated or lost, except in the case of parcels of trust land used in promoting the public interest or disposed of without impairing the public's interest in the remaining lands and waters.[195] In the late nineteenth and early twentieth centuries the federal government also asserted public trust powers as a means of protecting federal public lands.[196]

Recent decisions reflect a growing willingness to use the trust doctrine as a means of preventing the development and private exploitation of natural resources placed in the legal category of the public trust. In 1988, the U.S. Supreme Court expanded the common law doctrine in *Phillips Petroleum Co. v. Mississippi,*[197] holding that the public trust in the state of Mississippi includes title to all lands under waters influenced by the ebb and flow of the tide.[198]

The Supreme Court's decision is particularly significant for coastal zone protection. The court rejected Phillips' contention that "navigability -and not tidal influence- has become the *sine qua non* of the public trust interest in tidelands in this country."[199] The Court reviewed its earlier decisions, and reasoned that public trust extended to all lands beneath waters influenced by the ebb and flow of the tide, regardless of the navigability of the waters.[200] By rejecting a requirement of navigability, the court's decision gives additional strength to the states to regulate land use

---

[193] Illinois Central Railroad v. Illinois, 146 U.S. 387 (1892).

[194] Id., 146 U.S. at 459.

[195] Id., 146 U.S. at 452-53.

[196] See, for example, Knight v. United States Land Ass'n, 142 U.S. 161 (1891); Light v. United States, 220 U.S. 523 (1911). See also Wilkinson, The Public Trust Doctrine in Public Land Law, 14 U.C. Davis L. Rev. 269, 279-81 (1980).

[197] Phillips Petroleum Co. v. Mississippi, 484 U.S. 469 (1988).

[198] Id., 484 U.S. at 484.

[199] The dispute arose when the state granted oil and gas leases to land underlying the Bayou LaCroix and eleven small drainage streams in southwestern Mississippi. Record title to this land was held by Phillips Petroleum. The titles were traceable to prestatehood Spanish land grants. Phillips and its predecessors-in-interest paid taxes on the land for over 100 years. Despite Phillips' record title, Mississippi based its assertion of title on its claim that it owned all tidelands in the state and held them in public trust. Id., 484 U.S. at 472.

[200] Id., 484 U.S. at 479-80. The court reasoned that settled case law in Mississippi made clear the state's general public trust interest in tidelands, even though the issue of its claim to nonnavigable tidelands was one of first impression. Id. at 482.

activities further landward of the mean high water mark of the tide.[201] Coastal wetland areas, for instance, which may not be navigable, but which may be influenced by tidal waters, may now be subject to protection under a public trust theory.

---

[201] Commentators have noted that, in effect, the decision will "fortify the operation of the trust as a state tool for economic and environmental control of significant resources [and give strength to] legislatures and activists who choose to assert the public interest more forcefully in an age of ever-increasing property conflicts." See Note, Phillips Petroleum Co. v. Mississippi and the Public Trust Doctrine; Strengthening Sovereign Interest in Tidal Property, 38 Cath.U.L.Rev. 571, 597-98 (1989).

15

Endangered Species and
Sensitive Areas Protection

15.1 History and Policy of Endangered Species
and Sensitive Areas Protection

Wildlife and the natural areas they inhabit have had a profound impact on people since our earliest history as a species. Evidence of this impact exists from the earliest Paleolithic cave drawings to the most modern wildlife art.[1] Professor Rodgers said "[t]he human relationship to wildlife touches the deepest wellsprings of human emotion and behavior . . . [and] forges the best thinking about human ethics and morality."[2]

The indigenous peoples of North America evidently lived in harmony with nature, and incorporated natural themes into their artwork and verbal traditions. However, when colonists from Europe came to North America, they brought with them attitudes that humans were intended to dominate nature and to use nature specifically to benefit humans. These views of nature, often characterized as the "pioneer spirit," dominated American attitudes from Colonial times until the 1960s. A statement attributed to Colonial clergyman and author Cotton Mather manifests the early pioneer attitudes: "What is not useful [in nature] is vicious."

As a direct result of these attitudes, early U.S. federal laws were directed only to preserve economically valuable hunting and fishing resources. In the nineteenth century, regulation of wildlife was considered solely a state responsibility.[3] The first federal law to protect wildlife (however

---

[1] See Pfieffer, *The Creative Explosion: an Inquiry into the Origins of Art and Religion* (Cornell Univ. Press,1982). See also Wilson, *Biophilia* (Harvard Univ. Press, 1984).
[2] Rodgers, *Environmental Law*, 2d ed. (West,1994), hereafter, "Rodgers," at § 9.9.
[3] See Geer v. Connecticut, 161 U.S. 519 (1896).

indirectly) was the Lacey Act of 1900, which provides federal sanctions for interstate transportation of fish or wildlife taken in violation of federal, state, or foreign law.[4] Although it has been frequently amended, the Lacey Act remains a useful tool to prevent overexploitation of wildlife resources.[5] In 1918, the Migratory Bird Treaty Act[6] "g[ave] the Secretary of the Interior power to adopt regulations for the protection of migratory birds [and] concentrated on establishing refuges for wildlife."[7]

It wasn't until the 1960s, with the decade's heightened interest in environmental protection, that Congress began to provide the kind of emphasis on the protection of species that exists today. The Endangered Species Preservation Act of 1966 authorized the Secretary of the Interior to identify "the names of the species of native fish and wildlife found to be threatened with extinction," and to acquire lands to protect threatened wildlife.[8] Unfortunately, the 1966 Act applied only to "native" fish and wildlife, and was limited in that it applied only where "practicable."[9]

In 1969, Congress passed the Endangered Species Conservation Act.[10] The 1969 Act continued the basic approach of the 1966 Act, but also prohibited the importation of any endangered species into the United States.[11] The 1969 Act was the first attempt by Congress to enter the international arena of endangered species protection, but it suffered from the same major drawbacks as did the 1966 Act.

The Endangered Species Act of 1973 (ESA) was passed following Congress' finding that "various species of fish, wildlife, and plants in the United States have been rendered extinct as a consequence of economic growth and development untempered by adequate concern and conservation."[12] As will be discussed below, the 1973 Act is far more

---

[4] 16 U.S.C. §§ 3371-78, and 18 U.S.C. § 42.
[5] See Maine v. Taylor, 477 U.S. 131 (1986) [bait store owner imported out of state baitfish in violation of Maine law]. For a thorough discussion of the modern Lacey Act, see Littell, *Endangered and Other Protected Species: Federal Law and Regulation* (Bureau Nat'l Afairs, 1992) at Chapter 10.
[6] 16 U.S.C. 703 et seq. See Rholf, *the Endangered Species Act: a Guide to its Protection and Implementation* (Stanford Env'l Law Society, 1989).
[7] TVA v. Hill, 437 U.S. 153 (1978). For an interesting review of the influence of ecology and ecological thinking on the development of American law, see Bosselman and Tarlock, The Influence of Ecological Science on American Law: An Introduction, 69 Chi.-Kent L. Rev. 847 (1994).
[8] Pub. L. No. 89-669, 80 Stat. 926 (1966). See Tennessee Valley Authority v. Hill, 437 U.S. 153, 175 (1978).
[9] See Littell, *Endangered and Other Protected Species: Federal Law and Regulation* (Bureau Nat'l Afairs, 1992) at Chapter 2.
[10] Pub. L. No. 91-135, 83 Stat. 275 (1969).
[11] See Delbay Pharmaceuticals v. Dept. of Commerce, 409 F. Supp. 637 (D.D.C. 1976).
[12] 16 U.S.C. § 1531(a)(1). See Coggins, An Ivory Tower Perspective on Endangered Species, 8(1) Natural Resources & Env't 3 (1993).

expansive than either the 1966 or 1969 Acts. The ESA is probably the most important federal habitat-protection law because an endangered or threatened species has a "critical habitat" that cannot be developed or converted to human use under most circumstances. It is one of the few environmental laws which places an absolute prohibition on destroying a resource, although, as we will see, the prohibition may actually be something less than absolute. Of equal significance, the ESA prohibits the kind of "cost-benefit" analysis that characterizes most environmental laws.[13]

In addition to changes in how Americans regarded wildlife, attitudes about the relationship between humans and the land also began to change as early as the nineteenth century. Emergence of the "environmental ethic" (Aldo Leopold, Garrett Hardin, and others) led to increased public sensitivity to lost or degraded natural resources. During the 1870s and 1880s, George Perkins Marsh and Gifford Pinchot argued that advances in ecological theory could be used to allocate the earth's natural resources in a way that mutually benefitted both humans and nature. This view led to the development of the "conservationist" movement (i.e., careful management of the environment would ensure plentiful natural resources for the future). This view dominated policy decisions until the 1960s.[14]

Also during the 1870s and 1880s, John Muir and others advocated a form of natural morality that supported the preservation of large tracks of pristine wilderness. The "preservationist" movement resulted, and was largely responsible for developing the nation's system of national parks and wilderness areas, but the movement lacked the large public support enjoyed by the conservationists until the 1960s.

15.1.1 A Primer on "Habitats"

Discussion of the protection of wildlife or sensitive lands often begin with the term "habitat" (for example, several sections of the ESA speaks of the "critical habitat" of an endangered or threatened species). However, the legal and biological significance of "habitat" is often different. "Habitat" is often defined by nonbiologists as "the natural environment in which a plant or animal species lives." Many laws and regulations which protect natural resources speak in terms of fish and wildlife habitat.

---

[13] For an excellent review of the philosophy of endangered species protection, see Norton, *Why Preserve Natural Variety?* (Princeton Univ. Press, 1987).

[14] For an interesting discussion of the various ethics that have influenced American's relationship with the land, see Bosselman, Four Land Ethics: Order, Reform, Responsibility, Opportunity, 24 Envtl. L. 1439 (1994), and the sources cited therein.

"Critical habitat" must be set aside under the Federal Endangered Species Act.

The biologists, the concept of "ecosystem" has come to have a greater significance than does "habitat" in the traditional sense. Biologists have long argued that the concept is more meaningful than "habitat." An "ecosystem" includes all living (biotic) and nonliving (abiotic) components within an environment. Ecosystems are defined primarily by dominant plant species (e.g., "sphagnum bog," or "pine forest"). To conservationists, protecting ecosystems is especially important because it provides for the protection of numerous species (not just a particular target species), and protects environmental components on which they depend.[15]

A second biological issue related to habitats is how "big" is a habitat? It depends on the species. For some large, free-ranging species (e.g., wolves), the critical habitat may be many square miles. For smaller, stationary species (e.g., some plants), the habitat may be a few square meters. For migratory species, habitat is shifted during the season. In designing habitat "reserves" for protected species, it is often necessary to protect connecting "corridors" for movement between "patches" of habitat.

In recent years, biologists have discovered the importance of protecting "biodiversity,"[16] which can be defined as the number of different species of organisms (plants and animals) in an ecosystem. Healthy ecosystems consist of numerous groups of interactive and interdependent plants and animals. Biologists have determined that global (and often local) biodiversity is rapidly declining, which is a matter of great urgency for several reasons. First, rapidly increasing human populations are degrading the natural environment at an alarming rate. Second, science is finding new uses for biodiversity to relieve human suffering (e.g., new medicines) and abate environmental destruction. Third, biodiversity is being lost through increasing rates of extinction caused by destruction of natural habitats.

In response to losses of biodiversity in the United States, President Clinton and the U.S. Department of the Interior have launched a "National Biological Survey" under the auspices of the U.S. Fish and Wildlife Service (USFWS) to assess the condition of biodiversity in the United

---

[15] See Blair, Collins, and Knapp, Ecosystems as Functional Units in Nature. 14(3) Natural Resources & Env't 150 (2000); Ruhl, Ecosystem Management, the ESA, and the Seven Degrees of Relevance, 14(3) Natural Resources & Env't 156 (2000) [discussing the relationship between ecosystems and the ESA]; and Hodas, NEPA, Ecosystem Management and Environmental Accounting, 14(3) Natural Resources & Env't 185 (2000) [discussing the role of NEPA in ecosystem management].

[16] See the articles in Wilson (ed.), *Biodiversity* (Nat. Acad. Press, 1988); and Wilson, *the Diversity of Life* (W.W. Norton & Co., 1992).

States, and to recommend measures to reverse the trend in lost biodiversity.

15.1.2 Modern Approaches to Protecting Endangered Species and Sensitive Areas

The centerpiece of federal wildlife protection is the Endangered Species Act of 1973. There are, however, many others. Another federal statute that protects both land and, therefore, wildlife resources is the Coastal Zone Management Act, which places "stringent controls" (i.e., no access to federal flood insurance) on activities which adversely impact the coastal environment.[17] Still other federal laws that protect both wildlife and sensitive lands are the National Wildlife Refuge System Administration Act of 1966, which sets aside "refuges," but permits their use for "any purpose, including but not limited to hunting, fishing, public recreation, . . .," the Migratory Bird Conservation Act (and the associated "Stamp Act" and "Wetlands Act"), and the "National Environmental Policy Act" (NEPA), requires an Environmental Impact Statement (EIS) for federal actions that affect environmental resources.

Still other laws provide subsidies to states for the purchase of land: the Pittman-Robertson Program; the Dingell-Johnson Program; the Land and Water Conservation Fund Act; and the Lacey Act Amendments of 1981.

15.2 Overview of the Endangered Species Act

15.2.1 Section 2: Findings, Purposes, and Policy

In passing the ESA, Congress made a strong statement that plant and animal species are important, and that they are worthy of protection. ESA Section 2 contains the strong language used by Congress to show how seriously they regarded the plight of fish, wildlife, and plants:

(a) Findings
The Congress finds and declares that —
(1) various species of fish, wildlife, and plants in the United States have been rendered extinct as a consequence of economic growth and development untempered by adequate concern and conservation;
(2) other species of fish, wildlife, and plants have been so depleted in numbers that they are in danger of or threatened with extinction;

---

[17] 16 U.S.C. § 1451–1464. See Chapter 14.

(3) these species of fish, wildlife, and plants are of esthetic, ecological, educational, historical, recreational, and scientific value to the Nation and its people;

(4) the United States has pledged itself as a sovereign state in the international community to conserve to the extent practicable the various species of fish or wildlife and plants facing extinction, pursuant to - [treaties with foreign nations] and

(5) encouraging the States and other interested parties, through Federal financial assistance and a system of incentives, to develop and maintain conservation programs which meet national and international standards is a key to meeting the Nation's international commitments and to better safeguarding, for the benefit of all citizens, the Nation's heritage in fish, wildlife, and plants.[18]

ESA Section 2 also states the general purposes of the Act:

(b) Purposes
The purposes of this chapter are to provide a means whereby the ecosystems upon which endangered species and threatened species depend may be conserved, to provide a program for the conservation of such endangered species and threatened species, and to take such steps as may be appropriate to achieve the purposes of the treaties and conventions set forth in subsection (a) of this section.[19]

Finally, ESA Section 2(c) sets out the general policies of the ESA:

(1) It is further declared to be the policy of Congress that all Federal departments and agencies shall seek to conserve endangered species and threatened species and shall utilize their authorities in furtherance of the purposes of this chapter.

(2) It is further declared to be the policy of Congress that Federal agencies shall cooperate with State and local agencies to resolve water resource issues in concert with conservation of endangered species.[20]

15.2.2 Section 7: the "Jeopardy Provision"

Section 7 of the ESA states that every federal agency "shall . . . insure that any action [by the agency] is not likely to jeopardize the continued existence of any endangered species or threatened species, or result in the

---

[18] 16 U.S.C. § 1531(a).
[19] 16 U.S.C. § 1531(b).
[20] 16 U.S.C. § 1531(c).

destruction or adverse modification" of the critical habitat of the species.[21] It is important to note that Section 7 applies to all federal agency actions, but not to private individuals.

An "action" is defined in Department of the Interior regulations as "all activities or programs of any kind authorized, funded, or carried out, in whole or in part, by Federal agencies."[22] "Agency action" is defined in the statute as "any action authorized, funded, or carried out by" a federal agency. [23] Courts have construed "agency action" broadly.[24] In case of doubt whether a particular development project constitutes "federal agency action" or not, it is usually most prudent to assume that it does.[25]

If an agency has several "reasonable and prudent alternative" actions, it is allowed to pick the one that "best suits all of its interests, including political or business interests."[26]

The ESA Section 7 jeopardy provision has been quite controversial. Some members of Congress became concerned in the 1970s that several large, public projects might be thwarted due to ESA Section 7 concerns. In response, Congress created the "Endangered Species Committee" following the 1978 amendments. The Endangered Species Committee is composed of six members of the President's cabinet and subcabinet, plus one representative from each affected state. It has come to be known as the "God committee" because it can issue an "exemption" when a federal action may violate the Section 7 jeopardy provision.

The Committee has been convened only four times. The most significant was 1992 when the Committee voted 5-2 to allow logging in Oregon's old-growth forests, the critical habitat of the northern spotted owl.[27]

---

[21] 16 U.S.C. § 1536(a)(2). See Bennett v. Spear, 520 U.S. 154 (1997).

[22] 50 C.F.R. § 402.02.

[23] 16 U.S.C. § 1536(a)(2).

[24] See Conner v. Burford, 848 F.2d 1441 (9th Cir. 1988, *cert. denied sub nom.*, Sun Exploration & Production Co. v. Lujan, 489 U.S. 1012 (1989).

[25] But see Proffitt v. Dept. of the Interior, 825 F. Supp. 159 (W.D. Ky. 1993) [EPA technical assistance for an environmental assessment of a sewage treatment plant was *not* federal agency action].

[26] 50 C.F.R. § 402.14(h). See Southwest Center for Biodiversity v. U.S. Bureau of Reclamation, 143 F.3d 515 (9th Cir. 1998).

[27] See Thornton, The Search for a Conservation Planning Paradigm: Section 10 of the ESA, 8(1) Natural Resources & Env't (1993); and, *Endangered and Other Protected Species: Federal Law and Regulation* (Bureau Nat'l Afairs, 1992) at Chapter 6.

15.2.3 Section 7: the "Duty to Conserve"

Under Section 7 of the ESA, the Secretary of the Interior has an obligation to use programs which seek to conserve imperiled species "by carrying out programs for the conservation" of the imperiled species.[28] Of equal significance, however, is the mandate by ESA Section 4 that *all* federal agencies participate in conservation efforts:

> All other Federal agencies shall, in consultation with and with the assistance of the Secretary, utilize their authorities in furtherance of the purposes of this chapter by carrying out programs for the conservation of endangered species and threatened species listed pursuant to section 1533 [ESA Section 4]of this title.[29]

ESA Section 2(c)(1) reinforces this mandate in discussing congressional policy:

> It is further declared to be the policy of Congress that all Federal departments and agencies shall seek to conserve endangered species and threatened species and shall utilize their authorities in furtherance of the purposes of this chapter.[30]

Federal courts have interpreted this mandate strictly. One court held that the Department of the Interior must use "all methods which are necessary to bring any endangered . . . or threatened species to the point at which the [conservation measures under the Endangered Species Act] . . . are no longer necessary."[31] In addition, the Department of the Interior "must do far more than merely avoid the elimination of protected species. It must bring those species back from the brink [of extinction] so that they may be removed from the protected class, and it must use all methods necessary to do so."[32]

Some courts have given agencies broad discretion in choosing mitigation

---

[28] 16 U.S.C. § 1536(a)(1).

[29] Id. See Carson-Truckee Water Conservation District v. Clark, 741 F.2d 257, 261 (9th Cir. 1984), cert. denied sub nom. Nevada v. Hodel, 470 U.S. 1083 (1985) [all federal agencies must carry out programs for the conservation of endangered and threatened species].

[30] 16 U.S.C. § 1531(c)(1).

[31] National Wildlife Federation v. Hodel, 23 Env't Rep. (Cas) 1089, 1092 (E.D. Cal. 1985) [prohibiting the Secretary of the Interior from allowing hunters to use lead shot due to poisoning of bald eagles].

[32] Defenders of Wildlife v. Andrus, 428 F. Supp. 167, 170 (D.D.C. 1977) [disallowing pre-dawn game bird hunting since poor visibility might result in endangered species being shot by accident].

measures,[33] while others have imposed a more rigorous standard on the agencies.[34]

15.2.4 Section 7: the Consultation Requirement

As noted in section 15.2.3 above, Section 7(a) of the ESA requires federal agencies to act "in consultation with" the Secretary of the USFWS.[35] USFWS regulations state that the consultation process begins with a conference:

> (a) Each Federal agency shall confer with the Service[36] on any action which is likely to jeopardize the continued existence of any proposed species or result in the destruction or adverse modification of proposed critical habitat. The conference is designed to assist the Federal agency and any applicant[37] in identifying and resolving potential conflicts at an early stage in the planning process. . .
>
> (c) A conference between a Federal agency and the Service shall consist of informal discussions concerning an action that is likely to jeopardize the continued existence of the proposed species or result in the destruction or adverse modification of the proposed critical habitat at issue. Applicants may be involved in these informal discussions to the greatest extent practicable. During the conference, the Service will make advisory recommendations, if any, on ways to minimize or avoid adverse effects. If the proposed species is subsequently listed or the proposed critical habitat is designated prior to completion of the action, the Federal agency must review the action to determine whether formal consultation is required. . .

---

[33] See Hawksbill Sea Turtle v. FEMA, 11 F. Supp.2d 529 (D.V.I. 1996), aff'd 126 F.3d 461 (3d. Cir. 1997).

[34] See Sierra Club v. Glickman, 156 F.3d 606 (5th Cir. 1998).

[35] 16 U.S.C. §§ 1536(a)(1) and (2).

[36] The "service" is defined as "the U.S. Fish and Wildlife Service or the National Marine Fisheries Service, as appropriate." 50 CFR § 402.02.

[37] "Applicants" are defined as "any person, as defined in section 3(13) of the Act, who requires formal approval or authorization from a Federal agency as a prerequisite to conducting the action." 50 CFR § 402.02. A "person" under ESA § 3(13) is "an individual, corporation, partnership, trust, association, or any other private entity; or any officer, employee, agent, department, or instrumentality of the Federal Government, of any State, municipality, or political subdivision of a State, or of any foreign government; any State, municipality, or political subdivision of a State; or any other entity subject to the jurisdiction of the United States." 16 U.S.C. § 1532(13). This is the same definition of "person" that applies to ESA § 9 "takings" discussed in section 15.2.5.

(e) The conclusions reached during a conference and any recommendations shall be documented by the Service and provided to the Federal agency and to any applicant. The style and magnitude of this document will vary with the complexity of the conference. If formal consultation also is required for a particular action, then the Service will provide the results of the conference with the biological opinion.[38]

If it appears from informal consultation that a protected species may be present in the area of a "major construction activity" by a federal agency, then the agency must prepare a "biological assessment."[39] The purpose of the biological assessment is to:

[E]valuate the potential effects of the action on listed and proposed species and designated and proposed critical habitat and determine whether any such species or habitat are likely to be adversely affected by the action and is used in determining whether formal consultation or a conference is necessary.[40]

The biological assessment may then by used as part of an Environmental Impact Statement or an Environmental Assessment under NEPA.[41]

If the biological assessment indicates a likely effect on a protected species, then the agency must begin formal consultation. If the species is proposed to be listed as an endangered or threatened species but is not yet listed, then the agency is required only to "cofer" with the Service.[42] When requesting a formal consultation in writing, the agency must include:

(1) A description of the action to be considered;

(2) A description of the specific area that may be affected by the action;

(3) A description of any listed species or critical habitat that may be affected by the action;

(4) A description of the manner in which the action may affect any listed species or critical habitat and an analysis of any cumulative effects;

---

[38] 50 CFR § 402.10.

[39] 50 CFR § 402.12(b).

[40] 50 CFR § 402.12(a).

[41] See the discussion of NEPA in chapter 3. See Fund for Animals, Inc. v. Thomas, 127 F.3d 80 (D.C. Cir. 1997) [USFWS biological opinion used in Forest Service's Environmental Assessment adequately described the environmental consequences of "game baiting" practices].

[42] 16 U.S.C. § 1536(a)(4). See Enos v. Marsh, 769 F.2d 1363 (9th Cir. 1985).

(5) Relevant reports, including any environmental impact statement, environmental assessment, or biological assessment prepared; and

(6) Any other relevant available information on the action, the affected listed species, or critical habitat.[43]

The agency requesting formal consultation must provide the Service with the best scientific and commercial data available, or which can be obtained during the consultation, for an adequate review of the effects that an action may have upon listed species or critical habitat. The information may include the results of studies or surveys conducted by the agency or by a designated representative. The agency must also provide any applicant with the opportunity to submit information for consideration during the consultation.[44]

The formal consultation process concludes 90 days after it is initiated, unless it is extended to gather additional data.[45] The formal consultation process ends when the Service issues a "biological opinion," which contains:

(1) A summary of the information on which the opinion is based;

(2) A detailed discussion of the effects of the action on listed species or critical habitat; and

(3) The Service's opinion on whether the action is likely to jeopardize the continued existence of a listed species or result in the destruction or adverse modification of critical habitat (a "jeopardy biological opinion"); or, the action is not likely to jeopardize the continued existence of a listed species or result in the destruction or adverse modification of critical habitat (a "no jeopardy" biological opinion). A "jeopardy" biological opinion shall include reasonable and prudent alternatives, if any. If the Service is unable to develop such alternatives, it will indicate that to the best of its knowledge there are no reasonable and prudent alternatives.[46]

Federal agencies that do not follow the consultation requirements of ESA Section 7(a) can be sued. In *Natural Resources Defense Council v. Houston*, an environmental group sued the Bureau of Reclamation for renewing water supply contracts on the San Joaquin River without

---

[43] 50 CFR § 402.14(c).
[44] 50 CFR § 402.14(c).
[45] 50 CFR §§ 402.14(e) and (f).
[46] 50 CFR § 402.14(h).

consulting the National Marine Fisheries Service.[47] The district court granted a motion in favor of the environmental group and the Ninth Circuit affirmed, holding that the Bureau of reclamation had a legal obligation to request a formal consultation under the circumstances.[48]

15.2.5 Section 9: the "Taking" Provision

Without question, ESA Section 9 is the most contentious provision in the ESA. Simply stated, ESA Section 9 states that:

[I]t is unlawful for any person subject to the jurisdiction of the United States to — . . . (B) take any [endangered or threatened] species within the United States or the territorial sea of the United States.[49]

The meaning of the word "take" is made more clear by the ESA Section 3 definition: "The term 'take' means to harass, harm, pursue, hunt, shoot, wound, kill, trap, capture, or collect, or to attempt to engage in any such conduct."[50]

A controversy erupted in the early 1990s over what, exactly, was meant by the term "harm" in the definition of "take" in ESA Section 3. USFWS regulations define the term as:

Harm in the definition of "take" in the Act [ESA] means an act which actually kills or injures wildlife. Such act may include significant habitat modification or degradation where it actually kills or injures wildlife by significantly impairing essential behavioral patterns, including breeding, feeding or sheltering.[51]

A series of property owners challenged the USFWS definition, arguing that destruction of property that did not directly injure the animals themselves (red-cockaded woodpeckers and northern spotted owls) could not rise to the level of a "take" of a protected species in violation of ESA Section 9. The D.C. Circuit Court of Appeals initially agreed with the property owners, and held that "harm" to an endangered or threatened species requires a direct application of force against the animals, not the

---

[47] Natural Resources Defense Council v. Houston, 146 F.3d 1118 (9th Cir. 1998). See also Thomas v. Peterson, 753 F.2d 754 (9th Cir. 1985) [U.S. Forest Service violated ESA § 7 by allowing timber road in national forest without consultation with USFWS].
[48] Id.
[49] 16 U.S.C. § 1538(a)(1).
[50] 16 U.S.C. § 1532(19).
[51] 50 CFR § 17.3.

indirect effects of habitat modification.[52] However, in a controversial 6-3 decision, the U.S. Supreme Court reversed the D.C. Circuit, holding that modification of critical habitat is "harm" under the statute.[53] Subsequent federal courts have struggled with determinations of what is or is not "harm" to a protected species. Some courts still require direct evidence of physical injury to the protected species before ruling that an ESA Section 9 "take" has occurred,[54] while others require less evidence of direct injury.[55]

In another development, local governments have been held liable for takings committed by third parties if the local government could have prohibited the actions that caused the harm.[56] The local government may not be liable, however, if it has no discretion regarding the acts that caused the take of the protected species.[57]

A similar problem has arisen with the definition for "harass" within the ESA Section 3 meaning of "take." As it currently appears, "harass" is defined thus:

Harass in the definition of "take" in the Act means an intentional or negligent act or omission which creates the likelihood of injury to wildlife by annoying it to such an extent as to significantly disrupt normal behavioral patterns which include, but are not limited to, breeding, feeding, or sheltering.[58]

In the past, some federal courts have found that acts might not rise to the level of a "take" under ESA Section 9 because the ESA itself does not define "harass."[59]

---

[52] Babbitt v. Sweet Home Chapter of Communities for a Greater Oregon, 30 F.3d 190 (D.C. Cir. 1994).

[53] Babbitt v. Sweet Home Chapter of Communities for a Greater Oregon, 115 S.Ct. 2407 (1995).

[54] See Hawksbill Sea Turtle v. FEMA, 11 F. Supp.2d 529 (D.V.I. 1998) ["no direct evidence of Tree Boas that have died or been injured"]; and Coastside Habitat Coalition v. Prime Properties, Inc., No. C97-4025 CRB, 1998 WL 231024 (N.D. Cal. April 30, 1998) [applied an "imminent danger"for injunctive relief to be awarded].

[55] See United States v. Town of Plymouth, 6 F. Supp.2d 81 (D. Mass. 1998).

[56] See Strahan v. Coxe, 127 F.3d 155 (1st Cir. 1997); Loggerhead Turtle v. County Council of Volusia County, 148 F.3d 1231 (11th Cir. 1998); and United States v. Town of Plymouth, 6 F. Supp.2d 81 (D. Mass. 1998).

[57] See Strahan v. Linnon, No. 97-1787, 1998 U.S. App. LEXIS 16314 (First Cir. July 16, 1998).

[58] 50 CFR § 17.3.

[59] see United States v. Hayashi, 5 F.3d 1278 (9th Cir. 1993) [tuna fishermen were not guilty of a "take" of endangered porpoises under the MMPA by firing rifles in the water, because the Act does not define the term "harass"].

Furthermore, it is also unlawful for any person to "possess, sell, deliver, carry, transport, or ship, by any means whatsoever, any such species . . .".[60] This section is aimed primarily at halting the possession of animal "parts" when they are from an endangered or threatened species. Some defendants have argued that they are "innocent owners" of the parts in that they could not reasonably be expected to know that the parts are from a protected species. However, federal courts have used a "strict liability" approach, and held the defendants guilty of a "take" under the ESA regardless of their degree of fault.[61]

15.2.6 Section 9 Applies to Private Individuals

Unlike ESA Section 7, which applies only to agencies, ESA Section 9's prohibition against "taking" of protected species applies "to any person."[62] Courts have interpreted "person" loosely, and have applied the ESA Section 9 "taking" prohibition to a variety of activities. For example, private hunters have been prohibited from selling protected bald eagle feathers,[63] and from shooting a grizzly bear[64] or a Florida panther,[65] even if the act was allegedly done to protect private property. On the other hand, it is not a "taking" if the person has a good faith belief that they are protecting themselves, a family member, or another individual from bodily harm by an endangered species.[66]

As noted above, a local government may be liable for takings by third parties if it could have prevented them.[67]

---

[60] 16 U.S.C. § 1538(a)(1)(D).

[61] See United States v. One Handbag of Crocodilus Species, 856 F. Supp. 128 (E.D.N.Y., 1994) [An "innocent owner" of protected crocodile hide items is still liable under the ESA, since a "strict liability" approach is necessary to effect congressional intent].

[62] 16 U.S.C. § 1538(a)(1).

[63] United States v. Dion, 762 F.2d 674 (8th Cir. 1985), *rev'd in part on other grounds*, 476 U.S. 734 (1986).

[64] United States v. St. Onge, 676 F. Supp. 1041 (D. Mont. 1987). See also United States v. Clavette, 135 F.3d 1308 (9th Cir. 1998) [individual sentenced to 3 years' probation, fined $2,000, and ordered to pay $6,250 in restitution to the USFWS for killing grizzly bear in violation of ESA § 9].

[65] United States v. Billie, 667 F. Supp. 1485 (S.D. Fla. 1987).

[66] 16 U.S.C. § 1540(b)(3).

[67] See Strahan v. Coxe, 127 F.3d 155 (1st Cir. 1997); United States v. Town of Plymouth, 6 F. Supp.2d 81 (D. Mass. 1998); and Loggerhead Turtle v. County Council of Volusia County, 148 F.3d 1231 (11th Cir. 1998).

15.2.7 ESA Enforcement

Violations of ESA Section 9 are taken seriously.[68] Certainly the most formidable enforcement is criminal. ESA Section 11(b)(1) states that "any person who knowingly violates any provision of this chapter, of any permit or certificate issued hereunder, or of any regulation . . ." is subject to criminal sanctions.[69] The key challenge for the government is proving that the defendant "knowingly" violated the ESA.[70] A person who is convicted of violating the ESA can "be fined not more than $50,000 or imprisoned for not more than one year, or both."[71]

A frequently employed alternative to criminal enforcement are civil penalties. ESA Section 11(a) states that "any person who knowingly violates, and any person engaged in business as an importer or exporter of fish, wildlife, or plants who violates, any provision of this chapter, or any provision of any permit or certificate issued hereunder, or of any regulation . . . may be assessed a civil penalty by the Secretary of not more than $25,000 for each violation."[72] As with any civil trial, the burden of proof is lower in civil enforcement, and is an easier burden on the agency.

In some instances, the agency's only remedy is forfeiture of the protected species or parts, or of the equipment used in collecting or transporting the protected species. To obtain forfeiture, the agency must only make a *prima facie* showing of a violation, after which the burden shifts to the defendant to rebut the charge.[73]

15.3 The Listing Process Under ESA Section 4

The current "listing" process is found in Section 4 of the ESA, and is based on 1982 amendments which streamlined the process over earlier

---

[68] See Littell, *Endangered and Other Protected Species: Federal Law and Regulation* (Bureau Nat'l Afairs, 1992) at Chap. 7; and Sobeck, Enforcement of the Endangered Species Act, 8(1) Natural Resources & Env't 30 (1993).

[69] 16 U.S.C. § 1540(b)(1).

[70] See United States v. Billie, 667 F. Supp. 1485 (S.D. Fla. 1987); and United States v. Nguyen, 916 F.2d 1016 (5th Cir. 1990).

[71] 16 U.S.C § 1540(b)(1). Updated sentencing laws have increased the maximum fine to not more than $100,000, and increased jail sentences. See Sobeck, Enforcement of the Endangered Species Act, 8(1) Natural Resources & Env't 30 (1993).

[72] 16 U.S.C. § 1540(a)(1). Again, updated sentencing laws have increased the maximum civil penalties.

[73] 16 U.S.C. § 1540(e)(4)(A). See United States v. 2,507 Live Canary Winged Parakeets, 689 F. Supp 1107 (S.D. Fla. 1988).

versions.[74] The process begins when an interested person submits a petition to the Secretary of the Interior (or USFWS). Within 90 days, the Secretary of the Interior must publish a finding of whether or not the petition contains sufficient scientific and commercial information to support action, and must then begin a status review.[75] A finding may be postponed beyond the initial 90 days for budgetary reasons.[76] Within 12 months, the Secretary must publish one of three findings (usually in the Federal Register): (1) the action is not warranted, (2) the action is warranted, or (3) the action is warranted, but is precluded by other pending listing proposals.[77] If the action is "warranted but precluded," the Secretary must demonstrate that "expeditious progress" is taking place to list qualified species.[78]

If the Secretary finds that the action is warranted, a notice must be published along with the proposed regulation, and a public hearing held if one is requested. Within one year, the Secretary must adopt or withdraw the regulation. Once the species is "listed," the listing must be challenged in court since there is no administrative appeal.[79]

Emergency listings become effective as soon as published following a 240-day streamlined rulemaking procedure. These can occur if the Secretary finds that a species is in "imminent peril."[80] The Secretary's emergency listing of the desert tortoise successfully survived a legal challenge in *City of Las Vegas v. Lujan*.[81]

At this stage of the procedure, the Secretary must determine if a listed species is "endangered" (in danger of extinction), or "threatened" (likely to become an endangered species in the foreseeable future). The Secretary

---

[74] 16 U.S.C. § 1533. For excellent reviews of the listing process, see Littell, *Endangered and Other Protected Species: Federal Law and Regulation* (Bureau Nat'l Afairs, 1992) at Chapter 3; and Ruhl, Section 4 of the ESA - The Cornerstone of Species Protection Law, 8(1) Natural Resources & Env't 26 (1993).

[75] 16 U.S.C. § 1533(b)(3)(A), see Marbled Murrelet v. Babbitt, No. 91-522R (W.D. Wash, 1993) [the USFWS was ordered to issue a rule proposing critical habitat using the best scientific information then available, although the USFWS argued insufficient information].

[76] See Biodiversity Legal Foundation v. Babbitt, 146 F.3d 1249 (10th Cir. 1998).

[77] 16 U.S.C. § 1533(b)(3)(B). One court ruled that the 12 month period runs from the receipt of the petition, not from the date of the preliminary findings. Biodiversity Legal Foundation v. Babbitt, 63 F. Supp. 2d 31 (D.D.C. 1999) [23 month delay in making a preliminary finding exceeded the statutory maximum].

[78] 16 U.S.C. § 1533(b)(3)(B)(iii)(II).

[79] See Glover River Org. v. Dept. of Interior, 675 F.2d 251 (10th Cir. 1982); and Pacific Northwest Generating Co-op v. Brown, 822 F.2d 1479 (D. Or. 1993).

[80] 16 U.S.C. § 1533(b)(7).

[81] City of Las Vegas v. Lujan, 891 F.2d 927 (D.C. Cir. 1989).

must also determine the species,' "critical habitat" to assure protection,[82] and implement a "recovery plan."[83] All may be challenged in federal court,[84] and will be discussed more fully below.

Potential candidate species are placed into one of two categories while the listing process is completed. Species for which adequate data exist are listed as "category 1." About 600 U.S. species are in category 1. Species for which there are inadequate data are placed into "category 2," of which there are over 3,000 U.S. species.

Interestingly, it has been held that the ESA Section 4 listing process does not require the preparation of an EIS under NEPA because the designation does not alter the natural physical landscape (see chapter 3).[85] However, the federal courts are split as to whether an EIS is required prior to designating "critical habitat" for a listed species.[86]

Under ESA Section 4(b)(6)(B),[87] the Secretary's decision to list (or not to list) a species can be appealed to federal district court. However, the plaintiffs must demonstrate that they have standing, and that the case is ripe.[88]

A species may be "delisted" (i.e., removed from the list) if: (1) it is extinct; (2) it has recovered to the point that it no longer requires ESA protection; or (3) it was originally listed based on erroneous data.[89] For example, in 1998 The Bald Eagle was reclassified from "endangered" to "threatened."

As noted above, the ESA authorizes two levels of protection: "endangered" and "threatened."[90] While the statute specifically authorizes the protection of species determined to be *endangered*, Congress delegated to the Secretary the authority to protect *threatened* species.[91] In fact, the Secretary has authorized virtually identical protection for *threatened* species in the regulations.[92]

---

[82] 16 U.S.C. § 1333(b).
[83] 16 U.S.C. § 1533(f).
[84] See Trinity County Concerned Citizens v. Babbitt, No. 92-1194 (D.D.C. 1993) [unsuccessful challenge to Northern spotted owl critical habitat designation].
[85] Douglas County v. Babbitt, 48 F.3d 1495 (9th Cir. 1995).
[86] Compare Douglas County v. Babbitt, 48 F.3d 1495 (9th Cir. 1995) [no EIS required]; with Catron County. v. U.S. Fish & Wildlife Serv., 75 F.3d 1429 (10th Cir. 1996) [EIS required].
[87] 16 U.S.C. § 1533(b)(6)(B)(ii).
[88] See Glover River Org. v. Dept. of Interior, 675 F.2d 251 (10th Cir. 1982).
[89] 50 CFR § 424.11(d).
[90] See Sierra Club v. Clark, 755 F.2d 608 (8th Cir. 1985).
[91] 16 U.S.C. § 1533(d); see Christy v. Hodel, 857 F.2d 1324 (9th Cir. 1988), cert. denied sub nom. Christy v. Lujan, 109 S.Ct. 3176 (1989).
[92] 50 CFR §§ 17.31(a) and 17.71(a).

An interesting question has arisen as to the level of protection given to species that have been proposed for listing as either endangered or threatened, but for which the listing process is not yet complete. Under ESA Section 7(a)(4), government agencies must "confer with the Secretary" before taking an action that might jeopardize a candidate species.[93] However, this level of protection has not always worked to the advantage of the candidate species, and may even invite destruction of a species before it becomes listed.[94]

Under ESA Section 4(b)(7), the Secretary is authorized to issue emergency regulations "to prevent a significant risk to the well being" of a candidate species. This provision is rarely used, however, given the press of numerous candidate species.[95] In addition the USFWS regularly publishes a list of all candidate species, and monitors their status. Presumably, any candidate species that is threatened would justify emergency regulations. Despite these safeguards, candidate species do *not* enjoy the same level of protection as listed species.

Under ESA Section 4(a)(3),[96] the Secretary must designate critical habitat at the same time that a species is listed. "Critical habitat" is defined in ESA Section 3(5) as the "specific areas within the geographical area occupied by the species" that are essential to its conservation or that require special protection, or areas outside the area that the Secretary finds to be essential for its conservation.[97] Many critical habitat designations have been delayed (often because of a lack of scientific data), and there have been many legal challenges.[98] Courts have become increasingly impatient with attempts by the USFWS to postpone critical habitat designations.[99] As noted above, the federal courts are split as to whether an EIS is required prior to designating "critical habitat" for a listed species.

Under ESA Section 4(f), the Secretary must also create a recovery plan for each listed species, unless a recovery plan will not promote the

---

[93] 16 U.S.C. § 1536(a)(4); see Wilson v. Block, 708 F.2d 735 (D.C. Cir. 1983), cert. denied., 104 S.Ct. 371 (1983).
[94] See Sierra Club v. Clark, 755 F.2d 608 (8th Cir. 1985) [COE allowed to complete a harbor project that affected a candidate plant species].
[95] 16 U.S.C. § 1533(b)(7).
[96] 16 U.S.C. § 1533(a)(3).
[97] 16 U.S.C. § 1532(5).
[98] See, for example, Northern Spotted Owl v. Lujan, 758 F. Supp. 621 (E.D. Wash. 1991).
[99] See Conservation Council for Hawaii v. Babbitt, 2 F. Supp. 2d 1280 (D. Haw. 1998) [USFWS did not establish a rational basis for failing to designate critical habitat for 245 species of endangered and threatened plants].

conservation of the species.[100] The ESA amendments of 1988 added that the recovery plan must be prepared "without regard to taxonomic classification," because it was found that recovery plans were only being prepared for the so-called "warm fuzzies" (like bald eagles, wolves, etc.). Unfortunately, fewer than 400 recovery plans have been prepared to date (largely due to USFWS staffing and budget problems), and most of these are for "photogenic" species.

Plants receive considerably less protection under the ESA than do "fish and wildlife." Originally, plants received no protection at all, but the 1982 amendments added a new Section 9(a)(2), which makes it unlawful to "remove or reduce to possession,"or "maliciously damage or destroy," any protected plant species from land under federal jurisdiction.[101] However, the ESA's prohibition regarding plants does not extend to privately owned land!

ESA Section 11(g) contains a citizen suit provision that allows "any person" to commence a civil suit in federal district court on his or her own behalf to force compliance with ESA mandates.[102] A plaintiff is limited to injunctive or declaratory relief under ESA Section 11(g), although damages may be pursued under common law.[103] Under Section 11 of the ESA, a district court may award fees to "any party" where "appropriate."[104] However, the Ninth Circuit held in *Marbled Murrelet v. Babbitt* that a successful defendant in a citizen suit case may only collect attorney's fees if the plaintiff's action was "frivolous."[105]

Courts have been rather liberal in granting standing in citizen suit actions under the ESA. The U.S. Supreme Court in *Bennett v. Spear* held that the broadly worded standing requirements in ESA Section 11(g) allow "any person" to commence a civil suit, and allow a broad "zone of interest" because the overall subject matter of the ESA is the environment, a matter in which all persons have an interest.[106]

---

[100]   16 U.S.C. § 1533(f); see National Wildlife Federation v. National Park Svc., 669 F. Supp. 384 (D. Wyo. 1987); and Sierra Club v. Yeutter, 926 F.2d 429 (5th Cir. 1991).
[101]   16 U.S.C. § 1538(a)(2).
[102]   16 U.S.C. § 1540(g).
[103]   See Sierra Club v. Marsh, 816 F.2d 1376 (9th Cir. 1987).
[104]   16 U.S.C. § 1540(g)(4).
[105]   Marbled Murrelet v. Babbitt, 111 F.3d 1447 (9th Cir. 1997).
[106]   Bennett v. Spear, 520 U.S. 154 (1997). See also Lujan v. Defenders of Wildlife, 504 U.S. 555 (1992) [the desire to use or observe an animal species, even for purely esthetic purposes, is undeniably a cognizable interest for the purpose of standing]. Other courts have followed the liberal rules for standing: Animal Legal Defense Fund v. Glickman, 154 F.3d 426 (D.C. Cir. 1998) [merely "seeing with his own eyes the particular animal" whose condition caused esthetic injury is enough for standing]; Biodiversity Legal Foundation v.

15.4 Incidental Takes and Habitat Conservation Plans

The ESA's proscription against "taking" of protected species, even on private land, is often viewed as an "unfunded mandate" by private property owners who wish to develop their land for personal gain.[107] In a much anticipated move that appeased many private property owners, Congress amended the ESA in 1982 to add a new Section 10(a), which allows "incidental takings" of a protected species to take place.[108] These are generally perceived as an "exception" to Section 9 of the ESA.

ESA Section 10(a) actually arose from an historic Habitat Conservation Plan (HCP) prepared in cooperation between several groups of environmentalists and property owners on San Bruno Mountain near San Francisco.[109] The agreement allowed development of part of the critical habitat of several butterflies, but provided land and funding for the future recovery of the species (see further discussion below). Marked at the time as a model for the habitat conservation planning process, the success of the San Bruno Mountain HCP has been difficult to duplicate.[110]

Under ESA Section 10(a), a private person may obtain a permit from the Secretary of the Interior:

[U]nder such terms and conditions as [the Secretary] shall prescribe
  (A) any act otherwise prohibited by section 1538 of this title for scientific purposes or to
enhance the propagation or survival of the affected species, including, but not limited to, acts necessary for the establishment and maintenance of experimental populations pursuant to subsection (j) of this section; or

---

Babbitt, 63 F. Supp. 2d 31 (D.D.C. 1999) [status and mission of environmental group make it an acceptable plaintiff]; and Coalition for Sustainable Resources, Inc. v. U.S. Forest Service, 48 F. Supp. 2d 1303 (D. Wyo. 1999) [any harm claimed by plaintiff will suffice to show standing].

[107] See Thornton, The Search for a Conservation Planning Paradigm: Section 10 of the ESA. 8(1) Natural Resources & Env't 21 (1993).

[108] 16 U.S.C. § 1539(a). See Bean, *Reconciling Conflicts under the Endangered Species Act* (World Wildlife Fund, 1991). Some commentators have argued that ESA § 10(a) is politically motivated, and has violated congressional intent in the original ESA just to satisfy property owners. See Sheldon, Habitat Conservation Planning: Addressing the Achilles Heel of the Endangered Species Act, 6 N.Y.U. Env'l L.J. 279 (1998).

[109] See Friends of Endangered Species v. Jantzen, 760 F.2d 976 (9th Cir. 1985).

[110] See Thornton, The Search for a Conservation Planning Paradigm: Section 10 of the ESA. 8(1) Natural Resources & Env't 21 (1993).

(B) any taking otherwise prohibited by section 1538(a)(1)(B) of this title if such taking is incidental to, and not the purpose of, the carrying out of an otherwise lawful activity.[111]

However, an applicant for a permit under ESA Section 10(a) must submit a Habitat Conservation Plan that specifies:

(i) the impact which will likely result from such taking;
(ii) what steps the applicant will take to minimize and mitigate such impacts, and the funding that will be available to implement such steps;
(iii) what alternative actions to such taking the applicant considered and the reasons why such alternatives are not being utilized; and
(iv) such other measures that the Secretary may require as being necessary or appropriate for purposes of the plan.[112]

After opportunity for public comment, the Secretary can issue an ESA Section 10(a) permit if it is determined that:

(i) the taking will be incidental;
(ii) the applicant will, to the maximum extent practicable, minimize and mitigate the impacts of such taking;
(iii) the applicant will ensure that adequate funding for the plan will be provided;
(iv) the taking will not appreciably reduce the likelihood of the survival and recovery of the species in the wild;
(v) the measures, if any, required under subparagraph (A)(iv) will be met; [and] . . . has received such other assurances as [the Secretary] may require.[113]

In a related action, the Department of the Interior and the USFWS have instituted a "no surprises" rule, which states that a private landowner who has properly implemented an HCP will not be required at a later date to provide additional land, water, compensation, or face new restrictions from the federal government.[114]
Despite the uneven success of early HCPs, ESA Section 10(a) has become a popular mechanism to permit private property owners to develop

---

[111] Id. See Bennett v. Spear, 117 S.Ct. 1154 (1997).
[112] 16 U.S.C. § 1539(a)(2)(A).
[113] 16 U.S.C. § 1539(a)(2)(B).
[114] 50 CFR §§ 17 and 222. See Bosselman, The Statutory and Constitutional Mandate for a No Surprises Rule, 24 Ecol. L.Q. (1997).

at least some land, even though a protected species might be present. According to the USFWS, there were 225 HCPs in the United States as of February 1998 (over 18 million acres), with another 200 proposed.

It should also be noted that a federal court has held that an EIS must be prepared under NEPA Section 102(2)(c) if an ESA Section 10(a) "incidental take" is permitted.[115]

The San Bruno Mountain (California) HCP

The San Bruno Mountain HCP, the first in the nation, involved critical habitat for the endangered mission blue and San Bruno elfin butterflies, and the candidate callippe silverspot butterfly. The HCP was designed to protect the butterflies *and* ecological diversity.

When the HCP was approved in 1983, there were 3,400 acres of undeveloped land on San Bruno Mountain, 1,500 acres of which was privately owned. The HCP protected 90 percent of the critical habitat for the butterflies (800 private acres were donated and added to the 1,952 publicly owned acres), while limited development would occur on the remaining 10 percent.

Funding for long-term habitat protection (including monitoring and removal of exotic plants) came from an initial "interim fee" charged to landowners, plus a yearly fee to each residential or commercial unit that is formulated as a lien on the property.

The HCP was administered by San Mateo County, with consultation from USFWS and the California Department of Fish and Game. The HCP survived several legal challenges, the most significant of which was *Friends of Endangered Species v. Jantzen.*[116]

The North Key Largo (Florida) HCP.

Another early HCP was the North Key Largo HCP, which is located in one of the last relatively undisturbed tropical hardwood hammocks in the United States. This HCP was designed to protect several listed species, including two rodents, a butterfly, a snake, and the American crocodile.

The HCP preserved eighty-four percent of the area of North Key Largo, while permitting high-density development (3,500 units) on the remaining sixteen percent. The administration of the HCP was to be a collaborative effort by Monroe County, the State of Florida, and USFWS. However, a change in the composition of the Monroe County Board of Commissioners in 1986 resulted in the attempted withdrawal of the County from the HCP.

The current status of the North Key Largo HCP is "uncertain."

---

[115] Ramsey v. Kantor, 96 F.3d 434 (9th Cir. 1996).
[116] 760 F.2d 976 (9th Cir. 1985) [the HCP did not appreciably reduce the likelihood of survival of the species].

15.5 Other Federal Laws Protecting Species

The ESA is not the only federal law that has the effect of protecting species that are at risk. The Lacey Act of 1900,[117] discussed above, creates a federal violation for interstate transport of many species taken in violation of state, tribal, or other federal laws. The Wild Free-Roaming Horses and Burros Act of 1971 protects these species on public lands.[118]

In some cases, other environmental laws serve the purpose of protecting species or habitat. For example, the U.S. Supreme Court has allowed states to use their authority under Section 303 of the Clean Water Act to designate water quality criteria for a water body that protects specific aquatic species.[119]

Several other federal laws protect species in various ways.

15.5.1 The Marine Mammal Protection Act

The Marine Mammal Protection Act of 1972 (MMPA),[120] which protects marine mammals (such as dolphins and manatees) in a manner reminiscent of the ESA. For example, a person may not "take" a marine mammal (where "take" is more broadly defined than in the ESA), even if the animal is not actually harmed.[121] It is also unlawful to possess or import marine mammals that were taken unlawfully (even if they were taken in a foreign country), or if they were taken in an inhumane manner, or if they are taken when they are pregnant or nursing.[122] However, limited exemptions exist for the U.S. tunaboat industry.[123]

---

[117] 16 U.S.C. §§ 3371–3378; 18 U.S.C. § 42.

[118] 16 U.S.C. §§ 1331–1340.

[119] See PUD No. 1 v. Washington Dept. of Ecology, 114 S.Ct. 1900 (1994).

[120] 16 U.S.C §§ 1361–1407.

[121] See Katelnikoff v. Dept. of the Interior, 657 F. Supp. 659 (D. Alaska 1986); and Fed'n of Japan Salmon Fisheries Coop. Ass'n v. Baldridge, 679 F. Supp. 37 (D.D.C. 1987), aff'd sub nom. Kokechik Fisherman's Ass'n v. Sec. of Commerce, 839 F.2d 795 (D.C. Cir. 1988).

[122] See Animal Welfare Inst. v. Kreps, 561 F.2d 1002 (D.C. Cir. 1977); and Globe Fur and Dyeing Corp. v. United States, 467 F. Supp. 177 (D.D.C. 1978), aff'd, 612 F.2d 586 (D.C. Cir. 1988).

[123] 16 U.S.C. § 1374(h); see Earth Island Inst. v. Mosbacher, 746 F. Supp. 964 (N.D. Cal. 1990), aff'd, 929 F.2d 1449 (9th Cir. 1991) [dolphin kill limited to 20,500 animals]; and American Tunaboat Ass'n v. Brown, 67 F.3d 1414 (9th Cir. 1995) [failure by tuna fishery to reduce the yearly take of dolphins]), as well as for certain other commercial fisheries (16 U.S.C. § 1371(a)(4).

15.5.2 The Bald Eagle and Golden Eagle Protection Act

The Bald Eagle and Golden Eagle Protection Act of 1962[124] is designed to protect our national symbol and its closest relative. Interestingly, golden eagles were added to the Bald Eagle Protection Act of 1940 only because it became obvious that many juvenile bald eagles were being killed accidentally by hunters who thought they were golden eagles. Prohibited acts include "take" (defined broadly), as well as removing or destroying nests, eggs, or "parts" such as feathers.[125] There are limited exemptions to protect livestock, falconry, and Native American religious purposes, among others.

15.5.3 The Migratory Bird Treaty Act

The Migratory Bird Treaty Act of 1918[126] protects birds that migrate into and out of the United States by implementing treaties between the United States and several other countries (Great Britain on behalf of Canada, Mexico, Japan, and the "Soviet Union"). The Act does not prohibit all hunting of migratory birds, but rather limits the methods (e.g., trapping, snaring, poisoning, and the use of certain types of guns is prohibited.

15.6 State Protection of Endangered Species

Most states have the equivalent of the federal ESA, which serve to protect species (and habitats) at the state level. Many of these include species that do not qualify for federal endangered or threatened status, and most states require a designation of "critical habitat" for the species.

15.6.1 California

The California Endangered Species Act (CESA)[127] is administered by the California Department of Fish and Game (DFG), and parallels the main provisions of the Federal ESA.

The term "endangered species" is defined by the CESA as a species of plant, fish, or wildlife which is "in serious danger of becoming extinct

---

[124] 16 U.S.C. § 668.
[125] See Mountain States Legal Found'n v. Hodel, 799 F.2d 1423 (10th Cir. 1986).
[126] 16 U.S.C. §§ 703-711.
[127] California Fish & Game Code §§ 2050, et seq.

throughout all, or a significant portion of its range," and is limited to species or subspecies native to California. The CESA establishes a petitioning process for the listing of threatened or endangered species in which the California Fish and Game Commission is required to adopt regulations for the process and establish criteria for determining whether a species is endangered or threatened.[128]

The term "take" is defined as to "hunt, pursue, catch, capture, or kill, or attempt to hunt, pursue, catch, capture, or kill" a protected species.[129] The CESA prohibits the "taking" of listed species except as otherwise provided in State law, but unlike the Federal ESA in which the taking prohibition applies only to listed species, the CESA applies the take prohibitions to candidate species (those species petioned for listing).

All state "lead agencies" are required to consult with DFG to ensure that any action it undertakes is not likely to jeopardize the continued existence of any endangered or threatened species or result in destruction or adverse modification of essential habitat. A "lead agency" is defined under the California Environmental Quality Act as the public agency that has primary responsibility for carrying out or approving a project that may have a significant effect on the environment.[130]

California's Natural Community Conservation Planning Act (NCCP Act) was designed to protect endangered species and biodiversity by protecting endangered ecosystems.[131] The first of its kind in the United States, it has been cited with approval by Secretary of the Interior Bruce Babbitt. A hybrid between an "endangered species" and an "environmentally sensitive areas" protection statute, the Act requires the preparation of a "Natural Communities Conservation Plan" (NCCP). The NCCP process brings together state agencies (under the direction of the California Resources Agency), county and local governments, environmental groups, and private landowners. Once the Department of Fish & Game approves an NCCP, then it may authorize (by permit or agreement) developments that might otherwise have an adverse impact on listed or candidate species on the state list of endangered species.[132]

The first NCCP is for the Coastal Sage Scrub (CSS) community in southwestern California, which involves millions of acres of habitat in five counties. The CSS community supports many species of native plants and

---

[128] Calif. Code of Regulations, tit. 14 § 670.1(a) sets forth the required contents for such a petition.
[129] Calif. Fish and Game Code § 86.
[130] Calif. Pub. Res. Code °21067
[131] Calif. Fish & Game Code §§ 2800-2840.
[132] Calif. Fish & Game Code, §§ 2081, 2825(c) and 2835. See Bosselman, Planning to Prevent Species Endangerment, 44 Land Use L. & Zoning Dig. 4 (March, 1992).

animals, including the California gnatcatcher, cactus wren, and over 40 candidate species, but is seriously threatened by development for human uses. The NCCP is designed to provide for regional or areawide protection and perpetuation of biodiversity, while allowing compatible and appropriate land development and growth. Numerous regulatory agencies (local, state, and federal), conservation groups, and private landowners have participated in development of the NCCP.

15.6.2 Illinois

The Illinois Endangered Species Protection Act[133] establishes an Endangered Species Protection Board whose task is to define, identify and protect endangered species, and to supplement the federal ESA. The Illinois ESA states that:

It is unlawful for any person to possess, take, transport, sell, offer for sale, give or otherwise dispose of any animal or the product thereof of any animal species which occurs on the Illinois List, or to deliver, receive, carry, transport or ship in interstate or foreign commerce plants listed as endangered by the Federal government without a permit therefor issued by the Department as provided in Section 4 of this Act and to take plants on the Illinois list without the expressed written permission of the landowner or to sell or offer for sale plants or plant products of endangered species on the Illinois list.[134]

An "endangered species" is:
any species of plant or animal classified as endangered under the Federal Endangered Species Act of 1973 . . . plus such other species which the [Endangered Species] Board may list as in danger of extinction in the wild in Illinois due to one or more causes including but not limited to, the destruction, diminution or disturbance of habitat, overexploitation, predation, pollution, disease, or other natural or manmade factors affecting its prospects of survival.[135]

A "threatened species" is:
any species of plant or animal classified as threatened under the Federal Endangered Species Act of 1973 . . . plus such other species

---

[133] 520 ILCS § 10. The regulations are at 17 Il.Admin.Code Chap. 1, subchap. c: Endangered Species.
[134] 520 ILCS § 10/3.
[135] 520 ILCS § 10/2.

which the Board may list as likely to become endangered in the wild in Illinois within the foreseeable future.[136]

The term "take" is defined in the Illinois ESA at Section 10/2:

"Take" means, in reference to animals and animal products, to harm, hunt, shoot, pursue, lure, wound, kill, destroy, harass, gig, spear, ensnare, trap, capture, collect, or to attempt to engage in such conduct. "Take" means, in reference to plants and plant products, to collect, pick, cut, dig up, kill, destroy, bury, crush, or harm in any manner.[137]

"Essential Habitat" is:

the specific ecological conditions required by an endangered or threatened species for its survival and propagation, or physical examples of these conditions.[138]

All Illinois state and local agencies must further the purposes of the Illinois ESA by consulting with the Endangered Species Board concerning any actions that are likely to jeopardize the continued existence of any Illinois listed endangered or threatened species, or are likely to result in the destruction or adverse modification of the designated essential habitat of the species.[139]

"Any person" who violates any provision of the Illinois ESA is guilty of a Class A misdemeanor, which may include fines and imprisonment up to one year.[140] Moreover, property and equipment used in violations of the Illinois ESA, along with the protected species and their parts, may be searched and seized.[141]

15.6.3 Texas

The Texas Parks and Wildlife Department (TPWD) has authority to establish a list of endangered animals in the state. "Endangered species" are those species which the Executive Director of the TPWD has determined are "threatened with statewide extinction." "Threatened species" are those species that the TPW Commission has determined are

---

[136] Id.
[137] Id.
[138] Id.
[139] 520 ILCS § 10/11(b).
[140] 520 ILCS § 10/9.
[141] 520 ILCS § 10/8.

likely to become endangered in the future.[142]

The TPWD has also establish a list of threatened and endangered plant species for the state. An "endangered plant" is defined as one that is "in danger of extinction throughout all or a significant portion of its range," while a "threatened plant" is one which is likely to become endangered within the foreseeable future.[143]

TPWD regulations prohibit the taking, possession, transportation, or sale of any of the animal species designated by state law as endangered or threatened without the issuance of a permit. Commerce in threatened and endangered plants, and the collection of listed plant species from public land without a permit issued by TPWD are prohibited.

15.6.4 Florida

Several states permit the equivalent of a federal HCP by allowing "incidental taking" of listed species. The State of Florida issues permits for the development of critical habitat, but the activity must "clearly enhance the survival potential of the species."[144] An example of such an action occurred when the Florida Game and Freshwater Fish Commission issued a permit for the destruction of endangered rodent nests in Monroe County, Florida in exchange for revegetation, improvements, and an environmental easement on an adjacent 6-acre plot of habitat.[145]

15.7 CITES: The ESA in the International Arena

ESA Section 8(b) contains a little known section devoted to international issues on protected species:

International cooperation
(b) Encouragement of foreign programs.
In order to carry out further the provisions of this chapter, the Secretary, through the Secretary of State, shall encourage —
    (1) foreign countries to provide for the conservation of fish or wildlife and plants including endangered species and threatened species listed pursuant to section 1533 of this title;

---

[142] Texas Parks and Wildlife (TPW) Code Chap. 67-68; and Title 31, Sections 65.171 - 65.184, Texas Admin. Code.
[143] Texas Admin. Code Chapter 88 and Sections 69.01 - 69.14.
[144] Rule 39-27002(1), Fla. Admin. Code.
[145] This state "HCP" survived a legal challenge in Mangrove Chapter, Izaak Walton League v. Florida, 592 So.2d 1162 (Fla. 1st DCA 1992).

(2) the entering into of bilateral or multilateral agreements with foreign countries to provide for such conservation; and

(3) foreign persons who directly or indirectly take fish or wildlife or plants in foreign countries or on the high seas for importation into the United States for commercial or other purposes to develop and carry out with such assistance as he may provide, conservation practices designed to enhance such fish or wildlife or plants and their habitat.[146]

Beginning in 1969 (even before the ESA was passed), the United States had sponsored a multinational convention that aspired to create an international agreement on the conservation of endangered and other sensitive species. The result was the Convention on International Trade in Endangered Species (CITES). As of 1999, 145 countries, including the United States, have become members of CITES. ESA Section 8(b) served as enabling legislation for the U.S.'s mebership in CITES.

Under the CITES agreement, all member nations "shall not allow trade in specimens of species included in Appendices I, II and III except in accordance with the provisions of the present Convention."[147] Animal and plant species are placed in Articles I-III based on the following:

Fundamental Principles

1. Appendix I shall include all species threatened with extinction which are or may be affected by trade. Trade in specimens of these species must be subject to particularly strict regulation in order not to endanger further their survival and must only be authorized in exceptional circumstances.

2. Appendix II shall include:

(a) all species which although not necessarily now threatened with extinction may become so unless trade in specimens of such species is subject to strict regulation in order to avoid utilization incompatible with their survival; and

(b) other species which must be subject to regulation in order that trade in specimens of certain species referred to in sub-paragraph (a) of this paragraph may be brought under effective control.

3. Appendix III shall include all species which any Party identifies as being subject to regulation within its jurisdiction for the purpose of preventing or restricting exploitation, and as needing the co-operation of other Parties in the control of trade.

---

[146] 16 U.S.C. § 1537(b).
[147] CITES Article II(4).

Administration of CITES is described in Articles XI–XVIII. The governing body responsible for implementation and enforcement of CITES is the "Conference of the Parties," which meets at least once every 2 years. Day to day administration of CITES is put in the hands of a secretariat. Under Article XXIV, any member nation can withdrawn from CITES (called "Denunciation") 12 months after filing notification.

CITES Article VIII makes a fairly general statement regarding enforcement:

The Parties shall take appropriate measures to enforce the provisions of the present Convention and to prohibit trade in specimens in violation thereof. These shall include measures:

(a) to penalize trade in, or possession of, such specimens, or both; and

(b) to provide for the confiscation or return to the State of export of such specimens.

Member nations are free to adopt regulations that implement Article VIII, including regulations that are more strict than CITES.

The United States enforces CITES under ESA Section 9(c), which states:

(c) Violation of [CITES] Convention

(1) It is unlawful for any person subject to the jurisdiction of the United States to engage in any trade in any specimens contrary to the provisions of the Convention, or to possess any specimens traded contrary to the provisions of the Convention, including the definitions of terms in article I thereof.

(2) Any importation into the United States of fish or wildlife shall, if —

(A) such fish or wildlife is not an endangered species listed pursuant to section 1533 of this title but is listed in Appendix II to the Convention,

(B) the taking and exportation of such fish or wildlife is not contrary to the provisions of the Convention and all other applicable requirements of the Convention have been satisfied,

(C) the applicable [licensing] requirements . . . have been satisfied, and

(D) such importation is not made in the course of a commercial activity, be presumed to be an importation not in violation of any provision of this chapter or any regulation issued pursuant to this chapter.[148]

---

[148]   16 U.S.C. § 1538(c).

The United States has taken its responsibilities seriously, and has not hesitated to prosecute violators.[149]

Unfortunately, CITES has not yet lived up to its international potential. This is in large part the result of a tension between the investment in endangered species protection offered by developed nations, and the economic realities of developing nations. Professor Hill stated the problem in the following way: "The [CITES] Convention attempts to balance the vague intuitive notion that the preservation of species is good, against commercial demands for exploitation."[150] Nevertheless, CITES successes are that it has improved monitoring in the $5 billion per year trade in wildlife and wildlife products, and that it has raised international consciousness in the need for endangered species protection.[151]

15.8 Does the Endangered Species Act Work, and Is It Worth It?

In 1991, Representative Thomas Foley and Senator Mark Hatfield asked the National Academy of Sciences (NAS) to conduct a study of scientific issues in the ESA. The NAS reported in 1995 that the ESA had served as a workable and effective way to protect endangered species, at least from the scientific perspective.[152]

Unfortunately, statistics on ESA effectiveness are sobering. Of 711 species listed as of 1992, only 69 (9.7 percent) are recovering. Fourteen listed species (2.0 percent) are probably extinct, and 232 (32.6 percent) are declining despite ESA protection. In 1990, the USFWS had "delisted" 15 species; 9 are recovering, and 6 species became extinct despite the listings. Of the 2,944 category 2 species in 1990, 118 became extinct while waiting for the listing process to be completed.[153]

While some critics view the ESA as protecting a resource that seems to have only sentimental value, the truth may be different. Forty percent of prescription drugs come from plants, animals, or microorganisms, and

---

[149] See United States v. 3,210 Crusted Sides of Caiman Crocodilus Yacare, 636 F. Supp. 1281 (S.D. Fla. 1986); and United States v. 2,507 Live Canary Winged Parakeets, 689 F. Supp. 1106 (S.D. Fla. 1988). See Epstein, The Endangered Species Act Applies Extraterritorially, 5(1) the Transnational Lawyer 447 (1992);

[150] Hill, Convention on International Trade in Endangered Species: Fifteen Years Later, 13 Loy. L.A. Int'l & Comp. L.J. 231 (1990).

[151] See Balistrieri, CITES: The ESA and International Trade, 8(1) Natural Resources & Env't 33 (1993).

[152] National Research Council, *Science and the Endangered Species Act* (Nat. Acad. Press, 1995).

[153] Id.

most of these chemical compounds cannot be duplicated in the laboratory. For example, taxol, a cancer-fighting drug, is extracted from the Pacific yew tree; penicillin, an antibiotic that has saved millions of lives worldwide, is derived from a fungus; and digitalis, an important cardiac stimulant, comes from the common foxglove. Unfortunately, only about 5 percent of known plants have been studied for their pharmaceutical properties, and it is entirely possible that the next "miracle drug" will come from a listed species. In addition, threats to animals and plants from pesticides and other toxic or hazardous chemicals often serve as evidence of similar perils to human health (for example, damage caused by DDT in bald eagles suggested similar health threats to humans, and ultimately led to a ban on the use of the pesticide in the United States).[154]

Protecting endangered species as elements of biological diversity is becoming an increasingly strong emphasis among the scientific community. For example, among the many strains of existing crop plants are some that are (or may be) resistant to various diseases. These resistant strains have in the past and will continue in the future to provide critical protection against disastrous crop diseases that may lead to devastating famines. Likewise, many scientists feel that a possible response to the escalating health crisis involving antibiotic-resistant disease organisms in humans may be resolved by new strains of plants or animals that produce new antibiotics.

From a more practical economic perspective, it has been reported that wildlife-related recreation is a $50 billion industry yearly in the United States. Commercial and recreational fishing along the Pacific coast alone accounts for over 100,000 jobs, many of which are currently at risk due to overexploitation of Pacific salmon stocks. In the words of Eichenberg and, "destroying wildlife and habitats is as close as one can get to killing the goose that lays the golden egg."[155]

15.9 Federal Control of Wilderness Areas

Protection of endangered species and preservation of the wilderness areas they inhabit are inexorably intertwined. Although "critical habitat" is protected under the ESA, a more direct means of protecting land may be that of the Wilderness Act of 1964 (WA).[156]

In its introduction, the Wilderness Act makes a strong policy statement and establishes the National Wilderness Preservation System:

---

[154] Berry, Endangered Species and Other Endangered Laws, Environment & Development (Nov/Dec 1995).
[155] Id.
[156] 16 U.S.C. §§ 1131-1133.

In order to assure that an increasing population, accompanied by expanding settlement and growing mechanization, does not occupy and modify all areas within the United States and its possessions, leaving no lands designated for preservation and protection in their natural condition, it is hereby declared to be the policy of the Congress to secure for the American people of present and future generations the benefits of an enduring resource of wilderness. For this purpose there is hereby established a National Wilderness Preservation System to be composed of federally owned areas designated by Congress as "wilderness areas", and these shall be administered for the use and enjoyment of the American people in such manner as will leave them unimpaired for future use and enjoyment as wilderness . .
[157]

As of 1991, The National Wilderness Preservation System contained over 90 million acres (over 4 percent of all U.S. land). Approximately 39 million acres are in national parks, 34 million acres in national forests, 21 million acres in national wildlife refuges, and 106 million acres in Bureau of Land Management (BLM) lands.[158]

However, there are several sources of conflict within the Wilderness Act. First, despite the apparent contrary language in the introduction above, a variety of uses are specifically authorized by the Act, including logging, mining, cattle, pest control, fire management, and motorboat and snowmobile usage.[159] Since most area under Wilderness Act jurisdiction are under the regulatory control of other statutes, managers of wilderness areas may have little authority to control these uses.[160]

Another problem is that wilderness areas are frequently affected by factors from outside the area, such as acid rain, climate changes, and invasions of exotic species.

The question of whether or an EIS is required under NEPA (see Chapter 3) for activities in wilderness areas has received considerable attention. Federal courts have been in disagreement as to when an EIS is required. In

---

[157] 16 U.S.C. § 1131(a).
[158] Rodgers at § 9.9.
[159] 16 U.S.C.§ 1133(d). See Voyageurs Region National Park Ass'n v. Lujan, 966 F.2d 424 (8th Cir. 1992) [snowmobiles in wilderness area]; Friends of Boundary Waters Wilderness v. Robertson, 978 F.2d 1484 (8th Cir. 1992), cert. denied, 113 S.Ct. 2962 (1993) [motorized portaging denied because feasible alternative existed]; and Sierra Club v. Hodel, 848 F.2d 1068 (10th Cir. 1988) [road through wilderness area]. See also Rohlf and Honnold, Managing the Balance of Nature: The Legal Framework of Wilderness Management, 15 Ecol. L.Q. 249 (1988).
[160] See Rodgers at § 9.9.

Minnesota Public Interest Group v. Butz (Butz I),[161] the court held that logging would destroy the primitive character of the forest, and required the preparation of an EIS. However, in subsequent litigation the same court held that destruction of old growth forests was allowed by the Wilderness Act.[162]

Related to the Wilderness Act is the Federal Land Policy and Management Act (FLPMA),[163] which directs the BLM to administer over 170 million acres of public land in eleven western states. When it passed the FLPMA in 1976, Congress found that, among other things:

> [T]he public lands be managed in a manner that will protect the quality of scientific, scenic, historical, ecological, environmental, air and atmospheric, water resource, and archeological values; that, where appropriate, will preserve and protect certain public lands in their natural condition; that will provide food and habitat for fish and wildlife and domestic animals; and that will provide for outdoor recreation and human occupancy and use.[164]

With this statement, FLPMA attempted to reject the notion from the days of the Taylor Grazing Act of 1934[165] that all public land was available to the public for grazing and related activities. Instead, conservationists have argues that nonconsumptive uses of land (such as hiking, river rafting, and photography) are more consistent with the goals of the FLPMA than are activities such as cattle grazing.[166] The mandate for multiple uses (rather than a single use for cattle grazing) has been the source of considerable tension and litigation.[167]

---

[161] Minnesota Public Interest Group v. Butz (Butz I), 358 F. Supp. 584 (D. Minn. 1973), affirmed, 498 F.2d 1314 (8th Cir. 1974).

[162] Minnesota Public Interest Group v. Butz (Butz II), 401 F. Supp. 1276 (D.Minn. 1975), reversed, 541 F.2d 1292 (8th Cir. 1976).

[163] 43 U.S.C. § 1701-1785.

[164] 43 U.S.C. § 1701(a)(8).

[165] 43 U.S.C. § 315.

[166] See Wilderness Public Rights Fund v. Kleppe, 608 F.2d 1250 (9th Cir. 1979), cert. Denied, 446 U.S. 982 (1980). See also Coggins, Evans, and Lindeberg-Johnson, The Law of Public Rangeland Management I: The Extent and Distribution of Federal Power, 12 Env'l L. 536 (1982); and Coggins and Lindeberg-Johnson, The Law of Public Rangeland Management I: The Commons and the Taylor Act, 13 Env'l L. 1 (1982).

[167] See Coggins, The Law of Public Rangeland Management (IV): FLPMA, PRIA, and the Multiple Use Mandate, 14 Env'l L. 1 (1983).

15.10 State Protection of Environmentally Sensitive Areas

Many states have developed programs by which large tracts of public (or both public and private) land is designated as critical to the environmental health of the state, and is carefully regulated to avoid overexploitation. For example, the state of New York regulates over 6 million acres (60 percent of which is privately owned) in the Adirondack Park region.[168]

Additional tools for the protection of environmentally critical areas are found in a variety of federal and state laws that protect, for example, endangered species (and their protected "critical habitat"), coastal zones, wetlands, and forests. As discussed below, states have used a variety of techniques to regulate critical areas, but the unfortunate result has been a mixture of frequent failures and few successes.

Professor Malone noted that there are two basic mechanisms by which most states regulate environmentally critical areas.[169] Under the first approach, a state (or other governmental unit) simply adopts *ad hoc* legislation that designates the area of concern, and create the appropriate regulatory mechanism. This approach has been used with some success by California,[170] North Carolina,[171] and Massachusetts,[172] among others. A common thread in these more successful state programs is the emphasis on planning and on consistency requirements as a compliance tool. However, many states may find the difficulties of passing *ad hoc* legislation for each and every area of concern to be insurmountable.[173]

A second approach for critical area protection is based on Article 7, Part 2 of the American Law Institute's Model Land Development Code.[174] Under the Model Code approach, the state land planning agency (with authority from appropriate enabling legislation) develops a statewide plan that designates specific geographic areas within the state as "Areas of Critical State Concern" based, for example, on the presence of "historical, natural or environmental resources of regional or statewide importance,"[175] and promulgates appropriate regulations. Development within these areas is carefully monitored to assure that it is compatible with the particular

---

[168] *See* N.Y. Envtl. Conserv. L. § 9-0101.
[169] Malone, *Environmental Regulation of Land Use* (West, 1994) at § 13.01.
[170] Cal. Gov't. Code §§ 66800-66801 (Lake Tahoe).
[171] N.C. Gen. Stat. § 113A ff. (coastal zone).
[172] 1977 Mass. Acts ch. 831 (Martha's Vineyard).
[173] Much of the discussion in this section is based on Berry, Areas of Critical State Concern in American Planning Association, *Modernizing State Planning Statutes: The Growing Smart Working Papers* (1996).
[174] American Law Institute, Model Land Development Code(§§ 7-201 ff., *Areas of Critical State Concern*).
[175] Model Land Development Code § 7-201(3)(b).

historical, natural or environmental qualities of the area. Local governments draft plans that are consistent with the state plan, and then apply to a state land development agency for permission to develop within the area of critical state concern.

15.10.1 New York

One of the earliest (and still the largest) areas designated as of critical state concern is the New York Adirondack Park, which encompasses over 6 million acres of both public and private lands. Authority for development planning for the Park is in the 1971 Adirondack Park Agency Act of 1971.[176] The state legislature approved a regional land management plan in 1973 which set permissible densities for development on private lands, and set standards for permitted developments.[177] Applicants for new land uses and development with an impact of regional significance must obtain a permit from the Adirondack Park Agency.[178] Violators of any section of the act, Agency rules or regulations, or permit conditions are subject to fines of $500 per day, and the N.Y. Attorney General may seek injunctive relief.[179]

15.10.2 New Jersey

The state of New Jersey protects the ecologically important pinelands region under the Pinelands Protection Act of 1979.[180] The New Jersey act designates a "preservation area" that receives the highest level of protection, and a surrounding "protection area" that acts as a buffer. A Pinelands Commission was created by the act, which was responsible for preparing and adopting a comprehensive management plan, making periodic revisions, and identifying land management procedures for protection of the area. The commission is advised by a municipal council (composed of the mayors of municipalities within the area) to which the commission submits revisions of the management plan for review. Local

---

[176] *Supra*, note 2.
[177] N.Y. Exec. Law §§ 803-805. See Malone, supra note 13 at § 13.03[1].
[178] The Adirondack Park Agency consists of the state Commissioner of Environmental Conservation, the Secretary of State, the Commissioner of Commerce, and eight members appointed by the Governor and approved by the state senate. N.Y. Exec. Law § 803.
[179] *Id.* § 813.
[180] N.J. Stat. §§ 13.18A-1-29. The pinelands include many forest and wetland resources that are also protected under the federal National Parks and Recreation Act, 16 U.S.C. § 471i. See Malone, supra note 13 at § 13.03[3].

governments submit master plans and zoning ordinances that must be consistent with the comprehensive management plan.

15.10.3 Virginia

Another example of *ad hoc* legislation to protect critical areas is Virginia's Chesapeake Bay Preservation Act which protects the ecologically vulnerable tidewater region.[181] The Chesapeake Bay Local Assistance Board develops criteria for protection of the area, provides assistance to local governments in complying with the act and board regulations, and ensuring that local government comprehensive plans are consistent with the act. For their part, municipalities and counties within the tidewater area are required to develop comprehensive plans that establish preservation areas within their jurisdictions that comply with the board's criteria. The board is also authorized to develop administrative and legal actions to ensure compliance with the act.

15.10.4 Florida

Several states have adopted versions of the Model Code approach, but Florida's approach is closest.[182] The Florida Division of State Planning, in cooperation with local interests, recommends Areas of Critical State Concern to the state Administration Commission (the governor and cabinet) based on historical and environmental factors.[183] Once approved, all regional and state agencies must comply with the state plan,[184] and local governments have 6 months to prepare consistent comprehensive plans.[185] Developments of Regional Impact (DRI) within the critical areas may proceed only under local and regional plans and regulations.[186]

Florida's critical areas approach seems to have been relatively successful, having included a variety of wetlands and coastal resources, among others. However, it suffers a serious limitation in that no more than 5 percent of the state's land area may be designated as critical areas at one

---

[181] Va. Code §§ 10.1-2100-2115. See Malone, Supra note 13 at § 13.03[2].
[182] Florida Environmental Land and Water Management Act of 1972, Fla. Stat. §§ 380.012 ff. For discussion of other states, see Malone, *Environmental Regulation of Land Use* (West, 1994) at Chapt. 13.
[183] Fla. Stat. § 380.05(1)(a).
[184] Id., § 380.054(4).
[185] Id., § 380.05(5).
[186] Id., § 380.06(13).

time.[187] Such artificial limitations on protected land area may be popular among the regulated community, but they place artificial constraints on valid environmental planning.

The Florida/Model Code approach offers a useful model for protecting critical state areas, but whether it would work equally well in other states is certainly debatable. Florida's success is partly fueled by strong growth management and planning laws not present in most other states, and strong enforcement provisions that permit judicial review of inconsistent local plans and development projects.[188]

15.11 Local Protection of Endangered Species and Sensitive Lands

The primary mechanism for habitat protection at the local level remains zoning ordinances and performance standards. Such mechanisms are now quite common, and often rely on habitat-sensitive performance standards, and dedication of land for green ways and open space. Some local governments provide municipally owned and managed forest districts, or other mechanisms which protect wildlife habitat. Whatever the mechanism, it is critical that the planning process include adequate considerations of the habitat needs of local plants and animals if they are to succeed.

In designing effective local habitat protection measures, there are several important goals that should be considered:[189]

● The measures must provide proper *ecosystem components* to protect the plant and animal species targeted. For example, many animal species are dependent on particular plant species for survival (e.g., many songbird species require particular seed-bearing plants for food, trees for nesting and predator avoidance, etc.). Proper habitat design is critical to the long-term survival of most target species.

● The habitat must be of *sufficient size* to support the target species. Different species of animals often have very different requirements for feeding areas, breeding areas, etc. Where habitat areas are small and isolated from each other, care should be taken to create "corridors" which allow animals to move from one "patch" of habitat to another.

● Efforts must be made to include as much *potential habitat* as possible. It is often possible to augment protected habitat by permitting (or requiring) the inclusion of appropriate native plant species into lawns,

---

[187] Id., § 380.05(20).
[188] Id., §§ 380.05(13) and 380.07(2), respectively.
[189] Based on Berry, Endangered Species and Other Endangered Laws, Environment & Development (Nov/Dec 1995).

parks, etc. For example, several communities in Arizona and Florida are promoting the use of "xeriscaping," or the use of naturally occurring, drought-adapted plants, in landscaping. The use of these plants actually increases the available habitat area for target species.

- Proper procedures must be followed to *minimize human-induced, anthropogenic influences* on the habitats. Measures must be included to protect habitats from domestic pets (primarily dogs and cats, but also livestock), which often cause severe damage by destroying wildlife, or by competing with natural species for food. Even well-meaning intrusions by humans may frighten or otherwise interfere with the normal activities of animals and plants. Introduction of contamination from humans must, of course, be minimized by limiting the entry of runoff, garbage, and pollutants of all kinds into the habitat. Design and construction of effective, habitat-sensitive infrastructure is the most efficient way to achieve this goal.
- Proper procedures must be followed to *assure long-term protection* for the habitats. This can be accomplished most effectively by outright transfers of titles to the governing municipality coupled with strong municipal dedication to habitat monitoring and maintenance. Other techniques such as various forms of easements and covenants are effective so long as proper safeguards are taken.
- Residents must receive *proper education* of the goals of habitat protection, and the best methods for individual support of those goals.

Part

VIII

The International Perspective

16

International Environmental Law
and Policy

16.1 Introduction

International environmental law is the least familiar aspect of environmental law to most nonenvironmental lawyers. One reason is that it may not seem clear what, exactly, environmental law is. We will defer to Professor Guruswamy's definition that international environmental law "consists of international law dealing with the environment as found, primarily, in international agreements (also called treaties, conventions or pacts), together with the international mechanisms for implementing them; and secondarily, in international customary law (the common law of the international community)."[1] We have already referred to international environmental law at several points in this book; for example in the discussion of the extraterritorial reach of the National Environmental Policy Act (NEPA) in chapter 3, or the Convention on International Trade in Endangered Species (CITES) in chapter 15.

16.2 History of International Environmental Law

International environmental law is at once among the oldest and among the youngest areas within environmental law. It is among the oldest because the first international (i.e., multi-nation) environmental agreements took place in the early part of this century. The 1909 United States – United Kingdom Boundary Waters Treaty provided that water "shall not be polluted on either side [of the border between the U.S. and

---

[1] Guruswamy, International Environmental Law: Boundaries, Landmarks, and realities, 10(2) Natural Resources & Env't 43 (1995).

Canada] to the injury to health or property on the other side."[2] Other early international agreements were aimed primarily at economically important natural resources, such as the 1902 Convention for the Protection of Birds Useful in Agriculture, the 1911 Treaty for the Preservation and Protection of Fur Seals, the 1916 Convention for the Protection of Migratory Birds in the United States and Canada, the 1940 Washington Convention on Nature Protection and Wild Life Preservation, and numerous treaties dealing with fisheries and whaling.[3]

In the 1960s, international environmental issues began to emerge as powerful tools for reform of international environmental regulation. As was discussed in Chapter 3, the 1960s was a time of change at all levels, fueled by post World War II affluence, and focused on increased public apprehension over scientific evidence of rapid, irreversible environmental damage from anthropogenic pollution, the squandering of natural resources, and abrupt increases in human population growth rates. During the 1960s, there was a dramatic increase in the number of international agreements aimed specifically at protecting the environment rather than preserving economically valuable resources. International conventions addressed topics such as oil pollution of the oceans, and protection of threatened species from overexploitation.[4] The 1968 African Convention on the Conservation of Nature and Natural Resources is an excellent example. In the U.S., the National Environmental Policy Act (NEPA) was passed in 1969.[5]

Modern international environmental law was formed largely by the Stockholm Conference on the Human Environment in 1972.[6] An issue debated at length by the Stockholm Conference was the plight of the many poor less-developed countries (LDCs), for whom environmental protection was a luxury that overwhelming poverty made unaffordable. In the final analysis, the conference declared in its famous Principle 21, that:

> States have, in accordance with the Charter of the United Nations and the principles of international law, the sovereign right to exploit their

---

[2] For an excellent review of international environmental agreements in the early 1900s, see E.B. Weiss, International Environmental Law: Contemporary Issues and the Emergence of a New World Order, 81 Geo. L.J. 675 (1993).
[3] Id.
[4] Id.
[5] 42 U.S.C. §4331 et seq. See Chapter 3.
[6] See, generally, L. Sohn, The Stockholm Declaration on the Human Environment, 14 Harvard Int'l L.J. 23 (1973); and L. Guruswamy, International Environmental Law: Boundaries, Landmarks, and realities, 10(2) Natural Resources & Env't 43 (1995).

own resources pursuant to their own environmental policies, and the responsibility to ensure that activities within their jurisdiction or control do not cause damage to the environment of other States or of areas beyond the limits of national jurisdiction.[7]

This recognition that each country has both a right to its resources and an obligation to ensure that pollution does not affect other countries has driven international environmental legal efforts since the 1970s.

16.3 Sources of International Law

International environmental law since the 1970s has generally involved two broad areas of law; "customary" international law (the "common law" of the international community), and law that arises from agreements between two or more sovereign states.

16.3.1 "Customary" International Law

Customary international law is certainly the oldest and probably the weakest form of international environmental law. In fact, customary law is often called "soft law."[8] It is often based on little more than customary trade practices and habits, and is largely unenforceable. Soft laws are usually expressed as political or value statements, rather than legally binding provisions of an agreement.

Partly as a result of the 1972 Stockholm Conference discussed above, the United Nations General Assembly created a special subsidiary body known as the United Nations Environment Program (UNEP). The principle function of UNEP was supposed to be the promotion of regional conventions, and to generate environmental dialogue among member nations. What has happened in fact is that UNEP has become the primary negotiating body for draft resolutions sent to the U.N. General Assembly, such that UNEP has become a primary promoter of soft law. Regional organizations in Europe such as the Organization for Economic Cooperation and Development (OECD) have adopted a series of recommendations regarding transboundary pollution, but these do not become "hard" law until directives are passed by the affected governments.

Nongovernmental organizations (NGOs) have contributed to

---

[7] Stockholm Conference on the Human Environment, Principle 21.
[8] See Dupuy, Soft Law and the International Law of the Environment, 12 Mich. J. Int'l L. 420 (1991).

environmental soft laws as well. The International Law Association (ILA), an NGO, adopted the Montreal Rules of International Law Applicable to Transfrontier Pollution, which attempts to regulate through transboundary pollution through cooperation. Similarly, the Institute of International Law (IIL) has prepared and promoted resolutions on the Utilization of Non-Maritime International Waters, and Transboundary Air Pollution.[9]

In every case, international customary law (soft law) relies on the same message; namely, that it is in the best interest of a particular nation to abide by terms that it would wish its neighbors to follow. The extent to which these customary laws have been successful depends largely on the relationships between the countries involved. In any event, their lack of any real mechanism for enforcement places them largely beyond the scope of this book.

16.3.2 Bilateral and Multilateral Agreements

Bilateral and multilateral agreements are treaties, charters, conventions, protocols, etc. that create specific legally binding obligations on two or more signatory nations. These agreements use a variety of specific regulatory tools for designating responsibilities and authorizing enforcement.

Some of the best (and most successful) examples of bilateral agreements are those between the United States and Canada. Under the Great Lakes Water Quality Agreement of 1978 (GLWQA, amended in 1983 and 1987), the United States and Canada are committed to an ecosystem-wide system to restore and maintain the chemical, physical, and biological integrity of the great lakes. With over 90 percent of the freshwater surface in North America, the Great Lakes system had gradually degraded to the point that international cooperation became critically important. Under the GLWQA, an International Joint Commission (IJC) is created with membership made up of representatives from both nations. The IJC drafts regulations and makes recommendations on all actions affecting the Great Lakes, their tributaries, and adjacent riparian areas. Under the GLWQA, the entry of pollutants into the Great Lakes has decreased substantially, and water quality and ecosystems have improved dramatically.

In some situations, a nation (or group of nations) may enter into an agreement with a nongovernmental organization (NGO). For example, an agreement with the World Bank or the International Monetary Fund (both NGOs) may obligate a nation to clean up buried hazardous wastes in order to procure a loan.

The most popular model for the implementation of multinational

---

[9] Id.

agreements begins when an issue (or a group of related issued) is identified. One or more countries then calls for a "convention," which is attended by representatives of invited or interested countries, who negotiate the details of the agreement (including coverage, regulations, enforcement, and the like). This is then followed by a formal proposal called a "protocol," which is eventually ratified by every nation that wishes to be included.

We will examine several specific multinational agreements in section 6.3.

16.4 International Environmental Agreements

There are literally hundreds of international environmental agreements (or agreements with environmental components) effective throughout the world. However, a thorough review of all of them is beyond the scope of this book. We will examine a few international agreements in detail, and refer to a few others in general.

16.4.1 The North American Free Trade Agreement (NAFTA)

Implemented on January 1, 1994, the goal of the North American Free Trade Agreement (NAFTA) was to remove most barriers to trade and investment among the United States, Canada, and Mexico.[10] Under NAFTA, all nontariff barriers to agricultural trade between the United States and Mexico were eliminated. Many tariffs were eliminated immediately, with others being phased out over periods of 5 to 15 years.

NAFTA is not an environmental agreement. Rather, it is a trade agreement which contains certain safeguards that will have the effect of protecting the environment in all three signatory countries.

Although NAFTA encourages trading partners to adopt international and regional standards, the agreement explicitly recognizes each country's right to determine the necessary level of protection. Such flexibility permits each country to set more stringent standards, as long as they are scientifically based. In addition, NAFTA allows state and local governments to enact standards more stringent than those adopted at the national level, so long as these standards are scientifically defensible and are administered in a forthright and expeditious manner.

NAFTA imposes limitations on the development, adoption, and enforcement of sanitary and phytosanitary (SPS) measures. These are

---

[10] In the United States, NAFTA was implemented by the North American Free Trade Agreement Implementation Act, 19 U.S.C. § 3301.

measures taken to protect human, animal, or plant life or health from risks that may arise from animal or plant pests or diseases, or from food additives or contaminants. Disciplines contained in the NAFTA are designed to prevent the use of SPS measures as disguised restrictions on trade, while still safeguarding each country's right to protect consumers from unsafe products, or to protect domestic crops and livestock from the introduction of imported pests and diseases.

The NAFTA Committee on Sanitary and Phytosanitary Measures promotes the harmonization and equivalence of SPS measures, and facilitates technical cooperation, including consultations regarding disputes involving SPS measures. This committee meets periodically to review and resolve issues in the SPS area.

It is still too early to determine whether NAFTA's environmental goals will be met.

16.4.2 The Kyoto Protocol

The Kyoto Protocol was negotiated as a means of implementing the United Nations Framework Convention on Climate Change, to which the United States became a signatory in 1992, and by which the United States is legally bound. Unfortunately, it has become an example of the effect of the intrusion of the U.S. political process into international environmental issues.

The U.N. Framework Convention set an objective of stabilizing greenhouse gas concentrations in the atmosphere at a level that would prevent global warming, and anticipated that the parties would adopt protocols to the convention in order to achieve that objective. Such protocols must themselves be ratified by the participating states and meet their own standards for going into effect internationally before they can become legally binding. In this instance the Kyoto Protocol has been negotiated, sets binding targets for reduction of emissions of greenhouse gases by developed nations.

The President of the United States signed the Kyoto Protocol in 1998, and indicated his intent eventually to seek its ratification. However, as of this writing, the protocol has not been ratified by the United States or even submitted to the Senate for its consent. The protocol will not enter into force internationally until it has been ratified by at least 55 countries that accounted for at least 55 percent of the total carbon dioxide emissions in 1990. Both steps— ratification by the United States and entry into force internationally—are necessary for the Protocol to be legally binding on the United States.[11]

The Kyoto Protocol provides that it is open for signature from March 16, 1998, to March 15, 1999, and is subject to ratification, acceptance, or approval.[12] The United States initially delayed signing as a means of encouraging fuller participation in emissions reductions by developing states; but on November 12, 1998, it became the 58th nation (and the last major industrialized nation) to sign.

Signature in itself does not make the Protocol legally binding on the United States, but it has several consequences. First, it authenticates the text of the agreement (i.e., it represents the acknowledgment of the negotiating countries that the text expresses the agreement they have reached). Secondly, it at least begins the process by which the United States could become legally bound by the protocol (i.e., signature of a treaty is essentially a political statement of approval and represents "at least a moral obligation to seek (its) ratification").[13] The protocol cannot become legally binding until it is submitted to the U.S. Senate, the Senate gives its consent, the president signs and deposits the appropriate instruments of ratification with the United Nations, and the protocol gains sufficient ratifications to enter into force internationally.

The United States is not yet legally bound by the Kyoto Protocol, although signature of a treaty or protocol obligates a state "to refrain from acts that would defeat the object and purpose of the agreement."[14]

16.4.3 North American Agreement on Environmental Cooperation (NAAEC)

The North American Agreement on Environmental Cooperation (NAAEC) was signed in Washington, D.C. on September 9, 1993; in Ottawa, Canada on September 12, 1993; and in Mexico City on September 8, 1993. It went into force on January 1, 1994 (immediately after entry into force of NAFTA).

The NAAEC is an agreement between Canada, Mexico, and the United States to increase cooperation between the parties to foster the protection and improvement of the environment within the territories of the parties. The objectives of the agreement include: the promotion of sustainable development based on cooperation and mutually supportive environmental

---

[11] Ackerman, *Global Climate Change: Selected Legal Questions About the Kyoto Protocol* (Committee for the National Institute for the Environment, 98-349, 1999).

[12] Kyoto Protocol, Articl 23.

[13] American Law Institute, *Restatement of the Foreign Relations Law of the United States Third*, Vol. 1, § 312, Comment d (1987).

[14] Id., § 312(3).

and economic policies; the support of the environmental goals and objectives of NAFTA; the avoidance of new trade distortions and trade barriers; the promotion of economically efficient and effective environmental measures; public participation in the law making process; increased cooperation in the development and improvement of environmental law, policies, regulations, procedures and practices; enhanced compliance with and enforcement of environmental laws and regulations.

16.4.4 The Convention on Biological Diversity

The concept of "biological diversity," along with its values to humans, was discussed in Chapter 15. Interestingly, the conservation of biological diversity and the sustainable use of its components is not a new concept in the international arena. It was emphasized in the United Nations' 1972 Stockholm Conference on the Human Environment. In 1973, the first session of the Governing Council for the new UN Environment Programme (UNEP) identified the conservation of nature, wildlife and genetic resources as a priority.

The international community's concern over the loss of biological diversity worldwide inspired negotiations for a legally binding document whose primary goal would be a reversal of the current trend. Negotiations were also influenced by the growing recognition throughout the world of the need for a fair and equitable sharing of the benefits arising from the use of genetic resources.

The Convention on Biological Diversity (CBD) was signed at the Earth Summit in Rio de Janeiro in June, 1992. The objectives of the CBD are spelled out in Article I:

The objectives of this Convention, to be pursued in accordance with its relevant provisions, are the conservation of biological diversity, the sustainable use of its components and the fair and equitable sharing of the benefits arising out of the utilization of genetic resources, including by appropriate access to genetic resources and by appropriate transfer of relevant technologies, taking into account all rights over those resources and to technologies, and by appropriate funding.

In other words, the CBD seeks not only to protect biological diversity in the traditional sense. Its provisions on scientific and technical cooperation, access to financial and genetic resources, and the transfer of ecologically sound technologies are unique.[15] There is a clear attempt to ensure that the countries that supply the biological and genetic diversity to the rest of the conference will profit.

Article 25 of the CBD creates a multidisciplinary "subsidiary body for the provision of scientific, technical and technological advice" with membership open to all signatory countries. The purpose of the subsidiary body is to provide the signatory countries and other subsidiary bodies with timely advice relating to the implementation of the CBD. It reports regularly to the conference on all aspects of its work.

For these reasons, the CBD is a significant recent developments in international law, international relations, and the fields of environment and development.

16.4.5 The Basel Convention on Transboundary Movements of Hazardous Wastes

The 1989 Basel Convention on the Control of Transboundary Movements of Hazardous Wastes and Their Disposal was convened by the United Nations Environment Programme. Its goal is to stop the export of hazardous and toxic wastes to any country, unless the receiving country's government agrees in advance to accept the wastes.[16] In addition, both the exporting and importing countries must guarantee that they will follow prescribed methods for treatment and disposal of the wastes.

The Basel Convention requires that parties take all practical steps to ensure that the transboundary movement of hazardous wastes "is conducted in a manner which will protect human health and the environment against the adverse effects which may result from such movements."[17] The convention requires the exporting country to notify, or require the generator or exporter to notify, in writing, the competent authority of the countries concerned (including countries of export, import and transit) of any proposed transboundary movement of hazardous wastes or other wastes.[18] Countries may decide not to consent, partially or totally, to the import of hazardous wastes for disposal and may also decide to limit or ban the export of hazardous wastes or other wastes. The convention does not require a uniform definition of hazardous waste, and as long as either the exporting, importing or transit party considers the waste

---

[15] See Guruswamy, International Environmental Law: Boundaries, Landmarks, and realities, 10(2) Natural Resources & Env't 43 (1995).

[16] For an excellent review of the Basel Convention, see Rogus, the Basel Convention and the United States, 2 New England Int'l & Comp. L. Ann. 434 (1996).

[17] Basel Convention, Article 4, General Obligations.

[18] Id. at Article 6, Transboundary Movement between Parties.

hazardous, the waste qualifies as a "hazardous waste."[19]

Signatory countries to the Basel Convention are required to take appropriate measures to reduce the generation of hazardous wastes, and to reduce the transboundary movement of hazardous wastes to the minimum consistent with their environmentally sound and efficient management. In addition, both importing and exporting parties are bound to prevent planned transboundary movements if they have reason to believe that they will not be managed in an environmentally sound manner.[20] Parties may not trade with non-parties in wastes covered by the convention absent a separate agreement between them that satisfies standards set by the convention. Finally, a secretariat in Geneva is to organize periodic meetings of the parties and
perform other functions, such as compiling and transmitting information (including news of illicit trafficking) and cooperating with states in the provision of experts and equipment in emergencies.[21]

Unfortunately, because a Basel Convention resolution calls for a total ban on hazardous waste exports, the United States has failed to ratify the convention.

16.4.6 The Vienna Convention and the Montreal Protocol

A study by the U.S. National Academy of Sciences in 1976 led to a national ban on chlorofluorocarbons (CFCs) for nonessential aerosols.[22] Shortly after the United States banned CFCs, similar legislation followed in Canada and the Scandinavian countries. The issue of dangerous stratospheric ozone depletion was taken up by the United Nations, and a working group began to negotiate the key elements of an international treaty in 1981. This resulted in the Vienna Convention for the Protection of the Ozone Layer, signed in 1985 by 20 of the leading CFC-producing and consuming nations.

Unfortunately, the Vienna Convention did not establish specific controls or reduction targets, which was largely the result of a conflict between two groups of countries. One goup, headed by the United States, advocated a worldwide ban on CFCs, while another group, headed by the European Union nations, Japan and the former U.S.S.R. would not agree to more than a freeze in production capacity. Many countries felt that an early phase-out would have given the U.S. chemical manufacturers an advantage

---

[19] Id. at Article 3, National Definitions of Hazardous Wastes.
[20] Id. at Article 4, General Obligations.
[21] Id at Article 16, Secretariat. See Rogus, the Basel Convention and the United States, 2 New England Int'l & Comp. L. Ann. 434 (1996).
[22] See Chapter 11.

since many of them had already developed CFC substitutes, while most non-U.S. manufacturers had not. Due to this impasse, specific reduction measures were left to a later protocol.

Negotiations on the Montreal Protocol followed the signing of the Vienna Convention. On 16 September 1987, the Montreal Protocol on Substances that Deplete the Ozone Layer was adopted by 24 countries. The protocol became possible laregely as a result of a change in the position of West Germany, which led to a readjustment of the EU position. Under the protocol, CFC production is to be cut to half of 1986 levels by the year 1999, starting with a freeze in production and consumption within one year of the entry into force of the protocol.[23] In the case of halons, the Montreal Protocol provided only for a freeze on 1986 levels. No additional ozone depleting substances (ODS) have been included under the control measures.

The agreement therefore provided for trade controls of ODS with states who did not become party to the protocol. Unfortunately, the original protocol gave little specific financial support to developing countries, and China and India refused to sign. Following subsequent amendments, however, China (1991) as well as India (1992) have became parties to the Montreal Protocol.

The Montreal Protocol has twice been amended since coming into effect in 1989. The London Amendment in 1990 provided for additional reduction in CFCs, to be completely phased out by the year 2000. The amendment also submitted new substances like "other fully halogenated CFCs," carbon tetrachloride and methylchloroform to a control mechanism.[24] The London Amendment established a financial mechanism financed by the industrialized countries to allow for the incremental costs of acquiring and developing alternative technologies for developing countries. Two years later, the signatory countries adopted additional amendments at the annual meeting in Copenhagen, which once again moved forward the timetable for the reduction of ODS.

16.4.7 The International Organization for Standards (ISO)

An increasingly significant NGO is the International Organization for Standardization, also known as "ISO."[25] ISO is a private sector, international standards body founded in 1947 and based in Geneva,

---

[23] See Wood, The Multilateral Fund for the Implementation of the Montreal Protocol, 5(4) International Environmental Affairs 335 (1993).
[24] Montreal Protocol at Article 2C, Other Fully Halogenated CFCs.
[25] "ISO" is not an acronym, but is instead from the Greek word "*isos*" for "equal."

Switzerland. The primary purpose of ISO is to promote the international harmonization and development of manufacturing, product, and communications standards. More than 120 countries belong to ISO as full voting members, while several other countries serve as observer members. The United States is a full voting member and is officially represented by the American National Standards Institute (ANSI).

The role of ISO is to develop voluntary standards that cover many aspects of technology. ISO standards represent an international consensus on the state of the art in the technology concerned. The idea is that these standards contribute to making the development, manufacture, and supply of products and services more efficient, safer, and cleaner. In addition. the standards should make trade between countries easier and fairer, and should help to safeguard the quality of goods and services for consumers.[26]

ISO produces internationally harmonized standards through a structure of Technical Committees (TCs). The TCs usually divide into subcommittees, which are further subdivided in Working Groups where the actual writing of standards occurs.

To date, ISO has promulgated more than 8,000 internationally accepted standards for everything from paper sizes to film speeds.[27] ISO standards and enforcement are administered by experts from the affected industrial, technical, and business sectors, as well as selected individuals with relevant knowledge, such as representatives of government agencies and testing laboratories.

The two most recognizable standards are the ISO 9000 series and the ISO 14000 series. ISO's 9000 series adopted a series of Quality Management Standards (QMS) that have already had a significant impact on world trade, while the ISO 14000 series Environmental Management Standards (EMS) are expected to have a similar impact on trade and the environment.[28]

ISO 9000 Series

ISO Technical Committee 176 (ISO/TC176) was formed in 1979 to harmonize the increasing international activity in quality management and quality assurance standards. Subcommittee 1 was established to determine common terminology. It developed ISO 8402: Quality-Vocabulary, which

---

[26] See, generally, Rosenbaum, *ISO 14000 and the Law. Legal Guide for the Implementation of Environmental Management for the World Market* (AQA Press, 1998).

[27] U.S.E.P.A., *EPA Standards Network* (EPA, 1998).

[28] See Johnson, *ISO 14000: The Business Manager's Complete Guide to Environmental Management* (Wiley, 1997); Begley, Value of ISO 14000 Management Systems: Put to the Test, August 1997 Env'l Sci. & Tech. 364 (1997); Wilson, ISO 14000 Insight: A Variety of Drivers Influence an EMS Program, Sept. 1997 Pollution Engineering 53 (1997).

was published in 1986.[29] Subcommittee 2 was established to develop quality systems standards; the result being the ISO 9000 series, published in 1987, and revised in 1994. The United States had input into this development process through membership in ISO via ANSI. This input was channeled through a Technical Advisory Group (TAG). ASQ administers the United Sstates' TAG to ISO/TC176 on behalf of ANSI. Qualified U.S. experts participate in the meetings where these documents are drafted. ASQ continues to administer the U.S. TAG to ISO/TC176, and the United States continues to contribute to this process of developing international standards on quality assurance and quality management, and the generic supporting technologies necessary for full implementation.[30]

The ISO 9000 series is a set of five individual, but related, international standards on quality management and quality assurance. They are generic, not specific to any particular products. They can be used by manufacturing and service industries alike. These standards were developed to document the quality system elements to be implemented in order to maintain an efficient quality system within an organization or company. The ISO 9000 Series standards do not themselves specify the technology to be used for implementing quality system elements.

There are several benefits to implementing the ISO 9000 series in an organization or company, such as encouraging quality in products or services, and avoiding costly after-the-fact inspections, warranty costs, and rework. In addition, it is often possible to reduce the number of audits performed by customers. Increasingly, customers are accepting supplier quality system registration from an accredited third-party assessment based on these standards.[31]

ISO 1400 Series

ISO's EMS are a series of voluntary standards and guideline reference documents which include environmental management systems, eco-labeling, environmental auditing, life cycle assessment, environmental performance evaluation, and environmental aspects in product standards. Although the ISO 14000 series is based in many ways on the ISO 9000

---

[29] ASQ published ANSI/ASQ A8402-1994: Quality Systems Terminology. While this document is not an adoption of ISO 8402, it does contain many of the exact terms and definitions contained in ISO 8402.

[30] U.S. Standards Group on QEDS, Registrar Accreditation Board, *ANSI ASC Z-1 Committee on Quality Assurance Answers the Most Frequently Asked Questions About the ISO 9000* (ANSI/ASQ Q9000 Series, 1998).

[31] See Johnson, *The ISO 14000 EMS Audit Handbook* (St Lucie Press, 1997); and Welch, *Moving Beyond Environmental Compliance: A Handbook for Integrating Pollution Prevention with ISO 14000* (CRC Lewis Publ., 1998).

approach, its environmental scope is far greater.[32]

The idea behind EMS is to help companies and organizations to establish and meet policy goals with management oversight through objectives and targets, organizational structures and accountability, management controls, and review functions. EMS do not set requirements for environmental compliance or establish requirements for specific levels of pollution prevention or performance. The EMS specification document calls for environmental policies which include a commitment to both compliance with environmental laws and prevention of pollution.

In August 1991, ISO established a Strategic Advisory Group (SAGE) to assess the need for international environmental management standards and to recommend an overall strategic plan for such standards. SAGE was asked to incorporate several factors into environmental management standards, and these have become the basis of the ISO standards:

- Promote a common approach to environmental management similar to ISO 9000 Quality Management Standards;

- Enhance an organization's ability to attain and measure environmental performance; and

- Facilitate trade and remove trade barriers.

Based on the SAGE findings, ISO formed Technical Committee #207 (TC-207) in 1992 for EMS. Currently, nearly 50 countries have signed on to TC-207 as full voting members, with an additional 13 countries as observers. The United States, which is a voting member, participates in the process through a TAG. Under delegated authority from ANSI, the TAG is administered by the American Society for Testing and Materials (ASTM). TC-207 has six subcommittees, each of which contains several working groups, plus one working group on Environmental Aspects in Product Standards that reports directly to the full TC-207.

ISO assigns a document numbering system to each TC. Standards produced by TC-207 are assigned the 14000 designation. For example, the Environmental Management Systems Guidance Standard has become ISO 14000, and the EMS specification document became ISO 14001. Each subsequently completed ISO standard from this TC will has a 14000 designation.

---

[32] U.S.E.P.A., *EPA Standards Network* (EPA, 1998). See also Rosenbaum, *ISO 14000 and the Law. Legal Guide for the Implementation of Environmental Management for the World Market* (AQA Press, 1998).

ISO 14000 series elements to date (and publication timetable) are as follows:[33]

- Environmental Policy. An organization-based statement of policy and objectives which outlines the entity's commitment to environmental responsibility. Environmental policy sets the operational scope for the EMS and is, therefore, an instrumental element of the implementation process. There are no separate ISO guidelines on environmental policy.

- Environmental Management System (ISO14001, ISO14004). That part of the overall management system which includes organizational structure, responsibilities, practices, procedures, planning activities, processes and resources for developing, implementing, achieving, reviewing and maintaining the environmental policy. [Released in 1996.][34]

- Environmental Performance Evaluation (ISO14031). A process guide to measure, analyze, assess and describe the organization's environmental performance against agreed targets based on the entity's environmental policy objectives. [Released in 1998.]

- Environmental Auditing (ISO14010, ISO14011, ISO14012, ISO14015). Guidelines for general principles of environmental auditing, guidelines for auditing environmental management systems and the qualification criteria for environmental auditors. ISO14015 deals specifically with Environmental Site Assessment. [Released in 1996 (ISO14010, ISO14011, ISO14012); to be released in future (ISO14015)]

- Life Cycle Assessment (ISO14040, ISO14041, ISO14042, ISO14043). Guidelines to evaluate environmental attributes associated with a product, process, or service, including impacts along the entire continuum of a product's life from raw material extraction, through manufacturing processes, distribution and transportation, use, recycling, to final disposal. [Released in: 1997 (ISO14040); 1998 (ISO14041); and 1999 (ISO14042, ISO14043).]

- Environmental Labeling (ISO14020, 14021, 14022, 14023, 14024, 14025). Methods to identify products which meet specified

---

[33] Parto, *An Introduction To ISO 14000 - Who Needs It ?* (Univ. Waterloo, 1997).
[34] See Woodside, *ISO 14001 Implementation Manual* (McGraw-Hill, 1998); and Canadian Standards Association, *ISO 14001:1996, Environmental Management Systems — Specification with Guidance for Use* (CSA, 1996).

environmental requirements of a product class including procedures, terms and definitions, symbols, and testing techniques. [Released in: 1998 (ISO14020, ISO14021, ISO14024); and 1999 (ISO14022, ISO14023, ISO14025).]

●Environmental Aspects in Product Standards (Draft ISO Guide 64). Guide to raise environmental awareness in ISO's product standard writers. The guide recommends the use of life cycle thinking and recognized scientific methodologies to develop product standards which promote improved environmental aspects and account for environmental impacts. [Released in 1997.]

●Terms and Definitions (ISO14050). Terminology standard to ensure clarity and consistency in terms and definitions used by ISO and its member organizations. [Released in 1997.]

16.4.8 European Union Regulations and Directives

Protection of the environment has become one of the major challenges facing the European Community (EC).[35] The EC has been strongly criticized in the past for putting trade and economic development before environmental considerations, but it is now recognized that the European model of development cannot be based on the depletion of natural resources and the deterioration of the environment. Environmental action by the EC began in 1972 with four successive "action programmes," which developed strtegies and methodologies for approaching environmental problems. During this period, the EC adopted over 200 pieces of legislation, chiefly concerned with limiting pollution by introducing minimum standards for waste management, water pollution and air pollution.[36] The Treaty on European Union eventually conferred on the EC the status of a policy. With the subsequent Treaty of Amsterdam, the principle of sustainable development as one of the EC's major goals.[37]

The Fifth Community Action Programme on the Environment "Towards Sustainability" established the principles of an collective EC strategy of

---

[35] European Community, *Union Policy. Environment* (1998), hereafter "Union Policy." The European Union's official worldwide web site for the environment is found at http://europa.eu.int/scadplus/leg/en/s15000.htm. The information found on this particularly informative web site forms the basis of much of the following discussion.
[36] See Pinder (ed.), *The New Europe: Economy, Society and Environment* (John Wiley & Son Ltd., 1998).
[37] Union Policy.

voluntary action for the period 1992-2000, and marked the beginning of an EC approach which would consider numerous sources of pollution, including industry, energy, tourism, transport, and agriculture. This collective approach to environmental policy was confirmed by the European Commission (the governing body of the EC) in 1998, and all EC institutions are now obligated to address environmental considerations in all their other policies.[38]

The European Environment Agency (EEA)

The EC has available a series of environmental "instruments" to carry out its goals. For example, there is a financial instrument (the Life programme) and technical instruments such as the European Environment Agency (EEA).

The EEA was created by the European Union (EU) in 1993 with a mandate to coordinate and put to strategic use information of relevance to the protection and improvement of Europe's environment.[39] The EEA is based in Copenhagen, Denmark, and has a mandate to ensure a supply of objective, reliable and comprehensive environmental information, enabling member nations to take measures to protect their environment, to assess the result of such measures, and to insure that the public is properly informed about the state of the environment.[40] The geographical scope of the EEA's work is open to other countries that share the concerns of the EU, member states, and the objectives of the EEA.

The EEA executes its responsibilities in co-operation with the European Information and Observation Network (EIONET). EIONET was created and is operated by the EEA. EIONET consists of national networks, organized by the EEA to assist in information retrieval, identification of special environmental issues, and producing efficient and timely information on Europe's environment.[41]

The EEA has come to play an increasingly important role in recent years. Although its role is purely advisory, its work has become more and

---

[38] Id.
[39] See Ute, *Energy and Environment in the European Union : The Challenge of Integration* (Avebury Studies in Green Research, 1995).
[40] Council Regulation (EEC) No 1210/90 of 7 May 1990 on the establishment of the European Environment Agency and the European environment information and observation network. Amended by Council Regulation (EC) No 933/1999 of 29 April 1999.
[41] Id. The EEA's mission statement states: "The EEA aims to support sustainable development and to help achieve significant and measurable improvement in Europe's environment through the provision of timely, targeted, relevant and reliable information to policy making agents and the public."

more crucial for the adoption of new measures and for assessing the impact of decisions already adopted.[42]

Fifth Environmental Framework Programme

One of the most significant EC environmental measures is the fifth European environmental programme, which was adopted for the period 1993 to 2000. Among the principal points of the fifth Programme are:

- Sustainable growth.

- Integration of environmental matters into all other policies.

- Integration of decisions taken in the Rio Summit into EC policy.

- Recourse to preventive measures.

- Shared responsibility.

- Involvement of public authorities, traders, businesses and citizens.

The Programme is intended to guide the EC as well as national actions in the definition of environmental policy. In their resolution, member states noted that many forms of economic activity and development are not sustainable at present.[43]

Despite the number of environmental Directives and other measures, the Environmental Commissioner stated in 1998 that more than half the 166 European Union laws issued to date had been ignored by member states. As a result, the number of infringement proceedings has risen sharply. In particular, since August 1992, the European Commission opened proceedings against Portugal on dangerous wastes, PCBs, and PCTs; Italy on bathing water quality; Ireland on toxic waste; Germany on the disposal of used oils; and England on the quality of drinking water.[44]

Environmental Assessment

The Town and Country Planning (Assessment of Environmental Effects) Regulations of 1988 require an environmental statement to be produced in support of a specified range of major development proposals.[45]

---

[42] See http://www.eea.eu.int.
[43] Union Policy.
[44] The Institution of Civil Engineers, *the Environment File. European and UK Legislation* (1996-1999).
[45] Union Policy.

Environmental assessment is defined by an EC Department of the Environment circular as:

> [A] technique for drawing together, in a systematic way, expert quantitative analysis and qualitative assessment of a project's environmental effects, and presenting the results in a way which enables the importance of the predicted effects, and the scope for modifying or mitigating them, to be properly evaluated by the relevant decision-making body before a decision is given. Environmental assessment techniques can help both developers and public authorities with environmental responsibilities to identify likely effects at an early stage, and thus to improve the quality of both project planning and decision making.[46]

The information gathered by a developer and included in conjunction with a planning application, is referred to as an Environmental Statement (ES). The contents of the ES must include:

- The contents of the plan or program and its main objectives;

- The environmental characteristics of any area likely to be significantly affected by the plan or program;

- Any existing environmental problems which are relevant to the plan or program;

- The national, EC, or international environmental protection objectives which are relevant to the plan or program in question;

- The likely environmental effects of implementing the plan or program;

- Any alternative solution which might be considered.

The statement must also include a non-technical summary of this information.[47] The ES process bears many similarities to the Environmental Impact Statement (EIS) process under Section 102(2)(b) of the United States' National Environmental Policy Act (NEPA) discussed at length in chapter 3.

---

[46] Department of the Environment circular 15/88. See also Union Policy, and Brouwer (ed.), *Environment and Europe: European Union Environment Law and Policy and Its Impact on Industry* (Kluwer Law Int'l, 1995) .

[47] Union Policy. See also Pinder (ed.), *The New Europe : Economy, Society and Environment* (John Wiley & Son, 1998).

Waste Management

EC policy on waste management involves three complementary strategies:

● Eliminating waste at source by improving product design.

● Encouraging the recycling and re-use of waste.

● Reducing pollution caused by waste incineration.

The EC's approach has been to assign more responsibility to the producer. For example, a 1997 draft directive on end-of-life vehicles provides for the introduction of a system for collecting such vehicles at the manufacturer's expense.[48]

At the international level, this approach was also adopted at the first Conference of the Parties to the OSPAR Convention for the Protection of the Marine Environment of the North-East Atlantic.[49] One of the tasks of this conference was to negotiate the dismantling and disposal of offshore oil rigs and natural gas platforms . The parties to the convention adopted the position supported by the EC that the dumping of such installations at sea should be banned and that the costs of dismantling and disposing of such installations should be borne by their owners. The EC is a party to the Convention on the Control of Transboundary Movements of Hazardous Wastes and their Disposal (the Basel Convention, discussed in section 16.3.5 above). The EC ratified the amendment to the convention, banning exports of hazardous wastes from the OECD countries, the EC, and Lichtenstein to non-OECD countries, regardless of whether the waste is for disposal, recycling, or use.[50]

Water Pollution

A number of directives have been adopted by the member states to introduce water quality standards (primarily drinking water and bathing waters), and to monitor emissions of pollutants. The EC is a party to various international conventions aimed at protecting the marine environment (the OSPAR Convention, the Barcelona Convention for the Protection of the Mediterranean Sea against Pollution[51]) and watercourses (Helsinki Convention on the Protection and Use of Transboundary

---

[48] Union Policy.
[49] OSPAR (Oslo and Paris Commissions) refers to the Convention for the Protection of the Marine Environment of the North East Atlantic of 1992. There are delegations from the fifteen countries that border the North East Atlantic, the EC, and NGOs.
[50] Union Policy.
[51] EC Decision 77/585/EEC - CJL 240, 19.9.1997.

Watercourses and International Lakes; Convention on Cooperation for the Protection and Sustainable Use of the River Danube[52]).

The current proposals for directives are aimed at further improving the ecological quality of surface waters, at introducing EC action on fresh waters and surface waters, and at protecting EC estuaries, coastal waters, and groundwater.[53]

Air Pollution

Improving air quality is an EC priority, since the EC considers air pollution to be the main cause of global warming. To achieve a significant reduction in air pollution, the EC has attempted to combine national and international efforts to reduce emissions of the gases responsible. To this end, the United Nations Framework Convention of 1992 and the Kyoto Protocol of 1997 were adopted by the EC. The parties have undertaken to reduce their emissions of greenhouse gases by at least 5 percent of their 1990 levels during the period 2008–2012. To achieve this, the European Commission's strategy is to take action in all the economic sectors which produce polluting gases, primarily transport, energy, industry and agriculture.

The ECC is also a party to the Geneva Convention on Long-Range Transboundary Air Pollution[54] and to some of its protocols. EC legislation in this field is principally aimed at cutting emissions from industrial activities and road vehicles. Where transport is concerned, the strategy is:

● To reduce polluting emissions (catalytic converters, and road worthiness tests).

● To reduce the fuel consumption of private cars (in collaboration with car manufacturers).

● To promote clean vehicles (primarily through tax incentives).[55]

Nature Conservation

In Europe, some 1,000 plant species and more than 150 species of birds are severely threatened or on the brink of extinction. To combat this situation, EC legislation has introduced several measures to conserve wildlife (protection of certain species such as birds and seals) and natural habitats (protection of woodlands and watercourses). The EC is a party to a number of conventions, including the Bern Convention on the

---

[52] OJ L 342, 12.12.1997.
[53] Union Policy.
[54] EC Decision 81/462/EEC - OJ L 171, 27.6.1981.
[55] Union Policy.

Conservation of European Wildlife and Natural Habitats and the Bonn Convention on the Conservation of Migratory Species.[56]

International Cooperation

According to Article 130r of the Treaty establishing the European Union, one of the objectives of EC policy on the environment is to promote measures at the international level to consider regional or worldwide environmental problems. Under the treaty, the EC may cooperate with third countries and with competent international organizations. Although this recognition dates back only to the Treaty on European Union, the EC has been a party to international conventions on environmental conservation since the 1970s. The EC is a party to more than 30 conventions and agreements on the environment, and takes an active part in the negotiations leading to the adoption of these instruments.

The EC participates as an observer in the activities and negotiations taking place within the context of international bodies or programmes and, in particular, under the auspices of the United Nations. The EC is a full participant in the work of the United Nations Commission for Sustainable Development, which is the body responsible for adherence with the Conference on Environment and Development held at Rio de Janeiro in June, 1992.[57]

Some of these conventions are global in scope, while others are regional. Among the global conventions are the Vienna Convention for the Protection of the Ozone Layer,[58] its Montreal Protocol on substances which deplete the ozone layer[59] and the UN Conventions on Biological Diversity and on Climate Change.[60] The EC has also recently signed the Kyoto Protocol which provides for measures and commitments to reduce greenhouse emissions.[61]

Accession Strategies

An important environmental issue is the serious pollution problems in many countries that wish to join the EC, especially those in eastern and central Europe. The EC strategy is to assurance on the incorporation of the environmental EC *acquis*[62] into the legislation of the candidate countries.

A European Commission communication sets out the Union's pre-accession strategy for the central and aastern European countries

---

[56] Id.
[57] Id.
[58] OJ L 297, 31.10.1988.
[59] OJ L 297, 31.10.1988.
[60] OJ L 33, 7.2.1994.
[61] Union Policy.
[62] The existing body of EC law is known as the *acquis communautaire*.

(CEEC).[63] Its aim is to supplement the partnerships for accession and to help the applicant countries improve their national programs for the adoption of the EC *acquis*. The Commission focuses on the environmental issues which affect the ten CEEC applicants, except Cyprus, which will be dealt with in a separate document due to the island's special political situation.[64]

The EC acknowledges that enlargement of the European Union to include the CEEC is an environmental challenge on a scale which cannot be compared with previous accessions. There is a large gap between the levels of protection in the Member States of the Union and the CEEC. Full compliance with the EC's environmental *acquis* will probably only be achievable in the long term. However, the integration of these countries will provide for a considerable increase in biodiversity within Europe in view of their vast areas of uncultivated land.

The European Commission has drawn up a special strategy within the framework of its Agenda 2000. In the Commission's view, the applicant countries should define and start implementing realistic national strategies before accession which will guarantee gradual alignment in the long term. This strategy must include priority areas of action, key objectives to be attained by the date of accession, and timetables for the subsequent achievement of compliance. The communication therefore sets out details to be included by the applicant countries when drafting their national policies.

The main sector-specific challenges to the CEEC are:[65]

- Air pollution: This is largely due to emissions from stationary sources (power plants and district heating installations). The first step is to identify zones and agglomerations where EU limits are being exceeded. It is equally important to modernize refineries so that they comply with European standards;

- Waste management: Steps for the approximation of legislation have accelerated in some countries since 1997 (national investment programs, modernization of incinerators);

- Water pollution: Major investment programs to improve drinking water quality and the management of wastewater are under way in most CEEC countries;

---

[63] COM(98) 294 (final not published in the Official Journal).
[64] Union Policy.
[65] Id.

●Industrial pollution control and risk management: This area requires special attention on the part of the applicant countries since they have numerous heavily polluting industrial and energy production facilities;

●Nuclear safety and radiation protection: All of the CEEC countries have adopted a basic law, which must be supplemented by additional legislation in order to ensure full implementation (this legislation is also required in countries which do not produce nuclear power).

The Commission has defined a set of priority objectives which will help the applicant countries in drawing up their National Programmes for the Adoption of the *Acquis* (NPAA). These priorities must be determined on the basis of a detailed analysis of the environmental situation in each country.

The applicant countries must fill in the gaps in their legislation and administrative rules to improve the environment while at the same time improving the economy and competitiveness. A Commission staff working paper from 1997 identifies the main problems faced by the applicant countries and describes the steps to be taken.[66] When developing their NPAA, the applicant countries must consider:

●How programs for promoting energy efficiency, cleaner technologies and waste minimization and recycling can be integrated into the national economic and sectoral policies;

●How industrial and agricultural production can be guided towards sustainable development;

●How the environmental gains can be maintained during the transition period.

Since enlargement of the EC offers challenges and opportunities for the environment not only in the applicant countries, but for Europe and the entire planet, accession is seen by the EC as part of the process of sustainable development by integrating environmental issues into all policy areas.[67]

---

[66] European Commission, *Guide to the Approximation of the European Union Environmental Legislation* (1997).
[67] Union Policy.

16.5 Enforcement and Conflict in International Environmental Law

The problems facing international environmental agreements are the same as problems facing any legal system. [68] First, there must be acceptance within a nation of the need for protective laws. Second, there must be a mechanism for formulating effective laws once the nation agrees that they are necessary. And third, there must be some mechanism available to enforce the laws, even if it is at some rudimentary level.

In the United States, all three conditions are met for the most part. Like most other nations, we seem to agree that there is a need for internationally enforceable environmental laws, but we are often resistance to allowing international involvement in our affairs. Unfortunately, the many international agreements the United States has signed but not ratified is mute evidence to the difficulties we face whenever we attempt to enter the arena of international law.

Nevertheless, the United States has placed many regulations in its own environmental laws that apply to international environmental issues. For example, section 3017 of the Resource Conservation and Recovery Act (RCRA) states that "no person shall export any hazardous waste identified or listed . . . unless . . . (B) the government of the receiving country has consented to accept such hazardous waste . . . (D) the shipment conforms to the terms of the consent of the government of the receiving country."[69] If the U.S. and another country have entered into an international agreement establishing notice, export, and enforcement procedures for the transportation, treatment, storage, or disposal of hazardous wastes, then shipments to (or from) that country must conform to the terms of the agreement.[70] Any person who violates these terms is subject to whatever enforcement is prescribed by the agreement, as well as the general enforcement provisions of RCRA.[71] As we have noted throughout this book, there have been many other instances in which U.S. federal environmental laws have provisions that deal with international issues.

A second major problem with international environmental laws is the issue of enforcement. How is one country (or an international body) to enforce environmental when most nations are wary about involving foreign nations in their domestic affairs. In fact, the *Restatement of the*

---

[68] See, generally, Birnie and Boyle, *International Law and the Environment* (Oxford Univ. Press, 1993).
[69] 42 U.S.C. §6938(a).
[70] 42 U.S.C. §6938(b).
[71] See chapter 8.

Law on Foreign Relations states that "Under international law, a [nation] . . . is immune from the jurisdiction of courts of another nation state, except with respect to claims arising out of activities . . . by private persons."[72] This statement would seem to preclude any traditional form of enforcement, such as injunctive or declaratory relief discussed in Chapter 2.

Enforcement of most international environmental agreements is "soft" at best. The North American Free Trade Agreement (NAFTA), discussed in Section 16.3.1 above, is a good example. NAFTA recommends that disputes between the three countries (the U.S., Mexico, and Canada) on matters within the scope of the agreement will be resolved by cooperation and consultation.[73] If a signatory country wishes to challenge an environmental measure of another country, the complaining country will have the burden of proving that the environmental measure is not consistent with NAFTA.[74] If disputes cannot be resolved by the two countries, NAFTA establishes a Commission to assist parties in dispute settlement.[75] If the Commission cannot settle the dispute, it must establish an arbitral panel upon request of any disputing party.[76]

Still another problem area in international environmental law is the perception by many developing nations that they are the victims of exploitation by developed nations that have looted their countries of natural resources and left them in debt. In fact, the degree of debt in developing nations worldwide is staggering. Professor Sachs wrote the following about the crisis in Latin America at the end of the 1980s:

> Between 1981 and 1988 real per capita income declined in absolute terms in almost every country in South America. Many countries' living standards have fallen to levels of the 1950s and 1960s. Real wages in Mexico declined by about 50 percent between 1980 and

---

[72] American Law Institute, *Restatement of the Law on Foreign Relations* at §451. The *Restatements of the Law* have no legal force of their own, but they are considered authoritative, and are frequently cited with approval by courts, and by governmental entities when drafting legislation.

[73] NAFTA at Article 22: "The consulting Parties shall make every attempt to arrive at a mutually satisfactory resolution of the matter through consultations under this Article." See Rogus, the Basel Convention and the United States, 2 New England Int'l & Comp. L. Ann. 434 (1996).

[74] NAFTA at Article 23.

[75] Id. at Article 23: "If consulting parties fail to resolve the matter pursuant to Article 22, any such Party may request in writing a special session of the Council."

[76] Id. at Article 24: "If the matter has not been resolved within 60 days pursuant to Article 23, on the written request of any consulting Party and by 2/3 vote, an arbitral panel will convene."

1988. A decade of development has been wiped out throughout the debtor world.[77]

This level of poverty leads to political instability and violence, with a concomitant decrease in protection of human health and the environment.

Some economists have proposed a novel solution to at least part of the debt problem. These are generally known as "debt-for-nature" swaps, where the debtor country promises to protect the environment in return for purchases of the debt by outside groups. While these solutions are promising, they are largely untried and speculative.[78]

---

[77] Sachs, Making the Brady Plan Work, 68(3) Foreign Affairs 91 (1989).
[78] See V. Ferraro and M. Rosser, Global Debt and Third World Development pp. 332–355, in Klare and Thomas (eds.), *World Security: Challenges for a New Century* (St. Martin's Press, 1994).

A

Glossary of Environmental Terms

Abatement. Reducing the degree or intensity of, or eliminating, pollution.

Aboveground Storage Tank. A device situated so that the entire surface area of the tank is completely above the plane of the adjacent surrounding surface and the entire surface area of the tank (including the tank bottom) is able to be visually inspected.

Acidic. The condition of water or soil that contains a sufficient amount of acid substances to lower the pH below 7.0

Activated Carbon. A highly adsorbent form of carbon used to remove odors and toxic substances from liquid or gaseous emissions. In waste treatment it is used to remove dissolved organic matter from waste water. It is also used in motor vehicle evaporative control systems.

Acute Exposure. A single exposure to a toxic substance which results in severe biological harm or death. Acute exposures are usually characterized as lasting no longer than a day, as compared to longer, continuing exposure over a period of time.

Acute Toxicity. The ability of a substance to cause poisonous effects resulting in severe biological harm or death soon after a single exposure or dose. Also, any severe poisonous effect resulting from a single short-term exposure to a toxic substance.

Acutely Hazardous Waste. Commercial chemical products and manufacturing intermediates having the generic names listed in 40 CFR 261.33; off-specification commercial chemical products and manufacturing chemical intermediates which, if they met specifications, would have the generic names listed; and any residue or contaminated soil, water, or other debris resulting from the cleanup of a spill of any of these substances.

Administrative Order. A legal document signed by the EPA directing an individual, business, or other entity to take corrective action or refrain from an activity. It describes the violations and actions to be taken, and can be enforced in court. Such orders may be issued, for example, as a result of an administrative complaint whereby the respondent is ordered to pay a penalty for violating a statute.

Administrative Record. All documents which the EPA considered or relied on in selecting the response action at a Superfund site, culminating in the record of decision for remedial action or an action memorandum for removal actions.

Adsorption. An advanced method of treating waste in which activated carbon removes organic matter from wastewater .

Aeration. A process which promotes biological degradation of organic matter in water. The process may be passive (as when waste is exposed to air), or active (as when a mixing or bubbling device introduces the air).

Agricultural Pollution. Farming wastes, including runoff and leaching of pesticides and fertilizers; erosion and dust from plowing; improper disposal of animal manure and carcasses; crop residues, and debris.

Air Contaminant. Any particulate matter, gas, or combination thereof, other than water vapor.

Air Pollutant. Any substance in air that could, in high enough concentration, harm humans, other animals, vegetation, or material. Pollutants may include almost any natural or artificial composition of airborne matter capable of being airborne. They may be in the form of solid particles, liquid droplets, gases, or in combination thereof. Generally, they fall into two main groups: (1) those emitted directly from identifiable sources and (2) those produced in the air by interaction between two or more primary pollutants, or by reaction with normal atmospheric constituents, with or without photoactivation. Exclusive of pollen, fog, and dust, which are of natural origin, about 100 contaminants have been identified and fall into the following categories: solids, sulfur compounds, volatile organic chemicals, nitrogen compounds, oxygen compounds, halogen compounds, radioactive compounds, and odors.

Air Pollution. The presence of contaminant or pollutant substances in the air that do not disperse properly and interfere with human health or welfare, or produce other harmful environmental effects.

Air Pollution Control Device. Mechanism or equipment that cleans emissions generated by an incinerator by removing pollutants that would otherwise be released to the atmosphere.

Air Quality Control Region. Federally designated area that is required to meet and maintain federal ambient air quality standards. May include nearby locations in the same state or nearby states that share common air pollution problems.

Air Quality Criteria. The levels of pollution and lengths of exposure above which adverse health and welfare effects may occur.

Air Quality Standards. The level of pollutants prescribed by regulations that may not be exceeded during a given time in a defined area.

Air Stripping. A treatment system that removes volatile organic compounds (VOCs) from contaminated groundwater or surface water by forcing an airstream through the water and causing the compounds to evaporate.

Air Toxics. Any air pollutant for which a national ambient air quality standard (NAAQS) does not exist (i.e., excluding ozone, carbon monoxide, PM-10, sulfur dioxide, nitrogen oxide) that may reasonably be anticipated to cause cancer, developmental effects, reproductive dysfunctions, neurological disorders, heritable gene mutations, or other serious or irreversible chronic or acute health effects in humans.

Airborne Particulates. Total suspended particulate matter found in the atmosphere as solid particles or liquid droplets. Chemical composition of particulates varies widely, depending on location and time of year. Airborne particulates include: windblown dust, emissions from industrial processes, smoke from the burning of wood and coal, and motor vehicle or non-road engine exhausts, exhaust of motor vehicles.

Aliquot. A discrete sample used for analysis.

Alkaline. The condition of water or soil which contains a sufficient amount of alkali substance to raise the pH above 7.0.

Alternate Concentration Limit (ACL). An alternative to the concentration limit set by EPA or a state for a particular hazardous substance or waste. Proposing an ACL is a way of introducing site-specific considerations to the cleanup process. You must provide evidence to show that the ACL will not have adverse effects on human health and the environment. You must also include an analysis showing that concentrations of contaminants moving between the contamination source and receptors would present an acceptable level of risk to any person in contact with the water, soil, or air. Few ACLs have been permitted under RCRA. SARA has been even more stringent. EPA is currently debating the acceptable cancer risk rates for approval of ACLs.

Alternative Remedial Contract Strategy Contractors. Government contractors who provide project management and technical services to support remedial response activities at National Priorities List sites.

Ambient Air. Any unconfined portion of the atmosphere: open air, surrounding air.

Anaerobic. A life or process that occurs in, or is not destroyed by, the absence of oxygen.

Anaerobic Decomposition. Reduction of the net energy level and change in chemical composition of organic matter caused by microorganisms in an oxygenfree environment.

Animal Studies. Investigations using animals as surrogates for humans with the expectation that the results are pertinent to humans.

Anti-Degradation Clause. Part of federal air quality and water quality requirements prohibiting deterioration where pollution levels are above the legal limit.

Applicable or Relevant and Appropriate Requirements (ARARs). ARARs include the federal standards and more stringent state standards that are legally applicable or relevant and appropriate under the circumstances. ARARs include cleanup standards, standards of control, and other environmental protection requirements, criteria, or limitations.

RCRA has frequently been used as an ARAR for cleanup of Superfund sites.

Aqueous. Something made up of, similar to, or containing water; watery.

Aquifer. An underground geological formation, or group of formations, containing usable amounts of groundwater that can supply wells and springs.

Architectural Coatings. Coverings such as paint and roof tar that are used on exteriors of buildings.

Area Source. Any small source of non-natural air pollution that is released over a relatively small area but which cannot be classified as a point source. Such sources may include vehicles and other small engines, small businesses, and household activities.

Aromatic. A type of hydrocarbon, such as benzene or toluene, added to gasoline in order to increase octane. Some aromatics are toxic.

Artesian (Aquifer or well). Water held under pressure in porous rock or soil confined by impermeable geologic formations.

Asbestos. A mineral fiber that can pollute air or water and cause cancer or asbestosis when inhaled. EPA has banned or severely restricted its use in manufacturing and construction.

Asbestos Abatement. Procedures to control fiber release from asbestos-containing materials in a building or to remove them entirely, including removal, encapsulation, repair, enclosure, encasement, and operations and maintenance programs.

Asbestos-Containing Materials (ACM). Any waste that contains commercial asbestos and is generated by a source subject to an asbestos abatement project. It includes asbestos mill tailings, asbestos waste from control devices, friable asbestos from control devices, and bags or containers that previously contained commercial asbestos.

Asbestos Program Manager. A building owner or designated representative who supervises all aspects of the facility asbestos management and control program.

Asbestosis. A disease associated with inhalation of asbestos fibers. The disease makes breathing progressively more difficult and can be fatal.

Attainment Area. An area considered to have air quality as good as or better than the national ambient air quality standards as defined in the Clean Air Act. An area may be an attainment area for one pollutant and a non-attainment area for others.

Attenuation. The process by which a compound is reduced in concentration over time, through absorption, adsorption, degradation, dilution, and/or transformation.

Backfill. Earth used to fill a trench or an excavation.

Background Level. In toxic substances monitoring, the average presence in the environment, originally referring to naturally occurring phenomena.

Baffles. Fin-like devices installed vertically on the inside walls of liquid waste transport vehicles that are used to reduce the movement of the waste inside the tank.

Bench-Scale Tests. Laboratory testing of potential cleanup technologies.

Berm. An earthen mound used to direct the flow of runoff around or through a structure.

Best Available Control Measures (BACM). A term used to refer to the most effective measures (according to EPA guidance) for controlling small or dispersed particulates from sources such as roadway dust, soot and ash from woodstoves, and open burning of rush, timber, grasslands, or trash.

Best Available Control Technology (BACT). For any specific source, the necessary technology that would produce the greatest reduction of each pollutant regulated by the Clean Air Act, taking into account energy, environmental, economic, and other costs.

Best Demonstrated Available Technology (BDAT). The technology EPA establishes for a land-banned hazardous waste to reduce overall toxicity or mobility of toxic constituents in the waste. BDAT must be applied to such a waste prior to land disposal unless one can successfully demonstrate the validity of an equivalent treatment method.

Best Management Practice (BMP). Methods that have been determined to be the most effective, practical means of preventing or reducing pollution from nonpoint sources.

Bioassay. Study of living organisms to measure the effect of a substance, factor, or condition by comparing before-and-after exposure or other data.

Biochemical Oxygen Demand (BOD). A measure of the amount of oxygen consumed in the biological processes that break down organic matter in water. The greater the BOD, the greater the degree of pollution.

Biodegradable. Capable of decomposing rapidly under natural conditions.

Biological Oxygen Demand (BOD). An indirect measure of the concentration of biologically degradable material present in organic wastes. It usually reflects the amount of oxygen consumed in five days by biological processes breaking down organic waste.

Biomass. All of the living material in a given area; often refers to vegetation.

Bioremediation. A hazardous waste site remediation technique that utilizes microorganisms to metabolize hazardous organic constituents in waste to nonhazardous compounds. Environmental conditions are carefully controlled in an attempt to create optimum growth conditions for the organisms.

Biota. The animal and plant life of a given region.

Bog. A type of wetland that accumulates appreciable peat deposits. Bogs depend primarily on precipitation for their water source, and are usually acidic and rich in plant residue with a conspicuous mat of living green moss.

Boom. A floating device used to contain oil on a body of water.

Bottom Land Hardwoods. Forested freshwater wetlands adjacent to rivers in the southeastern United States, especially valuable for wildlife breeding, nesting, and habitat.

Brackish. Mixed fresh and salt water.

Brownfield. Abandoned, idled, or underused industrial and commercial property that has been taken out of productive use as a result of actual or perceived risks from environmental contamination.

Buffer Strip or Zone. Strips of grass or other erosion-resistant vegetation between a waterway and an area of more intensive land use.

By-Product. Material, other than the principal product, generated as a consequence of an industrial process.

Calibration. A check of the precision and accuracy of measuring equipment.

Cap. A layer of clay, or other impermeable material installed over the top of a closed landfill to prevent entry of rainwater and minimize leachate.

Capacity Assurance Plan. A statewide plan which supports a state's ability to manage the hazardous waste generated within its boundaries over a twenty-year period.

Carbon Absorber. An add-on control device that uses activated carbon to absorb volatile organic compounds from a gas stream. (The VOCs are later recovered from the carbon.)

Carbon Adsorption. A treatment system that removes contaminants from groundwater or surface water by forcing it through tanks containing activated carbon treated to attract the contaminants.

Carcinogen. Any substance that can cause or aggravate cancer.

CAS Registration Number. A number assigned by the Chemical Abstracts Service to identify a chemical.

Categorical Exclusion. A class of actions which either individually or cumulatively would not have a significant effect on the human environment and therefore would not require preparation of an environmental assessment or environmental impact statement under the National Environmental Policy Act (NEPA).

Categorical Pretreatment Standard. A technology-based effluent limitation for an industrial facility discharging into a municipal sewer system. Analogous in stringency to Best Availability Technology (BAT) for direct dischargers.

Cathodic Protection. A technique to prevent corrosion of a metal surface by making it the cathode of an electrochemical cell.

Chain-of-Custody. Procedures used to minimize the possibility of tampering with samples.

Characteristic Waste. A solid waste that is a hazardous waste because it exhibits one or more of the following hazardous characteristics: ignitability, corrosivity, reactivity, or toxicity.

Chemical Oxygen Demand (COD). A measure of the oxygen required to oxidize all compounds, both organic and inorganic, in water.

Chlorinated Hydrocarbons. These include a class of persistent, broad-spectrum insecticides that linger in the environment and accumulate in the food chain. Among them are DDT, aldrin, dieldrin, heptachlor, chlordane, lindane, endrin, mirex, hexachloride, and toxaphene. Other examples include TCE, used as an industrial solvent.

Chlorinated Solvent. An organic solvent containing chlorine atoms, e.g., methylene chloride and 1,1,1-trichloromethane, used in aerosol spray containers and in highway paint.

Chlorofluorocarbons (CFCs). A family of inert, nontoxic, and easily liquified chemicals used in refrigeration, air conditioning, packaging, insulation, or as solvents and aerosol propellants. Because CFCs are not destroyed in the lower atmosphere they drift into the upper atmosphere where their chlorine components destroy ozone.

Class I Area. Under the Clean Air Act, a Class I area is one in which visibility is protected more stringently than under the national ambient air quality standards; includes national parks, wilderness area, monuments, and other areas of special national and cultural significance.

Clay Lens. A naturally occurring, localized area of clay that acts as an impermeable layer to runoff infiltration.

Clean Air Act (CAA). The law that authorizes regulations governing releases of airborne contaminants from stationary and non-stationary sources. The regulations include National Ambient Air Quality Standards for specific pollutants.

Clean Water Act (CWA). The law that authorizes establishment of the regulatory program to restore and maintain the physical and biological integrity of the nation's waters. The CWA established, among other things, the National Pollutant Discharge Elimination System (NPDES) to regulate industrial and municipal point-source discharges.

Cleanup. Actions taken to deal with a release or threat of release of a hazardous substance that could affect humans and/or the environment. The term "cleanup" is sometimes used interchangeably with the terms remedial action, removal action, response action, or corrective action.

Closure. The procedure a landfill operator must follow when a landfill reaches its legal capacity for solid waste: ceasing acceptance of solid waste and placing a cap on the landfill site.

Closure Plan. A written plan (subject to approval by authorized regulatory agencies) which the owner/operator of a hazardous waste management facility must submit with the RCRA permit application or for interim status closure. The approved plan becomes part of the permit conditions subsequently imposed on the applicant. The plan identifies steps required to (1) completely or partially close a hazardous waste management unit at any point during its intended operating life, and (2) completely close the unit at the end of its intended operating life.

Coastal Zone. Lands and waters adjacent to the coast that exert an influence on the uses of the sea and its ecology, or whose uses and ecology are affected by the sea.

Combined Sewer Overflows. Discharge of a mixture of storm water and domestic waste when the flow capacity of a sewer system is exceeded during rainstorms.

Combined Sewers. A sewer system that carries both sewage and storm-water runoff. Normally, its entire flow goes to a waste treatment plant, but during a heavy storm, the volume of water may be so great as to cause overflows of untreated mixtures of storm water and sewage into receiving waters. Storm-water runoff may also carry toxic chemicals from industrial areas or streets into the sewer system.

Comment Period. Time provided for the public to review and comment on a proposed EPA action or rulemaking after publication in the Federal Register.

Composite Sample. Used to determine "average" loadings or concentrations of pollutants, such samples are collected at regular time intervals, pooled into one large sample, and can be developed on time or flow rate.

Comprehensive Environmental Response, Compensation, and Liability Act (CERCLA). Also known as Superfund, it is a program to identify sites where hazardous substances have been or might have been released into the environment and to ensure that they are cleaned up. CERCLA is primarily concerned with abandoned sites.

Comprehensive Environmental Response, Compensation, and Liability Information System (CERCLIS). EPA database which identifies hazardous waste sites that require investigation and possible remedial action to mitigate potential negative impacts on human health or the environment.

Concrete Aprons. A pad of non-erosive material designed to prevent scour holes developing at the outlet ends of culverts, outlet pipes, grade stabilization structures, and other water control devices.

Conditionally Exempt Small Quantity Generator (CESQG). Those who generate no more than 100 kilograms of hazardous waste per month. Other than the hazardous waste determination requirement in 40 CFR 262.11, CESQGs are exempt from RCRA provided they do not exceed certain quantity limits for hazardous waste storage or generation.

Conduit. Any channel or pipe for transporting the flow of water.

Consent Decree. A legal document, approved by a judge, that formalizes an agreement reached between EPA and potentially responsible parties (PRPs) through which PRPs will conduct all or part of a cleanup action at a Superfund site; cease or correct actions or processes that are polluting the environment; or otherwise comply with EPA initiated regulatory enforcement actions to resolve the contamination at the Superfund site involved. The consent decree describes the actions PRPs will take and may be subject to a public comment period.

Construction and Demolition Waste. Waste building materials, dredging materials, tree stumps, and rubble resulting from construction, remodeling, repair, and demolition of homes, commercial buildings, and other structures and pavements. May contain lead, asbestos, or other hazardous substances.

Contaminant. Any physical, chemical, biological, or radiological substance or matter that has an adverse affect on air, water, or soil.

Contamination. Introduction into water, air, and soil of microorganisms, chemicals, toxic substances, wastes, or wastewater in a concentration that makes the medium unfit for its next intended use. Also applies to surfaces of objects and buildings, and various household and agricultural use products.

Contingency Plan. A document setting out an organized, planned, and coordinated course of action to be followed in case of a fire, explosion, or release of hazardous waste constituents which could threaten human health or the environment.

Conveyance. Any natural or human-made channel or pipe in which concentrated water flows.

Corrective Action. Action to remedy releases from hazardous waste management units, solid waste management units, or any other sources or release(s) at or from a TSD facility. For hazardous waste management units, the owner or operator of a TSD facility must implement corrective action to ensure that these regulated units comply with the groundwater protection standard in the facility permit. Corrective action for solid waste management units and for releases beyond the facility boundary may be required in a corrective action permit if necessary to protect human health and the environment. Corrective action for any releases from unpermitted TSD facilities can be accomplished through an enforcement action pursuant to RCRA 3008(h).

Corrective Action Reporting System (CARS). EPA's national data-base of information on corrective action permits and enforcement actions.

Corrective Measures Implementation (CMI). The fourth and final step in the RCRA corrective action process. Includes designing, constructing, operating, maintaining, and monitoring selected corrective measures that have been approved by the regulatory agency. This stage combines activities that are often segregated under Superfund as remedial design (RD) and remedial assessment (RA).

Corrective Measures Study (CMS). The third step in the RCRA corrective action process. If the RCRA facility investigation (RFI) reveals a potential need for corrective measures, the agency requires the owner to perform a CMS to identify and recommend specific measures to correct the releases. Although analogous to the Superfund feasibility study (FS) stage, this study is usually less complicated.

Corrosion. The dissolution and wearing away of metal caused by a chemical reaction such as between water and the pipes, chemicals touching a metal surface, or contact between two metals.

Corrosive. A chemical agent that reacts with the surface of a material causing it to deteriorate or wear away.

Cost/Benefit Analysis. A quantitative evaluation of the costs which would be incurred versus the overall benefits to society of a proposed action such as the establishment of an acceptable dose of a toxic chemical.

Cost-Effective Alternative. An alternative control or corrective method identified after analysis as being the best available in terms of reliability, performance, and cost. Although costs are one important consideration, regulatory and compliance analysis does not require EPA to choose the least expensive alternative. For example, when selecting or approving a method for cleaning up a Superfund site the Agency balances costs with the long-term effectiveness of the methods proposed and the potential danger posed by the site.

Cost Recovery. A legal process by which potentially responsible parties who contributed to contamination at a Superfund site can be required to reimburse the Trust Fund for money spent during any cleanup actions by the federal government.

Cost Sharing. A publicly financed program through which society, as a beneficiary of environmental protection, shares part of the cost of pollution control with those who must actually install the controls. In the Superfund, the government may pay part of the cost of a cleanup action with those responsible for the pollution paying the major share.

Cover Material. Soil used to cover compacted solid waste in a sanitary landfill.

Criteria Pollutants. The 1970 amendments to the Clean Air Act required EPA to set National Ambient Air Quality Standards for certain pollutants known to be hazardous to human health. EPA has identified and set standards to protect human health and welfare for six pollutants: ozone, carbon monoxide, total suspended particulates, sulfur dioxide, lead, and nitrogen oxide. The term "criteria pollutants" derives from the requirement that EPA must describe the characteristics and potential health and welfare effects of these pollutants. It is on the basis of these criteria that standards are set or revised.

Culvert. A covered channel or a large-diameter pipe that directs water flow below the ground level.

Denuded. Land stripped of vegetation such as grass, or land that has had vegetation worn down due to impacts from the elements or humans.

Detention Ponds. A surface water impoundment constructed to hold and manage storm water runoff.

Dike. An embankment to confine or control water, often built along the banks of a river to prevent overflow of lowlands; a levee.

Discharge. A release or flow of surface water, storm water, or other liquid effluent from a conveyance or storage container. Can also apply to the release of chemical emissions into the air through designated venting mechanisms.

Disposal. The discharge, deposit, injection, dumping, spilling, leaking, or placing of any hazardous waste or hazardous substance into or on any land or water so that such waste or substance may enter the environment or be emitted into the air or discharged into any waters, including groundwater.

Downgradient. The direction that groundwater flows; similar to "downstream" for surface water.

Drip Guard. A device used to prevent drips of fuel or corrosive or reactive chemicals from contacting other materials or areas.

Ecological Risk Assessment. The application of a formal framework, analytical process, or model to estimate the effects of human actions(s) on a natural resource and to interpret the significance of those effects in light of the uncertainties identified in each component of the assessment process. Such analysis includes initial hazard identification, exposure and dose response assessments, and risk characterization.

Ecology. The relationship of living things to one another and their environment, or the study of such relationships.

Ecosystem. The interacting system of a living biological (biotic) community and its non-living (abiotic) environmental surroundings.

Effluent. Any discharge flowing from a conveyance. Generally refers to wastes discharged into surface waters.

Emergency Planning and Community Right-to-Know Act (EPCRA). The primary goal of EPCRA (also known as SARA Title III) is to facilitate public awareness and emergency response planning for chemical hazards. To fulfill this purpose, EPCRA requires that certain companies that manufacture, process, and use chemicals in specified quantities must file written reports, provide notification of spills/releases, and maintain toxic chemical inventories.

Emission Standard. The maximum amount of air polluting discharge legally allowed from a single source, mobile or stationary.

Emission. Pollution discharged into the atmosphere from smokestacks, other vents, and surface areas of commercial or industrial facilities; from residential chimneys; and from motor vehicle, locomotive, or aircraft exhausts.

Encapsulation. The treatment of asbestos-containing material with a liquid that covers the surface with a protective coating or embeds fibers in an adhesive matrix to prevent their release into the air.

Endangered Species. Animals, birds, fish, plants, and other living organisms threatened with extinction by human-made or natural changes in their environment and listed as endangered pursuant to Section 4 of the Endangered Species Act.

EPCRA Section 313 Water Priority Chemical. A chemical or chemical categories which are: (a) listed at 40 CFR 372.65 pursuant to Section 313 of the Emergency Planning and Community Right-to-Know Act (EPCRA); (b) present at or above threshold levels at a facility subject to EPCRA Section 313 reporting requirements; and (c) that meet at least one of the following criteria: (i) are listed in Appendix D of 40 CFR Part 122 on either Table II (organic priority pollutants), Table III (certain metals, cyanides, and phenols), or Table V (certain toxic pollutants and hazardous substances); (ii) are listed as a hazardous substance pursuant to Section 311 (b)(2)(A) of the Clean Water Act at 40 CFR 116.4; or (iii) are pollutants for which EPA has published acute or chronic water quality criteria.

Environmental Assessment. An environmental analysis prepared pursuant to the National Environmental Policy Act to determine whether a federal action would significantly affect the environment and thus require a more detailed environmental impact statement.

Environmental Audit. An independent assessment of the current status of a party's compliance with applicable environmental requirements or of a party's environmental compliance policies, practices, and controls.

Environmental Equity. Equal protection from environmental hazards of individuals, groups, or communities regardless of race, ethnicity, or economic status.

Environmental Exposure. Human exposure to pollutants originating from facility emissions. Threshold levels are not necessarily surpassed, but low level chronic pollutant exposure is one of the most common forms of environmental exposure.

Environmental Impact Statement (EIS). A document required of federal agencies by the National Environmental Policy Act for major projects or legislative proposals significantly affecting the environment. A tool for decision making, it describes the positive and negative effects of the undertaking and cites alternative actions.

Environmental Justice. The fair treatment of all races, cultures, incomes, and educational levels with respect to the development, implementation, and enforcement of environmental laws, regulations, and policies. Fair treatment implies that no population of people should be forced to shoulder a disproportionate share of the negative environmental impacts of pollution or environmental hazards due to a lack of political or economic strength.

EPA Hazardous Waste Number. A number assigned by EPA to waste that is hazardous by definition; to each hazardous waste listed in 40 CFR 261 Subpart D from specific and nonspecific sources identified by EPA (F, K, P, U); and to each characteristic waste identified in 40 CFR 261 Subpart C, including wastes with ignitable (D001), reactive (D002), corrosive (D003), and EP toxic (D004D017) characteristics.

EPA Identification Number. A number assigned by EPA to each generator; transporter; and treatment, storage, or disposal facility. Identification numbers are facility-specific, except for the transporter who has one number for all his operations.

Erosion. The wearing away of land surface by wind or water. Erosion occurs naturally from weather or runoff but can be intensified by land-clearing practices related to farming, residential or industrial development, road building, or timber-cutting.

Estuary. Regions of interaction between rivers and near-shore ocean waters, where tidal action and river flow mix fresh and salt water. Such areas include bays, mouths of rivers, salt marshes, and lagoons. These brackish water ecosystems shelter and feed marine life, birds, and wildlife.

Eutrophication. The slow aging process during which a lake, estuary, or bay evolves into a bog or marsh and eventually disappears. During the later stages of eutrophication the water body is choked by abundant plant life due to higher levels of nutritive compounds such as nitrogen and phosphorus. Human activities can accelerate the process.

Excavation. The process of removing earth, stone, or other materials.

Exposure Assessment. Identifying the pathways by which toxicants may reach individuals, estimating how much of a chemical an individual is likely to be exposed to, and estimating the number likely to be exposed.

Extraction Procedure (EP) Toxicity. One of the characteristics, along with ignitability, reactivity, and corrosivity, to make a waste a characteristic hazardous waste. The EP toxic list includes maximum concentrations for 14 constituents which, if exceeded, would make a waste hazardous. Effective September 1990, EP Toxicity was replaced by the Toxicity Characteristic.

Extremely Hazardous Substances. Chemicals identified by the EPA as toxic, and listed under SARA Title III. The list is subject to periodic revision.

Facility. All contiguous land, and structures, other appurtenances, and improvements on the land used for treating, storing, or disposing of hazardous waste. A facility may consist of several treatment, storage, or disposal operational units (e.g., one or more landfills, surface impoundments, or combinations of them). Under CERCLA 101(9), (1) any building, structure, installation, equipment, pipe or pipeline (including any pipe into a sewer or publicly owned treatment works), well, pit, pond, lagoon, impoundment, ditch, landfill, storage container, motor vehicle, rolling stock, or aircraft; or (2) any site or area where a hazardous substance has been deposited, stored, disposed of or placed, or has otherwise come to be located. Does not include any consumer product in consumer use or any vessel.

Fate and Transport Modeling. A mathematical process for simulating the behavior of contaminants in various environments to predict contaminant concentration and mobility. Models range from relatively simple analytical solutions to complex numerical models.

Feasibility Study. 1. Analysis of the practicability of a proposal; e.g., a description and analysis of potential cleanup alternatives for a site such as one on the National Priorities List. The feasibility study usually recommends selection of a cost-effective alternative. It usually starts as soon as the remedial investigation is under way; together, they are commonly referred to as the "RI/FS." 2. A small-scale investigation of a problem to ascertain whether a proposed research approach is likely to provide useful data.

Fen. A type of wetland that accumulates peat deposits. Fens are less acidic than bogs, deriving most of their water from groundwater rich in calcium and magnesium.

Fertilizer. Materials such as nitrogen and phosphorus that provide nutrients for plants. Commercially sold fertilizers may contain other chemicals or may be in the form of processed sewage sludge.

Filter Fabric. Textile of relatively small mesh or pore size that is used to (a) allow water to pass through while keeping sediment out (permeable), or (b) prevent both runoff and sediment from passing through (impermeable).

Filter Strip. Usually long, relatively narrow area of undisturbed or planted vegetation used to retard or collect sediment, organic matter, and other pollutants from runoff and wastewater.

Finding of No Significant Impact (FONSI). A document prepared by a federal agency showing why a proposed action would not have a significant impact on the environment and thus would not require preparation of an Environmental Impact Statement. A FONSI is based on the results of an environmental assessment.

First Flush. Individual sample taken during the first 30 minutes of a storm event. The pollutants in this sample can often be used as a screen for non-storm water discharges since such pollutants are flushed out of the system during the initial portion of the discharge.

Flange. A rim extending from the end of a pipe; can be used as a connection to another pipe.

Floodplain. Lowland and relatively flat areas adjoining inland and coastal waters and other flood prone areas such as offshore islands, including at a minimum that area subject to a one percent or greater chance of flooding in any given year. The base floodplain shall be used to designate the 100-year floodplain (one percent chance floodplain).

Flow Channel Liner. A covering or coating used on the inside surface of a flow channel to prevent the infiltration of water to the ground.

Flowmeter. A gauge that shows the speed of water moving through a conveyance.

Flow-Proportional Composite Sample. Combines discrete aliquots of a sample collected over time, based on the flow of the waste stream being sampled. There are two methods used to collect this type of sample. One collects a constant sample volume at time intervals which vary based on stream flow. The other collects aliquots at varying volumes based on stream flow, at constant time intervals.

Flow-Weighted Composite Sample. Means a composite sample consisting of a mixture of aliquots collected at a constant time interval, where the volume of each aliquot is proportional to the flow rate of the discharge.

Formaldehyde. A colorless, pungent, and irritating gas, CH_2O, used chiefly as a disinfectant and preservative and in synthesizing other compounds like resins.

Friable. Capable of being crumbled, pulverized, or reduced to powder by hand pressure.

Friable Asbestos. Any material containing more than one percent asbestos, and that can be crumbled or reduced to powder by hand pressure. (May include previously non-friable material which becomes broken or damaged by mechanical force.)

Fugitive Emissions. Emissions not caught by a capture system.

Gas Chromatograph/Mass Spectrometer. Highly sophisticated instrument that identifies the molecular composition and concentrations of various chemicals in water and soil samples.

General Permit. A permit issued under the NPDES program to cover a certain class or category of discharges.

Generator. Any person whose process produces a hazardous waste in excess of 100 kg/month or acutely hazardous waste in excess of 1 kg/month, or whose actions first cause a hazardous waste to become subject to regulation.

Geographic Information System (GIS). A computer system designed for storing, manipulating, analyzing, and displaying data in a geographic context.

Grab Sample. A discrete sample which is taken from a waste stream on a one-time basis with no regard to flow or time; instantaneous sample that is analyzed separately.

Grading. The cutting and/or filling of the land surface to a desired slope or elevation.

Granular Activated Carbon (GAC) Treatment. A filtering system often used in small water systems and individual homes to remove organics. GAC can be highly effective in removing elevated levels of radon from water.

Greenhouse Effect. The warming of the Earth's atmosphere attributed to a buildup of carbon dioxide or other gases; some scientists think that this buildup allows the sun's rays to heat the Earth, while infra-red radiation makes the atmosphere opaque to a counterbalancing loss of heat.

Groundwater. The supply of fresh water found beneath the Earth's surface, usually in aquifers, which supply wells and springs. Because groundwater is a major source of drinking water, there is growing concern over contamination from leaching agricultural or industrial pollutants or leaking underground storage tanks.

Habitat. The place where a population (e.g., human, animal, plant, microorganism) lives and its surroundings, both living and non-living.

Hazard Ranking System (HRS). The method EPA uses to determine which sites should be listed on the National Priorities List (NPL) under CERCLA. The HRS ranks sites by means of a mathematical rating scheme that combines the potential of a release to cause hazardous situations with the severity/magnitude of these potential impacts and the number of people who may be affected. Using the numerical scores from this scheme, EPA and the states list sites by priority and allocate resources for site investigation, enforcement, and cleanup. Sites receiving high HRS scores appear on the National Priorities List. Under SARA, the HRS must be revised by EPA to determine whether it is adequately identifying sites for the NPL (CERCLA 105(c)). Citizens may now petition EPA to conduct a preliminary assessment of a site near them. If the assessment indicates that a release may pose a threat to health or the environment, EPA will do an HRS scoring.

Hazardous Air Pollutants. Air pollutants which are not covered by ambient air quality standards but which, as defined in the Clean Air Act, may reasonably be expected to cause or contribute to irreversible illness or death. Such pollutants include asbestos, beryllium, mercury, benzene, coke oven emissions, radionuclides, and vinyl chloride.

Hazardous and Solid Waste Amendments of 1984 (HSWA). Amendments to RCRA which greatly increased the complexity of the RCRA regulatory program by imposing restrictions on land disposal of hazardous wastes, authorizing EPA to require corrective action for releases from hazardous waste management facilities, and instituting requirements for underground storage tanks containing petroleum and hazardous substances

Hazardous Substance (CERCLA definition). Under CERCLA 101(14), any element, compound, mixture, solution, or substance which, when released to the environment, may present substantial danger to public health/welfare or the environment. Also includes (1) any substance designated pursuant to Section 311(b)(2)(A) of the Federal Water Pollution Control Act; (2) any element, compound, mixture, solution, or substance designated pursuant to Section 102 of CERCLA; (3) any hazardous waste having the characteristics identified under or listed pursuant to Section 3001 of the Solid Waste Disposal Act; (4) any toxic pollutant listed under Section 307(a) of the Federal Water Pollution Control Act; (5) any hazardous air pollutant listed under Section 112 of the Clean Air Act; and (6) any imminently hazardous chemical substance or mixture so identified pursuant to the Toxic Substances Control Act. Excludes petroleum (including crude oil not otherwise specifically listed or designated as a hazardous substance under any of the above laws), natural gas, natural gas liquids, liquefied natural gas, or synthetic gas usable for fuel (or mixtures of natural gas and such synthetic gas). The definition of hazardous substances in CERCLA is broader than the definition of hazardous wastes under RCRA.

Hazardous Substance Superfund. The fund, largely financed by taxes on petroleum and chemicals, and by an "environmental tax" on corporations, which provides operating money for government-financed actions under CERCLA. The fund is a revolving fund in the sense that it enables the government to take action and then seek reimbursement later, or to clean up sites when responsible parties with sufficient cleanup funds cannot be found. Money recovered from PRPs is returned to the fund rather than to the U.S. Treasury.

Hazardous Waste (RCRA definition). A solid waste which because of its quantity, concentration or physical, chemical, or infectious characteristics may (1) cause or contribute to an increase in mortality or an increase in serious irreversible or incapacitating reversible illness; or (2) pose a substantial present or potential hazard to human health or the environment when improperly treated, stored, transported or disposed of, or otherwise managed. EPA hazardous waste regulations (40 CFR 261) specify that if a material qualifies as a solid waste, and does not qualify for an exemption, it is a hazardous waste if it is listed by 40 CFR Part 261, Subpart D, or if it exhibits any of the four hazardous waste characteristics (ignitability, reactivity, corrosivity and toxicity).

Hazardous Waste Constituent. A constituent that caused the waste to be listed as a hazardous waste under 40 CFR Part 261 Subpart D.

Hazardous Waste Generator. Any person whose act or process produces hazardous waste identified or listed in 40 CFR 261 or whose act first causes hazardous waste to become subject to regulation. Generators are subject to specific hazardous waste management regulations, which apply only to the particular site of generation. The regulations vary by the volume of waste annually generated but include reporting, testing, record-keeping, storage, and shipping and disposal requirements.

Hazardous Waste Management Unit. A contiguous area of land on or in which hazardous waste is placed, or the largest area in which there is significant likelihood of mixing hazardous waste constituents in the same area. A unit may be a surface impoundment, waste pile, land treatment area, landfill cell, incinerator, tank and its associated piping and underlying containment system, or container storage area. A container alone does not constitute a unit.

Heavy Metals. Metallic elements with high atomic weights, e.g., mercury, chromium, cadmium, arsenic, and lead; can damage living things at low concentrations and tend to accumulate in the food chain.

Holding Pond. A pond or reservoir, usually made of earth, built to store polluted runoff for a limited time.

Hydrogeology. The geology of groundwater, with particular emphasis on the chemistry and movement of water.

Hydrology. The science dealing with the properties, distribution, and circulation of water.

Ignitable. Capable of burning or causing a fire.

Illicit Connection. Any discharge to a municipal separate storm sewer that is not composed entirely of storm water except discharges authorized by an NPDES permit (other than the NPDES permit for discharges from the municipal separate storm sewer) and discharges resulting from fire fighting activities.

Indoor Air Pollution. Chemical, physical, or biological contaminants in indoor air.

Infiltration. 1. The penetration of water through the ground surface into sub-surface soil or the penetration of water from the soil into sewer or other pipes through defective joints, connections, or manhole walls. 2. A land application technique where large volumes of wastewater are applied to land, allowed to penetrate the surface and percolate through the underlying soil.

Inlet. An entrance into a ditch, storm sewer, or other waterway.

Interim Status. The period during which the owner/operator of an existing TSD facility is treated as having been issued a RCRA permit even though he/she has not yet received a final determination. An existing facility should have automatically qualified for interim status if the owner/operator filed both timely "notification" and the first part (Part A) of the RCRA permit application. Interim status continues until the permit is issued. Owners/operators of new facilities cannot by definition qualify for interim status, but need a RCRA permit prior to beginning construction of a hazardous waste management facility.

Intermediates. A chemical compound formed during the making of a product.

Interstitial Monitoring. The continuous surveillance of the space between the walls of an underground storage tank.

Irrigation. Human application of water to agricultural or recreational land for watering purposes.

Jute. A plant fiber used to make rope, mulch, netting, or matting.

Karst. A geologic formation of irregular limestone deposits with sinks, underground streams, and caverns.

Lagoon. A shallow pond where sunlight, bacterial action, and oxygen work to purify waste water.

Land Application Units. An area where wastes are applied onto or incorporated into the soil surface (excluding manure spreading operations) for treatment or disposal.

Land Ban. Phasing out of land disposal of most untreated hazardous wastes, as mandated by the 1984 amendments to RCRA.

Land Disposal. Includes, but is not limited to, placement in a landfill, surface impoundment, waste pile, injection well, land treatment facility, salt dome formation, salt bed formation, underground mine or cave, or concrete vault or bunker intended for disposal purposes. Land disposal facilities are a subset of TSD facilities. Groundwater monitoring is required at all land disposal facilities.

Land Treatment Units. An area of land where materials are temporarily located to receive treatment. Examples include sludge lagoons and stabilization ponds.

Landfills. An area of land or an excavation in which wastes are placed for permanent disposal, and which is not a land application unit, surface impoundment, injection well, or waste pile.

Large Quantity Generator. Person or facility generating more than 2200 pounds of hazardous waste per month. Such generators produce about 90 percent of the nation's hazardous waste, and are subject to all RCRA requirements.

Leachate. Liquid that has percolated through solid and/or hazardous waste and has extracted dissolved or suspended materials from the waste.

Leaching. The process by which soluble constituents are dissolved in a solvent such as water and carried down through the soil.

Lead Agency. The federal or state agency providing the On-Scene Coordinator (OSC) or the responsible official for a CERCLA response action. Includes the federal (EPA, Coast Guard, DOD, DOI, DOE, etc.) or state agency responsible for collecting data and performing assessments and other studies. The lead agency responsibility may shift during the stages of the Superfund process.

Leak-Detection System. A system capable of detecting the failure of a primary or secondary containment structure or the presence of a release of hazardous waste or accumulated liquid in the secondary containment structure. Detection is based on systemic operational controls (such as visual monitoring) or continuous automatic monitoring of any releases from the containment areas.

Level Spreader. A device used to spread out storm water runoff uniformly over the ground surface as sheet flow (i.e., not through channels). The purpose of level spreaders are to prevent concentrated, erosive flows from occurring and to enhance infiltration.

Liming. Treating soil with lime to neutralize acidity levels.

Limnology. The study of the physical, chemical, hydrological, and biological aspects of fresh water bodies.

Liner. 1. A relatively impermeable barrier designed to prevent leachate from leaking from a landfill. Liner materials include plastic and dense clay. 2. An insert or sleeve for sewer pipes to prevent leakage or infiltration.

Listed Waste. Wastes listed as hazardous under RCRA but which have not been subjected to the Toxic Characteristics Listing Process because the dangers they present are considered self-evident.

Liquid Level Detector. A device that provides continuous measures of liquid levels in liquid storage areas or containers to prevent overflows.

Local Emergency Planning Committee (LEPC). A committee appointed by the state emergency response commission, as required by SARA Title III, to formulate a comprehensive emergency plan for its jurisdiction.

Lowest Achievable Emission Rate. Under the Clean Air Act, the rate of emissions that reflects (a) the most stringent emission limitation in the implementation plan of any state for such source unless the owner or operator demonstrates such limitations are not achievable; or (b) the most stringent emissions limitation achieved in practice, whichever is more stringent. A proposed new or modified source may not emit pollutants in excess of existing new source standards.

Major Stationary Sources. Term used to determine the applicability of Prevention of Significant Deterioration and new source regulations. In a nonattainment area, any stationary pollutant source with potential to emit more than 100 tons per year is considered a major stationary source. In PSD areas the cutoff level may be either 100 or 250 tons, depending upon the source.

Manifest System. Tracking of hazardous waste from "cradle to grave" (generation through disposal) with accompanying documents known as manifests.

Marsh. A type of wetland that does not accumulate appreciable peat deposits and is dominated by herbaceous vegetation. Marshes may be either fresh or saltwater, tidal or non-tidal.

Material Safety Data Sheet (MSDS). Fact sheets required by OSHA on every commercial chemical that must be prepared by the chemical's manufacturer or importer and must include specific information, including its ingredients (one percent or more), known or suspected health risks associated with its use or exposure, proper safety precautions and waste disposal, and other information necessary to prevent or minimize a health and safety risk to employees or consumers.

Materials Management Practices. Practices used to limit the contact between significant materials and precipitation. These may include structural or nonstructural controls such as dikes, berms, sedimentation ponds, vegetation strips, spill response plans, and so on.

Material Storage Areas. Onsite locations where raw materials, products, final products, byproducts, or waste materials are stored.

Maximum Contaminant Level (MCL). The maximum permissible level of a contaminant in water delivered to any user of a public water system. MCLs are enforceable standards.

Maximum Contaminant Level Goal (MCLG). The maximum level of a contaminant in drinking water at which no known or anticipated adverse effect on human health would occur, and which includes an adequate margin of safety. MCLGs are nonenforceable health goals.

Mixed Funding Agreement. Allows EPA to reimburse parties for certain costs (with interest) of actions parties have agreed to perform, but EPA has agreed to finance. A mixed funding agreement can be used when some PRPs cannot currently pay for response costs, and other PRPs want to perform response actions and be reimbursed for the non-participating PRPs' shares (CERCLA 122(b)).

Mobile Source. Any non-stationary source of air pollution such as cars, trucks, motorcycles, buses, airplanes, locomotives.

Monitoring Well. 1. A well used to obtain water quality samples or measure groundwater levels. 2. Well drilled at a hazardous waste management facility or Superfund site to collect groundwater samples for the purpose of physical, chemical, or biological analysis to determine the amounts, types, and distribution of contaminants in the groundwater beneath the site.

Mulch. A natural or artificial layer of plant residue or other materials covering the land surface which conserves moisture, holds soil in place, aids in establishing plant cover, and minimizes temperature fluctuations.

Municipal Separate Storm Sewer System (MS4). All municipal separate storm sewers that are either: (a) located in an incorporated place (city) with a population of 100,000 or more as determined by the latest Decennial Census by the Bureau of Census (these cities are listed in Appendices F and G of 40 CFR Part 122); or (b) located in the counties with unincorporated urbanized populations of 100,000 or more, except municipal separate storm sewers that are located in the incorporated places, townships, or towns within such counties (these counties are listed in Appendices H and I of 40 CFR Part 122); or (c) owned or operated by a municipality other than those described in (a) or (b) and that are designated by the Director as part of the large or medium municipal separate storm sewer system.

National Ambient Air Quality Standards (NAAQS). Standards established by EPA that apply for outside air throughout the country.

National Contingency Plan (NCP). The basic policy directive for federal response actions under CERCLA Section 105. It sets forth the Hazard Ranking System and procedures and standards for responding to releases of hazardous Substances, pollutants, and contaminants. The plan is a regulation (40 CFR Part 300) subject to regular revision.

National Emissions Standards For Hazardous Air Pollutants (NESHAP). Emissions standards set by EPA for an air pollutant not covered by NAAQS that may cause an increase in fatalities or in serious, irreversible, or incapacitating illness. Primary standards are designed to protect human health, secondary standards to protect public welfare (e.g., building facades, visibility, crops, and domestic animals).

National Pollutant Discharge Elimination System (NPDES). A provision of the Clean Water Act which prohibits discharge of pollutants into waters of the United States unless a special permit is issued by EPA, a state, or, where delegated, a tribal government on an Indian reservation.

National Pollutant Discharge Elimination System Permit (NPDES Permit). A permit issued by the EPA or an approved state agency for the discharge of pollutants into navigable waters under the National Pollutant Discharge Elimination System.

National Priorities List (NPL). A list of sites designated as needing long-term remedial cleanup. The purpose of the list is to inform the public of the most serious hazardous waste sites in the nation. EPA revises the list periodically to add new sites or delete sites following cleanup. Sites on the list are generally slated for EPA enforcement or cleanup. Note that many elements of the CERCLA/SARA program apply to sites regardless of whether they are on the NPL.

National Response Center (NRC). The federal operations center that receives notifications of all releases of oil and hazardous substances into the environment. The Center, open 24 hours a day, is operated by the U.S. Coast Guard, which evaluates all reports and notifies the appropriate agency.

National Response Team (NRT). Representatives of 13 federal agencies who, as a team, coordinate federal responses to nationally significant incidents of pollution and provide advice and technical assistance to the responding agency(ies) before and during a response action.

Navigable Water. Water which by itself, or by uniting with other waters navigable, forms a continuous highway over which interstate or international commerce may be conducted in the customary mode of trade and travel on water.

New Source Performance Standards (NSPS). Uniform national EPA air emission and water effluent standards which limit the amount of pollution allowed from new sources or from modified existing sources.

New Source Review (NSR). Clean Air Act requirement that requires State Implementation Plans go include a permit review that applies to the construction and operation of new and modified major stationary sources in nonattainment areas to assure attainment of the national ambient air quality standards.

No Further Remedial Action Planned (NFRAP). Determination made by EPA following a preliminary assessment that a site does not pose a significant risk and so requires no further activity under CERCLA.

Non-Attainment Area. Area that does not meet one or more of the National Ambient Air Quality Standards for the criteria pollutants designated in the Clean Air Act.

Non-Binding Allocations of Responsibility (NBAR). Determination EPA may make of each PRP's share of responsibility for cleanup. Under SARA, the NBAR is an attempt to require EPA to provide more information to PRPs to encourage settlement.

Noncontact Cooling Water. Water used to cool machinery or other materials without directly contacting process chemicals or materials.

Non-Point Source. Diffuse pollution sources (i.e., without a single point of origin or not introduced into a receiving stream from a specific outlet). The pollutants are generally carried off the land by storm water. Common nonpoint sources are agriculture, forestry, urban, mining, construction, dams, channels, land disposal, saltwater intrusion, and city streets.

Notice Letter. EPA's formal notice to PRPs that CERCLA-related action is to be undertaken at a site for which those PRPs are considered responsible. Notice letters arc generally sent at least 60 days prior to scheduled obligation of funds for an RI/FS at a designated site. The intent is to give PRPs sufficient time to organize and to contact the government. A notice letter is sent again prior to implementing the remedy.

Notice of Intent (NOI). An application to notify the NPDES permitting authority of a facility's intention to be covered by a general permit; exempts a facility from having to submit an individual or group application.

Oil Sheen. A thin, glistening layer of oil on water.

Oil/Water Separator. A device installed, usually at the entrance to a drain, which removes oil and grease from water flows entering the drain.

On-Scene Coordinator (OSC). Under the NCP, a representative of EPA or the state who directs or coordinates operations at the scene of a removal action.

Opacity. The amount of light obscured by particulate pollution in the air; clear window glass has zero opacity, a brick wall is 100 percent opaque. Opacity is an indicator of changes in performance of particulate control systems.

Open Dump. A land disposal site at which solid (usually municipal) wastes are disposed of in a manner that does not protect the environment, renders them susceptible to open burning, and are exposed to the elements, vectors, and scavengers.

Open Dump Inventory. Under RCRA, EPA's Office of Solid Waste maintains an inventory of open dumps in the United States. The inventory includes all dump sites which do not comply with EPA's "Criteria for Classification of Solid Waste Disposal Facilities and Practices." (40 CFR Section 257).

Operator. The person responsible for the overall operation of a facility.

Organic Pollutants. Substances containing carbon which may cause pollution problems in receiving streams.

Organic Solvents. Liquid organic compounds capable of dissolving solids, gases, or liquids.

Outfall. The point, location, or structure where wastewater or drainage discharges from a sewer pipe, ditch, or other conveyance to a receiving body of water.

Owner. The person who owns a facility or part of a facility.

Ozone (O_3). Found in two layers of the atmosphere, the stratosphere and the troposphere. In the stratosphere (the atmospheric layer 7 to 10 miles or more above the earth's surface) ozone is a natural form of oxygen that provides a protective layer shielding the earth from ultraviolet radiation. In the troposphere (the layer extending up 7 to 10 miles from the earth's surface), ozone is a chemical oxidant and major component of photochemical smog. It can seriously impair the respiratory system and is one of the most widespread of all the criteria pollutants for which the Clean Air Act required EPA to set standards. Ozone in the troposphere is produced through complex chemical reactions of nitrogen oxides, which are among the primary pollutants emitted by combustion sources; hydrocarbons, released into the atmosphere through the combustion, handling and processing of petroleum products; and sunlight.

Ozone depletion. Destruction of the stratospheric ozone layer which shields the earth from ultraviolet radiation harmful to life. This destruction of ozone is caused by the breakdown of certain chlorine and/or-bromine containing compounds (chlorofluorocarbons or halons), which break down when they reach the stratosphere and then catalytically destroy ozone molecules.

Ozone Layer. The protective layer in the atmosphere, about 15 miles above the ground, that absorbs some of the sun's ultraviolet rays, thereby reducing the amount of potentially harmful radiation reaching the earth's surface.

Particulates. 1. Fine liquid or solid particles such as dust, smoke, mist, fumes, or smog found in air or emissions. 2. Very small solid suspended in water. They vary in size, shape, density, and electrical charge, can be gathered together by coagulation and flocculation.

Pathogens. Microorganisms that can cause disease in other organisms or in humans, animals, and plants (e.g., bacteria, viruses, or parasites) found in sewage, in runoff from farms or rural areas populated with domestic and wild animals, and in water used for swimming. Fish and shellfish contaminated by pathogens, or the contaminated water itself, can cause serious illness.

Permeability. The quality of a soil that enables water or air to move through it. Usually expressed in inches/hour or inches/day.

pH. An expression of the intensity of the basic or acid condition of a liquid. The pH may range from 0 to 14, where 0 is the most acid, 7 is neutral. Natural waters usually have a pH between 6.5 and 8.5.

Picocuries Per Liter (pCi/L). A unit of measure for levels of radon gas.

Plunge Pool. A basin used to slow flowing water, usually constructed to a design depth and shape. The pool may be protected from erosion by various lining materials.

Pneumatic Transfer. A system of hoses which uses the force of air or other gas to push material through; used to transfer solid or liquid materials from tank to tank.

Point Source. Any discernible, confined, and discrete conveyance, including but not limited to any pipe, ditch, channel, tunnel, conduit, well, discrete fissure, container, rolling stock, concentrated animal feeding operation, or vessel or other floating craft, from which pollutants are or may be discharged. This term does not include return flows from irrigated agriculture or agricultural storm water runoff.

Pollutant (Clean Water Act). Dredged spoil, solid waste, incinerator residue, filter backwash, sewage, garbage, sewage sludge, munitions, chemical wastes, biological materials, some radioactive materials, heat, wrecked or discarded equipment, rock, sand, cellar dirt, and industrial, municipal, and agricultural waste discharged into water.

Pollutant or Contaminant (CERCLA). Any element, substance, compound, or mixture, including disease-causing agents that, after release into the environment and upon exposure, ingestion, inhalation, or assimilation into any organism, either directly from the environment or indirectly by ingestion through food chains, or may reasonably be anticipated to cause death, disease, behavioral abnormalities, cancer, genetic mutation, physiological malfunctions, or physical deformations.

Pollution Prevention. Any source reduction or recycling activity that results in reduction of total volume of hazardous waste, reduction of toxicity of hazardous waste, or both, as long as that reduction is consistent with the goal of minimizing present and future risks to public health and the environment. Transfer of hazardous constituents from one environment medium to another does not constitute waste minimization.

Polychlorinated Biphenyls (PCBs). Halogenated organic compounds limited to the biphenyl molecule that have been chlorinated to varying degrees.

Polyvinyl Chloride (PVC). A plastic used in pipes because of its strength; does not dissolve in most organic solvents.

Porous Pavement. A human-made surface that will allow water to penetrate through and percolate into soil (as in porous asphalt pavement or concrete). Porous asphalt pavement is comprised of irregular shaped crushed rock precoated with asphalt binder. Water seeps through into lower layers of gravel for temporary storage, then filters naturally into the soil.

Potentially Responsible Parties (PRPs). Those identified by EPA as potentially liable under CERCLA for cleanup costs. PRPs may include generators and present or former owners/operators of certain facilities or real property where hazardous wastes have been stored, treated, or disposed of, as well as those who accepted hazardous waste for transport and selected the facility.

Precipitation. Any form of rain or snow.

Preventative Maintenance Program. A schedule of inspections and testing at regular intervals intended to prevent equipment failures and deterioration.

Process Waste Water. Water that comes into direct contact with or results from the production or use of any raw material, intermediate product, finished product, by-product, waste product, or wastewater.

Publicly Owned Treatment Works (POTW). Any device or system used in the treatment (including recycling and reclamation) of municipal sewage or industrial wastes of a liquid nature which is owned by a State or municipality. This includes sewers, pipes, or other conveyances if they convey wastewater to a POTW providing treatment.

Quality Assurance/Quality Control. A system of procedures, checks, audits, and corrective actions to ensure that all EPA research design and performance, environmental monitoring and sampling, and other technical and reporting activities are of the highest achievable quality.

Radon. A colorless naturally occurring, radioactive, inert gas formed by radioactive decay of radium atoms in soil or rocks.

Raw Material. Any product or material that is converted into another material by processing or manufacturing.

RCRA Facility Assessment (RFA). Usually the first step in the RCRA corrective action process. EPA conducts a comprehensive review of pertinent site information. This may be followed by visual site inspection and if necessary a sampling visit to make release determinations.

RCRA Facility Investigation (RFI). The second step in the RCRA corrective action process. If the RFA indicates a suspected release, the regulatory agency prescribes an RFI under a corrective action permit (RCRA 3004(u)) or enforcement action (RCRA 3008(h)). Such investigations can range from small specific activities to complex multimedia studies.

Reasonably Available Control Technology (RACT). Control technology that is both reasonably available, and both technologically and economically feasible. Usually applied to existing sources in nonattainment areas; in most cases it is less stringent than new source performance standards.

Receiving Waters. A river, lake, ocean, stream or other watercourse into which wastewater or treated effluent is discharged.

Record of Decision (ROD). Published by the government after completion of an RI/FS, the ROD identifies the remedial alternative chosen for implementation at a Superfund site. The ROD is part of the written administrative record. Judicial review of EPA cleanup decisions may be limited to the administrative record.

Recycle. The process of minimizing the generation of waste by recovering usable products that might otherwise become waste. Examples are the recycling of aluminum cans, wastepaper, and bottles.

Regulated Asbestos-Containing Material (RACM). Friable asbestos material or nonfriable ACM that will be or has been subjected to sanding, grinding, cutting, or abrading or has crumbled, pulverized, or reduced to powder in the course of demolition or renovation operations.

Release. Any spilling, leaking, pumping, pouring, emitting, emptying, discharging, injecting, escaping, leaching, dumping, or disposing into the environment (see CERCLA Section 101(22)). Includes the abandonment or discarding of barrels, containers, and other closed receptacles containing any hazardous substance, pollutant, or contaminant. Exclusions include (1) releases solely exposing workers in a workplace, with respect to a claim they may bring against the employer; (2) engine exhaust emissions from motor vehicles, rolling stock, aircraft, vessels, or pipeline pumping station engines; (3) nuclear releases subject to the Atomic Energy Act and financial requirements or the Nuclear Regulatory Commission (also excludes any release of source, byproduct, or special nuclear material from any processing site designated under Section 102(a) or 302(a) of the Uranium Mill Tailings Radiation Control Act); and (4) the normal application of fertilizer. Release also means substantial threat of release.

Remedial Action (RA). The actual construction or implementation phase of a Superfund site cleanup that follows remedial design.

Remedial Action Plan (RAP). This plan details the technical approach for implementing remedial response. It includes the methods to be followed during the entire remediation process--from developing the remedial design to implementing the selected remedy through construction.

Remedial Design. A phase of remedial action that follows the ROD, consent decree, and remedial investigation/feasibility study (RI/FS) and includes development or engineering drawings and specifications for a site cleanup.

Remedial Investigation/Feasibility Study (RI/FS). Extensive technical studies conducted by the government or by PRPs to investigate the scope of contamination (RI) and determine the remedial alternatives (FS) which, consistent with the NCP, may be implemented at a Superfund site. Government funded RI/FSs do not recommend a specific alternative for implementation. RI/FSs conducted by PRPs usually do recommend and technically support a remedial alternative. An RI/FS may include a variety of on- and off-site activities, such as monitoring, sampling, and analysis.

Remedial Response. Long-term action that stops or substantially reduces a release or threat of a release of hazardous substances that is serious but not an immediate threat to public health.

Remediation. Cleanup methods used to remove or contain a release or spill of hazardous substances from the environment.

Removal, Remove, or Removal Action. Under CERCLA 101(23), generally short-term actions taken to respond promptly to an urgent need. With regard to hazardous substances, the cleanup or removal of released substances from the environment; actions in response to the threat of release; actions that may be necessary to monitor, assess, and evaluate the release or threat; disposal of removed material; or other actions needed to prevent, minimize, or mitigate damage to public health or welfare or to the environment. Removal also includes, without being limited to, security fencing or other measures to limit access; provision of alternative water supplies; temporary evacuation and housing of threatened individuals not otherwise provided for; and any emergency assistance provided under the Disaster Relief Act.

Reportable Quantity (RQ). The quantity of a hazardous substance or oil that triggers reporting requirements under CERCLA, the Clean Water Act, or other environmental laws. Quantities are to be measured over a 24-hour period. If a substance is released in amounts exceeding its RQ, the release must be reported to the National Response Center, the State Emergency Response Commission, and/or other federal, state, and local authorities.

Residual. Amount of pollutant remaining in the environment after a natural or technological process has taken place, e.g., the sludge remaining after initial wastewater treatment, or particulates remaining in air after the air passes through a scrubbing or other pollutant removal process.

Response Action. Any remedial action, removal action, or cleanup at a site under CERCLA 101(25). Includes enforcement-related activities.

Retrofit. Addition of a pollution control device on an existing facility without making major changes to the generating plant.

Retention. The holding of runoff in a basin without release except by means of evaporation, infiltration, or emergency bypass.

Reverse Meniscus. The curved upper surface of a liquid in a container.

Rill Erosion. The formation of numerous, closely spread streamlets due to uneven removal of surface soils by storm water or other water.

Riparian Habitat. Areas adjacent to rivers and streams that have a high density, diversity, and productivity of plant and animal species relative to nearby uplands.

Risk Assessment. A qualitative and quantitative evaluation performed to define the risk posed to human health and/or the environment by the presence or potential presence and/or use of specific pollutants. Baseline risk assessments are performed as part of corrective action.

Route of Exposure. The avenue by which a chemical comes into contact with an organism (e.g., inhalation, ingestion, dermal contact, injection).

Runon. Storm water surface flow or other surface flow which enters property other than that where it originated.

Runoff. That part of precipitation, snow melt, or irrigation water that runs off the land into streams or other surface water. It can carry pollutants from the air and land into the receiving waters.

Runoff Coefficient. The fraction of total rainfall that will appear at the conveyance as runoff.

Sanitary Sewer. A system of underground pipes that carries sanitary waste or process wastewater to a treatment plant.

Sanitary Waste. Domestic sewage.

Scour. The clearing and digging action of flowing water, especially the downward erosion caused by stream water in sweeping away mud and silt from the stream bed and outside bank of a curved channel.

Scrubber. An air pollution device that uses a spray of water or reactant or a dry process to trap pollutants in emissions.

Sealed Gate. A device used to control the flow of liquid materials through a valve.

Secondary Containment. Structures, usually dikes or berms, surrounding tanks or other storage containers and designed to catch spilled material from the storage containers.

Sediment Trap. A device for removing sediment from water flows; usually installed at outfall points.

Sedimentation. The process of depositing soil particles, clays, sands, or other sediments that were picked up by flowing water.

Sediments. Soil, sand, and minerals washed from land into water, usually after rain. They pile up in reservoirs, rivers, and harbors, destroying fish-nesting areas and holes of water animals and cloud the water so that needed sunlight might not reach aquatic plants. Careless farming, mining, and building activities will expose sediment materials, allowing them to be washed off the land after rainfalls.

Septic Tank. A watertight, covered receptacle designed to receive or process, through liquid separation or biological digestion, the sewage discharged from a building sewer. The effluent from such a receptacle is distributed for disposal through the soil and settled solids and scum from the tank are pumped out periodically and hauled to a treatment facility.

Sheet Erosion. Erosion of thin layers of surface materials by continuous sheets of running water.

Sheetflow. Runoff which flows over the ground surface as a thin, even layer, not concentrated in a channel.

Shelf Life. The time for which chemicals and other materials can be stored before becoming unusable due to age or deterioration.

Significant Materials (Clean Water Act definition). Include, but are not limited to: raw materials; fuels; materials such as solvents, detergents and plastic pellets; finished materials such as metallic products; raw materials used in food processing or production; hazardous substances designated under CERCLA Section 101(14); any chemical the facility is required to report pursuant to EPCRA Section 313; fertilizers; pesticides; and waste products such as ashes, slag, and sludge that have a potential to be released with storm water discharges.

Significant Spills (Clean Water Act definition). Include, but are not limited to: releases of oil or hazardous substances in excess of reportable quantities under Section 311 of the Clean Water Act.

Site Inspection. The collection of information from a Superfund site to determine the extent and severity of hazards posed by the site. It follows and is more extensive than a preliminary assessment. The purpose is to gather information necessary to score the site, using the Hazard Ranking System, and to determine if the site presents an immediate threat that requires prompt removal action.

Slag. Non-metal containing waste leftover from the smelting and refining of metals.

Slide Gate. A device used to control the flow of water through storm water conveyances.

Sloughing. The movement of unstabilized soil layers down a slope due to excess water in the soils.

Sludge. A semi-solid residue from any of a number of air or water treatment processes. Sludge can be a hazardous waste.

Small Quantity Generator (SQG). A regulated facility that generates more than 100 kilograms and less than 1,000 kilograms (about 1 ton) of hazardous waste in a calendar month. However, even if a small quantity generator avoids the requirements of full generator status, the facility may still be subject to certain RCRA conditions (e.g., if the quantity of acutely hazardous wastes generated in a calendar month exceeds quantities specified under RCRA).

Sole Source Aquifer. An aquifer which is the sole or principal drinking water source for an area and which if contaminated would create a significant hazard to public health. Sole source aquifers may receive special protective status.

Solids Dewatering. A process for removing excess water from solids to lessen the overall weight of the wastes.

Solid Waste (RCRA definition). Garbage, refuse, sludge from a waste treatment plant, water supply treatment plant or air pollution control facility and other discarded material including solid, liquid, semi-solid, or contained gaseous materials resulting from industrial, commercial, mining and agriculture activities and from community activities, but does not include solids or dissolved materials in domestic sewage or solid or dissolved materials in irrigation return flows or industrial discharges that are point sources subject to permits under the federal Clean Water Act or source, special nuclear or byproduct material as defined by the Atomic Energy Act. EPA defines hazardous waste as a subset of solid waste.

Solid Waste Disposal Act (SWDA). The SWDA was amended in 1976 by The Resource Conservation and Recovery Act (RCRA). SWDA has since been amended by several public laws, including the Used Oil Recycling Act of 1980 (UORA). the Hazardous and Solid Waste Amendments of 1984 (HSWA), and the Medical Waste Tracking Act of 1988 (MWTA).

Solid Waste Management Unit (SWMU). Any unit in which wastes have been placed at any time, regardless of whether the unit was designed to accept solid or hazardous waste. Units include areas from which solid wastes have been routinely released.

Source Control. A practice or structural measure to prevent pollutants from entering storm water runoff or other environmental media.

Source Reduction. Reducing the amount of materials entering the waste stream by redesigning products or patterns of production or consumption (e.g., using returnable beverage containers). Synonymous with waste reduction.

Spent Solvent. A liquid solution that has been used and is no longer capable of dissolving solids, gases, or liquids.

Spill Guard. A device used to prevent spills of liquid materials from storage containers.

Spill Prevention Control and Countermeasures Plan (SPCCP). Plan required for both onshore and offshore facilities under 40 CFR 112.3 to prevent and respond to discharges of oil and hazardous substances as defined in the Clean Water Act.

Spoil. Dirt or rock removed from its original location, destroying the composition of the soil in the process, as in strip-mining, dredging, or construction.

State Emergency Response Commission (SERC). Commission appointed by each state governor according to the requirements of SARA Title III. The SERCs designate emergency planning districts, appoint local emergency planning committees, and supervise and coordinate their activities.

State Implementation Plan (SIP). EPA-approved state plans for the establishment, regulation, and enforcement of air pollution standards.

Stationary Source. A fixed-site producer of pollution, mainly power plants and other facilities using industrial combustion processes.

Stopcock Valve. A small valve for stopping or controlling the flow of water or other liquid through a pipe.

Storage (of Hazardous Waste). The holding of hazardous waste for a temporary period, at the end of which the hazardous waste is treated, disposed of, or stored elsewhere. Facilities are required to have a RCRA permit for storage of hazardous waste for more than 90 days; storage for less than 90 days does not require a RCRA permit.

Storm Drain. A slotted opening leading to an underground pipe or an open ditch for carrying surface runoff.

Storm Water. Runoff from a storm event, snow melt runoff, and surface runoff and drainage.

Storm Water Discharge Associated with Industrial Activity. [See 40 CFR Section 122.26(b)(14)]. The discharge from any conveyance which is used for collecting and conveying storm water and which is directly related to manufacturing, processing or raw materials storage areas at an industrial plant. The term does not include discharges from facilities or activities excluded from the NPDES program under 40 CFR Part 122. For the categories of industries identified in 40 CFR 122.26(b)(14)(i) through (x), the term includes, but is not limited to, storm water discharges from industrial plant yards; immediate access roads and rail lines used or traveled by carriers of raw materials, manufactured products, waste material, or by-products used or created by the facility; material handling sites; refuse sites; sites used for the application or disposal of process wastewaters (as defined at 40 CFR 401); sites used for the storage and maintenance of material handling equipment; sites used for residual treatment, storage, or disposal; shipping and receiving areas; manufacturing buildings; storage areas (including tank farms) for raw materials, and intermediate and finished products; and areas where industrial activity has taken place in the past and significant materials remain and are exposed to storm water. For the categories of industries identified in 40 CFR 122.26(b)(14)(xi), the term includes only storm water discharges from all the areas (except access roads and rail lines) that are listed in the previous sentence where material handling equipment or activities, raw materials, intermediate products, final products, waste material, by-products, or industrial machinery *are exposed to storm water.*

Subsoil. The bed or stratum of earth lying below the surface soil.

Sump. A pit or tank that catches liquid runoff for drainage or disposal.

Superfund. The program operated under the legislative authority of CERCLA that funds and carries out EPA hazardous waste emergency and long-term removal and remedial activities. These activities include establishing the National Priorities List, investigating sites for inclusion on the list, determining their priority, and conducting and/or supervising the cleanup and other remedial actions.

Superfund Amendments and Reauthorization Act of 1986 (SARA). 1986 amendments enacted to expand the scope of CERCLA, commonly known as SARA Title III or the Emergency Planning and Community Right-to-Know Act (EPCRA).

Surface Impoundment. A natural topographic depression, human-made excavation, or diked area formed primarily of earthen materials (may be lined with human-made materials) used for the treatment, storage, or disposal of liquid wastes. Examples of surface impoundments are holding, storage, settling, and aeration pits, ponds, and lagoons.

Surface Runoff. Precipitation, snow melt, or irrigation in excess of what can infiltrate the soil surface and be stored in small surface depressions; a major transporter of nonpoint source pollutants.

Surface Water. All water naturally open to the atmosphere (rivers, lakes, reservoirs, streams, wetlands impoundments, seas, estuaries, etc.) and subject to surface runoff; also refers to springs, wells, or other collectors which are directly influenced by surface water.

Suspended Loads. Sediment particles maintained in the water column by turbulence and carried with the flow of water.

Suspended Solids. Small particles of solid pollutants that float on the surface of, or are suspended in, sewage or other liquids. They resist removal by conventional means.

Swale. An elongated depression in the land surface that is at least seasonally wet, is usually heavily vegetated, and is normally without flowing water. Swales direct storm water flows into primary drainage channels and allow some of the storm water to infiltrate into the ground surface .

Swamp. A type of wetland dominated by woody vegetation but without appreciable peat deposits. Swamps may be fresh or salt water and tidal or non-tidal.

Tarp. A sheet of waterproof canvas or other material used to cover and protect materials, equipment, or vehicles.

Threatened Species. Animals, birds, fish, plants, and other living organisms threatened with extinction by human-made or natural changes in their environment and listed as threatened pursuant to Section 4 of the Endangered Species Act.

Tidal Marsh. Low, flat marshlands traversed by channels and tidal hollows, subject to tidal inundation; normally, the only vegetation present is salttolerant bushes and grasses.

Time Composite Sample. Prepared by collecting fixed volume aliquots at specified time intervals and combining into a single sample for analysis.

Topography. The physical features of a surface area including relative elevations and the position of natural and human-made features.

Total Dissolved Solids (TDS). All material that passes the standard glass river filter; now called total filtrable reside. Term is used to reflect salinity.

Total Suspended Solids (TSS). A measure of the suspended solids in wastewater, effluent, or water bodies, determined by tests for "total suspended nonfilterable solids."

Toxic Pollutants (Clean Water Act definition). Any pollutant listed as toxic under Section 501 (a)(1) of the Clean Water Act or, in the case of "sludge use or disposal practices," any pollutant identified in the regulations implementing Section 405(d) of the Clean Water Act.

Toxic Release Inventory (TRI). Under SARA Title III, Section 313, facilities that handle certain types of chemicals must report to EPA annually the quantities of these chemicals released to the environment during the year. These releases may be allowed under air or water permits, may be spills of waste or product materials, or may be fugitive smoke stack emissions or product/process tank/oil line losses. EPA is beginning to evaluate TRI data to determine whether facilities have environmentally significant releases that must be addressed under RCRA corrective action or Superfund authorities.

Toxic Waste. A waste that can produce injury if inhaled, swallowed, or absorbed through the skin.

Toxicity Characteristic Leachate Procedure (TCLP). The analytical method one must use to determine whether or not a waste is a characteristic hazardous waste based on toxicity. The TCLP is also necessary to comply with provisions of land disposal restrictions as well.

Toxicity Characteristic (TC) Rule. This rule replaced the Extraction Procedure (EP) toxicity test with the TC test to determine whether or not a waste is a characteristic waste based on toxicity. The TC test requires analysis of 25 organic compounds in addition to the eight metals and six pesticides that were subject to the EP test.

Transporter. A person transporting hazardous waste within the United States which requires a manifest. On-site movement of hazardous waste does not apply. Transporters must comply with 40 CFR Part 263.

Treatment. Any method, technique, or process, including neutralization, designed to change the physical, chemical, or biological character or composition of any hazardous waste so as to neutralize such waste, or so as to recover energy or material resources from the waste, or so as to render such waste nonhazardous, or less hazardous; safer to transport, store, or dispose of; or amenable for recovery, amenable for storage, or reduced in volume.

Treatment Standards (RCRA). Standards that hazardous wastes must meet prior to land disposal. A treatment standard generally expresses a treatment technology as concentration limits to give generators flexibility in choosing treatment options. Note that concentration limits are based upon the use of best demonstrated available technology (BDAT) for a particular waste or a similar waste.

Treatment, Storage, and Disposal (TSD) Facility. Site where a hazardous substance is treated, stored, or disposed of. TSD facilities are regulated by EPA and states under RCRA.

Tributary. A river or stream that flows into a larger river or stream.

Trichloroethylene (TCE). A stable, low boiling-point colorless liquid, toxic if inhaled. Used as a solvent or metal decreasing agent, and in other industrial applications.

Turbidity. 1. Haziness in air caused by the presence of particles and pollutants. 2. A cloudy condition in water due to suspended silt or organic matter.

Underground Storage Tank (RCRA definition). Any one or combination of tanks (including its connecting underground pipes) used to contain an accumulation of regulated substances, and the volume of which (including the volume of the underground pipes) is 10 percent or more beneath the surface of the ground. Regulated substances include hazardous chemical products regulated under CERCLA and petroleum products. Some tank uses are exempt from regulation, including septic tanks, residential/agricultural fuel or heating oil tanks, and wastewater collection systems.

Urea-Formaldehyde Foam Insulation. A material once used to conserve energy by sealing crawl spaces, attics, etc.; no longer used because emissions were found to be a health hazard.

Volatile Organic Compound (VOC). Any organic compound that participates in atmospheric photochemical reactions except those designated by EPA as having negligible photochemical reactivity.

Waste Minimization. Measures or techniques that reduce the amount of wastes generated during industrial production processes; term is also applied to recycling and other efforts to reduce the amount of waste going into the waste stream.

Waste Pile. Any noncontainerized accumulation of solid, nonflowing hazardous waste that is used for treatment or storage.

Waste Reduction. Using source reduction, recycling, or composting to prevent or reduce waste generation.

Water Table. The depth or level below which the ground is saturated with water.

Waters of the United States (Clean Water Act definition). "Waters of the United States" are defined in 33 CFR Section 328.3 as: (a) All waters, which are currently used, were used in the past, or may be susceptible to use in interstate or foreign commerce, including all waters which are subject to the ebb and flow of the tide; (b) All interstate waters, including interstate "wetlands;" (c) All other waters such as intrastate lakes, rivers, streams (including intermittent streams), mud flats, sand flats, "wetlands," sloughs, prairie potholes, wet meadows, playa lakes, or natural ponds, the use, degradation, or destruction of which would affect or could affect interstate or foreign commerce ....; (d) All impoundments of waters otherwise defined as waters of the United States under this definition; (e) Tributaries of waters identified in paragraphs (a) through (d) of this definition; (f) The territorial sea; and (g) "Wetlands" adjacent to waters (other than waters that are themselves wetlands) identified in paragraphs (a) through (f) of this definition.

Waste treatment systems, including treatment ponds or lagoons designed to meet the requirements of the Clean Water Act (other than cooling ponds as defined in 40 CFR 423.11 (m) which also meet the criteria of this definition) are not waters of the United States. This exclusion applies only to humanmade bodies of water which neither were originally created in waters of the United States (such as disposal area in wetlands) nor resulted from the impoundment of waters of the United States.

Waterway. A channel for the passage or flow of water.

Water Quality Standards. State-adopted and EPA-approved ambient standards for water bodies. The standards cover the use or the water body and the water quality criteria that must be met to protect the designated use or uses.

Weir. A device used to gauge the flow rate of liquid through a channel; is essentially a dam built across an open channel over which the liquid flows, usually through some type of notch.

Wet Well. A chamber used to collect water or other liquid and to which a pump is attached.

Wetlands. Those areas that are inundated or saturated by surface or groundwater at a frequency and duration sufficient to support, and that under normal circumstances do support, a prevalence of vegetation typically adapted for life in saturated soil conditions. Examples include: bogs, estuaries, fens, marshes, and swamps.

Wind Break. Any device designed to block wind flow and intended for protection against any ill effects of wind.

Environmental Acronyms

AAEE: American Academy of Environmental Engineers
AAPCO: American Association of Pesticide Control Officials
AARC: Alliance for Acid Rain Control
ACL: Alternate Concentration Limit
ACM: Asbestos-Containing Material
ACQR: Air Quality Control Region
ACS: American Chemical Society
ADR: Alternate Dispute Resolution
AEERL: Air and Energy Engineering Research Laboratory
AHERA: Asbestos Hazard Emergency Response Act
AIHC: American Industrial Health Council
ALAPO: Association of Local Air Pollution Control Officers
ALARA: As Low As Reasonably Achievable
ALJ: Administrative Law Judge
AMBIENS: Atmospheric Mass Balance of Industrially Emitted and
  Natural Sulfur
APA: Administrative Procedures Act
APCA: Air Pollution Control Association
APCD: Air Pollution Control District
APHA: American Public Health Association
APWA: American Public Works Association
AQCCT: Air-Quality Criteria and Control Techniques
AQCP: Air Quality Control Program
AQCR: Air-Quality Control Region
AQMP: Air-Quality Management Plan
ARARs: Applicable or Relevant and Appropriate
  Requirements
ASDWA: Association of State Drinking Water Administrators

ASHAA: Asbestos in Schools Hazard Abatement Act
ASIWCPA: Association of State and Interstate Water Pollution Control Administrators
ASTHO: Association of State and Territorial Health Officers
ASTM: American Society for Testing and Materials
ASTSWMO: Association of State and Territorial Solid Waste Management Officials
ATERIS: Air Toxics Exposure and Risk Information System
ATSDR: Agency for Toxic Substances and Disease Registry
AWRA: American Water Resources Association
AWWA: American Water Works Association

BACM: Best Available Control Measures
BACT: Best Available Control Technology
BADT: Best Available Demonstrated Technology
BART: Best Available Retrofit Technology
BAT: Best Available Technology
BATEA: Best Available Treatment Economically Achievable
BCT: Best Control Technology
BCPCT: Best Conventional Pollutant Control Technology
BDAT: Best Demonstrated Achievable Technology
BDCT: Best Demonstrated Control Technology
BDT: Best Demonstrated Technology
BMP: Best Management Practice
BO: Biological Opinion
BOD: Biochemical Oxygen Demand
BOD: Biological Oxygen Demand
BPT: Best Practicable Technology
BPWTT: Best Practical Wastewater Treatment Technology

CA: Corrective Action
CAA: Clean Air Act
CAAA: Clean Air Act Amendments
CAER: Community Awareness and Emergency Response
CARS: Corrective Action Reporting System
CAO: Corrective Action Order
CAP: Corrective Action Plan
CAP: Criteria Air Pollutant
CAS: Chemical Abstract Service
CATS: Corrective Action Tracking System
CDBG: Community Development Block Grant
CDD: Chlorinated dibenzo-p-dioxin
CDF: Chlorinated dibenzofuran

CEM: Continuous Emission Monitoring
CEMS: Continuous Emission Monitoring System
CEPP: Chemical Emergency Preparedness Plan
CEQ: Council on Environmental Quality
CERCLA: Comprehensive Environmental Response, Compensation, and Liability Act of 1980
CERCLIS: Comprehensive Environmental Response, Compensation, and Liability Information System
CESQG: Conditionally Exempt Small Quantity Generator
CFC: Chlorofluorocarbons
CFM: Chlorofluoromethanes
CFR: Code of Federal Regulations
CIAQ: Council on Indoor Air Quality
CIS: Chemical Information System
CMA: Chemical Manufacturers Association
CMI: Corrective Measures Implementation
CMS: Corrective Measures Study
COD: Chemical Oxygen Demand
CSO: Combined Sewer Overflow
CWA: Clean Water Act
CZMA: Coastal Zone Management Act
CZARA: Coastal Zone Management Act Reauthorization Amendments

DDT: Dichloro-Diphenyl-Trichloroethane
DNA: Deoxyribonucleic acid
DO: Dissolved Oxygen
DOW: Defenders Of Wildlife
DU: Ducks Unlimited
DWEL: Drinking Water Equivalent Level
DWS: Drinking Water Standard

EA: Endangerment Assessment
EA: Environmental Assessment
EDF: Environmental Defense Fund
EDTA: Ethylene Diamine Triacetic Acid
EHS: Extremely Hazardous Substance
EIR: Environmental Impact Report
EIS: Environmental Impact Statement
EIS/AS: Emissions Inventory System/Area Source
EIS/PS: Emissions Inventory System/Point Source
EL: Exposure Level
ELI: Environmental Law Institute

ELR: Environmental Law Reporter
EOP: End Of Pipe
EPA: Environmental Protection Agency
EPCRA: Emergency Preparedness and Community Right to Know Act
EPD: Emergency Planning District
EPTC: Extraction Procedure Toxicity Characteristic
ERC: Emergency Response Commission
ERC: Emissions Reduction Credit
ERNS: Emergency Response Notification System
ERT: Emergency Response Team
ESA: Endangered Species Act

FATES: FIFRA and TSCA Enforcement System
FDCA: Federal Food, Drug, and Cosmetic Act
FE: Fugitive Emissions
FEIS: Fugitive Emissions Information System
FERC: Federal Energy Regulatory Commission
FFFSG: Fossil-Fuel-Fired Steam Generator
FIFRA: Federal Insecticide, Fungicide, and Rodenticide Act
FIP: Final Implementation Plan
FLPMA: Federal Land Policy and Management Act
FOE: Friends Of the Earth
FOIA: Freedom Of Information Act
FONSI: Finding Of No Significant Impact
FR: Federal Register
FREDS: Flexible Regional Emissions Data System
FS: Feasibility Study
FSA: Food Security Act
FTTS: FIFRA/TSCA Tracking System
FWCA: Fish and Wildlife Coordination Act
FWPCA: Federal Water Pollution and Control Act
FWS: Fish and Wildlife Service

GAAP: Generally Accepted Accounting Principles
GAC: Granular Activated Carbon
GC/MS: Gas Chromatograph/ Mass Spectograph
GIS: Geographic Information Systems

HAP: Hazardous Air Pollutant
HAZMAT: Hazardous Materials
HAZOP: Hazard and Operability Study
HI: Hazard Index
HMTA: Hazardous Materials Transportation Act

HMTR: Hazardous Materials Transportation Regulations
HON: Hazardous Organic NESHAP
HRS: Hazard Ranking System
HSWA: Hazardous and Solid Waste Amendments
HW: Hazardous Waste

IAP: Indoor Air Pollution

LAER: Lowest Achievable Emission Rate
LDR: Land Disposal Restrictions
LEPC: Local Emergency Planning Committee
LERC: Local Emergency Response Committee
LLRW: Low Level Radioactive Waste

MCL: Maximum Contaminant Level
MCLG: Maximum Contaminant Level Goal
MOA: Memorandum of Agreement
MS4: Municipal Separate Storm Sewer System
MSDS: Material Safety Data Sheet
MSW: Municipal Solid Waste

NAAQS: National Ambient Air Quality Standard
NBAR: Non-Binding Allocations of Responsibility
NCP: National Contingency Plan
NEPA: National Environmental Policy Act of 1969
NESHAP: National Emissions Standards For Hazardous Air Pollutants
NFRAP: No Further Remedial Action Planned
NIOSH: National Institute of Occupational Safety and Health
NMFS: National Marine Fisheries Service
NOAA: National Oceanic and Atmospheric Administration
NOI: Notice of Intent
NPDES: National Pollutant Discharge Elimination System
NPL: National Priorities List
NRC: National Response Center
NRT: National Response Team
NSPS: New Source Performance Standards
NSR: New Source Review

ORM: Other Regulated Material
OSC: On-Scene Coordinator
OSHA: Occupational Safety and Health Administration

PCBs: Polychlorinated Biphenyls
PIGS: Pesticides in Groundwater Strategy
PIMS: Pesticide Incident Monitoring System
PM: Particulate Matter
PNA: Polynuclear Aromatic Hydrocarbons
POM: Particulate Organic Matter
POTW: Publicly Owned Treatment Works
PPA: Pollution Prevention Act
ppb: Parts Per Billion
ppm: Parts per Million
ppt: Parts Per Trillion
PRP: Potentially Responsible Party
PSD: Prevention of Significant Deterioration
PVC: Polyvinyl Chloride

QA/QC: Quality Assurance/Quality Control

RA: Remedial Action
RA: Risk Assessment
RACM: Reasonably Available Control Measures
RACM: Regulated Asbestos-Containing Material
RACT: Reasonably Available Control Technology
RAMS: Regional Air Monitoring System
RAP: Remedial Action Plan
RCRA: Resource Conservation and Recovery Act
RCRIS: Resource Conservation and Recovery Information System
RD/RA: Remedial Design/Remedial Action
RFA: RCRA Facility Assessment
RFI: RCRA Facility Investigation
RI: Remedial Investigation
RI/FS: Remedial Investigation/Feasibility Study
RNA: Ribonucleic Acid
ROD: Record Of Decision
RQ: Reportable Quantity
RRC: Regional Response Center
RRT: Regional Response Team
RUP: Restricted Use Pesticide

S&A: Sampling and Analysis
SAB: Science Advisory Board
SANE: Sulfur and Nitrogen Emissions
SARA: Superfund Amendments and Reauthorization Act of 1986

SCS: Soil Conservation Service
SDWA: Safe Drinking Water Act
SERC: State Emergency Planning Commission
SIC: Standard Industrial Classification
SIP: State Implementation Plan
SITE: Superfund Innovative Technology Evaluation
SLAMS: State/Local Air Monitoring Station
SMCL: Secondary Maximum Contaminant Level
SMCRA: Surface Mining Control and Reclamation Act
SOC: Synthetic Organic Chemicals
SOCMI: Synthetic Organic Chemicals Manufacturing Industry
SPCCP: Spill Prevention, Containment, and Countermeasures Plan
SPE: Secondary Particulate Emissions
SQG: Small Quantity Generator
SS: Suspended Solids
SSA: Sole Source Aquifer
SWDA: Solid Waste Disposal Act
SWMU: Solid Waste Management Unit

TAMS: Toxic Air Monitoring System
TCDD: Dioxin (Tetrachlorodibenzo-*p*-dioxin)
TCE: Trichloroethylene
TCLP: Toxicity Characteristic Leachate Procedure
TDS: Total Dissolved Solids
TMDL: Total Maximum Daily Load
TPQ: Threshold Planning Quantity
TQM: Total Quality Management
TRI: Toxic Release Inventory
TRIS: Toxic Chemical Release Inventory System
TSCA: Toxic Substances Control Act
TSDF: Treatment, Storage, and Disposal Facility
TSS: Total Suspended Solids
TVA: Tennessee Valley Authority
TWA: Time Weighted Average

UAO: Unilateral Administrative Order
UFFI: Urea-Formaldehyde Foam Insulation
UIC: Underground Injection Control
UST: Underground Storage Tank

VOC: Volatile Organic Compounds

WQS: Water Quality Standard
WRDA: Water Resources Development Act
WSRA: Wild and Scenic Rivers Act
WWF: World Wildlife Fund

About the Authors

James F. Berry is a biologist, college professor, and environmental attorney. He is a Professor of Biology at Elmhurst College in Elmhurst, IL, where he specializes in the fauna and ecology of wetlands and other sensitive areas, and on conservation biology. As an attorney, Dr. Berry specializes in environmental, land use, and natural resources law, and teaches courses on environmental law and regulation for the M.S.in Environmental Management program at the Stuart Graduate School of Business, Illinois Institute of Technology. He is licensed to practice law in all state courts in Florida and Illinois, as well as a number of federal district courts and courts of appeals. He is currently in private practice in Elmhurst, IL. Dr. Berry's biological education includes B.S. and M.S. degrees from Florida State University, and a Ph.D. from the University of Utah. He received his law degree from IIT Chicago-Kent College of Law, having completed the law school's Program in Environmental and Energy Law. He is a member of many scientific and legal professional organizations, and has published over 80 scientific and legal articles, book chapters, and coauthored *Wetlands: Guide to Science, Law, and Technology*, with Mark S. Dennison (1993).

Mark S. Dennison is an attorney and author of 16 books and more than 150 articles dealing primarily with environmental law and regulatory compliance issues. His books include: *Brownfields Redevelopment* (1998); *Wetland Mitigation* (1996); *Pollution Prevention Strategies and Technologies* (1995); *Environmental Reporting, Recordkeeping, and Inspections* (1995); *Storm Water Discharges* (1995); *Hazardous Waste Regulation Handbook* (1994); *Understanding Solid and Hazardous Waste Identification and Classification* (1993); *Wetlands: Guide to Science, Law, and Technology*, with James F. Berry (1993). Mr. Dennison was formerly editor-in-chief of the monthly newsletter, *Environmental Strategies for Real Estate* and editor of the quarterly journal, *Remediation*. He currently works in private practice in Westwood, New Jersey, specializing in environmental and real estate law. He is admitted to practice in New Jersey and New York. Mr. Dennison holds a B.A., *magna cum laude*, from the State University of New York (Oswego), an M.A. from Syracuse University, and a J.D. from New York Law School.